NUMERICAL MATHEMATICS AND SCIENTIFIC COMPUTATION

Series Editors

G. H. GOLUB Ch. SCHWAB
W. A. LIGHT E. SÜLI

NUMERICAL MATHEMATICS AND SCIENTIFIC COMPUTATION

*P. Dierckx: *Curve and surface fittings with splines*
*H. Wilkinson: *The algebraic eigenvalue problem*
*I. Duff, A. Erisman, and J. Reid: *Direct methods for sparse matrices*
*M. J. Baines: *Moving finite elements*
*J. D. Pryce: *Numerical solution of Sturm–Liouville problems*
K. Burrage: *Parallel and sequential methods for ordinary differential equations*
Y. Censor and S. A. Zenios: *Parallel optimization: theory, algorithms, and applications*
M. Ainsworth, J. Levesley, M. Marletta, and W. Light: *Wavelets, multilevel methods, and elliptic PDEs*
W. Freeden, T. Gervens, and M. Schreiner: *Constructive approximation on the sphere: theory and applications to geomathematics*
Ch. Schwab: *p- and hp- finite element methods: theory and applications to solid and fluid mechanics*
J. W. Jerome: *Modelling and computation for applications in mathematics, science, and engineering*
Alfio Quarteroni and Alberto Valli: *Domain decomposition methods for partial differential equations*
G. E. Karniadakis and S. J. Sherwin: *Spectral/hp element methods for CFD*
I. Babuška and T. Strouboulis: *The finite element method and its reliability*
B. Mohammadi and O. Pironneau: *Applied shape optimization for fluids*
S. Succi: *The lattice Boltzmann equation for fluid dynamics and beyond*
P. Monk: *Finite element methods for Maxwell's equations*
A. Bellen and M. Zennaro: *Numerical methods for delay differential equations*
M. Feistauer, J. Felcman, and I. Straškraba: *Mathematical and computational methods for compressible flow*

Monographs marked with an asterisk (*) appeared in the series 'Monographs in Numerical Analysis' which has been folded into, and is continued by, the current series

Mathematical and Computational Methods for Compressible Flow

M. Feistauer and J. Felcman
Charles University, Prague

I. Straškraba
Academy of Sciences of the Czech Republic

CLARENDON PRESS · OXFORD
2003

This book has been printed digitally and produced in a standard specification in order to ensure its continuing availability

OXFORD
UNIVERSITY PRESS

Great Clarendon Street, Oxford OX2 6DP

Oxford University Press is a department of the University of Oxford.
It furthers the University's objective of excellence in research, scholarship, and education by publishing worldwide in

Oxford New York

Auckland Cape Town Dar es Salaam Hong Kong Karachi
Kuala Lumpur Madrid Melbourne Mexico City Nairobi
New Delhi Shanghai Taipei Toronto

With offices in

Argentina Austria Brazil Chile Czech Republic France Greece
Guatemala Hungary Italy Japan South Korea Poland Portugal
Singapore Switzerland Thailand Turkey Ukraine Vietnam

Oxford is a registered trade mark of Oxford University Press in the UK and in certain other countries

Published in the United States by Oxford University Press Inc., New York

Reprinted 2006

ISBN 0-19-850588-4

PREFACE

Fluid dynamics is a very fascinating subject. The abundance of structures and processes in liquids and gases means that there exists a colourful pallet of methods and techniques used in their investigation. Mathematical models and methods play an important role in this area. The reason is that mathematics is a highly abstract and, thus, quite universal science. Particularly in connection with modern computers, mathematics offers efficient and powerful tools for the description and solution of many problems and processes in fluid dynamics from qualitative as well as quantitative point of view. The extensive application of mathematical and computational methods to the investigation of flow problems gave rise to (around 1950) computational fluid dynamics, commonly known under the acronym CFD.

The aim of this book is to provide the reader with a sufficiently detailed and extensive, mathematically precise and comprehensible guide through a wide spectrum of mathematical and computational methods used in CFD for the numerical simulation of compressible flow. The idea for writing this book came from Dr S. Adlung of OUP who had a number of discussions with M. Feistauer about mathematical and numerical methods in CFD. These discussions showed that a book collecting the most recent important results on the mathematical theory of compressible flow and its numerical solution was needed so we decided, together with Professor A. Novotný from the University of Toulon, to write a book on the mathematical theory and numerical methods for compressible flow. It was easy to say, but difficult to do. We eagerly collected the most up-to-date material, but found that the size of the book would be really huge. Therefore, we decided to split the material into two books, one on the mathematical theory of compressible flow (by A. Novotný and I. Straškraba) and the present book on mathematical and computational methods for compressible flow.

In contrast to other existing books presenting introductions to CFD, the present book aims to use the most recent numerical techniques applied on nonuniform unstructured meshes, including mesh adaptation, applicable directly to the numerical simulation of complicated, technically relevant compressible flow problems. Moreover, the emphasis is on mathematical and numerical analysis as well as on practical applications. We try to maintain a balance between the derivation of numerical schemes and their theoretical analysis, verification and applications.

We hope that the book will be useful to specialists – namely, pure and applied mathematicians, aerodynamicists, engineers, physicists and natural scientists. We also expect that the book will be suitable for advanced undergraduate, graduate and postgraduate students of mathematics and technical sciences.

As for references, there is a huge amount of literature on numerical and computational methods for compressible flow. We tried to quote the works relevant to the topics of the book, but it is clear that many significant references were unintentionally omitted. We apologize in advance to those authors whose contributions are not mentioned or do not receive the attention they deserve.

We have tried to avoid errors, but some may remain. Readers are welcome to send any correction electronically to the address `feist@karlin.mff.cuni.cz`. Errata will then be placed on the website

`http://www.karlin.mff.cuni.cz/katedry/knm/`

We are grateful to Professors I. Babuška, B. Cockburn, L. Demkowicz, J. E. Flaherty, P. Fraunie, T. Gallouët, P. Hansbo, R. Herbin, R. Jeltsch, C. Johnson, K. Kozel, M. Křížek, D. Kröner, M. Lukáčová, J. Neustupa, A. Novotný, J. T. Oden, M. R. Padula, P. Penel, W. Rachowicz, R. Rannacher, H.-G. Roos, C. Schwab, C.-W. Shu, T. Sonar, E. Süli, F. Toro, G. Warnecke and W. Wendland for valuable information, advice, comments, inspiring suggestions and stimulating discussions which helped us during the preparation of the manuscript.

Further, we are grateful to our friends and colleagues V. Dolejší, P. Knobloch, M. Křížek, K. Najzar, C. Schwab and V. Sobotíková and to our PhD students M. Bejček, T. Neustupa, A. Prachař, K. Švadlenka and J. Uchytil, who read parts of the manuscript and provided us with helpful suggestions. Moreover, we wish to thank V. Dolejší who provided us with many computational results and figures.

We are indebted to Mrs I. Marešová, who expertly typed in TEXa large part of the manuscript. Without her skilful, prompt and patient typing and retyping of many pages, the book would never have been completed. We are also very much obliged to Mrs S. Novotná for creating the database of literature which helped us work with the references.

We gratefully acknowledge sponsorship within the RiP program in the Mathematical Research Institute Oberwolfach (MFO), which is financed by the local German state Baden-Württemberg and was initiated by the Volkswagen Stiftung. The one-month stay in MFO helped us to complete the whole work successfully.

The work on the book was also partially supported by the Grant Agency of the Czech Republic (Projects No. 201/02/0684 and 101/01/0938), Grant Agency of Charles University (Project No. 275/2001/B-MAT/MFF) and by the Ministry of Education, Youth and Sports of the Czech Republic (Project No. MSM113200007). We gratefully acknowledge this support.

Our families gave us considerable support during work on the book. Our special thanks go to Jaroslava Feistauerová, Věra Felcmanová and Jana Hrušková.

Prague, December 2002

M. F.
J. F.
I. S.

CONTENTS

INTRODUCTION

In many areas of science and technology, one encounters the necessity to investigate the flow of liquids and gases in, for example, aviation and aeronautics, the car industry, the design of turbines, compressors and pumps, the chemical and food industry, medicine, biology, agriculture, meteorology, hydrology, oceanography and environmental protection. In the preface of the well-known book (Landau and Lifschitz, 1959) by L. D. Landau and E. M. Lifschitz, fluid dynamics is characterized as a branch of theoretical physics. The fundamental concepts and equations of fluid dynamics are connected with the names of Newton, Euler, Cauchy, Lagrange, Bernoulli, Huygens, d'Alembert, Kirchhoff, Helmholtz, Lamb, Stokes, Navier and others and belong to the area of classical rational mechanics. Therefore, it is natural that fluid dynamics uses extensive mathematical tools, particularly partial differential equations, and can also be considered as a mathematical science.

An image of flow can be obtained in two ways:

a) With the aid of *experiments*, which may give a realistic picture of real flow. In a number of cases the experimental investigation of a flow requires great cost, is lengthy and sometimes impossible, such as in the flow around space vehicles at re-entry or a loss-of-coolant accident in a nuclear reactor.

b) With the use of *mathematical models*. Most fluid dynamical models are represented by a system of partial differential equations expressing the fundamental laws of conservation of mass, momentum and energy, completed by constitutive relations and thermodynamical laws, together with boundary and initial conditions.

Theoretical investigation of the models consists in the study of the existence and uniqueness of a solution, its stability, continuous dependence on data, etc. This is important because of the validation of the adequacy and correctness of the model and to explain important qualitative features of flow processes.

From a practical point of view, quantitative behaviour of the flow is also important. Therefore, numerical and computational methods play a major role in modern mathematical modelling of flow. Mathematical and numerical methods and techniques applied by means of modern computers form a very efficient counterpart to experimental methods. The numerical and computer simulation of flow is the subject of *computational fluid dynamics*, abbreviated as CFD. The goal of this field of science is to obtain a realistic qualitative and quantitative image of flow with the aid of modern numerical methods, computational algorithms and specialized systems of computer programs and to obtain results comparable with experiments carried out, for example, in wind tunnels.

Modern CFD has become indispensable for design and optimization in aerodynamics. The growth of computer power allows the solution of more and more complex problems and the investigation of many different configurations rapidly and at acceptable cost. There are a number of commercial or research software packages available and widely used by engineers. CFD codes successively become design tools by industrial users. These codes are applied more and more as black boxes. Usually the detailed descriptions of the numerical algorithms hidden in commercial software are not available to the user. This means that it is often impossible to detect gross errors in the obtained result and diminishes the reliability of the computed data. Many times significant errors have been made when using CFD software packages without a solid background in fluid mechanics and numerical analysis. Thus, it should be necessary that users of fluid dynamical codes have a solid understanding of the basic principles of numerical methods and algorithms used in CFD. Moreover, progress in CFD is impossible without the development and analysis of new schemes based on the application of modern numerical techniques and approaches.

There are a number of monographs on CFD, such as (Ferziger and Perić, 1996), (Fletcher, 1991), (Hirsch, 1988), (Peyret and Taylor, 1983), (Roache, 1972), (Roache, 1998*a*), (Sod, 1985), (Wesseling, 2001), (Kovenya and Yanenko, 1981), (Kovenya *et al.*, 1990). Some of these books have a more or less engineering orientation, giving a useful survey on classical methods used in CFD, but without a deeper mathematical insight into stability, convergence and error estimate analysis. The exceptions are (Wesseling, 2001), (Sod, 1985), (Kovenya and Yanenko, 1981), (Kovenya *et al.*, 1990), which cover the mathematical treatment of these questions. However, all these books are concerned with finite difference or finite volume schemes applied on structured or uniform meshes.

The present book is concerned with mathematical and numerical methods for compressible flow. The aim is to provide the reader with a sufficient knowledge of modern numerical methods applied in CFD. We try to explain recent developments in CFD in a mathematically precise but comprehensible way. (For incompressible fluid dynamics there are several monographs, such as, for example, (Girault and Raviart, 1979), (Girault and Raviart, 1986), (Temam, 1977), covering the mathematical theory and numerical analysis of the incompressible Navier–Stokes equations.) We explain up-to-date finite volume and finite element techniques applied to the numerical solution of inviscid as well as viscous compressible flow on unstructured meshes and allowing the simulation of complex, technically relevant problems. Among others, we are concerned with finite volume methods using approximate Riemann solvers, finite element techniques, such as the streamline diffusion method and the discontinuous Galerkin method, and combined finite volume – finite element schemes. Many of the numerical methods and schemes explained in our book have not yet become part of commercial software, although they produce highly accurate and reliable results.

Our goal is to give a comprehensive insight into numerics for compressible flow, starting from the development of numerical schemes, their theoretical mathematical analysis, up to their verification on test problems and solution of practically relevant problems. In CFD one can see quite clearly two faces of numerical mathematics: *numerical mathematics as a science*, consisting of rigorous mathematical analysis of stability, convergence, error estimates etc., and *numerical mathematics as an art*, based on partial a priori knowledge of simulated processes, experience and intuition (see, for example, the preface of (Ralston, 1965)). Some important qualitative properties of schemes and algorithms used in CFD can often be proven mathematically rigorously only under restrictive assumptions, or if they are applied to simplified model problems. In this case we can consider numerical mathematics as a branch of science. Then these properties are heuristically extrapolated to complicated problems in CFD, on the basis of our experience or intuition and with the aid of numerous numerical experiments. In such a case numerical mathematics appears as an art. These two faces of numerical mathematics in CFD can also be traced in our book.

The contents of the book are as follows.

Chapter 1 gives an insight into the fundamental equations and relations describing compressible flow and explains some important physical aspects. Moreover, it contains a survey of mathematical concepts and results important for understanding the subsequent chapters.

Chapter 2 briefly describes the mathematical theory of compressible flow. Attention is paid to the investigation of qualitative properties of initial-boundary value problems for the inviscid Euler equations as well as the viscous Navier–Stokes system.

Chapter 3 is concerned with finite difference and finite volume methods for the numerical solution of the Euler equations. The main emphasis is on the finite volume method applied to the solution of multidimensional problems on unstructured meshes. We also pay attention to the finite difference method for the one-dimensional Euler equations, because it allows the reader to understand easily the main concepts and techniques for the solution of compressible inviscid multidimensional flow.

Finally, Chapter 4 is devoted to the theory of the finite element method and its application to compressible inviscid as well as viscous flow. The first part represents an introduction to the finite element method, which may be useful for readers not too familiar with this technique. Then we discuss conforming finite elements with streamline diffusion stabilization, combined finite volume–finite element method and, finally, the discontinuous Galerkin finite element method.

Of course, it was not possible to cover in detail all known models and methods used in CFD. No space was allocated, for example, to the low Mach number models, turbulence modelling, the use of multigrid, a detailed description of Krylov space methods, flows with chemical reactions, and fluid and structure interaction.

We proceed in such a way that we formulate the initial-boundary value problems of fluid dynamics, describe numerical schemes and discuss their theoretical analysis. The study is supplemented by a series of examples of the solution to various flow problems. They will allow the reader to achieve a sufficient understanding of the character of the models and methods discussed and their applicability.

1

FUNDAMENTAL CONCEPTS AND EQUATIONS

In this book we shall be concerned with the motion of compressible fluids, i.e. gases. They will be considered as continuous media or, simply, continua. This means that we assume that each point of the domain occupied by the fluid represents exactly one fluid particle. Moreover, we assume that functions describing the behaviour and motion of the fluid are continuous or even, according to the mathematical means used, sufficiently smooth. These assumptions are used mainly for the derivation of the equations and relations of fluid dynamics. In further considerations the strong assumptions are removed in the so-called weak formulations.

Depending on their compressibility and expansibility, fluids are divided into *liquids* (e.g. water, oil) and *gases* (e.g. air, steam). The characteristic feature of the gases is the substantial change of the volume under the influence of the pressure. This property is called the *compressibility* and the gases are therefore called *compressible fluids*. On the other hand, as a consequence of their expansibility, the gases fill up the whole space into which they can penetrate. Here we shall be concerned with the dynamics of gases only.

The behaviour of fluids is determined by intermolecular forces and the molecular mean free path characterizing heat phenomena. A fluid offers a relatively small resistance against forces causing it to change its shape. The property of resisting this change of shape is called *viscosity*. The viscosity force tries to reduce the difference between the velocities of neighbouring fluid layers. It resembles the friction force and, therefore, we sometimes encounter the notion of *internal friction* for viscosity phenomena. The friction forces also cause the fluid to adhere to the walls and bodies surrounding the fluid.

Very often the fluid viscosity is so small that it can be neglected. Then we speak of the model of *inviscid fluid*.

The viscous character and properties of the fluid are embodied in the relation between the stress tensor and the deformation velocity tensor (= analogous to the strain tensor in the elasticity theory). If this relation is linear, we speak of a *Newtonian fluid*, which is particularly the case of gases.

The viscosity of gases is very small and its effects can occur mainly in a neighbourhood of walls and bodies surrounded by the gas in the so-called *boundary layer*. Behind a body immersed in a moving fluid we can observe a more or less conspicuous *wake* linked to the boundary layer and containing vortices floating away from the body.

In compressible flow, if the velocity exceeds the speed of sound in some region, so-called *shock waves* or *contact discontinuities* can occur, representing jumps (slightly smeared due to the viscosity) in the velocity, density, pressure and other quantities.

Let us mention also that the fluid motion can be visualized. Flow visualization plays an important role in research, yielding qualitative insight and, recently, also quantitative results. This is nicely documented by the book (Van Dyke, 1988) which is a collection of beautiful photographs showing a number of fluid flows and fluid structures.

In this chapter, we introduce some basic mathematical concepts and notation. Then the governing equations and relations of gas dynamics will be presented. Finally, important mathematical results and advanced concepts will be covered.

1.1 Some mathematical concepts and notation

We assume the reader is familiar with the elements of linear algebra, mathematical analysis and the theory of the Lebesgue integral – see, for example, (Rudin, 1974).

1.1.1 *Basic notation*

By $\mathbb{R}$ and $\mathbb{N}$ we shall denote the set of all real numbers and the set of all positive integers, respectively. In the Euclidean space $\mathbb{R}^N$ $(N \geq 1)$ we shall use a Cartesian coordinate system with axes denoted by $x_1, \dots, x_N$. Points from $\mathbb{R}^N$ will be usually denoted by $x = (x_1, \dots, x_N)$, $y = (y_1, \dots, y_N)$, etc. In some physical situations we shall call elements of $\mathbb{R}^N$ vectors and treat them as columns. To emphasize this fact, the vectors will be usually denoted by bold letters. By $\boldsymbol{e}_1, \dots, \boldsymbol{e}_N$ we shall denote the *unit vectors* in the directions of the coordinate axes. If $\boldsymbol{a}$ is a (column) vector, then $\boldsymbol{a}^{\mathrm{T}}$ will denote the (row) vector transposed to $\boldsymbol{a}$. By $|\cdot|$ we shall denote the Euclidean norm in $\mathbb{R}^N$. Thus, $|x - y| = (\sum_{i=1}^N |x_i - y_i|^2)^{1/2}$ denotes the distance of two points $x, y \in \mathbb{R}^N$ and $|\boldsymbol{a}|$ denotes the magnitude of a vector $\boldsymbol{a} \in \mathbb{R}^N$.

Let us recall that an open set $\Omega \subset \mathbb{R}^N$ is *connected*, if its arbitrary two points can be connected with a piecewise linear curve in Ω. We usually denote by Ω a *domain*, i.e. an open and connected set in $\mathbb{R}^N$. We say that Ω is a bounded domain, if Ω is a domain and if it is bounded. We say that Ω is an unbounded domain, if Ω is a domain and if Ω is not bounded. We say that Ω is an exterior domain, if $\mathbb{R}^N \setminus \overline{\Omega}$ is a bounded domain. We also denote $B_r(x) := \{y \in \mathbb{R}^N;\ |x-y| < r\}$ as a ball in $\mathbb{R}^N$ with centre x and radius $r > 0$. We define the *diameter* of a set $M \subset \mathbb{R}^N$ as $\mathrm{diam}(M) = \sup\{|x-y|;\ x, y \in M\}$.

If $\mathcal{P}$ and $\mathcal{Q}$ are two sets and f is a mapping (function) defined on $\mathcal{P}$ with its values lying in $\mathcal{Q}$, we write $f : \mathcal{P} \to \mathcal{Q}$. For such a mapping and a subset $M \subset \mathcal{P}$ the symbol $f|_M$ denotes the restriction of f on the set M. In other words, if we write $g = f|_M$, then $g : M \to \mathcal{Q}$ and $g(x) = f(x)$ for all $x \in M$. Let us note that a one-to-one mapping f of a set $\mathcal{P}$ onto a set $\mathcal{Q}$ is called *bijection*.

By $f \circ g$ we denote the composition of functions f and g: $(f \circ g)(x) = f(g(x))$.

For $M \subset \mathbb{R}^N$ we denote by $C(M)$ (or $C^0(M)$) the linear space of all functions continuous on M. For $k \in \mathbb{N}$ and $\Omega \subset \mathbb{R}^N$ a domain, $C^k(\Omega)$ will denote the linear space of all functions which have continuous partial derivatives up to order k in Ω. Let $\partial\Omega$ and $\overline{\Omega}$ denote the boundary of the set Ω and its closure, respectively. The space $C^k(\overline{\Omega})$ is formed by all functions from $C^k(\Omega)$ whose all derivatives up to order k can be continuously extended onto $\overline{\Omega}$.

If $\mathcal{P}$ is some set, then by the symbol $\mathcal{P}^N$ we denote the Cartesian product $\mathcal{P} \times \mathcal{P} \times \cdots \times \mathcal{P}$ (N times). It means that $\mathcal{P}^N = \{(a_1, \ldots, a_N);\ a_i \in \mathcal{P},\ i = 1, \ldots, N\}$. For two sets $\mathcal{P}$ and $\mathcal{Q}$ we put $\mathcal{P} \times \mathcal{Q} = \{(x, y);\ x \in \mathcal{P},\ y \in \mathcal{Q}\}$.

If $m \in I\!N$ and $\boldsymbol{f} : \Omega \to I\!R^m$, i.e. $\boldsymbol{f} = (f_1, \ldots, f_m)$, $f_i : \Omega \to I\!R$ for $i = 1, \ldots, m$, then we write $\boldsymbol{f} \in [C^k(\Omega)]^m$, if $f_i \in C^k(\Omega)$ for all $i = 1, \ldots, m$. Similarly we define the space $[C^k(\overline{\Omega})]^m$. (Often we simply write $[C^k(\Omega)]^m = C^k(\Omega)^m$ etc.)

If M is some set, then the number of its elements will be denoted by card (M) or $\#M$.

Quantities describing fluid flow are functions of space and time. This means that we can write such a quantity as a function $f = f(x, t)$, where t is time and $x = (x_1, x_2, x_3)$ denotes points of a set $\Omega_t \subset I\!R^3$ occupied by the fluid at time t. Let $(0, T)$ $(0 < T)$ be a time interval during which we follow the fluid motion. Then the domain of definition of the function f is the set

$$\mathcal{M} = \{(x, t);\ x \in \Omega_t, t \in (0, T)\} \subset I\!R^4. \tag{1.1.1}$$

Let us assume that the set $\mathcal{M}$ is open. (If the domains $\Omega_t = \Omega$ are independent of t, then $\mathcal{M} = \Omega \times (0, T)$ is open.) If $t \in (0, T)$ is fixed, then $f(t) = f(\cdot, t)$ denotes the function '$x \to f(x, t)$' whose value at $x \in \Omega_t$ equals $f(x, t)$.

1.1.2 *Differential operators*

Let us consider a function $f = f(x, t)$ defined on the set $\mathcal{M}$ and having partial derivatives (in a classical or generalized sense) $\partial f/\partial x_1$, $\partial f/\partial x_2$ and $\partial f/\partial x_3$. Then we put

$$\operatorname{grad} f = \left(\frac{\partial f}{\partial x_1}, \frac{\partial f}{\partial x_2}, \frac{\partial f}{\partial x_3}\right)^{\mathrm{T}}. \tag{1.1.2}$$

If $\boldsymbol{f} = (f_1, f_2, f_3) : \mathcal{M} \to I\!R^3$ is a vector function with components which have first order partial derivatives (in a classical or generalized sense) with respect to x_1, x_2, x_3 in $\mathcal{M}$, then we set

$$\operatorname{div} \boldsymbol{f} = \sum_{i=1}^{3} \frac{\partial f_i}{\partial x_i}, \tag{1.1.3}$$

$$\operatorname{rot} \boldsymbol{f} = \left(\frac{\partial f_3}{\partial x_2} - \frac{\partial f_2}{\partial x_3}, \frac{\partial f_1}{\partial x_3} - \frac{\partial f_3}{\partial x_1}, \frac{\partial f_2}{\partial x_1} - \frac{\partial f_1}{\partial x_2}\right)^{\mathrm{T}}.$$

(The notation curl $\boldsymbol{f}$ is often used instead of rot $\mathbf{f}$.) The derivatives $\frac{\partial f}{\partial x_i}$ will also be denoted by $\partial f/\partial x_i$ or simply, $f_{x_i}, \partial_{x_i} f$ or $\partial_i f$. Similar notation is used for the derivative with respect to time: $\partial f/\partial t = f_t = \partial_t f$ and for higher order derivatives: $f_{x_i x_j}, \partial_{ij} f$, etc. Sometimes we shall use the Einstein summation convention over repeated indices. So, for example, $\operatorname{div} \boldsymbol{f} = \partial_i f_i$, $\sum_{j=1}^N a_{ij} x_j = a_{ij} x_j$, etc.

We define the *scalar product* in $\mathbb{R}^N$ by the relation

$$\boldsymbol{a} \cdot \boldsymbol{b} = \sum_{i=1}^{N} a_i\, b_i, \tag{1.1.4}$$

where $\boldsymbol{a} = (a_1, \dots, a_N)$ and $\boldsymbol{b} = (b_1, \dots, b_N) \in \mathbb{R}^N$. The *magnitude* of $\boldsymbol{a}$ is the number $|\boldsymbol{a}| = (\boldsymbol{a} \cdot \boldsymbol{a})^{\frac{1}{2}}$. For $\boldsymbol{a}$, $\boldsymbol{b} \in \mathbb{R}^3$ we define the *vector product*

$$\boldsymbol{a} \times \boldsymbol{b} = (a_2 b_3 - a_3 b_2,\, a_3 b_1 - a_1 b_3,\, a_1 b_2 - a_2 b_1)^{\mathrm{T}}. \tag{1.1.5}$$

The vector product can be written symbolically as a determinant:

$$\boldsymbol{a} \times \boldsymbol{b} = \det \begin{pmatrix} \boldsymbol{e}_1, \, \boldsymbol{e}_2, \, \boldsymbol{e}_3 \\ a_1, \, a_2, \, a_3 \\ b_1, \, b_2, \, b_3 \end{pmatrix}. \tag{1.1.6}$$

By $\boldsymbol{a} \otimes \boldsymbol{b}$ we denote the so-called *tensor product*:

$$\boldsymbol{a} \otimes \boldsymbol{b} = \begin{pmatrix} a_1 b_1, \, a_1 b_2, \, a_1 b_3 \\ a_2 b_1, \, a_2 b_2, \, a_2 b_3 \\ a_3 b_1, \, a_3 b_2, \, a_3 b_3 \end{pmatrix}. \tag{1.1.7}$$

In physics one often meets the concept of tensors (of order two) which can be expressed as 3×3 matrices. If $\mathbb{A} = (a_{ij})_{i,j=1}^3$ is a tensor function depending on $x \in \Omega_t$, then we define

$$\operatorname{div} \mathbb{A} = \left(\sum_{j=1}^{3} \frac{\partial a_{j1}}{\partial x_j}, \sum_{j=1}^{3} \frac{\partial a_{j2}}{\partial x_j}, \sum_{j=1}^{3} \frac{\partial a_{j3}}{\partial x_j} \right)^{\mathrm{T}}. \tag{1.1.8}$$

The *scalar product* of two tensors $\mathbb{A} = (a_{ij})_{i,j=1}^3$ and $\mathbb{B} = (b_{ij})_{i,j=1}^3$ is defined by

$$\mathbb{A} \cdot \mathbb{B} = \sum_{i,j=1}^{3} a_{ij} b_{ij}. \tag{1.1.9}$$

By δ_{ij} we denote the Kronecker delta: $\delta_{ii} = 1$, $\delta_{ij} = 0$ for $i \neq j$. Introducing the vector differential operator 'nabla'

$$\nabla = \left(\frac{\partial}{\partial x_1}, \frac{\partial}{\partial x_2}, \frac{\partial}{\partial x_3} \right)^{\mathrm{T}}, \tag{1.1.10}$$

we can express the operators grad, div and rot in the form

$$\operatorname{grad} f = \nabla f, \tag{1.1.11}$$

$$\operatorname{div} \boldsymbol{f} = \nabla \cdot \boldsymbol{f},$$
$$\operatorname{rot} \boldsymbol{f} = \nabla \times \boldsymbol{f}.$$

Let $\boldsymbol{f} : \mathcal{M} \to I\!R^3$ be a mapping whose components have first order partial derivatives with respect to x_1, x_2, x_3. We define the *Jacobi matrix* of the mapping $\boldsymbol{f}(\cdot, t)$ ($t \in (0, T)$ is fixed) as the matrix

$$\frac{D\boldsymbol{f}}{Dx}(x,t) = \begin{pmatrix} \frac{\partial f_1}{\partial x_1}(x,t), & \frac{\partial f_1}{\partial x_2}(x,t), & \frac{\partial f_1}{\partial x_3}(x,t) \\ \frac{\partial f_2}{\partial x_1}(x,t), & \frac{\partial f_2}{\partial x_2}(x,t), & \frac{\partial f_2}{\partial x_3}(x,t) \\ \frac{\partial f_3}{\partial x_1}(x,t), & \frac{\partial f_3}{\partial x_2}(x,t), & \frac{\partial f_3}{\partial x_3}(x,t) \end{pmatrix}. \tag{1.1.12}$$

The Jacobi matrix is also defined for a general differentiable mapping $\boldsymbol{f}$ of an open set $U \subset I\!R^N$ into $I\!R^m$ as

$$\frac{D\boldsymbol{f}}{Dx}(x) = \begin{pmatrix} \frac{\partial f_1}{\partial x_1}(x), & \frac{\partial f_1}{\partial x_2}(x), & \dots, & \frac{\partial f_1}{\partial x_N}(x) \\ \frac{\partial f_2}{\partial x_1}(x), & \frac{\partial f_2}{\partial x_2}(x), & \dots, & \frac{\partial f_2}{\partial x_N}(x) \\ \vdots & \vdots & \dots & \vdots \\ \frac{\partial f_m}{\partial x_1}(x), & \frac{\partial f_m}{\partial x_2}(x), & \dots, & \frac{\partial f_m}{\partial x_N}(x) \end{pmatrix}. \tag{1.1.13}$$

Let $f, f_i : \mathcal{M} \to I\!R$, $\boldsymbol{f} = (f_1, f_2, f_3)$ and let f, f_i have second order partial derivatives (in a classical or generalized sense) with respect to x_1, x_2, x_3. Then we set

$$\Delta f = \sum_{i=1}^{3} \frac{\partial^2 f}{\partial x_i^2}, \qquad \Delta \boldsymbol{f} = (\Delta f_1, \Delta f_2, \Delta f_3). \tag{1.1.14}$$

The symbol Δ is called the *Laplace operator*, or, briefly, *Laplacian*.

1.1.3 *Transformations of Cartesian coordinates*

Let us consider two Cartesian coordinate systems $(x_1, \dots, x_N)$ and $(x_1^*, \dots, x_N^*)$ in $I\!R^N$. Then the transition from x_i to x_i^* is realized by the relations

$$x_i^* = \sum_{j=1}^{N} a_{ij} x_j + c_i, \qquad i = 1, \dots, N, \tag{1.1.15}$$

where $\mathbb{A} = (a_{ij})_{i,j=1}^{N}$ is an *orthonormal matrix*. It means that

$$\mathbb{A}\,\mathbb{A}^{\mathrm{T}} = \mathbb{I}, \tag{1.1.16}$$

where $\mathbb{I}$ denotes the unit matrix and $\mathbb{A}^{\mathrm{T}} = (a_{ji})_{i,j=1}^{N}$ is the transpose of $\mathbb{A}$ (hence, in view of (1.1.16), $\mathbb{A}^{-1} = \mathbb{A}^{\mathrm{T}}$). Relations (1.1.15) can be written in the vector form

$$x^* = \mathbb{A}\, x + \mathbf{c},\ x = (x_1, \dots, x_N)^{\mathrm{T}},\ x^* = (x_1^*, \dots, x_N^*)^{\mathrm{T}},\ \mathbf{c} = (c_1, \dots, c_N)^{\mathrm{T}}. \quad (1.1.17)$$

The transformation inverse of (1.1.15) is

$$x_k = \sum_{i=1}^{N} a_{ik}\, x_i^* + c_k^*, \qquad k = 1, \dots, N, \quad (1.1.18)$$

where $c_k^* = -\sum_{i=1}^{N} a_{ik}\, c_i$. The vector form of (1.1.18) is

$$x = \mathbb{A}^{\mathrm{T}} x^* + \mathbf{c}^*, \qquad \mathbf{c}^* = (c_1^*, \dots, c_N^*)^{\mathrm{T}}. \quad (1.1.19)$$

1.1.4 *Hölder-continuous and Lipschitz-continuous functions*

A function $\boldsymbol{f} : M \to I\!R^N, M \subset I\!R^N$, is *$\mu$-Hölder-continuous with* $\mu \in (0,1]$, if there exists a constant L such that

$$|\boldsymbol{f}(x) - \boldsymbol{f}(y)| \le L|x - y|^{\mu} \qquad \forall x,\, y \in M. \quad (1.1.20)$$

If $\mu = 1$, we speak of a *Lipschitz-continuous* (or simply Lipschitz) function. If $\Omega \subset I\!R^N$ is an open set, then $C^{k,\mu}(\overline{\Omega})$ denotes the set of all functions whose derivatives of order k are μ-Hölder-continuous in $\overline{\Omega}$.

1.1.5 *Symbols 'o' and 'O'*

Let f be a function defined in $B(0) \setminus 0$, where $B(0)$ is a neighbourhood of zero, and let $\alpha \in I\!R$. We write $f(h) = o(h^{\alpha})$, if $\lim_{h \to 0} f(h)/h^{\alpha} = 0$. Further, we say that $f(h) = O(h^{\alpha})$, if there is a constant $c > 0$ such that $|f(h)| \le ch^{\alpha}$ for all $h \in B(0) \setminus \{0\}$.

1.1.6 *Measure and integral*

We expect that the reader is familiar with elements of differential and integral calculus, the theory of the (Lebesgue) measure and integral, substitution theorem, Fubini's theorem and the differentiations of an integral with respect to a parameter.

The Lebesgue measure of a set $M \subset I\!R^N$ will be denoted by $\mathrm{meas}_N(M)$ or simply by $|M|$. Hence, $|M|$ is the volume of M or the area of M or the length of M, if $N = 3$ or $N = 2$ or $N = 1$, respectively. If some statement $\mathcal{S}$ is valid in $M \setminus K$, where $|K| = 0$, we say that $\mathcal{S}$ holds *almost everywhere* (a.e.) in M. We can also write that $\mathcal{S}$ holds for *almost all* (a.a.) $x \in M$.

1.1.7 *Description of the boundary*

Let $\Omega \subset I\!R^N$ be a bounded domain. Its boundary $\partial\Omega$ is called *Lipschitz-continuous* (or simply Lipschitz), if there exist numbers $\alpha > 0$, $\beta > 0$, and a finite number of local Cartesian coordinate systems $x_1^r, \ldots, x_N^r$ and Lipschitz-continuous functions $a_r : \mathcal{M}_r = \{\hat{x}^r = (x_2^r, \ldots, x_N^r) \in \mathbb{R}^{N-1}; |\hat{x}^r| \leq \alpha\} \longrightarrow \mathbb{R}$ (called local maps), $r = 1, \ldots, R, (R \in \mathbb{N})$, such that

$$\begin{aligned} &\partial\Omega = \bigcup_{r=1}^{R} \Lambda_r, &(1.1.21)\\ &\Lambda_r = \{(x_1^r, \hat{x}^r);\ x_1^r = a_r(\hat{x}^r),\ |\hat{x}^r| < \alpha\},\\ &\{(x_1^r, \hat{x}^r);\ a_r(\hat{x}^r) < x_1^r < a_r(\hat{x}^r) + \beta,\ |\hat{x}^r| < \alpha\} \subset \Omega,\ r = 1, \ldots, R,\\ &\{(x_1^r, \hat{x}^r);\ a_r(\hat{x}^r) - \beta < x_1^r < a_r(\hat{x}^r),\ |\hat{x}^r| < \alpha\} \subset \mathbb{R}^n \setminus \overline{\Omega},\ r = 1, \ldots, R. \end{aligned}$$

We often speak of *Lipschitz domain* or domain with *Lipschitz-continuous boundary.* If $a_r \in C^k(\mathcal{M}_r)$ (or $C^{k,\mu}(\mathcal{M}_r)$, $\mu \in (0, 1)$) for all $r = 1, \ldots, R$, then we write $\partial\Omega \in C^k$ (or $\partial\Omega \in C^{k,\mu}$).

1.1.8 *Measure on the boundary of a domain*

If $\partial\Omega$ is Lipschitz-continuous, then it is possible to define an $(N-1)$-dimensional measure on $\partial\Omega$. Let us briefly describe its construction. It is known that the partial derivatives $\partial a_r/\partial x_j^r$, $j = 2, \ldots, N$, of the Lipschitz-continuous function $a_r : \mathcal{M}_r \to \mathbb{R}$ are defined almost everywhere on $\mathcal{M}_r$ and are measurable on $\mathcal{M}_r$. For any set $M \subset \Lambda_r$ let us construct the projection PM of M into $\mathcal{M}_r$. We say that the set M is measurable (on $\partial\Omega$), if PM is measurable with respect to the Lebesgue measure in $\mathbb{R}^{N-1}$, and define the measure of M as the Lebesgue integral

$$\tilde{\mu}(M) = \int_{\mathcal{M}_r} \chi(\hat{x}^r) \left[1 + \sum_{i=2}^{N} \left(\frac{\partial a_r(\hat{x}^r)}{\partial x_i^r}\right)^2\right]^{\frac{1}{2}} d\hat{x}^r, \tag{1.1.22}$$

where χ is the characteristic function of the set PM. (That is, $\chi = 1$ in PM and $\chi = 0$ outside PM.)

The decomposition (1.1.21) of $\partial\Omega$ implies the existence of disjoint measurable sets $M_r \subset \Lambda_r$ such that

$$\partial\Omega = \bigcup_{r=1}^{R} M_r. \tag{1.1.23}$$

Now we say that a set $M \subset \partial\Omega$ is measurable, if the sets $M \cap M_r$ are measurable and we define the measure of M by

$$\tilde{\mu}(M) = \sum_{r=1}^{R} \tilde{\mu}(M \cap M_r). \tag{1.1.24}$$

It is possible to prove that this definition is independent of the decomposition (1.1.23). The function $\tilde{\mu}$ is a σ-additive nonnegative measure.

We define the *surface integral* of a function f defined on a measurable subset M of $\partial\Omega$ as the integral induced by the measure $\tilde{\mu}$:

$$\int_M f(x)\,dS = \int f\chi\,d\tilde{\mu}, \tag{1.1.25}$$

where χ is the characteristic function of the set M.

If $\Omega \subset I\!R^2$ is a plane domain with a Lipschitz-continuous boundary $\partial\Omega$ formed by a piecewise smooth curve φ, then the integral (1.1.25) with $M = \partial\Omega$ is equal to the curvilinear integral.

In the sequel we shall use the notation $\text{meas}_{N-1} = \tilde{\mu}$ for the $(N-1)$-dimensional measure defined on the boundary $\partial\Omega$ of a domain $\Omega \subset I\!R^N$, respectively.

Let us note that the vector $\boldsymbol{n} = (n_1, \dots, n_N)$ with the components

$$\begin{aligned} n_1 &= -\frac{1}{b}, \\ n_i &= \frac{1}{b}\frac{\partial a_r}{\partial x_i^r}, \qquad i = 2, \dots, N, \end{aligned} \tag{1.1.26}$$

where

$$b = \left[1 + \sum_{i=2}^{N} \left(\frac{\partial a_r}{\partial x_i^r}\right)^2\right]^{\frac{1}{2}}, \tag{1.1.27}$$

is the vector of *unit outer normal* to $\partial\Omega$.

For more details see (Kufner *et al.*, 1977), Section 6.3.

1.1.9 *Green's theorem.*

Let $\Omega \subset I\!R^N$ be a bounded domain with a Lipschitz-continuous boundary and let u, $v \in C^1(\overline{\Omega})$. Then

$$\int_\Omega \frac{\partial u}{\partial x_i} v\,dx = \int_{\partial\Omega} uvn_i\,dS - \int_\Omega u\frac{\partial v}{\partial x_i}\,dx, \quad i = 1, \dots, N. \tag{1.1.28}$$

For proof, see, e.g. (Nečas, 1967), Theorem III.1.1. For $\boldsymbol{u} \in [C^1(\overline{\Omega})]^N$ we get from Green's theorem the identity

$$\int_\Omega \operatorname{div} \boldsymbol{u}\,dx = \int_{\partial\Omega} \boldsymbol{u}\cdot\boldsymbol{n}\,dS, \tag{1.1.29}$$

which is important in fluid dynamics. Later we shall state and use these formulae also for more general functions (see Section 1.3.3.2, Theorem 1.15).

1.1.10 *Lebesgue spaces*

Let $M \subset I\!R^N$ be a Lebesgue measurable set. We recall that we denote by $|M| = \text{meas}_N(M)$ the Lebesgue measure of this set. We say that two measurable functions defined on M are equivalent, if they differ at most on a set of zero Lebesgue measure. In such a situation we write $f_1 = f_2$ a.e. (almost everywhere) in M or $f_1(x) = f_2(x)$ for a.a. (almost all) $x \in M$. For $p \in [1, +\infty)$ the symbol $L^p(M)$ will denote the linear space of all (classes of equivalent) functions measurable on M for which

$$\int_M |u|^p \, dx < +\infty. \tag{1.1.30}$$

The space $L^p(M)$ equipped with the norm

$$\|u\|_{L^p(M)} = \left(\int_M |u|^p \, dx \right)^{1/p} \tag{1.1.31}$$

is a Banach space. Furthermore, we define the Banach space

$$\begin{aligned} L^\infty(M) = \{u;\ u \text{ is a measurable function on } M \text{ and } \|u\|_{L^\infty(M)} := \\ \operatorname{ess\,sup}_M |u| := \inf\{\sup_{x \in M \setminus Z} |u(x)|;\ Z \subset M,\ \operatorname{meas}_N(Z) = 0\} < \infty\}. \end{aligned} \tag{1.1.32}$$

We often use the notation $\|\cdot\|_{0,p,M}$ and $\|\cdot\|_{0,M}$ for the norm in $L^p(M)$ and in $L^2(M)$, respectively.

Similarly we define $L^p(M)$ for $M \subset \partial\Omega$, where $\Omega \subset I\!R^N$ is a domain with a Lipschitz-continuous boundary and M is measurable with respect to the $(N-1)$-dimensional measure $\tilde{\mu}$. In (1.1.30) and (1.1.31) we write $\int_M \ldots dS$ instead of $\int_M \ldots dx$ and in (1.1.32) we replace meas_N by $\operatorname{meas}_{N-1}$.

We say that $f \in L^p_{\mathrm{loc}}(M)$, if $f|_{M'} \in L^p(M')$ for any bounded $M' \subset M$ such that $\overline{M'} \subset M$.

1.1.10.1 *Basic properties of Lebesgue spaces* We shall list several important properties of spaces $L^p(M)$:

(i) If $1 \le p < \infty$, then $L^p(M)$ is a separable Banach space and if M is a domain, then $C_0^\infty(M)$ is dense in $L^p(M)$. On the contrary, the Banach space $L^\infty(M)$ is not separable.

(ii) If $1 \le p < \infty$, then to any $f \in [L^p(M)]^*$ (= dual of $L^p(M)$ – see 1.4.3) there exists a unique $u_f \in L^{p'}(M)$ which satisfies

$$\langle f, \varphi \rangle = \int_M u_f \varphi, \ \varphi \in L^p(M) \text{ and } \|f\|_{[L^p(M)]^*} = \|u_f\|_{L^{p'}(M)}.$$

(The symbol $\langle f, \varphi \rangle$ denotes the value of the functional f at the point φ.) Here p' is the so-called conjugate exponent, given by

$$\frac{1}{p'} + \frac{1}{p} = 1; \tag{1.1.33}$$

of course, if $p = 1$, then $p' = \infty$. Therefore, we identify $[L^p(M)]^*$ with $L^{p'}(M)$. For $p = p' = 2$ the above result is known as the Riesz theorem.

(iii) We have

$$[L^1(M)]^* \equiv L^\infty(M)$$

and

$$L^1(M) \subset [L^\infty(M)]^*.$$

(iv) If $1 < p < \infty$, then $L^p(M)$ is a *reflexive* Banach space. (Cf. 1.4.3.13.) Spaces $L^p(M)$ with $p = 1$ and $p = \infty$ are not reflexive.

(v) Let $1 \le p \le +\infty$, $f \in L^p(M)$ and $g \in L^{p'}(M)$. Then the Hölder inequality is valid:

$$\left| \int_M fg\,dx \right| \le \|f\|_{L^p(M)}\, \|g\|_{L^{p'}(M)}. \tag{1.1.34}$$

If $p = 2$, then (1.1.34) is called the *Cauchy inequality* – cf. 1.4.2.11.

(vi) The space $L^2(M)$ is a Hilbert space with the scalar product $\int_M fg dx$.

1.1.11 *One auxiliary result*

Lemma 1.1 *Let $\Omega \subset I\!R^N$ be an open set. Then we have:*

(i) If $f \in C^0(\Omega)$, then

$$\lim_{\mathrm{diam}(V(a))\to 0+} \frac{1}{|V(a)|} \int_{V(a)} f\,dx = f(a) \;\; \text{for all } a \in \Omega, \tag{1.1.35}$$

where $V(a) \subset \Omega$ denotes an open set containing a and

$$\mathrm{diam}(V(a)) := \sup\{|x - y|;\; x,\, y \in V(a)\}.$$

(ii) If $f \in C^0(\Omega)$, then $f = 0$ in Ω if and only if $\int_{\mathcal{V}} f\,dx = 0$ for any open and bounded set $\mathcal{V} \subset \overline{\mathcal{V}} \subset \Omega$.

Proof of assertion (ii) is easy and can be left to the reader. Proof of assertion (i) is a consequence of a more general result for which the reader is referred to (Rudin, 1974).

1.2 Governing equations and relations of gas dynamics

In this section, fundamental concepts, equations and relations of gas dynamics will be summarized. Details can be found in a number of books, such as, for example, (Feistauer, 1993), (Chorin and Marsden, 1993), (Shinbrot, 1973), (Landau and Lifschitz, 1959).

Let $(0,T) \subset I\!R$ be a time interval, during which we follow the fluid motion, and let $\Omega_t \subset I\!R^3$ denote the domain occupied by the fluid at time $t \in (0, T)$.

In our considerations we use the *fundamental hypothesis* that

exactly one fluid particle passes through each point $x \in \Omega_t$ at any time t. (1.2.1)

(We assume that the set $\mathcal{M}$ defined by (1.1.1) is open.)

1.2.1 *Description of the flow*

There are two possibilities for describing the fluid motion, as follows.

a) The *Lagrangian description* of the flow considers the motion of each individual fluid particle. The trajectories of the particles can be described by the equation

$$x = \varphi(X, t) \tag{1.2.2}$$

(i.e. $x_i = \varphi_i(X, t),\ i = 1, 2, 3$). X represents the *reference* determining the particle under consideration. Sometimes we use a more detailed description of the motion of fluid particles in the form

$$x = \varphi(X, t_0;\ t) \tag{1.2.3}$$

which determines, at time t, the position x of the particle passing through the point (given by the reference) X at time t_0. Then, of course,

$$X = \varphi(X, t_0;\ t_0),$$

provided the references are identical with the coordinates of particles at time t_0.

The components X_1, X_2, X_3 of the reference X are called *Lagrangian coordinates* in contrast to the *Eulerian coordinates* x_1, x_2, x_3.

The Lagrangian description in the form (1.2.3) is used, for example, if we study the flow of a piece of fluid formed by the same particles at each time instant and filling a domain $\mathcal{V}(t) \subset I\!R^3$ at time t.

The *velocity* and the *acceleration* of the fluid particle given by the reference X are defined as

$$\text{a)}\qquad \hat{\boldsymbol{v}}(X,t) = \frac{\partial \varphi}{\partial t}(X,t) \qquad \left(= \frac{\partial \varphi}{\partial t}(X, t_0;\ t)\right) \tag{1.2.4}$$

and

$$\text{b)}\qquad \hat{\boldsymbol{a}}(X,t) = \frac{\partial^2 \varphi}{\partial t^2}(X,t) \qquad \left(= \frac{\partial^2 \varphi}{\partial t^2}(X, t_0;\ t)\right),$$

respectively, provided the above derivatives exist.

b) The *Eulerian description* is based on the determination of the *velocity* $\boldsymbol{v}(x, t)$ of the fluid particle passing through the point x at time t. With respect to (1.2.2) and (1.2.4) we can write

$$\boldsymbol{v}\,(x,t) = \hat{\boldsymbol{v}}\,(X,t) = \frac{\partial \varphi}{\partial t}\,(X,t) \quad \text{where } x = \varphi(X,t). \tag{1.2.5}$$

Let us note that in our later considerations the velocity will also be denoted by $\boldsymbol{u}$.

Under the assumption that

$$\boldsymbol{v} \in \left[C^1(\mathcal{M})\right]^3, \tag{1.2.6}$$

the *acceleration* of the particle passing through the point x at time t is expressed as

$$\boldsymbol{a}(x,t) = \frac{\partial \boldsymbol{v}}{\partial t}(x,t) + \sum_{i=1}^{3} v_i(x,t)\frac{\partial \boldsymbol{v}}{\partial x_i}(x,t), \tag{1.2.7}$$

which can also be written in the form

$$\boldsymbol{a} = \frac{\partial \boldsymbol{v}}{\partial t} + (\boldsymbol{v}\cdot \mathrm{grad})\,\boldsymbol{v} = \frac{\partial \boldsymbol{v}}{\partial t} + (\boldsymbol{v}\cdot\nabla)\,\boldsymbol{v}. \tag{1.2.8}$$

(For simplicity we omit the variables x and t.)

Let us introduce the symbol

$$\frac{\mathrm{D}}{\mathrm{D}t} = \frac{\partial}{\partial t} + \boldsymbol{v}\cdot \mathrm{grad} = \frac{\partial}{\partial t} + \boldsymbol{v}\cdot\nabla \tag{1.2.9}$$

called the *material* (or total) *derivative* with respect to time. The partial derivative $\partial/\partial t$ is called the *local derivative* and the term $(\boldsymbol{v}\cdot\mathrm{grad})$ is referred to as the *convective derivative.* We see that the acceleration of a fluid particle is expressed in Eulerian coordinates as the material derivative of the velocity:

$$\boldsymbol{a} = \frac{\mathrm{D}\boldsymbol{v}}{\mathrm{D}t} := \frac{\partial \boldsymbol{v}}{\partial t} + (\boldsymbol{v}\cdot \mathrm{grad})\,\boldsymbol{v}. \tag{1.2.10}$$

1.2.1.1 *The transition from the Eulerian description to the Lagrangian description* This is equivalent to the determination of the paths of fluid particles on the basis of a given velocity field $\boldsymbol{v}(x,t)$. The trajectory of the fluid particle passing through a point $X \in \Omega_{t_0}$ at time $t_0 \in (0,T)$ is given as the solution of the initial value problem

$$\frac{dx}{dt} = \boldsymbol{v}(x,t), \qquad x(t_0) = X. \tag{1.2.11}$$

Theorems 10.1.1, 11.1.5 and 13.1.1 from (Kurzweil, 1986) immediately imply the following:

Theorem 1.2 *Under assumption* (1.2.6) *the following statements hold:*

1. For each $(X,t_0) \in \mathcal{M}$ *problem* (1.2.11) *has exactly one maximal solution* $\varphi(X,t_0;t)$ *(defined for* t *from a certain interval* $(\alpha_{X,t_0}, \beta_{X,t_0})$*).*

2. The mapping φ *has continuous first order partial derivatives with respect to* X_1, X_2, X_3, t_0, t *and continuous derivatives* $\partial^2\varphi/\partial t\partial X_i$, $\partial^2\varphi/\partial t_0\partial X_i$, $i = 1,2,3$, *in its domain of definition* $\{(X,t_0,t);\ (X,t_0)\in\mathcal{M},\ t\in(\alpha_{X,t_0},\beta_{X,t_0})\}$.

Let us recall that a solution of problem (1.2.11) is called maximal if any solution of the problem is its restriction.

1.2.1.2 *Derivatives along trajectories and streamlines* If $F \in C^1(\mathcal{M})$ represents some physical quantity transported by moving particles, then the function $\hat{F}(X,t) = F(\boldsymbol{\varphi}(X,t),t)$ represents the values of this quantity along the trajectory with equation $x = \boldsymbol{\varphi}(X,t)$. The *derivative* of this quantity *along the trajectory* is expressed with the aid of the chain rule in the form

$$\begin{aligned} \frac{\partial \hat{F}(X,t)}{\partial t} &= \frac{\partial F(\boldsymbol{\varphi}(X,t),t)}{\partial t} \\ &= \frac{\partial F}{\partial t}(\boldsymbol{\varphi}(X,t),t) + \frac{\partial \boldsymbol{\varphi}}{\partial t} \cdot \nabla F(\boldsymbol{\varphi}(X,t),t) \\ &= \frac{\partial F}{\partial t}(x,t) + \boldsymbol{v}(x,t) \cdot \nabla F(x,t) \\ &= \frac{\mathrm{D}F(x,t)}{\mathrm{D}t}. \end{aligned} \tag{1.2.12}$$

Therefore, the material derivative is sometimes called the derivative along the trajectory of a fluid particle.

A useful concept which helps us to make the description of the flow more vivid are streamlines. For a fixed t we define the *streamline* passing through a point $y \in \Omega_t$ as the curve in Ω_t with equation $x = x(\vartheta)$ determined by the conditions

$$\frac{dx}{d\vartheta}(\vartheta) = \boldsymbol{v}(x(\vartheta),t), \quad x(0) = y. \tag{1.2.13}$$

Here the parameter ϑ does not mean time. If $F \in C^1(\Omega_t)$, then by the chain rule we have

$$\begin{aligned} \frac{\partial}{\partial \vartheta} F(x(\vartheta),t) &= \nabla F(x(\vartheta),t) \cdot \frac{dx}{d\vartheta}(\vartheta) \\ &= \nabla F(x(\vartheta),t) \cdot \boldsymbol{v}(x(\vartheta),t). \end{aligned} \tag{1.2.14}$$

The expression

$$F_{\boldsymbol{v}}(x,t) = \nabla F(x,t) \cdot \boldsymbol{v}(x,t) \tag{1.2.15}$$

is called the *derivative* of the function F *along the streamline* at a point $x \in \Omega_t$ and time t. The streamlines are identical to the trajectories in the case of a steady (stationary) flow, when quantities describing the flow are independent of time. For general nonstationary flow the streamlines and trajectories are different.

1.2.2 *The transport theorem*

Let a function $F = F(x,t) : \mathcal{M} \to \mathbb{R}$ be the Eulerian representation of some physical quantity transported by fluid particles and let us consider a system of fluid particles filling a *bounded domain* $\mathcal{V}(t) \subset \Omega_t$ at time t. The total amount of the quantity given by the function F that is contained in the volume $\mathcal{V}(t)$ at time t equals the integral

$$\mathcal{F}(t) = \int_{\mathcal{V}(t)} F(x,t)\, dx. \tag{1.2.16}$$

In what follows we shall need to calculate the rate of change of the quantity F bound on the system of particles considered. In other words, we shall be interested in the derivative

$$\frac{d\mathcal{F}(t)}{dt} = \frac{d}{dt} \int_{\mathcal{V}(t)} F(x,t)\, dx. \tag{1.2.17}$$

Let us suppose that $F \in C^1(\mathcal{M})$ and $\boldsymbol{v} \in [C^1(\mathcal{M})]^3$ and let $\varphi = \varphi(X, t_0;\, t)$ be the mapping from Theorem 1.2. This mapping defines the change of the domain $\mathcal{V}(t)$ with time. Let $t_0 \in (0, T)$ be an arbitrary fixed time instant and $\mathcal{V}(t_0) \subset \Omega_{t_0}$. Then

$$\mathcal{V}(t) = \{\varphi(X, t_0;\, t);\ X \in \mathcal{V}(t_0)\} \tag{1.2.18}$$

(provided $\varphi(X, t_0;\, t)$ is defined for all $X \in \mathcal{V}(t_0)$). By $J(X,t)$ we shall denote the Jacobian of the mapping '$X \in \mathcal{V}(t_0) \longrightarrow \varphi(X, t_0;\, t) \in \mathcal{V}(t)$':

$$J(X,t) = \det \frac{D\varphi(X, t_0;\, t)}{DX} = \det \begin{pmatrix} \dfrac{\partial \varphi_1}{\partial X_1}, & \dfrac{\partial \varphi_1}{\partial X_2}, & \dfrac{\partial \varphi_1}{\partial X_3} \\ \dfrac{\partial \varphi_2}{\partial X_1}, & \dfrac{\partial \varphi_2}{\partial X_2}, & \dfrac{\partial \varphi_2}{\partial X_3} \\ \dfrac{\partial \varphi_3}{\partial X_1}, & \dfrac{\partial \varphi_3}{\partial X_2}, & \dfrac{\partial \varphi_3}{\partial X_3} \end{pmatrix} (X,\, t_0;\, t). \tag{1.2.19}$$

It is possible to prove the following technical result (see, for example, (Feistauer, 1993)).

Lemma 1.3 *Let $t_0 \in (0, T)$, $\mathcal{V}(t_0)$ be a bounded domain and let $\overline{\mathcal{V}(t_0)} \subset \Omega_{t_0}$. Then there exists an interval $(t_1, t_2) \ni t_0$ such that the following conditions are satisfied:*

a) The mapping '$t \in (t_1, t_2),\ X \in \mathcal{V}(t_0) \longrightarrow x = \varphi(X, t_0;\, t) \in \mathcal{V}(t)$' has continuous first order derivatives with respect to t, X_1, X_2, X_3 and continuous second order derivatives $\partial^2\varphi/\partial t\, \partial X_i$, $i = 1, 2, 3$.

b) The mapping '$X \in \mathcal{V}(t_0) \longrightarrow x = \varphi(X, t_0;\, t) \in \mathcal{V}(t)$' is a continuously differentiable one-to-one mapping of $\mathcal{V}(t_0)$ onto $\mathcal{V}(t)$ with the Jacobian (1.2.19) *which is continuous and bounded and satisfies the condition*

$$J(X,t) > 0 \qquad \forall\, X \in \mathcal{V}(t_0),\ \forall\, t \in (t_1, t_2).$$

c) The inclusion

$$\left\{(x,t);\ t \in [t_1, t_2],\ x \in \overline{\mathcal{V}(t)}\right\} \subset \mathcal{M}$$

holds and thus the mapping $\boldsymbol{v}$ has continuous and bounded first order derivatives on $\{(x,t);\ t \in (t_1, t_2),\ x \in \mathcal{V}(t)\}$.

d) $\boldsymbol{v}(\varphi(X, t_0;\, t),\, t) = \dfrac{\partial \varphi}{\partial t}(X, t_0;\, t)\ \forall\, X \in \mathcal{V}(t_0),\ \forall\, t \in (t_1, t_2)$.

The following lemma plays an important role in fluid dynamics.

Lemma 1.4 *Let conditions a)–d) from Lemma 1.3 be satisfied. Then the function $J = J(X,t)$ has a continuous and bounded partial derivative $\partial J/\partial t$ for $X \in \mathcal{V}(t_0)$, $t \in (t_1, t_2)$, and*

$$\frac{\partial J}{\partial t}(X,t) = J(X,t)\,\mathrm{div}\,\boldsymbol{v}(x,t), \qquad x = \varphi(X, t_0;\, t). \tag{1.2.20}$$

Proof can be found, for example, in (Feistauer, 1993), (Chorin and Marsden, 1993) or (Shinbrot, 1973).

Now we can prove the so-called *transport theorem* on the derivative of integral (1.2.16):

Theorem 1.5 *Let conditions a)–d) from Lemma 1.3 be satisfied and let the function $F = F(x,t)$ have continuous and bounded first order derivatives on the set $\{(x,t); t \in (t_1,t_2), x \in \mathcal{V}(t)\}$. Then for each $t \in (t_1,t_2)$ there exists a finite derivative*

$$\begin{aligned}\frac{d\mathcal{F}}{dt}(t) &= \frac{d}{dt}\int_{\mathcal{V}(t)} F(x,t)\,dx \\ &= \int_{\mathcal{V}(t)} \left[\frac{\partial F}{\partial t}(x,t) + \boldsymbol{v}(x,t)\cdot \mathrm{grad}\,F(x,t) + F(x,t)\,\mathrm{div}\,\boldsymbol{v}(x,t)\right] dx \\ &= \int_{\mathcal{V}(t)} \left[\frac{\partial F}{\partial t}(x,t) + \mathrm{div}(F\boldsymbol{v})(x,t)\right] dx.\end{aligned} \tag{1.2.21}$$

Proof By the substitution theorem, the integral $\mathcal{F}(t)$ can be written in the form

$$\mathcal{F}(t) = \int_{\mathcal{V}(t_0)} F(\varphi(X,t_0;t),\, t)\, J(X,t)\, dX.$$

Since t_0 is fixed and the integration domain $\mathcal{V}(t_0)$ does not depend on time t, we can apply the theorem on differentiation of an integral with respect to a parameter:

$$\begin{aligned}\frac{d\mathcal{F}}{dt}(t) = \int_{\mathcal{V}(t_0)} &\left[\left(\frac{\partial F}{\partial t}(\varphi(X,t_0;t),\, t) + \sum_{i=1}^{3} \frac{\partial F}{\partial x_i}(\varphi(X,t_0;t),\, t)\frac{\partial \varphi_i}{\partial t}(X,t_0;\, t)\right)\right. \\ &\left.\cdot J(X,t) + F(\varphi(X,t_0;\, t),\, t)\frac{\partial J}{\partial t}(X,t)\right] dX.\end{aligned}$$

The assumptions considered guarantee the correctness of the differentiation under the integral sign. In view of Lemma 1.4 and relation d) from Lemma 1.3, we get the identity

$$\frac{d\mathcal{F}}{dt}(t) = \int_{\mathcal{V}(t_0)} \left[\frac{\partial F}{\partial t}(\varphi(X,t_0;t),t) + \sum_{i=1}^{3}\frac{\partial F}{\partial x_i}(\varphi(X,t_0;t),t) v_i(\varphi(X,t_0;t),t)\right.$$

$$+ F(\varphi(X,t_0;t),t)\,\mathrm{div}\,\boldsymbol{v}(\varphi(X,t_0;t),t)\Big]\, J(X,t)\,dX.$$

If we use the inverse substitution, transforming the integral over $\mathcal{V}(t_0)$ onto the integral over $\mathcal{V}(t)$, we immediately obtain relation (1.2.21). □

In what follows, we shall introduce the mathematical formulation of fundamental physical laws: the law of conservation of mass, the law of conservation of momentum and the law of conservation of energy, called in brief *conservation laws* from which we will derive the fundamental differential equations of fluid dynamics: the continuity equation, the equations of motion and the energy equation.

1.2.3 *The continuity equation*

The *density of fluid* is a function

$$\rho : \mathcal{M} = \{(x,t);\ t \in (0,T),\ x \in \Omega_t\} \to (0,+\infty)$$

which allows us to determine the mass $m(\mathcal{V};t)$ of the fluid contained in any subdomain $\mathcal{V} \subset \Omega_t$:

$$m(\mathcal{V};t) = \int_{\mathcal{V}} \rho(x,t)\,dx. \tag{1.2.22}$$

Let $\rho \in C^1(\mathcal{M})$ and $\boldsymbol{v} \in [C^1(\mathcal{M})]^3$. Let us consider an arbitrary time instant $t_0 \in (0,T)$ and a moving piece of the fluid formed by the same particles at each instant and filling at time t_0 a bounded domain $\mathcal{V} \subset \overline{\mathcal{V}} \subset \Omega_{t_0}$, called a *control volume.* By $\mathcal{V}(t)$ we denote the domain occupied by this piece of fluid at time $t \in (t_1,t_2)$, where (t_1,t_2) is a sufficiently small time interval containing t_0 with properties from Lemma 1.3. Hence, $\mathcal{V}(t_0) = \mathcal{V}$ and conditions a)–d) from Lemma 1.3 are satisfied.

Since the domain $\mathcal{V}(t)$ is formed by the same particles at each time instant, the *conservation of mass* can be formulated in the following way:

The mass of the piece of fluid represented by the domain $\mathcal{V}(t)$ does not depend on time t.

This means that

$$\frac{dm(\mathcal{V}(t);t)}{dt} = 0, \qquad t \in (t_1,t_2). \tag{1.2.23}$$

With respect to (1.2.22),

$$m(\mathcal{V}(t);t) = \int_{\mathcal{V}(t)} \rho(x,t)\,dx. \tag{1.2.24}$$

Using the transport theorem 1.5, whose assumptions are satisfied for the function $F = \rho$, from (1.2.23) and (1.2.24), we get the identity

$$\int_{\mathcal{V}(t)} \left[\frac{\partial \rho}{\partial t}(x,t) + \mathrm{div}(\rho \boldsymbol{v})(x,t) \right] dx = 0, \quad t \in (t_1, t_2).$$

Now, if we substitute $t := t_0$ and take into account that $\mathcal{V}(t_0) = \mathcal{V}$, we conclude that

$$\int_{\mathcal{V}} \left[\frac{\partial \rho}{\partial t}(x,t_0) + \mathrm{div}(\rho \boldsymbol{v})(x,t_0) \right] dx = 0 \tag{1.2.25}$$

for an arbitrary $t_0 \in (0,T)$ and an arbitrary control volume $\mathcal{V}$ in Ω_{t_0}.

Using the continuity of the integrand in (1.2.25) and assertion ii) of Lemma 1.1 and writing t instead of t_0, we conclude that

$$\frac{\partial \rho}{\partial t}(x,t) + \mathrm{div}\,(\rho(x,t)\boldsymbol{v}(x,t)) = 0, \qquad t \in (0,T),\ x \in \Omega_t. \tag{1.2.26}$$

This equation is the differential form of the law of conservation of mass and is called the *continuity equation.*

1.2.4 *The equations of motion*

Basic dynamical equations describing fluid motion will be derived from the *law of conservation of momentum* which can be formulated in the following way:

The rate of change of the total momentum of a piece of fluid formed by the same particles at each time and occupying the domain $\mathcal{V}(t)$ at instant t is equal to the force acting on $\mathcal{V}(t)$.

Let $\rho \in C^1(\mathcal{M}), \boldsymbol{v} \in [C^1(\mathcal{M})]^3$. The total momentum of particles contained in $\mathcal{V}(t)$ is given by

$$\boldsymbol{\mathcal{H}}(\mathcal{V}(t)) = \int_{\mathcal{V}(t)} \rho(x,t)\boldsymbol{v}(x,t)\,dx. \tag{1.2.27}$$

Moreover, denoting by $\boldsymbol{\mathcal{F}}(\mathcal{V}(t))$ the force acting on the volume $\mathcal{V}(t)$, the law of conservation of momentum reads

$$\frac{d\boldsymbol{\mathcal{H}}(\mathcal{V}(t))}{dt} = \boldsymbol{\mathcal{F}}(\mathcal{V}(t)), \qquad t \in (t_1, t_2). \tag{1.2.28}$$

Using the transport theorem, we get

$$\int_{\mathcal{V}(t)} \left[\frac{\partial}{\partial t}(\rho(x,t)v_i(x,t)) + \mathrm{div}\,(\rho(x,t)v_i(x,t)\boldsymbol{v}(x,t)) \right] dx = \mathcal{F}_i(\mathcal{V}(t)),$$
$$i = 1,2,3,\ t \in (t_1, t_2).$$

Now, taking into account that $t_0 \in (0,T)$ is an arbitrary time instant and $\mathcal{V}(t_0) = \mathcal{V} \subset \overline{\mathcal{V}} \subset \Omega_{t_0}$, where $\mathcal{V}$ is an arbitrary control volume, we get the law of conservation of momentum in the form where we write t instead of t_0:

$$\int_{\mathcal{V}} \left[\frac{\partial}{\partial t}(\rho(x,t)v_i(x,t)) + \mathrm{div}\,(\rho(x,t)v_i(x,t)\boldsymbol{v}(x,t)) \right] dx = \mathcal{F}_i(\mathcal{V};t), \tag{1.2.29}$$

$$i = 1, 2, 3, \text{ for an arbitrary } t \in (0, T)$$
$$\text{and an arbitrary control volume } \mathcal{V} \text{ in } \Omega_t.$$

The vector $\boldsymbol{\mathcal{F}}(\mathcal{V}; t)$ with components $\mathcal{F}_i(\mathcal{V}; t)$ denotes the force acting on the volume $\mathcal{V}$ at time t.

Since we wish to rewrite (1.2.29) as a differential equation, it is necessary to specify the character of the vector $\boldsymbol{\mathcal{F}}(\mathcal{V}; t)$.

We distinguish two types of forces acting in fluids, the so-called volume forces and surface forces.

a) The *volume force* (also called outer or body force) $\boldsymbol{\mathcal{F}}_v(\mathcal{V}; t)$ acting at time t on the particles contained in a control volume $\mathcal{V} \subset \overline{\mathcal{V}} \subset \Omega_t$ is expressed by its density (related to the unit of mass) $\boldsymbol{f} \in C(\mathcal{M})^3$:

$$\boldsymbol{\mathcal{F}}_v(\mathcal{V}; t) = \int_{\mathcal{V}} \rho(x, t)\, \boldsymbol{f}(x, t)\, dx. \tag{1.2.30}$$

b) The *surface force* (or inner force) $\boldsymbol{F}_S$, by which the fluid contained outside the domain $\mathcal{V}$ acts on a set $S \subset \partial\mathcal{V}$, is expressed with the use of the *stress vector* $\boldsymbol{T}(x, t, \boldsymbol{n})$ characterizing the density of the surface force:

$$\boldsymbol{F}_S = \int_S \boldsymbol{T}(x, t, \boldsymbol{n}(x))\, dS. \tag{1.2.31}$$

Here $\boldsymbol{n}(x)$ is the unit outer normal to $\partial\mathcal{V}$ at x. We shall assume that $\boldsymbol{T} \in [C(\mathcal{M} \times \mathcal{S}_1)]^3$, where $\mathcal{S}_1$ is the surface of the unit sphere with centre at the origin. Then the total surface force acting at time t on the control volume $\mathcal{V}$ from outside has the form

$$\boldsymbol{\mathcal{F}}_s(\mathcal{V}; t) = \int_{\partial\mathcal{V}} \boldsymbol{T}(x, t, \boldsymbol{n}(x))\, dS. \tag{1.2.32}$$

The stress vector $\boldsymbol{T}(x, t, \boldsymbol{n})$ can be expressed with the aid of some of its values for certain normals. Let us choose the normals parallel to the coordinate axes and set

$$\tau_{ji} = T_i(x, t, \boldsymbol{e}_j), \qquad i, j = 1, 2, 3, \tag{1.2.33}$$
$$\boldsymbol{e}_1 = (1, 0, 0), \quad \boldsymbol{e}_2 = (0, 1, 0), \quad \boldsymbol{e}_3 = (0, 0, 1).$$

The quantities $\tau_{ji} = \tau_{ji}(x, t)$, $i, j = 1, 2, 3$, are called the *components of the stress tensor*

$$\mathcal{T} = \begin{pmatrix} \tau_{11}, \tau_{12}, \tau_{13} \\ \tau_{21}, \tau_{22}, \tau_{23} \\ \tau_{31}, \tau_{32}, \tau_{33} \end{pmatrix}. \tag{1.2.34}$$

Then

$$T_i(x, t, \boldsymbol{n}) = \sum_{j=1}^{3} n_j \tau_{ji}(x, t), \quad i = 1, 2, 3. \tag{1.2.35}$$

(See, for example, (Feistauer, 1993).)

1.2.5 *The equations of motion of general fluids*

Let us assume that ρ, v_i, $\tau_{ij} \in C^1(\mathcal{M})$ and $f_i \in C(\mathcal{M})$ $(i, j = 1,2,3)$. Expressing the total force acting on the fluid contained in a control volume $\mathcal{V}$ and substituting in (1.2.29), we obtain

$$\int_{\mathcal{V}} \left[\frac{\partial}{\partial t}(\rho(x,t)v_i(x,t)) + \operatorname{div}(\rho(x,t)v_i(x,t)\boldsymbol{v}(x,t))\right] dx \tag{1.2.36}$$

$$= \int_{\mathcal{V}} \rho(x,t) f_i(x,t)\, dx + \int_{\partial\mathcal{V}} \sum_{j=1}^{3} \tau_{ji}(x,t)\, n_j(x)\, dS, \quad i = 1,2,3,$$

for each $t \in (0,T)$ and an arbitrary control volume $\mathcal{V}$ in Ω_t.

Moreover, applying Green's theorem from Paragraph 1.1.9 and Lemma 1.1, we get the *equations of motion of a general fluid in differential conservative form*:

$$\frac{\partial}{\partial t}(\rho v_i) + \operatorname{div}(\rho v_i \boldsymbol{v}) = \rho f_i + \sum_{j=1}^{3} \frac{\partial \tau_{ji}}{\partial x_j}, \qquad i = 1,2,3. \tag{1.2.37}$$

This can be written as

$$\frac{\partial}{\partial t}(\rho\boldsymbol{v}) + \operatorname{div}(\rho\boldsymbol{v}\otimes\boldsymbol{v}) = \rho\boldsymbol{f} + \operatorname{div}\mathcal{T}. \tag{1.2.38}$$

1.2.6 *The law of conservation of the moment of momentum; symmetry of the stress tensor*

Let us assume that ρ, v_i, $\tau_{ji} \in C^1(\mathcal{M})$ and $f_i \in C(\mathcal{M})$. Similarly as above, consider a control volume $\mathcal{V} = \mathcal{V}(t)$ formed by the same fluid particles at each time instant $t \in (t_1, t_2)$. The law of conservation of the moment of momentum can be formulated in the following way:

The rate of change of the moment of momentum of the piece of fluid occupying the volume $\mathcal{V}(t)$ at any time t is equal to the sum of the moments of the volume and surface forces acting on this volume.

Hence,

$$\frac{d}{dt}\int_{\mathcal{V}(t)} x \times (\rho\boldsymbol{v})(x,t)\, dx \tag{1.2.39}$$

$$= \int_{\mathcal{V}(t)} x \times (\rho\boldsymbol{f})(x,t)\, dx + \int_{\partial\mathcal{V}(t)} x \times \boldsymbol{T}(x,t,\boldsymbol{n}(x))\,(x)\, dS.$$

It is possible to prove the following important result ((Feistauer, 1993)):

Theorem 1.6 *The law of conservation of the moment of momentum* (1.2.39) *is valid if and only if the stress tensor $\mathcal{T}$ is symmetric.*

1.2.7 *The Navier–Stokes equations*

The relations between the stress tensor and other quantities describing fluid flow, particularly the velocity and its derivatives, represent the so-called *rheological equations* of the fluid. The simplest rheological equation

$$\mathcal{T} = -p\,\mathbb{I}, \tag{1.2.40}$$

characterizes inviscid fluid. Here p is the pressure and $\mathbb{I}$ is the unit tensor:

$$\mathbb{I} = \begin{pmatrix} 1, 0, 0 \\ 0, 1, 0 \\ 0, 0, 1 \end{pmatrix}. \tag{1.2.41}$$

Besides the pressure forces, the friction shear forces also act in real fluids as a consequence of the *viscosity*. Therefore, in the case of viscous fluid, we add a contribution $\mathcal{T}'$ characterizing the shear stress to the term $-p\,\mathbb{I}$:

$$\mathcal{T} = -p\,\mathbb{I} + \mathcal{T}'. \tag{1.2.42}$$

In order to identify the viscous part $\mathcal{T}'$ of the stress tensor, we shall use *Stokes' postulates:*

1) $\mathcal{T} = -p\,\mathbb{I} + \mathcal{T}'$.
2) The tensor $\mathcal{T}'$ is a continuous function of the deformation velocity tensor,

$$\mathbb{D} = \mathbb{D}(\boldsymbol{v}) = (d_{ij})_{i,j=1}^{3}, \quad d_{ij} = \frac{1}{2}\left(\frac{\partial v_i}{\partial x_j} + \frac{\partial v_j}{\partial x_i}\right), \tag{1.2.43}$$

is independent of other kinematic variables and does not explicitly depend on the position in the fluid and on time either.
3) A fluid is an *isotropic* medium. This means that its properties are the same in all space directions.
4) If the deformation velocity tensor is zero, only the pressure force acts in the fluid. Hence, if $\mathbb{D} = 0$, then $\mathcal{T} = -p\,\mathbb{I}$.
5) The relation between $\mathcal{T}'$ and $\mathbb{D}$ is linear.

In mathematical language the above postulates can be formulated as follows:

1*) $\mathcal{T} = -p\,\mathbb{I} + \mathcal{T}'$.
2*) $\mathcal{T}' = f(\mathbb{D})$, f is continuous.
3*) The form of the mapping f is invariant with respect to the transformation of the Cartesian coordinate system: $\mathbb{S}\mathcal{T}'\mathbb{S}^{-1} = f(\mathbb{S}\,\mathbb{D}\,\mathbb{S}^{-1})$ for any orthonormal matrix $\mathbb{S}$.
4*) $f(0) = 0$.
5*) The mapping f is linear.

Then it is possible to show that the following representation holds true ((Feistauer, 1993)):

Theorem 1.7 *Under the above conditions 1*)–5*), the stress tensor has the form*

$$\mathcal{T} = (-p + \lambda \operatorname{div} \boldsymbol{v})\, \mathbb{I} + 2\mu\mathbb{D}(\boldsymbol{v}), \tag{1.2.44}$$

where λ, μ are constants or scalar functions of thermodynamical quantities.

If the stress tensor depends linearly on the velocity deformation tensor as in (1.2.44), the fluid is called *Newtonian*, which is the case of gases.

Let us assume that $\rho \in C^1(\mathcal{M})$ and $\partial \boldsymbol{v}/\partial t$ and $\partial^2 \boldsymbol{v}/\partial x_i \partial x_j \in C(\mathcal{M})$ $(i, j = 1, 2, 3)$ and substitute relation (1.2.44) into the general equations of motion (1.2.38). We get the so-called *Navier–Stokes equations*

$$\begin{aligned} &\frac{\partial(\rho \boldsymbol{v})}{\partial t} + \operatorname{div}(\rho \boldsymbol{v} \otimes \boldsymbol{v}) \\ &= \rho \boldsymbol{f} - \operatorname{grad} p + \operatorname{grad}(\lambda \operatorname{div} \boldsymbol{v}) + \operatorname{div}(2\mu\mathbb{D}(\boldsymbol{v})). \end{aligned} \tag{1.2.45}$$

1.2.8 *Properties of the viscosity coefficients*

Here μ and λ are called the first and the second *viscosity coefficients*, respectively, μ is also called *dynamical viscosity*. In the kinetic theory of gases the conditions

$$\mu \geq 0, \qquad 3\lambda + 2\mu \geq 0, \tag{1.2.46}$$

are derived. For monoatomic gases, $3\lambda + 2\mu = 0$. This condition is usually used even in the case of more complicated gases.

The viscosity coefficients can be functions of thermodynamical quantities. The most evident is the dependence on the absolute temperature θ. Often Sutherland's formula

$$\mu = \frac{c_1 \theta^{3/2}}{\theta + c_2} \tag{1.2.47}$$

derived in the kinetic theory of gases is used. (Here c_1 and c_2 are constants depending on the fluid.) Other relations and tables characterizing the dependence of μ on θ can be found, for example, in (Loitsianskii, 1973).

1.2.9 *The Reynolds number*

In the investigation of viscous flow the characteristic dimensionless *Reynolds number* plays an important role. It is defined as

$$Re = \frac{U^* L^* \rho^*}{\mu^*}, \tag{1.2.48}$$

where U^* is the characteristic velocity, L^* the characteristic length, ρ^* the characteristic density and μ^* the characteristic viscosity. The character of flow is different for small Reynolds numbers (laminar flow) and large Reynolds numbers (turbulent flow). The Reynolds number plays an important role in the similarity of flows. See, Section 1.2.23.

1.2.10 *Various forms of the Navier–Stokes equations*

For later use let us reformulate the Navier–Stokes equations (1.2.45), provided ρ and $\boldsymbol{v}$ are sufficiently regular and satisfy the continuity equation (1.2.26), and the viscosity coefficients μ and λ are constants.

First, taking into account (1.2.43) and putting $N = 3$, system (1.2.45) can be written in the form

$$(\rho\boldsymbol{v})_t + \operatorname{div}(\rho\boldsymbol{v} \otimes \boldsymbol{v}) \\ = \rho\boldsymbol{f} - \nabla p + \sum_{j=1}^{N} \mu(\boldsymbol{v}_{x_j} + \nabla v_j)_{x_j} + \lambda\nabla\operatorname{div}\boldsymbol{v}. \tag{1.2.49}$$

Further, we have

$$\sum_{j=1}^{N} \frac{\partial}{\partial x_j}(\nabla v_j) = \nabla\operatorname{div}\boldsymbol{v}, \quad \sum_{j=1}^{N} \frac{\partial}{\partial x_j}\boldsymbol{v}_{x_j} = \Delta\boldsymbol{v},$$

and (1.2.49) reads

$$(\rho\boldsymbol{v})_t + \operatorname{div}(\rho\boldsymbol{v} \otimes \boldsymbol{v}) = \rho\boldsymbol{f} - \nabla p + \mu\Delta\boldsymbol{v} + (\mu + \lambda)\nabla\operatorname{div}\boldsymbol{v}, \tag{1.2.50}$$

where, in view of (1.2.46),

$$\mu \geq 0, \quad \mu + \lambda \geq \mu/3 \geq 0. \tag{1.2.51}$$

Here μ is the *dynamical viscosity coefficient* and $\eta = \mu + \lambda$ is the so-called *bulk viscosity*. If we set $\lambda = -2\mu/3$, then $\eta = \mu/3$. In viscous gases,

$$\mu > 0, \quad \eta > 0. \tag{1.2.52}$$

Finally, using the continuity equation (1.2.26), we can express the Navier–Stokes equations in the *nonconservative form*

$$\rho(\boldsymbol{v}_t + (\boldsymbol{v} \cdot \nabla)\boldsymbol{v}) = \rho\boldsymbol{f} - \nabla p + \mu\Delta\boldsymbol{v} + (\mu + \lambda)\nabla\operatorname{div}\boldsymbol{v}. \tag{1.2.53}$$

1.2.11 *The energy equation*

Now we derive the energy equation representing the law of conservation of energy. Let us recall that the power of the force $\boldsymbol{F}$ acting on a particle passing through the point x at time t is

$$W(x,t) = \boldsymbol{F}(x,t) \cdot \boldsymbol{v}(x,t). \tag{1.2.54}$$

As in the preceding sections, we consider a piece of fluid represented by a control volume $\mathcal{V}(t)$ satisfying assumptions from Section 1.2.3. The law of conservation of energy can be formulated as follows:

The rate of change of the total energy of the fluid particles occupying the domain $\mathcal{V}(t)$ at time tis equal to the sum of powers of the volume force acting *on the volume $\mathcal{V}(t)$ and the surface force acting on the surface $\partial\mathcal{V}(t)$, and of the amount of heat transmitted to $\mathcal{V}(t)$.*

Let us denote by $\mathcal{E}(\mathcal{V}(t))$ the total energy of the fluid particles contained in the domain $\mathcal{V}(t)$ and by $Q(\mathcal{V}(t))$ the amount of heat transmitted to $\mathcal{V}(t)$ at time t. Taking into account the character of outer and inner forces acting on the domain $\mathcal{V}(t)$, determined by the density $\boldsymbol{f}$ of the volume force and the stress vector $\boldsymbol{T}$, we get the identity representing the law of conservation of energy:

$$\begin{aligned}\frac{d}{dt}\mathcal{E}(\mathcal{V}(t)) = &\int_{\mathcal{V}(t)} \rho(x,t)\boldsymbol{f}(x,t)\cdot\boldsymbol{v}(x,t)\,dx \\ &+ \int_{\partial\mathcal{V}(t)} \boldsymbol{T}(x,t,\boldsymbol{n}(x))\cdot\boldsymbol{v}(x,t)\,dS + Q(\mathcal{V}(t)).\end{aligned} \tag{1.2.55}$$

Further, we can write

$$\text{a) } \mathcal{E}(\mathcal{V}(t)) = \int_{\mathcal{V}(t)} E(x,t)\,dx, \tag{1.2.56}$$

$$\text{b) } E = \rho\left(e + \frac{|\boldsymbol{v}|^2}{2}\right),$$

$$\text{c) } Q(\mathcal{V}(t)) = \int_{\mathcal{V}(t)} \rho(x,t)q(x,t)\,dx - \int_{\partial\mathcal{V}(t)} \mathbf{q}(x,t)\cdot\boldsymbol{n}(x)\,dS.$$

Here E is the total energy, e is the specific internal energy (i.e. per unit mass) associated with molecular and atomic behaviour, $|\boldsymbol{v}|^2/2$ is the density of the kinetic energy, q represents the density of heat sources (related to unit of mass) and $\mathbf{q}$ is the heat flux. By virtue of the so-called *Fourier's law*,

$$\mathbf{q} = -k\,\operatorname{grad}\theta, \tag{1.2.57}$$

so that

$$\int_{\partial\mathcal{V}(t)} \mathbf{q}(x,t)\cdot\boldsymbol{n}(x)\,dS = -\int_{\partial\mathcal{V}(t)} k(x,t)\,\frac{\partial\theta(x,t)}{\partial n}\,dS, \tag{1.2.58}$$

where k is the heat conduction coefficient and θ is the absolute temperature. From the second law of thermodynamics (see Section 1.2.14) it can be proven that $k \geq 0$. Experiments show that k is a function of the absolute temperature: $k = k(\theta)$. We often suppose that k is constant.

Substituting (1.2.35) and (1.2.56), a–c) into (1.2.55), we get

$$\begin{aligned}&\frac{d}{dt}\int_{\mathcal{V}(t)} E(x,t)\,dx \\ &= \int_{\mathcal{V}(t)} \rho(x,t)\boldsymbol{f}(x,t)\cdot\boldsymbol{v}(x,t)\,dx\end{aligned} \tag{1.2.59}$$

$$+\int_{\partial\mathcal{V}(t)} \sum_{i,j=1}^{3} \tau_{ji}(x,t)\, n_j(x)\, v_i(x,t)\, dS$$

$$+\int_{\mathcal{V}(t)} \rho(x,t) q(x,t)\, dx - \int_{\partial\mathcal{V}(t)} \mathbf{q}(x,t)\cdot \boldsymbol{n}(x)\, dS.$$

As in the preceding considerations we assume some smoothness of functions describing the flow. That is, let ρ, u, v_i, τ_{ij}, $q_i \in C^1(\mathcal{M})$, and f_i, $q \in C(\mathcal{M})$ $(i,\ j = 1,2,3)$. By virtue of the transport theorem 1.5, Green's theorem and Lemma 1.1, we derive from (1.2.59) the *energy equation* written in the differential conservative form:

$$\begin{aligned}&\frac{\partial E}{\partial t} + \operatorname{div}(E\boldsymbol{v}) \\ &= \rho \boldsymbol{f}\cdot\boldsymbol{v} + \operatorname{div}(\mathcal{T}\boldsymbol{v}) + \rho q - \operatorname{div}\mathbf{q}.\end{aligned} \tag{1.2.60}$$

For a *Newtonian fluid* we have

$$\begin{aligned}\frac{\partial E}{\partial t} + \operatorname{div}(E\boldsymbol{v}) &= \rho\boldsymbol{f}\cdot\boldsymbol{v} - \operatorname{div}(p\boldsymbol{v}) + \operatorname{div}(\lambda\boldsymbol{v}\,\operatorname{div}\boldsymbol{v}) \\ &\quad + \operatorname{div}(2\mu\mathbb{D}(\boldsymbol{v})\boldsymbol{v}) + \rho q - \operatorname{div}\mathbf{q}.\end{aligned} \tag{1.2.61}$$

The system given by equations (1.2.26), (1.2.45), (1.2.61) forms the so-called *full system of equations of a Newtonian fluid.*

1.2.12 *Thermodynamical relations*

In order to complete the conservation law system, additional equations derived in thermodynamics have to be included.

The absolute temperature θ, the density ρ and the pressure p are called the *state variables.* All these quantities are positive functions. The gas is characterized by the equation of state

$$p = p(\rho,\theta) \tag{1.2.62}$$

and the relation

$$e = e(\rho,\theta). \tag{1.2.63}$$

On the basis of these equations it is possible to express p and θ as functions of e and ρ:

$$p = p(e,\rho), \tag{1.2.64}$$

$$\theta = \theta(e,\rho). \tag{1.2.65}$$

Very often we consider the so-called *perfect gas* (also called ideal gas) whose state variables satisfy the equation of state in the form

$$p = R\,\theta\,\rho. \tag{1.2.66}$$

$R > 0$ is the *gas constant,* which can be expressed in the form

$$R = c_p - c_v, \tag{1.2.67}$$

where c_p and c_v denote the *specific heat at constant pressure* and the *specific heat at constant volume*, respectively. From experiments we know that $c_p > c_v$, so that $R > 0$. We shall consider c_p and c_v to be constant, which is assumed for perfect gases. Experiments show that this is true for a relatively large range of temperature. The quantity

$$\gamma = \frac{c_p}{c_v} > 1 \tag{1.2.68}$$

is called the *Poisson adiabatic constant.* For example, for air, $\gamma = 1.4$.

The internal energy of a perfect gas is given by

$$e = c_v\theta. \tag{1.2.69}$$

Hence,

$$e = c_p\theta - R\theta = h - \frac{p}{\rho}, \tag{1.2.70}$$

where

$$h = c_p\theta \tag{1.2.71}$$

is the *enthalpy.*

1.2.13 *Entropy*

One of the important thermodynamical quantities is the entropy S, defined by the relation

$$\theta\, dS = de + p\, dV, \tag{1.2.72}$$

where $V = 1/\rho$ is the so-called specific volume. This identity is derived in thermodynamics under the assumption that the internal energy is a function of S and $V : e = e(S, V)$, which explains the meaning of the differentials in (1.2.72).

Theorem 1.8 *For a perfect gas we have*

$$\begin{aligned} S &= c_v \ln \frac{p/p_0}{(\rho/\rho_0)^\gamma} + \text{const} \\ &= c_v \ln \frac{\theta/\theta_0}{(\rho/\rho_0)^{\gamma-1}} + \text{const}, \end{aligned} \tag{1.2.73}$$

where p_0 *and* ρ_0 *are fixed (reference) values of pressure and density, respectively, and* $\theta_0 = p_0/(R\rho_0)$.

Proof Using (1.2.69) and the relation $V = 1/\rho$, we can write (1.2.72) in the form

$$\theta\, dS = c_v\, d\theta - \frac{p}{\rho^2}\, d\rho.$$

From this and (1.2.66)–(1.2.68) we obtain

$$dS = c_v \frac{d\theta}{\theta} - \frac{p}{\rho\theta}\frac{d\rho}{\rho} = c_v \frac{d\,(p/\rho)}{(p/\rho)} - R\frac{d\rho}{\rho} = c_v d\ln\frac{p/p_0}{(\rho/\rho_0)^\gamma} = c_v\, d\ln\frac{\theta/\theta_0}{(\rho/\rho_0)^{\gamma-1}},$$

which immediately yields (1.2.73). □

If the flow considered is a reversible process, which means that the system is in equilibrium with the surrounding medium at each time instant, the *first law of thermodynamics* is valid:

$$\delta Q = de + p\,dV,^{0} \tag{1.2.74}$$

where δQ is the elementary heat transmitted (related to unit of mass). This means that the heat transmitted to the system is equal to the sum of the energy increment and the elementary work performed on the system by the pressure force. From this and (1.2.72) we find

$$dS = \frac{\delta Q}{\theta}. \tag{1.2.75}$$

1.2.14 *The second law of thermodynamics*

In irreversible processes, equality (1.2.72) does not hold in general and is replaced by the inequality

$$dS \geq \frac{\delta Q}{\theta} \tag{1.2.76}$$

called the second law of thermodynamics. For a system of fluid particles occupying a domain $\mathcal{V}(t)$ at time t (cf. Section 1.2.3) we postulate the second law of thermodynamics mathematically in the form

$$\begin{aligned} &\frac{d}{dt}\int_{\mathcal{V}(t)} \rho(x,t)\, S(x,t)\, dx \\ &\geq \int_{\mathcal{V}(t)} \frac{\rho(x,t)q(x,t)}{\theta(x,t)}\, dx - \int_{\partial\mathcal{V}(t)} \frac{\mathbf{q}(x,t)\cdot \boldsymbol{n}(x)}{\theta(x,t)}\, dS, \end{aligned} \tag{1.2.77}$$

where q and $\mathbf{q}$ denote the density of heat sources and the heat flux (see (1.2.56), c). The left-hand side of (1.2.77) represents the rate of change of the entropy contained in the volume $\mathcal{V}(t)$, and the first and second integral on the right-hand side are called the *entropy production* and the *entropy flux*, respectively. Let $\rho,\ \theta,\ v_i,\ q_i \in C^1(\mathcal{M}),\ q,\ f_i \in C(\mathcal{M}),\ i = 1,2,3$. By virtue of the transport theorem and the continuity equation, from (1.2.77) we obtain the inequality

$$\rho\frac{\mathrm{D}S}{\mathrm{D}t} \geq \frac{\rho q}{\theta} - \operatorname{div}\left(\frac{\mathbf{q}}{\theta}\right). \tag{1.2.78}$$

1.2.15 *Dissipation form of the energy equation*

On the basis of the continuity equation (1.2.26), the Navier–Stokes equations (1.2.45) can be written in the *convective form*

$$\rho\frac{\partial \boldsymbol{v}}{\partial t} + \rho(\boldsymbol{v}\cdot\nabla)\boldsymbol{v} \tag{1.2.79}$$

[0] We use the symbol δQ, because the elementary heat transmitted depends on the way of the transition of the system from one state to another and, therefore, it cannot be expressed as the differential dQ. In theoretical thermodynamics δQ is a Pfaff 1-form and $1/\theta$ is its integration factor.

$$= \rho \boldsymbol{f} - \nabla p + \nabla(\lambda \mathrm{div} \boldsymbol{v}) + \mathrm{div}(2\mu \mathbb{D}(\boldsymbol{v})).$$

The scalar product of this equation with $\boldsymbol{v}$ yields

$$\rho \frac{\partial}{\partial t}\left(\frac{|\boldsymbol{v}|^2}{2}\right) + \rho(\boldsymbol{v}\cdot\nabla)\frac{|\boldsymbol{v}|^2}{2} \tag{1.2.80}$$
$$= \rho \boldsymbol{f}\cdot\boldsymbol{v} - \boldsymbol{v}\cdot\nabla p + \boldsymbol{v}\cdot\nabla(\lambda \mathrm{div}\boldsymbol{v}) + \boldsymbol{v}\cdot\mathrm{div}(2\mu\mathbb{D}(\boldsymbol{v})).$$

Further, the energy equation (1.2.61) and (1.2.56), b) imply that

$$\frac{\partial}{\partial t}\left[\rho\left(e + \frac{1}{2}|\boldsymbol{v}|^2\right)\right] + \mathrm{div}\left[\rho\left(e + \frac{1}{2}|\boldsymbol{v}|^2\right)\right] \tag{1.2.81}$$
$$= \rho \boldsymbol{f}\cdot\boldsymbol{v} - \mathrm{div}(p\boldsymbol{v}) + \mathrm{div}(\lambda\boldsymbol{v}\mathrm{div}\,\boldsymbol{v}) + \mathrm{div}(2\mu\mathbb{D}(\boldsymbol{v})\boldsymbol{v}) + \rho q - \mathrm{div}\,\boldsymbol{q}.$$

Taking into account that

$$\begin{aligned} \mathrm{div}(p\boldsymbol{v}) &= \boldsymbol{v}\cdot\nabla p + p\,\mathrm{div}\,\boldsymbol{v}, \\ \mathrm{div}(\lambda\boldsymbol{v}\mathrm{div}\,\boldsymbol{v}) &= \lambda(\mathrm{div}\,\boldsymbol{v})^2 + \boldsymbol{v}\cdot\nabla(\lambda\mathrm{div}\,\boldsymbol{v}), \\ \mathrm{div}(2\mu\mathbb{D}(\boldsymbol{v})\boldsymbol{v}) &= \boldsymbol{v}\cdot\mathrm{div}(2\mu\mathbb{D}(\boldsymbol{v})) + 2\mu\mathbb{D}(\boldsymbol{v})\cdot\mathbb{D}(\boldsymbol{v}), \end{aligned} \tag{1.2.82}$$

and the fact that the left-hand side of (1.2.80) can be transformed with the aid of (1.2.26) to the expression

$$\left(\partial(\rho|\boldsymbol{v}|^2)/\partial t + \mathrm{div}(\rho|\boldsymbol{v}|^2\boldsymbol{v})\right)/2,$$

we find that the energy equation can be written in the dissipation form

$$\frac{\partial}{\partial t}(\rho e) + \mathrm{div}(\rho e\boldsymbol{v}) + p\,\mathrm{div}\,\boldsymbol{v} \tag{1.2.83}$$
$$= \lambda(\mathrm{div}\,\boldsymbol{v})^2 + 2\mu\mathbb{D}(\boldsymbol{v})\cdot\mathbb{D}(\boldsymbol{v}) + \rho q - \mathrm{div}\,\boldsymbol{q}.$$

Here the quantity

$$D(\boldsymbol{v}) = \lambda(\mathrm{div}\boldsymbol{v})^2 + 2\mu\mathbb{D}(\boldsymbol{v})\cdot\mathbb{D}(\boldsymbol{v}) \tag{1.2.84}$$

is called *dissipation.*

Exercise 1.9 Prove that under assumption (1.2.46) the dissipation satisfies the condition $D(\boldsymbol{v}) \geq 0$.

If e, $\boldsymbol{q}$ and p are expressed by (1.2.69), (1.2.57) and (1.2.66), equation (1.2.83) can be written in the *temperature form*

$$\frac{\partial}{\partial t}(c_v\rho\theta) + \mathrm{div}(c_v\rho\theta\boldsymbol{v}) + R\rho\theta\mathrm{div}\boldsymbol{v} \tag{1.2.85}$$
$$= \mathrm{div}(k\nabla\theta) + D(\boldsymbol{v}) + \rho q.$$

1.2.16 *Entropy form of the energy equation*

Let the equation of state (1.2.66) be valid. From (1.2.72) and (1.2.69) it follows that

$$\theta\rho\frac{\mathrm{D}S}{\mathrm{D}t} = c_v\rho\frac{\mathrm{D}\theta}{\mathrm{D}t} - R\theta\frac{\mathrm{D}\rho}{\mathrm{D}t}. \tag{1.2.86}$$

Using the continuity equation, which implies that $\mathrm{D}\rho/\mathrm{D}t = \rho_t + \boldsymbol{v}\cdot\nabla\rho = -\rho\mathrm{div}\boldsymbol{v}$, and (1.2.66), we obtain

$$\begin{aligned}&\theta\partial(\rho S)/\partial t + \theta\mathrm{div}(\rho S\boldsymbol{v})\\&= \partial(c_v\rho\theta)/\partial t + \mathrm{div}(c_v\rho\theta\boldsymbol{v}) + R\rho\theta\mathrm{div}\,\boldsymbol{v}.\end{aligned}$$

This and (1.2.85) immediately give the entropy form of the energy equation

$$\frac{\partial(\rho S)}{\partial t} + \mathrm{div}(\rho S\boldsymbol{v}) = (\mathrm{div}(k\nabla\theta) + \rho q + D(\boldsymbol{v}))/\theta \tag{1.2.87}$$

or

$$\rho\frac{\mathrm{D}S}{\mathrm{D}t} = (D(\boldsymbol{v}) + \rho q + \mathrm{div}\,\boldsymbol{q})/\theta. \tag{1.2.88}$$

As a consequence of (1.2.88) and Exercise 1.9, we find that (provided the quantities describing the flow are sufficiently regular) the entropy inequality (1.2.78) is satisfied.

1.2.17 *Adiabatic flow*

If there is no heat transmission and heat exchange between fluid volumes, we speak of adiabatic flow. Hence, in adiabatic flow the heat sources and heat flux are zero, so that $q = 0$, $\mathbf{q} = 0$ and, with respect to (1.2.57), also $k = 0$.

It is known that heat conductivity and internal friction represent two faces of molecular transmission. Heat conductivity is related to the transmission of molecular kinetic energy and internal friction is conditioned by the transmission of molecular momentum. Therefore, it makes sense to speak of adiabatic flow particularly in the case of inviscid gas.

From (1.2.88) it follows that for *adiabatic inviscid flow*

$$\frac{\mathrm{D}S}{\mathrm{D}t} = \frac{\partial S}{\partial t} + \boldsymbol{v}\cdot\nabla S = 0. \tag{1.2.89}$$

If $x = \varphi(X,T)$ is the trajectory of a fluid particle and, hence, $\boldsymbol{v}(x,t) = \frac{\partial\varphi}{\partial t}(X,t)$, in view of (1.2.12) we have

$$\frac{\partial}{\partial t}S(\varphi(X,t),t) = 0.$$

Hence,

$$S(\varphi(X,t),t) = \mathrm{const.} \tag{1.2.90}$$

From this and (1.2.73) we see that we have proven

Theorem 1.10 *In adiabatic flow of an inviscid perfect gas*

$$S = \text{const} \quad \textit{along the trajectory of any fluid particle,} \tag{1.2.91}$$

$$p = \kappa \rho^\gamma \quad \textit{along the trajectory of any fluid particle,} \tag{1.2.92}$$

where κ is a constant dependent on the trajectory considered.

If condition (1.2.91) is satisfied, then we speak of *isentropic* flow. If $S = \text{const}$ in the whole flow field, then the flow is called *homoentropic.*

Remark 1.11 It is necessary to say that the statement $p = \kappa\rho^\gamma$ is not quite correct, although it is commonly used in physics. To be correct, we should write $p = cp_0(\rho/\rho_0)^\gamma$ where p_0 and ρ_0 are suitable reference values of the pressure and density and c is a dimensionless constant. If it makes sense to substitute $p := p_0$ and $\rho := \rho_0$, we see that $c = 1$. For the sake of simplicity, however, we often use the unprecise statement with $\kappa := p_0/\rho_0^\gamma$ mentioned above.

Summarizing the above results, we see that *the flow of a perfect gas, described by sufficiently smooth functions, satisfies the second law of thermodynamics. If the flow is moreover inviscid and adiabatic, then it is isentropic.*

In some special cases (e.g. in transonic inviscid flow) the assumptions on continuity or smoothness of functions describing the flow must be relaxed and, therefore, it is necessary to reformulate the conservation laws and the second law of thermodynamics in a suitable weak sense. We shall deal with these questions in detail in Chapters 2 and 3, where also the importance of the second law of thermodynamics will be made clear.

1.2.18 *Barotropic flow*

We say that the flow is barotropic if the pressure can be expressed as a function of the density:

$$p = p(\rho). \tag{1.2.93}$$

This means that $p(x,t) = p(\rho(x,t))$ for all $(x,t) \in \mathcal{M}$, or, more briefly, $p = p \circ \rho$. We assume that

$$p : (0, +\infty) \to (0, +\infty) \tag{1.2.94}$$
and there exists the continuous derivative
$$p' > 0 \text{ on } (0, +\infty).$$

From (1.2.92) it follows that in *adiabatic barotropic flow of an inviscid perfect gas* we have the relation

$$p = \kappa\rho^\gamma, \tag{1.2.95}$$

the constant κ being common for the whole flow field. Thus, the flow is homoentropic.

From the above considerations we conclude that the *viscous compressible barotropic flow* in a domain Ω and time interval $(0, T)$ is described by the following system considered in the space-time cylinder $Q_T = \Omega \times (0, T)$ (see Section 1.2.10):

$$(\rho \boldsymbol{v})_t + \operatorname{div}(\rho \boldsymbol{v} \otimes \boldsymbol{v}) + \nabla p(\rho) - \mu \Delta \boldsymbol{v} - (\mu + \lambda) \nabla (\operatorname{div} \boldsymbol{v}) = \rho \boldsymbol{f}, \quad (1.2.96)$$

$$\rho_t + \operatorname{div}(\rho \boldsymbol{v}) = 0, \quad (1.2.97)$$

$$p = p(\rho). \quad (1.2.98)$$

Here μ and λ are constant viscosity coefficients, $\mu, \mu + \lambda > 0$ and $\boldsymbol{f}$ denotes the density of the volume force. If we set $\mu = \lambda = 0$, we obtain the model of inviscid barotropic flow.

1.2.19 *Complete system describing the flow of a heat-conductive gas*

If we consider the flow of a real heat conductive gas, then the above system (1.2.96)–(1.2.98) is not sufficient and it is necessary to use the system consisting of the continuity equation, the Navier–Stokes equations, the energy equation and thermodynamical relations. That is, we use equations (1.2.68), b), (1.2.64) and (1.2.65). Then the complete system reads:

$$\rho_t + \operatorname{div}(\rho \boldsymbol{v}) = 0, \quad (1.2.99)$$

$$(\rho \boldsymbol{v})_t + \operatorname{div}(\rho \boldsymbol{v} \otimes \boldsymbol{v}) = \rho \boldsymbol{f} - \nabla p + \nabla(\lambda \operatorname{div} \boldsymbol{v}) + \operatorname{div}(2\mu \mathbb{D}(\boldsymbol{v})), (1.2.100)$$

$$E_t + \operatorname{div}(E \boldsymbol{v}) = \rho \boldsymbol{f} \cdot \boldsymbol{v} - \operatorname{div}(p \boldsymbol{v}) + \operatorname{div}(\lambda \boldsymbol{v} \operatorname{div} \boldsymbol{v}) \quad (1.2.101)$$

$$+ \operatorname{div}(2\mu \mathbb{D}(\boldsymbol{v}) \boldsymbol{v}) + \rho q + \operatorname{div}(k \nabla \theta), \quad (1.2.102)$$

$$e = \frac{E}{\rho} - \frac{1}{2} |\boldsymbol{v}|^2, \quad (1.2.103)$$

$$p = p(e, \rho), \quad (1.2.104)$$

$$\theta = \theta(e, \rho). \quad (1.2.105)$$

The velocity deformation tensor $\mathbb{D}(\boldsymbol{v})$ is expressed in (1.2.43). We simply call this system the *compressible Navier–Stokes equations* for a heat conductive gas. For a perfect gas, for which relations (1.2.66) and (1.2.79) are valid, we write equations (1.2.103)–(1.2.105) as

$$p = (\gamma - 1)(E - \rho |\boldsymbol{v}|^2 / 2), \quad \theta = (E/\rho - |\boldsymbol{v}|^2/2)/c_v. \quad (1.2.106)$$

If we set $\mu = \lambda = k = 0$, we obtain the model of inviscid compressible flow, described by the continuity equation, the Euler equations, the energy equation and thermodynamical relations. Since gases are light, usually it is possible to neglect the effect of the volume force. Neglecting heat sources also, we get the system

$$\rho_t + \operatorname{div}(\rho \boldsymbol{v}) = 0, \quad (1.2.107)$$

$$(\rho \boldsymbol{v})_t + \operatorname{div}(\rho \boldsymbol{v} \otimes \boldsymbol{v}) + \nabla p = 0, \quad (1.2.108)$$

$$E_t + \operatorname{div}((E+p)\boldsymbol{v}) = 0, \tag{1.2.109}$$

$$e = \frac{E}{\rho} - \frac{1}{2}|\boldsymbol{v}|^2, \quad p = p(e,\rho). \tag{1.2.110}$$

This system is simply called the *compressible Euler equations*. For a perfect gas equations (1.2.110) take the form

$$p = (\gamma - 1)(E - \rho|\boldsymbol{v}|^2/2). \tag{1.2.111}$$

1.2.20 *Speed of sound; Mach number*

A more general model than barotropic flow is obtained in thermodynamics under the assumption that the pressure is a function of the density and entropy: $p = p(\rho, S)$, where p is a continuously differentiable function and $\partial p/\partial \rho > 0$. For example, for a perfect gas, in view of Theorem 1.8, we have

$$p = f(\rho, S) = \kappa\rho^\gamma \exp(S/c_v), \qquad \kappa = \text{const} > 0. \tag{1.2.112}$$

(The adiabatic barotropic flow of an ideal perfect gas with $S = \text{const}$ is obviously a special case of this model.) Let us introduce the quantity

$$a = \sqrt{\frac{\partial f}{\partial \rho}} \tag{1.2.113}$$

which has the dimension m s^{-1} of velocity and is called the *speed of sound*. This terminology is based on the fact that a represents the speed of propagation of pressure waves of small intensity.

A further important characteristic of gas flow is the *Mach number*

$$M = \frac{|\boldsymbol{v}|}{a} \tag{1.2.114}$$

(which is obviously a dimensionless quantity). We say that the flow is *subsonic* or *sonic* or *supersonic* at a point x and time t, if

$$M(x,t) < 1 \qquad \text{or} \qquad M(x,t) = 1 \qquad \text{or} \qquad M(x,t) > 1, \tag{1.2.115}$$

respectively. We speak of *transonic flow* in the domain Ω_t, if there exist nonempty subsets Ω_t^1 and Ω_t^2 of Ω_t such that the flow is subsonic in Ω_t^1 and supersonic in Ω_t^2.

1.2.21 *Simplified models*

If the quantities describing the flow are independent of time and, hence, $\partial/\partial t \equiv 0$, we speak of *stationary* (or steady) flow. A solution of the governing equations independent of time is called a *steady-state* or *stationary* solution.

Sometimes the geometry of the domain occupied by the fluid and the character of the flow allow us to introduce a Cartesian coordinate system $x =$

(x_1, x_2, x_3) in $I\!R^3$ in such a way that the quantities describing the flow are independent of x_3, i.e. $\partial/\partial x_3 \equiv 0$, and the velocity component v_3 and the component f_3 of the volume force vanish. Then we can assume that

$$\Omega \subset I\!R^2,\ x = (x_1, x_2),\ \boldsymbol{v} = (v_1, v_2),\ \boldsymbol{f} = (f_1, f_2),\ \nabla = (\partial/\partial x_1, \partial/\partial x_2),$$

$$\mathbb{D}(\boldsymbol{v}) = (d_{ij}(\boldsymbol{v}))_{i,j=1}^2,\ d_{ij}(\boldsymbol{v}) = (\partial v_i/\partial x_j + \partial v_j/\partial v_i)/2,$$

etc. We easily find that the governing system of equations can again be written as (1.2.26) (the continuity equation), (1.2.49) with $N = 2$ or (1.2.50) or (1.2.53) (the Navier–Stokes equations) and (1.2.61) (the energy equation). Also relations from Pragaraphs 1.2.14–1.2.20 remain formally unchanged. In this way we get a two-dimensional (2D) model of the flow. We also speak of plane flow.

Similarly, we arrive at a 1D model, when the flow is described by quantities ρ, p, θ, $v = v_1$, etc., depending on $x = x_1 \in \Omega \subset I\!R$ only.

Hence, the flow problems can be formulated in $I\!R^N$, where $N = 1, 2$ or 3.

1.2.22 *Initial and boundary conditions*

The system of governing equations and relations has to be equipped with initial conditions determining the state at the initial time $t = 0$ and with boundary conditions which characterize the behaviour of the flow at the boundary $\partial\Omega$.

The *initial conditions* can be formulated, for example, as

$$\boldsymbol{v}(x, 0) = \boldsymbol{v}^0(x),\ \rho(x, 0) = \rho^0(x),\ \theta(x, 0) = \theta^0(x), \quad x \in \Omega, \tag{1.2.116}$$

with given initial data $\boldsymbol{v}^0$, ρ^0, θ^0.

The choice of the *boundary conditions* is more complex. First, let us assume that Ω is a fixed domain. For the velocity, the simplest choice is the Dirichlet (also called essential) condition

$$\boldsymbol{v}|_{\partial\Omega} = \boldsymbol{v}_D, \tag{1.2.117}$$

with a given function $\boldsymbol{v}_D$. On a fixed impermeable wall $\Gamma \subset \partial\Omega$ we assume the so-called *no-slip condition*

$$\boldsymbol{v}|_\Gamma = 0 \tag{1.2.118}$$

(i.e. $\boldsymbol{v}_D|_\Gamma = 0$), expressing the physical fact that a real fluid adheres to Γ.

If condition (1.2.117) is used, then on $\partial\Omega$ we distinguish *inflow through the inlet*

$$\Gamma_I(t) = \{x \in \partial\Omega; \boldsymbol{v}(x, t) \cdot \boldsymbol{n}(x) < 0\} \tag{1.2.119}$$

and *outflow through the outlet*

$$\Gamma_O(t) = \{x \in \partial\Omega; \boldsymbol{v}(x, t) \cdot \boldsymbol{n}(x) > 0\}, \tag{1.2.120}$$

where $\boldsymbol{n}$ denotes the unit outer normal to $\partial\Omega$. If $\Gamma_I(t) \neq \emptyset$, then it is necessary to prescribe the density on $\Gamma_I(t)$:

$$\rho(x,t) = \rho_D(x,t), \quad x \in \Gamma_I(t),\ t \in (0,T). \tag{1.2.121}$$

In the case of an *inviscid model*, when $\mu = \lambda = 0$ and $k = 0$, the boundary conditions (1.2.117) and (1.2.118) must be relaxed. There is no reason for the fluid to adhere to an impermeable wall Γ now and we consider the condition

$$(\boldsymbol{v} \cdot \boldsymbol{n})|_\Gamma = 0. \tag{1.2.122}$$

On the whole boundary, possibly with an inlet or an outlet, we use the boundary condition

$$(\boldsymbol{v} \cdot \boldsymbol{n})|_{\partial\Omega} = v_{nD} \tag{1.2.123}$$

with a given scalar function v_{nD}. On the inlet $\Gamma_I(t) \neq \emptyset$ we again prescribe the density, i.e. we use condition (1.2.121).

In practice, the condition prescribing the velocity (or normal component of the velocity in inviscid flow) seems to be often rather strong on the outlet. Therefore, one uses various 'soft' natural outlet boundary conditions. We shall mention these in further chapters.

There is also a model of viscous flow with the assumption that the fluid slips on the boundary. Then we use the *no-stick conditions*, also called *slip conditions*,

$$(\boldsymbol{v} \cdot \boldsymbol{n})|_{\partial\Omega} = 0, \quad \Big(\sum_{i,j=1}^{N} \tau_i(\partial_i v_j + \partial_j v_i) v_j\Big)\Big|_{\partial\Omega} = 0, \tag{1.2.124}$$

where τ_i are the components of the tangent unit vector to the boundary. Another possibility is that on the boundary $\partial\Omega$ the normal component of the stress tensor is prescribed:

$$\mathcal{T}_n|_{\partial\Omega} \equiv (\mathcal{T}\boldsymbol{n})|_{\partial\Omega} \equiv \left((-p + \lambda \operatorname{div} \boldsymbol{v})\boldsymbol{n} + 2\mu\mathbb{D}(\boldsymbol{v})\boldsymbol{n}\right)|_{\partial\Omega} = \boldsymbol{H} \tag{1.2.125}$$

with a given function $\boldsymbol{H}$. These conditions may appear in problems with a *free boundary*. For the definition of the boundary we use the so-called *kinematic condition*

$$\partial\Omega(t) = \{x; \eta(x,t) = 0\}, \tag{1.2.126}$$

where η satisfies

$$\frac{\partial\eta}{\partial t} + (\boldsymbol{v} \cdot \nabla)\eta = 0. \tag{1.2.127}$$

This equation means that the particle on the free boundary can move only along this boundary. Indeed, if

$$\boldsymbol{\xi}_0 \in \partial\Omega(0) \quad \text{and} \quad \frac{d\boldsymbol{\xi}}{dt} = \boldsymbol{v}(\boldsymbol{\xi},t), \quad t > 0,$$

then

$$\frac{d}{dt}\eta(\boldsymbol{\xi}(t),t) = \eta_t(\boldsymbol{\xi}(t),t) + \boldsymbol{v}(\boldsymbol{\xi}(t),t) \cdot \nabla\eta(\boldsymbol{\xi}(t),t) = 0,$$

which implies

$$\eta(\boldsymbol{\xi}(t), t) = \eta(\boldsymbol{\xi}_0, 0) = 0, \quad t > 0.$$

If the *heat conduction* is included in the model, then it is necessary to add boundary conditions characterizing the heat processes at the boundary. For example, we can assume that

$$\left(k\frac{\partial\theta}{\partial n}\right)\Big|_{\partial\Omega} = \beta(\theta - \chi)|_{\partial\Omega}, \tag{1.2.128}$$

where k is the heat conduction coefficient and β and χ are given functions defined on $\partial\Omega$. There are also other possibilities: either

$$\theta|_{\partial\Omega} = \chi \quad \text{(given temperature on the boundary)} \tag{1.2.129}$$

or

$$\left(k\frac{\partial\theta}{\partial n}\right)\Big|_{\partial\Omega} = q \quad \text{(given heat flux through the boundary).} \tag{1.2.130}$$

If the regularity of a solution up to the boundary is required, then initial and boundary data must satisfy certain *compatibility conditions* to be specified accordingly.

1.2.23 *Dimensionless form of gas dynamics equations*

In order to be able to carry out experiments on small models (e.g. measurements of flow past a model of an aircraft in a wind tunnel) and to transfer the results to the original real flow, we use the *dimensionless form* of gas dynamics equations.

Let us introduce the following positive *reference quantities*: a reference length L^*, a reference velocity U^* (scalar quantity), a reference density ρ^*, a reference volume force f^* (e.g. a gravitational constant g), a reference viscosity μ^* and a reference heat conduction coefficient k^*. All other reference quantities can be derived from these basic ones: we choose L^*/U^* for t, $\rho^* U^{*2}$ for both p and E, U^{*3}/L^* for heat sources q, U^{*2}/c_v for θ. We denote by primes the *dimensionless quantities*

$$\begin{aligned} &x_i' = x_i/L^*, \quad v_i' = v_i/U^*, \quad \boldsymbol{v}' = \boldsymbol{v}/U^*, \quad \rho' = \rho/\rho^*, \\ &p' = p/(\rho^* U^{*2}), \quad E' = E/(\rho^* U^{*2}), \quad \theta' = \frac{c_v\theta}{U^{*2}}, \quad q' = \frac{q\,L^*}{U^{*3}}, \\ &t' = t\,U^*/L^*, \quad \boldsymbol{f}' = \boldsymbol{f}/f^*, \quad \mu' = \frac{\mu}{\mu^*}, \quad \lambda' = \frac{\lambda}{\mu^*}, \quad k' = \frac{k}{k^*}. \end{aligned} \tag{1.2.131}$$

Let us multiply equation (1.2.99) by $L^*/(\rho^* U^*)$, (1.2.100) by $L^*/(\rho^* U^{*2})$ and (1.2.102) by $L^*/(\rho^* U^{*3})$. We get

$$\rho'_{t'} + \operatorname{div}(\rho' \boldsymbol{v}') = 0 \tag{1.2.132}$$

(i.e. the continuity equation remains unchanged),

$$(\rho' \boldsymbol{v}')_{t'} + \operatorname{div}(\rho' \boldsymbol{v}' \otimes \boldsymbol{v}') = \frac{1}{Fr^2}\,\rho' \boldsymbol{f}' - \nabla p' \tag{1.2.133}$$

$$+\frac{1}{Re}\left[\nabla(\lambda'\mathrm{div}\,\boldsymbol{v}') + \mathrm{div}(2\mu'\mathbb{D}(\boldsymbol{v}'))\right],$$

$$E'_{t'} + \mathrm{div}(E'\boldsymbol{v}') = \frac{1}{Fr^2}\rho' f' \cdot \boldsymbol{v}' - \mathrm{div}(p'\boldsymbol{v}') \tag{1.2.134}$$
$$+\frac{1}{Re}\left[\mathrm{div}(\lambda'\boldsymbol{v}'\mathrm{div}\,\boldsymbol{v}') + \mathrm{div}(2\mu'\mathbb{D}(\boldsymbol{v}')\boldsymbol{v}')\right]$$
$$+\rho' q' + \mathrm{div}\left(\frac{\gamma k'}{Re\,Pr}\nabla\theta'\right),$$

where

$$Fr = U^*/\sqrt{L^* f^*}, \quad Re = \rho^* U^* L^*/\mu^*, \quad Pr = c_p\,\mu^*/k^* \tag{1.2.135}$$

are the Froud, Reynolds and Prandtl number, respectively. In (1.2.132)–(1.2.134) the partial derivatives in the operators div, ∇ and in the velocity deformation tensor $\mathbb{D}$ are calculated with respect to x'. Further, multiplying equations in (1.2.106) by $1/(\rho^* U^{*2})$ and c_v/U^{*2}, respectively, we get their dimensionless form

$$p' = (\gamma - 1)\left(E' - \rho'|\boldsymbol{v}'|^2/2\right), \quad \theta' = \frac{E'}{\rho'} - |\boldsymbol{v}'|^2/2. \tag{1.2.136}$$

Supposing that $\mu' = \mu'(\theta')$, $\lambda' = \lambda'(\theta')$ and $k' = k'(\theta')$ are universal functions, we conclude that there are four similarity parameters Re, Fr, Pr, γ.

Let us consider two flows in geometrically similar domains Ω_1 and Ω_2 ('$\Omega_1 = L\,\Omega_2$', $L = \mathrm{const} > 0$). Let us suppose that for both flows the viscosity coefficients λ, μ and the heat conduction coefficient k are nonzero constants and choose μ and k as respective reference values. We call these flows *dynamically similar*, if they have the same Froud, Reynolds and Prandtl numbers. Then their dimensionless continuity equations, Navier–Stokes equations and energy equations are identical, provided γ and the ratio λ/μ are the same for both flows. Moreover, if their dimensionless boundary conditions become identical, then one flow field can be obtained by scaling the other one.

Exercise 1.12 Consider two samples of flow past a body. The velocity is assumed to be homogeneous at infinity and to have the direction of the axis x_1. In one case the body $\overline{\Omega}_1$ has the diameter $L_1 = 10\,\mathrm{m}$ and the far-field velocity $U_1^\infty = 10\,\mathrm{m\,s^{-1}}$. In the other case we consider a geometrically similar body $\overline{\Omega}_2$ of diameter $L = 1\,\mathrm{m}$. We neglect the volume forces and suppose that both flows have the same Reynolds and Prandtl numbers and viscosity coefficients. We choose the diameter of the body as L^* and the magnitude of the far-field velocity as U^*. Determine the far-field velocity U_2^∞ in order to have dynamically similar flows and to obtain one flow field by scaling the other one.

Further, let us mention the dimensionless form of the Euler equations, which is widely used in the computation of inviscid compressible flow.

Passing to the dimensionless quantities $x'_i, v'_i, \boldsymbol{v}', p', E', t'$ (see (1.2.131)) system (1.2.107)–(1.2.109) can be written in the form

$$\rho'_{t'} + \operatorname{div}(\rho' \boldsymbol{v}') = 0, \tag{1.2.137}$$

$$(\rho' \boldsymbol{v}')_{t'} + \operatorname{div}(\rho' \boldsymbol{v}' \otimes \boldsymbol{v}') + \nabla p' = 0, \tag{1.2.138}$$

$$E'_{t'} + \operatorname{div}((E' + p')\boldsymbol{v}') = 0. \tag{1.2.139}$$

In (1.2.137)–(1.2.139) the partial derivatives in operators div and ∇ are considered with respect to x'. For a perfect gas, for which relations (1.2.66) and (1.2.79) are valid, we write equations (1.2.110) in dimensionless form as

$$p' = (\gamma - 1)\left(E' - \rho' |\boldsymbol{v}'|^2/2\right). \tag{1.2.140}$$

In (1.2.114) the Mach number M was introduced. If we define the Mach number for the Euler equation in dimensionless form as

$$M' = \frac{|\boldsymbol{v}'|}{a'}, \tag{1.2.141}$$

where

$$a' = \sqrt{\gamma \frac{p'}{\rho'}}, \tag{1.2.142}$$

we see immediately that

$$M' = \frac{|\boldsymbol{v}/U^*|}{\sqrt{\gamma \frac{p\,\rho^*}{\rho^* U^{*2} \rho}}} = M. \tag{1.2.143}$$

Formally, system (1.2.137)–(1.2.139) has the same form and properties as system (1.2.107)–(1.2.109) and therefore the primes are omitted if we consider the Euler equations in dimensionless form.

1.3 Some advanced mathematical concepts and results

Here we summarize all important auxiliary results which will be used in our subsequent considerations. For proofs and further details, see, for example, (Kufner *et al.*, 1977). Note that the basic concepts used from functional analysis are explained in Section 1.4.

1.3.1 *Spaces of continuous, Hölder-continuous and continuously differentiable functions*

Let $\Omega \subset \mathbb{R}^N$ be a domain. We shall work with the linear spaces $C^k(\Omega)$ and $C^k(\overline{\Omega})$ defined in Section 1.1.1 and the linear spaces $C^{k,\mu}(\overline{\Omega})$ introduced in Section 1.1.4.

Let us put

$$C^\infty(\Omega) = \bigcap_{k=1}^{\infty} C^k(\Omega) \quad \text{and} \quad C^\infty(\overline{\Omega}) = \bigcap_{k=1}^{\infty} C^k(\overline{\Omega}). \tag{1.3.1}$$

A vector $\alpha = (\alpha_1, \dots, \alpha_N)$ with integer components $\alpha_i \geq 0$ is called a *multi-index of dimension* N. The number $|\alpha| = \sum_{i=1}^N \alpha_i$ is called the *length* of the multi-index α. In the following we shall often use the simplified notation

$$D^\alpha v = \frac{\partial^{|\alpha|} v}{\partial x_1^{\alpha_1} \dots \partial x_N^{\alpha_N}} \tag{1.3.2}$$

for partial derivatives of a function $v = v(x_1, \dots, x_N)$ of N real variables. We also denote

$$x^\alpha = x_1^{\alpha_1} \dots x_N^{\alpha_N}.$$

The linear space $C^k(\overline{\Omega})$, $k = 0, 1, \dots$, endowed with the norm

$$|u|_{C^k(\overline{\Omega})} := \sum_{|\alpha| \leq k} \sup_{x \in \Omega} |D^\alpha u(x)|$$

is a Banach space. It is well known that this space is separable and that it is not reflexive.

The linear space $C^{k,\mu}(\overline{\Omega})$, $k = 0, 1, \dots$, $\mu \in (0, 1]$ endowed with the norm

$$|u|_{C^{k,\mu}(\overline{\Omega})} = |u|_{C^{k,\mu}(\overline{\Omega})} + \sum_{|\alpha|=k} \sup_{x,y \in \Omega,\, x \neq y} \frac{|(D^\alpha u)(x) - (D^\alpha u)(y)|}{|x - y|^\mu}$$

is a Banach space. It is called the Hölder space. This space is neither separable nor reflexive.

1.3.2 *Distributions*

In this section, if not stated explicitly otherwise, Ω is a domain in $I\!R^N$. Let $C_0^\infty(\Omega)$ denote the linear space of all functions $v \in C^\infty(\overline{\Omega})$ whose *support*

$$\operatorname{supp} v = \overline{\{x;\ v(x) \neq 0\}} \tag{1.3.3}$$

is a compact (i.e. bounded and closed) subset of the domain Ω. In $C_0^\infty(\Omega)$ we introduce the *topology of locally uniform convergence*: we say that $v_n \to v$ in $C_0^\infty(\Omega)$ as $n \to \infty$, if

there exists a compact set $K \subset \Omega$ such that (1.3.4)

a) $\operatorname{supp} v_n \subset K$ for all $n = 1, 2, \dots$,

and

b) $D^\alpha v_n \longrightarrow D^\alpha v$ uniformly in K

as $n \longrightarrow \infty$ for all multiindices $\alpha = (\alpha_1, \dots, \alpha_n)$.

The space $C_0^\infty(\Omega)$ endowed with this topology is usually denoted by $\mathcal{D}(\Omega)$. By $\mathcal{D}'(\Omega)$ we denote the space of all continuous linear functionals defined on $\mathcal{D}(\Omega)$, i.e. $\mathcal{D}'(\Omega) = (C_0^\infty(\Omega))^*$. This means that $f : \mathcal{D}(\Omega) \to \mathbb{R}$ is an element of $\mathcal{D}'(\Omega)$, if the following conditions are satisfied (the symbol $\langle f, v\rangle$ denotes the value of the functional f at a point $v \in \mathcal{D}(\Omega)$):

$$\text{a)}\ \langle f, \alpha_1 v_1 + \alpha_2 v_2\rangle = \alpha_1\langle f, v_1\rangle + \alpha_2\langle f, v_2\rangle \qquad (1.3.5)$$
$$\forall\, \alpha_1, \alpha_2 \in \mathbb{R},\ \forall\, v_1, v_2 \in \mathcal{D}(\Omega),$$
$$\text{b)}\ v,\ v_n \in \mathcal{D}(\Omega),\ v_n \overset{n\to\infty}{\longrightarrow} v \text{ in } \mathcal{D}(\Omega) \Longrightarrow \langle f, v_n\rangle \to \langle f, v\rangle.$$

The space $\mathcal{D}'(\Omega)$ is endowed with the topology induced by the convergence

$$f_n \to f \ \text{ in } \mathcal{D}'(\Omega), \text{ if } \langle f_n, v\rangle \to \langle f, v\rangle,\ v \in \mathcal{D}(\Omega).$$

The elements of the space $\mathcal{D}'(\Omega)$ are called *generalized functions* or *distributions.*

Example 1.13 a) Let f be a *locally integrable* function in Ω, i.e. $f \in L^1_{\mathrm{loc}}(\Omega)$. (see Section 1.1.10). It can be shown that the mapping

$$v \in C_0^\infty(\Omega) \to \int_\Omega f v\, dx \qquad (1.3.6)$$

is an element of $\mathcal{D}'(\Omega)$. Hence, the space $L^1_{\mathrm{loc}}(\Omega)$ can be identified with a certain subspace in $\mathcal{D}'(\Omega)$. The distributions associated with functions f by (1.3.6) are called *regular distributions* and they will again be denoted by the symbol f. Therefore, we write

$$\langle f, v\rangle = \int_\Omega f v\, dx, \quad v \in \mathcal{D}(\Omega). \qquad (1.3.7)$$

b) For $a \in \Omega$ we define the functional $\delta_a : C_0^\infty(\Omega) \to \mathbb{R}$ attaining the value $v(a)$ for any $v \in C_0^\infty(\Omega)$. It is possible to prove that $\delta_a \in \mathcal{D}'(\Omega)$. Hence, we write

$$\langle \delta_a, v\rangle = v(a), \quad v \in C_0^\infty(\Omega). \qquad (1.3.8)$$

The generalized function δ_a is called the *Dirac distribution.*

The following assertion is sometimes useful.

Lemma 1.14 *Let f_1, $f_2 \in L^1_{\mathrm{loc}}(\Omega)$ and $\langle f_1, v\rangle = \langle f_2, v\rangle$ for all $v \in C_0^\infty(\Omega)$. Then $f_1 = f_2$ a.e. in Ω.*

1.3.2.1 *Derivatives of distributions* Let α be a multi-index. A distribution $f_\alpha \in \mathcal{D}'(\Omega)$ is called *the α-th distributional derivative* (or generalized derivative) of a distribution $f \in \mathcal{D}'(\Omega)$, if

$$\langle f_\alpha, v\rangle = (-1)^{|\alpha|}\langle f, D^\alpha v\rangle \quad \forall\, v \in \mathcal{D}(\Omega). \qquad (1.3.9)$$

It is obvious that every distribution has generalized derivatives of all orders. Moreover, Green's theorem implies that the *distributional derivatives of any function $f \in C^k(\overline{\Omega})$* (considered as a distribution in the sense of (1.3.7)) *up to order k are equal to the classical derivatives of f*. Therefore, we use the notation $D^\alpha f$ also for distributional derivatives and write

$$\langle D^\alpha f, v\rangle = (-1)^{|\alpha|}\langle f, D^\alpha v\rangle \qquad (1.3.10)$$
$$\text{for } f \in \mathcal{D}'(\Omega) \text{ and } v \in C_0^\infty(\Omega).$$

1.3.3 *Sobolev spaces*

Let $k \geq 0$ be an integer and $1 \leq p \leq \infty$. We denote by the symbol $W^{k,p}(\Omega)$ the space of all functions $u \in L^p(\Omega)$ such that their distributional derivatives up to the order k are also elements of the space $L^p(\Omega)$:

$$W^{k,p}(\Omega) = \{u;\ D^\alpha u \in L^p(\Omega) \text{ for all multiindices } \alpha \text{ with } |\alpha| = 0, \ldots, k\}. \tag{1.3.11}$$

(Since $L^p(\Omega) \subset L^1_{\text{loc}}(\Omega)$, each $u \in L^p(\Omega)$ can be considered as a distribution defined by (1.3.7).)

The space $W^{k,p}(\Omega)$ is equipped with the norm

$$\|u\|_{W^{k,p}(\Omega)} = \|u\|_{k,p,\Omega} = \left(\sum_{|\alpha|=0}^{k} \|D^\alpha u\|^p_{L^p(\Omega)}\right)^{1/p} = \left(\sum_{|\alpha|=0}^{k} \int_\Omega |D^\alpha u|^p \, dx\right)^{1/p}, \tag{1.3.12}$$

if $1 \leq p < \infty$, and

$$\|u\|_{W^{k,\infty}(\Omega)} = \|u\|_{k,\infty,\Omega} = \max_{|\alpha| \leq k} \|D^\alpha u\|_{L^\infty(\Omega)} = \max_{|\alpha| \leq k} \left\{ \operatorname*{ess\,sup}_{x \in \Omega} |D^\alpha u(x)| \right\}$$

for $p = \infty$.

Obviously,

$$L^p(\Omega) = W^{0,p}(\Omega) \supset W^{1,p}(\Omega) \supset W^{2,p}(\Omega) \supset \ldots. \tag{1.3.13}$$

Further, we define the space $W^{k,p}_0(\Omega)$ as the closure of the set $C^\infty_0(\Omega)$ in the space $W^{k,p}(\Omega)$:

$$W^{k,p}_0(\Omega) = \overline{C^\infty_0(\Omega)}^{W^{k,p}(\Omega)}. \tag{1.3.14}$$

For $p = 2$ we often use the notation $H^k(\Omega) = W^{k,2}(\Omega)$, $H^k_0(\Omega) = W^{k,2}_0(\Omega)$ and $\|\cdot\|_{k,\Omega} = \|\cdot\|_{k,2,\Omega}$.

In the following we assume that $\Omega \subset \mathbb{R}^N$ *is a domain with a Lipschitz-continuous boundary.*

1.3.3.1 *Fundamental properties of Sobolev spaces* a) For $1 \leq p \leq \infty$, $W^{k,p}(\Omega)$ is a Banach space. The space $H^k(\Omega) = W^{k,2}(\Omega)$ is a Hilbert space with the scalar product

$$(u, v)_{k,\Omega} = \int_\Omega \sum_{|\alpha|=0}^{k} D^\alpha u D^\alpha v \, dx, \quad u, v \in H^k(\Omega). \tag{1.3.15}$$

b) For $1 \leq p < \infty$, the space $W^{k,p}(\Omega)$ is separable.

c) For $1 < p < \infty$, the space $W^{k,p}(\Omega)$ is reflexive.

d) Let $1 \leq p < \infty$. Then $C^\infty(\overline{\Omega})$ is dense in $W^{k,p}(\Omega)$.

e) The spaces $W^{k,1}(\Omega)$ and $W^{k,\infty}(\Omega)$ are not reflexive and the space $W^{k,\infty}(\Omega)$ is not separable.

Let us note that the dual $(H_0^1(\Omega))^*$ of the space $H_0^1(\Omega)$ is usually denoted by $H^{-1}(\Omega)$.

Since $W_0^{k,p}(\Omega)$ is a closed subspace of $W^{k,p}(\Omega)$, the statements a)–d) remain valid, if $W^{k,p}(\Omega)$ and $C^\infty(\overline{\Omega})$ are replaced by $W_0^{k,p}(\Omega)$ and $C_0^\infty(\Omega)$, respectively. The spaces $W_0^{k,1}(\Omega)$ and $W_0^{k,\infty}(\Omega)$ are not reflexive, but $W_0^{k,\infty}(\Omega)$ is separable.

Proofs of all of these properties can be found, for example, in (Kufner *et al.*, 1977), Section 5.2.

1.3.3.2 *Theorem on traces*

Theorem 1.15 *Let $1 \leq p < \infty$ and Ω be a Lipschitz domain.*

(i) Then there exists a uniquely determined continuous linear mapping $\gamma_0^\Omega : W^{1,p}(\Omega) \to L^p(\partial\Omega)$ such that

$$\gamma_0^\Omega(u) = u|_{\partial\Omega} \quad \textit{for all } u \in C^\infty(\overline{\Omega}). \tag{1.3.16}$$

(ii) If $1 < p < \infty$, then Green's formula

$$\int_\Omega (u\partial_i v + v\partial_i u)\, dx = \int_{\partial\Omega} \gamma_0^\Omega(u)\gamma_0^\Omega(v) n_i \, dS, \tag{1.3.17}$$
$$u \in W^{1,p}(\Omega),\ v \in W^{1,p'}(\Omega)$$

holds.

For the proof see e.g. (Kufner *et al.*, 1977), Section 6.4 and (Nečas, 1967), Theorem III.1.1.

The function $\gamma_0^\Omega(u) \in L^p(\partial\Omega)$ is called the *trace* of the function $u \in W^{1,p}(\Omega)$ on the boundary $\partial\Omega$. For simplicity, when there is no confusion, the notation $u|_{\partial\Omega} = \gamma_0^\Omega(u)$ is used not only for $u \in C^\infty(\overline{\Omega})$ but also for $u \in W^{1,p}(\Omega)$. In the sequel, whenever it is not confusing, we omit the superscript Ω of γ_0 and write simply γ_0 instead of γ_0^Ω.

The continuity of the mapping γ_0 is equivalent to the existence of a constant $c > 0$ such that

$$\|u|_{\partial\Omega}\|_{L^p(\partial\Omega)} = \|\gamma_0(u)\|_{L^p(\partial\Omega)} \leq c\|u\|_{1,p,\Omega}, \quad u \in W^{1,p}(\Omega). \tag{1.3.18}$$

Note that the symbol c will often denote a positive *generic constant*, attaining, in general, different values in different places.

1.3.3.3 *The Friedrichs–Poincaré inequalities*

Lemma 1.16 (Friedrichs inequality) *Let $1 \le p < \infty$ and Ω be a bounded Lipschitz domain. Let a set $\Gamma \subset \partial\Omega$ be measurable with respect to the $(N-1)$-dimensional measure $\tilde{\mu} = \mathrm{meas}_{N-1}$ defined on $\partial\Omega$ and let $\mathrm{meas}_{N-1}(\Gamma) > 0$. Then there exists a constant $c = c(p, N, \Omega) > 0$ such that*

$$\|u\|_{1,\Omega} \le c\|\nabla u\|_{0,\Omega} \quad \textit{for all } u \in W^{1,2}(\Omega) \textit{ with } \gamma_0(u) = 0 \ \tilde{\mu}\textit{-a.e. on } \Gamma. \tag{1.3.19}$$

Lemma 1.17 (Poincaré inequality) *Let $1 \le p < \infty$ and Ω be a bounded Lipschitz domain. Then there exists a constant $c = c(p, N, \Omega) > 0$ such that*

$$\int_\Omega |u|^p\, dx \le c\left(\int_\Omega |\nabla u|^p\, dx + \left|\int_\Omega u\, dx\right|^p\right), \quad u \in W^{1,p}(\Omega). \tag{1.3.20}$$

1.3.3.4 *Imbedding theorems* a) Let $k \ge 0$ and $1 \le p \le \infty$ and Ω be a bounded Lipschitz domain. Then

$$W^{k,p}(\Omega) \hookrightarrow L^q(\Omega) \text{ where } \frac{1}{q} = \frac{1}{p} - \frac{k}{N}, \text{ if } k < \frac{N}{p}, \tag{1.3.21}$$

$$W^{k,p}(\Omega) \hookrightarrow L^q(\Omega) \text{ for all } q \in [1, \infty), \text{ if } k = \frac{N}{p},$$

$$W^{k,p}(\Omega) \hookrightarrow C^{0,k-N/p}(\overline{\Omega}), \quad \text{if } \frac{N}{p} < k < \frac{N}{p} + 1,$$

$$W^{k,p}(\Omega) \hookrightarrow C^{0,\alpha}(\overline{\Omega}) \text{ for all } \alpha \in (0,1), \text{ if } k = \frac{N}{p} + 1,$$

$$W^{k,p}(\Omega) \hookrightarrow C^{0,1}(\overline{\Omega}), \text{ if } k > \frac{N}{p} + 1.$$

b) Let $k > 0$, $1 \le p \le \infty$. Then

$$W^{k,p}(\Omega) \hookrightarrow\hookrightarrow L^q(\Omega) \text{ for all } q \in [1, p^*) \text{ with } \frac{1}{p^*} = \frac{1}{p} - \frac{k}{N}, \tag{1.3.22}$$

$$\text{if } k < \frac{N}{p},$$

$$W^{k,p}(\Omega) \hookrightarrow\hookrightarrow L^q(\Omega) \text{ for all } q \in [1, \infty), \text{ if } k = \frac{N}{p},$$

$$W^{k,p}(\Omega) \hookrightarrow\hookrightarrow C(\overline{\Omega}), \text{ if } k > \frac{N}{p}.$$

(We set $1/\infty := 0$.) The symbols $\hookrightarrow$ and $\hookrightarrow\hookrightarrow$ denote the continuous and compact imbedding, respectively (see Sections 1.4.7.4 and 1.4.7.5).

Statements (1.3.21) and (1.3.22) are called *Sobolev's imbedding theorems* and *Kondrashov's theorems on compact imbedding*, respectively. The compact imbedding $W^{1,2}(\Omega) \hookrightarrow\hookrightarrow L^2(\Omega)$ (i.e. the case $p = q = 2$, $k = 1$) is known as *Rellich's theorem.*

The statements mentioned can be *iterated*. For example,

$$W^{k+s,p}(\Omega) \hookrightarrow W^{s,q}(\Omega) \text{ where } \frac{1}{q} = \frac{1}{p} - \frac{k}{n}, \text{ if } k < \frac{N}{p},$$

etc.

1.3.3.5 *Characterization ofthe space $W_0^{1,p}$* The space $W_0^{1,p}(\Omega)$ with $1 \le p < \infty$, defined in (1.3.14), can be characterized as

$$W_0^{1,p}(\Omega) = \left\{ v \in W^{1,p}(\Omega); \gamma_0(v) = 0 \right\}. \tag{1.3.23}$$

1.3.3.6 *Sobolev–Slobodetskii spaces of functions with 'fractional derivatives'* If $k \ge 0$ is an integer, $\varepsilon \in (0,1)$ and $p \in [1,\infty)$, then $W^{k+\varepsilon,p}(\Omega)$ denotes the space of all functions $u \in W^{k,p}(\Omega)$ such that

$$I_\alpha(u) = \int_\Omega \int_\Omega \frac{|D^\alpha u(x) - D^\alpha u(y)|^p}{|x-y|^{N+p\varepsilon}}\, dx\, dy < \infty \quad \text{for } |\alpha| = k. \tag{1.3.24}$$

The space $W^{k+\varepsilon,p}(\Omega)$ equipped with the norm

$$\|u\|_{k+\varepsilon,p,\Omega} = \left(\|u\|_{k,p,\Omega}^p + \sum_{|\alpha|=k} I_\alpha(u) \right)^{1/p} \tag{1.3.25}$$

is a Banach space.

1.3.3.7 *Sobolev-Slobodetskii spaces on boundary* The Sobolev-Slobodetskii spaces can also be defined on the Lipschitz-continuous boundary $\partial\Omega$ of a domain Ω. Let $\partial\Omega$ be characterized with the use of local (sufficiently regular) maps $x_1{}' = a_r(x'), x' \in \mathcal{M}_r, a_r \in C^{k-1,1}(\mathcal{M}_r), r = 1, \dots, R$, from Section 1.1.7. We say that a function $u : \partial\Omega \to I\!R$ is an element of the space $W^{k+\varepsilon,p}(\partial\Omega)$, where $\varepsilon \in [0,1)$ and $p \ge 1$, if $u(a_r(x'), x') \in W^{k+\varepsilon,p}(\mathcal{M}_r)$ for all $r = 1, \dots, R$. The space of traces of all functions $u \in W^{1+\varepsilon,p}(\Omega)$ can be identified with the space $W^{1+\varepsilon-1/p,p}(\partial\Omega)$. Hence, by $H^{1/2}(\partial\Omega) = W^{1/2,2}(\partial\Omega)$ we denote the space of traces of all functions from $H^1(\Omega)$. Moreover, the operator of traces is a compact mapping of the space $W^{1,p}(\Omega)$ into $L^q(\partial\Omega)$, if either $p \in (1,N)$ and $1 \le q < p(N-1)/(N-p)$, or $p \ge N$ and $q \ge 1$.

1.3.4 *Functions with values in Banach spaces*

In the investigation of nonstationary problems we shall work with functions which depend on time and have values in Banach spaces. If $u(x,t)$ is a function of the space variable x and time t, then it is sometimes suitable to separate these variables and consider u as a function $u(t) = u(\cdot, t)$ which for each t under consideration attains a value $u(t)$ that is a function of x and belongs to a suitable space of functions depending on x. This means that $u(t)$ represents the mapping '$x \to [u(t)](x) = u(x,t)$'.

Let $a, b \in \mathbb{R}$, $a < b$, and let X be a Banach space with norm $\|\cdot\|$. By a *function defined on the interval* $[a, b]$ *with its values in the space* X we understand any mapping $u: [a, b] \to X$.

We say that a function $u: [a, b] \to X$ is *continuous at a point* $t_0 \in [a, b]$, if

$$\lim_{\substack{t \to t_0 \\ t \in [a,b]}} \|u(t) - u(t_0)\| = 0. \tag{1.3.26}$$

By the symbol $C([a, b]; X)$ we shall denote the space of all functions continuous on the interval $[a, b]$ (i.e. continuous at each $t \in [a, b]$) with values in X. The space $C([a, b]; X)$ equipped with the norm

$$\|u\|_{C([a,b]; X)} = \max_{t \in [a,b]} \|u(t)\| \tag{1.3.27}$$

is a Banach space.

1.3.4.1 *Some examples of spaces of functions with values in a Banach space*
Let X be a Banach space. For $p \in [1, \infty]$ we denote by $L^p(a, b; X)$ the space of (equivalent classes of) strongly measurable functions $u: (a, b) \to X$ such that

$$\|u\|_{L^p(a,b;X)} := \left[\int_a^b \|u(t)\|_X^p \, dt \right]^{1/p} < \infty, \quad \text{if } 1 \le p < \infty, \tag{1.3.28}$$

and

$$\begin{aligned} \|u\|_{L^\infty(a,b;X)} &:= \operatorname{ess} \sup_{t \in (a,b)} \|u(t)\|_X \\ &= \inf_{\text{meas}(N)=0} \; \sup_{t \in (a,b) \setminus N} \|u(t)\|_X < +\infty, \qquad \text{if } p = \infty. \end{aligned} \tag{1.3.29}$$

We speak of Bochner spaces. It can be proven that $L^p(a, b; X)$ is a Banach space. (The definition of a strongly measurable function $u: (a, b) \to X$ can be found in (Kufner *et al.*, 1977) or (Feistauer, 1993), Chapter 8.)

If the space X is reflexive, so is $L^p(a, b; X)$ for $p \in (1, \infty)$. Let $1 \le p < \infty$. Then the dual of $L^p(a, b; X)$ is $L^q(a, b; X^*)$, where $1/p + 1/q = 1$ and X^* is the dual of X (for $p = 1$ we set $q = \infty$). The duality between $L^q(a, b; X^*)$ and $L^p(a, b; X)$ becomes

$$\begin{aligned} \langle f, v \rangle &= \int_a^b \langle f(t), v(t) \rangle_{X^*,X} \, dt, \\ f &\in L^q(a, b; X^*), \quad v \in L^p(a, b; X). \end{aligned} \tag{1.3.30}$$

$\langle f(t), v(t) \rangle_{X^*,X}$ denotes the value of the functional $f(t) \in X^*$ at $v(t) \in X$.

If X is a separable Banach space, then $L^p(a, b; X)$ is also separable, provided $p \in [1, \infty)$. (See, for example, (Edwards, 1965), Section 8.18.1.)

We define analogously Sobolev spaces of functions with values in X:

$$W^{k,p}(a,b;X) = \Big\{ f \in L^p(a,b;X); \frac{d^j f}{dt^j} \in L^p(a,b;X),\ j = 1,\dots,k \Big\}, \tag{1.3.31}$$

where $k = 1, 2, \dots$ and $p \in [1, \infty]$. The norm of $f \in W^{k,p}(a,b;X)$ is defined by

$$\|f\|_{W^{k,p}(a,b;X)} = \Big(\sum_{j=1}^{k} \Big\| \frac{d^j f}{dt^j} \Big\|^p_{L^p(a,b;X)} \Big)^{1/p}. \tag{1.3.32}$$

These spaces have exactly the same properties as listed above for L^p-spaces.

We also define spaces of continuous and continuously differentiable functions on an interval I with values in X:

$$C(I;X) = C^0(I;X) = \tag{1.3.33}$$
$$\{ f : I \to X;\ f \text{ is bounded and continuous at each point of } I \},$$
$$C^k(I;X) = \Big\{ f \in C(I;X);\ \frac{d^j f}{dt^j} \in C(I;X) \text{ for all } j = 1,\dots,k \Big\}.$$

The norm of $f \in C^k(I;X), k = 0, 1, \dots$, is defined by

$$\|f\|_{C^k(I;X)} = \sup \Big\{ \Big\| \frac{d^j f}{dt^j}(t) \Big\|_{C(I;X)};\ j = 0,\dots,k,\ t \in I \Big\}. \tag{1.3.34}$$

These spaces are nonreflexive Banach spaces, separable, if X is.

1.4 Survey of concepts and results from functional analysis

In this section we give a survey of the basic concepts and results of functional analysis which are frequently used. All results are stated without proofs, which can be found in standard monographs (e.g. (Edwards, 1965), (Ljusternik and Sobolev, 1974), (Taylor, 1967), (Yosida, 1974), (Zeidler, 1989)).

1.4.1 *Linear vector spaces*

The set X is called a (real) *linear vector space* (linear space, in brief), if the addition of elements of X and multiplication by scalars is defined

$$u, v \in X \longrightarrow u + v \in X,$$
$$u \in X,\ \lambda \in I\!R \longrightarrow \lambda u \in X$$

so that for any u, v, $w \in X$ and λ, $\mu \in I\!R$ the following axioms are satisfied:

(i) $u + v = v + u$;

(ii) $u + (v + w) = (u + v) + w$;

(iii) in X there exists a uniquely determined element denoted by 0 and called the *zero element* such that $u + 0 = u$;

(iv) to each $u \in X$ there exists a uniquely determined element $(-u)$ such that $u + (-u) = 0$;

(v) $\lambda(u + v) = \lambda u + \lambda v$;

(vi) $(\lambda + \mu)u = \lambda u + \mu u$;

(vii) $(\lambda\mu)u = \lambda(\mu u)$;

(viii) $1u = u$;

(ix) $0u = 0$.

(The zero element in X is denoted by the same symbol as zero in $I\!R$ without the danger of misunderstanding.) In X we also define *subtraction*: $u - v = u + (-v)$.

1.4.1.1 *Convex sets* A subset M of a linear vector space X is called *convex*, if

$$u,\, v \in M,\ \lambda \in [0,1] \Longrightarrow \lambda u + (1 - \lambda)\, v \in M.$$

1.4.2 *Normed linear space*

1.4.2.1 *Definition* Let X be a linear vector space. A function $\|\cdot\| : X \to [0, \infty)$ is called a *norm* on X, if it satisfies the following axioms:

(i) $u \in X,\ \|u\| = 0 \Leftrightarrow u = 0$;

(ii) $\|\lambda u\| = |\lambda|\, \|u\|, \quad \lambda \in I\!R,\ u \in X$;

(iii) $\|u + v\| \le \|u\| + \|v\|, \quad u,\, v \in X$ (triangle inequality).

A linear space X equipped with a norm $\|\cdot\|$ is called a *normed linear space,* and it is a metric space with the *metric* (i.e. distance) $d(u, v) = \|u - v\|$ for $u, v \in X$.

1.4.2.2 *Balls* The ball of centre $a \in X$ with radius $\varepsilon > 0$ is defined as the set $B_\varepsilon(a) := \{u \in X;\ \|u - a\| < \varepsilon\}$. (We often refer to an ε-neighbourhood of a.)

1.4.2.3 *Bounded sets* A set $M \subset X$ (sequence $\{u_n\}_{n=1}^{\infty}$ with $u_n \in X$) is called *bounded* in X, if there exists a constant $K \ge 0$ such that $\|u\| \le K$ for all $u \in M$ ($\|u_n\| \le K$ for all $n = 1, 2, \ldots$).

1.4.2.4 *Open sets* We say that a set $M \subset X$ is *open*, if for every $a \in M$ there exists a ball with centre a contained in M.

1.4.2.5 *Sequences and limits* Let $u,\, u_n \in X$ for $n = 1, 2, \ldots$. We say that

$$\lim_{n\to\infty} u_n = u \text{ (in } X\text{) or, more simply, } u_n \to u \text{ as } n \to \infty,$$

if $\lim_{n\to\infty} \|u_n - u\| = 0$. In this situation, we speak of *strong convergence* of u_n to u in X.

1.4.2.6 *Closed sets, closure and boundary* A set $M \subset X$ is called *closed*, if the following holds:

$$u_n \in M \text{ for } n = 1, 2, \ldots,\ u_n \to u \text{ as } n \to \infty \implies u \in M.$$

The *closure* of a set $M \subset X$ is defined as the set

$$\overline{M} = \{u \in X;\ \text{there exists a sequence } \{u_n\} \subset M \text{ such that } u_n \to u\}.$$

A set M is closed if and only if $\overline{M} = M$. Moreover, M is closed if and only if $X \setminus M$ is open. The empty set $\emptyset$ is open as well as closed. The same is true for the set $M = X$. The boundary of a set $M \subset X$ is the set $\partial M = \overline{M} \cap \overline{(X \setminus M)}$.

1.4.2.7 *Continuity* Let f be a mapping from $\mathcal{D}(f) \subset X$ into Y. We say that f is continuous at a point $a \in \mathcal{D}(f)$ if the following implication holds:

$$x_n \in \mathcal{D}(f),\ x_n \to a \text{ as } n \to \infty \implies f(x_n) \to f(a) \text{ as } n \to \infty.$$

1.4.2.8 *Compactness* A set $M \subset X$ is called *compact* (*precompact* or *relatively compact*), if for every bounded sequence $\{u_n\}_{n=1}^{\infty} \subset M$ there exist a subsequence $\{u_{n_k}\}_{k=1}^{\infty}$ and an element $u \in M$ $(u \in X)$ such that $u_{n_k} \to u$ in X as $k \to \infty$.

1.4.2.9 *Connected sets* A set $M \subset X$ is called *connected*, if the following implication holds:

$$M = A \cup B,\ A \cap \bar{B} = \emptyset = \bar{A} \cap B \implies \text{either } A \text{ or } B \text{ is empty.}$$

1.4.2.10 *Banach space* Let X be a normed linear space. A sequence $\{u_n\} \subset X$ is called a *Cauchy* (or *fundamental*) *sequence* if the following holds: for any $\varepsilon > 0$ there exists $n_0 = n_0(\varepsilon)$ such that $\|u_m - u_n\| < \varepsilon$ for all integers m, $n > n_0$. We say that X is a *complete space*, if every Cauchy sequence $\{u_n\} \subset X$ is convergent: there exists a $u \in X$ such that $u_n \to u$ as $n \to \infty$. A complete normed linear space is called a *Banach space*.

1.4.2.11 *Hilbert space* A real scalar function $(\cdot, \cdot)$ defined on $X \times X$, where X is a linear vector space, is called a *scalar* (or *inner*) *product* on X, if for any u, v, $w \in X$ and $\lambda \in I\!R$ the following relations hold:

$$\begin{aligned}
&\text{(i)}\ (u + v, w) = (u, w) + (v, w);\\
&\text{(ii)}\ (\lambda u, v) = \lambda(u, v);\\
&\text{(iii)}\ (u, v) = (v, u);\\
&\text{(iv)}\ (u, u) > 0, \text{ provided } u \neq 0.
\end{aligned}$$

A scalar product induces the norm in X defined as $\|u\| = (u, u)^{1/2}$ for $u \in X$. The so-called *Cauchy inequality*

$$|(u, v)| \leq \|u\|\,\|v\| \quad \text{for } u,\ v \in X$$

can be proven. A linear vector space with a scalar product which is complete with respect to the induced norm is called a *Hilbert space*.

1.4.2.12 *Density, separability* A subset $M \subset X$ of a normed linear space X is called *dense* in X, if $\overline{M} = X$. The space X is called *separable*, if there exists a countable set $M \subset X$, dense in X. (M is countable, if all its elements can be ordered into a sequence.)

1.4.2.13 *Subspaces* Let X be a normed linear space with a norm $\|\cdot\|_X$ and let $M \subset X$ be a linear vector space. Then M equipped with the norm $\|\cdot\|_X$ is called a *subspace* of the space X. If M is closed in X, we speak of a *closed subspace* of X.

1.4.2.14 *Subspaces and separability* A closed subspace of a separable normed linear space is also a separable space.

1.4.2.15 *Subspaces of Banach spaces* A closed subspace of a Banach (Hilbert) space is also a Banach (Hilbert) space.

1.4.3 *Duals to Banach spaces, weak and weak-* topologies*

1.4.3.1 *Continuous linear functionals on normed linear spaces* A continuous linear functional defined on a normed linear space X is a continuous linear mapping $f : X \to I\!R$. For the value of f at a point $u \in X$, we use the notation $f(u)$ or $\langle f, u \rangle$.

1.4.3.2 *Dual spaces* The set of all continuous linear functionals on X forms a linear vector space, which is denoted by X^* (or X') and is called the *dual of* X. The space X^* together with the norm

$$\|f\|_{X^*} = \sup_{\substack{u \in X \\ u \neq 0}} \frac{|\langle f, u \rangle|}{\|u\|_X}, \qquad f \in X^*$$

is a normed linear space. The mapping $\langle \cdot, \cdot \rangle : X^* \times X \to I\!R$ is called the *duality* (mapping) between X and X^*.

It is known that X^* endowed with the norm $\|\cdot\|_{X^*}$ is a Banach space.

1.4.3.3 *Hahn–Banach theorem* Let the following assumptions be satisfied: X is a normed linear space, $M \subset X$ is a linear subspace, $\varphi : M \to I\!R$ is a linear mapping continuous on M in the norm $\|\cdot\|_X$ (it means that u_n, $u \in M$, $\|u_n - u\|_X \to 0 \Longrightarrow \varphi(u_n) \to \varphi(u)$). Then there exists $\phi \in X^*$ such that

$$\phi(u) = \varphi(u) \text{ for } u \in M \text{ and } \|\phi\|_{X^*} = \sup_{\substack{u \in M \\ u \neq 0}} \frac{|\varphi(u)|}{\|u\|_X}.$$

1.4.3.4 *Another calculation of the norm in X* One consequence of the Hahn–Banach theorem is the following formula:

$$\|x\|_X = \sup_{f \in X^*} \frac{\langle f, x \rangle}{\|f\|_{X^*}}.$$

1.4.3.5 *Weak convergence on X* Let X be a normed linear space and $\{u_n\} \subset X$. We say that

$$u_n \to u \in X \text{ weakly in } X \text{ as } n \to \infty,$$

if

$$\lim_{n\to\infty} f(u_n) = f(u) \quad \forall f \in X^*.$$

For weak convergence, sometimes, the notation $u_n \xrightarrow{w} u$ (or $u_n \rightharpoonup u$) is used.

1.4.3.6 *Strong convergence implies weak convergence* If $u_n \to u$ (strongly) in X, then $u_n \to u$ weakly in X.

1.4.3.7 *Weakly convergent sequences are bounded* If X is a Banach space, $\{u_n\} \subset X$, which is weakly convergent in X to $u \in X$, then it is bounded and $\|u\|_X \leq \liminf_{n\to\infty} \|u_n\|_X$.

1.4.3.8 *Weak-$*$ convergence in X^** Let X be a normed linear space, X^* be the dual of X and $\{f_n\}_{n=1}^{\infty} \subset X^*$. We say that

$$f_n \to f \text{ weak-}* \text{ (or weakly-}*) \text{ in } X^* \text{ as } n \to \infty,$$

if

$$\lim_{n\to\infty} f_n(u) = f(u), \quad u \in X.$$

1.4.3.9 *Strong convergence implies weak-$*$ convergence* If $f_n \to f$ in X^*, then $f_n \to f$ weak-$*$ in X^*.

1.4.3.10 *Weakly-$*$ convergent sequence is bounded* If X is a Banach space, then any sequence $\{f_n\} \subset X^*$ weakly-$*$ convergent in X^* to $f \in X^*$ is bounded and $\|f\|_{X^*} \leq \liminf \|f_n\|_{X^*}$.

1.4.3.11 *Reflexivity* If X is a Banach space and X^* is its dual, then we can consider the dual of X^*:

$$X^{**} = (X^*)^*.$$

Let $u \in X$. It is obvious that the mapping '$\varphi \in X^* \to \langle \varphi, u \rangle \in \mathbb{R}$' defines an element of X^{**}. It means that for each $u \in X$ there exists $Ju \in X^{**}$ such that

$$(Ju)(\varphi) := \langle \varphi, u \rangle, \quad \varphi \in X^*.$$

$J: X \to X^{**}$ is called the *canonical mapping* of the space X into X^{**}.

1.4.3.12 *Basic properties of the canonical mapping* The mapping J is linear, continuous and one-to-one. Its inverse J^{-1} is also continuous. (This means that J is an *isomorphism* between X and $J(X)$.) Further, $\|u\|_X = \|Ju\|_{X^{**}}$. ($J$ is an *isometric* mapping.)

1.4.3.13 *Reflexive Banach spaces* We say that a Banach space is *reflexive*, if $J(X) = X^{**}$. It means that for each $g \in X^{**}$ there exists a uniquely determined element $u_g \in X$ such that $g(\varphi) = \langle \varphi, u_g \rangle$ for all $\varphi \in X^*$ and $\|g\|_{X^{**}} = \|u_g\|_X$. For a reflexive Banach space X we simply write $X = X^{**}$.

1.4.3.14 *Subspaces of reflexive Banach spaces* Any closed subspace of a reflexive Banach space is also a reflexive Banach space.

1.4.3.15 Let X be a separable reflexive Banach space. Then X^* is also a reflexive and separable Banach space.

1.4.3.16 *A version of the Banach–Alaoglu theorem* Let X be a reflexive Banach space and let $\{u_n\} \subset X$ be a bounded sequence. Then there exists a subsequence $\{u_{n_k}\}_{k=1}^{\infty}$ weakly convergent in X.

1.4.3.17 *Another version of the Banach–Alaoglu theorem* Let X be a separable Banach space and let $\{f_n\} \subset X^*$ be a bounded sequence. Then there exists a subsequence $\{f_{n_k}\}_{k=1}^{\infty}$ weakly-$*$ convergent in X^*.

1.4.4 *Riesz representation theorem*

Let X be a Hilbert space. Then for each $\varphi \in X^*$ there exists a uniquely determined element $u_\varphi \in X$ such that

$$\langle \varphi, v \rangle = (u_\varphi, v) \quad \forall v \in X.$$

Moreover,

$$\|u_\varphi\|_X = \|\varphi\|_{X^*}.$$

1.4.4.1 *A corollary of the Riesz representation theorem* Any Hilbert space is a reflexive Banach space.

1.4.5 *Operators*

1.4.5.1 *Linear operators* Let X and Y be Banach spaces with norms $\|\cdot\|_X$ and $\|\cdot\|_Y$. A mapping $A : \mathcal{D}(A) \subset X \to Y$, where $\mathcal{D}(A)$ is a subspace of X, is called a *linear operator*, if it satisfies the conditions

$$\text{(i) } A(u+v) = A(u) + A(v), \quad u, v \in X$$
$$\text{(ii) } A(\lambda u) = \lambda A(u), \quad \lambda \in I\!R,\ u \in X.$$

We call $\mathcal{D}(A)$ the *domain* of A. The set

$$\mathcal{R}(A) = A(\mathcal{D}(A)) = \{y \in Y;\ y = Ax,\ x \in \mathcal{D}(A)\}$$

is called the *range* of A,

$$\mathcal{N}(A) = \{x \in X;\ Ax = 0\}$$

the *null space* (or *kernel*) of A, and

$$\mathcal{G}(A) = \{(x, Ax);\ x \in \mathcal{D}(A)\} \subset X \times Y$$

the graph of A.

1.4.5.2 *Bounded linear operators* We say that the linear operator A defined in Section 1.4.5.1 is bounded, if there exists $c \geq 0$ such that

$$\|Au\|_Y \leq c\|u\|_X, \quad u \in \mathcal{D}(A).$$

1.4.5.3 *Continuous operators* We say that a (not necessarily linear) mapping $A: X \to Y$ is *continuous*, if the implication

$$u_n \to u \text{ in } X \Longrightarrow A(u_n) \to A(u) \text{ in } Y$$

is valid.

1.4.5.4 *Continuous linear operators* A linear operator $A: X \to Y$ is continuous if and only if it is bounded.

The set $\mathcal{L}(X,Y)$ of all continuous linear operators $A: X \to Y$ equipped with the norm

$$\|A\| := \sup_{\substack{u \in X \\ u \neq 0}} \frac{\|Au\|_Y}{\|u\|_X}$$

forms a Banach space.

1.4.5.5 *Closed and densely defined linear operators* A linear operator is said to be a *closed linear operator*, if its graph is closed. A linear operator $A: \mathcal{D}(A) \subset X \to Y$ is called *densely defined* (in X), if $\mathcal{D}(A)$ is dense in X.

1.4.5.6 *Adjoint linear operators* For any densely defined linear operator $A: X \to Y$ one can define the so-called *adjoint operator* $A^*: Y^* \to X^*$ by setting

$$\mathcal{D}(A^*) = \{v \in Y^*; \text{ there exists } c \geq 0$$
$$\text{such that } |\langle v, Au\rangle| \leq c\|u\|_X, \text{ for all } u \in \mathcal{D}(A)\},$$
$$\langle A^*v, u\rangle = \langle v, Au\rangle, \quad v \in \mathcal{D}(A^*),\ u \in \mathcal{D}(A).$$

1.4.5.7 *Symmetric operators and selfadjoint operators* If $X = Y = H$ where H is a Hilbert space, one can identify H and H^* by using the Riesz representation theorem from Section 1.4.4). Then A is called the *symmetric operator*, if A^* is an extension of A, i.e. if $\mathcal{D}(A) \subset \mathcal{D}(A^*)$ and $A^*v = Av$ for $v \in \mathcal{D}(A)$.

The operator A is called the *selfadjoint operator*, if A is the symmetric operator and $\mathcal{D}(A) = \mathcal{D}(A^*)$.

A selfadjoint operator is closed. If A is a symmetric operator and $\mathcal{D}(A) = X$, then A is bounded and selfadjoint.

1.4.5.8 *Completely continuous linear operators* A completely continuous operator is defined as a continuous linear operator $A: X \to Y$ (X and Y being normed linear spaces) which satisfies the following condition: if $M \subset X$ is a bounded set, then $A(M)$ is precompact in Y. This means that any bounded sequence $\{u_n\}_{n=1}^{\infty} \subset X$ contains a subsequence $\{u_{n_k}\}_{k=1}^{\infty}$ such that the sequence $\{A(u_{n_k})\}_{k=1}^{\infty}$ is convergent in Y.

A completely continuous linear operator is also called a *compact* operator.

1.4.6 *Lax–Milgram lemma*

Let us assume that H is a Hilbert space with scalar product $(\cdot,\cdot)_H$ and let $a : H \times H \to I\!R$ be a bilinear form having the following properties:

(i) a is continuous, i.e. there exists a constant $M > 0$ such that
$|a(z,v)| \le M\|z\|_H\|v\|_H \quad \forall z, v \in H$;

(ii) a is H-*elliptic,* i.e. there exists a constant $\alpha > 0$ such that
$a(z,z) \ge \alpha\|z\|_H^2 \quad \forall z \in H$.

Let $\varphi \in H^*$. Then there exists exactly one solution $z \in H$ of the equation

$$a(z,v) = \langle \varphi, v\rangle \quad \forall v \in H.$$

The mapping $\varphi \to z$ assigning to $\varphi \in H^*$ the solution z of the above equation is an invertible continuous linear operator of H^* onto H.

1.4.7 *Imbeddings*

1.4.7.1 *Identity operator* Let X and Y be normed linear spaces with norms $\|\cdot\|_X$ and $\|\cdot\|_Y$, respectively, and let

$$X \subset Y$$

(in the sense of sets). Let us define the *identity operator* I from X into Y with the domain of definition X and the range $I(X) = X$ by the relation

$$Iu = u \text{ for } u \in X.$$

1.4.7.2 *Continuous imbedding* The identity operator I is linear. If I is continuous, we speak of the *continuous imbedding* of X into Y.

1.4.7.3 *Equivalence* The continuity of the imbedding I is equivalent to the existence of a constant $c > 0$ such that

$$\|u\|_Y \le c\|u\|_X \quad \forall u \in X.$$

1.4.7.4 *Notation* The fact of the continuous imbedding of X into Y is written in the form $X \hookrightarrow Y$.

1.4.7.5 *Compact imbedding* In the case when the imbedding operator I is completely continuous we speak of a *compact imbedding* and write $X \hookrightarrow\hookrightarrow Y$.

1.4.7.6 *Consequence of compact imbedding* Let $X \hookrightarrow\hookrightarrow Y$ and $\{u_n\} \subset X$:

a) If a sequence $\{u_n\}_{n=1}^{\infty}$ is bounded in X, then it is possible to extract a subsequence $\{u_{n_k}\}_{k=1}^{\infty}$ strongly convergent in Y.

b) If $u_n \to u$ weakly in X, then $u_n \to u$ strongly in Y.

1.4.7.7 *Dual imbedding* Let X and Y be normed linear spaces. Then we have:

$$\text{a)}\quad X \hookrightarrow Y \Longrightarrow Y^* \hookrightarrow X^*; \qquad \text{b)}\quad X \hookrightarrow\hookrightarrow Y \Longrightarrow Y^* \hookrightarrow\hookrightarrow X^*.$$

1.4.8 *Solution of nonlinear operator equations*

A number of problems can be transformed to an equation of the form

$$u = F(u),$$

for an unknown $u \in X$, where X is a Banach space and $F : X \to X$. Every solution of this equation is called a *fixed point* of the mapping F. Let us quote some fundamental theorems on the existence of a fixed point.

1.4.8.1 *The method of contractive mapping* A mapping $F : X \to X$ is called *contractive*, if there exists a constant $q \in [0, 1)$ such that

$$\|F(u) - F(v)\| \le q\|u - v\| \quad \forall u, v \in X.$$

1.4.8.2 *Banach theorem* Let X be a Banach space and $F : X \to X$ be a contractive mapping. Then there exists exactly one fixed point $u \in X$ of the mapping F. The fixed point u can be obtained as

$$u = \lim_{k\to\infty} u_k,$$

where $u_0 \in X$ is arbitrary and

$$u_{k+1} = F(u_k) \quad \text{for } k \ge 0.$$

1.4.8.3 *Brouwer theorem* Let $M \subset \mathbb{R}^N$ be a nonempty, bounded, closed, convex set and let $F : M \to M$ be a continuous mapping. Then there exists at least one fixed point $u \in M$ of the mapping F.

An extension of the Brouwer theorem to general Banach spaces is the Schauder theorem.

1.4.8.4 *Completely continuous (nonlinear) operator* Let X be a Banach space, $M \subset X$ and $F : M \to X$. We say that F is *completely continuous* (or compact) in M, if F is continuous and maps any bounded set $\widetilde{M} \subset M$ onto a precompact set.

1.4.8.5 *Schauder theorem* Let X be a Banach space, $M \subset X$ be a nonempty, bounded, closed, convex set and $F : M \to M$ be a completely continuous mapping. Then there exists at least one fixed point $u \in M$ of the mapping F.

2

BASIC FACTS FROM THE THEORY OF THE EULER AND NAVIER–STOKES EQUATIONS

In this chapter we shall study the basic qualitative properties of the Euler and Navier–Stokes equations of compressible flow. We start with the Euler equations as a nonlinear hyperbolic system. Then the existence of smooth solutions and the typical phenomenon of loss of smoothness and lifespan of smooth solutions will be discussed. This is followed by an introduction to the theory of weak solutions with such concepts as shock and rarefaction waves, Rankine–Hugoniot conditions and entropy admissible solutions. In particular, existence results for one, two and m scalar conservation laws are given, and the Riemann problem in one space dimension is studied.

In the second part of this chapter a short overview of existence theory for the Navier–Stokes equations of compressible flow is presented. No detailed arguments are given for the Euler and Navier–Stokes equations and we refer the reader in this respect to the recent monograph (Novotný and Straškraba, 2003).

2.1 Hyperbolic systems and the Euler equations

The Euler equations of (adiabatic) compressible fluid flow written in conservation form are as follows (see (1.2.107)–(1.2.109)):

$$\begin{aligned} &\rho_t + \operatorname{div}(\rho \boldsymbol{v}) = 0, \\ &(\rho \boldsymbol{v})_t + \operatorname{div}(\rho \boldsymbol{v} \otimes \boldsymbol{v}) + \nabla p = 0, \\ &E_t + \operatorname{div}\big(\boldsymbol{v}(E+p)\big) = 0. \end{aligned} \tag{2.1.1}$$

Here

$$E = \rho\Big(\frac{|\boldsymbol{v}|^2}{2} + e\Big) \tag{2.1.2}$$

is the total energy, $\rho(x,t)$ the density, $\boldsymbol{v}$ the velocity and e is the internal energy, which is to be specified by an appropriate constitutive equation, e.g.

$$e = e(\rho, p), \tag{2.1.3}$$

obtained from thermodynamical laws (cf. Section 1.2.19).

The nonlinear system (2.1.1) can be written in the form

$$\frac{\partial \boldsymbol{w}}{\partial t} + \sum_{j=1}^{N} \frac{\partial \boldsymbol{f}_j(\boldsymbol{w})}{\partial x_j} = 0. \tag{2.1.4}$$

We assume that $\boldsymbol{f}_j = (f_{j1}, \ldots, f_{jm})^{\mathrm{T}} : D \to \mathbb{R}^m$, $j = 1, \ldots, N$ $(m, N \in \mathbb{N})$, are continuously differentiable functions and $D \subset \mathbb{R}^m$ is an open set. We consider

(2.1.4) in a space-time cylinder $Q_T = \Omega \times (0,T)$, where $\Omega \subset \mathbb{R}^N$ is a domain occupied by a gas and $T > 0$. The vector functions $\boldsymbol{f}_j$ are called *fluxes* of the quantity $\boldsymbol{w} = (w_1, \dots, w_m)^{\mathrm{T}}$ in the directions x_j.

System (2.1.4) is called a *system of conservation laws.* Assuming that $\boldsymbol{w} \in C^1(Q_T)^m$, it can be written as a *quasilinear system* of the type

$$\mathbb{A}_0(\boldsymbol{w})\boldsymbol{w}_t + \sum_{j=1}^{N} \mathbb{A}_j(\boldsymbol{w})\boldsymbol{w}_{x_j} = 0 \tag{2.1.5}$$

with $m \times m$ matrices $\mathbb{A}_j(\boldsymbol{w}), j = 0, \dots, N$, which depend on the unknown function $\boldsymbol{w}$ in a generally nonlinear way. That is, in the case of (2.1.4), $\mathbb{A}_0(\boldsymbol{w}) = \mathbb{I}$ is the unit $m \times m$ matrix and $\mathbb{A}_j = D\boldsymbol{f}_j(\boldsymbol{w})/D\boldsymbol{w}$ is the Jacobi matrix of $\boldsymbol{f}_j, j = 1, \dots, N$.

It is a well known fact that even the simplest equations of the type (2.1.5) exhibit such nonlinear phenomena as nonexistence of global smooth solutions on a massive set of initial and/or boundary data.

To understand the main features of the Euler equations, it is desirable to present some facts from the theory of first order quasilinear systems of partial differential equations of the form (2.1.5). We start from the following definitions:

Definition 2.1 *A quasilinear system* (2.1.5) *is called symmetric in the domain* $D \subset \mathbb{R}^m$*, if the matrices* $\mathbb{A}_j(\boldsymbol{w}), j = 0, 1, \dots, N$*, are symmetric for all* $\boldsymbol{w} \in D$*, i.e.* $\mathbb{A}_j(\boldsymbol{w}) = \mathbb{A}_j(\boldsymbol{w})^{\mathrm{T}}$ *for all* $\boldsymbol{w} \in D$.

Definition 2.2 *Let* D *be an open set in* $\mathbb{R}^m$*, a definition region of matrices* $\mathbb{A}_j(\boldsymbol{w}), j = 0, \dots, N$*. We say that system* (2.1.5) *with* $\mathbb{A}_0(\boldsymbol{w}) = \mathbb{I}$ *is symmetrizable (in* D*), if for any* $\boldsymbol{w} \in D$ *there exists a positive definite symmetric matrix* $\tilde{\mathbb{A}}_0(\boldsymbol{w})$ *such that*

(i) for any subdomain D_1 *satisfying* $\overline{D}_1 \subset D$ *we have*

$$c^{-1}|\boldsymbol{z}|^2 \le (\tilde{\mathbb{A}}_0(\boldsymbol{w})\boldsymbol{z}, \boldsymbol{z})_{\mathbb{R}^m} \le c|\boldsymbol{z}|^2, \quad \tilde{\mathbb{A}}_0(\boldsymbol{w}) = \tilde{\mathbb{A}}_0(\boldsymbol{w})^{\mathrm{T}} \tag{2.1.6}$$

with a positive constant c *independent of* $\boldsymbol{w} \in D_1$ *and* $\boldsymbol{z} \in \mathbb{R}^m$*;*

(ii) the following relations hold:

$$\tilde{\mathbb{A}}_0(\boldsymbol{w})\mathbb{A}_j(\boldsymbol{w}) = \tilde{\mathbb{A}}_j(\boldsymbol{w}) \quad \textit{and} \quad \tilde{\mathbb{A}}_j(\boldsymbol{w}) = \tilde{\mathbb{A}}_j(\boldsymbol{w})^{\mathrm{T}},\ j = 1, \dots, N. \tag{2.1.7}$$

2.2 Existence of smooth solutions

In this section we introduce the method of characteristics, a breakdown of solution features, a local existence theorem for a symmetric quasilinear hyperbolic system and its consequences for the Euler equations.

2.2.1 *Hyperbolic systems and characteristics*

First, we consider system (2.1.5) and introduce the notion of hyperbolicity.

Definition 2.3 *System (2.1.5) is called hyperbolic in the region $D \subset \mathbb{R}^m$, if all solutions $\lambda_j = \lambda_j(\boldsymbol{w}, \boldsymbol{n}), j = 1, \dots, m$, of the m-th degree algebraic equation*

$$\det\Big(\lambda \mathbb{A}_0(\boldsymbol{w}) - \sum_{j=1}^{N} n_j \mathbb{A}_j(\boldsymbol{w})\Big) = 0 \tag{2.2.1}$$

are real for any $\boldsymbol{n} = (n_1, \dots, n_N)^{\mathrm{T}} \in \mathbb{R}^N$ and $\boldsymbol{w} \in D$. We call λ_j the generalized eigenvalues of system (2.1.5).

If moreover the generalized eigenvalues λ_j are simple, then the system is called strictly hyperbolic.

We say that system (2.1.5) is (strictly) diagonally hyperbolic, if in addition the matrix $\mathbb{P} := \sum_{j=1}^{N} n_j \mathbb{A}_j(\boldsymbol{w})$ is diagonalizable. This means that there exists a nonsingular matrix $\mathbb{T} = \mathbb{T}(\boldsymbol{w}, \boldsymbol{n})$ such that

$$\mathbb{T}^{-1}\mathbb{P}\,\mathbb{T} = \mathbb{A} = \mathbb{A}(\boldsymbol{w}, \boldsymbol{n}) = \mathrm{diag}(\lambda_1, \dots, \lambda_m) = \begin{pmatrix} \lambda_1 & & \emptyset \\ & \ddots & \\ \emptyset & & \lambda_m \end{pmatrix}. \tag{2.2.2}$$

Example 2.4 Write the equations of inviscid barotropic flow under zero volume forces in the form (cf. (1.2.96)–(1.2.98))

$$\begin{aligned} &\boldsymbol{v}_t + (\boldsymbol{v} \cdot \nabla)\boldsymbol{v} + \frac{1}{\rho}\nabla p(\rho) = 0, \quad \boldsymbol{v} = (v_1, v_2, v_3)^{\mathrm{T}}, \\ &\rho_t + \mathrm{div}\,(\rho \boldsymbol{v}) = 0 \end{aligned} \tag{2.2.3}$$

with the equation of state $p = p(\rho)\,(\rho > 0)$ assuming $p'(\rho) > 0$. In this case we can set $\boldsymbol{w} = (\boldsymbol{v}, \rho), D = \{(\boldsymbol{v}, \rho); \boldsymbol{v} \in \mathbb{R}^N, \rho > 0\}$. Then the corresponding determinant in (2.2.1) is equal to

$$\Big(\lambda - \sum_{k=1}^{3} n_k v_k\Big)^2 \Big[\Big(\lambda - \sum_{k=1}^{3} n_k v_k\Big)^2 - p'(\rho)\sum_{k=1}^{3} n_k^2\Big] \tag{2.2.4}$$

and the corresponding equation (2.2.1) has three real roots (one of them multiple)

$$\lambda_1 = \lambda_2 = \sum_{k=1}^{3} n_k v_k, \quad \lambda_{3,4} = \sum_{k=1}^{3} n_k v_k \pm \Big[p'(\rho)\sum_{k=1}^{3} n_k^2\Big]^{1/2}. \tag{2.2.5}$$

Hence, system (2.2.3) is hyperbolic in the domain D.

Definition 2.5 *Let $N = 1, \mathbb{A}_0 = \mathbb{I}, \mathbb{A}_1(\boldsymbol{w}) = \mathbb{A}(\boldsymbol{w})$ and let system (2.1.5) be hyperbolic. This means that the roots λ_j, $j = 1, \dots, m$, of the equation*

$$\det\big(\lambda \mathbb{I} - \mathbb{A}(\boldsymbol{w})\big) = 0 \tag{2.2.6}$$

are real. For a given smooth function $\boldsymbol{w}$ *satisfying* (2.1.5)*, curves in the* (x,t)*-plane parametrized by* $x=\xi_j(s)$*,* $t=t(s)=s, s\in J$*, where* J *is an open interval and*

$$\frac{d\xi_j(s)}{ds}=\lambda_j(\xi_j(s),s), \quad s\in J,\, j=1,\dots,m, \tag{2.2.7}$$

are called characteristics.

For example, for the so-called Burgers equation $u_t+uu_x=0$, we have $m=1$, $\lambda_1=\lambda(u)=u$, and $\xi_1=\xi(s)$ is given by

$$\frac{d\xi(s)}{ds}=u(\xi(s),s). \tag{2.2.8}$$

Of course, u must have certain regularity in order that ξ would be well defined as a solution of ordinary differential equation (2.2.8). Notice that the trajectory $\{(x,t);\, x=\xi(s),\, t=s\}$ in the (x,t)-plane is not determined uniquely unless we fix 'the initial' value of ξ at some point $s=s_0$. The concept of characteristics in some cases helps us to say something about possible solutions to one-dimensional equations. We illustrate this method later in Sections 2.2.3 and 2.2.5.

2.2.2 *Formulation of the hyperbolic problem*

Since the system (2.1.4) under consideration is nonstationary, it is to be equipped with the *initial condition*

$$\boldsymbol{w}(x,0)=\boldsymbol{w}^0(x), \quad x\in\Omega, \tag{2.2.9}$$

where $\boldsymbol{w}^0$ is a given vector-valued function. Moreover, if Ω has a nonempty boundary, it is necessary to add suitable *boundary conditions*

$$B(\boldsymbol{w})=0 \quad \text{on } \partial\Omega\times(0,T), \tag{2.2.10}$$

where B is some boundary operator. The choice of well-posed boundary conditions for systems of conservation laws is a delicate question, not completely satisfactorily solved (see, for example, the fundamental work (Bardos *et al.*, 1979) concerned with the boundary conditions for a scalar equation). We shall discuss the choice of boundary conditions for the Euler equations later, in Section 3.3.6 in connection with numerical approximations of the Euler equations. Therefore, in the sequel, we shall be concerned with the *Cauchy problem* for system (2.1.4) in the set $Q_T=\mathbb{R}^N\times(0,T)$ (i.e. $\Omega=\mathbb{R}^N$), equipped with the initial condition (2.2.9). This means that our problem is to find a solution of the *problem*

$$\text{a)}\quad \frac{\partial\boldsymbol{w}}{\partial t}+\sum_{j=1}^{N}\frac{\partial\boldsymbol{f}_j(\boldsymbol{w})}{\partial x_j}=0 \quad \text{in } Q_T=\mathbb{R}^N\times(0,T),$$

$$\text{b)}\quad \boldsymbol{w}(x,0)=\boldsymbol{w}^0(x), \quad x\in\mathbb{R}^N. \tag{2.2.11}$$

Now let us deal with the concept of a solution of problem (2.2.11) and investigate some important properties of this problem.

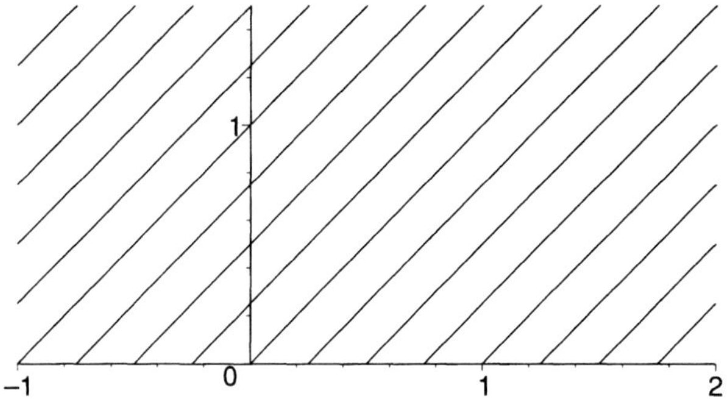

FIG. 2.2.1. Characteristics of a linear equation

Definition 2.6 *We say that a vector-valued function* $\boldsymbol{w}$ *is a classical solution of the Cauchy problem* (2.2.11), *if*

a) $\boldsymbol{w} \in C^1(\mathbb{R}^N \times (0,T))^m \cap C(\mathbb{R}^N \times [0,T))^m$,

b) $\boldsymbol{w}(x,t) \in D$ *for all* $(x,t) \in Q_T$,

c) $\boldsymbol{w}$ *satisfies* (2.2.11), *a) and b) for all* $(x,t) \in \mathbb{R}^N \times (0,T)$ *and* $x \in \mathbb{R}^N$, *respectively.*

The symbol D *denotes the domain of definition of the functions* $\boldsymbol{f}_j$, $j = 1,\dots,N$. *We say that* $\boldsymbol{w}$ *is a classical solution of* (2.2.11) *a), if* $\boldsymbol{w} \in C^1(\mathbb{R}^N \times (0,T))$ *and* (2.2.11), *a) holds.*

In what follows, we present a few simple examples that point out some specific properties of hyperbolic problems and illustrate the possibilities of the so-called *method of characteristics*.

2.2.3 *Linear scalar equation*

Let us consider the Cauchy problem

$$\text{a)}\ \frac{\partial w}{\partial t} + a\,\frac{\partial w}{\partial x} = 0 \quad \text{in } \mathbb{R} \times (0,\infty),$$
$$\text{b)}\ w(x,0) = w^0(x), \quad x \in \mathbb{R}, \tag{2.2.12}$$

where $a = \text{const} \in \mathbb{R}$ and $w = w(x,t) : \mathbb{R} \times [0,\infty) \to \mathbb{R}$. This problem can be solved by the *method of characteristics*. In coincidence with Definition 2.5, we define a characteristic of equation (2.2.12), a) as a parametrized curve

$$x = x(s), \quad t = s \quad (s \in (b,d)), \tag{2.2.13}$$

satisfying the condition

$$\frac{dx}{ds} = a. \tag{2.2.14}$$

Obviously, the characteristics are straight lines with the equation $x - at = c = \text{const}$. The angle α between the characteristic and the positive half-axis x satisfies the condition $\tan\alpha = 1/a$ (if $a = 0$, we put $\alpha = \pi/2$). See Fig. 2.2.1.

Theorem 2.7 *A function $w \in C^1(\mathbb{R}\times(0,\infty))$ is a classical solution of equation (2.2.11), a) if and only if w is constant along any characteristic.*

Proof 1) Let us consider any characteristic (2.2.13) satisfying (2.2.14). If w is a classical solution of (2.2.12), a), then

$$\frac{dw(x(s),s)}{ds} = \frac{\partial w(x(s),s)}{\partial t} + \frac{\partial w(x(s),s)}{\partial x}\frac{dx(s)}{ds}$$
$$= \left(\frac{\partial w}{\partial t} + a\,\frac{\partial w}{\partial x}\right)(x(s),\, s) = 0 \quad \forall s \in (b,d).$$

Thus, $w(x(s),s) = \text{const}$ for all $s \in (b,d)$. This means that w is constant along the characteristic (2.2.13).

2) Let $w \in C^1(\mathbb{R}\times(0,\infty))$ be constant along any characteristic and let $(x_0,t_0) \in \mathbb{R}\times(0,\infty)$. There exists a characteristic (defined on some interval $(b,d) \ni t_0$) such that $x(t_0) = x_0$. Then for each $s \in (b,d)$

$$0 = \frac{dw(x(s),s)}{ds} = \left(\frac{\partial w}{\partial t} + a\frac{\partial w}{\partial x}\right)(x(s),s).$$

For $s := t_0$ we have $(\partial w/\partial t + a\partial w/\partial x)\,(x_0,t_0) = 0$. □

From Theorem 2.7 we easily conclude that for $w^0 \in C^1(\mathbb{R})$, there exists a unique classical solution of problem (2.2.12). It can be expressed in the form

$$w(x,t) = w^0(x - at), \quad x \in \mathbb{R},\ t \in (0,\infty). \tag{2.2.15}$$

This solution is called a *travelling wave.*

2.2.4 *Solution of a linear system*

Our goal is to find a classical solution of a linear diagonally hyperbolic system with constant coefficients

$$\frac{\partial \boldsymbol{w}}{\partial t} + \mathbb{A}\,\frac{\partial \boldsymbol{w}}{\partial x} = 0 \quad \text{in } \mathbb{R}\times(0,\infty), \tag{2.2.16}$$

equipped with the initial condition (2.2.11), b). We assume that $\mathbb{A}$ is an $m\times m$ matrix which is diagonalizable and has real eigenvalues $\lambda_1,\dots,\lambda_m$. Hence, there exists a nonsingular matrix $\mathbb{T}$ such that

$$\mathbb{T}^{-1}\,\mathbb{A}\,\mathbb{T} = \mathbb{\Lambda} = \text{diag}\,(\lambda_1,\dots,\lambda_m). \tag{2.2.17}$$

The columns $\boldsymbol{r}_1,\dots,\boldsymbol{r}_m$ of $\mathbb{T}$ are eigenvectors of the matrix $\mathbb{T}$ and form a basis in $\mathbb{R}^m$. Therefore, $\mathbb{A}\,\boldsymbol{r}_s = \lambda_s\boldsymbol{r}_s$ and the solution can be sought in the form

$$\boldsymbol{w}(x,t) = \sum_{j=1}^{m} \mu_j(x,t)\,\boldsymbol{r}_j, \quad x \in \mathbb{R},\ t \in [0,\infty). \tag{2.2.18}$$

Similarly we can write

$$\boldsymbol{w}^0(x) = \sum_{j=1}^{m} \mu_j^0(x)\, \boldsymbol{r}_j, \quad x \in I\!R. \tag{2.2.19}$$

Substituting (2.2.18) into (2.2.16), we obtain

$$0 = \frac{\partial \boldsymbol{w}}{\partial t} + \mathbb{A}\, \frac{\partial \boldsymbol{w}}{\partial x} = \sum_{j=1}^{m} \left(\frac{\partial \mu_j}{\partial t} + \lambda_j\, \frac{\partial \mu_j}{\partial x} \right) \boldsymbol{r}_j,$$

which holds if and only if

$$\frac{\partial \mu_j}{\partial t} + \lambda_j \frac{\partial \mu_j}{\partial x} = 0, \quad j = 1, \ldots, m. \tag{2.2.20}$$

Moreover, each equation (2.2.20) is equipped with the initial condition

$$\mu_j(x, 0) = \mu_j^0(x), \quad x \in I\!R,\ j = 1, \ldots, m. \tag{2.2.21}$$

If $\boldsymbol{w}^0 \in C^1(I\!R)^m$, then, in view of Section 2.2.3, we find that every particular problem (2.2.20)–(2.2.21) ($j = 1, \ldots, m$) has a unique solution and this solution can be expressed in the form of the *travelling wave* (2.2.15):

$$\mu_j(x, t) = \mu_j^0(x - \lambda_j t).$$

Inserting these travelling waves into (2.2.18), we obtain the solution of problem (2.2.16) and (2.2.11), b) in the form

$$\boldsymbol{w}(x, t) = \sum_{j=1}^{m} \mu_j^0(x - \lambda_j t)\, \boldsymbol{r}_j. \tag{2.2.22}$$

2.2.5 *Nonlinear scalar equation*

Let $f \in C^2(I\!R)$, $a = f'$, $w_0 \in C^1(I\!R)$. Consider the equation

$$\frac{\partial w}{\partial t} + \frac{\partial f(w)}{\partial x} = 0 \quad \text{in } I\!R \times (0, T), \tag{2.2.23}$$

which can be rewritten for a classical solution in the form

$$\frac{\partial w}{\partial t} + a(w)\, \frac{\partial w}{\partial x} = 0 \quad \text{in } I\!R \times (0, T), \tag{2.2.24}$$

and equip this equation with the initial condition

$$w(x, 0) = w^0(x), \quad x \in I\!R. \tag{2.2.25}$$

In agreement with Definition 2.5, we consider the *characteristics* of equation (2.2.24) as curves given by the conditions

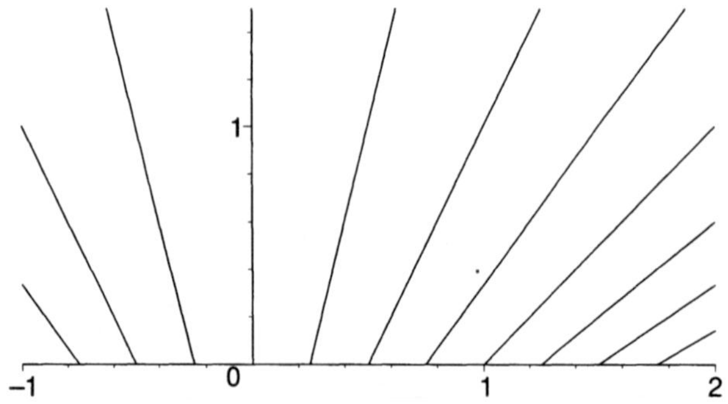

FIG. 2.2.2. Nonintersecting characteristics

$$\frac{dx}{ds} = a(w(x,s)), \;\; t = s. \tag{2.2.26}$$

As we see, the characteristics depend on the sought solution in this case. The classical solution w is constant along any characteristic, since

$$\begin{aligned}\frac{dw(x(s),s)}{ds} &= \frac{\partial w(x(s),s)}{\partial t} + \frac{\partial w(x(s),s)}{\partial x}\frac{dx(s)}{ds} \\ &= \left(\frac{\partial w}{\partial t} + a(w)\frac{\partial w}{\partial x}\right)(x(s),s) = 0.\end{aligned}$$

This and (2.2.26) imply that the characteristics are straight lines in the upper half-plane $t \geq 0$ starting from the axis x and determined by the conditions

$$\begin{aligned}&x - a(w(x,t))\,t = x_0 \in I\!R, \\ &w(x,t) = w(x_0,0) = w^0(x_0).\end{aligned} \tag{2.2.27}$$

This yields the implicit representation of the solution

$$w(x,t) = w^0(x - a(w(x,t))\,t), \quad x \in I\!R, \; t \geq 0. \tag{2.2.28}$$

The angle α between the characteristic passing through a point $(x_0, 0)$ and the axis x is given by the condition $\tan\alpha = 1/a(w^0(x_0))$.

Now let us investigate two cases:

I. Let $a' > 0$, $(w^0)' \geq 0$ in $I\!R$. Then $\tan\alpha = 1/a(w^0(x))$ is a nonincreasing function of x and exactly one characteristic passes through an arbitrary point (x,t), $x \in I\!R$, $t \geq 0$ (see Fig. 2.2.2). This implies the existence of a unique *global solution* $w \in C^1(I\!R \times (0,\infty))$.

II. If $a' > 0$ and $(w^0)' < 0$ in $I\!R$, then the function $\tan\alpha = 1/a(w^0(x))$ is increasing in $I\!R$ and, therefore, the characteristics passing through points $(x_1, 0)$ and $(x_2, 0)$ with $x_1 < x_2$ intersect each other at some point $P = (x_I, t_I)$ (see Fig. 2.2.3). Since $w^0(x_1) \neq w^0(x_2)$, the solution w becomes necessarily discontinuous at the finite time t_I. This means that the global classical solution does

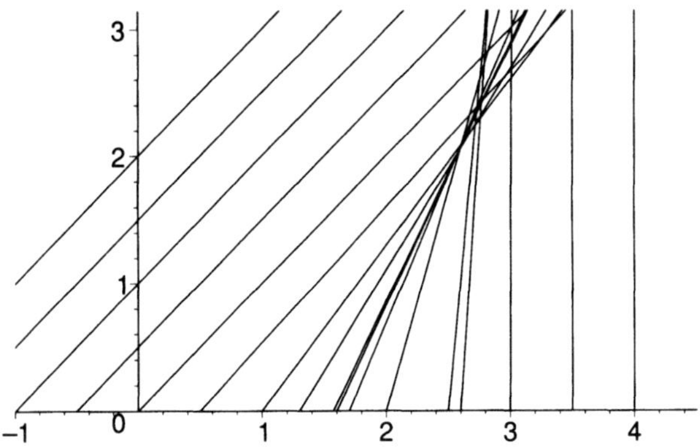

FIG. 2.2.3. Intersecting characteristics

not exist. In this case we can obtain only the *local existence* of a classical solution on the time interval $[0, T^*)$, where, as can be proven,

$$T^* = -1/\inf_{x \in \mathbb{R}} a'(w^0(x))\,(w^0)'(x). \tag{2.2.29}$$

The above example demonstrates an important property of nonlinear hyperbolic equations: *the possible rise of discontinuities of solutions in finite time, even in the case when the data are smooth.* These discontinuities appear in the solution of the Euler equations and are also observed in fluid dynamics. We call them *shock waves* or *contact discontinuities* according to their character.

From the above example it follows that the concept of a classical solution to nonlinear hyperbolic systems is rather restrictive. This situation is solved by introducing a weaker concept of solution, which does not exclude discontinuities. This will be the subject of Section 2.3.2.

2.2.6 *Symmetric hyperbolic systems*

In what follows we are going to introduce a local existence theorem for a first order symmetric quasilinear hyperbolic system (2.1.5) (see Definition 2.5) to be applied later to the Euler equations in several space variables. The solution to (2.1.5) is constructed with the aid of successive approximations defined as solutions of suitable linear problems. For the resolution of these auxiliary problems the theory of linear symmetric hyperbolic systems in the following form is applied.

Let us assume that $\mathbb{A}_j(x,t), j = 0, 1, \ldots, N$, and $\mathbb{B}(x,t)$ are symmetric $m \times m$ matrices in $\mathbb{R}^N \times (0,T)$ with some $T > 0$, $\boldsymbol{f}(x,t)$ and $\boldsymbol{v}^0(x)$ are m-dimensional vector functions defined in $\mathbb{R}^N \times (0,T)$ and $\mathbb{R}^N$, respectively. Let us consider the problem

$$\mathbb{A}_0(x,t)\frac{\partial \boldsymbol{v}}{\partial t}+\sum_{j=1}^{N}\mathbb{A}_j(x,t)\frac{\partial \boldsymbol{v}}{\partial x_j}+\mathbb{B}(x,t)\boldsymbol{v}=\boldsymbol{f}(x,t),\ x\in I\!R^N,\ t\in(0,T), \quad (2.2.30)$$

$$\boldsymbol{v}(x,0)=\boldsymbol{v}^0(x), \quad x\in I\!R^N. \quad (2.2.31)$$

Theorem 2.8 *Let*

$$\mathbb{A}_j\in\left[C([0,T];H^s(I\!R^N))\cap C^1(0,T;H^{s-1}(I\!R^N))\right]^{m\times m}, j=0,1,\dots,N,$$

$$\mathbb{A}_0(x,t) \textit{ invertible},\ \inf_{x,t}\|\mathbb{A}_0(x,t)\|_{I\!R^N\times I\!R^N}>0,\ \mathbb{B}\in C(0,T;H^{s-1}(I\!R^N))^{m\times m},$$

$$\boldsymbol{f}\in C(0,T;H^s(I\!R^N))^m,\ \boldsymbol{v}^0\in H^s(I\!R^N)^m,$$

where $s>\frac{N}{2}+1$ *is an integer. Then there exists a uniquely determined function*

$$\boldsymbol{v}\in\left[C([0,T);H^s(I\!R^N))\cap C^1(0,T;H^{s-1}(I\!R^N))\right]^m$$

satisfying (2.2.30) *and* (2.2.31) *pointwise.*

Proof One possible way to prove Theorem 2.8 is to regularize system (2.2.30), (2.2.31) with the help of a standard mollifier, solve the resulting system as an abstract differential equation with a uniformly continuous operator coefficient in a suitable Banach space, and solve this equation by a standard fixed point argument known from the theory of ordinary differential equations. Then a compactness argument using energy estimates is applied to let the regularization parameter tend to zero. A detailed description of the above steps can be found in the recent monograph (Novotný and Straškraba, 2003) □

2.2.7 *Quasilinear system*

Let us now consider a more general system than (2.1.5), namely

$$\mathbb{A}_0(\boldsymbol{u})\frac{\partial \boldsymbol{u}}{\partial t}+\sum_{j=1}^{N}\mathbb{A}_j(\boldsymbol{u})\frac{\partial \boldsymbol{u}}{\partial x_j}+\mathbb{B}\boldsymbol{u}=\boldsymbol{f}, \quad x\in I\!R^N,\ t\in(0,T), \quad (2.2.32)$$

$$\boldsymbol{u}(x,0)=\boldsymbol{u}^0(x), \quad x\in I\!R^N. \quad (2.2.33)$$

Here $\boldsymbol{u}=(u_1(x,t),\dots,u_m(x,t))^{\mathrm{T}}$, $\mathbb{A}_j(\boldsymbol{u}), j=1,\dots,N$, $\mathbb{B}(x,t)$ are given $m\times m$ matrices, $\boldsymbol{f}=\boldsymbol{f}(x,t), \boldsymbol{u}^0=\boldsymbol{u}^0(x)$ are given m-dimensional vector-valued functions. Assume for simplicity that

$\mathbb{A}_j(\boldsymbol{v}),\ j=0,1,\dots,N,$ are symmetric s times continuously differentiable for $\boldsymbol{v}\in D$ such that $\inf_{\boldsymbol{v}\in D}|\det\mathbb{A}_0(\boldsymbol{v})|>0,$ where $s\geq 1$ is an integer,
and D is a domain in $I\!R^m$; (2.2.34)
$\overline{\{\boldsymbol{u}^0(x);x\in I\!R^N\}}\subset D$;

$\mathbb{B},\boldsymbol{f}$ and $\boldsymbol{u}^0$ are s times continuously differentiable with respect to $x\in I\!R^N$ and $t\in(0,T_0)$. (2.2.35)

2.2.8 *Local existence for a quasilinear system*

We are now in a position to formulate the following local existence theorem for problem (2.2.32), (2.2.33).

Theorem 2.9 *Let assumptions* (2.2.34), (2.2.35) *be satisfied and* $s > \frac{N}{2} + 1$. *Then there exists* $T \in (0, T_0)$ *such that problem* (2.2.32), (2.2.33) *has a unique solution* $\boldsymbol{u} \in C^1(\mathbb{R}^N \times (0,T))$ *with values in* D.

Proof To prove Theorem 2.9, we define standard approximations by $\boldsymbol{u}^0(x,t) := \boldsymbol{u}^0(x)$ and $\boldsymbol{u}^{k+1}(x,t)$ for $k = 0, 1, \ldots$, as a solution of the system

$$\mathbb{A}_0(\boldsymbol{u}^k)\frac{\partial \boldsymbol{u}^{k+1}}{\partial t} + \sum_{j=1}^{N} \mathbb{A}_j(\boldsymbol{u}^k)\frac{\partial \boldsymbol{u}^{k+1}}{\partial x_j} + \mathbb{B}\boldsymbol{u}^{k+1} = \boldsymbol{f},\ x \in \mathbb{R}^N,\ t \in (0, T_0) \tag{2.2.36}$$

$$\boldsymbol{u}^{k+1}(x,0) = \boldsymbol{u}^0(x), \quad x \in \mathbb{R}^N. \tag{2.2.37}$$

Problem (2.2.36), (2.2.37) is solved with the aid of Theorem 2.8. Then estimates of $\boldsymbol{u}^k, k = 1, 2, \ldots$, are proven allowing us to establish the convergence of $\boldsymbol{u}^k$ to the sought solution $\boldsymbol{u}$. For details we refer to (Majda, 1984) or the monograph (Novotný and Straškraba, 2003). □

2.2.9 *Local existence for equations of inviscid barotropic flow*

To illustrate the application of the general Theorem 2.9, let us consider the equations of inviscid barotropic flow

$$\frac{\mathrm{D}p}{\mathrm{D}t} + \gamma p \operatorname{div} v = 0, \tag{2.2.38}$$

$$\frac{\mathrm{D}\boldsymbol{v}}{\mathrm{D}t} + \frac{1}{\rho}\nabla p = 0, \quad x \in \mathbb{R}^3,\ t \in (0,T), \tag{2.2.39}$$

$$p = \kappa \rho^\gamma, \tag{2.2.40}$$

$$p(x,0) = p^0(x),\ \boldsymbol{v}(x,0) = \boldsymbol{v}^0(x), \quad x \in \mathbb{R}^3, \tag{2.2.41}$$

with $\mathrm{D}/\mathrm{D}t = \frac{\partial}{\partial t} + \sum_{j=1}^{3} v_j \frac{\partial}{\partial x_j}$, the total derivative along fluid particle trajectories. Here $\rho = \left(p/\kappa\right)^{1/\gamma}$ is obtained by the inversion of the equation of state (2.2.40). We claim that system (2.2.38), (2.2.39) is symmetrized by the 4×4 matrix

$$\mathbb{A}_0(p) = \begin{pmatrix} (\gamma p)^{-1} & 0 \\ 0 & \left(\frac{p}{\kappa}\right)^{1/\gamma}\mathbb{I} \end{pmatrix}, \tag{2.2.42}$$

where $\mathbb{I}$ is the 3×3 unit matrix.

Indeed, since

$$\mathbb{A}_j(p, \boldsymbol{v}) = \begin{pmatrix} v_j & \gamma p & 0 & 0 \\ \frac{1}{\rho(p)} & v_j & 0 & 0 \\ 0 & 0 & v_j & 0 \\ 0 & 0 & 0 & v_j \end{pmatrix}, \quad j = 1, 2, 3, \tag{2.2.43}$$

by multiplication we get the matrices

$$\mathbb{A}_0(p)\mathbb{A}_j(p,\boldsymbol{v}) = \begin{pmatrix} \frac{v_j}{\gamma p} & 1 & 0 & 0 \\ 1 & \rho v_j & 0 & 0 \\ 0 & 0 & \rho v_j & 0 \\ 0 & 0 & 0 & \rho v_j \end{pmatrix}, \qquad (2.2.44)$$

which are symmetric.

Exercise 2.10 Derive (2.2.38) from the continuity equation (1.2.26) with the aid of (2.2.40) and prove (2.2.42)–(2.2.44).

Let us note that symmetrizing system (2.2.38), (2.2.39) by matrix (2.2.42) can be done by multiplying (2.2.38) by $(\gamma p)^{-1}$ and each equation in (2.2.39) by $\rho = (p/\kappa)^{1/\gamma}$. Put

$$\boldsymbol{u} = (p,\boldsymbol{v})^{\mathrm{T}},\ \mathbb{B} = 0,\ \boldsymbol{f} = 0,\ \boldsymbol{u}^0 = (p^0,\boldsymbol{v}^0)^{\mathrm{T}}. \qquad (2.2.45)$$

Then problem (2.2.38)–(2.2.41), after symmetrization by the matrix $\mathbb{A}_0(p)$ defined in (2.2.42), assumes the form (2.2.32), (2.2.33). Thus, Theorem 2.9 can be applied, giving us the following existence theorem for the original Euler system (2.2.38)–(2.2.41).

Theorem 2.11 *Let* $p^0 \in H^3(\mathbb{R}^3), \boldsymbol{v}^0 \in H^3(\mathbb{R}^3)^3$ *and* $d_0 := \inf_{x\in\mathbb{R}^N} p^0(x) > 0$. *Then there is* $T > 0$ *such that problem* (2.2.38)–(2.2.41) *has a unique solution* $(p,\boldsymbol{v}) \in C^1(\mathbb{R}^3 \times (0,T))^4 \cap C(\mathbb{R}^3 \times [0,T))^4$.

Proof Define

$$D := \{(p,\boldsymbol{v}) \in \mathbb{R}^4; d_0/2 < p < 2D_0\},$$

where $D_0 := \sup_{x\in\mathbb{R}^3} p^0(x)$. Then D is open in $\mathbb{R}^4$ and the elements of $\mathbb{A}_j(p,\boldsymbol{v})$, $j = 0,1,\ldots,N$, are infinitely differentiable with respect to $p,\boldsymbol{v}$ in D. Moreover, for $(p,\boldsymbol{v}) \in D$ we have

$$\det \mathbb{A}_0(p) = (\gamma p)^{-1}\left(\frac{p}{\kappa}\right)^{3/\gamma} \geq c_0 > 0, \qquad (2.2.46)$$

where c_0 is a constant dependent on d_0, D_0, κ and γ. Hence, all assumptions of Theorem 2.9 are satisfied and we conclude that there exists $T > 0$ such that the symmetrized problem has a unique solution $\boldsymbol{u} = (p,\boldsymbol{v}) \in C^1(\mathbb{R}^3 \times (0,T))$. Multiplying the symmetrized system by $\mathbb{A}_0(p)^{-1}$ given in (2.2.42), we obtain the original equations (2.2.38) and (2.2.39), which completes the proof. □

Let us note that also system (2.1.1) of the Euler equations for heat-conductive gas is symmetrizable, as shown in Section 4.3.2.

2.3 Weak solutions

In this section, we are concerned with the situation when a system of the type (2.1.4) does not possess a globally defined smooth solution despite the fact that the data may be smooth enough, e.g. so smooth that Theorem 2.9 holds true. Such a situation is typical even for very simple quasilinear hyperbolic equations as, for example, the so-called Burgers equation $u_t + uu_x = 0$. This fact focuses our attention on a weaker formulation of the solution to systems like (2.1.4) than the classical one. First, we summarize our knowledge about the blow up of solutions. Then we introduce weak solutions and comment on their typical structure composed of *shock waves* and *rarefaction waves.* The *Rankine–Hugoniot conditions* will be derived and the role of the so-called *entropy pairs* will be clarified. Moreover, we shall deal with the *Riemann problem* for a system of conservation laws, which plays an important role in both theoretical and numerical treatments of general initial–boundary value problems. Fundamental existence theorems for a single conservation law, two conservation laws and several conservation laws in one space dimension will be given.

2.3.1 *Blow up of classical solutions*

As we have already mentioned, nonlinear hyperbolic systems exhibit a typical nonlinear phenomenon: breakdown of classical solutions. Let us consider the situation as in Theorem 2.9, say for $T_0 = \infty$, where for data smooth enough we have the existence of a classical solution on some interval $(0, T)$ $(T > 0)$. We do not know how large (or small) T is, but due to the uniqueness we can consider a unique maximal classical solution on the interval $(0, T_{\max})$ with

$$T_{\max} = \sup\{T > 0; \text{the } C^1 \text{ solution from Theorem 2.9 exists on } (0,T)\}. \quad (2.3.1)$$

Without additional information we do not know whether

$$T_{\max} = \infty \quad (2.3.2)$$

or

$$T_{\max} < \infty. \quad (2.3.3)$$

If (2.3.2) holds, then no problem arises in this respect. If we have (2.3.3), then there are several possibilities. It is well known from the theory of nonlinear ordinary differential equations that the solution can tend to infinity in a finite time. A similar phenomenon is possible here. For example, if $\boldsymbol{w}$ is a maximal solution of (2.1.4) on $(0, T_{\max})$, then $x_0 \in \mathbb{R}^N$ can exist such that

$$\liminf_{t \uparrow T_{\max}} |\boldsymbol{w}(x_0, t)| = \infty. \quad (2.3.4)$$

On the other hand, it is also possible that

$$|w(x,t)| \leq C < \infty \quad \text{for all } x \text{ and } t \in [0, T_{\max}) \quad (2.3.5)$$

with a constant C independent of x and t, but

$$\liminf_{t \uparrow T_{\max}} \sum_{j=1}^{m} \Big(\sum_{k=1}^{N} \Big| \frac{\partial w_j}{\partial x_k}(x_0, t) \Big| + \Big| \frac{\partial w_j}{\partial t}(x_0, t) \Big| \Big) = \infty \quad (2.3.6)$$

for some x_0. This situation occurs generically. In fact, it can be shown that for any nontrivial initial data with compact support in $\mathbb{R}^N$ there exists a solution to (2.1.4) which develops a singularity of the type (2.3.6). A detailed discussion of this issue is given in (Majda, 1984), Chapter 3. The question arises whether relaxing the notion of a classical solution to a generalized one would allow us to continue the solution in time in some reasonable sense. This is our next subject.

2.3.2 *Generalized formulation for systems of conservation laws*

Let $M \subset \mathbb{R}^N$ be a measurable set. Recall that $L^\infty(M)^m$ denotes the space of all vector functions $\boldsymbol{f} = (f_1, \dots, f_m)$ with components $f_i \in L^\infty(M)$ for $i = 1, \dots, m$. The norm in $L^\infty(M)^m$ is defined by $\|f\|_{L^\infty(M)^m} = \max\{\|f_i\|_{L^\infty(M)};\ i = 1, \dots, m\}$. Further, we write $\boldsymbol{f} \in L^\infty_{\text{loc}}(M)^m$, if $\boldsymbol{f}|_{M\cap K} \in L^\infty(M \cap K)^m$ for every compact set $K \subset \mathbb{R}^N$.

By $C_0^\infty(\mathbb{R}^N \times [0,\infty))$ we denote the space of all infinitely differentiable functions whose supports are compact sets in $\mathbb{R}^N \times [0,\infty)$. Let us assume that $\boldsymbol{w}$ is a classical solution of problem (2.2.11) and $\boldsymbol{\varphi} \in [C_0^\infty(\mathbb{R}^N \times [0,\infty))]^m$. Then, using Green's theorem (from Section 1.1.9), we obtain the identity

$$0 = \int_0^\infty \int_{\mathbb{R}^N} \left(\frac{\partial \boldsymbol{w}}{\partial t} + \sum_{j=1}^N \frac{\partial \boldsymbol{f}_j(\boldsymbol{w})}{\partial x_j} \right) \cdot \boldsymbol{\varphi} \, dx dt = - \int_{\mathbb{R}^N} \boldsymbol{w}(x,0) \cdot \boldsymbol{\varphi}(x,0) \, dx$$

$$- \int_0^\infty \int_{\mathbb{R}^N} \left(\boldsymbol{w} \cdot \frac{\partial \boldsymbol{\varphi}}{\partial t} + \sum_{j=1}^N \boldsymbol{f}_j(\boldsymbol{w}) \cdot \frac{\partial \boldsymbol{\varphi}}{\partial x_j} \right) dx dt.$$

Hence, any classical solution of (2.2.11) satisfies the identity

$$\int_0^\infty \int_{\mathbb{R}^N} \left(\boldsymbol{w} \cdot \frac{\partial \boldsymbol{\varphi}}{\partial t} + \sum_{j=1}^N \boldsymbol{f}_j(\boldsymbol{w}) \cdot \frac{\partial \boldsymbol{\varphi}}{\partial x_j} \right) dx \, dt + \int_{\mathbb{R}^N} \boldsymbol{w}^0(x) \cdot \boldsymbol{\varphi}(x,0) \, dx = 0$$

$$\forall \boldsymbol{\varphi} \in [C_0^\infty(\mathbb{R}^N \times [0,\infty))]^m. \qquad (2.3.7)$$

Obviously, (2.3.7) also makes sense for $\boldsymbol{w} \in [L^\infty_{\text{loc}}(\mathbb{R}^N \times (0,\infty))]^m$. This leads us to the following definition:

Definition 2.12 *Let $\boldsymbol{w}^0 \in [L^\infty_{\text{loc}}(\mathbb{R}^N)]^m$. A vector function $\boldsymbol{w}$ is called a weak solution of problem* (2.2.11), *if $\boldsymbol{w} \in [L^\infty_{\text{loc}}(\mathbb{R}^N \times (0,\infty))]^m$, $\boldsymbol{w}(x,t) \in D$ for a.e. $(x,t) \in \mathbb{R}^N \times (0,\infty)$ and* (2.3.7) *holds.*

Exercise 2.13 Prove that any weak solution of class C^1 is a classical solution. Choosing test functions $\boldsymbol{\varphi} \in [C_0^\infty(\mathbb{R}^N \times (0,\infty))]^m$ in (2.3.7), show that *the weak solution satisfies equation* (2.2.11), a) *in the sense of distributions.* (Cf. Section 1.3.2.)

We often meet the so-called *piecewise smooth weak solutions* which admit discontinuities.

Definition 2.14 *We say that a function $\boldsymbol{w}$ is piecewise smooth, if there exists a finite number of smooth hypersurfaces Γ in $\mathbb{R}^N \times [0,\infty)$ outside of which the function $\boldsymbol{w}$ is of class C^1 and on which $\boldsymbol{w}$ and its first derivatives have one-sided*

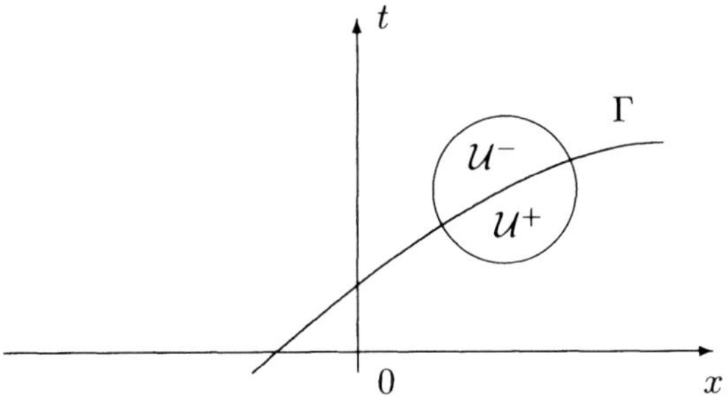

FIG. 2.3.4. Discontinuity of the solution

limits $\boldsymbol{w}^{\pm}(x,t)$ ($(x,t) \in \Gamma$), defined in the following way. Let $(x,t) \in \Gamma$ and let $\mathcal{U} = B_\varepsilon(x,t)$ be a ball with centre at (x,t) and a small radius $\varepsilon > 0$ such that $\mathcal{U} \setminus \Gamma$ has exactly two components $\mathcal{U}^{\pm}$ (see Fig. 2.3.4). Then we set

$$\boldsymbol{w}^{\pm}(x,t) = \lim_{\substack{(y,\vartheta)\to(x,t)\\ (y,\vartheta)\in\mathcal{U}^{\pm}}} \boldsymbol{w}(y,\vartheta), \tag{2.3.8}$$

provided this limit exists. Similarly we define one-sided limits of derivatives of $\boldsymbol{w}$). (If $N = 1$, then Γ is a curve in $\mathbb{R} \times [0,\infty)$.)

Let us show that the discontinuities on Γ of a piecewise smooth weak solution $\boldsymbol{w}$ are not arbitrary:

Theorem 2.15 *Let $\boldsymbol{w} : \mathbb{R}^N \times [0,\infty) \to \mathbb{R}^m$ be a piecewise smooth function. Then $\boldsymbol{w}$ is a solution of system (2.2.11), a) in the sense of distributions if and only if the following conditions are satisfied:*
a) $\boldsymbol{w}$ is a classical solution in any domain where $\boldsymbol{w}$ is of class C^1,
b) $\boldsymbol{w}$ satisfies the condition

$$(\boldsymbol{w}^+ - \boldsymbol{w}^-)n_t + \sum_{j=1}^{N} (\boldsymbol{f}_j(\boldsymbol{w}^+) - \boldsymbol{f}_j(\boldsymbol{w}^-))n_j = 0 \tag{2.3.9}$$

on smooth hypersurfaces Γ of discontinuity. Here $\boldsymbol{n} = (n_1, \ldots, n_N, n_t)$ denotes the normal to Γ.

Proof Let a piecewise smooth function $\boldsymbol{w}$ be a solution of (2.2.11), a) in the sense of distributions. Choosing suitable test functions φ, we easily find that $\boldsymbol{w}$ satisfies condition a) formulated above. Further, let $(x_0, t_0) \in \Gamma$. Let $\mathcal{U} = B_\varepsilon(x_0, t_0)$ be a ball with centre at (x_0, t_0) and with a sufficiently small radius $\varepsilon > 0$, as in Definition 2.14.

Suppose, for example, that $\boldsymbol{n}$ points from $\mathcal{U}^-$ to $\mathcal{U}^+$. Let $\boldsymbol{\varphi} \in [C_0^\infty(\mathcal{U})]^m$. Then, by virtue of (2.3.7),

$$0 = \int_{\mathcal{U}} \left(\boldsymbol{w} \cdot \frac{\partial \varphi}{\partial t} + \sum_{j=1}^{N} \boldsymbol{f}_j(\boldsymbol{w}) \cdot \frac{\partial \varphi}{\partial x_j} \right) dxdt$$

$$= \int_{\mathcal{U}^-} + \int_{\mathcal{U}^+} . \tag{2.3.10}$$

Applying Green's theorem (from Section 1.1.9) to the integrals over $\mathcal{U}^\pm$, from (2.3.10) we obtain

$$0 = -\int_{\mathcal{U}^-} \left(\frac{\partial \boldsymbol{w}}{\partial t} + \sum_{j=1}^{N} \frac{\partial \boldsymbol{f}_j(\boldsymbol{w})}{\partial x_j} \right) \cdot \varphi \, dxdt$$

$$- \int_{\mathcal{U}^+} \left(\frac{\partial \boldsymbol{w}}{\partial t} + \sum_{j=1}^{N} \frac{\partial \boldsymbol{f}_j(\boldsymbol{w})}{\partial x_j} \right) \cdot \varphi \, dxdt$$

$$- \int_{\Gamma \cap \mathcal{U}} \left(n_t \boldsymbol{w}^+ + \sum_{j=1}^{N} n_j \boldsymbol{f}_j(\boldsymbol{w}^+) \right) \cdot \varphi \, dS$$

$$+ \int_{\Gamma \cap \mathcal{U}} \left(n_t \boldsymbol{w}^- + \sum_{j=1}^{N} n_j \boldsymbol{f}_j(\boldsymbol{w}^-) \right) \cdot \varphi \, dS.$$

As $\boldsymbol{w}$ is a classical solution in $\mathcal{U}^\pm$, we have

$$\int_{\Gamma \cap \mathcal{U}} \left[n_t(\boldsymbol{w}^+ - \boldsymbol{w}^-) + \sum_{j=1}^{N} n_j(\boldsymbol{f}_j(\boldsymbol{w}^+) - \boldsymbol{f}_j(\boldsymbol{w}^-)) \right] \cdot \varphi \, dS = 0.$$

Since φ is arbitrary, (2.3.9) holds.

Now let $\boldsymbol{w}$ be a piecewise smooth function in $\mathbb{R}^N \times (0, \infty)$ satisfying (2.2.11), a) in the set $(\mathbb{R}^N \times (0,\infty)) \setminus \Gamma$ and condition (2.3.9) on Γ. Then for $\varphi \in C_0^\infty(\mathbb{R}^N \times [0,\infty))^m$ and a bounded domain $K \supset \operatorname{supp} \varphi$ with the Lipschitz-continuous boundary ∂K, it holds that

$$0 = \int_0^\infty \int_{\mathbb{R}^N} \left(\frac{\partial \boldsymbol{w}}{\partial t} + \sum_{s=1}^{N} \frac{\partial f_s(\boldsymbol{w})}{\partial x_s} \right) \cdot \varphi \, dxdt$$

$$= \int_K \left(\frac{\partial \boldsymbol{w}}{\partial t} + \sum_{s=1}^{N} \frac{\partial f_s(\boldsymbol{w})}{\partial x_s} \right) \cdot \varphi \, dxdt.$$

Using Green's theorem, and the relation $\varphi|_{\partial K} = 0$, we find that

$$0 = \int_{K \cap \Gamma} \left[(\boldsymbol{w}^+ - \boldsymbol{w}^-) n_t + \sum_{s=1}^{N} (\boldsymbol{f}_s(\boldsymbol{w}^+) - \boldsymbol{f}_s(\boldsymbol{w}^-)) n_s \right] \cdot \varphi \, dS$$

$$- \int_K \left(\boldsymbol{w} \cdot \frac{\partial \varphi}{\partial t} + \sum_{s=1}^{N} \boldsymbol{f}_s(\boldsymbol{w}) \cdot \frac{\partial \varphi}{\partial x_s} \right) dxdt - \int_{\mathbb{R}^N} \boldsymbol{w}^0(x) \varphi(x, 0) \, dx.$$

This together with (2.3.9) implies (2.3.7). □

Definition 2.16 *Relations* (2.3.9) *are called the Rankine–Hugoniot conditions.*

In the case $(n_1, \dots, n_N) \neq 0$, we can choose a normal $\boldsymbol{n} = (\boldsymbol{\nu}, -s)$ to Γ where $s \in I\!R$ and $\boldsymbol{\nu} \in I\!R^N$ is a unit vector. Then (2.3.9) can be written in the form

$$s[\boldsymbol{w}] = \sum_{j=1}^{N} \nu_j [\boldsymbol{f}_j(\boldsymbol{w})], \tag{2.3.11}$$

where

$$[\boldsymbol{w}] = \boldsymbol{w}^+ - \boldsymbol{w}^-, \quad [\boldsymbol{f}_j(\boldsymbol{w})] = \boldsymbol{f}_j(\boldsymbol{w}^+) - \boldsymbol{f}_j(\boldsymbol{w}^-). \tag{2.3.12}$$

The vector $\boldsymbol{\nu}$ and the number s can be interpreted as the direction and speed of propagation of the discontinuity Γ, respectively.

This is quite clear in the case $N = 1$. If we express the discontinuity Γ as a curve $x = \xi(t)$, we have $s = d\xi/dt$ and $\boldsymbol{n} = (1, -s)$. It is clear that s represents the speed of propagation of the discontinuity in dependence on time. Then (under the notation $\boldsymbol{f} = \boldsymbol{f}_1$), the Rankine–Hugoniot condition becomes

$$s[\boldsymbol{w}] = [\boldsymbol{f}(\boldsymbol{w})]. \tag{2.3.13}$$

2.3.3 *Examples of piecewise smooth weak solutions*

Theorem 2.15 can be used for the construction of weak solutions to some simple hyperbolic problems. Let us consider the Cauchy problem for the Burgers equation:

$$\begin{aligned} &\frac{\partial w}{\partial t} + w\,\frac{\partial w}{\partial x} = 0 \qquad \text{in } I\!R \times (0, \infty), \\ &w(x, 0) = w^0(x), \qquad\qquad x \in I\!R. \end{aligned} \tag{2.3.14}$$

The Burgers equation can be written as a conservation law equation (2.2.23) with the flux

$$f(w) = w^2/2. \tag{2.3.15}$$

a) Assume that the initial condition is defined by

$$w^0(x) = \begin{cases} 1, & x \le 0, \\ 1 - x, & 0 \le x \le 1, \\ 0, & 1 \le x. \end{cases} \tag{2.3.16}$$

The characteristic passing through a point $(x_0, 0)$ has the form $x = x_0 + w^0(x_0)t$ as follows from (2.2.27). Hence,

$$x = \begin{cases} x_0 + t, & x_0 \le 0, \\ x_0 + t(1 - x_0), & 0 \le x_0 \le 1, \\ x_0, & x_0 \ge 1, \end{cases} \tag{2.3.17}$$

see Fig. 2.3.5. For $t < 1$ the characteristics do not intersect each other and, therefore, in $I\!R \times (0,1)$ we obtain the continuous solution

$$w(x,t) = \left\{\begin{array}{ll} 1, & x \le t, \\ (1-x)\,/\,(1-t), & t \le x \le 1, \\ 0, & x \ge 1, \end{array}\right\} \; t < 1. \tag{2.3.18}$$

At time $t = 1$ the characteristics mutually intersect and the solution becomes discontinuous. For $t = 1$ the Rankine–Hugoniot condition yields the speed of propagation of the discontinuity $s = [(w^+)^2 - (w^-)^2]/[2(w^+ - w^-)] = (w^+ + w^-)/2 = 1/2$. Let us define the function

$$w(x,t) = \left\{\begin{array}{l} 1, \; x < (t+1)/2, \\ 0, \; x > (t+1)/2, \end{array}\right\} \; t \ge 1, \tag{2.3.19}$$

discontinuous on the line $x = (t+1)/2,\ t \ge 1$. Using Theorem 2.15, we easily show that (2.3.18)–(2.3.19) define a weak solution of problem (2.3.14)–(2.3.16). Passing through the discontinuity in the direction of the x axis, the function w jumps down.

b) Let

$$w^0(x) = \left\{\begin{array}{l} 0, \; x \le 0, \\ 1, \; x > 0. \end{array}\right. \tag{2.3.20}$$

The characteristics are shown in Fig. 2.3.6. We see that the solution obtained with the use of characteristics has the form

$$w(x,t) = \left\{\begin{array}{l} 0, \; x \le 0, \; t \ge 0, \\ 1, \; x > t, \; t \ge 0, \end{array}\right. \tag{2.3.21}$$

but it is not defined for $0 \le x \le t$. Taking $t = 0$, we obtain $s = 1/2$ for the speed of possible propagation of the discontinuity in the initial condition. We easily show that

$$w(x,t) = \left\{\begin{array}{l} 0, \; x < t/2, \\ 1, \; x > t/2, \end{array}\right\} \; t > 0 \tag{2.3.22}$$

is a weak solution discontinuous on the line $x = t/2$.

However, on the basis of Theorem 2.15, the continuous function

$$u(x,t) = \left\{\begin{array}{ll} 0, & x \le 0, \\ x/t, & 0 \le x \le t, \\ 1, & 0 \le t < x, \end{array}\right. \tag{2.3.23}$$

is also a weak solution. Obviously, the function u is a classical solution of the Burgers equation in the regions where u is smooth. In particular, if $0 < x < t$, we have

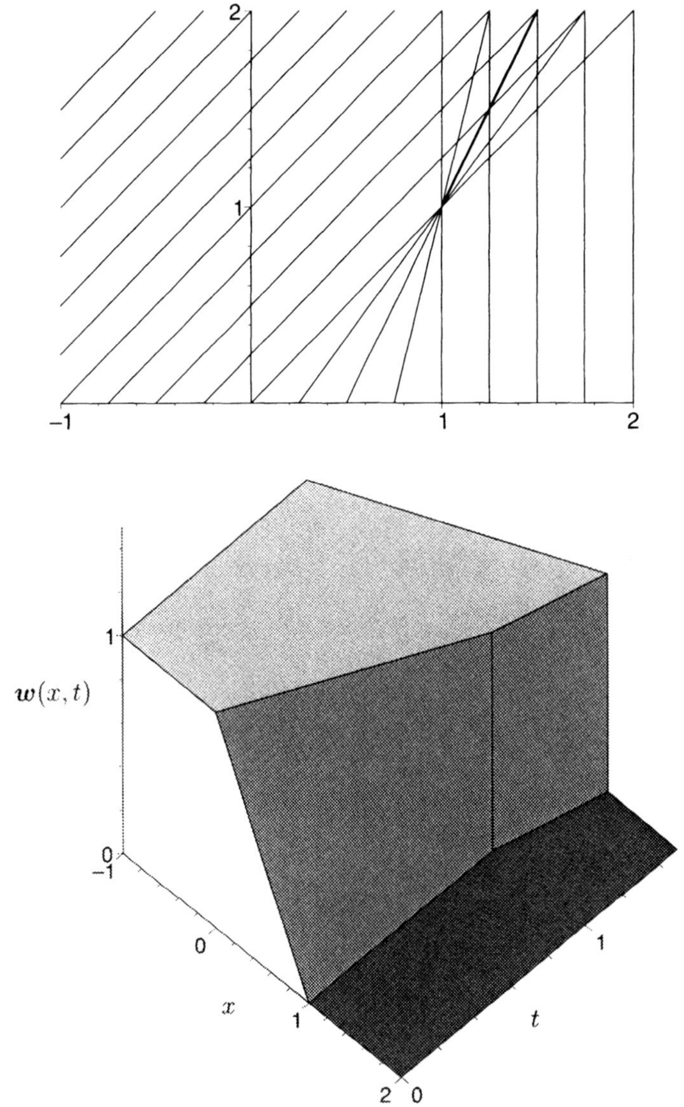

FIG. 2.3.5. Characteristics and discontinuity for initial condition (2.3.16)

$$\frac{\partial u}{\partial t} + \frac{\partial (u^2/2)}{\partial x} = -x/t^2 + x/t^2 = 0.$$

The Rankine–Hugoniot conditions are satisfied due to the continuity of u. The above considerations imply that the weak solution of problem (2.3.14), (2.3.20) is not unique.

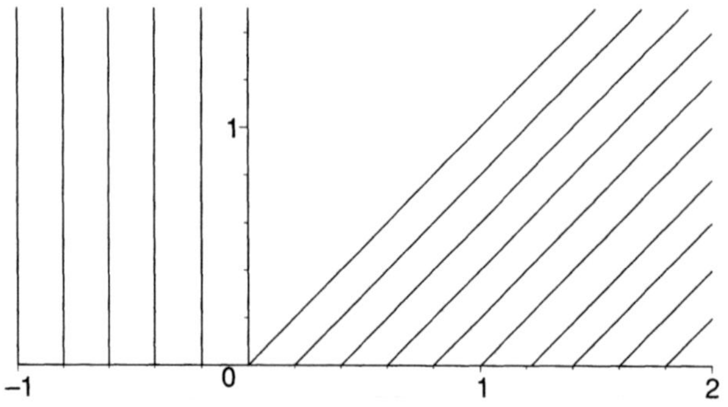

FIG. 2.3.6. Characteristics

2.3.4 *Entropy condition*

To distinguish physically admissible solutions from nonphysical ones, we introduce the so-called *entropy condition.* Let η, G_s : $D \to I\!R$ ($s = 1, \dots, N$) be sufficiently smooth functions satisfying the conditions

$$\nabla_w \eta(\boldsymbol{w})^{\mathrm{T}} \mathbb{A}_s(\boldsymbol{w}) = \nabla_w G_s(\boldsymbol{w}), \qquad \boldsymbol{w} \in D,\ s = 1, \dots, N, \tag{2.3.24}$$

where $\nabla_w \eta = (\frac{\partial \eta}{\partial w_1}, \dots, \frac{\partial \eta}{\partial w_m})^{\mathrm{T}}$ and $\mathbb{A}_s$ are the matrices defined by

$$\mathbb{A}_s(\boldsymbol{w}) = \frac{D\boldsymbol{f}_s(\boldsymbol{w})}{D\boldsymbol{w}} = \left(\frac{\partial f_{si}}{\partial w_j}\right)_{i,j=1}^{m}. \tag{2.3.25}$$

If $\boldsymbol{w}$ is a classical solution of system (2.2.11), a), then the function $\eta(\boldsymbol{w}) = \eta \circ \boldsymbol{w}$ satisfies the conservation law equation

$$\frac{\partial \eta(\boldsymbol{w})}{\partial t} + \sum_{s=1}^{N} \frac{\partial G_s(\boldsymbol{w})}{\partial x_s} = 0, \tag{2.3.26}$$

as follows from (2.2.11), a), (2.3.24) and the chain rule. On the other hand, a weak solution of (2.2.11), a) need not satisfy (2.3.26). If the weak solution is piecewise smooth, then it satisfies the Rankine–Hugoniot condition (2.3.9). Analogously, one can prove that if this $\boldsymbol{w}$ satisfies (2.3.26) in the sense of distributions, i.e.

$$\int_0^\infty \int_{I\!R^N} \left(\eta(\boldsymbol{w}) \frac{\partial \varphi}{\partial t} + \sum_{s=1}^{N} G_s(\boldsymbol{w}) \frac{\partial \varphi}{\partial x_s} \right) dx dt = 0,$$
$$\forall \varphi \in C_0^\infty(I\!R^N \times (0, \infty)),$$

then the jump condition

$$n_t[\eta(\boldsymbol{w})] + \sum_{s=1}^{N} n_s [G_s(\boldsymbol{w})] = 0 \tag{2.3.27}$$

holds. Conditions (2.3.9) and (2.3.27) are not, in general, compatible, as can be shown on the example of the Burgers equation (2.3.14) whose classical solution also satisfies the equation

$$\frac{\partial}{\partial t}\left(\frac{u^{p+1}}{p+1}\right)+\frac{\partial}{\partial x}\left(\frac{u^{p+2}}{p+2}\right)=0, \qquad p=1,2,\dots . \tag{2.3.28}$$

However, it is possible to show that a discontinuous piecewise smooth weak solution of (2.3.14) may not be a weak solution of the above equation (2.3.28) because it might not satisfy the corresponding Rankine–Hugoniot condition.

Definition 2.17 *Let D be a convex set. We say that a real-valued function η defined on D is convex, if*

$$\eta(x+\tau(y-x)) \le \eta(x)+\tau(\eta(y)-\eta(x)) \qquad \forall\, x,\, y \in D,\ \forall\, \tau \in [0,1].$$

Definition 2.18 *Let D be a convex set. A convex function $\eta : D \to I\!R$ ($\eta \in C^1(D)$) is called the entropy of system (2.2.11), a), if there exist functions $\boldsymbol{G} = (G_1, \dots, G_N) : D \to I\!R$, called entropy fluxes, such that the relations (2.3.24) hold. The pair $(\eta, \boldsymbol{G})$ is called the entropy–entropy flux pair.*

The concept of entropy allows us to specify physically admissible solutions of the hyperbolic system (2.2.11), a):

Definition 2.19 *We say that a weak solution $\boldsymbol{w}$ of (2.2.11) is an entropy solution, if for every entropy η of system (2.2.11) the condition*

$$\frac{\partial \eta(\boldsymbol{w})}{\partial t}+\sum_{j=1}^{N}\frac{\partial G_j(\boldsymbol{w})}{\partial x_j} \le 0 \tag{2.3.29}$$

is satisfied in the sense of distributions on $I\!R^N \times (0,\infty)$. This means that

$$\int_0^\infty \int_{I\!R^N}\left(\eta(\boldsymbol{w})\frac{\partial \varphi}{\partial t}+\sum_{j=1}^{N} G_j(\boldsymbol{w})\frac{\partial \varphi}{\partial x_j}\right) dx\, dt \ge 0, \tag{2.3.30}$$
$$\forall \varphi \in C_0^\infty(I\!R^N \times (0,\infty)),\ \varphi \ge 0.$$

Only entropy solutions are considered to be physically admissible. Inequality (2.3.30) is called *the entropy condition.* This condition is a generalization of the physical entropy condition in gas dynamics, which expresses the second law of thermodynamics (see Section 1.2.14). Therefore, we speak of physically admissible solutions of (2.2.11) even if no physical context is specified here. More about this issue can be found in Section 2.3.5.

It is obvious that every classical solution satisfies the entropy condition. Let us consider a piecewise smooth weak solution $\boldsymbol{w}$. Analogously to Theorem 2.15 one can show that $\boldsymbol{w}$ is an entropy solution if and only if the condition

$$n_t[\eta(\boldsymbol{w})] + \sum_{j=1}^{N} n_j[G_s(\boldsymbol{w})] \le 0 \qquad (2.3.31)$$

is satisfied on every hypersurface Γ of discontinuity. The normal

$$\boldsymbol{n} = (n_1, \dots, n_N, n_t)$$

to Γ is oriented as in Fig. 2.3.4, i.e. in the direction from $\mathcal{U}^-$ to $\mathcal{U}^+$.

In the one-dimensional case ($N = 1$) condition (2.3.31) has the form

$$s[\eta(\boldsymbol{w})] \ge [G(\boldsymbol{w})] \qquad \text{on } \Gamma, \qquad (2.3.32)$$

where s is the speed of propagation of the discontinuity Γ. Inequality (2.3.32) is called the *Lax shock entropy condition.*

In the case of a scalar equation

$$\frac{\partial w}{\partial t} + \sum_{j=1}^{N} \frac{\partial f_j(w)}{\partial x_j} = 0, \qquad (2.3.33)$$

every convex function $\eta \in C^1(D)$ is an entropy of this equation. Actually, the entropy fluxes G_j should satisfy

$$G_j' = \eta' \, f_j', \quad j = 1, \dots, N, \qquad (2.3.34)$$

and thus can be easily obtained by integration.

Example 2.20 Let us apply the Lax shock entropy condition (2.3.32) to problems from Section 2.3.3. In this case $D = I\!R$ and every convex function $\eta \in C^1(I\!R)$ is an entropy of equation (2.3.14), with flux G given by (2.3.34), i. e.,

$$G'(w) = \eta'(w) f'(w), \quad w \in I\!R. \qquad (2.3.35)$$

First, we shall be concerned with the solution (2.3.22) of problem (2.3.14) with w^0 defined by (2.3.20). Using, for example, the entropy $\eta(w) = w^2/2$ with flux $G(w) = w^3/3$ (see (2.3.35)) and taking into account that on the discontinuity, $s = 1/2$, $w^+ = 1$, $w^- = 0$, we find that

$$s[\eta(w)] - [G(w)] = \frac{1}{2}\left(\frac{(w^+)^2}{2} - \frac{(w^-)^2}{2}\right) - \left(\frac{(w^+)^3}{3} - \frac{(w^-)^3}{3}\right) = \frac{1}{4} - \frac{1}{3} < 0.$$

Hence, the piecewise smooth weak solution (2.3.22) *does not satisfy* the entropy condition.

On the other hand, the continuous weak solution (2.3.23) satisfies, of course, the entropy condition.

Let us show that the weak solution w given by (2.3.18)–(2.3.19) of problem (2.3.14) with w^0 from (2.3.16) satisfies the entropy condition. To this end, let us assume that η is a convex entropy of equation (2.3.14) with flux G satisfying (2.3.35). In this case, on the discontinuity, $s = 1/2 = (f(0) - f(1))/(0 - 1)$, $w^+ = 0$, $w^- = 1$. We need to verify inequality (2.3.32), i.e.

$$\frac{f(0) - f(1)}{0 - 1}\,(\eta(0) - \eta(1)) - (G(0) - G(1)) \geq 0. \tag{2.3.36}$$

Let us set

$$h(\xi) = \frac{f(\xi) - f(1)}{\xi - 1}\,(\eta(\xi) - \eta(1)) - (G(\xi) - G(1)) \tag{2.3.37}$$

for $\xi \in [0, 1)$. It is easy to see that $\lim_{\xi \to 1-} h(\xi) = 0$. Condition (2.3.36) is equivalent to the inequality $h(0) \geq 0$ and, hence, it is enough to show that $h'(\xi) \leq 0$ for $\xi \in (0, 1)$. By (2.3.37) and (2.3.35),

$$h'(\xi) = -\frac{1}{(\xi - 1)^2}\Big(f(\xi) - f(1) - (\xi - 1)\,f'(\xi)\Big)\Big(\eta(\xi) - \eta(1) - (\xi - 1)\,\eta'(\xi)\Big). \tag{2.3.38}$$

From the convexity of the functions f and η it follows that

$$f(\xi) - f(1) - (\xi - 1)\,f'(\xi) \geq 0 \tag{2.3.39}$$

and η satisfies a similar relation. This and (2.3.38) imply that $h'(\xi) \leq 0$, which was to be proven.

From the above results we see that passing in the direction of x, the 'jump down' is entropy admissible, whereas the 'jump up' violates the entropy condition.

Exercise 2.21 Prove (2.3.39) for convex $f \in C^1(\mathbb{R})$.
Hint: Use Definition 2.17 (for f) and the definition of the derivative f'.

2.3.5 *Entropy in fluid mechanics*

If the concept of an entropy solution of the inviscid flow system (2.1.1) is not to be void, then the existence of at least one entropy of the system has to be guaranteed. In Section 1.2.13 we introduced the physical entropy

$$S = c_v \ln \frac{p/p_0}{(\rho/\rho_0)^\gamma} + \text{const}$$

(where p_0, ρ_0 are fixed reference values of the pressure and density respectively) which can be rewritten in the form

$$S = c_v \left\{ \ln \left[\left(E - \sum_{i=1}^{N} (\rho v_i)^2/(2\rho) \right) / E_0 \right] - \gamma \ln(\rho/\rho_0) \right\} + \text{const}. \tag{2.3.40}$$

We can set const $= 0$. Putting $m_i = \rho v_i$, we can prove that the function

$$(\rho, m_1, \ldots, m_N, E) \to -\rho S \tag{2.3.41}$$

$$= -c_v \rho \left\{ \ln \left[\left(E - \sum_{i=1}^{N} m_i^2/(2\rho) \right) / E_0 \right] - \gamma \ln(\rho/\rho_0) \right\}$$

is strictly convex. After a lengthy calculation we find that the functions $G_i = -\rho v_i S$, $i = 1, \dots, N$, are entropy fluxes of system (2.1.1). This means that the physical entropy defines a mathematical one for the system of conservation laws of an inviscid gas. Using equation (1.2.89) and the continuity equation, we obtain the identity

$$\frac{\partial(\rho S)}{\partial t} + \sum_{j=1}^{N} \frac{\partial(\rho v_j S)}{\partial x_j} = 0 \tag{2.3.42}$$

for adiabatic flow. This means that a smooth solution of system (2.1.1) satisfies the entropy equation (2.3.42). For a weak solution (2.3.42) is replaced by the inequality

$$\frac{\partial(\rho S)}{\partial t} + \sum_{j=1}^{N} \frac{\partial(\rho v_j S)}{\partial x_j} \geq 0 \tag{2.3.43}$$

(in the sense of distributions), which corresponds to the entropy condition implied by the second law of thermodynamics (1.2.78) (where $q = 0$ and $\boldsymbol{q} = 0$ in the case of adiabatic flow). We see that the mathematical theory is in agreement with physical ideas, hypotheses and postulates.

2.3.6 *Method of artificial viscosity*

As follows from the above considerations, our goal is to find physically admissible weak entropy solutions of problem (2.2.11). For this purpose, we can use the method of *artificial viscosity* proposed in 1954 by P. D. Lax. This method is based on introducing an additional 'viscous' (dissipative) term $\varepsilon \Delta \boldsymbol{w}$ to the right-hand side of (2.2.11), a), where $\varepsilon > 0$ is a small parameter. Then we get the parabolic system

$$\frac{\partial \boldsymbol{w}}{\partial t} + \sum_{s=1}^{N} \frac{\partial \boldsymbol{f}(\boldsymbol{w})}{\partial x_s} = \varepsilon \Delta \boldsymbol{w} \tag{2.3.44}$$

and an entropy solution of (2.2.11) is sought on the basis of the following theorem as a limit of 'viscous' solutions of system (2.3.44) as $\varepsilon \to 0+$. For the existence of solutions of system (2.3.44) the theory of parabolic systems can be used (see (Serre, 1997), (Smoller, 1983)).

Theorem 2.22 *Let $\eta \in C^2(\mathbb{R})$ be the entropy of system* (2.2.11) *with entropy fluxes G_s and let $\{\boldsymbol{w}_\varepsilon\}_{\varepsilon>0}$ be a family of solutions of* (2.3.44) *satisfying the conditions*

$$a) \quad \frac{\partial \boldsymbol{w}_\varepsilon}{\partial t}, \frac{\partial \boldsymbol{w}_\varepsilon}{\partial x_s}, \frac{\partial^2 \boldsymbol{w}_\varepsilon}{\partial x_i \partial x_s} \tag{2.3.45}$$

are continuous in $\mathbb{R}^N \times (0, \infty)$,

b) $\|\boldsymbol{w}_\varepsilon\|_{[L^\infty(K)]^m} \le c(K)\ \forall\, \varepsilon > 0,\ \forall\, K \subset I\!R^N \times (0,\infty),\ K$ *compact,*

c) $\boldsymbol{w}_\varepsilon \to \boldsymbol{w}$ *a.e. in* $I\!R^N \times (0,\infty)$.

Then $\boldsymbol{w}$ *is a solution of system* (2.2.11), *a) in the sense of distributions and satisfies the entropy condition* (2.3.30).

The proof can be found in (Feistauer, 1993), Chapter 7 or (Godlewski and Raviart, 1991).

Lax's method of artificial viscosity is often used as a basis for the construction of sufficiently dissipative numerical schemes which give approximations of entropy solutions of nonlinear hyperbolic equations. The artificial viscosity is usually chosen in a more complicated form than in Theorem 2.22.

There is also a class of methods possessing the so-called *numerical viscosity* which again leads to physically admissible approximate solutions. In theory, the method of viscosity is used for the construction of a precompact family of approximations to obtain an exact solution as its accumulation point.

2.3.7 *Existence and uniqueness of weak entropy solutions for scalar conservation laws*

The theory of existence and qualitative properties of scalar conservation laws is now well developed. The following theorem is a typical representation of the existence and uniqueness results available.

Theorem 2.23 *Let* $f_j \in C^1(I\!R), j = 1,\dots,N$. *For any* $w^0 \in L^\infty(I\!R^N)$ *there exists a unique weak entropy solution* w *of the Cauchy problem for one scalar conservation law defined by*

$$\frac{\partial w}{\partial t} + \sum_{j=1}^{N} \frac{\partial f_j(w)}{\partial x_j} = 0, \quad x \in I\!R^N,\ t > 0, \tag{2.3.46}$$

$$w(x,0) = w^0(x), \quad x \in I\!R^N, \tag{2.3.47}$$

and

$$w \in C([0,\infty); L^1_{\text{loc}}(I\!R^N)), \quad \|w(\cdot,t)\|_{L^\infty(I\!R^N)} \le \|w^0\|_{L^\infty(I\!R^N)}. \tag{2.3.48}$$

The proof can be found in (Dafermos, 2000), where various typical methods are used. See also (Bressan, 2000). When using the viscosity method the scheme of Theorem 2.22 can be used. In the case of one conservation law (2.3.46), all hypotheses of Theorem 2.22 can be verified. The first version of this theorem was proven by Kruzhkov in his famous paper (Kruzhkov, 1970).

2.3.8 *Riemann problem*

The Riemann problem for a hyperbolic system

$$\frac{\partial \boldsymbol{w}}{\partial t} + \frac{\partial \boldsymbol{f}(\boldsymbol{w})}{\partial x} = 0, \quad (x,t) \in I\!R \times (0,\infty) \tag{2.3.49}$$

consists in finding its weak solution which satisfies the initial condition formed by two constant states $\boldsymbol{w}_L$, $\boldsymbol{w}_R \in D$:

$$\boldsymbol{w}(x,0) = \boldsymbol{w}^0(x) = \begin{cases} \boldsymbol{w}_L, & x < 0, \\ \boldsymbol{w}_R, & x > 0. \end{cases} \tag{2.3.50}$$

We again assume that D is an open subset of $\mathbb{R}^m$ and $\boldsymbol{f} \in C^1(D)^m$.

Theorem 2.24 *If the Riemann problem* (2.3.49)–(2.3.50) *has a unique piecewise smooth weak solution $\boldsymbol{w}$, then $\boldsymbol{w}$ can be written for $t > 0$ in the similarity form $\boldsymbol{w}(x,t) = \tilde{\boldsymbol{w}}(x/t)$, where $\tilde{\boldsymbol{w}} : \mathbb{R} \to \mathbb{R}^m$.*

Proof We easily find that for any fixed $\alpha > 0$ the function $\boldsymbol{w}(\alpha x, \alpha t)$ is also a weak solution of (2.3.49)–(2.3.50). Due to the uniqueness, we have $\boldsymbol{w}(\alpha x, \alpha t) = \boldsymbol{w}(x,t)$, which means that $\boldsymbol{w}$ is a homogeneous vector function of order 0. Hence, for any fixed x and t, taking $\alpha = 1/t$, we see that $\boldsymbol{w}(x,t) = \boldsymbol{w}(x/t, 1) =: \tilde{\boldsymbol{w}}(x/t)$. □

Note that sometimes we write $\tilde{\boldsymbol{w}}(x/t) = \boldsymbol{w}_{RS}(x/t; \boldsymbol{w}_L, \boldsymbol{w}_R)$ in order to emphasize the dependence on initial data $\boldsymbol{w}_L, \boldsymbol{w}_R$. The abbreviation *RS* stands for *Riemann solution.* We shall use this notation in Chapter 3 thus meeting demands of the traditional notation in the numerical treatment of the Riemann problem (cf. (3.2.70)). In what follows, let us investigate the Riemann problem in special cases.

2.3.9 *Linear Riemann problem*

The solution of the linear Riemann problem (2.2.16), (2.3.50) will be expressed with the aid of the results from Section 2.2.4. Let

$$\boldsymbol{w}_L = \sum_{i=1}^{m} \alpha_i \boldsymbol{r}_i, \qquad \boldsymbol{w}_R = \sum_{i=1}^{m} \beta_i \boldsymbol{r}_i. \tag{2.3.51}$$

Then

$$\boldsymbol{w}^0(x) = \sum_{i=1}^{m} \left[\beta_i H(x) + \alpha_i (1 - H(x))\right] \boldsymbol{r}_i,$$

where H is the Heaviside function: $H(x) = 1$ if $x > 0$ and $H(x) = 0$ if $x < 0$. Using (2.2.22), we write

$$\boldsymbol{w}(x,t) = \sum_{i=1}^{m} \left[\beta_i H(x - \lambda_i t) + \alpha_i (1 - H(x - \lambda_i t))\right] \boldsymbol{r}_i. \tag{2.3.52}$$

If $-\infty = \lambda_0 < \lambda_1 \le \lambda_2 \le \cdots \le \lambda_m < \lambda_{m+1} = \infty$ and $\lambda_i < \lambda_{i+1}$ for some $1 \le i \le m-1$, then $\boldsymbol{w}$ is constant in the domain $\Omega_i = \{(x,t); t > 0,\ \lambda_i < x/t < \lambda_{i+1}\}$:

$$\boldsymbol{w}(x,t) = \boldsymbol{w}_i \qquad \text{if } (x,t) \in \Omega_i, \tag{2.3.53}$$

$$\boldsymbol{w}_i = \sum_{k=1}^{i} \beta_k \boldsymbol{r}_k + \sum_{k=i+1}^{m} \alpha_k \boldsymbol{r}_k, \qquad i = 0, \ldots, m,$$

where we postulate $\sum_{k=r}^{s} = 0$ whenever $r > s$. The structure of the solution is illustrated in Fig. 3.1.10.

Let us show that the function $\boldsymbol{w}$ defined by (2.3.53) is a weak solution of problem (2.2.16), (2.3.50). According to Theorem 2.15 it is sufficient to verify the validity of the Rankine–Hugoniot conditions on every discontinuity $x/t = \lambda_i$ propagating with the speed $s_i = \lambda_i$. We obviously have

$$\begin{aligned}
&\boldsymbol{w}_i - \boldsymbol{w}_{i-1} = (\beta_i - \alpha_i)\boldsymbol{r}_i, \\
&\mathbb{A}(\boldsymbol{w}_i - \boldsymbol{w}_{i-1}) = (\beta_i - \alpha_i)\lambda_i \boldsymbol{r}_i = s_i(\boldsymbol{w}_i - \boldsymbol{w}_{i-1}),
\end{aligned}$$

which we wanted to prove.

2.3.10 *Nonlinear Riemann problem*

Now we shall be concerned with special solutions of the nonlinear Riemann problem (2.3.49)–(2.3.50).

By virtue of hyperbolicity, we have

$$\mathbb{A}(\boldsymbol{w}) = \frac{D\boldsymbol{f}(\boldsymbol{w})}{D\boldsymbol{w}} = \mathbb{T}\,\mathbb{\Lambda}\,\mathbb{T}^{-1}, \qquad \boldsymbol{w} \in D, \tag{2.3.54}$$

where $\mathbb{\Lambda} = \mathbb{\Lambda}(\boldsymbol{w}) = \operatorname{diag}(\lambda_1(\boldsymbol{w}), \ldots, \lambda_m(\boldsymbol{w}))$ and $\lambda_j = \lambda_j(\boldsymbol{w}) \in \mathbb{R}$, $j = 1, \ldots, m$, are the eigenvalues of the matrix $\mathbb{A} = \mathbb{A}(\boldsymbol{w})$. The columns of the matrix $\mathbb{T} = \mathbb{T}(\boldsymbol{w})$, denoted by $\boldsymbol{r}_s = \boldsymbol{r}_s(\boldsymbol{w}), s = 1, \ldots, m$, are the eigenvectors of $\mathbb{A}(\boldsymbol{w})$ associated with the eigenvalues $\lambda_s(\boldsymbol{w})$ and form a basis in $\mathbb{R}^m$.

Definition 2.25 *An eigenvector $\boldsymbol{r}_k = \boldsymbol{r}_k(\boldsymbol{w})$ of the matrix $\mathbb{A} = \mathbb{A}(\boldsymbol{w})$ is called genuinely nonlinear, if*

$$\nabla\lambda_k(\boldsymbol{w})^{\mathrm{T}} \cdot \boldsymbol{r}_k(\boldsymbol{w}) \neq 0 \qquad \forall\, \boldsymbol{w} \in D. \tag{2.3.55}$$

(We write $\nabla = \nabla_w = (\partial/\partial w_1, \ldots, \partial/\partial w_m)^{\mathrm{T}}$.) Further, we say that $\boldsymbol{r}_k$ is linearly degenerate, if

$$\nabla\lambda_k(\boldsymbol{w})^{\mathrm{T}} \cdot \boldsymbol{r}_k(\boldsymbol{w}) = 0 \qquad \forall\, \boldsymbol{w} \in D. \tag{2.3.56}$$

We call $\psi_k : D \to \mathbb{R}$ a k-Riemann invariant, if $\psi_k \in C^1(D)$ and

$$\nabla\psi_k(\boldsymbol{w})^{\mathrm{T}} \cdot \boldsymbol{r}_k(\boldsymbol{w}) = 0 \qquad \forall\, \boldsymbol{w} \in D. \tag{2.3.57}$$

It is obvious that if an eigenvector $\boldsymbol{r}_k$ is linearly degenerate, then the corresponding eigenvalue λ_k is k-Riemann invariant.

It can be proven (see (Godlewski and Raviart, 1996), Chapter 1, Lemma 3.1), at least locally, that there exist $(m-1)$ k-Riemann invariants whose gradients are linearly independent.

In the following we shall deal with the construction of solutions of the Riemann problem in the similarity form (cf. Theorem 2.24) for some special couples of initial states $\boldsymbol{w}_L$ and $\boldsymbol{w}_R$. It is obvious that every similarity solution is constant on any line $x/t = \text{const}$.

a) *Solution of the Riemann problem in the form of a rarefaction wave* Let us normalize a genuinely nonlinear vector $\boldsymbol{r}_k$ so that

$$\nabla\lambda_k(\boldsymbol{w})^{\mathrm{T}} \cdot \boldsymbol{r}_k(\boldsymbol{w}) = 1 \qquad \forall\, \boldsymbol{w} \in D. \tag{2.3.58}$$

Let us consider the initial value problem

$$\begin{aligned} d\tilde{\boldsymbol{w}}(\xi)/d\xi &= \boldsymbol{r}_k(\tilde{\boldsymbol{w}}(\xi)), \\ \tilde{\boldsymbol{w}}(0) &= \boldsymbol{w}_L, \end{aligned} \tag{2.3.59}$$

for an arbitrary $\boldsymbol{w}_L \in D$ and assume that $\tilde{\boldsymbol{w}} : [0, \xi_R] \to D$ ($\xi_R > 0$) is its solution. Let us set $\boldsymbol{w}_R = \tilde{\boldsymbol{w}}(\xi_R)$. Since

$$\frac{d}{d\xi}[\lambda_k(\tilde{\boldsymbol{w}}(\xi))] = \nabla\lambda_k(\tilde{\boldsymbol{w}}(\xi)) \cdot \boldsymbol{r}_k(\tilde{\boldsymbol{w}}(\xi)) = 1, \tag{2.3.60}$$

we have

$$\begin{aligned} \lambda_k(\tilde{\boldsymbol{w}}(\xi)) &= \xi + \text{const} \\ &= \xi + \lambda_k(\boldsymbol{w}_L) \end{aligned} \tag{2.3.61}$$

and, thus,

$$\lambda_k(\boldsymbol{w}_R) = \xi_R + \lambda_k(\boldsymbol{w}_L) > \lambda_k(\boldsymbol{w}_L). \tag{2.3.62}$$

For $x \in I\!R$ and $t > 0$ we now define

$$\boldsymbol{w}(x,t) = \begin{cases} \boldsymbol{w}_L, & x/t \le \lambda_k(\boldsymbol{w}_L), \\ \tilde{\boldsymbol{w}}(\frac{x}{t} - \lambda_k(\boldsymbol{w}_L)), & \lambda_k(\boldsymbol{w}_L) \le x/t \le \lambda_k(\boldsymbol{w}_R), \\ \boldsymbol{w}_R, & x/t \ge \lambda_k(\boldsymbol{w}_R), \end{cases} \tag{2.3.63}$$

$$\boldsymbol{w}(x,0) = \begin{cases} \boldsymbol{w}_L, \ x < 0, \\ \boldsymbol{w}_R, \ x > 0. \end{cases}$$

The vector-valued function (2.3.63) is called a *k-rarefaction wave* or *k-simple wave.*

Remark 2.26 In this case the constant state $\boldsymbol{w}_L$, identified with the function defined by

$$\boldsymbol{w}^L(x,t) = \boldsymbol{w}_L \tag{2.3.64}$$

on the set

$$\Omega_L = \{(x,t); x/t \le \lambda_k(\boldsymbol{w}_L)\}, \tag{2.3.65}$$

and the constant state $\boldsymbol{w}_R$, identified with the function defined by

$$w^R(x,t) = w_R \tag{2.3.66}$$

on the set

$$\Omega_R = \{(x,t); x/t \geq \lambda_k(w_R)\}, \tag{2.3.67}$$

are connected by the function

$$w_{rar} = \tilde{w}(x/t - \lambda_k(w_L)) \tag{2.3.68}$$

defined on the set

$$\Omega_{rar} = \{(x,t); \lambda_k(w_L) \leq x/t \leq \lambda_k(w_R)\}, \tag{2.3.69}$$

sometimes called the *rarefaction fan* . Very often the function w_{rar} from (2.3.68) is called the k-rarefaction wave, or even the set Ω_{rar} from (2.3.69) is called the k-rarefaction wave. It will be clear from the context which of these three meanings for the k-rarefaction wave is used.

Theorem 2.27 *Let r_k be a genuinely nonlinear eigenvector. Then for any left state $w_L \in D$ there exists a one-parameter family of right states $w_R = \tilde{w}(\xi)$, $\xi \in [0, \xi_R]$ $(\xi_R > 0)$, such that the states w_L and w_R can be connected by the k-rarefaction wave of the form* (2.3.63), *which is a solution of the Riemann problem* (2.3.49), (2.3.50).

Proof It is enough to prove that (2.3.63) is a solution of problem (2.3.49)–(2.3.50). We use Theorem 2.15. It is obvious that the function w defined by (2.3.63) is continuous in $\mathbb{R} \times [0, \infty) \setminus \{(0,0)\}$. In the regions, where $x/t < \lambda_k(w_L)$ or $x/t > \lambda_k(w_R)$, we have $w = $ const and, hence, w satisfies (2.3.49) in the classical sense. Let $\lambda_k(w_L) < x/t < \lambda_k(w_R)$. Then, in view of (2.3.61),

$$\begin{aligned}\lambda_k(w(x,t)) &= \lambda_k\big(\tilde{w}\big(\frac{x}{t} - \lambda_k(w_L)\big)\big) \\ &= \frac{x}{t} - \lambda_k(w_L) + \lambda_k(w_L) = x/t.\end{aligned}$$

Now, it follows from (2.3.63) and (2.3.59) that

$$\begin{aligned}&\frac{\partial w(x,t)}{\partial t} + \frac{\partial f(w(x,t))}{\partial x} \\ &= \frac{\partial w(x,t)}{\partial t} + \mathbb{A}(w(x,t))\frac{\partial w(x,t)}{\partial x} \\ &= -\frac{x}{t^2} r_k(w(x,t)) + \frac{1}{t}\mathbb{A}(w(x,t)) r_k(w(x,t)) \\ &= r_k(w(x,t))(-x/t^2 + \lambda_k(w(x,t))/t) = 0.\end{aligned}$$

□

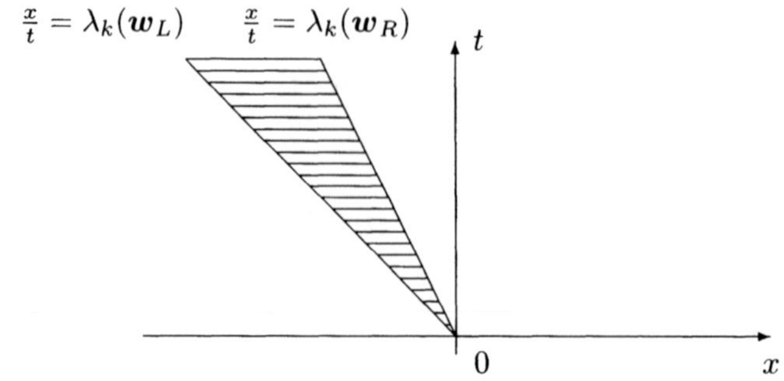

FIG. 2.3.7. Rarefaction fan

b) *Entropy discontinuity wave* Further, we shall be concerned with the possibility of writing the solution of the Riemann problem as a piecewise constant vector function with a discontinuity on a line $x = st$ (where $s \in \mathbb{R}$ is a constant and $t \in [0, \infty)$)

$$\boldsymbol{w}(x,t) = \begin{cases} \boldsymbol{w}_L, & x < st, \\ \boldsymbol{w}_R, & x > st. \end{cases} \tag{2.3.70}$$

This function is an entropy weak solution, if the speed of propagation of the discontinuity and the states $\boldsymbol{w}_L, \boldsymbol{w}_R$ satisfy conditions (2.3.13) and (2.3.32):

$$\begin{aligned} &\text{a)}\ \ s\,(\boldsymbol{w}_R - \boldsymbol{w}_L) = \boldsymbol{f}(\boldsymbol{w}_R) - \boldsymbol{f}(\boldsymbol{w}_L), \\ &\text{b)}\ \ s(\eta(\boldsymbol{w}_R) - \eta(\boldsymbol{w}_L)) \geq G(\boldsymbol{w}_R) - G(\boldsymbol{w}_L) \end{aligned} \tag{2.3.71}$$

for any (convex) entropy η with the corresponding flux G. Then we speak of an *entropy discontinuity wave.* Provided $\boldsymbol{w}_L$ is prescribed, relation (2.3.71), a) represents a system of m equations for $m+1$ unknowns (s and the components of $\boldsymbol{w}_R$). This leads us to the idea that the values $\boldsymbol{w}_R$ satisfying (2.3.71), a) for a fixed $\boldsymbol{w}_L$ form a curve in m-dimensional space. This is actually true under some conditions.

We distinguish two cases.

(i) Let us assume that the eigenvector $\boldsymbol{r}_k$ is genuinely nonlinear. In (Smoller, 1983), Theorems 17.11 and 17.14, the following result is proven:

Theorem 2.28 *If all eigenvalues λ_k are simple, then for every genuinely nonlinear eigenvector $\boldsymbol{r}_k$ and an arbitrary state $\boldsymbol{w}_L \in D$ there exists a parametrization of states $\boldsymbol{w}_R = \tilde{\boldsymbol{w}}_D(\xi)$, $\xi \in [-\xi_0, 0]$ $(\xi_0 > 0)$, such that the states $\boldsymbol{w}_L$ and $\boldsymbol{w}_R$ can be connected by an entropy discontinuity wave.*

The discontinuity wave (2.3.70) constructed to a genuinely nonlinear vector $\boldsymbol{r}_k$ is called a *k-shock wave.*

(ii) Another possibility is the *contact discontinuity* associated with a linearly degenerate eigenvector. Let $\boldsymbol{r}_k$ be a linearly degenerate eigenvector and let $\tilde{\boldsymbol{w}}$:

$[-\xi_1, \xi_1] \to D$ ($\xi_1 > 0$) be the solution of problem (2.3.59). We put $\boldsymbol{w}_R = \tilde{\boldsymbol{w}}(\xi_R)$ for $\xi_R \in [-\xi_1, \xi_1]$. Since

$$\frac{d}{d\xi}\{\lambda_k(\tilde{\boldsymbol{w}}(\xi))\} = \nabla\lambda_k(\tilde{\boldsymbol{w}}(\xi))^{\mathrm{T}} \cdot \boldsymbol{r}_k(\tilde{\boldsymbol{w}}(\xi)) = 0,$$

we have

$$\lambda_k(\tilde{\boldsymbol{w}}(\xi)) = \lambda_k(\boldsymbol{w}_L) = \lambda_k(\boldsymbol{w}_R), \qquad \xi \in [-\xi_1, \xi_1]. \tag{2.3.72}$$

Now we set $s = \lambda_k(\boldsymbol{w}_L) = \lambda_k(\boldsymbol{w}_R)$ and define the *k-contact discontinuity* (associated with the linearly degenerate eigenvector $\boldsymbol{r}_k$) as the function from (2.3.70).

One can easily show that $\boldsymbol{w}$ is a weak solution of (2.3.49)–(2.3.50). It suffices to verify the Rankine–Hugoniot condition (2.3.71), a):

$$-\lambda_k(\boldsymbol{w}_L)(\boldsymbol{w}_R - \boldsymbol{w}_L) + (\boldsymbol{f}(\boldsymbol{w}_R) - \boldsymbol{f}(\boldsymbol{w}_L)) = 0. \tag{2.3.73}$$

Actually, if $\xi \in [-\xi_1, , \xi_1]$, then

$$\begin{aligned}
&\frac{d}{d\xi}\left[\boldsymbol{f}(\tilde{\boldsymbol{w}}(\xi)) - \lambda_k(\boldsymbol{w}_L)\tilde{\boldsymbol{w}}(\xi)\right] \\
&= \left[\mathbb{A}(\tilde{\boldsymbol{w}}(\xi)) - \lambda_k(\tilde{\boldsymbol{w}}(\xi))\mathbb{I}\right] \frac{d\tilde{\boldsymbol{w}}(\xi)}{d\xi} \\
&= \left[\mathbb{A}(\tilde{\boldsymbol{w}}(\xi)) - \lambda_k(\tilde{\boldsymbol{w}}(\xi))\mathbb{I}\right] \boldsymbol{r}_k(\tilde{\boldsymbol{w}}(\xi)) = 0,
\end{aligned}$$

as follows from (2.3.72), (2.3.59) and the fact that $\boldsymbol{r}_k$ is the eigenvector of the matrix $\mathbb{A}$ associated with the eigenvalue λ_k. Substituting $\xi := 0$ and $\xi := \xi_R$ into the function $\boldsymbol{f}(\tilde{\boldsymbol{w}}(\xi)) - \lambda_k(\boldsymbol{w}_L)\tilde{\boldsymbol{w}}(\xi)$, we obtain the same values, which means that (2.3.73) holds.

Further, it is necessary to show that the k-contact discontinuity satisfies the entropy condition. Let η be the entropy with entropy flux G. This means that $\nabla_w \eta(\boldsymbol{w})^{\mathrm{T}}\, \mathbb{A}(\boldsymbol{w}) = \nabla_w G(\boldsymbol{w})$ (cf. (2.3.24) with $N = 1$). Then, by virtue of (2.3.59) and (2.3.72), for $\xi \in [-\xi_1, \xi_1]$ we have

$$\begin{aligned}
&\frac{d}{d\xi}\left[\lambda_k(\mathbf{w}_L)\, \eta(\tilde{\boldsymbol{w}}(\xi)) - G(\tilde{\boldsymbol{w}}(\xi))\right] \\
&= \left[\lambda_k(\boldsymbol{w}_L)\, \nabla_w \eta(\tilde{\boldsymbol{w}}(\xi))^{\mathrm{T}} - \nabla_w G(\tilde{\boldsymbol{w}}(\xi))^{\mathrm{T}}\right] \cdot \frac{d\tilde{\boldsymbol{w}}}{d\xi}(\xi) \\
&= \nabla_w \eta(\tilde{\boldsymbol{w}}(\xi))^{\mathrm{T}} \left[\lambda_k(\tilde{\boldsymbol{w}}(\xi))\, \mathbb{I} - \mathbb{A}(\tilde{\boldsymbol{w}}(\xi))\right] \boldsymbol{r}_k(\tilde{\boldsymbol{w}}(\xi)) = 0.
\end{aligned}$$

Substituting $\xi = 0$ and $\xi = \xi_R$ into $\lambda_k(\boldsymbol{w}_L)\eta(\tilde{\boldsymbol{w}}(\xi)) - G(\tilde{\boldsymbol{w}}(\xi))$, we get equality in (2.3.71), b). (Note that we do not need the convexity of the entropy η in this case.)

Remark 2.29 The function (2.3.70) is discontinuous across the line

$$S_{\text{disc}} = \{(x,t);\ x - st = 0,\ t \in (0,T)\}. \tag{2.3.74}$$

This line is sometimes called the k-shock wave or k-contact discontinuity and s is called the speed of the k-shock wave or k-contact discontinuity.

The above results can be formulated in the following way:

Theorem 2.30 *For every linearly degenerate eigenvector $\boldsymbol{r}_k$ and an arbitrary left state $\boldsymbol{w}_L \in D$ there exists a one-parameter family of right states $\boldsymbol{w}_R = \tilde{\boldsymbol{w}}(\xi)$, $\xi \in [-\xi_1, \xi_1]$ $(\xi_1 > 0)$, such that the states $\boldsymbol{w}_L$ and $\boldsymbol{w}_R$ can be connected by the k-contact discontinuity of the form* (2.3.70). *This solution satisfies the entropy condition* (2.3.71), *b), in which the equality holds.*

Remark 2.31 In cases when it is not necessary to emphasize that solutions of the Riemann problem from Theorems 2.27, 2.28 and 2.30 are associated with the eigenvector $\boldsymbol{r}_k$, we simply speak of the rarefaction (or simple) wave, shock wave and contact discontinuity, respectively.

The following *theorem on the solvability of the Riemann problem* demonstrates the importance of the given particular solutions of the Riemann problem.

Theorem 2.32 *Let us assume that for each $\boldsymbol{w} \in D$ all eigenvalues $\lambda_k(\boldsymbol{w})$ of the matrix $\mathbb{A}(\boldsymbol{w})$ are simple and that every eigenvector $\boldsymbol{r}_k$ is either genuinely nonlinear or linearly degenerate. Then to any $\boldsymbol{w}_L \in D$ there exists its neighbourhood $B(\boldsymbol{w}_L) \subset D$ such that the following statement holds: for any $\boldsymbol{w}_R \in B(\boldsymbol{w}_L)$ the Riemann problem* (2.3.49), (2.3.50) *has a unique solution. This solution consists of at most $m+1$ constant states separated by simple waves or entropy shock waves or contact discontinuities. There is exactly one solution of this structure.*

For the proof see, for example, (Godlewski and Raviart, 1996), Chapter 1, Theorem 6.1 or (Smoller, 1983), Theorem 17.18.

For the application of this theorem to the 1D Euler equations, see Section 3.1.6.

2.3.11 *Existence result for the 2×2 Euler system of barotropic flow*

Recent developments in weak compactness techniques enable us to prove the existence of weak solutions for a large class of 2×2 conservation laws with one space variable. Here we briefly describe a representative result for the Euler system of barotropic flow. We consider system (1.2.96)–(1.2.98) with $N = 1$ and $\lambda = \mu = 0$ written in the conservative form

$$\rho_t + m_x = 0, \tag{2.3.75}$$

$$m_t + \left(\frac{m^2}{\rho} + p(\rho)\right)_x = 0, \quad x \in I\!R,\, t > 0.$$

We denote $m = \rho v$, the momentum of the gas, $\rho \geq 0$, the density, $p = p(\rho) \geq 0$, the pressure (a given function of ρ). We shall consider resolution of system (2.3.75) in the physical region

$$D := \{(\rho, m); \rho \geq 0, |m| \leq C\rho\}$$

with a constant C depending on initial data and the equation of state for p. In the case $\rho = 0$ we define $m^2/\rho = 0$ so that there is no doubt about the interpretation of this function when the denominator is zero.

Next, we assign to (2.3.75) the initial data

$$(\rho, m)\big|_{t=0} = (\rho^0, m^0), \tag{2.3.76}$$

measurable functions with values in the region D, and we are interested in the resolution of problem (2.3.75), (2.3.76) in the sense of distributions.

In what follows we keep sufficiently general assumptions on the function $p(\cdot)$ which include the gamma law $p(\rho) = \kappa\rho^\gamma$ ($\kappa > 0, \gamma > 1$ constants), to admit more general pressure laws, especially those having higher unbounded derivatives near the vacuum. Our main sources here are (Chen and Le Floch, 2000) and (Dafermos, 2000).

Assume the following:

$$p(\cdot) \in C^4(0, \infty); \tag{2.3.77}$$

there exist $\gamma \in (1,3)$, $C > 0$, $\delta > 0$ such that (2.3.78)
$p(\rho) = \kappa\rho^\gamma\big(1 + P(\rho)\big)$, $|P^{(n)}(\rho)| \le C\rho^{1-n}$ for $n = 0, 1, 2, 3, 4$ and $\rho \in (0, \delta)$
(clearly, $p(0) := p(0+) = 0$, $p'(0) := p'(0+) = 0$, so that $p(\cdot) \in C^1([0, \infty))$);

$$p'(\rho) > 0,\ 2p'(\rho) + \rho p''(\rho) > 0 \text{ for } \rho > 0. \tag{2.3.79}$$

Note that the first assumption in (2.3.79) implies strict hyperbolicity of system (2.3.75) away from the vacuum (while at the vacuum the two characteristic speeds may coincide and the system be nonstrictly hyperbolic). The second inequality in (2.3.79) guarantees genuine nonlinearity (also away from the vacuum).

In order to formulate the existence result, we define the notion of admissible solution to (2.3.75), (2.3.76). First we recall that the *entropy–entropy flux pair* for (2.3.75) is a couple (η, G) satisfying the conditions

$$\begin{aligned} \frac{\partial G}{\partial \rho} &= \frac{\partial \eta}{\partial m}\Big(p'(\rho) - \frac{m^2}{\rho^2}\Big), \\ \frac{\partial G}{\partial m} &= \frac{\partial \eta}{\partial \rho} + \frac{2m}{\rho}\frac{\partial \eta}{\partial m} \end{aligned} \tag{2.3.80}$$

(cf. (2.3.24)). The compatibility condition (in fact the Cauchy–Riemann condition of interchangeability of partial derivatives) now, after a short computation, yields the following wave equation for η:

$$\frac{\partial^2 \eta}{\partial \rho^2} - \Big(p'(\rho) - \frac{m^2}{\rho^2}\Big)\frac{\partial^2 \eta}{\partial m^2} + \frac{2m}{\rho}\frac{\partial^2 \eta}{\partial \rho \partial m} = 0. \tag{2.3.81}$$

The pairs (η, G) satisfying (2.3.80), (2.3.81) are called *weak entropy–entropy flux pairs* because of their singularities near the vacuum, i.e. near the line $\{\rho = 0\}$ in the state space D. The analysis of weak entropy–entropy flux pairs is presented in the recent monograph (Novotný and Straškraba, 2003) and appeared originally in (Chen and Wang, 2001).

Definition 2.33 *A couple* $(\rho, m) \in L^\infty(I\!R)^2$ *is called a weak entropy solution to the problem* (2.3.75), (2.3.76), *if*

$$\int_0^\infty \int_{I\!R} \Big(\rho\varphi_t + m\varphi_x\Big)\, dx\, dt + \int_{I\!R} \rho^0(x)\varphi(x,0)\, dx = 0, \tag{2.3.82}$$

$$\int_0^\infty \int_{I\!R} \Big(m\varphi_t + \Big(\frac{m^2}{\rho} + p(\rho)\Big)\varphi_x\Big)\, dx\, dt + \int_{I\!R} m^0(x)\varphi(x,0)\, dx = 0$$

hold for any $\varphi \in C_0^\infty(I\!R \times [0,\infty))$, *and if for any (convex) entropy function* $\eta \in C^2(D)$ *satisfying* (2.3.81) *and the corresponding entropy flux* $G \in C^2(D)$ *satisfying* (2.3.80) *we have*

$$\int_0^\infty \int_{I\!R} \Big(\eta(\rho,m)\varphi_t + G(\rho,m)\varphi_x\Big)\, dx\, dt \geq 0 \; for\; all \; \varphi \in C_0^\infty(I\!R \times (0,\infty)),\; \varphi \geq 0. \tag{2.3.83}$$

Now we are in a position to state the existence theorem.

Theorem 2.34 *Let assumptions* (2.3.77)–(2.3.79) *be satisfied and let*

$$(\rho^0, m^0) \in L^\infty(I\!R)^2$$

be such that $0 \leq \rho^0(x) \leq C_0, m^0(x) \leq C_0\rho^0(x)$ *for a.a.* $x \in I\!R$ *and some* $C_0 > 0$. *Then there exists a weak entropy solution* $(\rho, m) \in L^\infty(I\!R \times (0,\infty))^2$ *of the Cauchy problem* (2.3.75), (2.3.76) *satisfying*

$$0 \leq \rho(x,t) \leq K(C_0),\; m(x,t) \leq K(C_0)\rho(x,t) \; for \; a.a. \; (x,t) \in I\!R \times (0,\infty) \tag{2.3.84}$$

with a constant $K(C_0) > 0$ *depending only on* C_0.

Further, let

$$\{(\rho_\varepsilon, m_\varepsilon)\}_{\varepsilon\in(0,1]} \subset L^\infty(I\!R \times (0,\infty))^2$$

be a family of couples satisfying

$$0 \leq \rho_\varepsilon(x,t) \leq K,\; |m_\varepsilon(x,t)| \leq K\rho_\varepsilon(x,t)$$

for a.a. $(x,t) \in I\!R \times (0,\infty)$ *and all* $\varepsilon \in (0,1]$, *and let for any entropy pair* (η, G) *described above, the family*

$$\{\partial_t\eta(\rho_\varepsilon, m_\varepsilon) + \partial_x G(\rho_\varepsilon, m_\varepsilon)\}_{\varepsilon\in(0,1]}$$

be relatively compact in $H^{-1}_{\mathrm{loc}}(I\!R \times (0,\infty))$. *Then the family* $\{(\rho_\varepsilon, m_\varepsilon)\}_{\varepsilon\in(0,1]}$ *is compact in* $L^r_{\mathrm{loc}}(I\!R \times (0,\infty))$ *for any fixed, but arbitrary,* $r \in [1,\infty)$.

Note that by compactness of a family $\{z_\varepsilon\}$ in $H^{-1}_{\mathrm{loc}}(I\!R \times (0,\infty))$ we mean that $\{z_\varepsilon|\Omega\}$ is compact in $H^{-1}(\Omega)$ for any fixed bounded domain $\Omega \subset I\!R \times (0,\infty)$. The second part of Theorem 2.34 is important for the convergence of suitable approximations to the exact solution. In fact, the family $\{(\rho_\varepsilon, m_\varepsilon)\}$ is thus constructed in the course of the proof.

The proof is very technical and uses the *method of compensated compactness* and the representation of weak limits by *Young measures*. For details we refer to (Chen and Le Floch, 2000), (Chen and Wang, 2001) and (Novotný and Straškraba, 2003). For the concept of Young measures and measure-valued solutions to conservation laws, see (Feistauer, 1993), Chapter 7, (Málek *et al.*, 1996) or (Kröner, 1997).

2.3.12 *Global existence results for general 1D systems*

Let us again consider problem (2.2.11). In 1965, J. Glimm in (Glimm, 1965) formulated a scheme which allowed him to prove an existence theorem for problem (2.2.11), where $N = 1$, with small data. Let $TV_{\mathbb{R}}(\boldsymbol{w})$ mean the *total variation* of a function $\boldsymbol{w} : \mathbb{R} \to \mathbb{R}^m$ defined by

$$TV_{\mathbb{R}}(\boldsymbol{w}) = \sup\Big\{\sum_{j=1}^{k-1} |\boldsymbol{w}(x_{j+1}) - \boldsymbol{w}(x_j)|;\ a < x_1 < \cdots < x_k < b\Big\}.$$

The following theorem holds true.

Theorem 2.35 *Let us assume that* $N = 1$, *system* (2.2.11), *a) is strictly hyperbolic and all eigenvectors of the matrix* $\mathbb{A} = D\boldsymbol{f}(\boldsymbol{w})/D\boldsymbol{w}$ *are either genuinely nonlinear or linearly degenerate in a neighbourhood of a constant state* $\overline{\boldsymbol{w}}$. *Then there exist two positive constants* δ_1 *and* δ_2 *such that for initial data satisfying*

$$\|\boldsymbol{w}^0 - \overline{\boldsymbol{w}}\|_{L^\infty(\mathbb{R})^m} \le \delta_1, \quad TV_{\mathbb{R}}\boldsymbol{w}^0 \le \delta_2,$$

the Cauchy problem (2.2.11) *has a global weak entropy solution* $\boldsymbol{w}(x,t)$ *in* $\mathbb{R} \times [0,\infty)$ *satisfying entropy inequality* (2.3.29) *in the sense of distributions for any entropy–entropy flux pair and*

$$\begin{aligned}
&\|\boldsymbol{w}(\cdot,t) - \overline{\boldsymbol{w}}\|_{L^\infty(\mathbb{R})^m} \le C_0\|\boldsymbol{w}^0 - \overline{\boldsymbol{w}}\|_{L^\infty(\mathbb{R})^m}, \quad t \in [0,\infty),\\
&TV_{\mathbb{R}}(\boldsymbol{w}(\cdot,t)) \le C_0 TV_{\mathbb{R}}(\boldsymbol{w}^0), \quad t \in [0,\infty),\\
&\|\boldsymbol{w}(\cdot,t_1) - \boldsymbol{w}(\cdot,t_2)\|_{L^1(\mathbb{R})^m} \le C_0|t_1 - t_2|\, TV_{\mathbb{R}}(\boldsymbol{w}^0), \quad t_1,t_2 \in [0,\infty),
\end{aligned}$$

for some constant $C_0 > 0$. *In addition,* $\boldsymbol{w} \in C([0,\infty); L^1_{\mathrm{loc}}(\mathbb{R}))^m$ *and the initial condition* (2.2.11), *b) is thus naturally satisfied.*

In the original proof in (Glimm, 1965), an approximate family $\boldsymbol{w}_h$ is constructed by means of a sequence of solutions of auxiliary *Riemann problems* (see Section 2.3.8). Then the structure of Riemann solutions is used to derive a priori bounds of the L^∞-norm and the variation of approximations. Finally, a compactness argument is applied. The details of the proof are presented in (Chen and Wang, 2001) and (Novotný and Straškraba, 2003)

Recent developments have led to another method using the the so-called *front tracking approximations.* A detailed proof using the method of front tracking approximations is given in (Bressan, 2000) (proof of Theorem 7.1).

2.4 Nonstationary Navier–Stokes equations of compressible flow

In this section we summarize some theoretical results for the nonstationary compressible Navier–Stokes equations. We distinguish solutions local in time, global in time, for small or large data, and weak solutions, strong solutions or classical solutions.

In general, regardless of the concrete equations, by *classical solutions* we mean functions that are continuous together with all derivatives appearing in the equations, in the domain where the equations are considered, and such that the equations are satisfied pointwise (cf. Definition 2.6).

For *strong solutions* the minimal requirement is that the unknown functions and all terms present in the equations are integrable over the domain of definition, and the equations are satisfied almost everywhere.

Both classical and strong solutions are sometimes called in short *regular solutions.*

Weak solutions are usually defined from case to case. The minimal requirement here is that the unknown functions are integrable over the domain of definition and the equations are satisfied *in the sense of distributions* (see Section 1.3.2).

2.4.1 *Results for the full system of compressible Navier–Stokes equations*

The full system of compressible Navier–Stokes equations for a heat-conductive gas in a bounded domain Ω with Dirichlet boundary conditions and initial data reads (see (1.2.99)–(1.2.105))

$$
\begin{aligned}
&\rho_t + \operatorname{div}(\rho \boldsymbol{v}) = 0, \quad x \in \Omega \subset I\!\!R^N,\ t \in (0,T)\ (N = 1,2,3),\\
&(\rho \boldsymbol{v})_t + \operatorname{div}(\rho \boldsymbol{v} \otimes \boldsymbol{v}) - \operatorname{div}(\mu \nabla \boldsymbol{v}) - \nabla\big((\mu+\lambda)\operatorname{div}\boldsymbol{v}\big) + \nabla p(\rho,\theta) = \rho \boldsymbol{f}(x,t),\\
&c_v \rho(\theta_t + \boldsymbol{v}\cdot\nabla\theta) - k\Delta\theta + p(\rho,\theta)\operatorname{div}\boldsymbol{v} - 2\mu(\mathbb{D}(\boldsymbol{v})\cdot\mathbb{D}(\boldsymbol{v})) - \lambda(\operatorname{div}\boldsymbol{v})^2 = 0,\\
&\mathbb{D}(\boldsymbol{v}) = \big(d_{ij}(\boldsymbol{v})\big)_{i,j=1}^N,\ d_{ij}(\boldsymbol{v}) := \frac{1}{2}\big(\partial_i v_j + \partial_j v_i\big), \qquad (2.4.1)\\
&p(\rho,\theta) = \theta q(\rho),\ q = q(\rho) > 0\ (\rho > 0, \theta > 0),\\
&(\rho,\boldsymbol{v},\theta)(x,0) = (\rho^0,\boldsymbol{v}^0,\theta^0)(x),\ x \in \Omega,\ \inf_x\{\rho^0,\theta^0\} > 0,\\
&\boldsymbol{v}(x,t) = 0,\ \theta(x,t) = \overline{\theta}(x,t),\ x \in \partial\Omega,\ t \geq 0.
\end{aligned}
$$

We make the following assumptions.

(i) $3\lambda + 2\mu \geq 0, \mu > 0, k > 0$ are constants;
(ii) $e = c_v\theta$, where $c_v = \text{const} > 0$ is the specific heat at constant volume;
(iii) $q'(\rho) > 0$ for $\rho > 0$.

Exercise 2.36 Derive the third equation in system (2.4.1) from equation (1.2.85) with $q = 0$, the first equation in (2.4.1) and thermodynamical relation (ii).

In the stationary case system (2.4.1) is reduced to

$$\operatorname{div}(\rho \boldsymbol{v}) = 0, \quad x \in \Omega \subset I\!\!R^N\ (N = 1,2,3), \qquad (2.4.2)$$

$$\operatorname{div}(\rho \boldsymbol{v} \otimes \boldsymbol{v}) - \operatorname{div}(\mu \nabla \boldsymbol{v}) - \nabla\big((\mu + \lambda)\operatorname{div} \boldsymbol{v}\big) + p(\rho, \theta) = \rho \boldsymbol{f}(x),$$
$$c_v \rho \boldsymbol{v} \cdot \nabla\theta - k\Delta\theta + p(\rho,\theta)\operatorname{div}\boldsymbol{v} - 2\mu(\mathbb{D}(\boldsymbol{v}) \cdot \mathbb{D}(\boldsymbol{v})) - \lambda(\operatorname{div}\boldsymbol{v})^2 = 0$$

and equipped with the corresponding boundary conditions. The solution of system (2.4.2) is called a *stationary solution.* In what follows, a special stationary solution of the type $(\rho, 0, \theta)$ called the *rest state* or *equilibrium* plays an important role. It is clear that if we put in (2.4.2) $\boldsymbol{v} = 0$ and $\boldsymbol{f} = \nabla F$ with a given potential F, then system (2.4.2) becomes

$$\begin{aligned} \nabla p(\rho,\theta) &= \rho \nabla F, \\ \Delta\theta &= 0. \end{aligned} \tag{2.4.3}$$

If we search for a solution satisfying the boundary condition

$$\theta = \overline{\theta} = \text{const} \quad \text{on } \partial\Omega, \tag{2.4.4}$$

then by Green's theorem, assuming that θ and Ω are smooth enough, we get

$$0 = \int_\Omega \Delta(\theta - \overline{\theta})(\theta - \overline{\theta})\, dx = \int_{\partial\Omega} \frac{d(\theta - \overline{\theta})}{dn}(\theta - \overline{\theta})\, dS - \int_\Omega |\nabla(\theta - \overline{\theta})|^2\, dx.$$

Consequently, since $\theta = \overline{\theta}$ on $\partial\Omega$, we get $\theta = \overline{\theta} + c$ in Ω for some constant c. Moreover, we find that $c = 0$ and, thus, $\theta = \overline{\theta}$ in Ω. Further, seeking solutions of (2.4.3) with $\rho > 0$ only, by integration of the first equation in (2.4.3) we get

$$\int_1^\rho \frac{1}{r}\frac{\partial p(r, \overline{\theta})}{\partial r}\, dr = F + K \tag{2.4.5}$$

with a constant K. The constant K can be fixed by prescribing the total mass of the gas contained in Ω, i.e. by the condition

$$\int_\Omega \rho\, dx = M \tag{2.4.6}$$

with a given $M > 0$. Obviously, it follows from (2.4.5) that $\mathcal{F}(\rho) = F + K$, where $\mathcal{F}$ is an increasing function in $(0, \infty)$. This implies that $\rho = \mathcal{F}^{-1}(F + K)$, where $\mathcal{F}^{-1}$ is the inverse of $\mathcal{F}$. Now we can already see that the constant K is uniquely determined by a given M from (2.4.6).

Let us now formulate the following global existence result.

Theorem 2.37 *Let $\Omega \subset \mathbb{R}^3$ be a bounded domain of class $C^{2,\alpha}$ $(\alpha \in (0,1])$ and let assumptions (i),(ii),(iii) hold. Let $\boldsymbol{f} = \nabla F$, $F \in H^4(\Omega)$, $(\rho^0, \boldsymbol{v}^0, \theta^0) \in H^3(\Omega)^5$, $\boldsymbol{v}^0|_{\partial\Omega} = 0$, $\theta^0|_{\partial\Omega} = \overline{\theta}$, $\boldsymbol{v}_t(0)|_{\partial\Omega} = 0$, $\theta_t(0)|_{\partial\Omega} = 0$, where $\boldsymbol{v}_t(x,0)$ and $\theta_t(x,0)$ are computed from the partial differential equations in (2.4.1), in which we put $\rho := \rho^0, \theta := \theta^0, \boldsymbol{v} := \boldsymbol{v}^0$. In addition, we assume that there is a rest state $(\widehat{\rho}, 0, \overline{\theta})$ with $\overline{\theta} =$ const > 0.*

Then there exist positive constants ε_0, β *and* $C_0 = C_0(\overline{\rho}, \overline{\theta}, \|F\|_{H^4(\Omega)})$ *(where* $\overline{\rho} := \frac{1}{|\Omega|}\int_\Omega \rho^0\,dx$*) such that if*

$$\|(\rho^0 - \widehat{\rho}, \boldsymbol{v}^0, \theta^0 - \overline{\theta})\|_{H^3(\Omega)^5} \leq \varepsilon_0, \tag{2.4.7}$$

then the initial-boundary value problem (2.4.1) *has a unique solution* $(\rho, \boldsymbol{v}, \theta)$ *global in time satisfying*

$$\begin{aligned}&(\rho, \boldsymbol{v}, \theta) \in \big(C^0 \cap L^\infty\big)([0,\infty); H^3(\Omega))^5 \cap C^1(0,\infty; H^2(\Omega) \times H^1(\Omega)^4),\\ &\inf_{x,t} \rho(x,t) > 0, \inf_{x,t} \theta(x,t) > 0,\end{aligned}$$

and

$$\sup_{x\in\Omega} |(\rho(x,t) - \widehat{\rho}(x), \boldsymbol{v}(x,t), \theta(x,t) - \overline{\theta})| \leq C_0 \exp(-\beta t).$$

The proof of Theorem 2.37 can be found in (Matsumura and Padula, 1992).

For a generalization of these results, see (Hoff, 1995) (where the global solution for small data allows discontinuities) and (Matsumura and Yamagata, 2001).

The solution from Theorem 2.37 is a strong solution with smooth first order partial derivatives because of the Sobolev imbedding $H^3(\Omega) \hookrightarrow C^1(\Omega)$ (see (1.3.21)).

Note that the existence of a regular solution on some maximal time interval $(0, T_{\max})$ can be proven even without the assumption that ε_0 in (2.4.7) is small enough. However, without additional information about this solution, we do not a priori know how large or small $T_{\max}$ can be. In general, we can give only a very pessimistic lower estimate of $T_{\max}$ in terms of the data. From the point of view of applications, such results are not very useful.

2.4.2 *Results for equations of barotropic flow*

Since no results on the global solvability of problem (2.4.1) with completely large data have been obtained up to now, we restrict ourselves to equations of barotropic flow (see (1.2.96)–(1.2.98))

$$\begin{aligned}&(\rho\boldsymbol{v})_t + \operatorname{div}(\rho\boldsymbol{v}\otimes\boldsymbol{v}) - \mu\Delta\boldsymbol{v} - (\lambda+\mu)\nabla\operatorname{div}\boldsymbol{v} + \nabla p(\rho) = \rho\boldsymbol{f},\\ &\rho_t + \operatorname{div}(\rho\boldsymbol{v}) = 0, \quad x \in \Omega \subset I\!R^N,\ t \in (0,T)\ (T>0).\end{aligned} \tag{2.4.8}$$

We impose the initial conditions

$$\rho|_{t=0} = \rho^0, \quad (\rho\boldsymbol{v})|_{t=0} = \boldsymbol{m}^0, \tag{2.4.9}$$

where

$$\rho^0 \geq 0 \text{ a.e. in } \Omega, \quad \rho^0 \in L^\infty(\Omega),$$

$$\frac{|\boldsymbol{m}^0|^2}{\rho^0} \in L^1(\Omega) \text{ (we set } |\boldsymbol{m}^0(x)|^2/\rho^0(x) = 0, \text{ if } \rho^0(x) = 0),$$

and the Dirichlet boundary condition

$$\boldsymbol{v}(x,t) = 0, \quad x \in \partial\Omega, \quad t > 0. \tag{2.4.10}$$

The unknown functions are $\boldsymbol{v} = \boldsymbol{v}(x,t) = (v_1(x,t), \dots, v_N(x,t))^{\mathrm{T}}$ (or $\boldsymbol{m} = \rho\boldsymbol{v}$) and $\rho = \rho(x,t)$.

As an example of a local existence result, we formulate the following theorem.

Theorem 2.38 *Let $\mu > 0, \lambda \geq -2\mu/3$ be constants, $\Omega \subset I\!R^N (N \leq 3)$ a bounded domain with $\partial\Omega \in C^3$, $\rho^0 \in H^2(\Omega)$, $\inf_{x\in\Omega} \rho^0 > 0$, $\boldsymbol{v}^0 \in H^2(\Omega) \cap H_0^1(\Omega)$, $\boldsymbol{f} \in L^\infty(0,T; H^1(\Omega))^N \cap W^{1,2}(0,T; L^2(\Omega))^N$ ($T > 0$ given) and $p(\cdot) \in C^2((0,\infty))$.*

Then for any $r_0 > 0$ there exists a $T^* = T^*(r_0, \overline{p}) \in (0,T]$ and positive constants $\alpha(r_0,\overline{p}), \beta(r_0,\overline{p})$ such that there exists a unique solution of problem (2.4.8)–(2.4.10) on $\Omega \times (0,T^*)$ satisfying

$$\begin{aligned}
&\rho \in L^\infty(0,T^*; H^2(\Omega)) \cap W^{1,\infty}(0,T^*; H^1(\Omega)), \\
&0 < \alpha(r_0,\overline{p}) \leq \rho(x,t) \leq \beta(r_0,\overline{p}) < \infty \quad \forall (x,t) \in \Omega \times (0,T^*), \\
&\boldsymbol{v} \in L^2(0,T^*; H^3(\Omega))^N \cap W^{1,2}(0,T^*; H_0^1(\Omega))^N.
\end{aligned}$$

Theorem 2.38 is taken from (Salvi and Straškraba, 1993), Theorem 1, and its proof can be found therein. Note that a similar statement can be made for the equations of heat-conductive gas flow (2.4.1). A corresponding theorem can be found, for example, in (Tani, 1977). The history of local existence theorems for compressible flow started in 1962 with the paper (Nash, 1962) and were continued by (Solonnikov, 1980). The history of the global existence of regular solutions for small data started with the famous papers (Matsumura and Nishida, 1982), (Matsumura and Nishida, 1983). Since then many variants have appeared. For further results in this regard we refer to (Valli and Zajączkowski, 1986), (Valli, 1992), (Hoff, 1997), (Novotný and Straškraba, 2003) and references therein.

For large data the existence of a global smooth solution is not known in the heat-conductive, as well as in the barotropic, case. Nevertheless, when we restrict ourselves to weak solutions, recent developments yield important global existence results.

Definition 2.39 *By a weak solution of* (2.4.8)–(2.4.10) *we call a couple* $(\boldsymbol{v}, \rho)$ *such that*

$$\rho,\ p(\rho),\ \rho|\boldsymbol{v}|^2,\ |\nabla\boldsymbol{v}| \ \in\ L^1_{\mathrm{loc}}(\Omega \times (0,\infty))$$

and, putting $Q_T = \Omega \times (0,T)$, for any $T > 0$ and any $\boldsymbol{\varphi} \in C^1(0,T; C_0^\infty(\Omega))^N$, $\psi \in C^1(0,T; C^\infty(\Omega))$ such that $\boldsymbol{\varphi}(x,T) \equiv 0$, $\psi(x,T) \equiv 0$, the following integral identities hold:

$$\int_{Q_T} \Big(\rho\boldsymbol{v} \cdot \boldsymbol{\varphi}_t + \rho\big((\boldsymbol{v}\cdot\nabla)\boldsymbol{\varphi}\cdot\boldsymbol{v}\big) - \mu\nabla\boldsymbol{v}\cdot\nabla\boldsymbol{\varphi} - (\lambda+\mu)\mathrm{div}\,\boldsymbol{v}\,\mathrm{div}\,\boldsymbol{\varphi}$$

$$+p(\rho)\operatorname{div}\boldsymbol{\varphi} + \rho\boldsymbol{f}\cdot\boldsymbol{\varphi}\Big)(x,t)\,dx\,dt + \int_\Omega \rho^0(x)\boldsymbol{v}^0(x)\boldsymbol{\varphi}(x,0)\,dx = 0, \quad (2.4.11)$$

$$\int_{Q_T} \big(\rho\psi_t + \rho(\boldsymbol{v}\cdot\nabla)\psi\big)\,dx\,dt + \int_\Omega \rho^0(x)\psi(x,0)\,dx = 0.$$

Theorem 2.40 *Let $\mu > 0, \lambda \geq -\frac{2}{3}\mu$ be constants, $\Omega \subset I\!R^N$ $(N \leq 3)$ a bounded domain with $\partial\Omega \in C^{2,\alpha}$ $(\alpha > 0)$, $p(\rho) = \kappa\rho^\gamma$, $\kappa > 0$, $\gamma > 3/2$.*

Then, for any

$$\rho^0 \in L^\gamma(\Omega),\ \rho^0 \geq 0,\ \frac{|\boldsymbol{m}^0|^2}{\rho^0} \in L^1(\Omega),\ \boldsymbol{f} \in L^\infty(\Omega\times(0,\infty))^N,$$

there exists a weak solution to (2.4.8)–(2.4.10) such that

$$\rho \in L^\infty(0,\infty; L^\gamma(\Omega)) \cap L^{\frac{5}{3}\gamma-1}(\Omega\times(0,T))$$
$$\cap\, C([0,T]; L^\gamma_{\text{weak}}(\Omega)) \cap C([0,T]; L^\beta(\Omega)),$$

for any $T > 0$ and $\beta \in [1,\gamma)$,

$$\boldsymbol{v} \in L^2(0,\infty; W_0^{2,1}(\Omega))^N,\ \ \rho|\boldsymbol{v}|^2 \in L^\infty(0,\infty; L^1(\Omega)),$$
$$\rho\boldsymbol{v} \in C([0,T]; L^{\frac{2\gamma}{\gamma+1}}_{\text{weak}}(\Omega))^N,$$

and, in the sense of distributions on $(0,T)$, the energy inequality

$$\frac{d}{dt}\int_\Omega \left(\frac{1}{2}\rho|\boldsymbol{v}|^2 + \frac{\kappa}{\gamma-1}\rho^\gamma\right)dx + \int_\Omega \left(\mu|\nabla\boldsymbol{v}|^2 + (\lambda+\mu)|\operatorname{div}\boldsymbol{v}|^2\right)dx \leq 0$$

holds true. In addition, the so-called renormalized continuity equation

$$b(\rho)_t + \operatorname{div}(b(\rho)\boldsymbol{v}) + \big(\rho b'(\rho) - b(\rho)\big)\operatorname{div}\boldsymbol{v} = 0$$

holds in the sense of distributions, i.e. in $\mathcal{D}'(Q_T)$, for any function $b \in C^1(I\!R)$.

Here $L^\delta_{\text{weak}}(\Omega)$ means the linear space $L^\delta(\Omega)$ endowed with the weak topology of $L^\delta(\Omega)$, induced by the usual L^δ-norm.

A strategy for the proof of Theorem 2.40, in a slightly weaker form, was first given in (Lions, 1993). Complete proof was then published in the monograph (Lions, 1998). The result of the present Theorem 2.40 is proven in (Feireisl *et al.*, 2001). A detailed proof elaborated for a wider readership is presented in the recent monograph (Novotný and Straškraba, 2003).

For the barotropic case, existence and uniqueness results analogous to Theorem 2.37 can be derived. We do not paraphrase it here but instead refer the reader to (Novotný and Straškraba, 2003) or (Salvi and Straškraba, 1993).

2.5 Existence results for stationary compressible Navier–Stokes equations

In this section we review several results on the existence and uniqueness of solutions to system (2.4.2). We adopt here the notions of classical, strong and regular solution from the beginning of Section 2.4.

2.5.1 *Existence of a regular solution for small data*

The existence of a regular solution to (2.4.2) for general large data (regardless of how smooth) is not known. The only existence results for classical or strong solutions of system (2.4.2) have been obtained for small data. We formulate a typical result in this direction for the problem given by (2.4.2) and the boundary conditions

$$\boldsymbol{v} = 0, \quad \theta = \overline{\theta} = \text{const} > 0 \quad \text{on } \partial\Omega. \tag{2.5.1}$$

The second boundary condition in (2.5.1) means that a constant temperature on the boundary is maintained. We also assume that

$$\boldsymbol{f} = \nabla F + \boldsymbol{g},$$

where a potential F and a nonpotential volume force $\boldsymbol{g}$ are given. For later use, we define the following problem: for a given total mass $M > 0$, find a density function $\widehat{\rho} \in L^1(\Omega)$ satisfying

$$\nabla \int_1^{\widehat{\rho}} \frac{1}{r} \frac{\partial p(r, \overline{\theta})}{\partial r}\, dr = \nabla F \text{ in } \mathcal{D}'(\Omega), \quad \int_\Omega \widehat{\rho}\, dx = M > 0, \ \widehat{\rho} > 0. \tag{2.5.2}$$

(Cf. (2.4.5)–(2.4.6).) Note that if $\widehat{\rho}$ satisfies (2.5.2), then the triple $(\rho, \boldsymbol{v}, \theta) := (\widehat{\rho}, 0, \overline{\theta})$ is a *rest state* for system (2.4.2), with $\boldsymbol{f}$ replaced by ∇F only, i.e. a solution with zero velocity, total prescribed mass M and the potential force on the right-hand side of the momentum equation. The existence of a solution to (2.5.2) is known under fairly general assumptions and has been studied for example in (Beirão da Veiga, 1987), (Matsumura and Padula, 1992) (see also (Novotný and Straškraba, 2003)).

Theorem 2.41 *Let $\Omega \subset \mathbb{R}^N (N = 2, 3)$ be a bounded domain with a boundary of class C^4 and*

$$\boldsymbol{f} = \nabla F + \boldsymbol{g}, \text{ where } F \in C^3(\overline{\Omega}), \ \boldsymbol{g} \in H^1(\Omega)^N \text{ are given.}$$

Let, in addition, p have the form

$$p(\rho, \theta) = \theta q(\rho)$$

with $q \in C^1(0, \infty), q'(r) > 0$ for $r > 0$, and suppose that there exists a rest state $(\widehat{\rho}, 0, \overline{\theta})$, where $\widehat{\rho} \in H^2(\Omega)$ and satisfies (2.5.2), and $\overline{\theta} = \text{const} > 0$.

Then there exist constants $\varepsilon_0 > 0$ and $C_0 > 0$ such that if

$$\|\boldsymbol{g}\|_{H^1(\Omega)^N} < \varepsilon_0,$$

then there exists a unique regular solution $(\rho, \boldsymbol{v}, \theta)$ of problem (2.4.2), (2.5.1) such that

$$\rho \in H^2(\Omega), \ \boldsymbol{v} \in H^3(\Omega)^N \cap H_0^1(\Omega)^N, \ \theta - \overline{\theta} \in H^2(\Omega) \cap H_0^1(\Omega)$$

and

$$\|\rho - \widehat{\rho}\|_{H^2(\Omega)} + \|\boldsymbol{v}\|_{H^3(\Omega)^N} + \|\theta - \overline{\theta}\|_{H^2(\Omega)} \leq C_0 \|\boldsymbol{g}\|_{H^1(\Omega)^N}.$$

The proof of Theorem 2.41 can be found for example in (Novotný and Padula, 1993).

A modification of Theorem 2.41 for the barotropic case can be found in (Novotný, 1993). Since the existence of weak solutions for large data is not known for system (2.4.2) in its full generality, we comment on results for barotropic flow only.

2.5.2 *Existence of weak solutions for barotropic flow*

Barotropic steady equations are of the form (cf. (2.4.8))

$$\begin{aligned}&\operatorname{div}(\rho \boldsymbol{v}\otimes \boldsymbol{v}) - \mu\Delta \boldsymbol{v} - (\lambda+\mu)\nabla \operatorname{div} \boldsymbol{v} + \nabla p(\rho) = \rho \boldsymbol{f}(x),\\ &\operatorname{div}(\rho \boldsymbol{v}) = 0, \quad x \in \Omega \subset I\!R^N.\end{aligned} \tag{2.5.3}$$

If regular solutions are considered, a similar result as in Theorem 2.41 holds also for problem (2.5.3), (2.4.10). We shall not paraphrase it here and refer the reader to (Novotný and Straškraba, 2003).

Now, let us consider weak solutions of (2.5.3), (2.4.10).

Definition 2.42 *Let $\boldsymbol{f} \in L^\infty(\Omega)^N$. By a weak solution of system* (2.5.3) *with the Dirichlet boundary condition* (2.4.10) *we mean a couple* $(\rho, \boldsymbol{v})$ *such that the following relations hold:*

$$\rho,\ p(\rho),\ \rho|\boldsymbol{v}| \in L^1(\Omega), \quad \boldsymbol{v} \in W_0^{1,2}(\Omega)^N, \tag{2.5.4}$$

$$\int_\Omega \Big(\rho v_i v_j \partial_j \varphi_i + \mu v_i \Delta \varphi_i + (\mu+\lambda) v_j \partial_i \partial_j \varphi_i + p(\rho)\partial_i \varphi_i + \rho f_i \varphi_i\Big)\, dx = 0$$

$$\textit{for all } \boldsymbol{\varphi} = (\varphi_1, \ldots, \varphi_N) \in C_0^\infty(\Omega)^N,$$

$$\int_\Omega \rho v_j \partial_j \varphi\, dx = 0 \quad \textit{for any } \varphi \in C_0^\infty(\Omega).$$

Here the Einstein summation convention over repeated indices is used.

Theorem 2.43 *Assume that*

$$\Omega \textit{ is a bounded domain in } I\!R^3,\ \Omega \in C^{2,\alpha},\ \textit{for some } \alpha \in (0,1];$$

$$p(\rho) = \kappa \rho^\gamma,\ \textit{where } \kappa > 0,\ \gamma > 5/3;$$

$$\boldsymbol{f} \in L^\infty(\Omega)^3.$$

Then for any $M > 0$, there exists a weak solution of system (2.5.3) *with the Dirichlet boundary condition* (2.4.10) *such that*

$$\rho \in L^{s(\gamma)}(\Omega),\ \textit{with } s(\gamma) = \begin{cases} 3(\gamma-1) & \textit{if } 3/2 < \gamma < 3,\\ 2\gamma & \textit{if } \gamma \geq 3,\end{cases}$$

$$\int_\Omega \rho\, dx = M.$$

The theory of weak stationary solutions to the barotropic equations is studied thoroughly in the fundamental monograph (Lions, 1998). The version of the existence result contained in Theorem 2.43 is taken from (Novo and Novotný, 2002) and appears also in (Novotný and Straškraba, 2003).

From a practical point of view it would be useful to investigate initial-boundary value problems for viscous compressible flow with a general nonhomogeneous boundary condition for the velocity, allowing to consider the inlet or outlet. Some results of this type are presented in (Valli and Zajączkowski, 1986).

3

FINITE DIFFERENCE AND FINITE VOLUME METHODS FOR NONLINEAR HYPERBOLIC SYSTEMS AND THE EULER EQUATIONS

In modern technologies one often encounters the necessity to solve compressible flow with a complicated structure. There are several conceivable models of compressible flow. Let us mention, for example, the model of inviscid, stationary, irrotational or rotational subsonic flow using the stream function formulation, and the models of transonic flow based on the small perturbation equation or full potential equation. There exists an extensive literature about the finite difference or finite element methods for the numerical solution of these models. (For a survey of mathematical and numerical methods for these models, see (Feistauer, 1998).) In a number of problems, the potential models are not sufficiently accurate, particularly in high speed (transonic or hypersonic) flow, because of the appearance of the so-called strong shocks with large entropy and vorticity production. This leads to the necessity of using the complete system of conservation laws consisting of the continuity equation, the Euler equations of motion and the energy equation (called the Euler equations in brief), which has been widely used during the last few decades for the modelling of flows in aeronautics, the aviation industry and steam or gas turbine design. Successively, the Euler equations have begun to be applied also to low Mach number problems on the one hand and to problems with chemical reactions on the other. These models neglect, of course, viscosity, but in many situations they give good results, reliable from the point of view of comparisons with experiments.

The system of the Euler equations is nonlinear, first order and hyperbolic in the case of unsteady flow. The investigation of nonlinear hyperbolic systems belongs to the most difficult area of partial differential equations. A survey of theoretical results for the compressible Euler equations was the subject of Chapter 2 of this book.

There exists a very wide literature on numerical schemes for the Euler equations, but theoretical aspects (such as stability, convergence, error estimates) are mostly omitted and the complete theory of these schemes is still missing because of a number of difficult obstacles. Therefore, the qualitative properties, as the order of approximation, stability, convergence and error estimates, are investigated in the sequel for simplified situations, e.g. for 1D systems, linearized problems, scalar equations and problems in the whole space without boundary conditions (the Cauchy problem). The results obtained are then applied or extended in a suitable (often quite heuristical) way to more complex problems.

From the literature devoted to various methods and techniques applied to hyperbolic problems, conservation laws and the Euler equations, we mention only the fundamental monographs (Godlewski and Raviart, 1991), (Godlewski and Raviart, 1996), (Kröner, 1997), (Le Veque, 1992), (Toro, 1997), (Wesseling, 2001) and (Feistauer, 1993).

This chapter is devoted to the numerical simulation of inviscid flow using the complete system of conservation law equations, which belongs to important subjects in CFD. In the framework of nonlinear hyperbolic equations of the conservation laws we develop and discuss here the most common and efficient numerical schemes for the solution of one- and multidimensional flows and present their analysis.

In the first section of this chapter we give a survey of some basic results concerning the hyperbolic systems and the Euler equations (unless they have already been treated in Chapter 2). Then we deal with problems with one space variable. In this case the most important concepts, methods and techniques are explained. The results obtained are then extended to the solution of multidimensional problems by the finite volume method on general nonuniform and unstructured meshes. Special attention is paid to the derivation of multidimensional Riemann solvers and to adaptive mesh refinement techniques. Some examples of the solution of test problems as well as some complicated, technically relevant problems are presented.

3.1 Further properties of the Euler equations

In this section we shall be concerned with some special properties of the Euler equations important for the construction of numerical schemes which were not introduced in Chapter 2.

3.1.1 *The Euler equations*

Let us consider the unsteady flow of an inviscid gas in a domain $\Omega \subset I\!R^N$ ($1 \leq N \leq 3$) and time interval $(0,T)$ ($0 < T \leq +\infty$). It is governed by the continuity equation, the Euler equations of motion and the energy equation, to which we add closing thermodynamical relations. See, for example, Section 1.2.19. We assume that the flow is adiabatic, i.e. we neglect the heat transfer. Moreover, because the gas is light, we neglect the outer volume force. We shall be concerned with the flow of a perfect gas, for which the equation of state has the form

$$p = R\rho\theta, \tag{3.1.1}$$

where R is the gas constant. The system of governing equations (see (1.2.107)–(1.2.110)) considered in the space-time cylinder $Q_T = \Omega \times (0,T)$ can be written in the form

$$\frac{\partial \rho}{\partial t} + \sum_{s=1}^{N} \frac{\partial(\rho v_s)}{\partial x_s} = 0, \tag{3.1.2}$$

$$\frac{\partial(\rho v_i)}{\partial t} + \sum_{s=1}^{N} \frac{\partial(\rho v_i v_s + \delta_{is} p)}{\partial x_s} = 0, \quad i = 1, \dots, N, \tag{3.1.3}$$

$$\frac{\partial E}{\partial t} + \sum_{s=1}^{N} \frac{\partial((E+p)v_s)}{\partial x_s} = 0, \tag{3.1.4}$$

$$p = (\gamma - 1)(E - \rho|\boldsymbol{v}|^2/2). \tag{3.1.5}$$

Here v_s are the components of the velocity vector $\boldsymbol{v} = (v_1, \dots, v_N)^{\mathrm{T}}$ in the directions x_s $(s = 1, \dots, N)$, ρ is the density, p is the pressure, E is the total energy, i.e.

$$E = \rho(c_v\theta + |\boldsymbol{v}|^2/2), \tag{3.1.6}$$

and θ is the absolute temperature. For a perfect gas we assume that the specific heat c_v at constant volume is a constant. (Note that relation (3.1.5) can be obtained from the definition (3.1.6) of the total energy, the equation of state (3.1.1) and the relations $\gamma = c_p/c_v$ and $R = c_p - c_v$. Moreover, $c_p, c_v > 0, \gamma > 1, R > 0$. Cf. Section 1.2.12.)

System (3.1.2)–(3.1.5) can be written as

$$\frac{\partial \boldsymbol{w}}{\partial t} + \sum_{s=1}^{N} \frac{\partial \boldsymbol{f}_s(\boldsymbol{w})}{\partial x_s} = 0, \tag{3.1.7}$$

where

$$\boldsymbol{w} = (w_1, \dots, w_m)^{\mathrm{T}} = (\rho, \rho v_1, \dots, \rho v_N, E)^{\mathrm{T}} \in \mathbb{R}^m, \quad m = N + 2, \tag{3.1.8}$$

is the so-called *state vector*, and

$$\begin{aligned} \boldsymbol{f}_s(\boldsymbol{w}) &= (f_{s1}(\boldsymbol{w}), \dots, f_{sm}(\boldsymbol{w}))^{\mathrm{T}} \\ &= (\rho v_s, \rho v_1 v_s + \delta_{1s} p, \dots, \rho v_N v_s + \delta_{Ns} p, (E+p)v_s)^{\mathrm{T}} \\ &= \Bigg(w_{s+1}, w_2 w_{s+1}/w_1 + \delta_{1s}(\gamma - 1)\Big(w_m - \sum_{i=2}^{m-1} w_i^2/(2w_1)\Big), \dots, \\ &\quad w_{m-1} w_{s+1}/w_1 + \delta_{m-2,s}(\gamma - 1)\Big(w_m - \sum_{i=2}^{m-1} w_i^2/(2w_1)\Big), \\ &\quad w_{s+1}\Big(\gamma w_m - (\gamma - 1)\sum_{i=2}^{m-1} w_i^2/(2w_1)\Big)/w_1 \Bigg)^{\mathrm{T}} \end{aligned} \tag{3.1.9}$$

is the *flux* of the quantity $\boldsymbol{w}$ in the direction x_s. Often, $\boldsymbol{f}_s, s = 1, \dots, N$, are called *inviscid Euler fluxes*. Usually, system (3.1.2)–(3.1.5) (i.e. (3.1.7)) is called the system of the Euler equations, or simply Euler equations. The functions $\rho, v_1, \dots, v_N, p$ are called *primitive variables*, whereas $w_1 = \rho$, $w_2 = \rho v_1, \dots, w_{m-1} = \rho v_N$, $w_m = E$ are *conservative variables*. Sometimes $\rho, v_1, \dots, v_N, \theta$ are called *physical variables.* (We also distinguish characteristic and entropy variables – see Sections 3.2.22.5, 3.5.5 and 4.3.2.) The domain of definition of the vector-valued functions $\boldsymbol{f}_s$ is the open set $D \subset I\!R^m$ of vectors $\boldsymbol{w} = (w_1, \dots, w_m)^{\mathrm{T}}$ such that the corresponding density and pressure are positive:

$$D = \Big\{ \boldsymbol{w} \in I\!R^m; w_1 = \rho > 0,\ w_s = \rho v_{s-1} \in I\!R \text{ for } s = 2, \dots, m-1, \qquad (3.1.10)$$
$$w_m - \sum_{i=2}^{m-1} w_i^2/(2w_1) = p/(\gamma - 1) > 0 \Big\}.$$

(Cf. (1.2.68).) Obviously, $\boldsymbol{f}_s \in C^1(D)^m$.

If $\boldsymbol{\vartheta} : M \to D$ is a vector function defined in a set M with values in D and $\varphi : D \to I\!R$, then $\varphi(\boldsymbol{\vartheta})$ means the composite function $\varphi \circ \boldsymbol{\vartheta}$, i.e. $(\varphi(\boldsymbol{\vartheta}))(y) = \varphi(\boldsymbol{\vartheta}(y))$ for $y \in M$.

Differentiation in (3.1.7) and the use of the chain rule lead to a *first order quasilinear system* of partial differential equations

$$\frac{\partial \boldsymbol{w}}{\partial t} + \sum_{s=1}^{N} \mathbb{A}_s(\boldsymbol{w}) \frac{\partial \boldsymbol{w}}{\partial x_s} = 0, \qquad (3.1.11)$$

where $\mathbb{A}_s(\boldsymbol{w})$ are $m \times m$ matrices defined for $\boldsymbol{w} \in D$ by

$$\mathbb{A}_s(\boldsymbol{w}) = \frac{D\boldsymbol{f}_s(\boldsymbol{w})}{D\boldsymbol{w}} = \left(\frac{\partial f_{si}(\boldsymbol{w})}{\partial w_j} \right)_{i,j=1}^{m} \qquad (3.1.12)$$
$$= \text{the Jacobi matrix of the mapping } \boldsymbol{f}_s.$$

For $\boldsymbol{w} \in D$ and $\boldsymbol{n} = (n_1, \dots, n_N)^{\mathrm{T}} \in I\!R^N$ we denote

$$\boldsymbol{\mathcal{P}}(\boldsymbol{w}, \boldsymbol{n}) = \sum_{s=1}^{N} \boldsymbol{f}_s(\boldsymbol{w}) n_s, \qquad (3.1.13)$$

which is the flux of the quantity $\boldsymbol{w}$ in the direction $\boldsymbol{n}$. The Jacobi matrix $D\boldsymbol{\mathcal{P}}(\boldsymbol{w}, \boldsymbol{n})/D\boldsymbol{w}$ can be expressed in the form

$$\frac{D\boldsymbol{\mathcal{P}}(\boldsymbol{w}, \boldsymbol{n})}{D\boldsymbol{w}} = \mathbb{P}(\boldsymbol{w}, \boldsymbol{n}) = \sum_{s=1}^{N} \mathbb{A}_s(\boldsymbol{w}) n_s. \qquad (3.1.14)$$

Now we shall investigate some properties of the system (3.1.7) of the Euler equations.

Lemma 3.1 *The vector-valued functions* $\boldsymbol{f}_s$ *defined by* (3.1.9) *are homogeneous mappings of order 1:*

$$\boldsymbol{f}_s(\alpha \boldsymbol{w}) = \alpha \boldsymbol{f}_s(\boldsymbol{w}), \quad \alpha > 0. \qquad (3.1.15)$$

Moreover, we have

$$\boldsymbol{f}_s(\boldsymbol{w}) = \mathbb{A}_s(\boldsymbol{w}) \boldsymbol{w}. \qquad (3.1.16)$$

Similarly,

$$\boldsymbol{\mathcal{P}}(\alpha \boldsymbol{w}, \boldsymbol{n}) = \alpha \boldsymbol{\mathcal{P}}(\boldsymbol{w}, \boldsymbol{n}), \quad \alpha \neq 0, \qquad (3.1.17)$$

$$\mathcal{P}(\boldsymbol{w},\boldsymbol{n}) = \mathbb{P}(\boldsymbol{w},\boldsymbol{n})\boldsymbol{w}. \tag{3.1.18}$$

Proof Relation (3.1.15) immediately follows from (3.1.9). Since $\boldsymbol{f}_s \in C^1(D)^m$, the expression $(D\boldsymbol{f}_s(\boldsymbol{w})/D\boldsymbol{w})\boldsymbol{w} = \mathbb{A}_s(\boldsymbol{w})\boldsymbol{w}$ is the derivative of $\boldsymbol{f}_s$ in the direction $\boldsymbol{w}$ at the point $\boldsymbol{w}$. By the definition of the derivative and (3.1.15),

$$\begin{aligned}\mathbb{A}_s(\boldsymbol{w})\boldsymbol{w} &= \lim_{\alpha\to 0}\frac{\boldsymbol{f}_s(\boldsymbol{w}+\alpha\boldsymbol{w}) - \boldsymbol{f}_s(\boldsymbol{w})}{\alpha} \\ &= \lim_{\alpha\to 0}\frac{(1+\alpha)\boldsymbol{f}_s(\boldsymbol{w}) - \boldsymbol{f}_s(\boldsymbol{w})}{\alpha} = \boldsymbol{f}_s(\boldsymbol{w}).\end{aligned} \tag{3.1.19}$$

Relations (3.1.17) and (3.1.18) are consequences of the definitions of $\mathcal{P}$ and $\mathbb{P}$ and the above results. □

The analysis of further properties of the Euler equations will be carried out separately for the case of three-dimensional flow ($N = 3$) and two-dimensional flow ($N = 2$). The case $N = 1$ will be treated as an exercise.

3.1.2 *Diagonalization of the Jacobi matrix*

Lemma 3.2 *Let $N = 3$. Then, with the notation $u = v_1$, $v = v_2$, $w = v_3$, $\gamma_1 = \gamma - 1$, we have*

$$\mathbb{A}_1(\boldsymbol{w}) \tag{3.1.20}$$
$$= \begin{pmatrix} 0 & 1 & 0 & 0 & 0 \\ \frac{\gamma_1}{2}|\boldsymbol{v}|^2 - u^2 & (3-\gamma)u & -\gamma_1 v & -\gamma_1 w & \gamma_1 \\ -uv & v & u & 0 & 0 \\ -uw & w & 0 & u & 0 \\ u\left(\gamma_1|\boldsymbol{v}|^2 - \gamma\frac{E}{\rho}\right) & \gamma\frac{E}{\rho} - \gamma_1 u^2 - \frac{\gamma_1}{2}|\boldsymbol{v}|^2 & -\gamma_1 uv & -\gamma_1 uw & \gamma u \end{pmatrix}.$$

Proof First, we express the vector-valued function $\boldsymbol{f}_1(\boldsymbol{w})$ with the aid of the variables $w_1, \dots, w_5$:

$$\begin{aligned}\boldsymbol{f}_1(\boldsymbol{w}) = \Big(&w_2, w_2^2/w_1 + (\gamma-1)\left(w_5 - (w_2^2+w_3^2+w_4^2)/(2w_1)\right), \\ &w_2w_3/w_1, w_2w_4/w_1, w_2\left(\gamma w_5 - (\gamma-1)(w_2^2+w_3^2+w_4^2)/(2w_1)\right)/w_1\Big)^{\mathrm{T}}.\end{aligned} \tag{3.1.21}$$

Then, by differentiation of $\boldsymbol{f}_1$, the elements of the matrix $\mathbb{A}_1(\boldsymbol{w})$ are calculated and, by (3.1.8), expressed with the aid of the variables ρ, u, v, w, E. □

Lemma 3.3 *Let $N = 3$. Then the matrix $\mathbb{A}_1(\boldsymbol{w})$ has the eigenvalues*

$$\tilde\lambda_1 = u - a, \quad \tilde\lambda_2 = \tilde\lambda_3 = \tilde\lambda_4 = u, \quad \tilde\lambda_5 = u + a, \tag{3.1.22}$$

where

$$a = \sqrt{\gamma p/\rho} \tag{3.1.23}$$

is the speed of sound. The corresponding eigenvectors have the form

$$\boldsymbol{r}_1(\boldsymbol{w}) = \left(1, u-a, v, w, \frac{|\boldsymbol{v}|^2}{2} + \frac{a^2}{\gamma-1} - ua\right)^{\mathrm{T}}, \tag{3.1.24}$$

$$\boldsymbol{r}_2(\boldsymbol{w}) = \left(1, u, v, w, \frac{|\boldsymbol{v}|^2}{2}\right)^{\mathrm{T}},$$

$$\boldsymbol{r}_3(\boldsymbol{w}) = \left(1, u, v-a, w, \frac{|\boldsymbol{v}|^2}{2} - va\right)^{\mathrm{T}},$$

$$\boldsymbol{r}_4(\boldsymbol{w}) = \left(1, u, v, w-a, \frac{|\boldsymbol{v}|^2}{2} - wa\right)^{\mathrm{T}},$$

$$\boldsymbol{r}_5(\boldsymbol{w}) = \left(1, u+a, v, w, \frac{|\boldsymbol{v}|^2}{2} + \frac{a^2}{\gamma - 1} + ua\right)^{\mathrm{T}}.$$

Proof Let us set

$$\mathbb{M}(\boldsymbol{w}, \lambda) = \mathbb{A}_1(\boldsymbol{w}) - \lambda \mathbb{I}. \tag{3.1.25}$$

The eigenvalues of the matrix $\mathbb{A}_1(\boldsymbol{w})$ are the solutions of the equation

$$\det\mathbb{M}(\boldsymbol{w}, \lambda) = 0 \tag{3.1.26}$$

with respect to λ. We denote

$\mathbb{M}_1$ = the matrix obtained from $\mathbb{M}$ by omitting the first row and first column,
$\mathbb{M}_2$ = the matrix obtained from $\mathbb{M}$ by omitting the first row and second column,
$\mathbb{M}_{11}$ = the matrix obtained from $\mathbb{M}_1$ by omitting the second row and first column,
$\mathbb{M}_{12}$ = the matrix obtained from $\mathbb{M}_1$ by omitting the second row and second column,
$\mathbb{M}_{21}$ = the matrix obtained from $\mathbb{M}_2$ by omitting the second row and first column,
$\mathbb{M}_{22}$ = the matrix obtained from $\mathbb{M}_2$ by omitting the second row and second column.

Then the expansion of $\det\mathbb{M}$ with respect to the first row gives

$$\det\mathbb{M} = -\lambda \det\mathbb{M}_1 - \det\mathbb{M}_2. \tag{3.1.27}$$

In a similar way we find that

$$\begin{aligned} \det\mathbb{M}_1 &= -v\det\mathbb{M}_{11} - (\lambda - u)\det\mathbb{M}_{12}, \\ \det\mathbb{M}_2 &= uv\det\mathbb{M}_{21} - (\lambda - u)\det\mathbb{M}_{22}. \end{aligned} \tag{3.1.28}$$

We calculate

$$\begin{aligned} \det\mathbb{M}_{11} &= -v(u-\lambda)^2(\gamma - 1), \\ \det\mathbb{M}_{12} &= -\frac{1}{2}(u-\lambda)\Big[(v^2 - w^2)(\gamma - 1) - u^2(3+\gamma) + 6u\lambda \end{aligned} \tag{3.1.29}$$

$$
\begin{aligned}
&\quad - 2(\lambda^2 - a^2)\Big],\\
\det\mathbb{M}_{21} &= \det\mathbb{M}_{11},\\
\det\mathbb{M}_{22} &= \frac{1}{2}(u-\lambda)\Big[-2u^3 - u^2\lambda(\gamma-3) - 2v^2u(1-\gamma)\\
&\quad - w^2\lambda(\gamma-1) + 2a^2u - v^2\lambda(\gamma-1)\Big].
\end{aligned}
$$

Now, from (3.1.27)–(3.1.29), after elementary but tedious calculation, we obtain

$$\det\mathbb{M} = (u-\lambda)^3(\lambda - u - a)(\lambda - u + a),$$

which already yields assertion (3.1.22).

The verification that the vectors (3.1.24) are eigenvectors of the matrix $\mathbb{A}_1(\boldsymbol{w})$ can be carried out by substitution into the system

$$(\mathbb{A}_1(\boldsymbol{w}) - \tilde{\lambda}_i(\boldsymbol{w})\mathbb{I})\boldsymbol{r}_i(\boldsymbol{w}) = 0. \tag{3.1.30}$$

□

The eigenvectors $\boldsymbol{r}_1(\boldsymbol{w}), \ldots, \boldsymbol{r}_5(\boldsymbol{w})$ of the matrix $\mathbb{A}_1(\boldsymbol{w})$ are linearly independent for each $\boldsymbol{w} \in D$ and, hence, they form a basis in the space $I\!R^5$. Denoting by $\tilde{\mathbb{T}}(\boldsymbol{w})$ the matrix having the vectors $\boldsymbol{r}_1(\boldsymbol{w}), \ldots, \boldsymbol{r}_5(\boldsymbol{w})$ as its columns, it is clear that $\tilde{\mathbb{T}}(\boldsymbol{w})$ is nonsingular and (3.1.30) (where $i = 1, \ldots, 5$) implies that

$$\mathbb{A}_1(\boldsymbol{w})\tilde{\mathbb{T}}(\boldsymbol{w}) = \tilde{\mathbb{T}}(\boldsymbol{w})\tilde{\Lambda}(\boldsymbol{w}),$$

where

$$\tilde{\Lambda}(\boldsymbol{w}) = \operatorname{diag}\left(\tilde{\lambda}_1(\boldsymbol{w}), \ldots, \tilde{\lambda}_5(\boldsymbol{w})\right),$$

which is the diagonal matrix with diagonal entries $\tilde{\lambda}_1(\boldsymbol{w}), ..., \tilde{\lambda}_5(\boldsymbol{w})$. Hence,

$$\tilde{\mathbb{T}}^{-1}(\boldsymbol{w})\mathbb{A}_1(\boldsymbol{w})\tilde{\mathbb{T}}(\boldsymbol{w}) = \tilde{\Lambda}(\boldsymbol{w}). \tag{3.1.31}$$

This means that the matrix $\mathbb{A}_1$ is diagonalizable. We have

$$
\tilde{\mathbb{T}}(\boldsymbol{w}) \tag{3.1.32}
$$

$$
= \begin{pmatrix}
1 & 1 & 1 & 1 & 1\\
u-a & u & u & u & u+a\\
v & v & v-a & v & v\\
w & w & w & w-a & w\\
\frac{|\boldsymbol{v}|^2}{2} + \frac{a^2}{\gamma-1} - ua & \frac{|\boldsymbol{v}|^2}{2} & \frac{|\boldsymbol{v}|^2}{2} - va & \frac{|\boldsymbol{v}|^2}{2} - wa & \frac{|\boldsymbol{v}|^2}{2} + \frac{a^2}{\gamma-1} + ua
\end{pmatrix}.
$$

It is possible to verify that its inverse has the form

$$
\tilde{\mathbb{T}}^{-1}(\boldsymbol{w}) \tag{3.1.33}
$$

$$= \frac{1}{a^2}\begin{pmatrix} \frac{1}{2}\left(\frac{(\gamma-1)|\boldsymbol{v}|^2}{2}+ua\right) & -\frac{a+u(\gamma-1)}{2} & -\frac{v(\gamma-1)}{2} & -\frac{w(\gamma-1)}{2} & \frac{\gamma-1}{2} \\ a^2-a(v+w)-(\gamma-1)\frac{|\boldsymbol{v}|^2}{2} & u(\gamma-1) & a+v(\gamma-1) & a+w(\gamma-1) & 1-\gamma \\ va & 0 & -a & 0 & 0 \\ wa & 0 & 0 & -a & 0 \\ \frac{1}{2}\left(\frac{(\gamma-1)|\boldsymbol{v}|^2}{2}-ua\right) & \frac{a-u(\gamma-1)}{2} & -\frac{v(\gamma-1)}{2} & -\frac{w(\gamma-1)}{2} & \frac{\gamma-1}{2} \end{pmatrix}.$$

3.1.3 *Rotational invariance of the Euler equations in 3D*

Any $\boldsymbol{n} = (n_1, n_2, n_3)^{\mathrm{T}} \in I\!R^3$ can be expressed in spherical coordinates in the form

$$\boldsymbol{n} = r(\cos\alpha\cos\beta, \sin\alpha\cos\beta, \sin\beta)^{\mathrm{T}}, \tag{3.1.34}$$

where $r = |\boldsymbol{n}|$, $\alpha \in [0, 2\pi)$ and $\beta \in [-\pi/2, \pi/2]$. Then we define the nonsingular matrix

$$\mathbb{Q} = \mathbb{Q}(\boldsymbol{n}) = \begin{pmatrix} 1 & 0 & 0 & 0 & 0 \\ 0 & \cos\alpha\cos\beta & \sin\alpha\cos\beta & \sin\beta & 0 \\ 0 & -\sin\alpha & \cos\alpha & 0 & 0 \\ 0 & -\cos\alpha\sin\beta & -\sin\alpha\sin\beta & \cos\beta & 0 \\ 0 & 0 & 0 & 0 & 1 \end{pmatrix}. \tag{3.1.35}$$

If we define a new Cartesian coordinate system $\tilde{x}_1, \tilde{x}_2, \tilde{x}_3$ by

$$\begin{pmatrix} \tilde{x}_1 \\ \tilde{x}_2 \\ \tilde{x}_3 \end{pmatrix} = \mathbb{Q}_0(\boldsymbol{n}) \begin{pmatrix} x_1 \\ x_2 \\ x_3 \end{pmatrix} + \tilde{\sigma}, \tag{3.1.36}$$

where $\tilde{\sigma} \in I\!R^3$ and

$$\mathbb{Q}_0(\boldsymbol{n}) = \begin{pmatrix} \cos\alpha\cos\beta & \sin\alpha\cos\beta & \sin\beta \\ -\sin\alpha & \cos\alpha & 0 \\ -\cos\alpha\sin\beta & -\sin\alpha\sin\beta & \cos\beta \end{pmatrix}, \tag{3.1.37}$$

then the transformation of the state vector $\boldsymbol{w}$ yields the state vector

$$\boldsymbol{q} = \mathbb{Q}(\boldsymbol{n})\boldsymbol{w}. \tag{3.1.38}$$

We consider the transformed state vector $\boldsymbol{q}$ as a function of $\tilde{x} = (\tilde{x}_1, \tilde{x}_2, \tilde{x}_3)$ and time t:

$$\boldsymbol{q} = \boldsymbol{q}(\tilde{x}, t) = \mathbb{Q}(\boldsymbol{n})\boldsymbol{w}(\mathbb{Q}_0^{-1}(\boldsymbol{n})(\tilde{x} - \tilde{\sigma}), t). \tag{3.1.39}$$

Theorem 3.4 *Let $\boldsymbol{w} \in D$ and $\boldsymbol{n} \in I\!R^3$, $|\boldsymbol{n}| = 1$. Then*

$$\boldsymbol{\mathcal{P}}(\boldsymbol{w}, \boldsymbol{n}) = \mathbb{Q}^{-1}(\boldsymbol{n})\boldsymbol{f}_1(\mathbb{Q}(\boldsymbol{n})\boldsymbol{w}) \tag{3.1.40}$$

and

$$\mathbb{P}(\boldsymbol{w}, \boldsymbol{n}) = \mathbb{Q}^{-1}(\boldsymbol{n})\mathbb{A}_1(\mathbb{Q}(\boldsymbol{n})\boldsymbol{w})\mathbb{Q}(\boldsymbol{n}). \tag{3.1.41}$$

Proof The matrix $\mathbb{Q}_0(\boldsymbol{n})$ is orthonormal, as can be verified, and thus $\mathbb{Q}_0^{-1}(\boldsymbol{n}) = \mathbb{Q}_0^{\mathrm{T}}(\boldsymbol{n})$. With the aid of this fact, it is possible to show that also $\mathbb{Q}^{-1}(\boldsymbol{n}) = \mathbb{Q}^{\mathrm{T}}(\boldsymbol{n})$. Using this result, we can obtain relations (3.1.40)–(3.1.41). □

Theorem 3.5 (On the transformation of the Euler equations) *A vector-valued function* $\boldsymbol{w} \in C^1(Q_T)^m$ *satisfies the Euler equations* (3.1.7) *if and only if the function* $\boldsymbol{q} = \boldsymbol{q}(\tilde{x}, t)$ *given by* (3.1.39) *satisfies the transformed system of the Euler equations*

$$\frac{\partial \boldsymbol{q}}{\partial t} + \sum_{s=1}^{N} \frac{\partial \boldsymbol{f}_s(\boldsymbol{q})}{\partial \tilde{x}_s} = 0. \tag{3.1.42}$$

We shall omit the proof for $N = 3$, because it is rather complicated and lengthy. If $N = 2$, see Exercise 3.9.

As we see, under the transformation of the Cartesian frame of reference, the system of the Euler equations remains formally unchanged. This property is called the *rotational invariance of the Euler equations.*

3.1.4 *Hyperbolicity of the Euler equations in 3D*

Theorem 3.6 *The 3D Euler equations* (3.1.2)–(3.1.5) *form a diagonally hyperbolic system in the sense of Definition 2.3.*

Proof If $\boldsymbol{n} \in I\!R^3$, $\boldsymbol{n} = 0$, then the matrix $\mathbb{P}(\boldsymbol{w}, \boldsymbol{n}) = 0$ is diagonalizable and has the only eigenvalue $\lambda = 0$. Let $\boldsymbol{n} \in I\!R^3$, $\boldsymbol{n} \neq 0$. It can be expressed with the aid of spherical coordinates in the form (3.1.34):

$$\boldsymbol{n} = r(\cos\alpha\cos\beta, \sin\alpha\cos\beta, \sin\beta)^{\mathrm{T}}, \quad \alpha \in [0, 2\pi), \beta \in [-\pi/2, \pi/2], r = |\boldsymbol{n}| > 0. \tag{3.1.43}$$

By virtue of (3.1.41),

$$\mathbb{P}(\boldsymbol{w}, \boldsymbol{n}) = r\mathbb{Q}^{-1}(\boldsymbol{n})\mathbb{A}_1(\mathbb{Q}(\boldsymbol{n})\boldsymbol{w})\mathbb{Q}(\boldsymbol{n}) \tag{3.1.44}$$

and consequently, since $\det\mathbb{Q} = \det\mathbb{Q}^{-1} = 1$,

$$\begin{aligned}\det(\mathbb{P}(\boldsymbol{w}, \boldsymbol{n}) - \mu\mathbb{I}) &= \det\left(r\mathbb{Q}^{-1}(\boldsymbol{n})\left(\mathbb{A}_1(\mathbb{Q}(\boldsymbol{n})\boldsymbol{w}) - \frac{\mu}{r}\mathbb{I}\right)\mathbb{Q}(\boldsymbol{n})\right) \\ &= r^5 \det\left(\mathbb{A}_1(\mathbb{Q}(\boldsymbol{n})\boldsymbol{w}) - \frac{\mu}{r}\mathbb{I}\right).\end{aligned}$$

From this and Lemma 3.3 we see that the matrix $\mathbb{P}(\boldsymbol{w}, \boldsymbol{n})$ has real eigenvalues.

It remains to show that the matrix $\mathbb{P}(\boldsymbol{w}, \boldsymbol{n})$ is diagonalizable. By (3.1.31) and (3.1.44),

$$\mathbb{P}(\boldsymbol{w}, \boldsymbol{n}) = r\mathbb{Q}^{-1}(\boldsymbol{n})\tilde{\mathbb{T}}(\mathbb{Q}(\boldsymbol{n})\boldsymbol{w})\tilde{\Lambda}(\mathbb{Q}(\boldsymbol{n})\boldsymbol{w})\tilde{\mathbb{T}}(\mathbb{Q}(\boldsymbol{n})\boldsymbol{w})^{-1}\mathbb{Q}(\boldsymbol{n}). \tag{3.1.45}$$

Under the notation

$$\Lambda(\boldsymbol{w}, \boldsymbol{n}) = r\tilde{\Lambda}(\mathbb{Q}(\boldsymbol{n})\boldsymbol{w}) = \operatorname{diag}\left(r\tilde{\lambda}_1(\mathbb{Q}(\boldsymbol{n})\boldsymbol{w}), \ldots, r\tilde{\lambda}_5(\mathbb{Q}(\boldsymbol{n})\boldsymbol{w})\right), \tag{3.1.46}$$

$$\mathbb{T}(\boldsymbol{w}, \boldsymbol{n}) = \mathbb{Q}^{-1}(\boldsymbol{n})\tilde{\mathbb{T}}(\mathbb{Q}(\boldsymbol{n})\boldsymbol{w}),$$

we see that the matrix $\mathbb{P}(\boldsymbol{w}, \boldsymbol{n})$ satisfies the relation $\mathbb{T}^{-1}\mathbb{P}\mathbb{T} = \Lambda = \Lambda(\boldsymbol{w}, \boldsymbol{n}) = \operatorname{diag}(\lambda_1, \ldots, \lambda_5)$ and, hence, $\mathbb{T}$ diagonalizes the matrix $\mathbb{P}$. □

Remark 3.7 From (3.1.45) and (3.1.46) it follows that the eigenvalues of the matrix $\mathbb{P}(\boldsymbol{w}, \boldsymbol{n})$ have the form

$$\lambda_i(\boldsymbol{w}, \boldsymbol{n}) = r\tilde{\lambda}_i(\mathbb{Q}(\boldsymbol{n})\boldsymbol{w}), \quad i = 1, \dots, 5. \tag{3.1.47}$$

By the transformation (3.1.36) of the Cartesian coordinates, in view of the relation (3.1.38), the component u appearing in (3.1.22) is transformed to $\boldsymbol{v} \cdot \boldsymbol{n}/|\boldsymbol{n}|$ (= the second component of the vector $\mathbb{Q}(\boldsymbol{n})\boldsymbol{w}$ divided by ρ). From this, (3.1.47), and (3.1.22), taking into account that $r = |\boldsymbol{n}|$, we immediately find that

$$\begin{aligned}
\lambda_1(\boldsymbol{w}, \boldsymbol{n}) &= \boldsymbol{v} \cdot \boldsymbol{n} - a|\boldsymbol{n}|, \\
\lambda_2(\boldsymbol{w}, \boldsymbol{n}) &= \lambda_3(\boldsymbol{w}, \boldsymbol{n}) = \lambda_4(\boldsymbol{w}, \boldsymbol{n}) = \boldsymbol{v} \cdot \boldsymbol{n}, \\
\lambda_5(\boldsymbol{w}, \boldsymbol{n}) &= \boldsymbol{v} \cdot \boldsymbol{n} + a|\boldsymbol{n}|
\end{aligned} \tag{3.1.48}$$

for $\boldsymbol{w} \in D$, $\boldsymbol{n} \in I\!R^3$.

3.1.5 *The 2D case*

Similar results to the above can be obtained for 2D flow. We summarize the main results without proofs.

Lemma 3.8 *If $N = 2$, then*

$$\begin{aligned}
\boldsymbol{f}_1(\boldsymbol{w}) = \Big(& w_2, w_2^2/w_1 + (\gamma - 1)[w_4 - (w_2^2 + w_3^2)/(2w_1)], \\
& w_2 w_3/w_1, w_2[\gamma w_4 - (\gamma - 1)(w_2^2 + w_3^2)/(2w_1)]/w_1\Big)^{\mathrm{T}}
\end{aligned} \tag{3.1.49}$$

and, with the notation $\boldsymbol{v} = (u, v)$,

$$\mathbb{A}_1(\boldsymbol{w}) \tag{3.1.50}$$

$$= \begin{pmatrix}
0 & 1 & 0 & 0 \\
\frac{\gamma-1}{2}|\boldsymbol{v}|^2 - u^2 & (3-\gamma)u & (1-\gamma)v & \gamma - 1 \\
-uv & v & u & 0 \\
u((\gamma-1)|\boldsymbol{v}|^2 - \gamma\frac{E}{\rho}) & \gamma\frac{E}{\rho} - (\gamma-1)u^2 - \frac{\gamma-1}{2}|\boldsymbol{v}|^2 & (1-\gamma)uv & \gamma u
\end{pmatrix}.$$

The matrix $\mathbb{A}_1(\boldsymbol{w})$ has the eigenvalues

$$\tilde{\lambda}_1 = u - a, \quad \tilde{\lambda}_2 = \tilde{\lambda}_3 = u, \quad \tilde{\lambda}_4 = u + a \tag{3.1.51}$$

and the corresponding eigenvectors

$$\begin{aligned}
\boldsymbol{r}_1(\boldsymbol{w}) &= (1, u - a, v, |\boldsymbol{v}|^2/2 + a^2(\gamma - 1) - ua)^{\mathrm{T}}, \\
\boldsymbol{r}_2(\boldsymbol{w}) &= (1, u, v, |\boldsymbol{v}|^2/2)^{\mathrm{T}}, \\
\boldsymbol{r}_3(\boldsymbol{w}) &= (1, u, v - a, |\boldsymbol{v}|^2/2 - va)^{\mathrm{T}},
\end{aligned} \tag{3.1.52}$$

$$\boldsymbol{r}_4(\boldsymbol{w}) = (1, u+a, v, |\boldsymbol{v}|^2/2 + a^2/(\gamma-1) + ua)^{\mathrm{T}}.$$

We have

$$\tilde{\mathbb{T}}^{-1}(\boldsymbol{w})\mathbb{A}_1(\boldsymbol{w})\tilde{\mathbb{T}}(\boldsymbol{w}) = \tilde{\mathbb{A}}(\boldsymbol{w}), \tag{3.1.53}$$

where

$$\tilde{\mathbb{A}}(\boldsymbol{w}) = \mathrm{diag}(\tilde{\lambda}_1, \dots, \tilde{\lambda}_4), \tag{3.1.54}$$

$$\tilde{\mathbb{T}}(\boldsymbol{w}) = \begin{pmatrix} 1 & 1 & 1 & 1 \\ u-a & u & u & u+a \\ v & v & v-a & v \\ \frac{|\boldsymbol{v}|^2}{2} + \frac{a^2}{\gamma-1} - ua & \frac{|\boldsymbol{v}|^2}{2} & \frac{|\boldsymbol{v}|^2}{2} - va & \frac{|\boldsymbol{v}|^2}{2} + \frac{a^2}{\gamma-1} + ua \end{pmatrix} \tag{3.1.55}$$

and

$$\tilde{\mathbb{T}}^{-1}(\boldsymbol{w}) = \frac{1}{a^2}\begin{pmatrix} \frac{1}{2}\left(\frac{(\gamma-1)|\boldsymbol{v}|^2}{2} + ua\right) & -\frac{a+u(\gamma-1)}{2} & -\frac{v(\gamma-1)}{2} & \frac{\gamma-1}{2} \\ a^2 - va - (\gamma-1)\frac{|\boldsymbol{v}|^2}{2} & u(\gamma-1) & a+v(\gamma-1) & 1-\gamma \\ va & 0 & -a & 0 \\ \frac{1}{2}\left(\frac{(\gamma-1)|\boldsymbol{v}|^2}{2} - ua\right) & \frac{a-u(\gamma-1)}{2} & -\frac{v(\gamma-1)}{2} & \frac{\gamma-1}{2} \end{pmatrix}. \tag{3.1.56}$$

The *rotational invariance* of the Euler equations is represented by the relations

$$\mathcal{P}(\boldsymbol{w}, \boldsymbol{n}) = \sum_{s=1}^{2} \boldsymbol{f}_s(\boldsymbol{w})n_s = \mathbb{Q}^{-1}(\boldsymbol{n})\boldsymbol{f}_1(\mathbb{Q}(\boldsymbol{n})\boldsymbol{w}), \tag{3.1.57}$$

$$\mathbb{P}(\boldsymbol{w}, \boldsymbol{n}) = \sum_{s=1}^{2} \mathbb{A}_s(\boldsymbol{w})n_s = \mathbb{Q}^{-1}(\boldsymbol{n})\mathbb{A}_1(\mathbb{Q}(\boldsymbol{n})\boldsymbol{w})\mathbb{Q}(\boldsymbol{n}),$$

$$\boldsymbol{n} = (n_1, n_2) \in I\!R^2, |\boldsymbol{n}| = 1, \boldsymbol{w} \in D,$$

where

$$\mathbb{Q}(\boldsymbol{n}) = \begin{pmatrix} 1 & 0 & 0 & 0 \\ 0 & n_1 & n_2 & 0 \\ 0 & -n_2 & n_1 & 0 \\ 0 & 0 & 0 & 1 \end{pmatrix}. \tag{3.1.58}$$

Exercise 3.9 Prove Theorem 3.5 in the case $N = 2$.
Hint: Use (3.1.57) and the similar relations

$$\begin{aligned} n_2\boldsymbol{f}_1(\boldsymbol{q}) + n_1\boldsymbol{f}_2(\boldsymbol{q}) &= \mathbb{Q}(\boldsymbol{n})\boldsymbol{f}_2(\mathbb{Q}^{-1}(\boldsymbol{n})\boldsymbol{q}), \\ n_2\mathbb{A}_1(\boldsymbol{q}) + n_1\mathbb{A}_2(\boldsymbol{q}) &= \mathbb{Q}(\boldsymbol{n})\mathbb{A}_2(\mathbb{Q}^{-1}(\boldsymbol{n})\boldsymbol{q})\mathbb{Q}^{-1}(\boldsymbol{n}), \quad \boldsymbol{q} \in D. \end{aligned} \tag{3.1.59}$$

Now we come to the hyperbolicity of the Euler equations.

Theorem 3.10 *The 2D Euler equations form a diagonally hyperbolic system in the sense of Definition 2.3.*

Proof The proof is carried out in a similar way as for Theorem 3.6. The eigenvalues of the matrix $\mathbb{P}(\boldsymbol{w}, \boldsymbol{n})$ have the form

$$\begin{aligned} \lambda_1(\boldsymbol{w}, \boldsymbol{n}) &= \boldsymbol{v} \cdot \boldsymbol{n} - a|\boldsymbol{n}|, \\ \lambda_2(\boldsymbol{w}, \boldsymbol{n}) &= \lambda_3(\boldsymbol{w}, \boldsymbol{n}) = \boldsymbol{v} \cdot \boldsymbol{n}, \\ \lambda_4(\boldsymbol{w}, \boldsymbol{n}) &= \boldsymbol{v} \cdot \boldsymbol{n} + a|\boldsymbol{n}| \end{aligned} \tag{3.1.60}$$

and the diagonalization of $\mathbb{P}(\boldsymbol{w}, \boldsymbol{n})$ is realized by the matrix

$$\mathbb{T}(\boldsymbol{w}, \boldsymbol{n}) = \mathbb{Q}^{-1}(\boldsymbol{n})\tilde{\mathbb{T}}(\mathbb{Q}(\boldsymbol{n})\boldsymbol{w})$$

and its inverse

$$\mathbb{T}^{-1}(\boldsymbol{w}, \boldsymbol{n}) = \tilde{\mathbb{T}}^{-1}(\mathbb{Q}(\boldsymbol{n})\boldsymbol{w})\mathbb{Q}(\boldsymbol{n})$$

. □

Exercise 3.11 Derive the Jacobi matrix $\mathbb{A}(\boldsymbol{w}) = \mathbb{A}_1(\boldsymbol{w}) = D\boldsymbol{f}(\boldsymbol{w})/D\boldsymbol{w}$ and its eigenvalues and eigenvectors for 1D flow ($N = 1$). In this case the Euler equations have the form

$$\frac{\partial \boldsymbol{w}}{\partial t} + \frac{\partial \boldsymbol{f}(\boldsymbol{w})}{\partial x} = 0, \tag{3.1.61}$$

where $x = x_1$ and, if we write $u = v_1$,

$$\boldsymbol{w} = (\rho, \rho u, E)^{\mathrm{T}}, \quad \boldsymbol{f}(\boldsymbol{w}) = (\rho u, \rho u^2 + p, (E + p)u)^{\mathrm{T}}. \tag{3.1.62}$$

To this system, equation (3.1.5) with $|\boldsymbol{v}| = |u|$ is added.

Solution:

$$\mathbb{A}(\boldsymbol{w}) = \begin{pmatrix} 0 & 1 & 0 \\ \frac{1}{2}(\gamma - 3)u^2 & (3 - \gamma)u & \gamma - 1 \\ \frac{1}{2}(\gamma - 2)u^3 - \frac{a^2 u}{\gamma - 1} & \frac{3 - 2\gamma}{2}u^2 + \frac{a^2}{\gamma - 1} & \gamma u \end{pmatrix}, \tag{3.1.63}$$

$$\begin{aligned} &\lambda_1 = u - a, \quad \lambda_2 = u, \quad \lambda_3 = u + a, \\ &\boldsymbol{r}_1(\boldsymbol{w}) = \left(1, u - a, \frac{u^2}{2} + \frac{a^2}{\gamma - 1} - ua\right)^{\mathrm{T}}, \\ &\boldsymbol{r}_2(\boldsymbol{w}) = \left(1, u, u^2/2\right)^{\mathrm{T}}, \\ &\boldsymbol{r}_3(\boldsymbol{w}) = \left(1, u + a, \frac{u^2}{2} + \frac{a^2}{\gamma - 1} + ua\right)^{\mathrm{T}}. \end{aligned} \tag{3.1.64}$$

Exercise 3.12 Verify that the diagonalization matrices $\mathbb{T}, \mathbb{T}^{-1}$ of the matrix $\mathbb{A} = D\boldsymbol{f}/D\boldsymbol{w}$ have the form

$$\mathbb{T}(\boldsymbol{w}) = \begin{pmatrix} 1 & 1 & 1 \\ u - a & u & u + a \\ \frac{u^2}{2} + \frac{a^2}{\gamma - 1} - ua & \frac{u^2}{2} & \frac{u^2}{2} + \frac{a^2}{\gamma - 1} + ua \end{pmatrix}, \tag{3.1.65}$$

$$\mathbb{T}^{-1}(\boldsymbol{w}) = \frac{1}{a^2}\begin{pmatrix} \frac{1}{2}\left(\frac{\gamma-1}{2}u^2 + ua\right) & -\frac{a+u(\gamma-1)}{2} & \frac{\gamma-1}{2} \\ a^2 - \frac{\gamma-1}{2}u^2 & u(\gamma-1) & 1-\gamma \\ \frac{1}{2}\left(\frac{\gamma-1}{2}u^2 - ua\right) & \frac{a-u(\gamma-1)}{2} & \frac{\gamma-1}{2} \end{pmatrix}. \qquad (3.1.66)$$

Remark 3.13 The eigenvalues of the matrix $\mathbb{A}(\boldsymbol{w})$ are simple in the 1D case ($N = 1$) and the system of the Euler equations is strictly diagonally hyperbolic according to Definition 2.3 (in contrast to the case $N = 2$ or 3, where the Euler equations are only diagonally hyperbolic).

As we shall see later, efficient numerical schemes are based on the solution of the Riemann problem for the system of the Euler equations with one space variable $x = x_1$, i.e. we set $\partial/\partial x_2 = \partial/\partial x_3 = 0$ in (3.1.7). Therefore, in what follows we shall be concerned with the Euler equations with one space variable in the form

$$\frac{\partial \boldsymbol{w}}{\partial t} + \frac{\partial \boldsymbol{f}(\boldsymbol{w})}{\partial x} = 0 \quad \text{in } Q_T = I\!R \times (0,T), \qquad (3.1.67)$$

where $\boldsymbol{f} := \boldsymbol{f}_1$ and $x := x_1$ and $\boldsymbol{w} \in I\!R^m$, $\boldsymbol{f}_1 \in I\!R^m$ are defined by (3.1.8)–(3.1.9) for $N \in \{1,2,3\}$. Then

$$\mathbb{A}(\boldsymbol{w}) = \frac{D\boldsymbol{f}(\boldsymbol{w})}{D\boldsymbol{w}} \qquad (3.1.68)$$

is given by (3.1.20), (3.1.50) and (3.1.63) for $N = 3, 2$ and 1, respectively.

According to the definition of the Riemann problem introduced in Section 2.3.8, our goal is to find an entropy weak solution $\boldsymbol{w} : Q_T \to I\!R^m$ of system (3.1.67), equipped with the initial condition formed by two given constant states $\boldsymbol{w}_L, \boldsymbol{w}_R \in I\!R^m$:

$$\boldsymbol{w}(x,0) = \begin{cases} \boldsymbol{w}_L, & x < 0, \\ \boldsymbol{w}_R, & x > 0. \end{cases} \qquad (3.1.69)$$

3.1.6 *Solution of the Riemann problem for 1D Euler equations*

In this case $N = 1$, $m = 3$, $\boldsymbol{w} = (w_1, w_2, w_3)^{\mathrm{T}} = (\rho, \rho u, E)^{\mathrm{T}}$, $\boldsymbol{f}(\boldsymbol{w}) = (\rho u, \rho u^2 + p, (E+p)\,u)^{\mathrm{T}}$. The matrix $\mathbb{A}(\boldsymbol{w})$ has simple eigenvalues $\lambda_1 = u - a$, $\lambda_2 = u$, $\lambda_3 = u + a$, where $a = \sqrt{\gamma p/\rho}$. The corresponding eigenvectors are defined by (3.1.64).

Exercise 3.14 Prove that the vectors $\boldsymbol{r}_1(\boldsymbol{w})$ and $\boldsymbol{r}_3(\boldsymbol{w})$ are genuinely nonlinear and $\boldsymbol{r}_2(\boldsymbol{w})$ is linearly degenerate for all $\boldsymbol{w} \in D$. (See Definition 2.25.)

It follows from Exercise 3.14 that Theorem 2.32 on the solvability of the Riemann problem can be applied to the 1D Euler equations.

In this section we shall try to express the solution of the Riemann problem for the 1D Euler equations with the aid of analytical tools. This means that we want to find a solution $\boldsymbol{w} = \boldsymbol{w}(x,t)$ defined for all $x \in I\!R$ and $t \geq 0$, satisfying equation (3.1.67) equipped with condition (3.1.69). In view of 2.3.10, the solvability of this problem is guaranteed by Theorem 2.32. Of course, it makes sense to assume $\boldsymbol{w}_L \neq \boldsymbol{w}_R$. Otherwise the solution of the Riemann problem is $\boldsymbol{w}(x,t) = \boldsymbol{w}_L = \boldsymbol{w}_R$ for all $x \in I\!R$ and $t \geq 0$.

The exact solution to the Riemann problem is an invaluable reference solution that is useful in assessing the performance of numerical methods and checking the correctness of programs in the early stages of development. The Riemann problem for 1D Euler equations includes the so-called shock-tube problem, a basic physical problem in gas dynamics. We use this problem in Section 3.5, for testing higher order finite volume methods, and in Section 3.7, where the realization of its numerical solution is discussed. It is not possible to express the exact analytical solution of the Riemann problem in a closed form, because its evaluation requires the solution of a nonlinear algebraic equation. However, it is possible to apply *iterative* schemes whereby the solution can be computed numerically to any desired accuracy.

First we shall study in detail the solutions of the Riemann problem for couples of initial states $\boldsymbol{w}_L = (\rho_L, \rho_L u_L, E_L)$ and $\boldsymbol{w}_R = (\rho_R, \rho_R u_R, E_R)$ introduced in Theorem 2.27 (rarefaction wave) and Theorem 2.28 (entropy discontinuity wave). We shall denote by p_L, a_L and p_R, a_R the pressure and speed of sound of states $\boldsymbol{w}_L$ and $\boldsymbol{w}_R$, respectively, see (3.1.5) and (3.1.23).

3.1.6.1 *k-rarefaction wave, $k = 1, 3$* The k-rarefaction wave is constructed with the use of a genuinely nonlinear eigenvector $\boldsymbol{r}_k$ defined in (3.1.64) and k-Riemann invariants defined in (2.3.57). From Exercise 3.14 we see that it makes sense to consider $k = 1, 3$.

Exercise 3.15 Let $k \in \{1, 3\}$. Verify that

$$\psi_1^{(k)}(\boldsymbol{w}) = \frac{p}{\rho^\gamma}, \tag{3.1.70}$$

$$\psi_2^{(k)}(\boldsymbol{w}) = u + (2 - k)\,\frac{2a}{\gamma - 1} \tag{3.1.71}$$

are the k-Riemann invariants and that for fixed $k = 1, 3$ the gradients $\nabla\psi_1^{(k)}$, $\nabla\psi_2^{(k)}$ are linearly independent.

Lemma 3.16 *Across a k-rarefaction wave the k-Riemann invariant is constant.*

Proof Let us consider for $k = 1, 3$ the function $\boldsymbol{w}$ defined by relations (2.3.63) with $\lambda_k(\boldsymbol{w}_R)$ given by (2.3.62) and $\tilde{\boldsymbol{w}}$ satisfying (2.3.59) for $\xi \in [0, \xi_R]$. Then

$$\frac{d}{d\xi}\,\psi_i^{(k)}(\tilde{\boldsymbol{w}}) = \nabla\psi_i^{(k)}(\tilde{\boldsymbol{w}}) \cdot \boldsymbol{r}_k(\tilde{\boldsymbol{w}}) \equiv 0, \quad i = 1, 2,$$

which implies

$$\psi_i^{(k)}(\tilde{\boldsymbol{w}})\Big|_{[0,\xi_R]} = \psi_i^{(k)}(\boldsymbol{w}_L) = \text{const.} \tag{3.1.72}$$

Hence, taking into account (2.3.63), we have

$$\psi_i^{(k)}(\boldsymbol{w}(x,t)) = \psi_i^{(k)}\,(\boldsymbol{w}_L)$$

in the set Ω_{rar} defined by (2.3.69), i.e. for x, t such that $x \in I\!R, t > 0$ and $\lambda_k(\boldsymbol{w}_L) \le x/t \le \lambda_k(\boldsymbol{w}_R)$. This together with (3.1.72) completes the proof. □

Lemma 3.16 allows us to express the relation between the ratios ρ_R/ρ_L and p_R/p_L for the k-rarefaction wave, $k = 1, 3$.

a) *Isentropic law across the k-rarefaction wave, k* $= 1, 3$ As the k-Riemann invariant $\psi_1^{(k)}$ defined in (3.1.70) is constant across a k-rarefaction wave, the *isentropic law*

$$p = C\rho^\gamma, \tag{3.1.73}$$

where C is a constant, can be used across the k-rarefaction wave. C is evaluated at the initial left data state by applying the isentropic law, namely

$$p_L = C\,\rho_L^\gamma,$$

and so the constant C is

$$C = p_L/\rho_L^\gamma,$$

from which we write

$$\frac{\rho_R}{\rho_L} = \left(\frac{p_R}{p_L}\right)^{1/\gamma}. \tag{3.1.74}$$

b) *Pressure ratio p_R/p_L for the k-rarefaction wave, k* $= 1, 3$ For $k = 1, 3$, a k-rarefaction wave is identified by the condition

$$(2-k)\,p_R < (2-k)\,p_L. \tag{3.1.75}$$

To prove this, by evaluating the constant k-Riemann invariant $\psi_2^{(k)}$ (see (3.1.71)) across a k-rarefaction wave on the left- and right-hand data states, we write

$$u_L + (2-k)\,\frac{2a_L}{\gamma - 1} = u_R + (2-k)\,\frac{2a_R}{\gamma - 1}. \tag{3.1.76}$$

Substitution of ρ_R from (3.1.74) into the definition of a_R gives

$$a_R = a_L\left(\frac{p_R}{p_L}\right)^{\frac{\gamma-1}{2\gamma}}. \tag{3.1.77}$$

For $k = 1, 3$ we have for eigenvalues λ_k from (3.1.64)

$$\lambda_k = u - (2-k)\,a \tag{3.1.78}$$

and from (2.3.62) we get

$$u_L - (2-k)\,a_L = \lambda_k(\boldsymbol{w}_L) < \lambda_k(\boldsymbol{w}_R) = u_R - (2-k)\,a_R,$$

which implies

$$u_L - u_R < (2-k)(a_L - a_R). \tag{3.1.79}$$

Using (3.1.76), we can write

$$u_L - u_R = (2-k)\frac{2}{\gamma-1}(a_R - a_L) < (2-k)(a_L - a_R)$$

and therefore

$$(2-k)\,a_R < (2-k)\,a_L,$$

which together with (3.1.77) yields (3.1.75).

c) *Relation between the velocity and pressure initial data for the k-rarefaction wave, $k = 1, 3$* For the k-rarefaction wave, $k = 1, 3$, it holds that

$$u_R = u_L + F_{kL}(p_R), \tag{3.1.80}$$

where

$$F_{kL}(p) = (2-k)\frac{2a_L}{\gamma-1}\left[1 - \left(\frac{p}{p_L}\right)^{\frac{\gamma-1}{2\gamma}}\right]. \tag{3.1.81}$$

According to (3.1.75), we define the function F_{kL} for p satisfying

$$(2-k)\,p < (2-k)\,p_L. \tag{3.1.82}$$

Similarly to (3.1.80), it holds that

$$u_L = u_R + F_{kR}(p_L), \tag{3.1.83}$$

where

$$F_{kR}(p) = (2-k)\frac{2a_R}{\gamma-1}\left[1 - \left(\frac{p}{p_R}\right)^{\frac{\gamma-1}{2\gamma}}\right]. \tag{3.1.84}$$

The function F_{kR} is defined according to (3.1.75) for p satisfying

$$(2-k)\,p_R < (2-k)\,p. \tag{3.1.85}$$

In order to derive relation (3.1.80), we substitute a_R from (3.1.77) into equation (3.1.76):

$$\begin{aligned} u_R &= u_L + (2-k)\frac{2}{\gamma-1}(a_L - a_R) \\ &= u_L + (2-k)\frac{2a_L}{\gamma-1}\left[1 - \left(\frac{p_R}{p_L}\right)^{\frac{\gamma-1}{2\gamma}}\right], \end{aligned}$$

which is the sought relation (3.1.80). The derivation of (3.1.83) is analogous.

d) *Explicit form of the k-rarefaction wave* Let us consider the k-rarefaction wave solution $\boldsymbol{w}$ defined in (2.3.63). Its explicit form can be obtained with the help of characteristics defined in (2.2.7).

Lemma 3.17 *For $k = 1, 3$ let the curves in the (x, t)-plane given by $x = x(s) := \xi_k(s)$, $t = t(s) := s$, satisfy*

$$\frac{d\xi_k(s)}{ds} = \lambda_k(\xi_k(s), s), \quad s > 0,\ k = 1, 3. \tag{3.1.86}$$

Then ξ_k are straight lines along which the k-rarefaction wave $\boldsymbol{w}$ defined in (2.3.63) is constant.

Proof Let $\boldsymbol{\ell}_k$ be a left eigenvector of $\mathbb{A}(\boldsymbol{w})$, i.e.

$$\boldsymbol{\ell}_k^{\mathrm{T}} \mathbb{A}(\boldsymbol{w}) = \lambda_k \boldsymbol{\ell}_k^{\mathrm{T}}.$$

Let ξ_k be the corresponding characteristic. Then we see that

$$\begin{aligned}\boldsymbol{\ell}_k \cdot \frac{d}{dt}\boldsymbol{w}(\xi_k(t), t) &= \boldsymbol{\ell}_k \cdot \left(\lambda_k \frac{\partial \boldsymbol{w}}{\partial x} + \frac{\partial \boldsymbol{w}}{\partial t}\right) \\ &= \boldsymbol{\ell}_k \cdot \left(\mathbb{A}(\boldsymbol{w})\frac{\partial \boldsymbol{w}}{\partial x} + \frac{\partial \boldsymbol{w}}{\partial t}\right) = 0. \end{aligned} \tag{3.1.87}$$

We know from Lemma 3.16 that the k-Riemann invariants $\psi_i^{(k)}$, $i = 1, 2$, are constant across a k-rarefaction wave. Therefore,

$$0 = \frac{d}{dt}\psi_i^{(k)}(\boldsymbol{w}(\xi_k(t), t)) = \nabla\psi_i^{(k)} \cdot \left(\frac{\partial \boldsymbol{w}}{\partial x}\frac{d\xi_k}{dt} + \frac{\partial \boldsymbol{w}}{\partial t}\right) \tag{3.1.88}$$

for $i = 1, 2$, and (3.1.87) and (3.1.88) imply that

$$\begin{pmatrix} \boldsymbol{\ell}_k^{\mathrm{T}} \\ \nabla^{\mathrm{T}}\psi_1^{(k)} \\ \nabla^{\mathrm{T}}\psi_2^{(k)} \end{pmatrix} \left(\frac{\partial \boldsymbol{w}}{\partial x}\frac{d\xi_k}{dt} + \frac{\partial \boldsymbol{w}}{\partial t}\right) = 0. \tag{3.1.89}$$

We know from Exercise 3.15 that k-Riemann invariants $\nabla\psi_1^{(k)}$, $\nabla\psi_2^{(k)}$ are linearly independent and this is also true for $\boldsymbol{\ell}_k$, $\nabla\psi_1^{(k)}$, $\nabla\psi_2^{(k)}$. The last statement follows if we multiply

$$\alpha_0 \boldsymbol{\ell}_k + \alpha_1 \nabla\psi_1^{(k)} + \alpha_2 \nabla\psi_2^{(k)} = 0 \tag{3.1.90}$$

by the right eigenvector $\boldsymbol{r}_k$ of $\mathbb{A}(\boldsymbol{w})$. From the definition of the k-Riemann invariant we obtain $\alpha_0\, \boldsymbol{\ell}_k \cdot \boldsymbol{r}_k = 0$. Since the matrix $\mathbb{A}$ is diagonalizable in the form $\mathbb{A} = \mathbb{T}\Lambda\mathbb{T}^{-1}$ with $\Lambda = \mathrm{diag}(\lambda_1, \lambda_2, \lambda_3)$ and the columns of $\mathbb{T}$ are the right eigenvectors $\boldsymbol{r}_j$ and the rows of $\mathbb{T}^{-1}$ are the left eigenvectors $\boldsymbol{\ell}_j$, we have $\boldsymbol{\ell}_k \cdot \boldsymbol{r}_j = \delta_{kj}$. Hence, $\boldsymbol{\ell}_k \cdot \boldsymbol{r}_k \neq 0$. This implies that $\alpha_0 = 0$, and thus we have shown that $\boldsymbol{\ell}_k$, $\nabla\psi_1^{(k)}$, $\nabla\psi_2^{(k)}$ are linearly independent. Then from (3.1.89) we get

$$\frac{d\xi_k}{dt}(t) = \lambda_k(\boldsymbol{w}(\xi_k(t), t)) = \mathrm{const.} \tag{3.1.91}$$

□

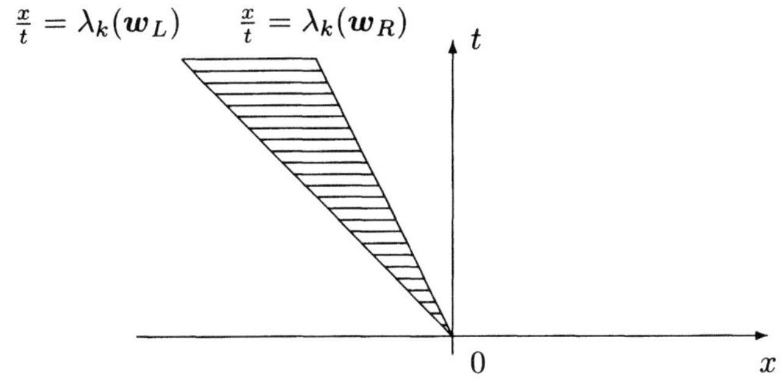

FIG. 3.1.1. k-rarefaction wave, $k = 1, 3$

Now we are ready to write the explicit definition of $\boldsymbol{w}$ in (3.3.79). The k-rarefaction wave is enclosed by two lines

$$L_1 = \left\{(x,t); \frac{x}{t} = \lambda_k(\boldsymbol{w}_L) = u_L - (2-k)\, a_L,\, t \geq 0\right\}, \tag{3.1.92}$$

$$L_2 = \left\{(x,t); \frac{x}{t} = \lambda_k(\boldsymbol{w}_R) = u_R - (2-k)\, a_R,\, t \geq 0\right\}, \tag{3.1.93}$$

see Fig. 3.1.1. We find the solution $\boldsymbol{w}$ inside the fan (the hatched region) from Fig. 3.1.1 in terms of ρ, u and p. This is easily obtained by considering the characteristic ξ_k through the origin $(0,0)$ and a general point (x,t) inside the fan. It follows from (3.1.91) and (3.1.78) that the characteristic $x = \xi_k(t)$ is a straight line $x = C\,t$ with slope

$$\frac{dx}{dt} = C = u - (2-k)\, a, \tag{3.1.94}$$

where u and a are respectively the sought velocity and speed of sound at (x,t). Also, the use of the k-Riemann invariant (3.1.71), which is constant in the fan, yields

$$u_L + (2-k)\,\frac{2a_L}{\gamma - 1} = u + (2-k)\,\frac{2a}{\gamma - 1} \tag{3.1.95}$$

and similarly

$$u_R + (2-k)\,\frac{2a_R}{\gamma - 1} = u + (2-k)\,\frac{2a}{\gamma - 1}. \tag{3.1.96}$$

The simultaneous solution of equations (3.1.94), (3.1.95) (or (3.1.96)) for u and a, the use of the definition of the speed of sound a and the isentropic law (3.1.73) give the resulting expression for ρ, u and p in the fan. Let us present them for $k = 1$ using the left state data $\boldsymbol{w}_L$ and (3.1.75) and for $k = 3$ using the right state data $\boldsymbol{w}_R$ and (3.1.75):

a) Case $k = 1$, $p_R < p_L$:

$$\begin{aligned}
\rho &= \rho_L \left[\frac{2}{\gamma+1} + \frac{\gamma-1}{(\gamma+1)\,a_L} \left(u_L - \frac{x}{t} \right) \right]^{\frac{2}{\gamma-1}}, \\
u &= \frac{2}{\gamma+1} \left[a_L + \frac{\gamma-1}{2} u_L + \frac{x}{t} \right], \\
p &= p_L \left[\frac{2}{\gamma+1} + \frac{\gamma-1}{(\gamma+1)\,a_L} \left(u_L - \frac{x}{t} \right) \right]^{\frac{2\gamma}{\gamma-1}}.
\end{aligned} \tag{3.1.97}$$

b) Case $k = 3$, $p_L < p_R$:

$$\begin{aligned}
\rho &= \rho_R \left[\frac{2}{\gamma+1} - \frac{\gamma-1}{(\gamma+1)\,a_R} \left(u_R - \frac{x}{t} \right) \right]^{\frac{2}{\gamma-1}}, \\
u &= \frac{2}{\gamma+1} \left[-a_R + \frac{\gamma-1}{2} u_R + \frac{x}{t} \right], \\
p &= p_R \left[\frac{2}{\gamma+1} - \frac{\gamma-1}{(\gamma+1)\,a_R} \left(u_R - \frac{x}{t} \right) \right]^{\frac{2\gamma}{\gamma-1}}.
\end{aligned} \tag{3.1.98}$$

3.1.6.2 *k-shock wave, $k = 1, 3$* First we pay attention to condition (2.3.71) b), which is one of two conditions imposed on the speed s of a k-shock wave in order that (2.3.70) is an entropy weak solution. Liu has introduced in (Liu, 1975) another entropy condition, which further implies the shock inequality

$$\lambda_k(\boldsymbol{w}_R) < s < \lambda_k(\boldsymbol{w}_L), \tag{3.1.99}$$

and can be used instead of (2.3.71), b). The idea in deriving condition (3.1.99) is as follows. Let $(x_0, t_0) \in S_{\text{disc}}$, where S_{disc} is defined in (2.3.74), i.e. $x_0/t_0 = s$. We assume that the eigenvalues satisfy

$$\lambda_1(\boldsymbol{w}_R) < \cdots < \lambda_k(\boldsymbol{w}_R) < s < \lambda_{k+1}(\boldsymbol{w}_R) < \cdots < \lambda_m(\boldsymbol{w}_R) \tag{3.1.100}$$

and

$$\lambda_1(\boldsymbol{w}_L) < \cdots < \lambda_j(\boldsymbol{w}_L) < s < \lambda_{j+1}(\boldsymbol{w}_L) < \cdots < \lambda_m(\boldsymbol{w}_L), \tag{3.1.101}$$

where, as we solve the 1D Riemann problem, $m = 3$. The values $\boldsymbol{w}_R$ and $\boldsymbol{w}_L$ of $\boldsymbol{w}$ on S have to respect k conditions given by the initial values on the right-hand side (see relation (3.1.100)), $m - j$ conditions given by the initial values on the left-hand side (see (3.1.101)) and are related by m jump conditions (see (2.3.71), a)). Now we have in all $2m + 1$ parameters (components of $\boldsymbol{w}_L$, $\boldsymbol{w}_R$ and s) and $k+m-j+m = 2m+k-j$ conditions. A necessary condition imposed on $\boldsymbol{w}_L$, $\boldsymbol{w}_R$ is the equal number of parameters and conditions, i.e. $2m + 1 = 2m + k - j$, or

$$j = k - 1. \tag{3.1.102}$$

This means in particular that

$$\lambda_k(\boldsymbol{w}_R) < s < \lambda_{k+1}(\boldsymbol{w}_R),$$

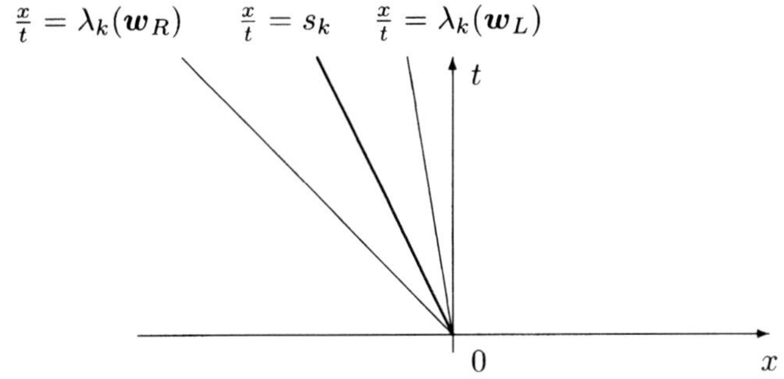

FIG. 3.1.2. k-shock wave, $k = 1, 3$

$$\lambda_{k-1}(\boldsymbol{w}_L) < s < \lambda_k(\boldsymbol{w}_L), \tag{3.1.103}$$

from where we get (3.1.99). More details can be found in (Chorin and Marsden, 1993), (Smoller, 1994).

Theorem 3.18 *Let system* (3.1.67) *be hyperbolic and genuinely nonlinear. Assume that there exists a strictly convex entropy function* η *and that* $\boldsymbol{w}$ *is a solution of* (3.1.67) *with a weak shock, i.e. the jump is small enough. Then* (2.3.71), *b) is equivalent to* (3.1.99).

For the proof see (Smoller, 1994), page 401.

Condition (3.1.99) defines the wedge, where the k-shock wave is positioned, see Fig. 3.1.2. In order to be able to find the solution of the Riemann problem analytically, we need the relations for density, velocity, pressure and speed of sound across the k-shock wave. To this end, we transform the equations to a frame of reference moving with the shock. Denoting the k-shock

$$S_k = \{(x, t); x = s_k t,\ t \in [0, \infty)\} \tag{3.1.104}$$

and using the transformation

$$\tilde{x} = x - s_k t, \tag{3.1.105}$$

in the new frame the shock speed is zero and the state $\boldsymbol{w} = (\rho, \rho u, E)^{\mathrm{T}}$ is transformed into the state $\tilde{\boldsymbol{w}}$

$$\tilde{\boldsymbol{w}} = (\rho, \rho\tilde{u}, \tilde{E})^{\mathrm{T}}, \tag{3.1.106}$$

where the relative velocity $\tilde{u}$ is

$$\tilde{u} = \frac{d\tilde{x}}{dt} = \frac{dx}{dt} - s_k = u - s_k \tag{3.1.107}$$

and for the energy we have

$$\tilde{E} = \frac{p}{\gamma - 1} + \frac{1}{2}\rho\tilde{u}^2. \tag{3.1.108}$$

The Rankine–Hugoniot conditions, see (2.3.71), a), in the new frame give

$$\rho_L \tilde{u}_L = \rho_R \tilde{u}_R, \tag{3.1.109}$$

$$\rho_L \tilde{u}_L^2 + p_L = \rho_R \tilde{u}_R^2 + p_R, \tag{3.1.110}$$

$$(\tilde{E}_L + p_L)\,\tilde{u}_L = (\tilde{E}_R + p_R)\,\tilde{u}_R. \tag{3.1.111}$$

a) *Relation of pressure and density ratio across the k-shock wave* By using the definition of total energy E, relation (3.1.111) can be written as

$$\left(\frac{p_L}{\gamma - 1} + \frac{1}{2}\rho_L \tilde{u}_L^2 + p_L\right)\tilde{u}_L = \left(\frac{p_R}{\gamma - 1} + \frac{1}{2}\rho_R \tilde{u}_R^2 + p_R\right)\tilde{u}_R,$$

or, equivalently,

$$\left(\frac{p_L}{\rho_L(\gamma - 1)} + \frac{1}{2}\tilde{u}_L^2 + \frac{p_L}{\rho_L}\right)\rho_L \tilde{u}_L = \left(\frac{p_R}{\rho_R(\gamma - 1)} + \frac{1}{2}\tilde{u}_R^2 + \frac{p_R}{\rho_R}\right)\rho_R \tilde{u}_R. \tag{3.1.112}$$

Substituting (3.1.109) into (3.1.112), provided $\tilde{u}_L \neq 0$, we obtain

$$\frac{p_L}{\rho_L(\gamma - 1)} + \frac{1}{2}\tilde{u}_L^2 + \frac{p_L}{\rho_L} = \frac{p_R}{\rho_R(\gamma - 1)} + \frac{1}{2}\tilde{u}_R^2 + \frac{p_R}{\rho_R},$$

from which we get

$$\frac{p_L}{\rho_L(\gamma - 1)} - \frac{p_R}{\rho_R(\gamma - 1)} = \frac{1}{2}(\tilde{u}_R^2 - \tilde{u}_L^2) - \frac{\rho_R p_L - \rho_L p_R}{\rho_L \rho_R}. \tag{3.1.113}$$

Using (3.1.109) in (3.1.110), we get

$$\begin{aligned}\rho_L \tilde{u}_L^2 &= \rho_R \tilde{u}_R \frac{\rho_R \tilde{u}_R}{\rho_R} + p_R - p_L \\ &= \rho_L \tilde{u}_L \frac{\rho_L \tilde{u}_L}{\rho_R} + p_R - p_L.\end{aligned}$$

After some manipulation we find that

$$\tilde{u}_L^2 = \frac{\rho_R}{\rho_L}\frac{p_R - p_L}{\rho_R - \rho_L}. \tag{3.1.114}$$

In a similar way we obtain

$$\tilde{u}_R^2 = \frac{\rho_L}{\rho_R}\frac{p_R - p_L}{\rho_R - \rho_L}. \tag{3.1.115}$$

Substitution of (3.1.114)–(3.1.115) into (3.1.113) gives

$$\frac{p_L}{\rho_L(\gamma - 1)} - \frac{p_R}{\rho_R(\gamma - 1)} = \frac{1}{2}(p_L - p_R)\frac{(\rho_L + \rho_R)}{\rho_L \rho_R} - \frac{\rho_R p_L - \rho_L p_R}{\rho_L \rho_R}$$

$$= \frac{1}{2}(p_L + p_R)\, \frac{\rho_L - \rho_R}{\rho_L \rho_R}.$$

Performing some algebraic operations, we obtain

$$\frac{\rho_L}{\rho_R} = \frac{\frac{p_L}{p_R} + \frac{\gamma-1}{\gamma+1}}{\frac{\gamma-1}{\gamma+1}\frac{p_L}{p_R} + 1}. \tag{3.1.116}$$

This establishes a useful relation between the density ratio ρ_L/ρ_R and the pressure ratio p_L/p_R across the shock wave.

b) *Speed of sound ratio across the k-shock wave, $k = 1, 3$* Across the k-shock wave, $k = 1, 3$, it holds that

$$(2-k)\, a_L < (2-k)\, a_R. \tag{3.1.117}$$

This is obtained in two steps. First we establish conditions (3.1.99) and then we use the Rankine–Hugoniot conditions, namely (3.1.109) and (3.1.111). From (3.1.99) we have

$$u_R - (2-k)\, a_R < s_k < u_L - (2-k)\, a_L$$

or, using (3.1.107),

$$(2-k)\, a_L < \tilde{u}_L, \tag{3.1.118}$$

$$(2-k)\, a_R > \tilde{u}_R. \tag{3.1.119}$$

The Rankine–Hugoniot condition (3.1.111) can be rewritten in the form (3.1.112) and, using the definition of a, we get

$$\frac{2}{\gamma-1}\, a_L^2 + \tilde{u}_L^2 = \frac{2}{\gamma-1}\, a_R^2 + \tilde{u}_R^2. \tag{3.1.120}$$

Remark 3.19 It follows from (3.1.118) and (3.1.119) that for $k = 1$ $\tilde{u}_L$ is positive and, due to (3.1.109), $\tilde{u}_R$ is positive as well. Similarly, if $k = 3$, then $\tilde{u}_R$ in (3.1.119) is negative and the same holds for $\tilde{u}_L$.

Due to Remark 3.19, it follows from (3.1.118)–(3.1.119) that

$$(2-k)\, a_L^2 < (2-k)\, \tilde{u}_L^2,$$
$$(2-k)\, a_R^2 > (2-k)\, \tilde{u}_R^2.$$

This together with (3.1.120) gives

$$\begin{aligned}
(2-k)\left(\frac{2}{\gamma-1}\, a_L^2 + a_L^2\right) &< (2-k)\left(\frac{2}{\gamma-1}\, a_L^2 + \tilde{u}_L^2\right) \\
&= (2-k)\left(\frac{2}{\gamma-1}\, a_R^2 + \tilde{u}_R^2\right) \\
&< (2-k)\left(\frac{2}{\gamma-1}\, a_R^2 + a_R^2\right)
\end{aligned}$$

and so (3.1.117) holds true.

c) *Relative velocity across the k-shock wave, $k = 1, 3$* Across the k-shock wave, $k = 1, 3$, it holds that

$$\tilde{u}_L > \tilde{u}_R. \tag{3.1.121}$$

This means that in the direction of the relative velocity $\tilde{u}$ (see Remark 3.19) the velocity decreases by jump.

Relation (3.1.121) follows immediately from (3.1.120), (3.1.117) and Remark 3.19.

d) *Density across the k-shock wave, $k = 1, 3$* Across the k-shock wave, $k = 1, 3$, it holds that

$$(2 - k)\,\rho_L < (2 - k)\,\rho_R. \tag{3.1.122}$$

This means that in the direction of the relative velocity $\tilde{u}$ (see Remark 3.19) the density increases by jump.

The relation (3.1.122) follows from (3.1.109) and (3.1.121).

e) *Pressure across the k-shock wave, $k = 1, 3$* Across the k-shock wave, $k = 1, 3$, it holds that

$$(2 - k)\,p_L < (2 - k)\,p_R. \tag{3.1.123}$$

This means that the pressure jumps up passing the k-shock wave in the direction of relative velocity $\tilde{u}$ (see Remark 3.19). Relation (3.1.122) is a consequence of relations (3.1.117), i.e.

$$(2 - k)\,a_L = (2 - k)\sqrt{\gamma\frac{p_L}{\rho_L}} < (2 - k)\sqrt{\gamma\frac{p_R}{\rho_R}} = (2 - k)\,a_R$$

and relation (3.1.122).

f) *Relation of the shock speed s to the density and pressure ratios across the k-shock wave, $k = 1, 3$* For the speed s it holds that

$$s = s_k := u_R + a_R\sqrt{\frac{\gamma + 1}{2\gamma}\frac{p_L}{p_R} + \frac{\gamma - 1}{2\gamma}} \tag{3.1.124}$$

$$= u_L - a_L\sqrt{\frac{\gamma + 1}{2\gamma}\frac{p_R}{p_L} + \frac{\gamma - 1}{2\gamma}}. \tag{3.1.125}$$

This is established by introducing the Mach numbers

$$M_R = \frac{u_R}{a_R}, \quad M_{s_k} = \frac{s_k}{a_R}. \tag{3.1.126}$$

Manipulation of equations (3.1.115), (3.1.116) and (3.1.126) leads to expressions for the density and pressure ratios across the shock as functions of the relative Mach number $M_R - M_{s_k}$, namely

$$\frac{\rho_L}{\rho_R} = \frac{(\gamma + 1)\,(M_R - M_{s_k})^2}{(\gamma - 1)\,(M_R - M_{s_k})^2 + 2}, \tag{3.1.127}$$

$$\frac{p_L}{p_R} = \frac{2\gamma(M_R - M_{s_k})^2 - (\gamma - 1)}{\gamma + 1}. \tag{3.1.128}$$

In terms of the pressure ratio (3.1.128) we first note the following relation:

$$M_R - M_{s_k} = -\sqrt{\frac{\gamma+1}{2\gamma}\frac{p_L}{p_R} + \frac{\gamma-1}{2\gamma}}.$$

This leads to an expression for the shock speed as a function of the pressure ratio across the shock, namely (3.1.124). In a similar way, by introducing the Mach number

$$M_L = \frac{u_L}{a_L}, \tag{3.1.129}$$

relation (3.1.125) can be derived.

g) *Relation of velocity and pressure across the k-shock wave, $k = 1, 3$* For the k-shock wave, $k = 1, 3$, it holds that

$$u_R = u_L + F_{kL}(p_R), \tag{3.1.130}$$

$$u_L = u_R + F_{kR}(p_L), \tag{3.1.131}$$

where

$$F_{kL}(p) = -(2-k)\,(p - p_L)\left(\frac{\frac{2}{(\gamma+1)\,\rho_L}}{p + \frac{\gamma-1}{\gamma+1}\,p_L}\right)^{1/2} \tag{3.1.132}$$

for p satisfying

$$(2-k)\,p_L < (2-k)\,p, \tag{3.1.133}$$

and

$$F_{kR}(p) = (2-k)\,(p_R - p)\left(\frac{\frac{2}{(\gamma+1)\,\rho_R}}{p + \frac{\gamma-1}{\gamma+1}\,p_R}\right)^{1/2} \tag{3.1.134}$$

for p satisfying

$$(2-k)\,p < (2-k)\,p_R. \tag{3.1.135}$$

To derive (3.1.130)–(3.1.131), we introduce the mass flux $Q = (2-k)\,\rho_L \tilde{u}_L$. In view of (3.1.109) and Remark 3.19,

$$Q = (2-k)\,\rho_R \tilde{u}_R > 0. \tag{3.1.136}$$

From equation (3.1.110) we have

$$(\rho_L \tilde{u}_L)\,\tilde{u}_L + p_L = (\rho_R \tilde{u}_R)\,\tilde{u}_R + p_R.$$

This and (3.1.136) yield

$$\frac{1}{2-k}\,Q\,\tilde{u}_L + p_L = \frac{1}{2-k}\,Q\,\tilde{u}_R + p_R$$

and, thus,

$$Q = (2-k)\,\frac{p_R - p_L}{\tilde{u}_L - \tilde{u}_R}. \tag{3.1.137}$$

By equation (3.1.107),

$$\tilde{u}_L - \tilde{u}_R = u_L - u_R$$

and so Q becomes

$$Q = (2-k)\,\frac{p_R - p_L}{u_L - u_R}, \tag{3.1.138}$$

from which we obtain

$$u_R = u_L - (2-k)\,\frac{p_R - p_L}{Q}. \tag{3.1.139}$$

We are now close to having related u_R to data on the left-hand side. We seek to express the right-hand side of (3.1.139) purely in terms of p_R and $\boldsymbol{w}_L$, which means that we need to express Q as a function of p_R and the data $\boldsymbol{w}_L$. We substitute the relations

$$\tilde{u}_L = \frac{Q}{(2-k)\,\rho_L}, \qquad \tilde{u}_R = \frac{Q}{(2-k)\,\rho_R}$$

obtained from (3.1.136) into equation (3.1.137) to produce

$$Q^2 = \frac{p_R - p_L}{\frac{1}{\rho_L} - \frac{1}{\rho_R}}. \tag{3.1.140}$$

The density ρ_R is related to the pressure p_R via (3.1.116), which can be rewritten as

$$\rho_R = \rho_L \left(\frac{\frac{\gamma-1}{\gamma+1} + \frac{p_R}{p_L}}{\frac{\gamma-1}{\gamma+1}\frac{p_R}{p_L} + 1} \right). \tag{3.1.141}$$

Substitution of ρ_R into (3.1.140) yields

$$\begin{aligned}
Q^2 &= \frac{p_R - p_L}{\frac{1}{\rho_L} - \frac{1}{\rho_R}} = \frac{p_R - p_L}{\frac{1}{\rho_L}\left(1 - \frac{\frac{\gamma-1}{\gamma+1}\frac{p_R}{p_L}+1}{\frac{\gamma-1}{\gamma+1}+\frac{p_R}{p_L}}\right)} \\
&= \frac{p_R - p_L}{\frac{1}{\rho_L}\frac{\frac{\gamma-1}{\gamma+1}+\frac{p_R}{p_L}-\frac{\gamma-1}{\gamma+1}\frac{p_R}{p_L}-1}{\frac{\gamma-1}{\gamma+1}+\frac{p_R}{p_L}}} = \frac{(p_R - p_L)\left(\frac{\gamma-1}{\gamma+1} + \frac{p_R}{p_L}\right)}{\frac{1}{\rho_L}\left(-\frac{2}{\gamma+1} + \frac{p_R}{p_L}\frac{2}{\gamma+1}\right)} \\
&= \frac{(p_R - p_L)\left(\frac{\gamma-1}{\gamma+1} + \frac{p_R}{p_L}\right)}{\frac{1}{\rho_L}\frac{2}{\gamma+1}\left(-1 + \frac{p_R}{p_L}\right)} = \frac{(p_R - p_L)\left(\frac{\gamma-1}{\gamma+1} + \frac{p_R}{p_L}\right)}{\frac{1}{\rho_L}\frac{2}{\gamma+1}\frac{p_R-p_L}{p_L}}
\end{aligned}$$

$$= \frac{p_R + \frac{\gamma-1}{\gamma+1} p_L}{\frac{2}{(\gamma+1)\,\rho_L}}.$$

It follows from this equality that

$$Q = \left(\frac{p_R + \frac{\gamma-1}{\gamma+1} p_L}{\frac{2}{(\gamma+1)\,\rho_L}} \right)^{1/2}. \tag{3.1.142}$$

Hence the velocity u_R in (3.1.139) satisfies

$$u_R = u_L - (2-k)\,(p_R - p_L) \left(\frac{\frac{2}{(\gamma+1)\,\rho_L}}{p_R + \frac{\gamma-1}{\gamma+1} p_L} \right)^{1/2},$$

which is the sought relation (3.1.130).

In a similar way we obtain (3.1.131). We write (3.1.139) as

$$u_L = u_R + (2-k)\,\frac{p_R - p_L}{Q}. \tag{3.1.143}$$

Further, we express Q as a function of p_L and the data $\boldsymbol{w}_R$. To this end, we use (3.1.140) again and substitute ρ_L obtained from (3.1.116) into (3.1.140). We get

$$Q^2 = \frac{p_R - p_L}{\frac{1}{\rho_L} - \frac{1}{\rho_R}} = \frac{p_R - p_L}{\frac{1}{\rho_R}\left(\frac{\frac{\gamma-1}{\gamma+1}\frac{p_L}{p_R} + 1}{\frac{p_L}{p_R} + \frac{\gamma-1}{\gamma+1}} - 1 \right)}$$

$$= \frac{p_R - p_L}{\frac{1}{\rho_R}\,\frac{\frac{\gamma-1}{\gamma+1}\frac{p_L}{p_R} + 1 - \frac{p_L}{p_R} - \frac{\gamma-1}{\gamma+1}}{\frac{p_L}{p_R} + \frac{\gamma-1}{\gamma+1}}} = \frac{(p_R - p_L)\left(\frac{\gamma-1}{\gamma+1} + \frac{p_L}{p_R}\right)}{\frac{1}{\rho_R}\left(\frac{2}{\gamma+1} - \frac{p_L}{p_R}\,\frac{2}{\gamma+1}\right)}$$

$$= \frac{(p_R - p_L)\left(\frac{\gamma-1}{\gamma+1} + \frac{p_L}{p_R}\right)}{\frac{1}{\rho_R}\,\frac{2}{\gamma+1}\left(1 - \frac{p_L}{p_R}\right)} = \frac{(p_R - p_L)\left(\frac{\gamma-1}{\gamma+1} + \frac{p_L}{p_R}\right)}{\frac{2}{(\gamma+1)\,\rho_R}\left(\frac{p_R - p_L}{p_R}\right)}$$

$$= \frac{p_L + \frac{\gamma-1}{\gamma+1} p_R}{\frac{2}{(\gamma+1)\,\rho_R}}.$$

This gives the sought relation for Q

$$Q = \left(\frac{p_L + \frac{\gamma-1}{\gamma+1} p_R}{\frac{2}{(\gamma+1)\,\rho_R}} \right)^{1/2}. \tag{3.1.144}$$

Substitution into (3.1.143) gives

$$u_L = u_R + (2-k)\,(p_R - p_L) \left(\frac{\frac{2}{(\gamma+1)\,\rho_R}}{p_L + \frac{\gamma-1}{\gamma+1} p_R} \right)^{1/2},$$

which is the sought relation (3.1.131).

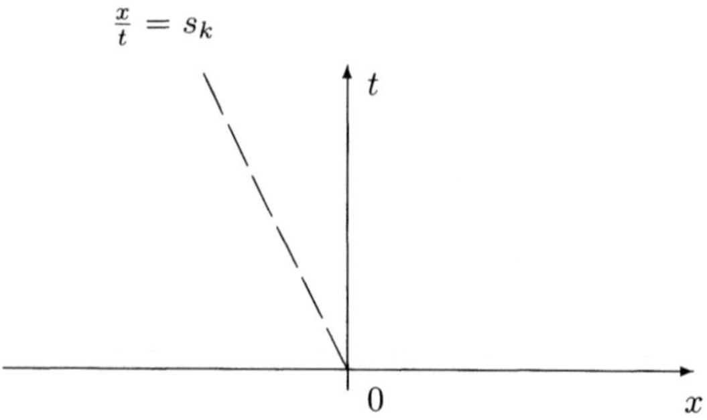

FIG. 3.1.3. k-contact discontinuity, $k = 2$

3.1.6.3 *k-contact discontinuity, $k = 2$* The k-contact discontinuity, defined in (2.3.70), is associated with a linearly degenerate vector $\boldsymbol{r}_k$, $k = 2$. Its position is given by the value

$$s = s_2 := \lambda_k(\boldsymbol{w}_L) = \lambda_k(\boldsymbol{w}_R), \tag{3.1.145}$$

as stated in (2.3.70), see Fig. 3.1.3.

The behaviour of the pressure and the velocity across the k-contact discontinuity will be investigated with the use of k-Riemann invariants defined in (2.3.57).

Exercise 3.20 Verify that for $k = 2$

$$\psi_1^{(k)}(\boldsymbol{w}) = p,$$
$$\psi_2^{(k)}(\boldsymbol{w}) = u$$

are the k-Riemann invariants and that the gradients $\nabla\psi_1^{(k)}$, $\nabla\psi_2^{(k)}$ are linearly independent.

Velocity u and pressure p across the k-contact discontinuity, $k = 2$ Across the k-contact discontinuity, $k = 2$, the velocity u and the pressure p are continuous, i.e.

$$u_L = u_R, \tag{3.1.146}$$
$$p_L = p_R. \tag{3.1.147}$$

Indeed, expressing u in terms of components of $\boldsymbol{w}$, we have $u = w_2/w_1$. Hence, if $\tilde{\boldsymbol{w}}$ is the solution of problem (2.3.59), then

$$\frac{d}{d\xi} u(\tilde{\boldsymbol{w}}(\xi)) = \nabla_w u \frac{d\tilde{\boldsymbol{w}}}{d\xi}(\xi) = \nabla_w u \cdot \boldsymbol{r}_k(\tilde{\boldsymbol{w}}(\xi)) = 0,$$

which is a consequence of the fact that u is a k-Riemann invariant with $k = 2$. Similarly, for p we have

$$p = (\gamma - 1)\left(w_3 - \frac{1}{2} w_1 \left(\frac{w_2}{w_1}\right)^2\right)$$

and

$$\frac{d}{d\xi}\, p(\tilde{\boldsymbol{w}}(\xi)) = \nabla_w p\, \frac{d\tilde{\boldsymbol{w}}}{d\xi}(\xi) = \nabla_w p \cdot \boldsymbol{r}_k(\tilde{\boldsymbol{w}}(\xi)) = 0.$$

Solving (2.3.59) with the right-hand side vector $\boldsymbol{r}_k$, $k = 2$, defined in (3.1.64) we have

$$\frac{d\rho}{d\xi} = 1, \qquad \rho(0) = \rho_L, \tag{3.1.148}$$

$$\frac{d(\rho u)}{d\xi} = u, \qquad (\rho u)(0) = \rho_L u_L, \tag{3.1.149}$$

$$\frac{dE}{d\xi} = \frac{1}{2} u^2, \qquad E(0) = E_L. \tag{3.1.150}$$

The solution of (3.1.148) is given by

$$\rho(\xi) = \xi + \rho_L. \tag{3.1.151}$$

For the solution of (3.1.149) we have

$$\begin{aligned} \frac{d\rho}{d\xi} u + \rho \frac{du}{d\xi} &= u, \\ \frac{du}{d\xi} &= 0, \\ u(\xi) &= u_L. \end{aligned} \tag{3.1.152}$$

Further, the pressure is determined by solving (3.1.150) with the use of (3.1.152):

$$\frac{dE}{d\xi} = \frac{1}{2} u^2 = \frac{1}{2} u_L^2,$$

from where we get $E(\xi) = \frac{1}{2} u_L^2 \xi + E_L$ and using (3.1.5)

$$E(\xi) = \frac{p(\xi)}{\gamma - 1} + \frac{1}{2}(\xi + \rho_L)\, u_L^2 = \frac{1}{2} u_L^2 \xi + \frac{p_L}{\gamma - 1} + \frac{1}{2} \rho_L u_L^2.$$

Finally,

$$p(\xi) = p_L. \tag{3.1.153}$$

For $\xi := \xi_R$ (see Section 2.3.10 b) (ii)) we obtain from (3.1.152) and (3.1.153) relations (3.1.146) and (3.1.147), respectively.

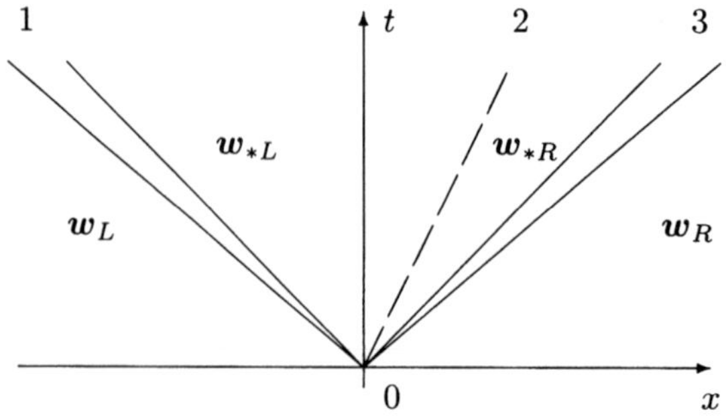

FIG. 3.1.4. Structure of the solution of the Riemann problem

3.1.6.4 *Solution strategy* Now we are able to investigate the general Riemann problem. Up to now we have considered special states $\boldsymbol{w}_L$, $\boldsymbol{w}_R$ that could be connected by a k-rarefaction wave ($k = 1, 3$), k-shock wave ($k = 1, 3$) or k-contact discontinuity ($k = 2$). Now we shall consider arbitrary states $\boldsymbol{w}_L$ and $\boldsymbol{w}_R$.

Remark 3.21 Physically, in the context of the Euler equations, the Riemann problem is a generalization of the so-called *shock-tube problem*: two stationary gases ($u_L = u_R = 0$) in a tube are separated by a diaphragm. The rupture of the diaphragm generates a nearly *centred* wave system that typically consists of a rarefaction wave, a contact discontinuity and a shock wave. This physical problem is reasonably well approximated by solving the shock-tube problem for the Euler equations. In the Riemann problem the velocities u_L and u_R are allowed to be nonzero, but the structure of the solution is the same as that of the shock-tube problem.

We shall seek the exact solution of the Riemann problem in the class consisting of a 1-shock or 1-rarefaction wave, or a 2-contact discontinuity followed by a 3-shock or 3-rarefaction wave. The structure of the solution is as depicted in Fig. 3.1.4.

We choose the convention of representing the k-shock wave and k-rarefaction wave $k = 1, 3$ by a pair of rays emanating from the origin and the 2-contact discontinuity by a dashed line. For given $\boldsymbol{w}_L$, $\boldsymbol{w}_R$, the unknown region between the 1- and 3-waves, the *Star Region*, is divided by the 2-contact discontinuity into two subregions, *Star Left* ($\boldsymbol{w}_{*L}$) and *Star Right* ($\boldsymbol{w}_{*R}$). The waves connect four constant states. From left to right there are $\boldsymbol{w}_L$ (left data state), $\boldsymbol{w}_{*L}$ between the 1-(shock or rarefaction) wave and the 2-contact discontinuity, $\boldsymbol{w}_{*R}$ between the 2-contact discontinuity and the 3-(shock or rarefaction) wave and $\boldsymbol{w}_R$ (right data state). Analysing the structure of eigenvalues $\lambda_1 < \lambda_2 < \lambda_3$ (see (3.1.64)) together with conditions for the speed s of the k-shock wave, $k = 1, 3$ (see (3.1.99)) and behaviour of the velocity u, the pressure p and the speed of

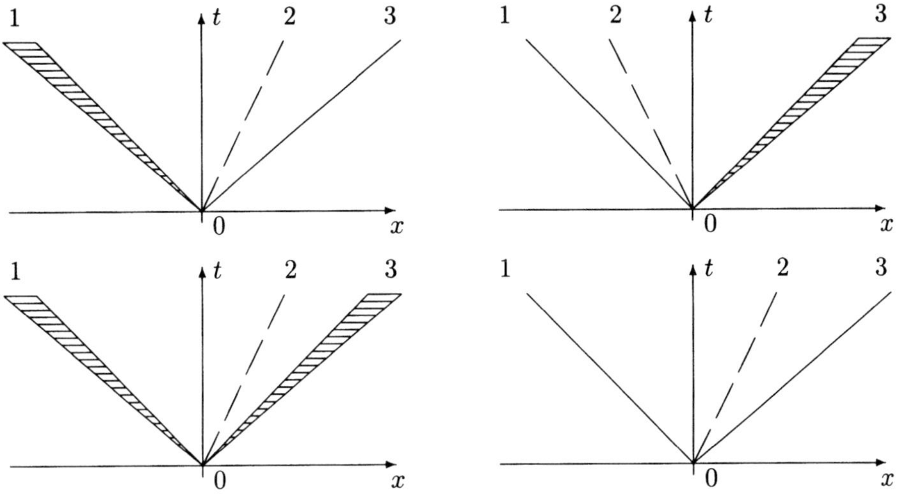

FIG. 3.1.5. Possible wave patterns

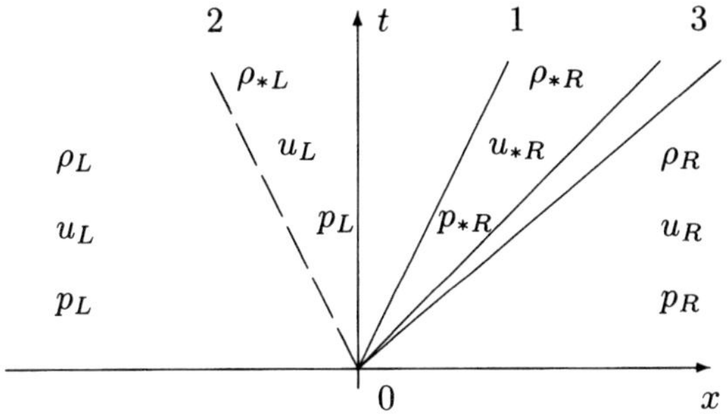

FIG. 3.1.6. Contradictory wave patterns

sound a across the waves, we find that there can be four possible wave patterns, which are shown in Fig. 3.1.5. Here the k-shock wave, k-rarefaction wave and k-contact discontinuity are marked by the corresponding k, the hatched wedge denotes the k-rarefaction wave, the dashed line the k-contact discontinuity and the full line the k-shock wave.

Exercise 3.22 Show that the situation in Fig. 3.1.6 leads to a contradiction with (3.1.145) and (3.1.99).
Hint: In the situation in Fig. 3.1.6 relation (3.1.145) gives

$$u_L = s_2, \tag{3.1.154}$$

where s_2 is the speed of the contact discontinuity. Applying (3.1.99) to the situation in Fig. 3.1.6 yields

$$u_{*R} - a_{*R} < s_1 < u_L - a_{*L}. \tag{3.1.155}$$

In Fig. 3.1.6, speed s_1 of the shock wave and s_2 satisfy

$$s_2 < s_1. \tag{3.1.156}$$

By the substitution of s_1 from (3.1.156) into (3.1.155) and using (3.1.154) we come to the contradiction with the positiveness of the speed of sound a_{*L}.

An analysis based on the eigenstructure of the Euler equations, see (3.1.146) and (3.1.147), reveals that both the pressure p and the velocity u in the *Star Region* are constant, i.e.

$$p_{*L} = p_{*R} =: p_*, \tag{3.1.157}$$

$$u_{*L} = u_{*R} =: u_*. \tag{3.1.158}$$

In what follows, we shall determine the unknown values ρ_{*L}, ρ_{*R}, u_*, p_* and then, using (3.1.8) and (3.1.5), the states $\boldsymbol{w}_{*L} = (\rho_{*L}, \rho_{*L}u_*, E_{*L})^{\mathrm{T}}$ and $\boldsymbol{w}_{*R} = (\rho_{*R}, \rho_{*R}u_*, E_{*R})$. This, together with formulae for speeds s_k (see (3.1.145), (3.1.124)) and formulae for the k-rarefaction wave fan (see (3.1.97), (3.1.98)) give the complete solution of the Riemann problem.

The solution strategy is based on computation of pressure p_*. Then the solution for u_* follows and the remaining unknowns are found by using standard gas dynamics relations. We describe these steps in detail.

a) *Equation for the pressure* p_* Adopting the notation from Fig. 3.1.4, the relation between the velocity and pressure for the 1-rarefaction wave (see (3.1.80)) or 1-shock wave (see (3.1.130)) gives

$$u_* = u_L + F_{1L}(p_*), \tag{3.1.159}$$

where the function $F_{1L}(p)$ is defined by (3.1.81) for $p \le p_L$ (1-rarefaction wave) (see (3.1.82)) and for $p > p_L$ (1-shock wave) the function $F_{1L}(p)$ is defined by (3.1.132). If we put all this together, the definition of the function F_{1L} reads

$$F_{1L}(p) = \begin{cases} \dfrac{2a_L}{\gamma - 1}\left[1 - \left(\dfrac{p}{p_L}\right)^{\frac{\gamma-1}{2\gamma}}\right] & \text{if } p \le p_L, \\[2ex] -(p - p_L)\left(\dfrac{\frac{2}{(\gamma+1)\rho_L}}{p + \frac{\gamma-1}{\gamma+1}p_L}\right)^{1/2} & \text{if } p > p_L. \end{cases} \tag{3.1.160}$$

Analogously, the relation between the velocity and pressure for the 3-rarefaction wave (see (3.1.83)) or 3-shock wave (see (3.1.131)) gives

$$u_* = u_R + F_{3R}(p_*), \tag{3.1.161}$$

where the function $F_{3R}(p)$ is defined by (3.1.84) for $p_R \geq p$ (3-rarefaction wave) (see (3.1.85)) and for $p_R < p$ (3-shock wave) the function $F_{3R}(p)$ is defined by (3.1.134). All together, the definition of F_{3R} reads

$$F_{3R}(p) = \begin{cases} -\dfrac{2a_R}{\gamma - 1}\left[1 - \left(\dfrac{p}{p_R}\right)^{\frac{\gamma-1}{2\gamma}}\right] & \text{if } p \leq p_R, \\ (p - p_R)\left(\dfrac{\frac{2}{(\gamma+1)\rho_R}}{p + \frac{\gamma-1}{\gamma+1}p_R}\right)^{1/2} & \text{if } p > p_R. \end{cases} \tag{3.1.162}$$

From (3.1.159), (3.1.161) we derive the implicit equation for p_*

$$u_L + F_{1L}(p_*) = u_R + F_{3R}(p_*), \tag{3.1.163}$$

which can be rewritten as

$$F(p_*) = 0, \tag{3.1.164}$$

where

$$F(p) = F_{3R}(p) - F_{1L}(p) + u_R - u_L. \tag{3.1.165}$$

A detailed analysis of the pressure function $F(p)$ reveals a particularly simple behaviour and that for *physically relevant data* there exists a unique solution to the equation $F(p) = 0$.

b) *Numerical solution for pressure p_** Given data $\boldsymbol{w}_L$, $\boldsymbol{w}_R$, the pressure function $F(p)$ behaves as shown in Fig. 3.1.7. It is monotone and concave as we shall demonstrate. The first derivative of F_{1L} with respect to p is

$$F'_{1L}(p) = \begin{cases} -\dfrac{1}{\rho_L a_L}\left(\dfrac{p}{p_L}\right)^{-\frac{\gamma+1}{2\gamma}} & \text{if } p \leq p_L, \\ \left(\dfrac{\frac{2}{(\gamma+1)\rho_L}}{\frac{\gamma-1}{\gamma+1}p_L + p}\right)^{1/2}\left(\dfrac{p - p_L}{2\left(\frac{\gamma-1}{\gamma+1}p_L + p\right)} - 1\right) & \text{if } p > p_L. \end{cases} \tag{3.1.166}$$

By inspection, $F'_{1L} < 0$ and analogously we find that $F'_{3R} > 0$. As $F' = F'_{3R} - F'_{1L}$, the function $F(p)$ is monotone as claimed.

The second derivative of the function F_{1L} is

$$F''_{1L}(p) = \begin{cases} \dfrac{(\gamma+1)a_L}{2\gamma^2 p_L^2}\left(\dfrac{p}{p_L}\right)^{-\frac{3\gamma+1}{2\gamma}} & \text{if } p \leq p_L, \\ \dfrac{1}{4}\left(\dfrac{\frac{2}{(\gamma+1)\rho_L}}{\frac{\gamma-1}{\gamma+1}p_L + p}\right)^{1/2}\left(\dfrac{4\frac{\gamma-1}{\gamma+1}p_L + p + 3p_L}{\left(\frac{\gamma-1}{\gamma+1}p_L + p\right)^2}\right) & \text{if } p > p_L \end{cases} \tag{3.1.167}$$

and it holds that $F''_{1L} > 0$. Analogously $F''_{3R} < 0$ and $F''(p) < 0$, i.e. the function $F(p)$ is concave as anticipated.

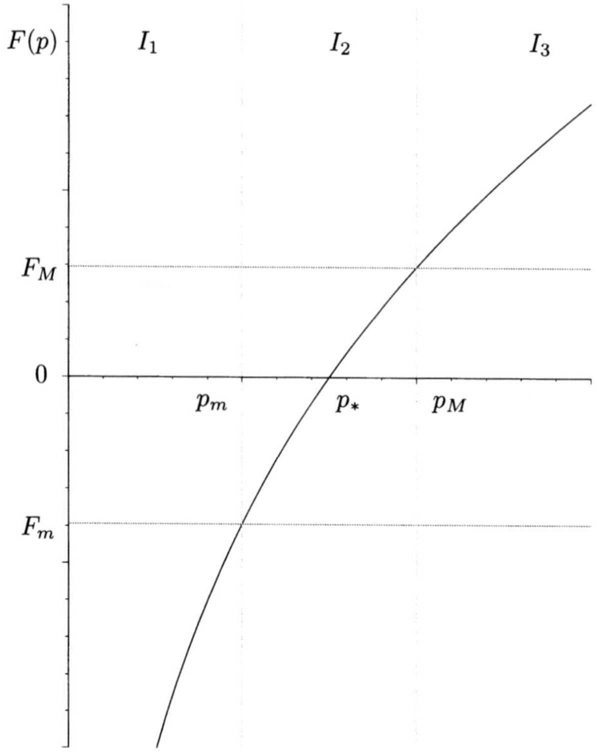

FIG. 3.1.7. Pressure function $F(p)$

From equations (3.1.166), (3.1.167) it can be seen that $F'_{1L} \to 0$, $F'_{3R} \to 0$ as $p \to \infty$ and $F''_{1L} \to 0$, $F''_{3R} \to 0$ as $p \to \infty$. This behaviour of F_{1L}, F_{3R}, and thus of $F(p)$, has implications when devising iteration schemes to find the zero p_* of $F(p)$.

The velocity difference $\Delta u = u_R - u_L$ and the pressure values p_L, p_R are the most important parameters of $F(p)$. With reference to Figs 3.1.7–3.1.9 we define

$$
\begin{aligned}
p_m &= \min(p_L, p_R), \qquad & p_M &= \max(p_L, p_R), \\
F_m &= F(p_m), & F_M &= F(p_M).
\end{aligned}
$$

For given p_L, p_R it is the velocity difference Δu which determines the value of p_*. Three intervals I_1, I_2, I_3 can be identified:

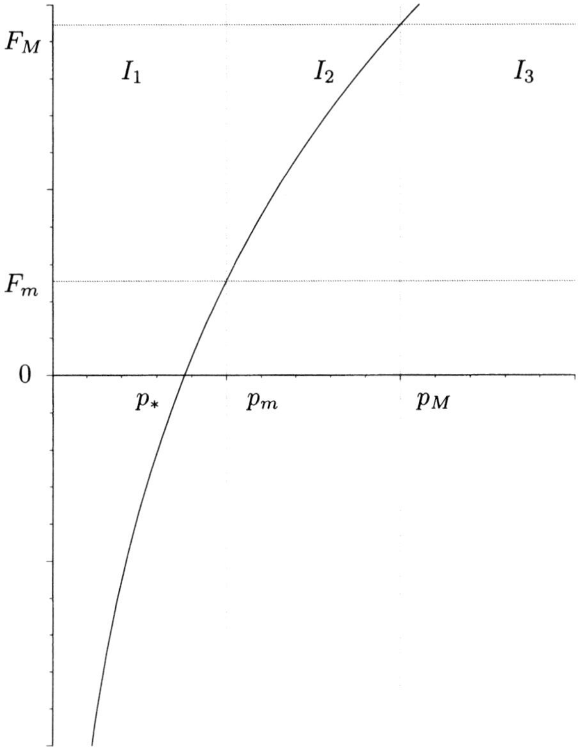

FIG. 3.1.8. Pressure function $F(p)$ for large Δu

$$\begin{aligned} p_* &\in I_1 = (0, p_m) \quad \text{if } F_m > 0, \\ p_* &\in I_2 = [p_m, p_M] \text{ if } F_m \le 0 \text{ and } F_M \ge 0, \\ p_* &\in I_3 = (p_M, \infty) \text{ if } F_M < 0. \end{aligned} \tag{3.1.168}$$

For sufficiently large Δu as in Fig. 3.1.8, the solution p_* lies in I_1 and thus $p_* < p_L$, $p_* < p_R$; so both the 1- and 3-waves are rarefaction waves.

For Δu, for which the graph of $F(p)$ is as in Fig. 3.1.7, the solution p_* lies between p_L and p_R. If $p_L < p_R$, we have in the solution of the Riemann problem the 1-shock wave, 2-contact discontinuity and 3-rarefaction wave. If $p_R < p_L$, the solution is formed by the 1-rarefaction wave, 2-contact discontinuity and 3-shock wave.

For sufficiently small values of Δu, corresponding to the graph of $F(p)$ in Fig. 3.1.9, p_* lies in I_3, i.e. $p_* > p_L$, $p_* > p_R$, which means that both the 1- and 3-waves are shock waves. The interval, where p_* lies, is identified by noting the signs of F_m and F_M; see (3.1.168).

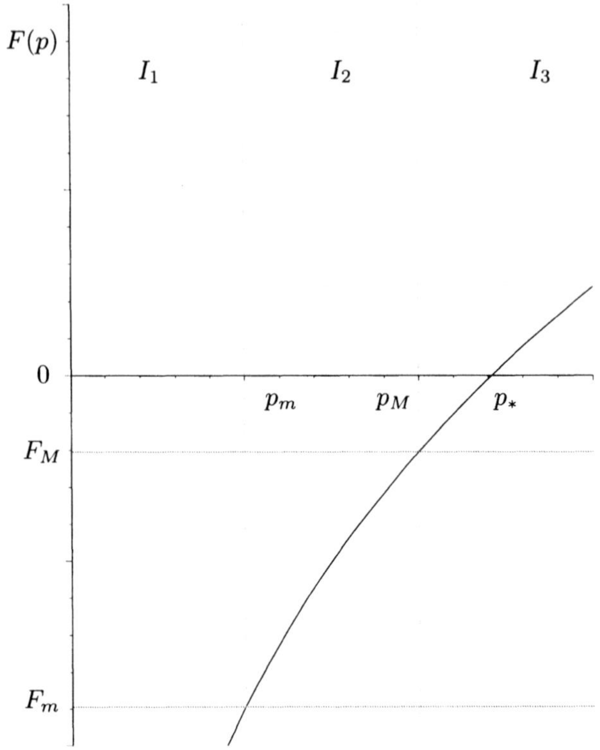

FIG. 3.1.9. Pressure function $F(p)$ for small Δu

Another observation on the behaviour of $F(p)$ is the following: in I_1 both $F'(p)$ and $F''(p)$ vary rapidly; this may lead to numerical difficulties when searching for the root of $F(p) = 0$. For large p the function $F(p)$ is asymptotically linear. There exists a unique positive p_* for pressure, provided Δu is *sufficiently small.* The critical value of Δu can be found analytically in terms of initial data. Clearly, for a positive solution for pressure p_* we require

$$F(0) < 0 \tag{3.1.169}$$

(see Figs 3.1.7–3.1.9).

Direct evaluation of $F(p)$ gives the *pressure positivity condition*

$$u_R - u_L < \frac{2}{\gamma - 1}(a_L + a_R). \tag{3.1.170}$$

c) *Iterative scheme for finding the pressure* p_* Given the particularly simple behaviour of the pressure function $F(p)$ in (3.1.165) and the availability of ana-

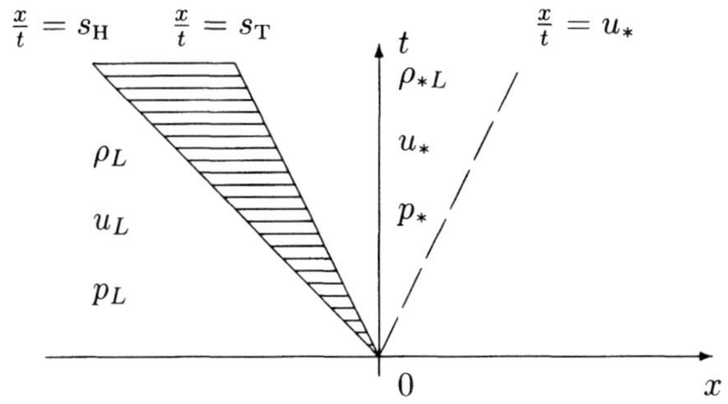

FIG. 3.1.10. Solution for the density ρ_{*L} in the case of the 1-rarefaction wave

lytic expressions for the derivative of $F(p)$, the Newton method to find the root of $F(p) = 0$ can be used.

d) *Solution for u_** Once (3.1.164) is solved for p_* the solution for u_* follows as in (3.1.159) or (3.1.161). Combining this, we can also write

$$u_* = \frac{1}{2}(u_L + u_R) + \frac{1}{2}(F_{1L}(p_*) + F_{3R}(p_*)), \tag{3.1.171}$$

where F_{1L} and R_{3R} are defined in (3.1.160) and (3.1.162), respectively.

e) *Solution for the density ρ_{*L} in the case of the 1-rarefaction wave* Having solved (3.1.164) for p_* and (3.1.171) for u_*, we use the fact that for $p_* \leq p_L$, the states $\boldsymbol{w}_L$ and $\boldsymbol{w}_{*L}$ are connected by the 1-rarefaction wave, while for $p_* > p_L$ the states $\boldsymbol{w}_L$ and $\boldsymbol{w}_{*L}$ are connected by the 1-shock wave as given by relations (3.1.75) and (3.1.123). In the case of the 1-rarefaction wave the situation is depicted in Fig. 3.1.10. For ρ_{*L} we use the relation (3.1.74) which in the notation of Fig. 3.1.10 reads

$$\rho_{*L} = \rho_L \left(\frac{p_*}{p_L}\right)^{1/\gamma}. \tag{3.1.172}$$

The 1-rarefaction wave is enclosed by the *Head* and *Tail* which are the lines $\frac{x}{t} = s_{\mathrm{H}}$, $\frac{x}{t} = s_{\mathrm{T}}$, where

$$s_{\mathrm{H}} = \lambda_1(\boldsymbol{w}_L) = u_L - a_L, \tag{3.1.173}$$

$$s_{\mathrm{T}} = \lambda_1(\boldsymbol{w}_{*L}) = u_* - a_{*L} \tag{3.1.174}$$

(see Fig. 3.1.10) and

$$a_{*L} = \sqrt{\gamma \frac{p_*}{\rho_{*L}}}. \tag{3.1.175}$$

The density ρ, velocity u and pressure p inside the 1-rarefaction wave are given by relations (3.1.97).

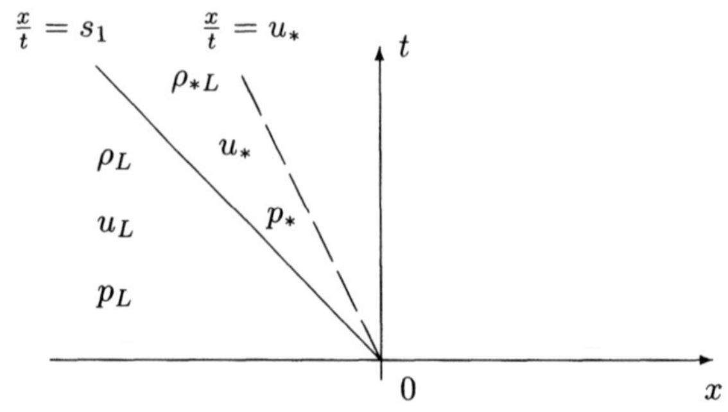

FIG. 3.1.11. Solution for density ρ_{*L} in the case of the 1-shock wave

f) *Solution for density ρ_{*L} in the case of the 1-shock wave* If the solution p_* of (3.1.164) satisfies $p_* > p_L$, it follows from (3.1.123) that $\boldsymbol{w}_L$ and $\boldsymbol{w}_{*L}$ are connected by the 1-shock wave. The corresponding situation is depicted in Fig. 3.1.11. For ρ_{*L} we use relation (3.1.116) which in the notation of Fig. 3.1.11 reads

$$\rho_{*L} = \rho_L \frac{\frac{\gamma-1}{\gamma+1}\frac{p_L}{p_*} + 1}{\frac{p_L}{p_*} + \frac{\gamma-1}{\gamma+1}}. \tag{3.1.176}$$

The speed s of the 1-shock wave is given by (3.1.125), i.e.

$$s = s_1 := u_L - a_L \sqrt{\frac{\gamma+1}{2\gamma}\frac{p_*}{p_L} + \frac{\gamma-1}{2\gamma}}. \tag{3.1.177}$$

g) *Solution for density ρ_{*R} in the case of the 3-rarefaction wave* The 3-rarefaction wave, see Fig. 3.1.12, is identified by the condition $p_* \leq p_R$ (cf. (3.1.75)). The pressure p_* and velocity u_* in the *Star Region* are known. The density ρ_{*R} is found from the isentropic law (3.1.74) as

$$\rho_{*R} = \rho_R \left(\frac{p_*}{p_R}\right)^{1/\gamma}, \tag{3.1.178}$$

from which the speed of sound follows as in (3.1.77)

$$a_{*R} = a_R \left(\frac{p_*}{p_R}\right)^{\frac{\gamma-1}{2\gamma}}. \tag{3.1.179}$$

Head and *Tail* are given respectively by

$$s_{\mathrm{H}} = \lambda_3(\boldsymbol{w}_R) = u_R + a_R, \tag{3.1.180}$$

$$s_{\mathrm{T}} = \lambda_3(\boldsymbol{w}_{*R}) = u_* + a_{*R} \tag{3.1.181}$$

(cf. Fig. 3.1.12). The density ρ, velocity u and pressure p inside the 3-rarefaction wave are given by relations (3.1.98).

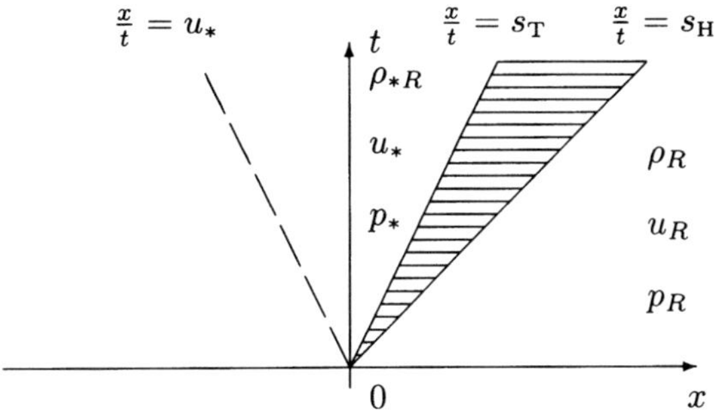

FIG. 3.1.12. Solution for density ρ_{*R} in the case of the 3-rarefaction wave

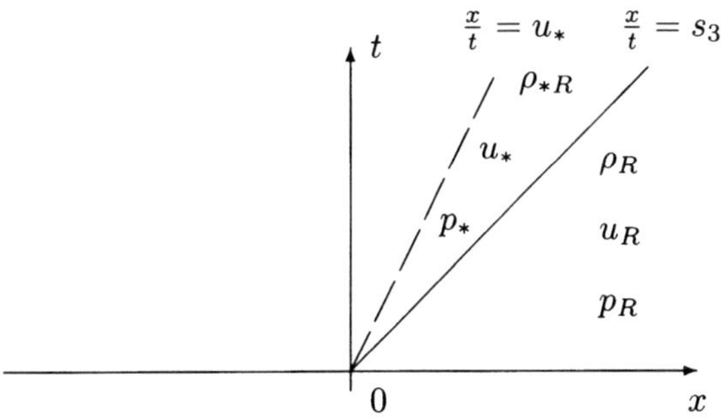

FIG. 3.1.13. Solution for ρ_{*R} in the case of the 3-shock wave

h) *Solution for ρ_{*R} in the case of the 3-shock wave* The 3-shock wave, see Fig. 3.1.13, is identified by the condition $p_* > p_R$ (see (3.1.123)). We know the pressure p_* and the velocity u_*. The density ρ_{*R} is found to be

$$\rho_{*R} = \rho_R \frac{\frac{p_*}{p_R} + \frac{\gamma-1}{\gamma+1}}{\frac{\gamma-1}{\gamma+1}\frac{p_*}{p_R} + 1} \tag{3.1.182}$$

as follows from (3.1.116). The speed s of the 3-shock wave is given by (3.1.124), i.e.

$$s = s_3 := u_R + a_R\sqrt{\frac{\gamma+1}{2\gamma}\frac{p_*}{p_R} + \frac{\gamma-1}{2\gamma}}. \tag{3.1.183}$$

Exercise 3.23 Solve the Riemann problem with initial data

$$u_L = 0,\ u_R = 0,$$

$$\rho_L = 1,\ \rho_R = 1.101463,$$

$$p_L = 2,\ p_R = 1,$$

and with the Poisson constant $\gamma = 1.4$. Draw the density distribution in the (x,t)-plane for $x \in [-3,3]$, $t \in [0,1]$.
Hint: Use the Newton method to find the root of equation $F(p) = 0$ with the initial approximation $p_*^{(0)} := \frac{1}{2}(p_L + p_R)$.

Show that the solution can be expressed in the following way:

$$\rho(x,t) = \begin{cases} \rho_L, & x < s_{\mathrm{H}}t, \\ \rho_L \left[\frac{2}{\gamma+1} + \frac{\gamma-1}{(\gamma+1)\,a_L}\left(-\frac{x}{t}\right)\right]^{\frac{2}{\gamma-1}}, & s_{\mathrm{H}}t < x < s_{\mathrm{T}}t, \\ \rho_{*L}, & s_{\mathrm{T}}t < x < u_*t, \\ \rho_{*R}, & u_*t < x < s_3t, \\ \rho_R, & s_3t < x, \end{cases} \tag{3.1.184}$$

$$u(x,t) = \begin{cases} 0, & x < s_{\mathrm{H}}t, \\ \frac{2}{\gamma+1}\left[a_L + \frac{x}{t}\right], & s_{\mathrm{H}}t < x < s_{\mathrm{T}}t, \\ u_*, & s_{\mathrm{T}}t < x < s_3t, \\ 0, & s_3t < x, \end{cases} \tag{3.1.185}$$

$$p(x,t) = \begin{cases} p_L, & x < s_{\mathrm{H}}t, \\ p_L \left[\frac{2}{\gamma+1} + \frac{\gamma-1}{(\gamma+1)\,a_L}\left(-\frac{x}{t}\right)\right]^{\frac{2\gamma}{\gamma-1}}, & s_{\mathrm{H}}t < x < s_{\mathrm{T}}t, \\ p_*, & s_{\mathrm{T}}t < x < s_3t, \\ p_R, & s_3t < x, \end{cases} \tag{3.1.186}$$

where $p_* = 1.5$ is the root of $F(p)$, defined in (3.1.165), $u_* = 0.337$ is given by relation (3.1.171), $s_{\mathrm{H}} = -1.673$ from (3.1.173), $\rho_{*L} = 0.814$ from (3.1.172), $s_{\mathrm{T}} = -1.269$ from (3.1.174), $s_3 = 1.348$ from (3.1.183) and $\rho_{*R} = 1.469$ from (3.1.182). The structure of the solution corresponds to the wave patterns in Fig. 3.1.5, upper left diagram. The density distribution is shown in Fig. 3.7.1 in Section 3.7.1.

3.1.7 *Solution of the Riemann problem for the split 3D Euler equations*

For the purpose of using the Riemann problem solution in connection with numerical methods of the Godunov type, see Section 3.2.12, it is useful to solve the Riemann problem for the split 3D Euler equations:

$$\frac{\partial \boldsymbol{w}}{\partial t} + \frac{\partial \boldsymbol{f}(\boldsymbol{w})}{\partial x} = 0 \quad \text{in } Q_T = I\!R \times (0,T), \tag{3.1.187}$$

$$\boldsymbol{w}(x,0) = \begin{cases} \boldsymbol{w}_L, & x < 0, \\ \boldsymbol{w}_R, & x > 0, \end{cases} \tag{3.1.188}$$

where $\boldsymbol{f} := \boldsymbol{f}_1$, $x = x_1$ and $\boldsymbol{w}$, $\boldsymbol{f}_1$ are defined by (3.1.8)–(3.1.9), with $N = 3$, $m = 5$.

Exercise 3.24 Prove that the eigenvectors $\boldsymbol{r}_1(\boldsymbol{w})$ and $\boldsymbol{r}_5(\boldsymbol{w})$ of the matrix $\mathbb{A}(\boldsymbol{w}) = D\boldsymbol{f}(\boldsymbol{w})/D\boldsymbol{w}$ are genuinely nonlinear and the eigenvectors $\boldsymbol{r}_2(\boldsymbol{w})$, $\boldsymbol{r}_3(\boldsymbol{w})$, $\boldsymbol{r}_4(\boldsymbol{w})$ of the matrix $\mathbb{A}(\boldsymbol{w})$ are linearly degenerate for all $\boldsymbol{w} \in D$. See Definition 2.25 and Lemmas 3.2–3.3.

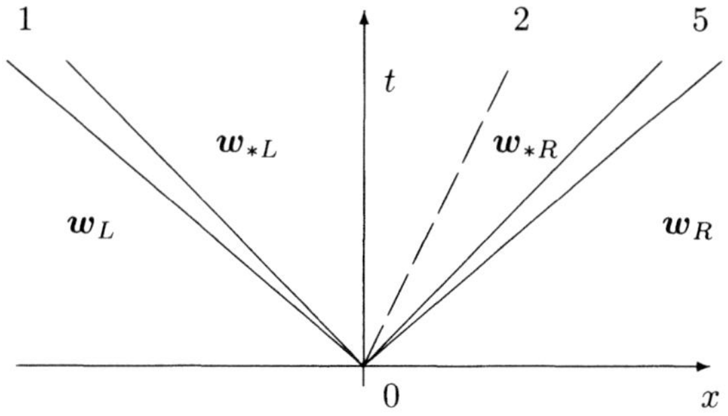

FIG. 3.1.14. Structure of the solution of the Riemann problem for the split 3D Euler equation

The structure of the similarity solution (cf. Theorem 2.24) of the Riemann problem (3.1.187)–(3.1.188) is shown in Fig. 3.1.14, where the notation for the velocity components introduced in Lemma 3.2 is used. Similarly as for the Riemann problem for 1D Euler equations (see Section 3.1.6.4), using the notation from Section 2.3.10, we choose the convention of representing the k-shock wave and k-rarefaction wave, $k = 1, 5$, by a pair of rays emanating from the origin. The multiple eigenvalue $\tilde{\lambda}_2 = \tilde{\lambda}_3 = \tilde{\lambda}_4$ will be associated with a discontinuity wave (having the same position as the contact discontinuity for 1D Euler equations, i.e. the line $x/t = \tilde{\lambda}_i$, $i = 2, 3, 4$), which will be called a *middle wave* (see (Toro, 1997), Section 3.2.4) and represented by a dashed line in Fig. 3.1.14. The region between the 1- and 5-waves, the *Star Region*, is divided by the middle wave into two subregions, *Star Left* ($\boldsymbol{w}_{*L}$) and *Star Right* ($\boldsymbol{w}_{*R}$). Analysing the behaviour of k-Riemann invariants defined in (2.3.57) in a similar way as in Section 3.1.6.3 we come to the conclusion that both the pressure p and the velocity component u are constant in the *Star Region*.

Exercise 3.25 Verify that for $k = 2, 3, 4$

$$\psi_1^{(k)}(\boldsymbol{w}) = p, \tag{3.1.189}$$

$$\psi_2^{(k)}(\boldsymbol{w}) = u \tag{3.1.190}$$

are the k-Riemann invariants.

The behaviour of the velocity components v, w across the 1- and 5-rarefaction waves is investigated similarly as in Section 3.1.6.1, where the Riemann problem for 1D Euler equations is considered.

Exercise 3.26 Let $k \in \{1, 5\}$. Verify that

$$\psi_1^{(k)}(\boldsymbol{w}) = v, \tag{3.1.191}$$

$$\psi_2^{(k)}(\boldsymbol{w}) = w \tag{3.1.192}$$

are the k-Riemann invariants and that for fixed $k = 1, 5$ the gradients $\nabla\psi_1^{(k)}$, $\nabla\psi_2^{(k)}$ are linearly independent.

Lemma 3.27 *Across a k-rarefaction wave, $k \in \{1, 5\}$, the k-Riemann invariant is constant.*

Proof See the proof of Lemma 3.16. □

From the previous lemma we see immediately that across the 1- and 5-rarefaction waves there is no change in the velocity components v, w. In fact, this is also true for the 1- and 5-shock waves. Consider a k-shock wave of speed $s = s_k$. By transforming to a frame of reference in which the shock speed is zero (cf. (3.1.105)) and applying the Rankine–Hugoniot conditions (2.3.71) we obtain for $k = 1$

$$\rho_L \tilde{u}_L = \rho_{*L} \tilde{u}_{*L} \tag{3.1.193}$$

$$\rho_L \tilde{u}_L^2 + p_L = \rho_{*L} \tilde{u}_{*L}^2 + p_{*L} \tag{3.1.194}$$

$$\rho_L \tilde{u}_L \tilde{v}_L = \rho_{*L} \tilde{u}_{*L} \tilde{v}_{*L} \tag{3.1.195}$$

$$\rho_L \tilde{u}_L \tilde{w}_L = \rho_{*L} \tilde{u}_{*L} \tilde{w}_{*L} \tag{3.1.196}$$

$$(\tilde{E}_L + p_L) \tilde{u}_L = (\tilde{E}_{*L} + p_{*L}) u_{*L}. \tag{3.1.197}$$

Substitution of condition (3.1.193) into equations (3.1.195) and (3.1.196) gives directly

$$\begin{aligned} \tilde{v}_L &= \tilde{v}_{*L}, \\ \tilde{w}_L &= \tilde{w}_{*L}. \end{aligned}$$

Since by definition $\tilde{v} = v + s_k$, $\tilde{w} = w + s_k$ we have

$$\begin{aligned} v_{*L} &= v_L, \\ w_{*L} &= w_L. \end{aligned} \tag{3.1.198}$$

A similar analysis for the 5-shock wave gives an equivalent result:

$$\begin{aligned} v_{*R} &= v_R, \\ w_{*R} &= w_R. \end{aligned} \tag{3.1.199}$$

Hence, the velocity components v and w remain constant across the k-(rarefaction, shock) waves, $k = 1, 5$, irrespective of their type.

Therefore, finding the solution of the Riemann problem for the split 3D Euler equations is fundamentally the same as finding the solution for the corresponding 1D Euler equations. The solution for the variables v and w consists of single jump discontinuities across the middle wave from the values v_L, w_L of the left data state to the values v_R, w_R of the right data state. Summarizing the above considerations, the structure of the solution of the Riemann problem for the split 3D Euler equations is as drawn in Fig. 3.1.15.

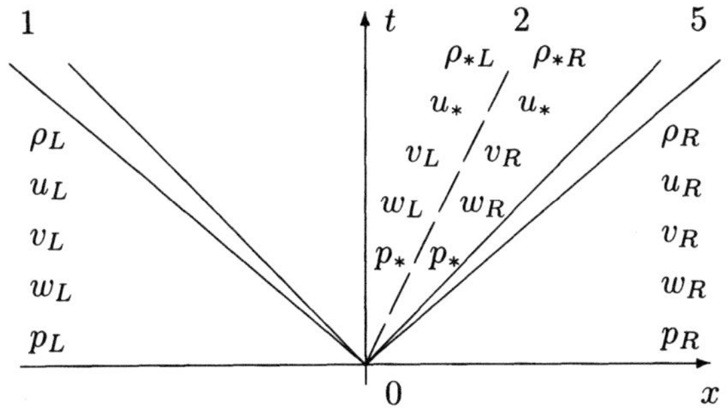

FIG. 3.1.15. Structure of the solution $\boldsymbol{w}(x,t) = \boldsymbol{w}_{\mathrm{RS}}(x/t; \boldsymbol{w}_L, \boldsymbol{w}_R)$ of the Riemann problem for the split 3D Euler equations in primitive variables

Remark 3.28 Finding the solution of the Riemann problem for the split 2D Euler equations is analogous. The solution for the corresponding one-dimensional Euler equations is completed by the solution for the variable v. It consists of single jump discontinuity across the middle wave from the value v_L of the left data state to the value v_R of the right data state.

3.2 Numerical methods for hyperbolic systems with one space variable

First we shall be concerned with numerical schemes for the solution of hyperbolic equations with one space variable. For this simplified situation we shall explain the derivation of basic schemes and their analysis. In further sections these ideas, results, methods and techniques will be extended to multidimensional problems. Let us consider the Cauchy problem

$$\frac{\partial \boldsymbol{w}}{\partial t} + \frac{\partial \boldsymbol{f}(\boldsymbol{w})}{\partial x} = 0 \quad \text{in } I\!R \times (0, \infty) \tag{3.2.1}$$

$$\boldsymbol{w}(x, 0) = \boldsymbol{w}^0(x), \quad x \in I\!R, \tag{3.2.2}$$

where $\boldsymbol{f} = (f_1, \ldots, f_m)^{\mathrm{T}}$, $f_i \in C^1(I\!R^m)$, $\boldsymbol{w} = (w_1, \ldots, w_m)^{\mathrm{T}}$. (For simplicity we assume that the domain of definition of the flux $\boldsymbol{f}$ is $D = I\!R^m$.) We assume that system (3.2.1) is diagonally hyperbolic, i.e. the Jacobi matrix $D\boldsymbol{f}/D\boldsymbol{w}$ is diagonalizable and has real eigenvalues for each $\boldsymbol{w} \in I\!R^m$. For simplicity of treatment we consider (3.2.1) for all $x \in I\!R$ and therefore do not need boundary conditions. Let us describe the *finite difference* discretization of problem (3.2.1)–(3.2.2). We denote $Z = \{0, \pm1, \pm2, \ldots\}$, $Z^+ = \{0, 1, \ldots\}$ and define the *mesh* in $I\!R \times [0, +\infty)$ formed by the mesh points

$$(x_i, t_k) = (ih, k\tau), \quad i \in Z, \ k \in Z^+, \tag{3.2.3}$$

where $h > 0$ and $\tau > 0$ is the *mesh size* in the direction x and t, respectively. τ is also called the *time step*. We set $x_{i\pm\frac{1}{2}} = x_i \pm h/2$. By $\boldsymbol{w}_i^k$ we denote the *approximate solution* at (x_i, t_k) approximating the value $\boldsymbol{w}(x_i, t_k)$. In order to derive the discretization of problem (3.2.1)–(3.2.2), let us assume that $\boldsymbol{w} \in C^1(I\!R \times [0,\infty))^m$ is its classical solution and integrate (3.2.1) over the set $\left(x_{i-\frac{1}{2}}, x_{i+\frac{1}{2}}\right) \times (t_k, t_{k+1})$. Using Fubini's theorem, we obtain

$$\int_{x_{i-\frac{1}{2}}}^{x_{i+\frac{1}{2}}} (\boldsymbol{w}(x,t_{k+1}) - \boldsymbol{w}(x,t_k))dx + \int_{t_k}^{t_{k+1}} \left(\boldsymbol{f}(\boldsymbol{w}(x_{i+\frac{1}{2}},t)) - \boldsymbol{f}(\boldsymbol{w}(x_{i-\frac{1}{2}},t))\right)dt = 0. \tag{3.2.4}$$

Now we want to approximate this relation with the aid of the values of the approximate solution. We write

$$\int_{x_{i-\frac{1}{2}}}^{x_{i+\frac{1}{2}}} \boldsymbol{w}(x,t_k)\,dx \approx h\boldsymbol{w}_i^k. \tag{3.2.5}$$

This means that the value $\boldsymbol{w}_i^k$ can be interpreted as an approximation of the *integral average* of the function $\boldsymbol{w}(.,t_k)$ on the interval $(x_{i-\frac{1}{2}}, x_{i+\frac{1}{2}})$. The values of the flux $\boldsymbol{f}(\boldsymbol{w}(x_{i\pm\frac{1}{2}},t))$ of the quantity $\boldsymbol{w}$ at the points $x_{i\pm\frac{1}{2}}$ and time t must also be approximated with the aid of the values $\boldsymbol{w}_j^n, j \in Z, n \in Z^+$, of the approximate solution. That is, we write

$$\int_{t_k}^{t_{k+1}} \boldsymbol{f}(\boldsymbol{w}(x_{i\pm\frac{1}{2}},t))\,dt \approx \tau\left[\vartheta\boldsymbol{g}_{i\pm\frac{1}{2}}^{k+1} + (1-\vartheta)\boldsymbol{g}_{i\pm\frac{1}{2}}^{k}\right], \tag{3.2.6}$$

where

$$\begin{aligned}\boldsymbol{g}_{i+\frac{1}{2}}^k &= \boldsymbol{g}(\boldsymbol{w}_{i-\ell+1}^k, \dots, \boldsymbol{w}_{i+\ell}^k),\\ \boldsymbol{g}_{i-\frac{1}{2}}^k &= \boldsymbol{g}(\boldsymbol{w}_{i-\ell}^k, \dots, \boldsymbol{w}_{i+\ell-1}^k), \quad \vartheta \in [0,1],\end{aligned} \tag{3.2.7}$$

and ℓ is a positive integer. The function $\boldsymbol{g} = \boldsymbol{g}(\boldsymbol{v}_1, \dots, \boldsymbol{v}_{2\ell}) \in I\!R^m$, $\boldsymbol{v}_1, \dots, \boldsymbol{v}_{2\ell} \in I\!R^m$, is called the *numerical flux* and is used to approximate the flux $\boldsymbol{f}(\boldsymbol{w}(x_{i-\frac{1}{2}}, t_k))$ of the quantity $\boldsymbol{w}$ at the points $(x_{i-\frac{1}{2}}, t_k)$ with the aid of the values of the approximate solution.

In the sequel, we shall always assume that the *numerical flux is a continuous function.*

The described discretization process leads to the following numerical scheme for the solution of the Cauchy problem (3.2.1)–(3.2.2):

$$\boldsymbol{w}_i^{k+1} = \boldsymbol{w}_i^k - \frac{\tau}{h}\left\{\vartheta\left(\boldsymbol{g}_{i+\frac{1}{2}}^{k+1} - \boldsymbol{g}_{i-\frac{1}{2}}^{k+1}\right) + (1-\vartheta)\left(\boldsymbol{g}_{i+\frac{1}{2}}^{k} - \boldsymbol{g}_{i-\frac{1}{2}}^{k}\right)\right\}, \tag{3.2.8}$$
$$i \in Z,\ k \in Z^+.$$

This is equipped with the initial conditions

$$\text{a) } \boldsymbol{w}_i^0 = \boldsymbol{w}^0(x_i) \quad \text{or} \quad \text{b) } \boldsymbol{w}_i^0 = \frac{1}{h}\int_{x_{i-\frac{1}{2}}}^{x_{i+\frac{1}{2}}} \boldsymbol{w}^0(x)\,dx, \quad i \in Z. \tag{3.2.9}$$

Condition a) can be used if $\boldsymbol{w}^0 \in C(\mathbb{R})^m$, whereas b) makes sense in the more general situation when $\boldsymbol{w}^0 \in L^1_{\text{loc}}(\mathbb{R})^m$.

Provided the approximate solution $\boldsymbol{w}_i^k$, $i \in Z$, is known at a time level t_k, from (3.2.8) the approximate solution $\boldsymbol{w}_i^{k+1}$, $i \in Z$, at time t_{k+1} is computed. If $\vartheta = 0$, (3.2.8) represents an explicit formula for computing $\boldsymbol{w}_i^{k+1}$. We speak of an *explicit method.* If $\vartheta \in (0,1]$, we get an *implicit method* requiring the solution of a nonlinear system at each time level. Usually one uses explicit schemes. This scheme is called *linear*, if the numerical flux $\boldsymbol{g}$ depends linearly on its variables, provided the scheme is applied to a linear system, i.e. system (3.2.1) where $\boldsymbol{f}(\boldsymbol{w}) = \mathbb{A}\boldsymbol{w}$ and $\mathbb{A}$ is a constant diagonalizable $m \times m$ matrix with real eigenvalues. Very often $\ell = 1$, i.e. $\boldsymbol{g} = \boldsymbol{g}(\boldsymbol{v}_1, \boldsymbol{v}_2)$. Then we speak of a *three-point scheme*, because in (3.2.8) only the values of the approximate solution at points x_{i-1}, x_i and x_{i+1} appear. As we see, the derived scheme (3.2.8) can be written in a simple form

$$\boldsymbol{w}_i^{k+1} = \boldsymbol{\Phi}\left(\boldsymbol{w}_{i-\ell}^k, \ldots, \boldsymbol{w}_{i+\ell}^k, \boldsymbol{w}_{i-\ell}^{k+1}, \ldots, \boldsymbol{w}_{i+\ell}^{k+1}\right), \quad i \in Z,\ k \in Z^+, \tag{3.2.10}$$

with a mapping $\boldsymbol{\Phi} : \mathbb{R}^{2\ell+1} \times \mathbb{R}^{2\ell+1} \to \mathbb{R}^m$. The formulation of the scheme in the form (3.2.10) is more general than (3.2.8). If scheme (3.2.10) can be expressed in form (3.2.8) with the aid of a numerical flux, we call (3.2.10) *conservative.*

3.2.1 *Example of a nonconservative scheme*

Let us write system (3.2.1) in the form

$$\frac{\partial \boldsymbol{w}}{\partial t} + \mathbb{A}(\boldsymbol{w})\frac{\partial \boldsymbol{w}}{\partial x} = 0, \quad \mathbb{A}(\boldsymbol{w}) = D\boldsymbol{f}(\boldsymbol{w})/D\boldsymbol{w}, \tag{3.2.11}$$

and discretize this system at a mesh point (x_i, t_k) in such a way that the derivative with respect to time is approximated by the forward difference, whereas the derivative with respect to space is approximated by the central difference:

$$\begin{aligned} \frac{\partial \boldsymbol{w}}{\partial t}(x_i, t_k) &\approx \frac{\boldsymbol{w}_i^{k+1} - \boldsymbol{w}_i^k}{\tau} \\ \frac{\partial \boldsymbol{w}}{\partial x}(x_i, t_k) &\approx \frac{\boldsymbol{w}_{i+1}^k - \boldsymbol{w}_{i-1}^k}{2h}. \end{aligned} \tag{3.2.12}$$

Then we obtain the nonconservative scheme

$$\boldsymbol{w}_i^{k+1} = \boldsymbol{w}_i^k - \frac{\tau}{2h}\mathbb{A}(\boldsymbol{w}_i^k)(\boldsymbol{w}_{i+1}^k - \boldsymbol{w}_{i-1}^k). \tag{3.2.13}$$

In what follows, we shall deal mainly with conservative schemes, because nonconservative schemes usually do not yield correct solutions. This fact was theoretically analysed, for example, in (Hou and Le Floch, 1994). (However, in Section 4.3 we shall pay attention to a finite element analogy of nonconservative schemes, which yield accurate numerical solutions, provided a suitable stabilization is used.)

3.2.2 *Semidiscretization in space*

Another way to construct numerical schemes for the solution of problem (3.2.1)–(3.2.2) is based on the semidiscretization in space, also called the *method of lines.* In this case we leave time t continuous and integrate equation (3.2.1) over the interval $(x_{i-\frac{1}{2}}, x_{i+\frac{1}{2}})$ with respect to x:

$$\int_{x_{i-\frac{1}{2}}}^{x_{i+\frac{1}{2}}} \frac{\partial \boldsymbol{w}(x,t)}{\partial t}\, dx + \boldsymbol{f}(\boldsymbol{w}(x_{i+\frac{1}{2}},t)) - \boldsymbol{f}(\boldsymbol{w}(x_{i-\frac{1}{2}},t)) = 0, \qquad (3.2.14)$$

$$i \in Z,\ t \in (0, +\infty).$$

Approximating $\boldsymbol{w}(x_i,t)$ by $\boldsymbol{w}_i(t)$ and again using numerical flux $\boldsymbol{g}$ for the approximation of the flux $\boldsymbol{f}$, we arrive at the system of ordinary differential equations

$$\frac{d\boldsymbol{w}_i}{dt} + (\boldsymbol{g}_{i+\frac{1}{2}} - \boldsymbol{g}_{i-\frac{1}{2}})/h = 0, \quad i \in Z, \qquad (3.2.15)$$

with

$$\begin{aligned} \boldsymbol{g}_{i-\frac{1}{2}} &= \boldsymbol{g}(\boldsymbol{w}_{i-\ell}, \dots, \boldsymbol{w}_{i+\ell-1}), \\ \boldsymbol{g}_{i+\frac{1}{2}} &= \boldsymbol{g}(\boldsymbol{w}_{i-\ell+1}, \dots, \boldsymbol{w}_{i+\ell}), \end{aligned} \qquad (3.2.16)$$

and equipped with initial conditions

$$\boldsymbol{w}_i(0) = \boldsymbol{w}_i^0, \quad i \in Z, \qquad (3.2.17)$$

defined by (3.2.9). Problem (3.2.15), (3.2.17) can be written symbolically in the form

$$\frac{d\boldsymbol{W}_h}{dt} = \boldsymbol{\Phi}_h(\boldsymbol{W}_h), \quad \boldsymbol{W}_h(0) = \boldsymbol{W}_h^0. \qquad (3.2.18)$$

It can be solved numerically by some method for the solution of systems of ordinary differential equations (see, for example, (Ralston, 1965), Chapter 5, or (Schatzman, 2002), Section 15). There are packages of programs for the solution of these systems ((Hindmarsh, 1983)) but they are not quite suitable for the solution of system (3.2.18) obtained by the discretization of the Euler equations. Their efficiency is relatively low, because they do not use some special features of the Euler equations.

In order to obtain a *time discretization* of system (3.2.18), we define a time mesh in the interval $(0,T)$ formed by points $0 = t_0 < t_1 < t_2 < \dots$ and denote by $\tau_k = t_{k+1} - t_k$ the time step between t_k and t_{k+1}. By $\boldsymbol{W}_h^k$ we denote the approximation of $\boldsymbol{W}_h(t_k)$.

Using, for example, the well-known *Euler forward method*

$$\boldsymbol{W}_h^{k+1} = \boldsymbol{W}_h^k + \tau_k \boldsymbol{\Phi}_h(\boldsymbol{W}_h^k), \qquad (3.2.19)$$

we obtain the explicit method (3.2.8) with $\vartheta = 0$.

Another possibility is to use the *Runge–Kutta methods* for the approximation of system (3.2.18). As an example let us introduce two Runge–Kutta schemes. Assume that $\boldsymbol{W}_h^k$ is known. Then

$$\begin{aligned} \boldsymbol{W}^{(0)} &= \boldsymbol{W}_h^k, \\ \boldsymbol{W}^{(1)} &= \boldsymbol{W}^{(0)} + \tau_k \boldsymbol{\Phi}_h(\boldsymbol{W}^{(0)}), \\ \boldsymbol{W}^{(2)} &= \boldsymbol{W}^{(1)} + \tau_k \boldsymbol{\Phi}_h(\boldsymbol{W}^{(1)}), \\ \boldsymbol{W}_h^{k+1} &= \frac{1}{2}(\boldsymbol{W}^{(0)} + \boldsymbol{W}^{(2)}), \end{aligned} \tag{3.2.20}$$

(second order method) or

$$\begin{aligned} \boldsymbol{W}^{(0)} &= \boldsymbol{W}_h^k, \\ &\vdots \\ \boldsymbol{W}^{(r)} &= \boldsymbol{W}^{(0)} + \alpha_r \tau_k \boldsymbol{\Phi}_h(\boldsymbol{W}^{(r-1)}), \quad r = 1, \ldots, 4, \\ &\vdots \\ \boldsymbol{W}_h^{k+1} &= \boldsymbol{W}^{(4)}, \\ \alpha_1 &= \frac{1}{4}, \quad \alpha_2 = \frac{1}{3}, \quad \alpha_3 = \frac{1}{2}, \quad \alpha_4 = 1 \end{aligned} \tag{3.2.21}$$

(fourth order method).

3.2.3 *Space-time nonuniform grid*

The above considerations can also be carried out on a *space-time nonuniform grid.* Let us consider a partition formed by points $\{x_{i+\frac{1}{2}}\}_{i\in Z}$ such that $x_{i-\frac{1}{2}} < x_{i+\frac{1}{2}}$ for all $i \in Z$ and set $x_i = (x_{i+\frac{1}{2}} + x_{i-\frac{1}{2}})/2$, $\mathcal{I}_i = (x_{i-\frac{1}{2}}, x_{i+\frac{1}{2}})$, $h_i = x_{i+\frac{1}{2}} - x_{i-\frac{1}{2}}$. Let $0 = t_0 < t_1 < \ldots$, $\tau_k = t_{k+1} - t_k$. Then the process represented by (3.2.4)–(3.2.8) can be modified to the scheme

$$\begin{aligned} \boldsymbol{w}_i^{k+1} &= \boldsymbol{w}_i^k - \frac{\tau_k}{h_i}\left\{\vartheta\left(\boldsymbol{g}_{i+\frac{1}{2}}^{k+1} - \boldsymbol{g}_{i-\frac{1}{2}}^{k+1}\right) + (1-\vartheta)\left(\boldsymbol{g}_{i+\frac{1}{2}}^{k} - \boldsymbol{g}_{i-\frac{1}{2}}^{k}\right)\right\}, \\ & i \in Z, \ k \in Z^+, \end{aligned} \tag{3.2.22}$$

where $\boldsymbol{g}_{i\pm\frac{1}{2}}^k$ are again defined by (3.2.7). Similarly as in (3.2.5), the value $\boldsymbol{w}_i^k$ can again be interpreted as an approximation of the *integral average* of the function $\boldsymbol{w}(\cdot, t_k)$ on the interval $\mathcal{I}_i$.

Similarly one can define the method of lines over a space nonuniform grid and then apply the Runge–Kutta schemes.

3.2.4 *Qualitative properties of numerical schemes for conservation laws*

For reasons of simplicity, in our further treatment we shall be concerned with the case of uniform meshes leading to scheme (3.2.8). In the investigation of numerical schemes we try to answer the following questions:

a) *consistency* of the method with equation (3.2.1) and *accuracy* of the scheme,
b) *stability*,
c) *convergence* of approximate solutions to the exact one and *error estimates*,
d) *computation* of approximate solutions (i.e. numerical realization of the method).

The following concept is very important.

Definition 3.29 *We say that the method* (3.2.8) *is consistent, if*

$$\boldsymbol{g}(\boldsymbol{v}, \ldots, \boldsymbol{v}) = \boldsymbol{f}(\boldsymbol{v}) \quad \forall \boldsymbol{v} \in I\!R^m. \tag{3.2.23}$$

The consistency is closely related to the convergence of the method as follows from the well-known *Lax–Wendroff convergence theorem* formulated below ((Lax and Wendroff, 1960)). Given an approximate solution $\boldsymbol{w}_i^k$, $i \in Z$, $k \in Z^+$, obtained from scheme (3.2.8) (using constant steps h and τ), equipped with the initial condition (3.2.9), b), we define the piecewise constant function $\boldsymbol{w}_{h\tau}(x,t)$:

$$\boldsymbol{w}_{h\tau}(x,t) = \boldsymbol{w}_i^k \text{ for } x \in (x_{i-\frac{1}{2}}, x_{i+\frac{1}{2}}),\ t \in (t_k, t_{k+1}),\ i \in Z,\ k \in Z^+. \tag{3.2.24}$$

Theorem 3.30 *Let scheme* (3.2.8) *be consistent and let* $\boldsymbol{w}^0 \in L^\infty(I\!R)^m$. *Furthermore, let us consider sequences* $h_n \to 0+$, $\tau_n \to 0+$ *as* $n \to \infty$ *and assume that the vector functions* $\boldsymbol{w}^n = \boldsymbol{w}_{h_n\tau_n}$, $n = 1, 2, \ldots$, *associated with the approximate solutions by* (3.2.24) *satisfy the following conditions:*

$$\begin{aligned} &a)\ \|\boldsymbol{w}^n\|_{L^\infty(I\!R\times(0,\infty))^m} \le c \quad \forall n = 1, 2, \ldots, \\ &b)\ \boldsymbol{w}^n \to \boldsymbol{w} \text{ a.e. in } I\!R \times (0,\infty) \text{ as } n \to \infty. \end{aligned} \tag{3.2.25}$$

Then $\boldsymbol{w}$ *is a weak solution of problem* (3.2.1)–(3.2.2).

Proof The proof can be found in (Godlewski and Raviart, 1991) or (Feistauer, 1993). □

3.2.5 *Order of the scheme*

Let us write scheme (3.2.8) in the form (3.2.10). Let $\boldsymbol{w} \in C^1(I\!R \times [0,\infty))^m$ be a classical solution of problem (3.2.1)–(3.2.2). Substituting $\boldsymbol{w}$ into (3.2.10) leads to the relation

$$\begin{aligned} &\boldsymbol{w}(x_i, t_{k+1}) \\ &= \boldsymbol{\Phi}\Big(\boldsymbol{w}(x_{i-\ell}, t_k), \ldots, \boldsymbol{w}(x_{i+\ell}, t_k), \boldsymbol{w}(x_{i-\ell}, t_{k+1}), \ldots, \boldsymbol{w}(x_{i+\ell}, t_{k+1})\Big) + \tau\boldsymbol{\varepsilon}_i^k, \\ &\quad i \in Z,\ k \in Z^+. \end{aligned} \tag{3.2.26}$$

The quantity $\boldsymbol{\varepsilon}_i^k$ is called a (*local relative*) *truncation error* of the scheme.

Definition 3.31 *We say that scheme (3.2.8), i.e. (3.2.10), is of order p in time and q in space, if, under the assumption that the exact solution* $\boldsymbol{w}$ *of (3.2.1)–(3.2.2) and the flux* $\boldsymbol{f}$ *are sufficiently smooth, there exist constants* $M, \tau_0, h_0 > 0$ *such that*

$$|\varepsilon_i^k| \le M(\tau^p + h^q), \quad i \in Z, \; k \in Z^+, \; \tau \in (0, \tau_0), \; h \in (0, h_0). \qquad (3.2.27)$$

We simply write $\varepsilon_i^k = O(\tau^p + h^q)$.

Usually the investigation of the truncation error is based on the use of the Taylor formula, as we shall see in the sequel (cf., for example, Section 3.2.8 and 3.2.10).

3.2.6 *Stability of the scheme*

Assumption (3.2.25), a) is closely related to the concept of the stability of the scheme which can be formulated in the following way: *If the initial condition* $\mathbf{w}^0 = \{\boldsymbol{w}_i^0\}_{i \in Z}$ *is bounded (in some norm), then the approximate solution* $\mathbf{w}^k = \{\boldsymbol{w}_i^k\}_{i \in Z}$ *remains bounded at every time level* $t_k \in [0, T]$ $(T < \infty)$. Quite often the *von Neumann stability criterion* is applied. Let us assume that the approximate solution $\boldsymbol{w}_j^k$ is expanded in the Fourier series

$$\begin{aligned} \boldsymbol{w}_j^k &= \sum_{n=-\infty}^{+\infty} \boldsymbol{F}_k(n) \exp(injh) \qquad (3.2.28) \\ &= \sum_{n=-\infty}^{+\infty} \boldsymbol{F}_k(n) \exp(inx), \quad j \in Z, \; k \in Z^+, \\ x &= jh, \quad i^2 = -1. \end{aligned}$$

Definition 3.32 *We say that the scheme is von Neumann stable, if there exists a constant* $c > 0$ *independent of* n, k, h, τ *such that the condition*

$$|\boldsymbol{F}_k(n)| \le c|\boldsymbol{F}_0(n)| \quad \forall n \in Z, \; \forall k \in Z^+, \; \forall h, \tau > 0, \qquad (3.2.29)$$

is satisfied for any initial condition $\boldsymbol{w}_j^0$, $j \in Z$, *expressed in the form (3.2.28) with* $k = 0$. *Moreover, we call the scheme weakly von Neumann stable, if*

$$|\boldsymbol{F}_k(n)| \le c(1 + c^*\tau)^k |\boldsymbol{F}_0(n)| \quad \forall n \in Z, \; \forall k \in Z^+, \; \forall h, \tau > 0, \qquad (3.2.30)$$

with constants $c, c^* > 0$ *independent of* n, k, h, τ.

The von Neumann stability criterion is easily applied to the case when a linear scheme is used for the numerical solution of a linear problem (3.2.1)–(3.2.2), when $\boldsymbol{f}(\boldsymbol{w}) = \mathbb{A}\boldsymbol{w}$. Then (3.2.10) (where we write j instead of i) has the form

$$\boldsymbol{w}_j^{k+1} = \sum_{s=-\ell}^{\ell} \left[\mathbb{C}_s \boldsymbol{w}_{j+s}^k + \tilde{\mathbb{C}}_s \boldsymbol{w}_{j+s}^{k+1} \right], \quad j \in Z, \; k \in Z^+, \qquad (3.2.31)$$

where $\mathbb{C}_s, \tilde{\mathbb{C}}_s$ are $m \times m$ matrices (dependent on h and τ). Substituting (3.2.28) into (3.2.31), we get

$$\begin{aligned}
&\sum_{n=-\infty}^{\infty} \boldsymbol{F}_{k+1}(n) \exp(inx) \\
&= \sum_{s=-\ell}^{\ell} \sum_{n=-\infty}^{\infty} \left[\mathbb{C}_s \boldsymbol{F}_k(n) + \tilde{\mathbb{C}}_s \boldsymbol{F}_{k+1}(n)\right] \exp(in(x+sh)) \\
&= \sum_{n=-\infty}^{\infty} \exp(inx) \sum_{s=-\ell}^{\ell} \left[\mathbb{C}_s \boldsymbol{F}_k(n) + \tilde{\mathbb{C}}_s \boldsymbol{F}_{k+1}(n)\right] \exp(insh).
\end{aligned}$$

Due to the linear independence of the system $\{\exp(inx)\}_{n=-\infty}^{+\infty}$, we find that

$$\begin{aligned}
&\left(\mathbb{I} - \sum_{s=-\ell}^{\ell} \tilde{\mathbb{C}}_s \exp(insh)\right) \boldsymbol{F}_{k+1}(n) \\
&= \sum_{s=-\ell}^{\ell} \mathbb{C}_s \exp(insh) \boldsymbol{F}_k(n) \quad \forall n \in Z, \ \forall h, \tau > 0.
\end{aligned} \tag{3.2.32}$$

(Here $\mathbb{I}$ denotes the identity $m \times m$ matrix.) If the matrix

$$\mathbb{I} - \sum_{s=-\ell}^{\ell} \tilde{\mathbb{C}}_s \exp(insh)$$

is nonsingular, then (3.2.32) implies that

$$\boldsymbol{F}_{k+1}(n) = \mathbb{G} \boldsymbol{F}_k(n), \quad n \in Z, \ h, \tau > 0, \tag{3.2.33}$$

where

$$\mathbb{G} = \mathbb{G}(n,h) = \left(\mathbb{I} - \sum_{s=-\ell}^{\ell} \tilde{\mathbb{C}}_s \exp(insh)\right)^{-1} \sum_{s=-\ell}^{\ell} \mathbb{C}_s \exp(insh) \tag{3.2.34}$$

is the so-called *amplification matrix*. In the scalar case ($m = 1$) we speak of an *amplification factor*. Let us notice that (3.2.33)–(3.2.34) can be simply derived, if we use the ansatz

$$w_j^k := \boldsymbol{F}_k(n) \exp(inx) \quad (x = jh), \quad n \in Z, \ k \in Z^+. \tag{3.2.35}$$

Now we can derive the criterion that guarantees the stability condition (3.2.29) or (3.2.30).

Theorem 3.33 *Let the amplification matrix* $\mathbb{G}(n,h)$ *be diagonalizable:*

$$\mathbb{G}(n,h) = \mathbb{T}(n,h)\mathbb{B}(n,h)\mathbb{T}^{-1}(n,h), \tag{3.2.36}$$

where $\mathbb{T}(n,h)$ *and* $\mathbb{T}^{-1}(n,h)$ *are nonsingular matrices with entries uniformly bounded by a constant independent of* n *and* h *and* $\mathbb{B}$ *is a diagonal matrix. (Its diagonal entries are eigenvalues* $\mu_1,\ldots,\mu_n$ *of the matrix* $\mathbb{G}$*.) Then the scheme is von Neumann stable if and only if*

$$\sigma(\mathbb{G}(n,h)) \le 1 \quad \forall n \in Z,\ \forall h > 0, \tag{3.2.37}$$

where $\sigma(\mathbb{G})$ *denotes the spectral radius of the matrix* $\mathbb{G}$*. Further, if there exists a constant* c^* *independent of* n, k, h, τ *and*

$$\sigma(\mathbb{G}(n,h)) \le 1 + c^*\tau, \quad \forall n \in Z,\ \forall h, \tau > 0, \tag{3.2.38}$$

then the scheme is weakly von Neumann stable.

Proof In view of (3.2.33),

$$\boldsymbol{F}_k(n) = \mathbb{G}^k \boldsymbol{F}_0(n), \quad k \in Z^+,\ n \in Z. \tag{3.2.39}$$

By (3.2.36),

$$\mathbb{G}^k = \mathbb{T}\mathbb{B}^k\mathbb{T}^{-1}.$$

If $\mathbb{T} = (t_{ij})_{i,j=1}^m$ and $\mathbb{T}^{-1} = (\vartheta_{ij})_{i,j=1}^m$, then

$$(\mathbb{G}^k)_{ij} = \sum_{r=1}^m t_{ir}\vartheta_{rj}\mu_r^k.$$

From this, (3.2.39) and the boundedness of the entries t_{ij} and ϑ_{ij} we see that conditions (3.2.37) and (3.2.38) imply inequalities (3.2.29) and (3.2.30), respectively. On the other hand, let us assume that $|\mu_r| > 1$ for an eigenvalue μ_r of the matrix $\mathbb{G}$. If we choose $\boldsymbol{F}_0$ as an eigenvector of $\mathbb{G}$ associated with the eigenvalue μ_r, i.e. $\mathbb{G}\boldsymbol{F}_0 = \mu_r\boldsymbol{F}_0$, then

$$\boldsymbol{F}_k = \mu_r^k \boldsymbol{F}_0$$

and $|\boldsymbol{F}_k| = |\mu_r|^k|\boldsymbol{F}_0| \to \infty$ as $k \to \infty$. Hence, (3.2.29) does not hold. □

3.2.7 *Stability in a nonlinear case*

The von Neumann method is usually applied to the investigation of the stability of numerical schemes for the solution of the general nonlinear problem (3.2.1) in a heuristic way based on the application of the von Neumann stability criterion to the linearized system

$$\frac{\partial \boldsymbol{u}}{\partial t} + \mathbb{A}(\boldsymbol{w}_i^k)\frac{\partial \boldsymbol{u}}{\partial x} = 0,\text{[1]} \quad i \in Z,\ k \in Z^+, \tag{3.2.40}$$

where $\mathbb{A}(\boldsymbol{w}) = D\boldsymbol{f}(\boldsymbol{w})/D\boldsymbol{w}$ and $\boldsymbol{w}_i^k$ is the approximate solution of (3.2.1) obtained with the aid of scheme (3.2.8). Then this scheme is used for the approximation of system (3.2.40) and its stability is analysed with the aid of the von

[1]Here and in the sequel, $\boldsymbol{u}, \boldsymbol{v}, \boldsymbol{w}$ will represent elements of the state space $I\!R^m$. It will be clear when the symbol $\boldsymbol{v}$ is used in the sense of the velocity.

Neumann method. If we establish the stability of scheme (3.2.8) applied to system (3.2.40), we say that the method (3.2.8) used for the solution of system (3.2.1) is *linearly von Neumann stable.* Often we speak of *linear L^2-stability* in this case. This approach is also called the *method of frozen coefficients.* Nevertheless, the linear von Neumann stability need not guarantee the stability of the scheme applied to a nonlinear problem and it is necessary to be careful in the use of the linear stability. Later we shall also be concerned with L^∞-stability.

In what follows, some well-known methods for the solution of system (3.2.1) will be derived and analysed.

3.2.8 *Lax–Friedrichs scheme*

One of the widely used methods is the Lax–Friedrichs scheme with the numerical flux

$$\boldsymbol{g}_{\mathrm{LF}}(\boldsymbol{u},\boldsymbol{v}) = \frac{1}{2}(\boldsymbol{f}(\boldsymbol{u})+\boldsymbol{f}(\boldsymbol{v})) - \frac{h}{2\tau}(\boldsymbol{v}-\boldsymbol{u}). \tag{3.2.41}$$

It is used in an explicit form of (3.2.8), i.e. with $\vartheta = 0$:

$$\boldsymbol{w}_i^{k+1} = \frac{1}{2}(\boldsymbol{w}_{i+1}^k + \boldsymbol{w}_{i-1}^k) - \frac{\tau}{2h}(\boldsymbol{f}_{i+1}^k - \boldsymbol{f}_{i-1}^k), \tag{3.2.42}$$

where $\boldsymbol{f}_{i\pm1}^k := \boldsymbol{f}(\boldsymbol{w}_{i\pm1}^k)$. Under the assumption that the exact solution $\boldsymbol{w}$ as well as the flux $\boldsymbol{f}$ have continuous third derivatives, we find that the truncation error is of order $O(\tau+h)$, provided $\tau/h = \lambda = $ const. Actually, then $\boldsymbol{f}(\boldsymbol{w}) \in C^3(\mathbb{R}\times[0,\infty))^m$ and by the Taylor formula we have

$$\begin{aligned}
&\text{a)}\ \ \boldsymbol{w}(x_i,t_{k+1}) = \boldsymbol{w}(x_i,t_k) + \tau\frac{\partial \boldsymbol{w}}{\partial t}(x_i,t_k) + \frac{\tau^2}{2}\frac{\partial^2 \boldsymbol{w}}{\partial t^2}(x_i,t_k) + O(\tau^3), \qquad (3.2.43)\\
&\text{b)}\ \ \boldsymbol{w}(x_{i\pm1},t_k) = \boldsymbol{w}(x_i,t_k) \pm h\frac{\partial \boldsymbol{w}}{\partial x}(x_i,t_k) + \frac{h^2}{2}\frac{\partial^2 \boldsymbol{w}}{\partial x^2}(x_i,t_k) + O(h^3),\\
&\text{c)}\ \ \boldsymbol{f}(\boldsymbol{w}(x_{i\pm1},t_k)) = \boldsymbol{f}(\boldsymbol{w}(x_i,t_k)) \pm h\frac{\partial \boldsymbol{f}(\boldsymbol{w})}{\partial x}(x_i,t_k)\\
&\qquad\qquad + \frac{h^2}{2}\frac{\partial^2 \boldsymbol{f}(\boldsymbol{w})}{\partial x^2}(x_i,t_k) + O(h^3).
\end{aligned}$$

Substituting into (3.2.42) and using equation (3.2.1) at the point (x_i,t_k), we get

$$\begin{aligned}
\boldsymbol{w}(x_i,t_{k+1}) = {} & \frac{1}{2}\left(\boldsymbol{w}(x_{i+1},t_k) + \boldsymbol{w}(x_{i-1},t_k)\right)\\
& - \frac{\tau}{2h}\left(\boldsymbol{f}(\boldsymbol{w}(x_{i+1},t_k)) - \boldsymbol{f}(\boldsymbol{w}(x_{i-1},t_k))\right) + \tau O(\tau+h).
\end{aligned}$$

Hence, the accuracy of the Lax–Friedrichs scheme is of first order in time and space.

3.2.9 *Stability of the Lax–Friedrichs scheme*

Let us investigate the linear von Neumann stability of the Lax–Friedrichs scheme.

a) First we consider a *linear scalar equation*

$$\frac{\partial w}{\partial t} + a\frac{\partial w}{\partial x} = 0 \quad \text{in } I\!R \times (0, \infty) \tag{3.2.44}$$

with constant $a \in I\!R$. Then scheme (3.2.42) reads

$$w_j^{k+1} = (w_{j+1}^k + w_{j-1}^k)/2 - a\lambda(w_{j+1}^k - w_{j-1}^k)/2, \tag{3.2.45}$$

where $\lambda := \tau/h$. Substituting the ansatz $w_j^k = F_k \exp(injh)$ $(i^2 = -1)$ into (3.2.45), we get

$$\begin{aligned} F_{k+1} &= F_k \left\{[\exp(inh) + \exp(-inh)]/2 - a\lambda[\exp(inh) - \exp(-inh)]/2\right\} \\ &= F_k \left[\cos nh - ia\lambda \sin nh\right]. \end{aligned}$$

The amplification factor $G = \cos nh - ia\lambda \sin nh$ satisfies the stability condition

$$|G|^2 = \cos^2 nh + a^2\lambda^2 \sin^2 nh = 1 - \sin^2 nh(1 - a^2\lambda^2) \le 1 \tag{3.2.46}$$

for each $h > 0$ and $n \in Z$ if and only if $|a|\lambda = |a|\tau/h \le 1$, i.e.

$$\tau \le h/|a|. \tag{3.2.47}$$

This inequality is called the *Courant–Friedrichs–Lewy* (CFL) *stability condition.* Due to the bound imposed on the time step by the CFL condition, we say that the Lax–Friedrichs scheme is *conditionally stable.*

b) Now let us consider a linear system

$$\frac{\partial \boldsymbol{w}}{\partial t} + \mathbb{A}\frac{\partial \boldsymbol{w}}{\partial x} = 0 \quad \text{in } I\!R \times (0, \infty), \tag{3.2.48}$$

where $\mathbb{A}$ is an $m \times m$ diagonalizable matrix with real eigenvalues: $\mathbb{A} = \mathbb{T}\mathbb{\Lambda}\mathbb{T}^{-1}$, $\mathbb{\Lambda} = \operatorname{diag}(\lambda_1, \dots, \lambda_m)$, $\lambda_1, \dots, \lambda_m \in I\!R$. Then, similarly as above, the ansatz $\boldsymbol{w}_j^k = \boldsymbol{F}_k \exp(injh)$ $(i^2 = -1)$ substituted into (3.2.42) (where we write j instead of i) leads to the relation

$$\boldsymbol{F}_{k+1} = \mathbb{G}\boldsymbol{F}_k \tag{3.2.49}$$

with the amplification matrix

$$\mathbb{G} = \cos nh\mathbb{I} - i\lambda \sin nh\mathbb{A}. \tag{3.2.50}$$

Using the diagonalization of the matrix $\mathbb{A}$, we find that

$$\begin{aligned} \mathbb{G} &= \mathbb{T}\mathbb{B}\mathbb{T}^{-1}, \\ \mathbb{B} &= \cos nh\mathbb{I} - i\lambda \sin nh\mathbb{\Lambda}. \end{aligned} \tag{3.2.51}$$

This implies that $\mathbb{G}$ and $\mathbb{B}$ have the same eigenvalues

$$\mu_j = \cos nh - i\lambda\lambda_j \sin nh, \quad j = 1, \dots, m. \tag{3.2.52}$$

Since the matrix $\mathbb{T}$ is independent of n, h, Theorem 3.33 can be applied. We get the stability of the scheme under the condition $|\mu_j| \le 1$, $j = 1, \dots, m$, i.e.

$\tau \le h/|\lambda_j|$, $j = 1, \dots, m$ (cf. (3.2.46)–(3.2.47)). Hence, in this case we have obtained the CFL stability condition in the form

$$\tau \le h/\sigma(\mathbb{A}), \tag{3.2.53}$$

where $\sigma(\mathbb{A})$ is the spectral radius of the matrix $\mathbb{A}$. Using the well-known estimate $\sigma(\mathbb{A}) \le \|\mathbb{A}\|_\infty = \max_{i=1,\dots,m} \sum_{j=1}^{m} |a_{ij}|$ for $\mathbb{A} = (a_{ij})_{i,j=1}^{m}$, we easily get a simple sufficient condition for the von Neumann stability in the form

$$\tau \le h/\|\mathbb{A}\|_\infty. \tag{3.2.54}$$

c) Finally, if the nonlinear system (3.2.1) is solved by the Lax–Friedrichs scheme, the linear von Neumann stability condition has the form

$$\tau \le h/\sigma(\mathbb{A}(\boldsymbol{w}_i^k)), \quad i \in Z,\ k \in Z^+,$$

as follows from the considerations in b). Usually, at each time level t_k, $k \in Z^+$, one uses in general a different time step τ_k satisfying the condition

$$\tau_k \le h/\sigma(\mathbb{A}(\boldsymbol{w}_i^k)), \quad i \in Z. \tag{3.2.55}$$

Then we set $t_{k+1} = t_k + \tau_k$. Very often, in order to suppress the effect of a possible nonlinear instability, we use the time step τ_k following from the more restrictive condition

$$\tau_k \le \mathrm{CFL}h/\sigma(\mathbb{A}(\boldsymbol{w}_i^k)), \quad i \in Z, \tag{3.2.56}$$

where CFL $\in (0, 1)$ is a suitable number.

Remark 3.34 The Lax–Friedrichs scheme can be quite easily obtained from the simple scheme of the form

$$\boldsymbol{w}_i^{k+1} = \boldsymbol{w}_i^k - \frac{\tau}{2h}\left(\boldsymbol{f}(\boldsymbol{w}_{i+1}^k) - \boldsymbol{f}(\boldsymbol{w}_{i-1}^k)\right), \tag{3.2.57}$$

with the aid of the approximation $\boldsymbol{w}(x_i, t_k) \approx (\boldsymbol{w}(x_{i+1}, t_k) + \boldsymbol{w}(x_{i-1}, t_k))/2$. Scheme (3.2.57) has the truncation error of order $O(\tau + h^2)$ under the assumption that $\partial^2\boldsymbol{w}/\partial t^2$, $\partial^3\boldsymbol{w}/\partial x^3$ and the third derivatives of $\boldsymbol{f}$ are continuous. However, the von Neumann stability criterion (applied to a linearized equation) is not satisfied for this scheme, even if $\tau > 0$ is chosen arbitrarily small. We say that scheme (3.2.57) is unconditionally unstable. On the other hand, we can show as an exercise that the implicit version to (3.2.57)

$$\boldsymbol{w}_i^{k+1} = \boldsymbol{w}_i^k - \frac{\tau}{2h}\left(\boldsymbol{f}(\boldsymbol{w}_{i+1}^{k+1}) - \boldsymbol{f}(\boldsymbol{w}_{i-1}^{k+1})\right) \tag{3.2.58}$$

is linearly von Neumann stable for arbitrary $h, \tau > 0$. Schemes with this property are called unconditionally stable. However, the transition from t_k to t_{k+1} requires the solution of a nonlinear algebraic system. Therefore, explicit schemes are usually preferred.

Now let us consider the scalar equation

$$\frac{\partial w}{\partial t} + \frac{\partial f(w)}{\partial x} = 0 \tag{3.2.59}$$

and assume that $f \in C^3(I\!R)$ and let $w \in C^3(I\!R \times [0,\infty))$ be its classical solution. We set $a = f'$. The order of accuracy of the Lax–Friedrichs scheme can be treated in the following way.

Theorem 3.35 *The Lax–Friedrichs scheme* (3.2.42) *with* $\lambda = \tau/h = \text{const}$ *approximates the equation*

$$\frac{\partial w}{\partial t} + \frac{\partial f(w)}{\partial x} = \frac{h}{2}\frac{\partial}{\partial x}\left[(1/\lambda - \lambda a^2(w))\frac{\partial w}{\partial x}\right] \tag{3.2.60}$$

with a truncation error of second order in time and space.

Proof We shall use relations (3.2.43). Moreover, under the notation

$$\begin{aligned} &w_x = \partial w/\partial x, \quad w_t = \partial w/\partial t, \quad f(w)_x = \partial f(w)/\partial x,\\ &w_{xx} = \partial^2 w/\partial x^2, \text{ etc.,}\end{aligned}$$

we can write

$$\begin{aligned} f(w)_x &= a(w)w_x,\\ f(w)_{xt} &= a'(w)w_t w_x + a(w)w_{xt},\\ f(w)_{xx} &= a'(w)w_x^2 + a(w)w_{xx}.\end{aligned}$$

From (3.2.59) we successively get

$$\begin{aligned} w_t &= -a(w)w_x,\\ w_{tx} &+ f(w)_{xx} = 0, \quad w_{tt} + f(w)_{xt} = 0,\\ w_{xt} &= w_{tx} = -a'(w)w_x^2 - a(w)w_{xx},\\ w_{tt} &= -a'(w)w_t w_x - a(w)w_{xt}\\ &= a(w)(2a'(w)w_x^2 + a(w)w_{xx}) = (a^2(w)w_x)_x.\end{aligned}$$

Now let us substitute the exact solution w into scheme (3.2.42) and use (3.2.43). We obtain

$$\begin{aligned} &\tau w_t(x_i,t_k) + \frac{1}{2}\tau^2 w_{tt}(x_i,t_k) + O(\tau^3)\\ &= \frac{1}{2}h^2 w_{xx}(x_i,t_k) + O(h^3) - \lambda(hf(w)_x(x_i,t_k) + O(h^3)).\end{aligned}$$

This and the above relations yield

$$\begin{aligned} (w_t + f(w)_x)(x_i,t_k) &= h(w_{xx}(x_i,t_k)/\lambda - \lambda w_{tt}(x_i,t_k))/2 + O(\tau^2 + h^2)\\ &= \frac{h}{2}\left[(1/\lambda - \lambda a^2(w))w_x\right]_x (x_i,t_k) + O(\tau^2 + h^2),\end{aligned}$$

which we wanted to prove. □

Remark 3.36 Note that equation (3.2.60) is of the parabolic type and admits a well-posed initial value problem provided the right-hand side of (3.2.60) has the character of a diffusion term.[2] This is true, provided $1/\lambda - \lambda a^2(w) \geq 0$, i.e. $|a(w)|\tau/h \leq 1$. In this way we again obtain the CFL condition – see (3.2.47). The right-hand side of (3.2.60) is called the numerical viscosity or numerical dissipation. Its presence causes dissipation which is the reason that the Lax–Friedrichs scheme gives physically admissible solutions.

3.2.10 *Lax–Wendroff scheme*

Let us assume that $\boldsymbol{w}$ and $\boldsymbol{f}$ have continuous fourth order derivatives. Then the same is true for $\boldsymbol{f}(\boldsymbol{w})$. By the Taylor formula

$$\boldsymbol{w}(x_i, t_{k+1}) = \boldsymbol{w}(x_i, t_k) + \tau \frac{\partial \boldsymbol{w}(x_i, t_k)}{\partial t} + \frac{\tau^2}{2} \frac{\partial^2 \boldsymbol{w}(x_i, t_k)}{\partial t^2} + O(\tau^3). \tag{3.2.61}$$

Further, from (3.2.1) it follows that

$$\begin{aligned} \frac{\partial^2 \boldsymbol{w}}{\partial t^2} &= -\frac{\partial^2 \boldsymbol{f}(\boldsymbol{w})}{\partial x \partial t} = -\frac{\partial}{\partial x}\left(\frac{\partial \boldsymbol{f}(\boldsymbol{w})}{\partial t}\right) \\ &= -\frac{\partial}{\partial x}\left(\frac{D\boldsymbol{f}(\boldsymbol{w})}{D\boldsymbol{w}}\frac{\partial \boldsymbol{w}}{\partial t}\right) = \frac{\partial}{\partial x}\left(\mathbb{A}(\boldsymbol{w})\frac{\partial \boldsymbol{f}(\boldsymbol{w})}{\partial x}\right), \end{aligned} \tag{3.2.62}$$

where $\mathbb{A}(\boldsymbol{w}) = D\boldsymbol{f}(\boldsymbol{w})/D\boldsymbol{w}$. The application of the Taylor formula yields the approximations

$$\frac{\partial \boldsymbol{f}(\boldsymbol{w})}{\partial x}(x_i, t_k) = \frac{\boldsymbol{f}(\boldsymbol{w}(x_{i+1}, t_k)) - \boldsymbol{f}(\boldsymbol{w}(x_{i-1}, t_k))}{2h} + O(h^2), \tag{3.2.63}$$

$$\begin{aligned} \frac{\partial}{\partial x}\left(\mathbb{A}(\boldsymbol{w})\frac{\partial \boldsymbol{f}(\boldsymbol{w})}{\partial x}\right)(x_i, t_k) &= \frac{1}{h^2}\left\{\mathbb{A}\left(\frac{\boldsymbol{w}(x_{i+1}, t_k) + \boldsymbol{w}(x_i, t_k)}{2}\right)\boldsymbol{f}(\boldsymbol{w}(x_{i+1}, t_k))\right. \\ &- \left[\mathbb{A}\left(\frac{\boldsymbol{w}(x_{i+1}, t_k) + \boldsymbol{w}(x_i, t_k)}{2}\right) + \mathbb{A}\left(\frac{\boldsymbol{w}(x_i, t_k) + \boldsymbol{w}(x_{i-1}, t_k)}{2}\right)\right]\boldsymbol{f}(\boldsymbol{w}(x_i, t_k)) \\ &\left. + \mathbb{A}\left(\frac{\boldsymbol{w}(x_i, t_k) + \boldsymbol{w}(x_{i-1}, t_k)}{2}\right)\boldsymbol{f}(\boldsymbol{w}(x_{i-1}, t_k))\right\} + O(h^2). \end{aligned}$$

Now using equation (3.2.1), relations (3.2.61)–(3.2.63) and approximating the exact solution $\boldsymbol{w}(x_\alpha, t_\beta)$ by the approximate values $\boldsymbol{w}_\alpha^\beta$, we obtain the Lax–Wendroff scheme

$$\boldsymbol{w}_i^{k+1} = \boldsymbol{w}_i^k - \lambda(\boldsymbol{f}_{i+1}^k - \boldsymbol{f}_{i-1}^k)/2 \tag{3.2.64}$$

[2]Note that, in general, the diffusion term has the form $\mathrm{div}(\mathbb{K}\nabla u)$, where $\mathbb{K}$ is a positive definite matrix.

$$+\lambda^2\left[\mathbb{A}^k_{i+\frac{1}{2}}(\boldsymbol{f}^k_{i+1}-\boldsymbol{f}^k_i)-\mathbb{A}^k_{i-\frac{1}{2}}(\boldsymbol{f}^k_i-\boldsymbol{f}^k_{i-1})\right]\Big/2,$$

where

$$\lambda=\tau/h,\quad \boldsymbol{f}^k_i=\boldsymbol{f}(\boldsymbol{w}^k_i),\quad \mathbb{A}^k_{i+\frac{1}{2}}=\mathbb{A}((\boldsymbol{w}^k_{i+1}+\boldsymbol{w}^k_i)/2). \tag{3.2.65}$$

The Lax–Wendroff scheme can be written in the form (3.2.8) with $\vartheta=0$ and the numerical flux

$$\boldsymbol{g}_{\mathrm{LW}}(\boldsymbol{u},\boldsymbol{v})=(\boldsymbol{f}(\boldsymbol{u})+\boldsymbol{f}(\boldsymbol{v}))/2-\lambda\mathbb{A}((\boldsymbol{u}+\boldsymbol{v})/2)(\boldsymbol{f}(\boldsymbol{v})-\boldsymbol{f}(\boldsymbol{u}))/2. \tag{3.2.66}$$

It is a *three-point scheme* and has *second order of accuracy* in space and time.

3.2.11 *Stability of the Lax–Wendroff scheme*

Let us apply the von Neumann method to the investigation of the stability of scheme (3.2.64) used for the solution of the linear scalar equation (3.2.59). Then scheme (3.2.64) reads

$$w^{k+1}_j=w^k_j-\lambda a(w^k_{j+1}-w^k_{j-1})/2+\lambda^2a^2(w^k_{j+1}-2w^k_j+w^k_{j-1})/2,\quad j\in Z. \tag{3.2.67}$$

The representation $w^k_j=F_k\exp(injh)$ ($i^2=-1$) leads to

$$\begin{aligned}&F_{k+1}\exp(injh)\\&=F_k\Big\{\exp(injh)-\lambda a[\exp(in(j+1)h)-\exp(in(j-1)h)]/2\\&\quad+\lambda^2a^2[\exp(in(j+1)h)-2\exp(injh)+\exp(in(j-1)h)]/2\Big\}\end{aligned}$$

and, hence, $F_{k+1}=GF_k$, where

$$G=1-i\lambda a\sin nh+\lambda^2a^2(\cos nh-1).$$

Now we apply Theorem 3.33. Since

$$\begin{aligned}|G|^2&=[1+\lambda^2a^2(\cos nh-1)]^2+\lambda^2a^2\sin^2 nh\\&=1-\lambda^2a^2(2-2\cos nh-\lambda^2a^2+2\lambda^2a^2\cos nh-\lambda^2a^2\cos^2 nh-sin^2nh)\\&=1-\lambda^2a^2(1-\lambda^2a^2)(1-\cos nh)^2,\end{aligned}$$

it is obvious that $|G|\le 1$ if and only if $1-\lambda^2a^2\ge 0$, i.e. $\lambda|a|\le 1$, which is the CFL stability condition (3.2.47). Hence, the Lax–Wendroff scheme is conditionally stable.

Exercise 3.37 Using the results from 4.2.10 and a similar approach as in part b) of Section 3.2.9, show that the stability of the Lax–Wendroff method (3.2.64) applied to the linear system (3.2.48) is equivalent to condition (3.2.53).

In what follows we will explain a widely used class of *methods of the Godunov type*.

3.2.12 *The Godunov method*

The methods of the Godunov type are based on the approximate solution of the *Riemann problem*

$$\frac{\partial \boldsymbol{w}}{\partial t} + \frac{\partial \boldsymbol{f}(\boldsymbol{w})}{\partial x} = 0, \quad x \in I\!R,\ t > 0, \tag{3.2.68}$$

$$\boldsymbol{w}(x,0) = \begin{cases} \boldsymbol{w}_L, & x < 0 \\ \boldsymbol{w}_R, & x > 0. \end{cases} \tag{3.2.69}$$

According to Theorem 2.32, this problem has a unique weak entropy solution

$$\boldsymbol{w}(x,t) = \boldsymbol{w}_{\mathrm{RS}}(x/t; \boldsymbol{w}_L, \boldsymbol{w}_R) \tag{3.2.70}$$

for some couples of initial states $\boldsymbol{w}_L, \boldsymbol{w}_R \in I\!R^m$. Then *the Godunov method* ((Godunov, 1959)) is defined as the scheme (3.2.8) with $\vartheta = 0$ and the numerical flux $\boldsymbol{g} = \boldsymbol{g}_{\mathrm{G}}$, where

$$\boldsymbol{g}_{\mathrm{G}}(\boldsymbol{u}, \boldsymbol{v}) = \boldsymbol{f}(\boldsymbol{w}_{\mathrm{RS}}(0; \boldsymbol{u}, \boldsymbol{v})). \tag{3.2.71}$$

Hence, the Godunov method reads

$$\boldsymbol{w}_i^{k+1} = \boldsymbol{w}_i^k - \frac{\tau}{h}\left[\boldsymbol{f}(\boldsymbol{w}_{\mathrm{RS}}(0; \boldsymbol{w}_i^k, \boldsymbol{w}_{i+1}^k)) - \boldsymbol{f}(\boldsymbol{w}_{\mathrm{RS}}(0; \boldsymbol{w}_{i-1}^k, \boldsymbol{w}_i^k))\right]. \tag{3.2.72}$$

The function $\boldsymbol{g}_{\mathrm{G}}$ is called the *Godunov numerical flux* or the *exact Riemann solver.*

The Godunov scheme assumes the construction of an exact solution of the Riemann problem, which is in general difficult. This drawback is avoided by the use of an *approximation* of the Godunov numerical flux, called an *approximate Riemann solver* or a *Riemann numerical flux.* We will denote it by $\boldsymbol{g}_{\mathrm{R}}(\boldsymbol{u}, \boldsymbol{v})$. The resulting methods are called the *methods of the Godunov type.*

Now we shall be concerned with the derivation of the methods of the Godunov type in several special cases.

3.2.13 *Riemann solver for a scalar equation*

First let us consider a linear scalar Riemann problem

$$\frac{\partial w}{\partial t} + a\frac{\partial w}{\partial x} = 0, \quad x \in I\!R,\ t > 0, \tag{3.2.73}$$

$$w(x,0) = \begin{cases} w_L, & x < 0 \\ w_R, & x > 0, \end{cases}$$

with $a = \text{const} \in I\!R$, $w_L, w_R \in I\!R$. Hence, the flux $f(w) = aw$. Then the unique weak solution (obtained with the aid of the method of characteristics) has the form

$$w(x,t) = \begin{cases} w_L, & x - at < 0 \\ w_R, & x - at > 0, \end{cases} \tag{3.2.74}$$

and $f(w(0,t)) = a^+ w_L + a^- w_R$, where $a^+ = \max(a,0)$, $a^- = \min(a,0)$. We see that in this case we obtain the Godunov numerical flux

$$g_{\mathrm{R}}(u,v) = g_{\mathrm{G}}(u,v) = a^+ u + a^- v = \{a(u+v) - |a|(v-u)\}/2. \tag{3.2.75}$$

This can be written in the form

$$g_{\mathrm{R}}(u,v) = f^+(u) + f^-(v), \tag{3.2.76}$$

where $f^+(w) = a^+ w,\ f^-(w) = a^- w$. Hence,

$$(f^+)' \geq 0, \quad (f^-)' \leq 0. \tag{3.2.77}$$

This motivates the *flux splitting method* for a general nonlinear scalar problem in which the flux f is split into *forward* and *backward flux* f^+ and f^-, respectively,

$$f(w) = f^+(w) + f^-(w), \tag{3.2.78}$$

satisfying (3.2.77). Then the approximate Riemann solver is defined in the form (3.2.76). On the basis of these considerations we come to the following method.

3.2.14 *Engquist–Osher scheme*

Let us set $a(u) = f'(u)$ and

$$\begin{aligned} f^+(u) &= \frac{1}{2} f(0) + \int_0^u a^+(q)\, dq, \\ f^-(u) &= \frac{1}{2} f(0) + \int_0^u a^-(q)\, dq. \end{aligned} \tag{3.2.79}$$

Since $a = a^+ + a^-$, $a^+ \geq 0$, $a^- \leq 0$ and

$$f(u) = f(0) + \int_0^u a(q)\, dq,$$

conditions (3.2.78) and (3.2.77) are satisfied. Then, substituting (3.2.79) into (3.2.76), we obtain the Engquist–Osher numerical flux, which, by virtue of $|a| = a^+ - a^-$, can be written in the following forms:

$$\begin{aligned} g_{\mathrm{EO}}(u,v) &= f(0) + \int_0^u a^+(q)\, dq + \int_0^v a^-(q)\, dq \\ &= f(u) + \int_u^v a^-(q)\, dq \end{aligned} \tag{3.2.80}$$

$$= f(v) - \int_u^v a^+(q)\,dq$$
$$= \frac{1}{2}\left\{f(u) + f(v) - \int_u^v |a(q)|\,dq\right\}.$$

With the aid of this numerical flux we obtain from (3.2.8), where $\vartheta := 0$, the three-point scheme

$$w_i^{k+1} = w_i^k - \frac{\lambda}{2}\left(f(w_{i+1}^k) - f(w_{i-1}^k)\right) + \frac{\lambda}{2}\left(\int_{w_i^k}^{w_{i+1}^k} |a(q)|\,dq - \int_{w_{i-1}^k}^{w_i^k} |a(q)|\,dq\right) \tag{3.2.81}$$

with $\lambda = \tau/h$.

If this scheme is applied to the linear equation (3.2.73), it becomes

$$w_i^{k+1} = w_i^k - \lambda a \begin{cases} (w_i^k - w_{i-1}^k), & \text{if } a > 0, \\ (w_{i+1}^k - w_i^k), & \text{if } a < 0. \end{cases}$$

We see that this scheme is obtained if the time derivative is approximated by the forward difference

$$\frac{\partial w}{\partial t}(x_i, t_k) \approx \frac{w_i^{k+1} - w_i^k}{\tau},$$

while the space derivative is approximated by the backward difference

$$\frac{\partial w}{\partial x}(x_i, t_k) \approx (w_i^k - w_{i-1}^k)/h, \quad \text{if } a > 0,$$

or the forward difference

$$\frac{\partial w}{\partial x}(x_i, t_k) \approx (w_{i+1}^k - w_i^k)/h, \quad \text{if } a < 0.$$

This means that we use the finite difference oriented against the direction of the speed a of the propagation of disturbances. This approach, called *upwinding*, is often used in CFD.

3.2.15 *Riemann numerical flux for a linear system*

Let us consider the Riemann problem

$$\text{a)} \quad \frac{\partial \boldsymbol{w}}{\partial t} + \mathbb{A}\frac{\partial \boldsymbol{w}}{\partial x} = 0, \quad x \in I\!R,\ t > 0, \tag{3.2.82}$$

$$\text{b)}\ \ \boldsymbol{w}(x,0) = \begin{cases} \boldsymbol{w}_L, & x < 0, \\ \boldsymbol{w}_R, & x > 0, \end{cases}$$

where $\mathbb{A}$ is an $m \times m$ diagonalizable matrix with real eigenvalues $\lambda_1, \dots, \lambda_m$ and $\boldsymbol{w}_L, \boldsymbol{w}_R \in I\!R^m$ are constant states. Hence, there exists a nonsingular matrix $\mathbb{T}$ such that

$$\mathbb{A} = \mathbb{T}\mathbb{\Lambda}\mathbb{T}^{-1}, \quad \mathbb{\Lambda} = \operatorname{diag}(\lambda_1, \dots, \lambda_m). \tag{3.2.83}$$

First we solve the Riemann problem (3.2.82). Obviously, the columns of $\mathbb{T}$, let us denote them by $\boldsymbol{r}_1, \dots, \boldsymbol{r}_m$, are the eigenvectors of the matrix $\mathbb{A} : \mathbb{A}\boldsymbol{r}_s = \lambda_s \boldsymbol{r}_s$, and form a basis in $I\!R^m$. Hence, we can write

$$\boldsymbol{w}(x,t) = \sum_{s=1}^{m} \mu_s(x,t)\boldsymbol{r}_s, \quad x \in I\!R,\ t \geq 0, \tag{3.2.84}$$

$$\boldsymbol{w}_L = \sum_{s=1}^{m} \alpha_s \boldsymbol{r}_s, \quad \boldsymbol{w}_R = \sum_{s=1}^{m} \beta_s \boldsymbol{r}_s. \tag{3.2.85}$$

Substituting (3.2.84) into (3.2.82), a) we get

$$0 = \sum_{s=1}^{m} \left(\frac{\partial \mu_s}{\partial t} + \lambda_s \frac{\partial \mu_s}{\partial x} \right) \boldsymbol{r}_s,$$

which holds if and only if

$$\frac{\partial \mu_s}{\partial t} + \lambda_s \frac{\partial \mu_s}{\partial x} = 0, \quad x \in I\!R,\ t > 0,\ s = 1, \dots, m. \tag{3.2.86}$$

This equation is equipped with the initial condition following from (3.2.82), b) and (3.2.85):

$$\mu_s(x,0) = \begin{cases} \alpha_s, & x < 0, \\ \beta_s, & x > 0. \end{cases} \tag{3.2.87}$$

In virtue of (3.2.74), problem (3.2.86)–(3.2.87) has a unique weak solution in the form

$$\mu_s(x,t) = \begin{cases} \alpha_s, & x - \lambda_s t < 0, \\ \beta_s, & x - \lambda_s t > 0, \end{cases} \quad s = 1, \dots, m. \tag{3.2.88}$$

On the basis of these results we can already construct the solution of problem (3.2.82). Let us assume that $-\infty = \lambda_0 < \lambda_1 \leq \lambda_2 \leq \dots \leq \lambda_m < \lambda_{m+1} = +\infty$ and let $\lambda_s < \lambda_{s+1}$ for some $s \geq 0$. From (3.2.84) and (3.2.88) it follows that $\boldsymbol{w}$ is constant in the domain $\Omega_s = \{(x,t); t > 0, \lambda_s < x/t < \lambda_{s+1}\}$ (see Fig. 3.2.1):

$$\boldsymbol{w}(x,t) = \boldsymbol{w}_s, \quad \text{if } (x,t) \in \Omega_s, \tag{3.2.89}$$

$$\boldsymbol{w}_s = \sum_{n=1}^{s} \beta_n \boldsymbol{r}_n + \sum_{n=s+1}^{m} \alpha_n \boldsymbol{r}_n, \quad s = 0, \dots, m.$$

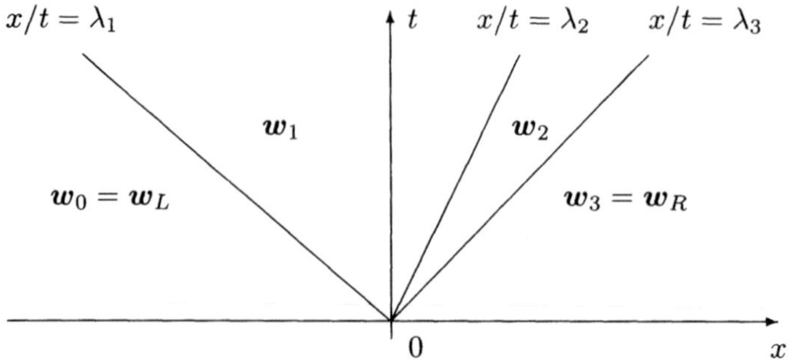

FIG. 3.2.1. Solution of the linear Riemann problem for $m = 3$

In order to verify that the vector function (3.2.89) is really a weak solution of problem (3.2.82), we use Theorem 2.15. Obviously, since $\boldsymbol{w}$ is constant in each domain Ω_s, it satisfies here system (3.2.82), a). Now it is sufficient to verify the Rankine–Hugoniot conditions (2.3.13) on every discontinuity $x/t = \lambda_s$, $s = 1, \dots, m$, propagating with the speed $s_s = \lambda_s$. Actually we have

$$\boldsymbol{w}_s - \boldsymbol{w}_{s-1} = (\beta_s - \alpha_s)\boldsymbol{r}_s$$

and, hence,

$$\mathbb{A}(\boldsymbol{w}_s - \boldsymbol{w}_{s-1}) = (\beta_s - \alpha_s)\lambda_s \boldsymbol{r}_s = s_s(\boldsymbol{w}_s - \boldsymbol{w}_{s-1}),$$

which we wanted to prove. Finally we can construct the Godunov numerical flux $\boldsymbol{g}_G(\boldsymbol{w}_L, \boldsymbol{w}_R) = \boldsymbol{f}(\boldsymbol{w}(0,t)) = \mathbb{A}\boldsymbol{w}(0,t)$, where $\boldsymbol{w}$ is expressed in the form (3.2.89). To this end, we define the matrices

$$\mathbb{\Lambda}^{\pm} = \operatorname{diag}(\lambda_1^{\pm}, \dots, \lambda_m^{\pm}), \quad |\mathbb{\Lambda}| = \operatorname{diag}(|\lambda_1|, \dots, |\lambda_m|) \tag{3.2.90}$$

and

$$\mathbb{A}^{\pm} = \mathbb{T}\mathbb{\Lambda}^{\pm}\mathbb{T}^{-1}, \quad |\mathbb{A}| = \mathbb{T}|\mathbb{\Lambda}|\mathbb{T}^{-1}, \tag{3.2.91}$$

where $\mathbb{T}$ is the matrix from (3.2.83). Thus,

$$\mathbb{A} = \mathbb{A}^{+} + \mathbb{A}^{-}, \quad |\mathbb{A}| = \mathbb{A}^{+} - \mathbb{A}^{-}. \tag{3.2.92}$$

Obviously the matrices $\mathbb{A}^{\pm}$ have eigenvalues $\lambda_s^{\pm}$ associated with eigenvectors $\boldsymbol{r}_s$, $s = 1, \dots, m$. Let $\lambda_1 \le \lambda_2 \le \cdots \le \lambda_s \le 0 < \lambda_{s+1} \le \cdots \le \lambda_m$. Then, we can write

$$\mathbb{A}^{+}\boldsymbol{w}_L = \mathbb{A}^{+}\left(\sum_{n=1}^{m} \alpha_n \boldsymbol{r}_n\right) = \sum_{n=1}^{m} \alpha_n \lambda_n^{+} \boldsymbol{r}_n = \sum_{n=s+1}^{m} \alpha_n \lambda_n \boldsymbol{r}_n, \tag{3.2.93}$$

$$\mathbb{A}^- \boldsymbol{w}_R = \mathbb{A}^- \left(\sum_{n=1}^{m} \beta_n \boldsymbol{r}_n \right) = \sum_{n=1}^{m} \beta_n \lambda_n^- \boldsymbol{r}_n = \sum_{n=1}^{s} \beta_n \lambda_n \boldsymbol{r}_n.$$

Moreover, in view of (3.2.89),

$$\boldsymbol{f}(\boldsymbol{w}(0,t)) = \mathbb{A}\boldsymbol{w}(0,t) = \mathbb{A}\boldsymbol{w}_s = \sum_{n=1}^{s} \beta_n \lambda_n \boldsymbol{r}_n + \sum_{n=s+1}^{m} \alpha_n \lambda_n \boldsymbol{r}_n. \tag{3.2.94}$$

From (3.2.93) and (3.2.94) it follows that the Godunov numerical flux (i.e. the exact Riemann solver) can be expressed in the form (writing $\boldsymbol{u} := \boldsymbol{w}_L$, $\boldsymbol{v} := \boldsymbol{w}_R$)

$$\boldsymbol{g}_{\mathrm{R}}(\boldsymbol{u}, \boldsymbol{v}) = \boldsymbol{g}_{\mathrm{G}}(\boldsymbol{u}, \boldsymbol{v}) = \mathbb{A}^+ \boldsymbol{u} + \mathbb{A}^- \boldsymbol{v}. \tag{3.2.95}$$

This can be rewritten in the form

$$\boldsymbol{g}_{\mathrm{R}}(\boldsymbol{u}, \boldsymbol{v}) = \{\mathbb{A}\boldsymbol{u} + \mathbb{A}\boldsymbol{v} - |\mathbb{A}|(\boldsymbol{v} - \boldsymbol{u})\}/2. \tag{3.2.96}$$

We see that (3.2.95) and (3.2.96) represent the multidimensional analogy to (3.2.75).

By (3.2.92) we have in (3.2.95)

$$\mathbb{A}^+ \boldsymbol{u} + \mathbb{A}^- \boldsymbol{v} = \mathbb{A}\boldsymbol{u} + \mathbb{A}^-(\boldsymbol{v} - \boldsymbol{u}). \tag{3.2.97}$$

The Godunov numerical flux can be expressed in the form

$$\boldsymbol{g}_{\mathrm{R}}(\boldsymbol{u}, \boldsymbol{v}) = \mathbb{A}\boldsymbol{u} + \int_{\boldsymbol{u}}^{\boldsymbol{v}} \mathbb{A}^- \, d\boldsymbol{w}, \tag{3.2.98}$$

where

$$\int_{\boldsymbol{u}}^{\boldsymbol{v}} \mathbb{A}^- \, d\boldsymbol{w} = \left(\int_{\boldsymbol{u}}^{\boldsymbol{v}} \boldsymbol{a}_1^- \, d\boldsymbol{w}, \ldots, \int_{\boldsymbol{u}}^{\boldsymbol{v}} \boldsymbol{a}_m^- \, d\boldsymbol{w} \right)^{\mathrm{T}}. \tag{3.2.99}$$

In (3.2.99) $\boldsymbol{a}_n^-$ denotes the n-th row of the matrix $\mathbb{A}^-$, $n = 1, \ldots, m$. By

$$\int_{\boldsymbol{u}}^{\boldsymbol{v}} \boldsymbol{b}(\boldsymbol{w}) \, d\boldsymbol{w} \tag{3.2.100}$$

we denote the curvilinear integral of the vector function $\boldsymbol{b} : \mathbb{R}^m \to \mathbb{R}^m$ along a curve in $\mathbb{R}^m$ with the initial and terminal points $\boldsymbol{u}$ and $\boldsymbol{v}$, respectively. As the vector field $\boldsymbol{a}_n^-$, $n = 1, \ldots, m$, is constant, it is potential and the integral in (3.2.98) is path independent and does not depend on the choice of a curve connecting $\boldsymbol{u}$ and $\boldsymbol{v}$.

Exercise 3.38 Show that the Godunov numerical flux for the linear system can be expressed in the form

$$\boldsymbol{g}_{\mathrm{R}}(\boldsymbol{u}, \boldsymbol{v}) = \mathbb{A}\boldsymbol{v} - \int_{\boldsymbol{u}}^{\boldsymbol{v}} \mathbb{A}^+ \, d\boldsymbol{w} \tag{3.2.101}$$

or

$$\boldsymbol{g}_{\mathrm{R}}(\boldsymbol{u}, \boldsymbol{v}) = \frac{1}{2} \left\{ \mathbb{A}\boldsymbol{u} + \mathbb{A}\boldsymbol{v} - \int_{\boldsymbol{u}}^{\boldsymbol{v}} |\mathbb{A}| \, d\boldsymbol{w} \right\}. \tag{3.2.102}$$

3.2.16 *Riemann solver for a nonlinear hyperbolic system*

In the case of a nonlinear hyperbolic system (3.2.1) we construct an approximate Riemann solver in a heuristic way using an analogy with the previous considerations. We now have

$$\mathbb{A}(\boldsymbol{w}) = D\boldsymbol{f}(\boldsymbol{w})/D\boldsymbol{w} = \mathbb{T}(\boldsymbol{w})\mathbb{\Lambda}(\boldsymbol{w})\mathbb{T}^{-1}(\boldsymbol{w}) \tag{3.2.103}$$

and define the matrices $\mathbb{\Lambda}^{\pm} = \mathbb{\Lambda}^{\pm}(\boldsymbol{w})$, $|\mathbb{\Lambda}| = |\mathbb{\Lambda}(\boldsymbol{w})|$, $\mathbb{A}^{\pm} = \mathbb{A}^{\pm}(\boldsymbol{w})$, $|\mathbb{A}| = |\mathbb{A}(\boldsymbol{w})|$ by (3.2.90) and (3.2.91). Under the assumption that there exist vector functions $\boldsymbol{f}^{+}(\boldsymbol{w})$ and $\boldsymbol{f}^{-}(\boldsymbol{w})$ such that

$$\begin{aligned} \boldsymbol{f}(\boldsymbol{w}) &= \boldsymbol{f}^{+}(\boldsymbol{w}) + \boldsymbol{f}^{-}(\boldsymbol{w}), \\ \frac{D\boldsymbol{f}^{\pm}(\boldsymbol{w})}{D\boldsymbol{w}} &= \mathbb{A}^{\pm}(\boldsymbol{w}), \end{aligned} \tag{3.2.104}$$

it is natural to introduce an approximate Riemann solver as

$$\boldsymbol{g}_{\mathrm{R}}(\boldsymbol{u},\boldsymbol{v}) = \boldsymbol{f}^{+}(\boldsymbol{u}) + \boldsymbol{f}^{-}(\boldsymbol{v}) \tag{3.2.105}$$

(cf. (3.2.75), (3.2.76), (3.2.95)). In analogy to (3.2.80), (3.2.98), (3.2.101) and (3.2.102) we symbolically write

$$\begin{aligned} \boldsymbol{g}_{\mathrm{R}}(\boldsymbol{u},\boldsymbol{v}) &= \boldsymbol{f}(\boldsymbol{u}) + \int_{\boldsymbol{u}}^{\boldsymbol{v}} \mathbb{A}^{-}(\boldsymbol{w})\, d\boldsymbol{w} \\ &= \boldsymbol{f}(\boldsymbol{v}) - \int_{\boldsymbol{u}}^{\boldsymbol{v}} \mathbb{A}^{+}(\boldsymbol{w})\, d\boldsymbol{w} \\ &= \frac{1}{2}\left\{\boldsymbol{f}(\boldsymbol{u}) + \boldsymbol{f}(\boldsymbol{v}) - \int_{\boldsymbol{u}}^{\boldsymbol{v}} |\mathbb{A}(\boldsymbol{w})|\, d\boldsymbol{w}\right\}. \end{aligned} \tag{3.2.106}$$

The integrals in (3.2.106) do not make sense in general. However, these formulae can be used for the definition of numerical schemes in such a way that the integrals are evaluated with the aid of some numerical quadratures or computed with the use of suitable integration paths in the state space. The resulting formulae are called *flux vector splitting schemes* (of the Godunov type).

3.2.17 *Flux vector splitting schemes for the Euler equations with one space variable*

The Euler equations describing 1D flow can be written in the form (3.2.1) with $m = 3$, $\boldsymbol{w} = (\rho, \rho u, E)^{\mathrm{T}}$, $\boldsymbol{f}(\boldsymbol{w}) = (\rho u, \rho u^2 + p, (E+p)u)^{\mathrm{T}}$ and $p = (\gamma - 1)(E - \rho u^2/2)$, see Section 3.1. By virtue of Lemma 3.1 we have

$$\boldsymbol{f}(\boldsymbol{w}) = \mathbb{A}(\boldsymbol{w})\boldsymbol{w}, \tag{3.2.107}$$

for $\mathbb{A}(\boldsymbol{w}) = D\boldsymbol{f}(\boldsymbol{w})/D\boldsymbol{w}$.

The simplest method is the *Steger–Warming scheme* ((Steger and Warming, 1981)) with the numerical flux $\boldsymbol{g}_{\mathrm{SW}}$ obtained on the basis of the relation

$$\boldsymbol{f}(\boldsymbol{w}) = \mathbb{A}^{+}(\boldsymbol{w})\boldsymbol{w} + \mathbb{A}^{-}(\boldsymbol{w})\boldsymbol{w} \tag{3.2.108}$$

and written in the form (3.2.105) with $\boldsymbol{f}^{\pm}(\boldsymbol{w}) = \mathbb{A}^{\pm}(\boldsymbol{w})\boldsymbol{w}$, i.e.

$$\boldsymbol{g}_{\mathrm{SW}}(\boldsymbol{u},\boldsymbol{v}) = \mathbb{A}^{+}(\boldsymbol{u})\boldsymbol{u} + \mathbb{A}^{-}(\boldsymbol{v})\boldsymbol{v}. \tag{3.2.109}$$

The Steger–Warming numerical flux can be derived from (3.2.106) with the aid of the 'quadrature formula'

$$\int_{\boldsymbol{u}}^{\boldsymbol{v}} |\mathbb{A}(\boldsymbol{w})|\, d\boldsymbol{w} \approx |\mathbb{A}(\boldsymbol{v})|\boldsymbol{v} - |\mathbb{A}(\boldsymbol{u})|\boldsymbol{u}, \tag{3.2.110}$$

which is precise (in a scalar case), if $\mathbb{A}$ is constant. Hence, the accuracy of this integration is very low. This explains why the Steger–Warming scheme is rather diffusive (i.e. yields strongly smeared discontinuities), as follows from numerical experiments.

Other schemes can be obtained on the basis of the quadrature formula

$$\int_{\boldsymbol{u}}^{\boldsymbol{v}} |\mathbb{A}(\boldsymbol{w})|\, d\boldsymbol{w} \approx \left|\mathbb{A}\left(\frac{\boldsymbol{u}+\boldsymbol{v}}{2}\right)\right| (\boldsymbol{v}-\boldsymbol{u}) \tag{3.2.111}$$

and the approximation

$$\frac{1}{2}(\boldsymbol{f}(\boldsymbol{u}) + \boldsymbol{f}(\boldsymbol{v})) \approx \boldsymbol{f}\left(\frac{\boldsymbol{u}+\boldsymbol{v}}{2}\right) = \mathbb{A}\left(\frac{\boldsymbol{u}+\boldsymbol{v}}{2}\right)\frac{\boldsymbol{u}+\boldsymbol{v}}{2} \tag{3.2.112}$$

following from (3.2.107). Substituting (3.2.111) and (3.2.112) into (3.2.106) and using (3.2.92), we derive the *Vijayasundaram scheme* ((Vijayasundaram, 1986), (Dervieux and Vijayasundaram, 1983)) with the numerical flux

$$\boldsymbol{g}_{\mathrm{V}}(\boldsymbol{u},\boldsymbol{v}) = \mathbb{A}^{+}\left(\frac{\boldsymbol{u}+\boldsymbol{v}}{2}\right)\boldsymbol{u} + \mathbb{A}^{-}\left(\frac{\boldsymbol{u}+\boldsymbol{v}}{2}\right)\boldsymbol{v}. \tag{3.2.113}$$

Then the Vijayasundaram scheme reads

$$\begin{aligned}\boldsymbol{w}_i^{k+1} &= \boldsymbol{w}_i^k - \lambda\Big\{\left(\mathbb{A}^{k+}_{i+\frac12}\boldsymbol{w}_i^k + \mathbb{A}^{k-}_{i+\frac12}\boldsymbol{w}_{i+1}^k\right) \\ &\quad - \left(\mathbb{A}^{k+}_{i-\frac12}\boldsymbol{w}_{i-1}^k + \mathbb{A}^{k-}_{i-\frac12}\boldsymbol{w}_i^k\right)\Big\}, \quad \lambda = \tau/h,\end{aligned} \tag{3.2.114}$$

$$\begin{aligned}\mathbb{A}^{k\pm}_{i-\frac12} &= \mathbb{A}^{\pm}\left(\frac{\boldsymbol{w}_{i-1}^k + \boldsymbol{w}_i^k}{2}\right), \\ \mathbb{A}^{k\pm}_{i+\frac12} &= \mathbb{A}^{\pm}\left(\frac{\boldsymbol{w}_i^k + \boldsymbol{w}_{i+1}^k}{2}\right).\end{aligned}$$

We can see a *partial upwinding* in this scheme. The numerical flux is computed at the point $x_{i+\frac12}$ ($x_{i-\frac12}$) with the use of the value $\boldsymbol{w}_i^k$ or $\boldsymbol{w}_{i+1}^k$ ($\boldsymbol{w}_{i-1}^k$ or $\boldsymbol{w}_i^k$)

corresponding to the mesh point located in the upwind direction from the point $x_{i+\frac{1}{2}}$ ($x_{i-\frac{1}{2}}$) with respect to the propagation speed given by the eigenvalues λ_s. On the other hand, the matrices $\mathbb{A}^{\pm}$ are evaluated with the use of a centred representation. (In the Steger–Warming scheme we speak of a *full upwinding.*)

Further, if we use (3.2.106) and (3.2.111), we arrive at the *Van Leer scheme* with the numerical flux

$$g_{\mathrm{VL}}(\boldsymbol{u},\boldsymbol{v}) = \frac{1}{2}\left\{\boldsymbol{f}(\boldsymbol{u}) + \boldsymbol{f}(\boldsymbol{v}) - \left|\mathbb{A}\left(\frac{\boldsymbol{u}+\boldsymbol{v}}{2}\right)\right|(\boldsymbol{v}-\boldsymbol{u})\right\}. \tag{3.2.115}$$

3.2.18 *The Roe scheme*

A very popular scheme is the Roe scheme (Roe, 1981), based on the approximation of the Riemann problem (3.2.68)–(3.2.69). By introducing the Jacobi matrix $\mathbb{A}(\boldsymbol{w})$ in (3.2.103) and using the chain rule, equation (3.2.68) can be written in the nonconservative form

$$\frac{\partial \boldsymbol{w}}{\partial t} + \mathbb{A}(\boldsymbol{w})\frac{\partial \boldsymbol{w}}{\partial x} = 0. \tag{3.2.116}$$

Roe's approach replaces the Jacobian matrix $\mathbb{A}(\boldsymbol{w})$ in (3.2.106) by a constant matrix

$$\tilde{\mathbb{A}} = \tilde{\mathbb{A}}(\boldsymbol{u},\boldsymbol{v}), \tag{3.2.117}$$

which is the function of the data states $\boldsymbol{u} = \boldsymbol{w}_L$, $\boldsymbol{v} = \boldsymbol{w}_R$ introduced in (3.2.69). The *Roe matrix* $\tilde{\mathbb{A}}(\boldsymbol{w}_L,\boldsymbol{w}_R)$ is required to satisfy the following properties:

Property (A). The matrix $\tilde{\mathbb{A}}(\boldsymbol{w}_L,\boldsymbol{w}_R)$ is diagonalizable. That is, $\tilde{\mathbb{A}}(\boldsymbol{w}_L,\boldsymbol{w}_R)$ is written in the form

$$\tilde{\mathbb{A}}(\boldsymbol{w}_L,\boldsymbol{w}_R) = \tilde{\mathbb{T}}(\boldsymbol{w}_L,\boldsymbol{w}_R)\tilde{\mathbb{\Lambda}}(\boldsymbol{w}_L,\boldsymbol{w}_R)\tilde{\mathbb{T}}^{-1}(\boldsymbol{w}_L,\boldsymbol{w}_R) \tag{3.2.118}$$

where

$$\tilde{\mathbb{\Lambda}}(\boldsymbol{w}_L,\boldsymbol{w}_R) = \operatorname{diag}\left(\lambda_1(\boldsymbol{w}_L,\boldsymbol{w}_R),\ldots,\lambda_m(\boldsymbol{w}_L,\boldsymbol{w}_R)\right) \tag{3.2.119}$$

and the columns

$$\boldsymbol{r}_1(\boldsymbol{w}_L,\boldsymbol{w}_R),\ldots,\boldsymbol{r}_m(\boldsymbol{w}_L,\boldsymbol{w}_R) \tag{3.2.120}$$

of $\tilde{\mathbb{T}}(\boldsymbol{w}_L,\boldsymbol{w}_R)$ are eigenvectors of $\tilde{\mathbb{A}}(\boldsymbol{w}_L,\boldsymbol{w}_R)$ corresponding to eigenvalues $\lambda_1(\boldsymbol{w}_L,\boldsymbol{w}_R),\ldots,\lambda_m(\boldsymbol{w}_L,\boldsymbol{w}_R)$.

Property (B). Consistency with the exact Jacobian:

$$\tilde{\mathbb{A}}(\boldsymbol{w},\boldsymbol{w}) = \mathbb{A}(\boldsymbol{w}). \tag{3.2.121}$$

Property (C). Conservation across discontinuities:

$$\boldsymbol{f}(\boldsymbol{w}_R) - \boldsymbol{f}(\boldsymbol{w}_L) = \tilde{\mathbb{A}}(\boldsymbol{w}_L,\boldsymbol{w}_R)(\boldsymbol{w}_R - \boldsymbol{w}_L). \tag{3.2.122}$$

Property (A) on diagonalization is an obvious requirement; the approximate problem should at the very least preserve the mathematical character of the original non-linear system. Property (B) ensures consistency with the conservation laws. Adopting the notation (3.2.91), (3.2.92) for $\tilde{\mathbb{A}}, \tilde{\mathbb{T}}, \tilde{\Lambda}$, let us approximate $\mathbb{A}^-(\boldsymbol{w})$ in (3.2.106) by $\tilde{\mathbb{A}}^-(\boldsymbol{w}_L, \boldsymbol{w}_R)$ and denote the corresponding numerical flux $\boldsymbol{g}_{\text{Roe}}$. We get

$$\begin{aligned} \boldsymbol{g}_{\text{Roe}}(\boldsymbol{w}_L, \boldsymbol{w}_R) &= \boldsymbol{f}(\boldsymbol{w}_L) + \tilde{\mathbb{A}}^-(\boldsymbol{w}_L, \boldsymbol{w}_R) \int_{\boldsymbol{w}_L}^{\boldsymbol{w}_R} d\boldsymbol{w} \\ &= \boldsymbol{f}(\boldsymbol{w}_L) + \tilde{\mathbb{A}}^-(\boldsymbol{w}_L, \boldsymbol{w}_R)(\boldsymbol{w}_R - \boldsymbol{w}_L). \end{aligned} \tag{3.2.123}$$

Using Property (A), the data difference $\boldsymbol{w}_R - \boldsymbol{w}_L$ can be written

$$\boldsymbol{w}_R - \boldsymbol{w}_L = \sum_{n=1}^{m} \gamma_n \boldsymbol{r}_n(\boldsymbol{w}_L, \boldsymbol{w}_R). \tag{3.2.124}$$

Substituting (3.2.124) into (3.2.123), we get the Roe numerical flux in the form

$$\begin{aligned} \boldsymbol{g}_{\text{Roe}}(\boldsymbol{w}_L, \boldsymbol{w}_R) &= \boldsymbol{f}(\boldsymbol{w}_L) + \sum_{n=1}^{m} \gamma_n \tilde{\mathbb{A}}^-(\boldsymbol{w}_L, \boldsymbol{w}_R) \boldsymbol{r}_n(\boldsymbol{w}_L, \boldsymbol{w}_R) \\ &= \boldsymbol{f}(\boldsymbol{w}_L) + \sum_{\lambda_n \le 0} \gamma_n \lambda_n(\boldsymbol{w}_L, \boldsymbol{w}_R) \boldsymbol{r}_n(\boldsymbol{w}_L, \boldsymbol{w}_R). \end{aligned} \tag{3.2.125}$$

Formula (3.2.125) represents a computable quantity, if the matrix $\tilde{\mathbb{A}}(\boldsymbol{w}_L, \boldsymbol{w}_R)$, its eigenvalues $\lambda_1(\boldsymbol{w}_L, \boldsymbol{w}_R), \dots, \lambda_m(\boldsymbol{w}_L, \boldsymbol{w}_R)$ and eigenvectors $\boldsymbol{r}_1(\boldsymbol{w}_L, \boldsymbol{w}_R), \dots,$ $\boldsymbol{r}_m(\boldsymbol{w}_L, \boldsymbol{w}_R)$ are available.

A breakthrough in constructing $\tilde{\mathbb{A}}(\boldsymbol{w}_L, \boldsymbol{w}_R)$ resulted from Roe's idea of introducing a parameter vector

$$\boldsymbol{z} = (z_1, \dots, z_m)^{\mathrm{T}}, \tag{3.2.126}$$

such that both the vector of conservative variables $\boldsymbol{w}$ and the flux $\boldsymbol{f}(\boldsymbol{w})$ could be expressed in terms of $\boldsymbol{z}$. This means that

$$\boldsymbol{w} = \overline{\boldsymbol{w}}(\boldsymbol{z}), \quad \boldsymbol{f}(\boldsymbol{w}) = \boldsymbol{f}(\overline{\boldsymbol{w}}(\boldsymbol{z})) = \overline{\boldsymbol{f}}(\boldsymbol{z}). \tag{3.2.127}$$

Further, we are interested in such a parameter $\boldsymbol{z}$ that both $\overline{\boldsymbol{w}}$ and $\overline{\boldsymbol{f}}$ are quadratic functions of $\boldsymbol{z}$, i.e.

$$\overline{\boldsymbol{w}}(\boldsymbol{z}) = \left(\sum_{k,\ell=1}^{m} \alpha_{k\ell}^1 z_k z_\ell, \dots, \sum_{k,\ell=1}^{m} \alpha_{k\ell}^m z_k z_\ell \right)^{\mathrm{T}}, \tag{3.2.128}$$

$$\overline{\boldsymbol{f}}(\boldsymbol{z}) = \left(\sum_{k,\ell=1}^{m} \beta_{k\ell}^1 z_k z_\ell, \dots, \sum_{k,\ell=1}^{m} \beta_{k\ell}^m z_k z_\ell \right)^{\mathrm{T}}. \tag{3.2.129}$$

Lemma 3.39 *The quadratic dependence of* $\boldsymbol{w}$ *and* $\boldsymbol{f}(\boldsymbol{w})$ *on* $\boldsymbol{z}$ *given by* (3.2.128)–(3.2.129) *implies*

$$\boldsymbol{w}_R - \boldsymbol{w}_L = \frac{D\overline{\boldsymbol{w}}}{D\boldsymbol{z}}\left(\frac{\boldsymbol{z}_L + \boldsymbol{z}_R}{2}\right)(\boldsymbol{z}_R - \boldsymbol{z}_L), \tag{3.2.130}$$

$$\boldsymbol{f}(\boldsymbol{w}_R) - \boldsymbol{f}(\boldsymbol{w}_L) = \frac{D\overline{\boldsymbol{f}}}{D\boldsymbol{z}}\left(\frac{\boldsymbol{z}_L + \boldsymbol{z}_R}{2}\right)(\boldsymbol{z}_R - \boldsymbol{z}_L), \tag{3.2.131}$$

where $\boldsymbol{z}_L, \boldsymbol{z}_R$ *are such that*

$$\begin{aligned} \boldsymbol{w}_L &= \overline{\boldsymbol{w}}(\boldsymbol{z}_L) \\ \boldsymbol{w}_R &= \overline{\boldsymbol{w}}(\boldsymbol{z}_R). \end{aligned} \tag{3.2.132}$$

Proof For the i-th element of the vector on the left-hand side of (3.2.130) we have

$$\begin{aligned} (w_R - w_L)_i &= w_{Ri} - w_{Li} = (\overline{w}(\boldsymbol{z}_R))_i - (\overline{w}(\boldsymbol{z}_L))_i \\ &= \sum_{k,\ell=1}^{m} \alpha^i_{k\ell}(z_{Rk}z_{R\ell} - z_{Lk}z_{L\ell}). \end{aligned} \tag{3.2.133}$$

In order to evaluate the i-th element of the right-hand side vector in (3.2.130), we multiply the i-th row of the matrix $\frac{D\overline{\boldsymbol{w}}}{D\boldsymbol{z}}\left(\frac{1}{2}(\boldsymbol{z}_L + \boldsymbol{z}_R)\right)$ by the vector $(\boldsymbol{z}_R - \boldsymbol{z}_L)$. In view of (3.2.128), the i-th row $\tilde{\boldsymbol{b}}_i$ of the matrix $D\overline{\boldsymbol{w}}/D\boldsymbol{z}$ has the form

$$\begin{aligned} \tilde{\boldsymbol{b}}_i(z) &= \left(\frac{\partial}{\partial z_1}\sum_{k,\ell=1}^{m} \alpha^i_{k\ell} z_k z_\ell, \dots, \frac{\partial}{\partial z_m}\sum_{k,\ell=1}^{m} \alpha^i_{k\ell} z_k z_\ell\right) \\ &= \left(\sum_{k,\ell=1}^{m} \alpha^i_{k\ell}(\delta_{\ell 1} z_k + \delta_{k1} z_\ell), \dots, \sum_{k,\ell=1}^{m} \alpha^i_{k\ell}(\delta_{\ell m} z_k + \delta_{km} z_\ell)\right). \end{aligned} \tag{3.2.134}$$

Evaluating $\tilde{\boldsymbol{b}}_i$ at $\frac{1}{2}(\boldsymbol{z}_l + \boldsymbol{z}_R)$ and multiplying it by the vector $(\boldsymbol{z}_R - \boldsymbol{z}_L)$, we get for the i-th element of the right-hand side vector in (3.2.130)

$$\begin{aligned} &\tilde{\boldsymbol{b}}_i\left(\frac{1}{2}(\boldsymbol{z}_L + \boldsymbol{z}_R)\right)\cdot(\boldsymbol{z}_R - \boldsymbol{z}_L) \\ &= \sum_{j=1}^{m}\left(\sum_{k,\ell=1}^{m} \alpha^i_{k\ell}\left(\delta_{\ell j}\frac{z_{Rk} + z_{Lk}}{2} + \delta_{kj}\frac{z_{L\ell} + z_{R\ell}}{2}\right)\right)(z_{Rj} - z_{Lj}) \\ &= \sum_{k,\ell=1}^{m} \alpha^i_{k\ell}\left(\frac{z_{Rk} + z_{Lk}}{2}(z_{R\ell} - z_{L\ell}) + \frac{z_{L\ell} + z_{R\ell}}{2}(z_{Rk} - z_{Lk})\right) \end{aligned} \tag{3.2.135}$$

$$= \sum_{k,\ell=1}^{m} \alpha^i_{k\ell} \left(z_{Rk} z_{R\ell} - z_{Lk} z_{L\ell} \right).$$

By comparison of (3.2.135) with (3.2.133) for $i = 1, \dots, m$ we obtain (3.2.130). The proof of (3.2.131) is quite analogous. □

Let us substitute $(\boldsymbol{z}_R - \boldsymbol{z}_L)$ from (3.2.130) into (3.2.131). We get

$$\boldsymbol{f}(\boldsymbol{w}_R) - \boldsymbol{f}(\boldsymbol{w}_L) = \frac{D\overline{\boldsymbol{f}}}{D\boldsymbol{z}} \left(\frac{\boldsymbol{z}_L + \boldsymbol{z}_R}{2} \right) \left[\frac{D\overline{\boldsymbol{w}}}{D\boldsymbol{z}} \left(\frac{\boldsymbol{z}_L + \boldsymbol{z}_R}{2} \right) \right]^{-1} (\boldsymbol{w}_R - \boldsymbol{w}_L), \tag{3.2.136}$$

provided $\left(\frac{D\overline{\boldsymbol{w}}}{D\boldsymbol{z}} \right)^{-1}$ exists. By the chain rule we have

$$\frac{D\overline{\boldsymbol{f}}}{D\boldsymbol{z}}(\boldsymbol{z}) = \frac{D\boldsymbol{f}}{D\boldsymbol{w}}(\overline{\boldsymbol{w}}(\boldsymbol{z})) \frac{D\overline{\boldsymbol{w}}}{D\boldsymbol{z}}(\boldsymbol{z}) = \mathbb{A}(\overline{\boldsymbol{w}}(\boldsymbol{z})) \frac{D\overline{\boldsymbol{w}}}{D\boldsymbol{z}}(\boldsymbol{z}). \tag{3.2.137}$$

The substitution of (3.2.137), where $\boldsymbol{z} = \frac{1}{2}(\boldsymbol{z}_L + \boldsymbol{z}_R)$, into (3.2.136) yields

$$\boldsymbol{f}(\boldsymbol{w}_R) - \boldsymbol{f}(\boldsymbol{w}_L) = \mathbb{A} \left(\overline{\boldsymbol{w}} \left(\frac{\boldsymbol{z}_L + \boldsymbol{z}_R}{2} \right) \right) (\boldsymbol{w}_R - \boldsymbol{w}_L). \tag{3.2.138}$$

In order to satisfy Property (C) we set

$$\tilde{\mathbb{A}}(\boldsymbol{w}_L, \boldsymbol{w}_R) = \mathbb{A} \left(\overline{\boldsymbol{w}} \left(\frac{\boldsymbol{z}_L + \boldsymbol{z}_R}{2} \right) \right), \tag{3.2.139}$$

where $\boldsymbol{z}_L$ and $\boldsymbol{z}_R$ are from (3.2.132).

Now we shall describe how to choose the parameter vector $\boldsymbol{z}$ resulting in the quadratic dependence in (3.2.128). It is defined by

$$\boldsymbol{z} = (z_1, \dots, z_m)^{\mathrm{T}} = \sqrt{\rho}(1, v_1, \dots, v_{m-2}, H)^{\mathrm{T}} \tag{3.2.140}$$

where v_i are the components of the velocity and

$$H = \frac{E + p}{\rho}. \tag{3.2.141}$$

The quadratic dependence in (3.2.128) and (3.2.129) then reads

$$\overline{\boldsymbol{w}}(\boldsymbol{z}) = \left(z_1^2, z_1 z_2, \dots, z_1 z_{m-1}, \frac{z_1 z_m}{\gamma} + \frac{\gamma - 1}{2\gamma}(z_2^2 + \cdots + z_{m-1}^2) \right)^{\mathrm{T}}, \tag{3.2.142}$$

and

$$\overline{\boldsymbol{f}}(\boldsymbol{z}) = \left(z_1 z_2, z_2^2 + \frac{\gamma - 1}{\gamma} \left[z_1 z_m - \frac{1}{2}(z_2^2 + \cdots + z_{m-1}^2) \right], z_2 z_3, \dots, z_2 z_m \right)^{\mathrm{T}}. \tag{3.2.143}$$

In what follows, we shall pay attention to the 3D case, i.e. $N = 3$, when $\mathbb{A} = \mathbb{A}_1 = D\boldsymbol{f}_1/D\boldsymbol{w}$, where $\boldsymbol{f}_1$ is defined by (3.1.9). The cases $N = 1$, $N = 2$ will be let to the reader as an exercise. The Jacobian matrix $D\overline{\boldsymbol{w}}/D\boldsymbol{z}$ has the form

$$\frac{D\overline{\boldsymbol{w}}}{D\boldsymbol{z}}(\boldsymbol{z}) = \begin{pmatrix} 2z_1 & 0 & 0 & 0 & 0 \\ z_2 & z_1 & 0 & 0 & 0 \\ z_3 & 0 & z_1 & 0 & 0 \\ z_4 & 0 & 0 & z_1 & 0 \\ \frac{z_5}{\gamma} & \frac{\gamma-1}{\gamma}z_2 & \frac{\gamma-1}{\gamma}z_3 & \frac{\gamma-1}{\gamma}z_4 & \frac{z_1}{\gamma} \end{pmatrix} \tag{3.2.144}$$

and due to the fact that $z_1 = \sqrt{\rho} > 0$, its inverse in (3.2.136) exists. Let us introduce the so-called *Roe's averages* $\hat{\rho}, \hat{u}, \hat{v}, \hat{w}, \hat{H}$ such that $\frac{1}{2}(\boldsymbol{z}_L + \boldsymbol{z}_R)$ in (3.2.138) can be written in the form

$$\frac{1}{2}(\boldsymbol{z}_L + \boldsymbol{z}_R) = \sqrt{\hat{\rho}}(1, \hat{u}, \hat{v}, \hat{w}, \hat{H})^{\mathrm{T}}. \tag{3.2.145}$$

We adopt the notation used in Lemma 3.2, i.e. $u = v_1$, $v = v_2$, $w = v_3$. Actually, since $\frac{1}{2}(\boldsymbol{z}_L + \boldsymbol{z}_R)$ expressed in terms of $\rho_L, \rho_R, u_L, u_R, \dots$ has the form

$$\frac{1}{2}(\boldsymbol{z}_L + \boldsymbol{z}_R) = \frac{1}{2}\left(\sqrt{\rho_L} + \sqrt{\rho_R}, \sqrt{\rho_L}u_L + \sqrt{\rho_R}u_R, \dots, \sqrt{\rho_L}H_L + \sqrt{\rho_R}H_R\right)^{\mathrm{T}} \tag{3.2.146}$$

(cf. (3.2.140)) we see that (3.2.145) holds, if we set

$$\begin{aligned} \sqrt{\hat{\rho}} &= \frac{1}{2}\left(\sqrt{\rho_L} + \sqrt{\rho_R}\right), \\ \hat{u} &= \frac{\sqrt{\rho_L}u_L + \sqrt{\rho_R}u_R}{\sqrt{\rho_L} + \sqrt{\rho_R}}, \\ \hat{v} &= \frac{\sqrt{\rho_L}v_L + \sqrt{\rho_R}v_R}{\sqrt{\rho_L} + \sqrt{\rho_R}}, \\ \hat{w} &= \frac{\sqrt{\rho_L}w_L + \sqrt{\rho_R}w_R}{\sqrt{\rho_L} + \sqrt{\rho_R}}, \\ \hat{H} &= \frac{\sqrt{\rho_L}H_L + \sqrt{\rho_R}H_R}{\sqrt{\rho_L} + \sqrt{\rho_R}}. \end{aligned} \tag{3.2.147}$$

Using (3.2.142) and (3.2.145), we are able to express the argument of the matrix $\mathbb{A}$ in (3.2.139) as

$$\overline{\boldsymbol{w}}\left(\frac{1}{2}(\boldsymbol{z}_L + \boldsymbol{z}_R)\right) = (\hat{\rho}, \hat{\rho}\hat{u}, \hat{\rho}\hat{v}, \hat{\rho}\hat{w}, \hat{E}), \tag{3.2.148}$$

where we define

$$\hat{E} = \frac{1}{\gamma}\hat{\rho}\hat{H} + \frac{\gamma - 1}{2\gamma}\hat{\rho}|\hat{\boldsymbol{v}}|^2 \tag{3.2.149}$$

with

$$|\hat{\boldsymbol{v}}|^2 = \hat{u}^2 + \hat{v}^2 + \hat{w}^2. \tag{3.2.150}$$

Denoting

$$\hat{\boldsymbol{w}} = (\hat{\rho}, \hat{\rho}\hat{u}, \hat{\rho}\hat{v}, \hat{\rho}\hat{w}, \hat{E})^{\mathrm{T}} \tag{3.2.151}$$

and substituting (3.2.148) into (3.2.139), we see that the Roe matrix $\tilde{\mathbb{A}}$ has the form

$$\tilde{\mathbb{A}}(\boldsymbol{w}_L, \boldsymbol{w}_R) = \mathbb{A}_1(\hat{\boldsymbol{w}}), \tag{3.2.152}$$

where $\mathbb{A}_1$ is defined in (3.1.20) as a function of $\boldsymbol{w}$ given by (3.1.8). Lemma 3.3 provides us with the eigenvalues and eigenvectors of $\mathbb{A}_1(\hat{\boldsymbol{w}})$. Now we are able to diagonalize the matrix $\tilde{\mathbb{A}}(\boldsymbol{w}_L, \boldsymbol{w}_R)$ by (3.1.31)–(3.1.33), to compute $\tilde{\mathbb{A}}^-(\boldsymbol{w}_L, \boldsymbol{w}_R)$ according to (3.2.91), (3.2.90), (3.2.83) and to evaluate the Roe numerical flux from (3.2.123). However, this procedure is quite expensive and, therefore, Roe proposed the following method.

The eigenvalues of the matrix $\tilde{\mathbb{A}}(\boldsymbol{w}_L, \boldsymbol{w}_R) = \mathbb{A}_1(\hat{\boldsymbol{w}})$ can be expressed in the form

$$\hat{\lambda}_1 = \hat{u} - \hat{a}, \quad \hat{\lambda}_2 = \hat{\lambda}_3 = \hat{\lambda}_4 = \hat{u}, \quad \hat{\lambda}_5 = \hat{u} + \hat{a}, \tag{3.2.153}$$

where we set

$$\hat{a}^2 = (\gamma - 1)\left(\hat{H} - \frac{1}{2}|\hat{\boldsymbol{v}}|^2\right). \tag{3.2.154}$$

The corresponding eigenvectors are

$$\begin{aligned}
\hat{\boldsymbol{r}}_1 &= \left(1, \hat{u} - \hat{a}, \hat{v}, \hat{w}, \hat{H} - \hat{u}\hat{a}\right)^{\mathrm{T}}, \qquad (3.2.155)\\
\hat{\boldsymbol{r}}_2 &= \left(1, \hat{u}, \hat{v}, \hat{w}, \frac{1}{2}|\hat{\boldsymbol{v}}|^2\right)^{\mathrm{T}},\\
\hat{\boldsymbol{r}}_3 &= (0, 0, 1, 0, \hat{v})^{\mathrm{T}},\\
\hat{\boldsymbol{r}}_4 &= (0, 0, 0, 1, \hat{w})^{\mathrm{T}},\\
\hat{\boldsymbol{r}}_5 &= \left(1, \hat{u} + \hat{a}, \hat{v}, \hat{w}, \hat{H} + \hat{u}\hat{a}\right)^{\mathrm{T}}.
\end{aligned}$$

Exercise 3.40 Prove (3.2.155) using Lemma 3.3 and relation (3.2.154).
Hint: Subtract $\boldsymbol{r}_3(\boldsymbol{w})$ from $\boldsymbol{r}_2(\boldsymbol{w})$ and $\boldsymbol{r}_4(w)$ from $\boldsymbol{r}_2(w)$ defined by (3.1.24).

Now we can define the eigenvalues $\lambda_n(\boldsymbol{w}_L, w_R)$ and vectors $\boldsymbol{r}_n(\boldsymbol{w}_L, \boldsymbol{w}_R)$ from (3.2.119) and (3.2.120) as $\lambda_n(\boldsymbol{w}_L, \boldsymbol{w}_R) = \hat{\lambda}_n$, $\boldsymbol{r}_n(\boldsymbol{w}_L, \boldsymbol{w}_R) = \hat{\boldsymbol{r}}_n$, $n = 1, \ldots, 5$.

In order to determine completely the Roe numerical flux $\boldsymbol{g}_{\mathrm{Roe}}(\boldsymbol{w}_L, \boldsymbol{w}_R)$ in (3.2.125), we need, in addition, the coefficients γ_n. They are obtained from (3.2.124).

When written in full, these equations read

$$\gamma_1 + \gamma_2 + \gamma_5 = \Delta w_1, \tag{3.2.156}$$

$$\gamma_1(\hat{u} - \hat{a}) + \gamma_2 \hat{u} + \gamma_5(\hat{u} + \hat{a}) = \Delta w_2, \tag{3.2.157}$$

$$\gamma_1 \hat{v} + \gamma_2 \hat{v} + \gamma_3 + \gamma_5 \hat{v} = \Delta w_3, \tag{3.2.158}$$

$$\gamma_1\hat{w} + \gamma_2\hat{w} + \gamma_4 + \gamma_5\hat{w} = \Delta w_4, \tag{3.2.159}$$

$$\gamma_1(\hat{H} - \hat{u}\hat{a}) + \frac{1}{2}\gamma_2|\hat{\boldsymbol{v}}|^2 + \gamma_3\hat{v} + \gamma_4\hat{w} + \gamma_5(\hat{H} + \hat{u}\hat{a}) = \Delta w_5. \tag{3.2.160}$$

Here the right-hand side terms of equations (3.2.156)–(3.2.160) are known: they are jumps Δw_n in the conservative quantity w_n, namely

$$\Delta w_n = (w_R)_n - (w_L)_n.$$

The substitution of equation (3.2.156) into (3.2.158) and (3.2.159) gives directly

$$\gamma_3 = \Delta w_3 - \hat{v}\Delta w_1, \quad \gamma_4 = \Delta w_4 - \hat{w}\Delta w_1. \tag{3.2.161}$$

Then one solves (3.2.156), (3.2.157) and (3.2.160) for unknowns $\gamma_1, \gamma_2, \gamma_5$. Computationally, it is convenient to arrange the solution as follows:

$$\gamma_2 = \frac{\gamma - 1}{\hat{a}^2}\left[\Delta w_1(\hat{H} - \hat{u}^2) + \hat{u}\Delta w_2 - \overline{\Delta w_5}\right], \tag{3.2.162}$$

$$\gamma_1 = \frac{1}{2\hat{a}}\left[\Delta w_1(\hat{u} + \hat{a}) - \Delta w_2 - \hat{a}\gamma_2\right], \tag{3.2.163}$$

$$\gamma_5 = \Delta w_1 - (\gamma_1 + \gamma_2). \tag{3.2.164}$$

Here

$$\overline{\Delta w_5} = \Delta w_5 - (\Delta w_3 - \hat{v}\Delta w_1)\hat{v} - (\Delta w_4 - \hat{v}\Delta w_1)\hat{w}. \tag{3.2.165}$$

3.2.18.1 *An algorithm* To compute the Roe numerical flux $\boldsymbol{g}_{\text{Roe}}(\boldsymbol{w}_L, \boldsymbol{w}_R)$ by (3.2.125) we apply the following procedure:

1. Compute the Roe average values for $\hat{u}, \hat{v}, \hat{w}, \hat{H}$ and $\hat{a}$ according to (3.2.147), (3.2.154).
2. Compute the eigenvalues $\hat{\lambda}_n$ according to (3.2.153).
3. Compute the eigenvectors $\hat{\boldsymbol{r}}_n$ according to (3.2.155).
4. Compute the coefficients γ_n according to (3.2.161)–(3.2.164).
5. Compute the Roe numerical flux from (3.2.125).

Exercise 3.41 Show that instead of (3.2.125) the numerical flux $\boldsymbol{g}_{\text{Roe}}(\boldsymbol{w}_L, \boldsymbol{w}_R)$ can be expressed by the formula

$$\boldsymbol{g}_{\text{Roe}}(\boldsymbol{w}_L, \boldsymbol{w}_R) = \boldsymbol{f}(\boldsymbol{w}_R) - \sum_{\lambda_n \geq 0} \gamma_n\lambda_n\boldsymbol{r}_n, \tag{3.2.166}$$

or alternatively

$$\boldsymbol{g}_{\text{Roe}}(\boldsymbol{w}_L, \boldsymbol{w}_R) = \frac{1}{2}\left(\boldsymbol{f}(\boldsymbol{w}_L) + \boldsymbol{f}(\boldsymbol{w}_R)\right) - \frac{1}{2}\sum_{n=1}^{m} \gamma_n|\lambda_n|\boldsymbol{r}_n. \tag{3.2.167}$$

Hint: Use the relations in (3.2.106).

Exercise 3.42 Show that (3.2.125) can be derived by integrating (3.2.68) over the control volume $(x_L, 0) \times (0, T^*)$, where $x_L < 0 < T^*$ and $|x_L|$ is sufficiently large such that the initial data $\boldsymbol{w}_L$, $\boldsymbol{w}_R$ in (3.2.69) are not perturbed in $(-\infty, x_L) \times (0, T^*)$.
Hint:

1. Show that the result of the integration can be written as

$$\int_{x_L}^{0} \boldsymbol{w}(x, T^*)\, dx + x_L \boldsymbol{w}_L + T^* \boldsymbol{g}_{\mathrm{G}}(\boldsymbol{w}_L, \boldsymbol{w}_R) - T^* \boldsymbol{f}(\boldsymbol{w}_L) = 0, \qquad (3.2.168)$$

   where $\boldsymbol{g}_{\mathrm{G}}(\boldsymbol{w}_L, , \boldsymbol{w}_R)$ is the Godunov numerical flux defined in (3.2.71).
2. Approximate problem (3.2.116), (3.2.69) by a linear one

$$\frac{\partial \tilde{\boldsymbol{w}}}{\partial t} + \tilde{\mathbb{A}} \frac{\partial \tilde{\boldsymbol{w}}}{\partial x} = 0 \qquad (3.2.169)$$

$$\tilde{\boldsymbol{w}}(x, 0) = \begin{cases} \boldsymbol{w}_L, & x < 0 \\ \boldsymbol{w}_R, & x > 0 \end{cases} \qquad (3.2.170)$$

   where $\tilde{\mathbb{A}} = \tilde{\mathbb{A}}(\boldsymbol{w}_L, \boldsymbol{w}_R)$ is a constant matrix.
3. Choose $x_L < 0$, $|x_L|$ sufficiently large, such that both the solutions of (3.2.68), (3.2.69) and (3.2.169), (3.2.170) satisfy

$$\boldsymbol{w}(x_L, t) = \boldsymbol{w}_L = \tilde{\boldsymbol{w}}(x_L, t), \quad t \in [0, T^*). \qquad (3.2.171)$$

4. Show by integration of (3.2.169) over $(x_L, 0) \times (0, T^*)$ that

$$\int_{x_L}^{0} \tilde{\boldsymbol{w}}(x, T^*)\, dx + x_L \boldsymbol{w}_L + T^* \tilde{\boldsymbol{g}}_{\mathrm{G}}(\boldsymbol{w}_L, \boldsymbol{w}_R) - T^* \tilde{\mathbb{A}} \boldsymbol{w}_L = 0 \qquad (3.2.172)$$

   where $\tilde{\boldsymbol{g}}_{\mathrm{G}}$ is the Godunov numerical flux for linear problem (3.2.169), (3.2.170).
5. Apply (3.2.94) and express the Godunov numerical flux $\tilde{\boldsymbol{g}}_{\mathrm{G}}$ as

$$\tilde{\boldsymbol{g}}_{\mathrm{G}}(\boldsymbol{w}_L, \boldsymbol{w}_R) = \tilde{\mathbb{A}} \boldsymbol{w}_L + \sum_{\lambda_n \le 0} \lambda_n \gamma_n \boldsymbol{r}_n, \qquad (3.2.173)$$

   where λ_n and $\boldsymbol{r}_n$ are the eigenvalues and eigenvector of $\tilde{\mathbb{A}}$, respectively, and the coefficients γ_n satisfy

$$\boldsymbol{w}_R - \boldsymbol{w}_L = \sum_{n=1}^{m} \gamma_n \boldsymbol{r}_n. \qquad (3.2.174)$$

6. Approximate $\int_{x_L}^{0} \boldsymbol{w}(x, T^*)\, dx$ in (3.2.168) by $\int_{x_L}^{0} \tilde{\boldsymbol{w}}(x, T^*)\, dx$ obtained from equation (3.2.172). This leads to the approximation

$$\boldsymbol{g}_{\mathrm{G}}(\boldsymbol{w}_L, \boldsymbol{w}_R) \approx \boldsymbol{f}(\boldsymbol{w}_L) + \sum_{\lambda_n \le 0} \lambda_n \gamma_n \boldsymbol{r}_n =: \boldsymbol{g}_{\mathrm{Roe}}(\boldsymbol{w}_L, \boldsymbol{w}_R). \qquad (3.2.175)$$

3.2.19 *Consistency of the schemes*

The schemes by Van Leer and Roe are consistent for any system (3.2.1). Concerning the Steger–Warming and Vijayasundaram schemes we get the consistency for systems with flux $\boldsymbol{f}$ homogeneous of order 1, implying the property (3.2.107). Thus, the Steger–Warming and Vijayasundaram schemes applied to the 1D Euler equations are consistent. All these three schemes are *first order accurate* both in space and time.

The evaluation of the integrals in (3.2.106) along suitable integration paths constructed with the aid of the Riemann invariants was used for the derivation of the well-known *Osher–Solomon scheme* ((Osher and Solomon, 1982)), which is very robust and gives good results in a number of flow problems. This scheme has a differentiable numerical flux in contrast to the above flux vector splitting schemes whose numerical fluxes are only locally Lipschitz-continuous. We will present the derivation of the Osher–Solomon scheme in Section 3.4 directly for 3D flow.

3.2.20 *Linear stability of flux vector splitting schemes*

The Steger–Warming, Vijayasundaram, Van Leer and Roe schemes become identical if they are applied to the linear system (3.2.82), a). Then they read

$$\boldsymbol{w}_j^{k+1} = \boldsymbol{w}_j^k - \lambda\left(|\mathbb{A}|\boldsymbol{w}_j^k + \mathbb{A}^-\boldsymbol{w}_{j+1}^k - \mathbb{A}^+\boldsymbol{w}_{j-1}^k\right), \quad \lambda = \tau/h. \tag{3.2.176}$$

Using the ansatz $\boldsymbol{w}_j^k = \boldsymbol{F}_k \exp(injh)$, then by virtue of (3.2.92) and (3.2.83), we obtain the amplification matrix

$$\begin{aligned} \mathbb{G} &= \mathbb{I} - \lambda\left[|\mathbb{A}|(1-\cos nh) + i\mathbb{A}\sin nh\right] \\ &= \mathbb{T}\left\{\mathbb{I} - \lambda\left[|\mathbb{\Lambda}|(1-\cos nh) + i\mathbb{\Lambda}\sin nh\right]\right\}\mathbb{T}^{-1}. \end{aligned}$$

This implies that the eigenvalues of the matrix $\mathbb{G}$ have the form

$$\mu_j = 1 - \lambda\left[|\lambda_j|(1-\cos nh) + i\lambda_j \sin nh\right],$$

where λ_j are eigenvalues of $\mathbb{A}$. Now we easily find that $|\mu_j| \le 1$ if and only if $\lambda|\lambda_j| \le 1$. This means that scheme (3.2.176) is stable if and only if $\tau\sigma(\mathbb{A})/h \le 1$, which is again the CFL condition (3.2.53).

3.2.21 *Higher order schemes*

In the previous sections we have described one second order scheme, i.e. the Lax–Wendroff scheme, and several flux vector splitting methods of the Godunov type which are of first order only. Numerical experiments show that a good scheme for the numerical solution of nonlinear hyperbolic problems should possess an adequate rate of dissipation (in the form of numerical viscosity) in order to produce numerical solutions approximating physically admissible entropy solutions. This is satisfied, e.g. by the first order methods of the Godunov type. The second order Lax–Wendroff method produces solutions with nonphysical behaviour near discontinuities, because this method does not contain a sufficient amount of dissipation (cf. Section 3.2.21.2). On the other hand, the first order schemes yield solutions with discontinuities rather smeared. There are two ways to cure these problems:

a) Adaptive mesh refinement of the computational mesh in the vicinity of discontinuities. This will be treated in detail in Section 3.6 directly for multidimensional problems.

b) The construction of higher order schemes with a sufficient rate of dissipation.

In the following part of this section we shall be concerned with subject b).

There are basically three types of finite difference techniques for increasing the order of accuracy and leading to schemes not suffering from spurious oscillations:

(i) second order MUSCL-type TVD schemes using a flux limiter,
(ii) ENO schemes,
(iii) WENO schemes.

These methods are usually developed on the example of the scalar Cauchy problem

$$\begin{aligned} &\frac{\partial w}{\partial t} + \frac{\partial f(w)}{\partial x} = 0 \quad \text{in } I\!R \times (0, \infty), \\ &w(x, 0) = w^0(x), \quad x \in I\!R, \end{aligned} \tag{3.2.177}$$

where $f \in C^1(I\!R)$, and then heuristically applied to nonlinear systems.

3.2.21.1 *κ-schemes* The order of accuracy of the three-point scheme

$$w_i^{k+1} = w_i^k - \frac{\tau}{h}\left(g_{i+\frac{1}{2}}^k - g_{i-\frac{1}{2}}^k\right), \quad i \in Z, \ k \in Z^+, \tag{3.2.178}$$

with

$$g_{i+\frac{1}{2}}^k = g(w_i^k, w_{i+1}^k), \tag{3.2.179}$$

where g is a numerical flux, can be increased in such a way that we set

$$g_{i+\frac{1}{2}}^k = g\left(w_{i+\frac{1}{2}}^-, w_{i+\frac{1}{2}}^+\right) \tag{3.2.180}$$

(the superscript k denoting the time level is omitted for simplicity) and

$$\begin{aligned} w_{i+\frac{1}{2}}^- &= w_i + \frac{1+\kappa}{4}(w_{i+1} - w_i) + \frac{1-\kappa}{4}(w_i - w_{i-1}), \\ w_{i+\frac{1}{2}}^+ &= w_{i+1} + \frac{1+\kappa}{4}(w_i - w_{i+1}) + \frac{1-\kappa}{4}(w_{i+1} - w_{i+2}), \quad \kappa \in [-1, 1]. \end{aligned} \tag{3.2.181}$$

If $\kappa = -1$, we obtain the fully one-sided upwind scheme, $\kappa = 0$ – the Fromm scheme, $\kappa = \frac{1}{3}$ – the upwind biased scheme, $\kappa = 1$ – the central scheme, equivalent to the unconditionally unstable scheme (3.2.57). For $\kappa \neq 1$, the above scheme is no longer a three-point formula. We see that this new scheme can be written in the general form (3.2.8) (with $m = 1$, $\vartheta = 0$) and, hence, in the form (3.2.10), i.e.

$$w_i^{k+1} = \Phi(w_{i-\ell}^k, \ldots, w_i^k, \ldots, w_{i+\ell}^k), \quad i \in Z. \tag{3.2.182}$$

Exercise 3.43 Investigate the order of accuracy of the scheme defined by (3.2.178), (3.2.180), (3.2.181) with the Lax–Friedrichs or Engquist–Osher numerical flux, applied to a linear equation.

3.2.21.2 *Gibbs phenomenon* Numerical experiments unfortunately show that κ-schemes produce numerical approximations of discontinuous solutions to problem (3.2.177) containing nonphysical *spurious oscillations*, *wiggles*, *overshoots* and *undershoots* in the vicinity of discontinuities. This is a common fault of higher order methods (i.e. methods of order $p > 1$ in space), applied to problems with solutions containing discontinuities (e.g. in the case of nonlinear conservation laws) or steep gradients in boundary or internal layers (in the case of convection–diffusion problems with dominating convection). We speak of the *Gibbs phenomenon* (or Gibbs effect).

The goal is, of course, to construct higher order schemes that do not suffer from this drawback. To this end, A. Harten, J. Hyman and P. D. Lax ((Harten *et al.*, 1976)) introduced the concepts of monotone and TVD schemes for the numerical solution of conservation laws.

3.2.21.3 *Monotone and TVD schemes* We introduce some important concepts.

Definition 3.44 *A numerical scheme for the solution of problem* (3.2.177), *written in the form* (3.2.182), *is called monotone if* Φ *is nondecreasing in all its arguments:*

$$\alpha_j \leq \alpha_j^*, \quad j = 1, \ldots, 2\ell + 1 \Rightarrow \Phi(\alpha_1, \ldots, \alpha_{2\ell+1}) \leq \Phi(\alpha_1^*, \ldots, \alpha_{2\ell+1}^*). \tag{3.2.183}$$

Further, for $w^k = \{w_i^k\}_{i \in Z}$ *we call*

$$TV(w^k) = \sum_{i=-\infty}^{+\infty} |w_{i+1}^k - w_i^k| \tag{3.2.184}$$

the total variation of the sequence w^k. *We say that* (3.2.182) *is a TVD scheme (total variation diminishing), if*

$$TV(w^{k+1}) \leq TV(w^k), \quad k \in Z^+. \tag{3.2.185}$$

In the sequel, we shall briefly characterize the main properties of monotone and TVD schemes. (For proofs, see, for example, (Godlewski and Raviart, 1991), Chapter 4.) First we introduce a criterion for verifying the TVD property:

Theorem 3.45 *Let us assume that scheme (3.2.182) can be written in the form*

$$w_i^{k+1} = w_i^k + \lambda A_{i+\frac{1}{2}}^k (w_{i+1}^k - w_i^k) + \lambda B_{i-\frac{1}{2}}^k (w_{i-1}^k - w_i^k), \tag{3.2.186}$$

where $\lambda = \tau/h$, *the coefficients* $A_{i+\frac{1}{2}}^k$ *and* $B_{i-\frac{1}{2}}^k$ *depend on* w_j^k, $j = i, i \pm 1, i \pm 2, \ldots$, *and satisfy the conditions*

$$\begin{aligned} &a)\ A_{i+\frac{1}{2}}^k \geq 0, \quad B_{i+\frac{1}{2}}^k \geq 0, \\ &b)\ 1 - \lambda\left(A_{i+\frac{1}{2}}^k + B_{i+\frac{1}{2}}^k\right) \geq 0, \quad i \in Z. \end{aligned} \tag{3.2.187}$$

Then (3.2.182) *is a TVD scheme.*

Theorem 3.46 *If scheme* (3.2.182) *is conservative, consistent and monotone, then it is TVD and* L^∞*-stable:*

$$\|w^{k+1}\|_\infty := \sup_{i\in Z} |w_i^{k+1}| \leq \|w^k\|_\infty, \quad k \in Z^+. \tag{3.2.188}$$

Remark 3.47 This result can still be improved. Let us assume that the initial condition from (3.2.177) satisfies the condition $\|w^0\|_{L^\infty(\mathbb{R})} \leq M$. Defining the initial condition for the approximate solution as $w_i^0 = \int_{x_{i-1/2}}^{x_{i+1/2}} w^0(x)\,dx/h$, we have $\|w^0\|_\infty \leq M$. Then inequality (3.2.188) can be proven under a weaker assumption on the monotonicity of the scheme. That is, it is sufficient to suppose that the function $\Phi(\alpha_1, \ldots, \alpha_{2\ell+1})$ from (3.2.182) is nondecreasing in each variable $\alpha_j \in [-M, M]$, $j = 1, \ldots, 2\ell + 1$.

Example 3.48 a) The Lax–Friedrichs scheme (3.2.42) (for $m = 1$) is monotone if

$$1 - \lambda f'(w_{i+1}) \geq 0, \quad 1 + \lambda f'(w_{i-1}) \geq 0,$$

i.e. if the CFL condition

$$\lambda \sup_{i\in Z} |f'(w_i)| \leq 1 \tag{3.2.189}$$

is satisfied.

b) The Engquist–Osher scheme (3.2.81) is monotone if

$$\begin{gathered} 1 - \lambda f'(w_i) \geq 0, \\ \lambda\left(\mp f'(w_{i\pm 1}) + |f'(w_{i\pm 1})|\right)/2 \geq 0. \end{gathered}$$

The first inequality is the CFL condition, the second one is satisfied automatically.

With respect to Remark 3.47, if $\|w^0\|_{L^\infty(\mathbb{R})} \leq M$ and the initial condition w_i^0 is defined in the same way as in this remark, then it is sufficient to require the condition

$$\lambda \max_{w\in[-M,M]} |f'(w)| \leq 1, \tag{3.2.190}$$

guaranteeing the monotonicity of the Lax–Friedrichs and Engquist–Osher schemes. This leads us to the following result.

Theorem 3.49 *Let the numerical flux $g = g(u, v)$ of the three-point scheme (3.2.178) be Lipschitz-continuous:*

$$|g(u,v) - g(u^*,v^*)| \leq L(|u-u^*| + |v-v^*|), \quad u,v,u^*,v^* \in I\!R, \tag{3.2.191}$$

nondecreasing with respect to u and nonincreasing with respect to v and let the CFL condition $2\lambda L \leq 1$ be satisfied. Then the scheme is monotone and TVD.

Proof Scheme (3.2.178) can be written as (3.2.182) with

$$\Phi(w_{i-1}, w_i, w_{i+1}) = w_i - \lambda(g(w_i, w_{i+1}) - g(w_{i-1}, w_i)), \quad \lambda = \tau/h > 0$$

(the superscript k is omitted). The monotonicity properties of the numerical flux g immediately imply that Φ is nondecreasing with respect to w_{i-1} and w_{i+1}. Let $w_{i-1}, w_{i+1}, w_i, w_i^* \in I\!R$, $w_i^* \geq w_i$. Then,

$$\begin{aligned}
&\Phi(w_{i-1}, w_i^*, w_{i+1}) - \Phi(w_{i-1}, w_i, w_{i+1}) \\
&= (w_i^* - w_i) - \lambda(g(w_i^*, w_{i+1}) - g(w_i, w_{i+1})) + \lambda(g(w_{i-1}, w_i^*) - g(w_{i-1}, w_i)) \\
&\geq (w_i^* - w_i) - \lambda|g(w_i^*, w_{i+1}) - g(w_i, w_{i+1})| - \lambda|g(w_{i-1}, w_i^*) - g(w_{i-1}, w_i)| \\
&\geq (w_i^* - w_i)(1 - 2\lambda L),
\end{aligned}$$

as follows from (3.2.191). Now, by virtue of the CFL condition $2\lambda L \leq 1$, we have the implication

$$w_i^* \geq w_i \Rightarrow \Phi(w_{i-1}, w_i^*, w_{i+1}) \geq \Phi(w_{i-1}, w_i, w_{i+1}),$$

which means that Φ is nondecreasing with respect to w_i. The TVD property follows from Theorem 3.45. Scheme (3.2.178) can be written in the form (3.2.186) with

$$A^k_{i+\frac{1}{2}} = \begin{cases} -\dfrac{g(w_i^k, w_{i+1}^k) - g(w_i^k, w_i^k)}{w_{i+1}^k - w_i^k}, & w_i^k \neq w_{i+1}^k, \\ 0, & w_i^k = w_{i+1}^k, \end{cases}$$

$$B^k_{i-\frac{1}{2}} = \begin{cases} \dfrac{g(w_{i-1}^k, w_i^k) - g(w_i^k, w_i^k)}{w_{i-1}^k - w_i^k}, & w_i^k \neq w_{i-1}^k, \\ 0, & w_i^k = w_{i-1}^k. \end{cases}$$

Obviously, $A^k_{i+\frac{1}{2}}, B^k_{i-\frac{1}{2}} \geq 0$ and $\left|A^k_{i+\frac{1}{2}}\right|, \left|B^k_{i-\frac{1}{2}}\right| \leq L$. This and the condition $2\lambda L \leq 1$ imply condition (3.2.187), b). □

Note that if g is consistent, then the monotonicity implies TVD, as follows from Theorem 3.46. Unfortunately, the concept of monotonicity does not play any role in the case of higher order schemes, since by virtue of (Harten *et al.*, 1976), *monotone schemes are only first order accurate.* Therefore there is a chance to get an accurate scheme with good properties, producing approximate solutions without oscillations, if there are no oscillations in the initial condition, provided the scheme is TVD. This is guaranteed by the result of (Harten *et al.*, 1983).

Theorem 3.50 *A TVD scheme is monotonicity preserving: if $TV(w^k) < \infty$, then*

$$w_i^k \le w_{i+1}^k \quad \forall i \in Z \quad \Rightarrow \quad w_i^{k+1} \le w_{i+1}^{k+1} \quad \forall i \in Z. \tag{3.2.192}$$

This implies that oscillations cannot arise in the numerical solution obtained by a TVD method, unless they are in the initial condition.

Now let us describe how to construct a TVD method. One possibility is to start from a κ-scheme (see (3.2.180)) with $\kappa = -1$ and to modify the increments $w_i - w_{i-1}$ and $w_{i+1} - w_i$ in a suitable way with the aid of a flux (slope) limiter.

Definition 3.51 *We call a function $\psi : I\!R \to I\!R$ a flux limiter, if it is continuous and satisfies the following conditions:*

$$\begin{aligned} \alpha &\le \psi(r) \le M, \\ -M &\le \psi(r)/r \le 2 + \alpha, \quad r \in I\!R, \end{aligned} \tag{3.2.193}$$

with constants $\alpha \in [-2, 0]$ and $M \in (0, \infty)$.

The concept of the flux limiter is applied in the following way to the construction of TVD higher order schemes in the form (3.2.178), (3.2.180). The values $w_{i+\frac{1}{2}}^{\pm}$ are defined by the formulae

$$\begin{aligned} w_{i+\frac{1}{2}}^{-} &= w_i^k + \psi(R_i^k)(w_i^k - w_{i-1}^k)/2, \\ w_{i+\frac{1}{2}}^{+} &= w_{i+1}^k + \psi(1/R_{i+1}^k)(w_{i+1}^k - w_{i+2}^k)/2, \\ R_i^k &= \frac{w_{i+1}^k - w_i^k}{w_i^k - w_{i-1}^k}. \end{aligned} \tag{3.2.194}$$

The analysis of the order and TVD properties of this scheme were first investigated in (Sweby, 1984). See also (Godlewski and Raviart, 1991).

In the case of a system of equations, the states $\boldsymbol{w}_{i+\frac{1}{2}}^{\pm}$ are obtained in such a way that formulae (3.2.194) are applied to each component w_s of the vector $\boldsymbol{w}$ separately:

$$\begin{aligned} w_{s,i+\frac{1}{2}}^{-} &= w_{s,i}^k + \psi(R_{s,i}^k)(w_{s,i}^k - w_{s,i-1}^k)/2, \\ w_{s,i+\frac{1}{2}}^{+} &= w_{s,i+1}^k + \psi(1/R_{s,i+1}^k)(w_{s,i+1}^k - w_{s,i+2}^k)/2, \\ R_{s,i}^k &= \frac{w_{s,i+1}^k - w_{s,i}^k}{w_{s,i}^k - w_{s,i-1}^k}. \end{aligned} \tag{3.2.195}$$

3.2.21.4 *Examples of flux limiters*

a) Van Leer: $\psi_{\mathrm{VL}}(r) = (r + |r|)/(1 + |r|)$ $(M = 2,\ \alpha = 0)$;
b) Van Albada: $\psi_{\mathrm{VA}}(r) = (r^2 + r)(r^2 + 1)$ $(M = 2,\ \alpha = -\frac{1}{2})$;
c) Charkravarthy–Osher: $\psi_{\mathrm{CO}}(r) = \max(0, \min(\phi r, 1), \min(r, \phi))$, where the constant ϕ is chosen from the interval $[1, 2]$;

d) $\psi_\kappa = \left(\frac{1-\kappa}{2} + \frac{1+\kappa}{2} r\right) \frac{2r}{r^2+1}, \ \kappa \in [-1,1]$;

e) Minmod limiter is the Charkravarthy–Osher limiter with $\phi = 1$.

3.2.22 *ENO and WENO schemes*

In what follows, we shall briefly explain basic principles of the *ENO methods* (essentially non-oscillatory methods) and *WENO methods* (weighted essentially non-oscillatory methods). We start from the scalar problem (3.2.177) and carry out its semidiscretization in space over a grid in $\mathbb{R}$ formed by intervals $\mathcal{I}_i = (x_{i-\frac{1}{2}}, x_{i+\frac{1}{2}})$, $x_{i-\frac{1}{2}} < x_{i+\frac{1}{2}}$, $i \in Z$. Integration over $\mathcal{I}_i$ of equation (3.2.177) yields

$$\frac{d\overline{w}(x_i,t)}{dt} = -\frac{1}{h_i}\left(f(w(x_{i+\frac{1}{2}},t)) - f(w(x_{i-\frac{1}{2}},t))\right), \tag{3.2.196}$$

where we set $x_i = (x_{i+\frac{1}{2}} + x_{i-\frac{1}{2}})/2$, $h_i = x_{i+\frac{1}{2}} - x_{i-\frac{1}{2}}$ and $\overline{w}(x_i,t)$ denotes the cell average of $w(\cdot,t)$ over $\mathcal{I}_i$:

$$\overline{w}(x_i,t) = \frac{1}{h_i}\int_{x_{i-\frac{1}{2}}}^{x_{i+\frac{1}{2}}} w(x,t)\,dx. \tag{3.2.197}$$

Let us set $h = \sup_{j\in Z} h_j$ and let $h \in (0,\infty)$. We approximate (3.2.196) by the following conservative scheme for unknown approximations $w_i(t)$ of the cell averages $\overline{w}(x_i,t)$:

$$\frac{dw_i(t)}{dt} = -\frac{1}{h_i}\left(g_{i+\frac{1}{2}}(t) - g_{i-\frac{1}{2}}(t)\right), \tag{3.2.198}$$

where

$$g_{i+\frac{1}{2}}(t) = g\left(w^-_{i+\frac{1}{2}}(t), w^+_{i+\frac{1}{2}}(t)\right), \tag{3.2.199}$$

$g = g(u,v)$ is a numerical flux and $w^\pm_{i+\frac{1}{2}}$ are obtained by the ENO and WENO reconstruction. (Compare with (3.2.178) and (3.2.180).)

First we shall be concerned with the ENO reconstruction.

3.2.22.1 *Cell averages reconstruction* For a function $v \in L^1_{\mathrm{loc}}(\mathbb{R})$ we denote by $\overline{v}_i$ its *cell averages*

$$\overline{v}_i = \frac{1}{h_i}\int_{x_{i-\frac{1}{2}}}^{x_{i+\frac{1}{2}}} v(x)\,dx, \quad i \in Z. \tag{3.2.200}$$

For $i \in Z$ and $r, s \in Z^+$ let $S(i)$ denote a stencil

$$S(i) = \{\mathcal{I}_{i-r}, \ldots, \mathcal{I}_{i+s}\} \tag{3.2.201}$$

formed by the interval $\mathcal{I}_i$, r intervals $\mathcal{I}_j$ to the left and s intervals $\mathcal{I}_j$ to the right of $\mathcal{I}_i$. Our goal is to find a polynomial $p(x)$, called a *cell averages reconstruction*, such that

$$\frac{1}{h_j} \int_{\mathcal{I}_j} p(x)\, dx = \overline{v}_j, \quad j = i - r, \ldots, i + s. \tag{3.2.202}$$

In order to construct the polynomial p, we proceed in the following way. Let $i_0 \in Z$ and

$$V(x) = \int_{x_{i_0 - \frac{1}{2}}}^{x} v(\xi)\, d\xi \tag{3.2.203}$$

be the primitive function of v. Clearly,

$$V\left(x_{i+\frac{1}{2}}\right) = \sum_{j=i_0}^{i} \int_{\mathcal{I}_j} v(\xi)\, d\xi = \sum_{j=i_0}^{i} \overline{v}_j h_j, \quad i \in Z. \tag{3.2.204}$$

Hence, if we know the averages $\overline{v}_j$ of the function v, we can easily evaluate its primitive function at the points $x_{i+\frac{1}{2}}$, $i \in Z$. Let us denote $k = r + s + 1$. Then the set $\left\{x_{i-r-\frac{1}{2}}, x_{i-r+\frac{1}{2}}, \ldots, x_{i+s+\frac{1}{2}}\right\}$ consists of $k + 1$ points. It is well known that there exists exactly one *interpolation polynomial* $P(x)$ of degree $\leq k$ such that

$$P\left(x_{j+\frac{1}{2}}\right) = V\left(x_{j+\frac{1}{2}}\right), \quad j = i - r - 1, \ldots, i + s. \tag{3.2.205}$$

Now we easily prove the following result:

Lemma 3.52 *If we set* $p(x) = P'(x)$, *then the degree of* p *is* $\leq k - 1$ *and* $p(x)$ *satisfies conditions* (3.2.202).

Proof It is necessary to verify (3.2.202). Actually, in view of (3.2.205), (3.2.204) and (3.2.200),

$$\begin{aligned}
&\frac{1}{h_j} \int_{\mathcal{I}_j} p(x)\, dx = \frac{1}{h_j} \int_{\mathcal{I}_j} P'(x)\, dx \\
&= \frac{1}{h_j} \left(P(x_{j+\frac{1}{2}}) - P(x_{j-\frac{1}{2}})\right) \\
&= \frac{1}{h_j} \left(V(x_{j+\frac{1}{2}}) - V(x_{j-\frac{1}{2}})\right) \\
&= \frac{1}{h_j} \left(\int_{x_{i_0-\frac{1}{2}}}^{x_{j+\frac{1}{2}}} v(x)\, dx - \int_{x_{i_0-\frac{1}{2}}}^{x_{j-\frac{1}{2}}} v(x)\, dx\right) \\
&= \frac{1}{h_j} \int_{x_{j-\frac{1}{2}}}^{x_{j+\frac{1}{2}}} v(x)\, dx = \overline{v}_j, \quad j = i - r, \ldots, i + s.
\end{aligned}$$

□

Furthermore, using the approximation properties of P and $p = P'$ (see, for example (Schatzman, 2002), Section 4.3 or (Ralston, 1965), Sections 3.2 and 4.1), we conclude that provided $v \in C^{k+1}(\mathcal{I})$, where $\mathcal{I} = \left[x_{i-r-\frac{1}{2}}, x_{i+s+\frac{1}{2}}\right]$,

$$P(x) = V(x) + O(h^{k+1}), \quad p(x) = v(x) + O(h^k), \quad x \in \mathcal{I}. \tag{3.2.206}$$

From the above considerations we know how to construct the approximation polynomial $p(x)$ from given cell averages $\overline{v}_j$, provided i, k and r are given. Then we can calculate the approximation $v_{i+\frac{1}{2}}$ of the value $v(x_{i+\frac{1}{2}})$:

$$v_{i+\frac{1}{2}} = p(x_{i+\frac{1}{2}}) = v(x_{i+\frac{1}{2}}) + O(h^k). \tag{3.2.207}$$

Of course, for a given k, we have some freedom in the choice of r. We could choose, say, stencil (3.2.201) with $s = k - r - 1$ and fixed $r \geq 0$ for all $i \in Z$. It would work well, if the function v were sufficiently regular. However, the generic functions for the solution of conservation laws are piecewise smooth with isolated discontinuities. If a discontinuity point of v is an element of the interval $\left[x_{i-r-\frac{1}{2}}, x_{i+s+\frac{1}{2}}\right]$, then the Gibbs phenomenon appears as spurious oscillations and under/overshoots of the polynomial $p(x)$. This motivates us to use the ENO strategy using the *adaptive stencil* in order to avoid including the discontinuity in the stencil, if possible.

3.2.22.2 *ENO strategy* The interpolation polynomial $P(x)$ with degree $\leq k$ and property (3.2.205) (where $k = r + s + 1$) and its derivative $p(x) = P'(x)$ can be expressed with the aid of *divided differences* $V\left[x_{i-\frac{1}{2}}, \ldots, x_{i+j-\frac{1}{2}}\right]$ of the function V:

$$V\left[x_{i-\frac{1}{2}}\right] = V\left(x_{i-\frac{1}{2}}\right), \quad \text{if } j = 0 \tag{3.2.208}$$

(divided difference of 0-th degree),

$$V\left[x_{i-\frac{1}{2}}, \ldots, x_{i+j-\frac{1}{2}}\right] = \frac{V\left[x_{i+\frac{1}{2}}, \ldots, x_{i+j-\frac{1}{2}}\right] - V\left[x_{i-\frac{1}{2}}, \ldots, x_{i+j-\frac{3}{2}}\right]}{x_{i+j-\frac{1}{2}} - x_{i-\frac{1}{2}}}, \quad \text{if } j \geq 1$$

(divided difference of j-th degree). We can note that

$$V\left[x_{i-\frac{1}{2}}, x_{i+\frac{1}{2}}\right] = \frac{V_{i+\frac{1}{2}} - V_{i-\frac{1}{2}}}{x_{i+\frac{1}{2}} - x_{i-\frac{1}{2}}} = \overline{v}_i. \tag{3.2.209}$$

This means that the divided differences of V of degree $j \geq 1$ can be expressed in terms of $\overline{v}_m$, $m \in Z$, avoiding the computation of V. The *Newton form* of the interpolation polynomial $P(x)$ reads

$$P(x) = \sum_{j=0}^{k} V\left[x_{i-r-\frac{1}{2}}, \dots, x_{i-r+j-\frac{1}{2}}\right] \prod_{m=0}^{j-1} \left(x - x_{i-r+m-\frac{1}{2}}\right). \tag{3.2.210}$$

(See (Schatzman, 2002), Section 4.2.) Differentiation of (3.2.210) yields

$$p(x) = \sum_{j=1}^{k} V\left[x_{i-r-\frac{1}{2}}, \dots, x_{i-r+j-\frac{1}{2}}\right] \sum_{m=0}^{j-1} \prod_{\substack{\ell=0 \\ \ell\neq m}}^{j-1} \left(x - x_{i-r+\ell-\frac{1}{2}}\right). \tag{3.2.211}$$

There is an important property of the divided difference that, provided V is sufficiently smooth ((Schatzman, 2002), Section 4.6.1),

$$V\left[x_{i-\frac{1}{2}}, \dots, x_{i+j-\frac{1}{2}}\right] = \frac{V^{(j)}(\xi)}{j!}, \tag{3.2.212}$$

with some $\xi \in \left[x_{i-\frac{1}{2}}, x_{i+j-\frac{1}{2}}\right]$. However, if the derivative of V is discontinuous at a point of the interval $\mathcal{I}$, then

$$V\left[x_{i-\frac{1}{2}}, \dots, x_{i+j-\frac{1}{2}}\right] = O(h^{-j+1}). \tag{3.2.213}$$

From these results we see that the divided differences of the function V represent a measure of its smoothness. This leads us to the idea, which is the basis of the ENO strategy, to choose adaptively such a stencil (3.2.201) for which the corresponding divided differences are minimal in comparison with other possible stencils.

3.2.22.3 *Algorithm of 1D ENO reconstruction* Let averages $\overline{v}_i$, $i \in Z$, of a function v be given. Then the ENO reconstruction of the values $v^-_{i+\frac{1}{2}}$ and $v^+_{i-\frac{1}{2}}$ approximating $v(x^-_{i+\frac{1}{2}})$ and $v(x^+_{i-\frac{1}{2}})$, respectively, is obtained as follows:

1. Set $\tilde{S}_2(i) = \left\{x_{i-\frac{1}{2}}, x_{i+\frac{1}{2}}\right\}$ and $S(i) = \{\mathcal{I}_i\}$.
2. For $\ell = 2, \dots, k$, assuming that

$$\tilde{S}_\ell(i) = \left\{x_{j+\frac{1}{2}}, x_{j+\ell-\frac{1}{2}}\right\}$$

with some $j \in Z$ is known, we add one of the neighbouring points, $x_{j-\frac{1}{2}}$ or $x_{j+\ell+\frac{1}{2}}$, to the stencil, according to the ENO procedure:
– If

$$\left|V\left[x_{j-\frac{1}{2}}, \dots, x_{j+\ell-\frac{1}{2}}\right]\right| \leq \left|V\left[x_{j+\frac{1}{2}}, \dots, x_{j+\ell+\frac{1}{2}}\right]\right|,$$

add $x_{j-\frac{1}{2}}$ to the stencil $\tilde{S}_\ell(i)$ to obtain

$$\tilde{S}_{\ell+1}(i) = \left\{x_{j-\frac{1}{2}}, \dots, x_{j+\ell-\frac{1}{2}}\right\}.$$

– Else add $x_{j+\ell+\frac{1}{2}}$ to $\tilde{S}_\ell(i)$ to obtain

$$\tilde{S}_{\ell+1}(i) = \left\{x_{j+\frac{1}{2}}, \ldots, x_{j+\ell+\frac{1}{2}}\right\}.$$

3. Write the resulting stencil $\tilde{S}_{k+1}(i)$ as $\left\{x_{i-r-\frac{1}{2}}, \ldots, x_{i+s+\frac{1}{2}}\right\}$, with $s = k - r - 1$, and define $S(i)$ by (3.2.201).
4. Construct the polynomial $p_i(x)$ on the interval $\mathcal{I}_i$ as $p_i = p|_{\mathcal{I}_i}$, where $p(x)$ is the polynomial of degree $\leq k$ satisfying (3.2.202), and set

$$v^-_{i+\frac{1}{2}} = p_i\left(x_{i+\frac{1}{2}}\right), \quad v^+_{i-\frac{1}{2}} = p_i\left(x_{i-\frac{1}{2}}\right).$$

Now we finish the derivation of the ENO scheme for the space semidiscretization of problem (3.2.177) by defining the values $w^{\pm}_{i+\frac{1}{2}}(t)$, $t > 0$. Using the notation $v(x) = w(x,t)$ for a fixed $t > 0$ and $\overline{v}_i$ for the cell averages of $v(x)$ on $\mathcal{I}_i$, $i \in Z$, we apply the ENO algorithm 3.2.22.3 yielding the values $v^+_{i-\frac{1}{2}}$, $v^-_{i+\frac{1}{2}}$ for each interval $\mathcal{I}_i$, $i \in Z$. Then we define

$$w^+_{i-\frac{1}{2}}(t) = v^+_{i-\frac{1}{2}}, \quad w^-_{i+\frac{1}{2}}(t) = v^-_{i+\frac{1}{2}}. \tag{3.2.214}$$

Finally, (3.2.199) and (3.2.214) substituted into (3.2.198) yield a system of ordinary differential equations of the form similar to (3.2.15) equipped with initial condition (3.2.17). This system is usually solved by some Runge–Kutta scheme. See, for example, (3.2.19), (3.2.20) or (3.2.21).

3.2.22.4 *The WENO method* The WENO method tries to improve the properties of the ENO strategy using the following idea: instead of using only one of the candidate stencils for the calculation of the values $v^-_{i+\frac{1}{2}}$ $(i \in Z)$, one uses the convex combination of these values $v^{(r)-}_{i+\frac{1}{2}}$ obtained for k candidate stencils

$$S_r(i) = \{\mathcal{I}_{i-r}, \ldots, \mathcal{I}_{i-r+k-1}\}, \quad r = 0, \ldots, k-1, \tag{3.2.215}$$

$$v^-_{i+\frac{1}{2}} = \sum_{r=0}^{k-1} \omega_r v^{(r)-}_{i+\frac{1}{2}}, \tag{3.2.216}$$

with suitable weights ω_r satisfying the conditions

$$\omega_r \geq 0, \quad \sum_{r=0}^{k-1} \omega_r = 1. \tag{3.2.217}$$

The weights ω_r are chosen in such a way that ω_r is small for the stencil $S_r(i)$ giving $p_i^{(r)}$ with high oscillations and ω_r is larger for less oscillating $p_i^{(r)}$. The values $v^+_{i-\frac{1}{2}}$ are defined in a similar way. For a description of the weights ω_r, see (Shu, 1999).

3.2.22.5 *Application of the ENO or WENO method to a system* There are basically two ways:

a) The easiest way is to apply ENO or WENO schemes to the system of equations (3.2.1) in a component-by-component fashion. For the 1D Euler equations the ENO/WENO reconstruction is applied to the *conservative variables* $\rho, \rho u, E$ or the *primitive variables* ρ, u, p or the *physical variables* ρ, u, θ.

b) Another possibility is to apply the ENO/WENO reconstruction to the so-called *characteristic variables.* We proceed in the following way. At the point $x_{i+\frac{1}{2}}$ we compute the average state $\boldsymbol{w}_{i+\frac{1}{2}} = (\boldsymbol{w}_i + \boldsymbol{w}_{i+1})/2$ and the Jacobi matrix $\mathbb{A}(\boldsymbol{w}_{i+\frac{1}{2}}) = (D\boldsymbol{f}/D\boldsymbol{w})(\boldsymbol{w}_{i+\frac{1}{2}})$. Using its diagonalization

$$\mathbb{A}(\boldsymbol{w}_{i+\frac{1}{2}}) = \mathbb{T}(\boldsymbol{w}_{i+\frac{1}{2}})\mathbb{\Lambda}(\boldsymbol{w}_{i+\frac{1}{2}})\mathbb{T}^{-1}(\boldsymbol{w}_{i+\frac{1}{2}}), \tag{3.2.218}$$
$$\mathbb{\Lambda}(\boldsymbol{w}_{i+\frac{1}{2}}) = \operatorname{diag}\left(\lambda_1(\boldsymbol{w}_{i+\frac{1}{2}}), \dots, \lambda_m(\boldsymbol{w}_{i+\frac{1}{2}})\right),$$

where $\lambda_s(\boldsymbol{w}_{i+\frac{1}{2}})$ are eigenvalues of $\mathbb{A}(\boldsymbol{w}_{i+\frac{1}{2}})$, we define a new vector variable

$$\boldsymbol{q} = (q_1, \dots, q_m) = \mathbb{T}^{-1}(\boldsymbol{w}_{i+\frac{1}{2}})\boldsymbol{w}. \tag{3.2.219}$$

The components q_i and $\boldsymbol{q}$ are called *characteristic variables.* With the aid of ENO and WENO reconstruction applied to variables q_i, the values $\boldsymbol{q}^{\pm}_{i+\frac{1}{2}}$ are computed and then the transformation back to physical space gives the states

$$\boldsymbol{w}^{\pm}_{i+\frac{1}{2}} = \mathbb{T}(\boldsymbol{w}_{i+\frac{1}{2}})\boldsymbol{q}^{\pm}_{i+\frac{1}{2}} \tag{3.2.220}$$

which are substituted in the numerical flux similarly as in (3.2.199). It is clear that the characteristic approach is much more costly than the componentwise application of ENO/WENO, but numerical experiments prove its robustness.

More details about ENO and WENO scheme can be found in (Shu, 1999), (Casper and Atkins, 1993), (Abgrall, 1994), (Harten, 1989), (Harten, 1991), (Harten and Osher, 1987), (Harten *et al.*, 1987), (Hu and Shu, 1998), (Jiang and Shu, 1996), (Liu *et al.*, 1994), (Liu and Osher, 1998), (Nessyahu and Tadmor, 1990), (Osher and Shu, 1988), (Osher and Shu, 1989), (Osher and Shu, 1991), (Shu, 1990), (Shu and Osher, 1988), (Shu and Osher, 1989), (Sonar, 1997*a*), (Sonar, 1997*b*), (Friedrich, 1998), (Fürst and Kozel, 1999), (Fürst and Kozel, 2001). There are very few theoretical results on ENO or WENO, but in practice these schemes are robust and stable.

3.3 The finite volume method for the multidimensional Euler equations

In this section we shall be concerned with the numerical solution of 2D and 3D Euler equations. This subject belongs to the most important parts of CFD. During the last few decades, a number of efficient methods and techniques for the solution of the Euler equations have been developed. There is an extensive literature on this topic. Many methods are based on the use of generalized body fitted coordinates and the application of 1D finite difference schemes from Section 3.2 via dimensional splitting. (See, for example, (Fletcher, 1991), (Kovenya and Yanenko, 1981), (Kovenya *et al.*, 1990), (Wesseling, 2001).) Here we shall pay attention to the *finite volume method* (FVM) on unstructured meshes. The finite volume method is now one of the most popular methods because of its flexibility, adaptability and applicability to the solution of flow problems in domains with a complicated geometry. We shall derive some basic finite volume schemes and explain their analysis representing the stability, monotonicity and convergence. We shall also discuss the numerical treatment of boundary conditions for the Euler equations.

Let us consider the flow of an inviscid perfect gas in a bounded domain $\Omega \subset I\!R^N$ and time interval $(0,T)$ with $T > 0$. Here $N = 2$ or 3 for 2D or 3D flow, respectively. Our goal is to solve numerically the Euler equations

$$\frac{\partial \boldsymbol{w}}{\partial t} + \sum_{s=1}^{N} \frac{\partial \boldsymbol{f}_s(\boldsymbol{w})}{\partial x_s} = 0 \quad \text{in } Q_T = \Omega \times (0,T) \tag{3.3.1}$$

(Q_T is called a space-time cylinder), equipped with the initial condition

$$\boldsymbol{w}(x,0) = \boldsymbol{w}^0(x), \quad x \in \Omega, \tag{3.3.2}$$

with a given vector function $\boldsymbol{w}^0$ and boundary conditions

$$B(\boldsymbol{w}(x,t)) = 0 \quad \text{for } (x,t) \in \partial\Omega \times (0,T). \tag{3.3.3}$$

Here B is a suitable boundary operator. The specification of the boundary conditions and their approximation will be given later (see Section 3.3.6). The state vector $\boldsymbol{w} = (\rho, \rho v_1, \ldots, \rho v_N, E)^{\mathrm{T}} \in I\!R^m$, $m = N + 2$ (i.e. $m = 4$ or 5 for 2D or 3D flow, respectively), the fluxes $\boldsymbol{f}_s$, $s = 1, \ldots, N$, are m-dimensional mappings specified in (3.1.9) and defined in the set D – see (3.1.10). The Jacobi matrices $\mathbb{A}_s(\boldsymbol{w}) = D\boldsymbol{f}_s(\boldsymbol{w})/D\boldsymbol{w}$ can be expressed with the aid of the matrix $\mathbb{A}_1(\boldsymbol{w})$ (see (3.1.20) and (3.1.50) for $N = 3$ and $N = 2$, respectively) and formula (3.1.41). Let us recall that for each $\boldsymbol{w} \in D$ and $\boldsymbol{n} = (n_1, \ldots, n_N)^{\mathrm{T}} \in I\!R^N$ with $|\boldsymbol{n}| = 1$ the mapping

$$\boldsymbol{\mathcal{P}}(\boldsymbol{w}, \boldsymbol{n}) = \sum_{s=1}^{N} n_s \boldsymbol{f}_s(\boldsymbol{w}) \tag{3.3.4}$$

has the Jacobi matrix

$$\mathbb{P}(\boldsymbol{w}, \boldsymbol{n}) = D\boldsymbol{\mathcal{P}}(\boldsymbol{w}, \boldsymbol{n})/D\boldsymbol{w} = \sum_{s=1}^{N} n_s \mathbb{A}_s(\boldsymbol{w}), \tag{3.3.5}$$

with eigenvalues $\lambda_i = \lambda_i(\boldsymbol{w}, \boldsymbol{n})$:

$$\lambda_1 = \boldsymbol{v} \cdot \boldsymbol{n} - a, \quad \lambda_2 = \cdots = \lambda_{m-1} = \boldsymbol{v} \cdot \boldsymbol{n}, \quad \lambda_m = \boldsymbol{v} \cdot \boldsymbol{n} + a, \tag{3.3.6}$$

where $\boldsymbol{v} = (v_1, \ldots, v_N)^{\mathrm{T}}$ is the velocity and $a = \sqrt{\gamma p/\rho}$ is the local speed of sound. The matrix $\mathbb{P}(\boldsymbol{w}, \boldsymbol{n})$ is diagonalizable with the aid of the matrices

$\mathbb{T} = \mathbb{T}(\boldsymbol{w}, \boldsymbol{n})$ and $\mathbb{T}^{-1} = \mathbb{T}^{-1}(\boldsymbol{w}, \boldsymbol{n})$ determined by (3.1.46), (3.1.32), (3.1.33) and (3.1.35) for $N = 3$ and (3.1.46), (3.1.55), (3.1.56) and (3.1.58) for $N = 2$:

$$\mathbb{P}(\boldsymbol{w}, \boldsymbol{n}) = \mathbb{T}\,\mathbb{A}\,\mathbb{T}^{-1}, \quad \mathbb{A} = \operatorname{diag}(\lambda_1, \ldots, \lambda_m). \tag{3.3.7}$$

The mapping $\boldsymbol{\mathcal{P}}(\boldsymbol{w}, \boldsymbol{n})$ is called the *flux of the quantity* $\boldsymbol{w}$ *in the direction* $\boldsymbol{n}$.

Now let us deal with the finite volume discretization of system (3.3.1). First we describe the construction of a finite volume mesh.

3.3.1 *Finite volume mesh*

Let $\Omega \subset I\!R^N$ be a domain occupied by the fluid. If $N = 2$, then by Ω_h we denote a polygonal approximation of Ω. This means that the boundary $\partial\Omega_h$ of Ω_h consists of a finite number of closed simple piecewise linear curves. For $N = 3$, Ω_h will denote a polyhedral approximation of Ω. The system $\mathcal{D}_h = \{D_i\}_{i \in J}$, where $J \subset Z^+ = \{0, 1, \ldots\}$ is an index set and $h > 0$, will be called a *finite volume mesh* in Ω_h, if D_i, $i \in J$, are *closed polygons* or *polyhedrons*, if $N = 2$ or 3, respectively, with mutually disjoint interiors such that

$$\overline{\Omega}_h = \bigcup_{i \in J} D_i. \tag{3.3.8}$$

The elements $D_i \in \mathcal{D}_h$ are called *finite volumes*. Two finite volumes D_i, $D_j \in \mathcal{D}_h$ are either disjoint or their intersection is formed by a common part of their boundaries ∂D_i and ∂D_j. If $\partial D_i \cap \partial D_j$ contains at least one straight segment or a plane manifold, if $N = 2$ or 3, respectively, then we call D_i and D_j *neighbouring finite volumes* (or simply neighbours). For two neighbours D_i, $D_j \in \mathcal{D}_h$ we set

$$\Gamma_{ij} = \partial D_i \cap \partial D_j = \Gamma_{ji}. \tag{3.3.9}$$

Obviously, Γ_{ij} is formed by a finite number β_{ij} of straight segments ($N = 2$) or plane manifolds ($N = 3$) $\Gamma_{ij}^\alpha = \Gamma_{ji}^\alpha$:

$$\Gamma_{ij} = \bigcup_{\alpha=1}^{\beta_{ij}} \Gamma_{ij}^\alpha. \tag{3.3.10}$$

See Fig. 3.3.1. We will call Γ_{ij}^α *faces* of D_i.

Further, we introduce the following *notation*:

$$\begin{aligned}
|D_i| &= N\text{-dimensional measure of } D_i \\
&= \text{area of } D_i \text{ if } N = 2, \text{ or volume of } D_i \text{ if } N = 3, \\
|\Gamma_{ij}^\alpha| &= (N-1)\text{-dimensional measure of } \Gamma_{ij}^\alpha \\
&= \text{the length of } \Gamma_{ij}^\alpha \text{ if } N = 2, \text{ or area of } \Gamma_{ij}^\alpha \text{ if } N = 3, \\
\boldsymbol{n}_{ij}^\alpha &= ((n_{ij}^\alpha)_1, \ldots, (n_{ij}^\alpha)_N)^{\mathrm{T}} = \text{unit outer normal to } \partial D_i \text{ on } \Gamma_{ij}^\alpha, \\
h_i &= \operatorname{diam}(D_i),
\end{aligned} \tag{3.3.11}$$

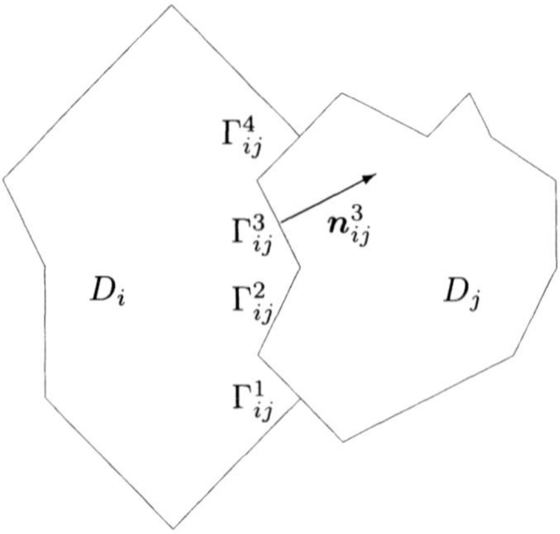

FIG. 3.3.1. Neighbouring finite volumes in 2D, $\Gamma_{ij} = \bigcup_{\alpha=1}^4 \Gamma_{ij}^\alpha$

$h = \sup_{i \in J} h_i$,
$|\partial D_i| = (N-1)$-dimensional measure of ∂D_i,
$s(i) = \{j \in J; j \neq i, D_j$ is a neighbour of $D_i\}$.

Clearly, $\boldsymbol{n}_{ij}^\alpha = -\boldsymbol{n}_{ji}^\alpha$.

The straight segments or plane manifolds for $N = 2$ or 3, respectively, that form the intersections of $\partial\Omega_h$ with finite volumes D_i adjacent to $\partial\Omega_h$ will be denoted by S_j and numbered by negative indexes j forming an index set $J_B \subset Z^- = \{-1, -2, \dots\}$. Hence, $J \cap J_B = \emptyset$ and $\partial\Omega_h = \bigcup_{j \in J_B} S_j$. For a finite volume D_i adjacent to the boundary $\partial\Omega_h$, i.e. if $S_j \subset \partial\Omega_h \cap \partial D_i$ for some $j \in J_B$, we set

$$\begin{aligned} &\gamma(i) = \{j \in J_B; S_j \subset \partial D_i \cap \partial\Omega_h\}, \\ &\Gamma_{ij} = \Gamma_{ij}^1 = S_j, \quad \beta_{ij} = 1 \quad \text{for } j \in \gamma(i). \end{aligned} \tag{3.3.12}$$

If D_i is not adjacent to $\partial\Omega_h$, then we put $\gamma(i) = \emptyset$. By $\boldsymbol{n}_{ij}^\alpha$ we again denote the unit outer normal to ∂D_i on Γ_{ij}^α. Then, putting

$$S(i) = s(i) \cup \gamma(i), \tag{3.3.13}$$

we have

$$\partial D_i = \bigcup_{j \in S(i)} \bigcup_{\alpha=1}^{\beta_{ij}} \Gamma_{ij}^\alpha, \tag{3.3.14}$$

$$\partial D_i \cap \partial\Omega_h = \bigcup_{j \in \gamma(i)} \bigcup_{\alpha=1}^{\beta_{ij}} \Gamma_{ij}^\alpha,$$

$$|\partial D_i| = \sum_{j\in S(i)} \sum_{\alpha=1}^{\beta_{ij}} |\Gamma_{ij}^\alpha|.$$

3.3.1.1 *Finite volumes in 2D* In practical computations one uses several types of finite volumes meshes:

a) *Triangular mesh* In this case $\mathcal{D}_h$ is a triangulation of the domain Ω_h with the usual properties from the finite element method (Ciarlet, 1979) (see Section 4.1): $D_i \in \mathcal{D}_h$ are closed triangles satisfying conditions (3.3.8) and

$$\begin{aligned} &\text{if } D_i,\ D_j \in \mathcal{D}_h,\ D_i \neq D_j, \text{ then either } D_i \cap D_j = \emptyset \\ &\text{or } D_i \cap D_j \text{ is a common vertex of } D_i \text{ and } D_j \\ &\text{or } D_i \cap D_j \text{ is a common side of } D_i \text{ and } D_j. \end{aligned} \tag{3.3.15}$$

The triangulation satisfying (3.3.15) is called *conforming*. Then, under the above notation, Γ_{ij} consists of only one straight segment and, thus, we have $\beta_{ij} = 1$ and simply write $\partial D_i = \bigcup_{j\in S(i)} \Gamma_{ij}$. See Fig. 3.3.2, a).

b) *Quadrilateral mesh* Now $\mathcal{D}_h$ consists of closed convex quadrilaterals D_i with properties (3.3.8) and (3.3.15). See Fig. 3.3.2, b).

c) *Dual finite volume mesh over a triangular grid* Let $\mathcal{T}_h = \{K_i\}_{i\in I}$ be a triangulation of the domain Ω_h formed by closed triangles K_i having properties (3.3.8) and (3.3.15), where we write K_i instead of D_i, $\mathcal{T}_h$ instead of $\mathcal{D}_h$ and I instead of J. Let $\sigma_h = \{P_i; i \in J\}$ be the set of all vertices of all triangles $K \in \mathcal{T}_h$ ($J \subset Z^+$ is a suitable index set). For each $i \in J$ we define the *dual finite volume* D_i *associated with the vertex* P_i as a closed polygon obtained in the following way. We join the centre of gravity of every triangle $K \in \mathcal{T}_h$ that contains the vertex P_i to the midpoint of every side of K containing P_i. (See Fig. 3.3.2, c).) If $P_i \in \partial\Omega_h$, then we complete the obtained contour by the straight segments joining P_i with the midpoints of boundary sides (i.e. sides which are subsets of $\partial\Omega_h$) that contain P_i. In this way we get the boundary ∂D_i of the finite volume D_i. It is obvious that (3.3.8) holds. Moreover, $\beta_{ij} \leq 2$.

d) *Barycentric finite volumes over a triangular grid* Similarly as in c), let $\mathcal{T}_h$ be a triangulation of the domain Ω_h. By $\mathcal{Q}_h = \{Q_i; i \in J\}$ (where $J \subset Z^+$ is a suitable index set) we denote the set of all midpoints of sides of all triangles $K \in \mathcal{Q}_h$. Now we proceed in the following way. We join the centre of gravity of each triangle $K \in \mathcal{T}_h$ with its vertices. Then each midpoint $Q_i \in \Omega_h \cap \mathcal{Q}_h$ is an interior point of a quadrilateral D_i with boundary formed by the above straight segments. If $Q_i \in \partial\Omega_h$ is the midpoint of a straight segment S_i with end points $P_1, P_2 =$ vertices of a triangle $K \in \mathcal{T}_h$ adjacent to $\partial\Omega_h$, then the associated finite volume D_i is a triangle with boundary formed by S_i and the straight segments connecting the centre of gravity of K with the points P_1 and P_2. See Fig. 3.3.2, d). In this case $\beta_{ij} = 1$.

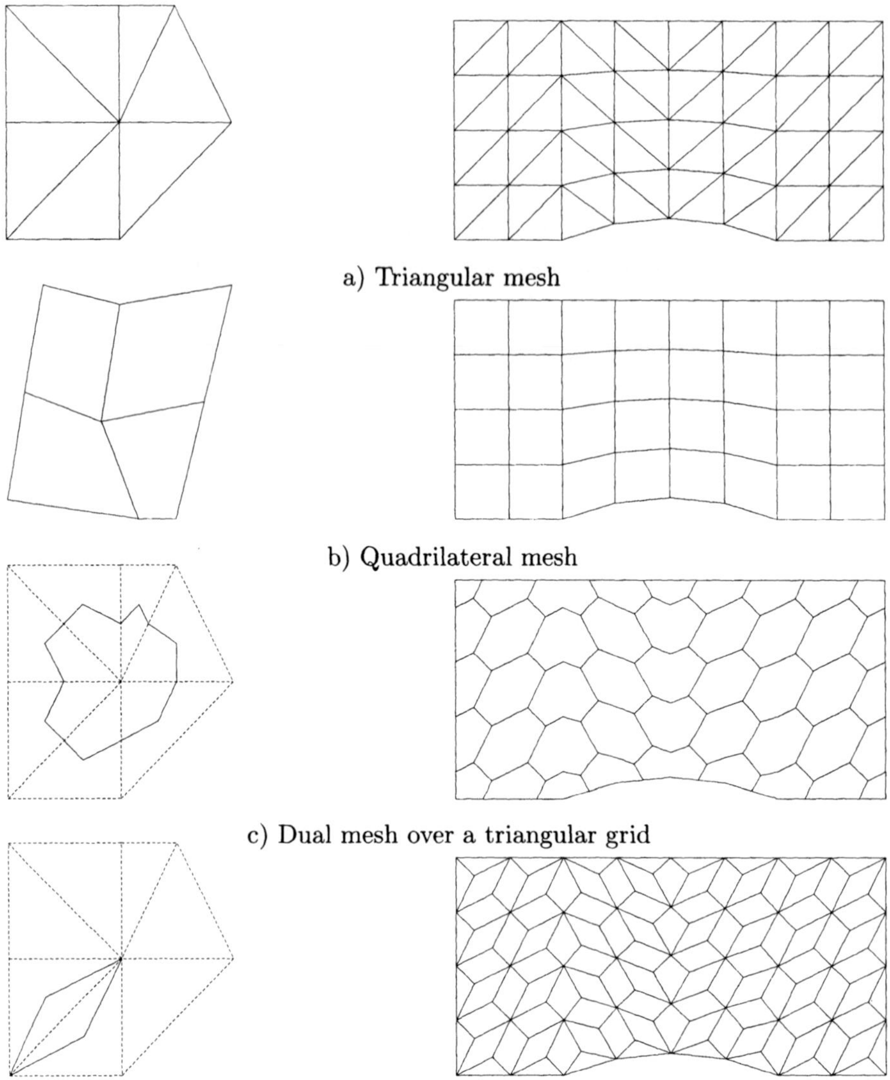

a) Triangular mesh

b) Quadrilateral mesh

c) Dual mesh over a triangular grid

d) Barycentric mesh over a triangular grid

FIG. 3.3.2. Finite volume meshes in 2D

3.3.1.2 *Finite volumes in 3D* In this case we construct a partition $\mathcal{D}_h = \{D_i\}_{i \in J}$ of Ω_h formed by *finite volumes* D_i which are *closed polyhedrons.* We again use a suitable index set $J \subset Z^+$ and assume (3.3.8). In 3D, the finite volumes D_i are usually chosen in the following way:

a) *Tetrahedral mesh* In this case $\mathcal{D}_h$ is a partition of the domain Ω_h formed by a finite number of closed tetrahedra with properties from the finite element method: $D_i \in \mathcal{D}_h$ satisfy (3.3.8) and

$$\begin{aligned} &\text{if } D_i,\ D_j \in \mathcal{D}_h,\ D_i \neq D_j, \text{ then either } D_i \cap D_j = \emptyset \\ &\text{or } D_i \cap D_j \text{ is a common vertex of } D_i \text{ and } D_j \\ &\text{or } D_i \cap D_j \text{ is a common edge of } D_i \text{ and } D_j \\ &\text{or } D_i \cap D_j \text{ is a common face of } D_i \text{ and } D_j. \end{aligned} \tag{3.3.16}$$

If condition (3.3.16) is satisfied, we call the mesh $\mathcal{D}_h$ *conforming.*

b) *Hexahedral mesh* Here $\mathcal{D}_h$ consists of closed convex hexahedra.

c) *Dual finite volumes over a tetrahedral mesh* Let $\mathcal{T}_h$ be a tetrahedral grid in Ω_h formed by a finite number of closed tetrahedra K_i with properties (3.3.8) and (3.3.16) (where we write K_i instead of D_i, $\mathcal{T}_h$ instead of $\mathcal{D}_h$ and I instead of J). Denoting by $\sigma_h = \{P_i; i \in J\}$ the set of all vertices of all elements $K \in \mathcal{T}_h$, we associate each $P_i \in \sigma_h$ with a *dual finite volume* D_i constructed in the following way. For each tetrahedron $K \in \mathcal{T}_h$ such that $P_i \in K$ we denote

C_K – centre of gravity of K,

P_i^j, $j = 1,2,3$ – midpoint of the closed edge $\mathcal{E}_j$ of K such that $P_i \in \mathcal{E}_j$,

Q_i^k, $k = 1,2,3$ – centre of gravity of the closed face of K containing P_i^j, P_i^ℓ, $j \neq \ell$, $k \in \{1,2,3\} \setminus \{j,\ell\}$.

(See Figs 3.3.3–3.3.12.) Further we denote by D_i^K the closed hexahedron with edges

$$\begin{aligned} &\overline{C_K Q_i^j},\ j = 1,2,3, \\ &\overline{P_i P_i^j},\ j = 1,2,3, \\ &\overline{P_i^j Q_i^k},\ j = 1,2,3,\ k \in \{1,2,3\} \setminus \{j\} \end{aligned}$$

(see the shaded hexahedron in Fig. 3.3.9). One can find that the quadruples $\{C_K, P_i^j, Q_i^k, Q_i^\ell\}$, $k,\ell = 1,2,3$, $k \neq \ell$, $j \in \{1,2,3\} \setminus \{k,\ell\}$ are coplanar and for $I_K = \{i \in J; P_i \text{ is the vertex of } K\}$ we have

$$\bigcup_{i \in I_K} D_i^K = K. \tag{3.3.17}$$

We define the dual finite volume D_i as

$$D_i = \bigcup_{K \ni P_i} D_i^K. \tag{3.3.18}$$

d) *Barycentric finite volumes over a tetrahedral mesh* $\mathcal{T}_h$ from c) is constructed as follows. We denote by $\mathcal{S}_h$ the set of all faces of all tetrahedra $K \in \mathcal{T}_h$.

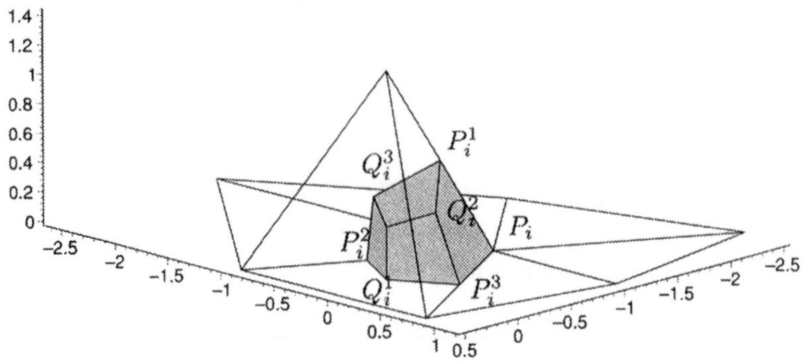

FIG. 3.3.3. A part D_i^K (shaded hexahedron) of the dual finite volume D_i

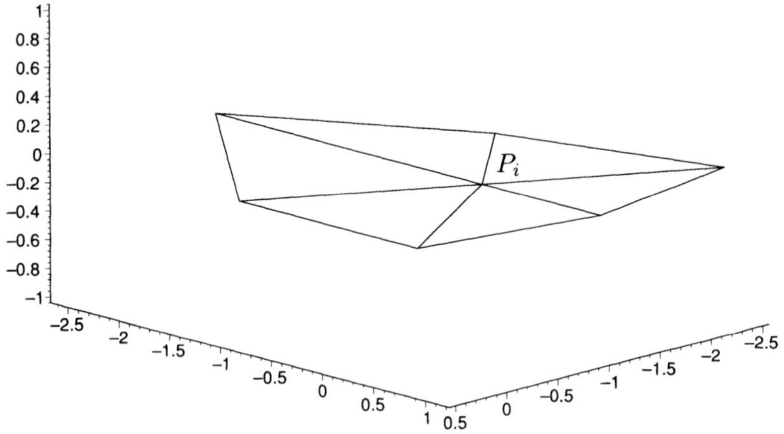

FIG. 3.3.4. A part of $\partial\Omega_h$ with the vertex P_i

We introduce a numbering of faces $S \in \mathcal{S}_h$ in such a way that $\mathcal{S}_h = \{S_i; i \in J\}$, where $J \subset Z^+$ is a suitable index set. The *barycentric finite volume* D_i is a closed polyhedron defined in the following way. We join the centre of gravity of every tetrahedron $K \in \mathcal{T}_h$ to its vertices. Then around the face $S_i \in \mathcal{S}_h$, we obtain a closed hexahedron containing S_i formed by a union of two tetrahedra with common face S_i. If $S_j \in \partial\Omega_h$ is a face with vertices P_1^K, P_2^K, P_3^K of tetrahedron $K \in \mathcal{T}_h$ adjacent to $\partial\Omega_h$, then we denote by D_j the tetrahedron with boundary formed by the face S_j and triangles with triples of vertices (C_K, P_1^K, P_2^K), (C_K, P_1^K, P_3^K) and (C_K, P_2^K, P_3^K). See Figs 3.3.13–3.3.14.

It is possible to find that in all cases a)–d), relations (3.3.8)–(3.3.10) and (3.3.14) hold. In the cases a), b), d) we have $\beta_{ij} = 1$; in the case c) of dual finite volumes β_{ij} is the number of tetrahedra that have the edge connecting P_i and P_j in common. (Cf. Figs 3.3.15–3.3.16.)

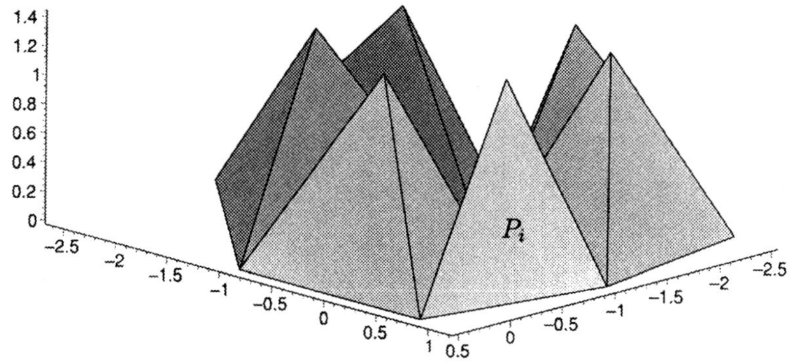

FIG. 3.3.5. Tetrahedra with one face on $\partial\Omega_h$

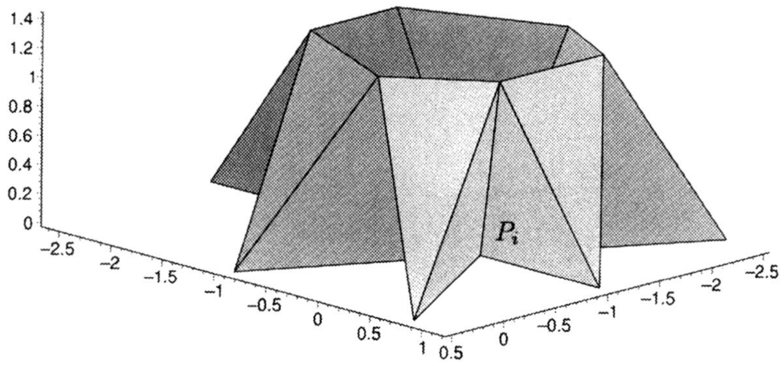

FIG. 3.3.6. Tetrahedra with just one edge on $\partial\Omega_h$

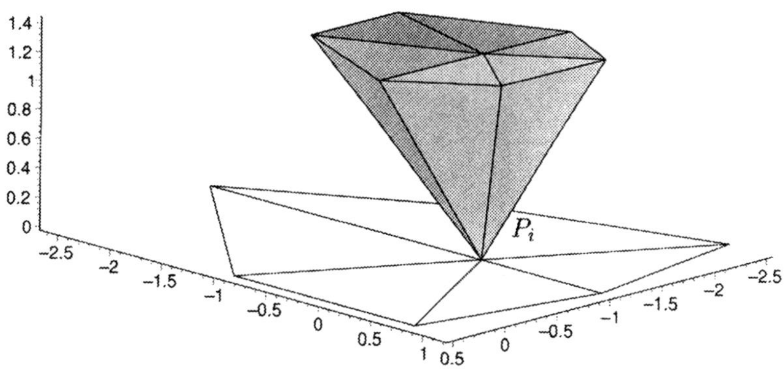

FIG. 3.3.7. Tetrahedra with just one vertex on $\partial\Omega_h$

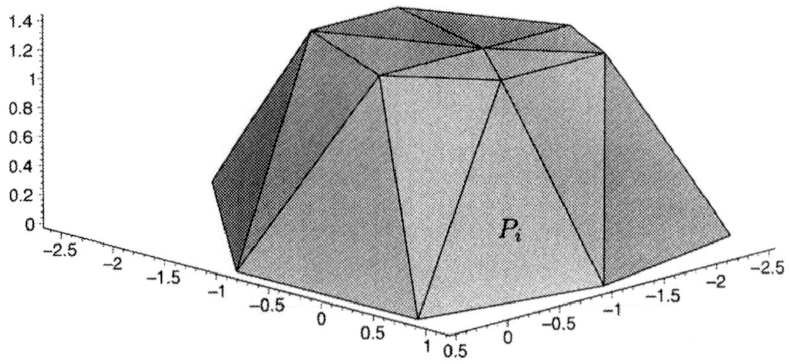

FIG. 3.3.8. Tetrahedra with common vertex P_i

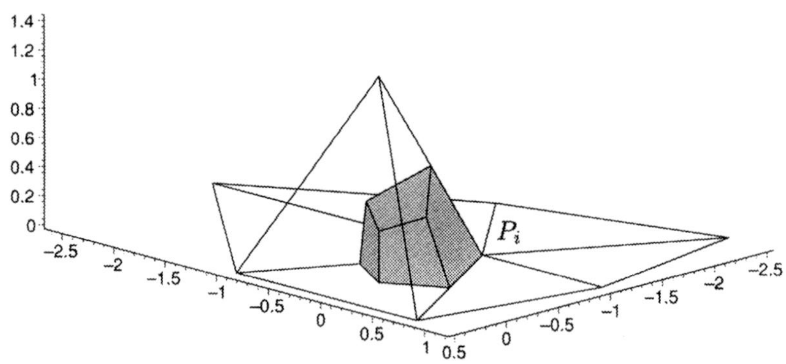

FIG. 3.3.9. The construction of the dual finite volume D_i

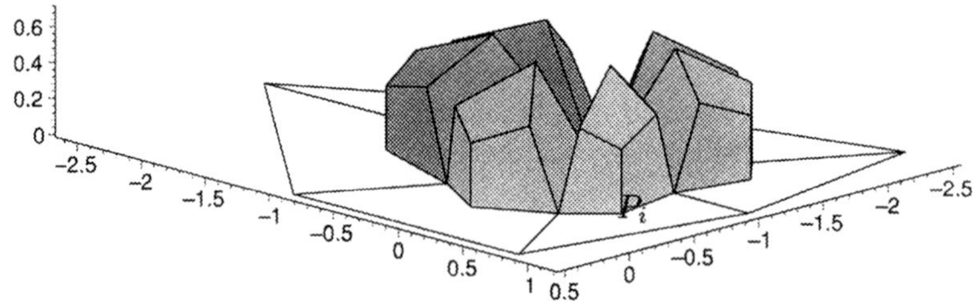

FIG. 3.3.10. The construction of the dual finite volume D_i

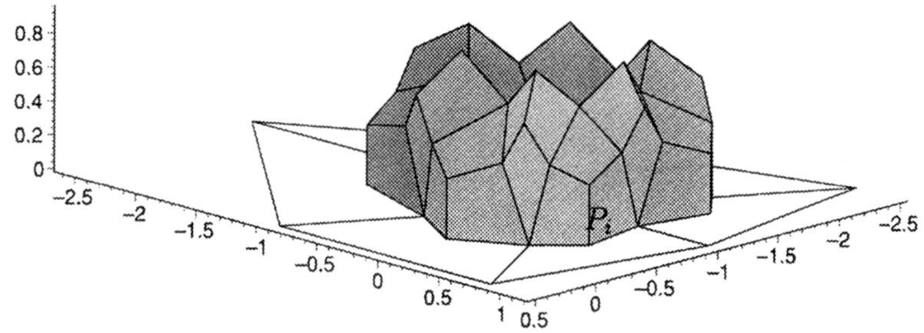

FIG. 3.3.11. The construction of the dual finite volume D_i

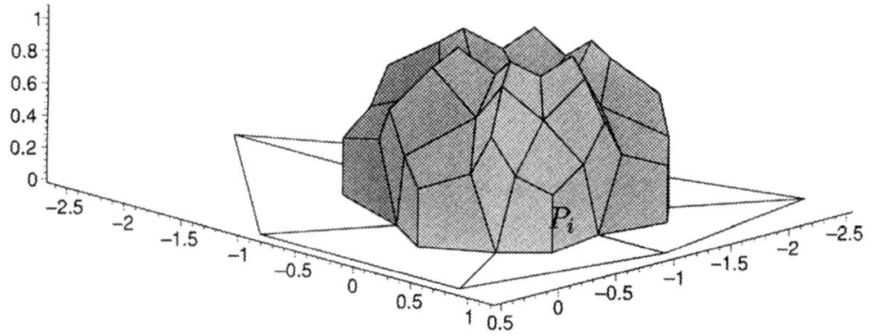

FIG. 3.3.12. The construction of the dual finite volume D_i

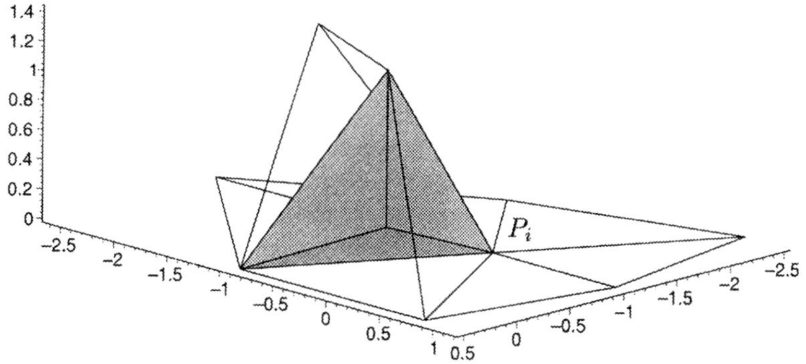

FIG. 3.3.13. Barycentric finite volume $D_i \in \mathcal{D}_h$ (union of two tetrahedra)

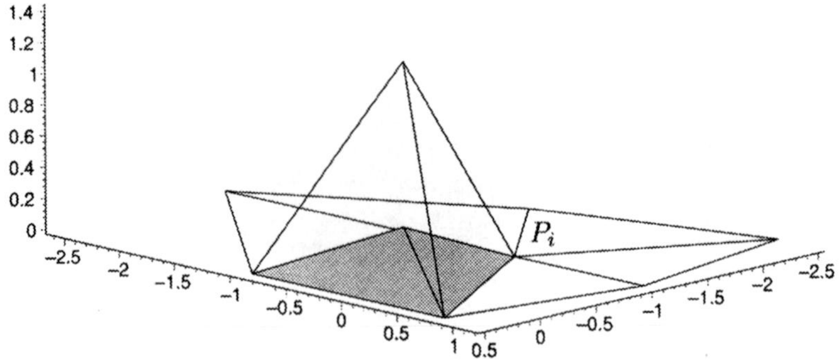

FIG. 3.3.14. Barycentric finite volume $D_i \in \mathcal{D}_h$ with one face on the $\partial\Omega_h$ (tetrahedron)

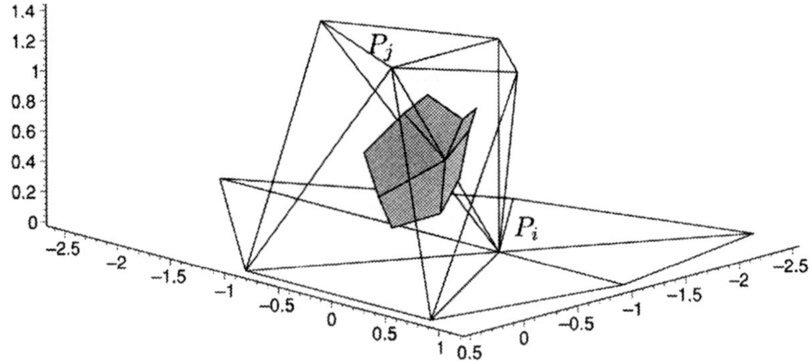

FIG. 3.3.15. Dual finite volume D_i: the construction of Γ_{ij}^{α} if $P_j \in \Omega_h$

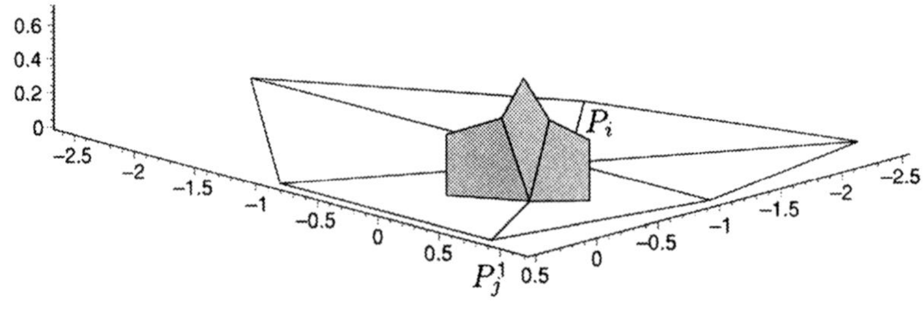

FIG. 3.3.16. Dual finite volume D_i: the construction of Γ_{ij}^{α} if $P_i, P_j \in \partial\Omega_h$

Remark 3.53 In practical applications of the finite volume method as well as the finite element method various types of meshes (also called grids) are used:

structured–unstructured,
uniform–nonuniform,
isotropic–anisotropic,
regular–irregular.

Mesh terminology is fairly consistent in the literature (see, for example, (Simpson, 1994)) for triangular or quadrilateral meshes in 2D and tetrahedral or hexahedral meshes in 3D.

A mesh is called *structured*, if the same number of elements meet at any vertex of the mesh. (For example, six triangles meet at each vertex.) In an *unstructured mesh* there is no restriction placed on the number of elements which can meet at any vertex. In a *uniform mesh* all elements are of the same form. Thus, the length scales of all elements are translation invariant. In an *isotropic mesh* all length scales of all elements are essentially the same. If some length scales of some elements are clearly different, we speak of an *anisotropic mesh*. In such a mesh, elements may be rather stretched. A mesh is *regular*, if for two arbitrary different (closed) elements their intersection is either empty or it consists of a common vertex or a common edge or (in 3D) a common face.

3.3.2 *Derivation of a general finite volume scheme*

In order to derive a finite volume scheme, we can proceed similarly as in Section 3.2. Let us assume that $\boldsymbol{w} : \overline{\Omega} \times [0,T] \to I\!R^m$ is a classical (i.e. C^1-) solution of system (3.3.1), $\mathcal{D}_h = \{D_i\}_{i\in J}$ is a finite volume mesh in a polyhedral approximation Ω_h of Ω. Let us construct a partition $0 = t_0 < t_1 < \dots$ of the time interval $[0,T]$ and denote by $\tau_k = t_{k+1} - t_k$ the time step between t_k and t_{k+1}. Integrating equation (3.3.1) over the set $D_i \times (t_k, t_{k+1})$ and using Green's theorem on D_i, we get the identity

$$\int_{D_i} \boldsymbol{w}(x,t)\,dx\Big|_{t=t_k}^{t_{k+1}} + \int_{t_k}^{t_{k+1}} \left(\int_{\partial D_i} \sum_{s=1}^{N} \boldsymbol{f}_s(\boldsymbol{w}) n_s \, dS \right) dt = 0.$$

Moreover, taking into account (3.3.14), we can write

$$\int_{D_i} (\boldsymbol{w}(x,t_{k+1}) - \boldsymbol{w}(x,t_k))\,dx \qquad (3.3.19)$$
$$+ \int_{t_k}^{t_{k+1}} \left(\sum_{j\in S(i)} \sum_{\alpha=1}^{\beta_{ij}} \int_{\Gamma_{ij}^{\alpha}} \sum_{s=1}^{N} \boldsymbol{f}_s(\boldsymbol{w}) n_s \, dS \right) dt = 0.$$

Now we shall approximate the integral averages $\int_{D_i} \boldsymbol{w}(x,t_k)dx/|D_i|$ of the quantity $\boldsymbol{w}$ over the finite volume D_i at time instant t_k by $\boldsymbol{w}_i^k$:

$$\boldsymbol{w}_i^k \approx \frac{1}{|D_i|} \int_{D_i} \boldsymbol{w}(x,t_k)\,dx, \qquad (3.3.20)$$

called the value of *the approximate solution* on D_i at time t_k. Further, we approximate the flux $\sum_{s=1}^{N} \boldsymbol{f}_s(\boldsymbol{w})(n_{ij}^\alpha)_s$ of the quantity $\boldsymbol{w}$ through the face Γ_{ij}^α in the direction $\boldsymbol{n}_{ij}^\alpha$ with the aid of a *numerical flux* $\boldsymbol{H}(\boldsymbol{w}_i^\ell, \boldsymbol{w}_j^\ell, \boldsymbol{n}_{ij}^\alpha)$, depending on the value of the approximate solution $\boldsymbol{w}_i^\ell$ on the finite volume D_i, the value $\boldsymbol{w}_j^\ell$ on D_j, and on the normal $\boldsymbol{n}_{ij}^\alpha$ at suitable time instants t_ℓ:

$$\sum_{s=1}^{N} \boldsymbol{f}_s(\boldsymbol{w})(n_{ij}^\alpha)_s \approx \boldsymbol{H}(\boldsymbol{w}_i^\ell, \boldsymbol{w}_j^\ell, \boldsymbol{n}_{ij}^\alpha). \tag{3.3.21}$$

We choose, for example, $\ell = k$ or $\ell = k+1$. If $\Gamma_{ij}^\alpha \subset \partial\Omega_h$ (i.e. the finite volume D_i is adjacent to $\partial\Omega_h$, $j \in \gamma(i)$, $\alpha = 1$ and $\Gamma_{ij}^1 = \Gamma_{ij}$), then there is no neighbour D_j of D_i adjacent to the face Γ_{ij} from the exterior of Ω_h and it is necessary to specify $\boldsymbol{w}_j^\ell$ on the basis of boundary conditions – see Section 3.3.6. In such a way we arrive at the approximation

$$\begin{aligned} &\int_{t_k}^{t_{k+1}} \left(\int_{\Gamma_{ij}^\alpha} \sum_{s=1}^{N} \boldsymbol{f}_s(\boldsymbol{w})(n_{ij}^\alpha)_s \, dS \right) dt \\ &\approx \tau_k \left[\vartheta \boldsymbol{H}(\boldsymbol{w}_i^{k+1}, \boldsymbol{w}_j^{k+1}, \boldsymbol{n}_{ij}^\alpha) + (1-\vartheta)\boldsymbol{H}(\boldsymbol{w}_i^k, \boldsymbol{w}_j^k, \boldsymbol{n}_{ij}^\alpha) \right] |\Gamma_{ij}^\alpha|, \quad \vartheta \in [0,1]. \end{aligned} \tag{3.3.22}$$

Using (3.3.19), (3.3.20) and (3.3.22), we obtain the following *finite volume scheme*:

$$\begin{aligned} \boldsymbol{w}_i^{k+1} = \boldsymbol{w}_i^k - \frac{\tau_k}{|D_i|} \sum_{j \in S(i)} \sum_{\alpha=1}^{\beta_{ij}} &\Big[\vartheta \boldsymbol{H}(\boldsymbol{w}_i^{k+1}, \boldsymbol{w}_j^{k+1}, \boldsymbol{n}_{ij}^\alpha) \\ &+ (1-\vartheta)\boldsymbol{H}(\boldsymbol{w}_i^k, \boldsymbol{w}_j^k, \boldsymbol{n}_{ij}^\alpha) \Big] |\Gamma_{ij}^\alpha|, \quad D_i \in \mathcal{D}_h, \; t_k \in [0,T), \; \vartheta \in [0,1]. \end{aligned} \tag{3.3.23}$$

If $\vartheta \in (0,1]$, then formula (3.3.23) is *implicit* and requires the solution of a complicated nonlinear system with respect to unknown values $\boldsymbol{w}_i^{k+1}$, $i \in J$. This is the reason that one usually prefers to use the *explicit scheme* with $\vartheta = 0$:

$$\begin{aligned} \boldsymbol{w}_i^{k+1} &= \boldsymbol{w}_i^k - \frac{\tau_k}{|D_i|} \sum_{j \in S(i)} \sum_{\alpha=1}^{\beta_{ij}} \boldsymbol{H}(\boldsymbol{w}_i^k, \boldsymbol{w}_j^k, \boldsymbol{n}_{ij}^\alpha) |\Gamma_{ij}^\alpha|, \\ &D_i \in \mathcal{D}_h, \; t_k \in [0,T). \end{aligned} \tag{3.3.24}$$

Method (3.3.23) is equipped with *initial conditions* $\boldsymbol{w}_i^0$, $i \in J$, defined by

$$\boldsymbol{w}_i^0 = \frac{1}{|D_i|} \int_{D_i} \boldsymbol{w}^0(x) \, dx, \tag{3.3.25}$$

under the assumption that the function $\boldsymbol{w}^0$ from (3.3.2) is locally integrable: $\boldsymbol{w}^0 \in L^1_{\mathrm{loc}}(\Omega)^m$.

Definition 3.54 *We define a finite volume approximate solution of* (3.3.1) *as piecewise constant vector-valued functions* $\mathbf{w}_h^k, k = 0, 1, \ldots,$ *defined a.e. in* Ω_h *so that* $\mathbf{w}_h^k|_{\overset{\circ}{D}_i} = \boldsymbol{w}_i^k$ *for all* $i \in J$*, where* $\overset{\circ}{D}_i$ *is the interior of* D_i*, i.e.* $\overset{\circ}{D}_i = D_i \setminus \partial D_i$*, and* $\boldsymbol{w}_i^k$ *are obtained from the formula* (3.3.23). *The function* $\mathbf{w}_h^k$ *is the approximate solution at time* $t = t_k$*. The vector* $\boldsymbol{w}_i^k$ *is the value of the approximate solution on the finite volume* D_i *at time* t_k*.*

3.3.3 *Properties of the numerical flux*

In what follows, we shall assume that the numerical flux $\boldsymbol{H}$ has the following properties:

1. $\boldsymbol{H}(\boldsymbol{u}, \boldsymbol{v}, \boldsymbol{n})$ is defined and continuous on $D \times D \times \mathcal{S}_1$, where D is the domain of definition of the fluxes $\boldsymbol{f}_s$ and $\mathcal{S}_1$ is the unit sphere in $\mathbb{R}^N$: $\mathcal{S}_1 = \{\boldsymbol{n} \in \mathbb{R}^N; |\boldsymbol{n}| = 1\}$.
2. $\boldsymbol{H}$ is *consistent*:

$$\boldsymbol{H}(\boldsymbol{u}, \boldsymbol{u}, \boldsymbol{n}) = \boldsymbol{\mathcal{P}}(\boldsymbol{u}, \boldsymbol{n}) = \sum_{s=1}^{N} \boldsymbol{f}_s(\boldsymbol{u}) n_s, \quad \boldsymbol{u} \in D, \ \boldsymbol{n} \in \mathcal{S}_1. \tag{3.3.26}$$

3. $\boldsymbol{H}$ is *conservative*:

$$\boldsymbol{H}(\boldsymbol{u}, \boldsymbol{v}, \boldsymbol{n}) = -\boldsymbol{H}(\boldsymbol{v}, \boldsymbol{u}, -\boldsymbol{n}), \quad \boldsymbol{u}, \boldsymbol{v} \in D, \ \boldsymbol{n} \in \mathcal{S}_1. \tag{3.3.27}$$

If $\boldsymbol{H}$ satisfies conditions (3.3.26) and (3.3.27), the *method* (3.3.23) is called *consistent* and *conservative*, respectively. (Note that the conservativity of the scheme means that the flux from the finite volume D_i into D_j through Γ_{ij}^α has the same magnitude, but opposite sign, as the flux from D_j into D_i.)

3.3.4 *Construction of some numerical fluxes*

One possible way to construct a numerical flux $\boldsymbol{H}$ is to use an analogy with the 1D case, replacing the 1D flux $\boldsymbol{f}(\boldsymbol{w})$ from Section 3.2 by the N-dimensional flux $\boldsymbol{\mathcal{P}}(\boldsymbol{w}, \boldsymbol{n})$ in the direction $\boldsymbol{n} \in \mathcal{S}_1$ defined in (3.3.4). In this way we obtain the generalization of the Lax–Friedrichs scheme (cf. Section 3.2.8) and flux vector splitting schemes of the Godunov type (cf. Section 3.2.17).

a) The *Lax–Friedrichs numerical flux* is defined by

$$\boldsymbol{H}_{\mathrm{LF}}(\boldsymbol{u}, \boldsymbol{v}, \boldsymbol{n}) = \frac{1}{2}\left(\boldsymbol{\mathcal{P}}(\boldsymbol{u}, \boldsymbol{n}) + \boldsymbol{\mathcal{P}}(\boldsymbol{v}, \boldsymbol{n}) - \frac{1}{\lambda}(\boldsymbol{v} - \boldsymbol{u})\right), \quad \boldsymbol{u}, \boldsymbol{v} \in D, \ \boldsymbol{n} \in \mathcal{S}_1. \tag{3.3.28}$$

Here $\lambda > 0$ is independent of $\boldsymbol{u}, \boldsymbol{v}$, but depends, in general, on Γ_{ij}^α in scheme (3.3.23). (See the theoretical analysis in Section 3.3.8.)

To obtain flux vector splitting schemes, we use relations (3.3.4)–(3.3.7). On the basis of (3.3.7) (similarly as in (3.2.90)–(3.2.91)) we define the matrices

$$\mathbb{\Lambda}^{\pm} = \mathrm{diag}(\lambda_1^{\pm}, \ldots, \lambda_m^{\pm}), \quad |\mathbb{\Lambda}| = \mathrm{diag}(|\lambda_1|, \ldots, |\lambda_m|), \tag{3.3.29}$$

$$\mathbb{P}^{\pm} = \mathbb{T}\mathbb{A}^{\pm}\mathbb{T}^{-1}, \quad |\mathbb{P}| = \mathbb{T}|\mathbb{A}|\mathbb{T}^{-1}, \tag{3.3.30}$$

depending on $\boldsymbol{w} \in D$ and $\boldsymbol{n} \in \mathcal{S}_1$. Now we define the following schemes:

b) *The Steger–Warming scheme* has the numerical flux

$$\boldsymbol{H}_{\mathrm{SW}}(\boldsymbol{u},\boldsymbol{v},\boldsymbol{n}) = \mathbb{P}^{+}(\boldsymbol{u},\boldsymbol{n})\boldsymbol{u} + \mathbb{P}^{-}(\boldsymbol{v},\boldsymbol{n})\boldsymbol{v}, \quad \boldsymbol{u},\boldsymbol{v} \in D,\ \boldsymbol{n} \in \mathcal{S}_1. \tag{3.3.31}$$

c) *The Vijayasundaram scheme*:

$$\boldsymbol{H}_{\mathrm{V}}(\boldsymbol{u},\boldsymbol{v},\boldsymbol{n}) = \mathbb{P}^{+}\left(\frac{\boldsymbol{u}+\boldsymbol{v}}{2},\boldsymbol{n}\right)\boldsymbol{u} + \mathbb{P}^{-}\left(\frac{\boldsymbol{u}+\boldsymbol{v}}{2},\boldsymbol{n}\right)\boldsymbol{v}. \tag{3.3.32}$$

d) *The Van Leer scheme*:

$$\boldsymbol{H}_{\mathrm{VL}}(\boldsymbol{u},\boldsymbol{v},\boldsymbol{n}) = \frac{1}{2}\left\{\boldsymbol{\mathcal{P}}(\boldsymbol{u},\boldsymbol{n}) + \boldsymbol{\mathcal{P}}(\boldsymbol{v},\boldsymbol{n}) - \left|\mathbb{P}\left(\frac{\boldsymbol{u}+\boldsymbol{v}}{2},\boldsymbol{n}\right)\right|(\boldsymbol{v}-\boldsymbol{u})\right\}. \tag{3.3.33}$$

3.3.5 *Another construction of the multidimensional numerical flux*

The above schemes can also be derived on the basis of the following general approach, based on the rotational invariance (3.1.40) and (3.1.57) of the 2D and 3D Euler equations. Let $\Gamma = \Gamma_{ij}^{\alpha}$ be a part of the interface between the finite volumes D_i and D_j with normal $\boldsymbol{n} = \boldsymbol{n}_{ij}^{\alpha}$ pointing from D_i into D_j.

In $I\!R^N$ ($N = 2$ or 3) we introduce a new Cartesian coordinate system $\tilde{x}_1, \ldots, \tilde{x}_N$ with the origin at the midpoint of the face Γ, the coordinate $\tilde{x}_1$ oriented in the direction of the normal $\boldsymbol{n}$ and $\tilde{x}_2, \ldots, \tilde{x}_N$ tangent to Γ. The rotational invariance of the Euler equations implies that these equations transformed into this new coordinate system have the form

$$\frac{\partial \boldsymbol{q}}{\partial t} + \sum_{s=1}^{N} \frac{\partial \boldsymbol{f}_s(\boldsymbol{q})}{\partial \tilde{x}_s} = 0, \tag{3.3.34}$$

see Theorem 3.5, where

$$\boldsymbol{q} = \mathbb{Q}(\boldsymbol{n})\boldsymbol{w} \tag{3.3.35}$$

with matrix $\mathbb{Q}(\boldsymbol{n})$ defined in (3.1.58) or (3.1.35) for $N = 2$ or 3, respectively. Now we neglect the tangential derivatives $\partial/\partial\tilde{x}_s$, $s > 1$, and get the system with one space variable $\tilde{x}_1$

$$\frac{\partial \boldsymbol{q}}{\partial t} + \frac{\partial \boldsymbol{f}_1(\boldsymbol{q})}{\partial \tilde{x}_1} = 0. \tag{3.3.36}$$

In analogy to (3.1.40) or (3.1.57) we define the numerical flux for the multidimensional Euler system (3.3.1) in the form

$$\boldsymbol{H}(\boldsymbol{u},\boldsymbol{v},\boldsymbol{n}) = \mathbb{Q}^{-1}(\boldsymbol{n})\boldsymbol{g}_{\mathrm{R}}(\mathbb{Q}(\boldsymbol{n})\boldsymbol{u}, \mathbb{Q}(\boldsymbol{n})\boldsymbol{v}), \tag{3.3.37}$$

where $\boldsymbol{g}_{\mathrm{R}} = \boldsymbol{g}_{\mathrm{R}}(\boldsymbol{q}_1, \boldsymbol{q}_2)$ is a numerical flux (approximate Riemann solver) for the system (3.3.36) with one space variable $\tilde{x}_1$.

Obviously, the consistency of $\boldsymbol{g}_{\mathrm{R}}$ implies that the numerical flux $\boldsymbol{H}$ defined by (3.3.37) is consistent. The proof of the conservativity of numerical fluxes for the multidimensional Euler equations is difficult. However, the test of the conservativity can be included in numerical codes.

Exercise 3.55 Derive schemes (3.3.28), (3.3.31)–(3.3.33) with the aid of definition (3.3.37).

The multidimensional Roe numerical flux $\boldsymbol{H}_{\mathrm{Roe}}(\boldsymbol{u}, \boldsymbol{v}, \boldsymbol{n})$ is defined also by (3.3.37) with $\boldsymbol{g}_{\mathrm{R}} := \boldsymbol{g}_{\mathrm{Roe}}$ (see (3.2.125)).

There are also other finite volume numerical schemes based on approximate Riemann solvers. An extensive treatment of various approximate Riemann solvers can be found, for example, in (Toro, 1997). A number of numerical experiments with the solution of complicated technically relevant problems (see Section 3.7) showed that the *Osher–Solomon scheme* appears as one of the most robust methods. Its derivation is complicated and, therefore, a separate Section 3.4 is devoted to its description.

3.3.6 *Boundary conditions*

The choice of appropriate boundary conditions is a very important and delicate question in the numerical simulation of fluid flow. The imposition of boundary conditions is, fundamentally, a physical problem, but it must correspond to the mathematical character of the solved equations. Great care is required in their numerical implementation. Usually two types of boundaries are considered: *reflective* and *transparent* or *transmissive*. The reflective boundaries usually consist of fixed walls. Transmissive or transparent boundaries arise from the need to replace unbounded or rather large physical domains by bounded or sufficiently small computational domains. The corresponding boundary conditions are devised so that they allow the passage of waves without any effect on them. For 1D problems the objective is reasonably well attained. For multidimensional problems this is a substantial area of current research, usually referred to *open-end* boundary conditions, *transparent* boundary conditions, *far-field* boundary conditions, *radiation* boundary conditions or *non-reflecting* boundary conditions. Useful publications dealing with boundary conditions are (Bayliss and Turkel, 1980), (Hedstrom, 1979), (Roe, 1989), (Giles, 1990), (Gustafsson, 1982), (Gustafsson and Fern, 1987), (Gustafsson and Fern, 1988), (Gustafsson and Kreiss, 1979), (Hagstrom and Hariharan, 1988), (Kröner, 1991), (Karni, 1992), (Godlewski and Raviart, 1996), Chapter V. A rigorous mathematical theory of boundary conditions to conservation laws was developed only for a scalar equation in (Bardos *et al.*, 1979).

In the numerical solution of the compressible Euler equations, one uses various more or less heuristic approaches. Here we explain one possibility used often in practice. Let $D_i \in \mathcal{D}_h$ be a finite volume adjacent to the boundary $\partial\Omega_h$, i.e. ∂D_i is formed by faces $\Gamma = \Gamma_{ij}^1 \subset \partial\Omega_h$ $(j \in \gamma(i))$ and let $\boldsymbol{n} = \boldsymbol{n}_{ij}^1$ be a unit outer normal to ∂D_i on Γ. (See Section 3.3.1.) In order to be able to compute the numerical flux $\boldsymbol{H}(\boldsymbol{w}_i^k, \boldsymbol{w}_j^k, \boldsymbol{n})$, it is necessary to specify the value $\boldsymbol{w}_j^k$. We describe here how to determine boundary conditions for the Euler equations. (Another approach will be explained in Section 3.4 in connection with the Osher–Solomon scheme.)

Similarly as in Section 3.3.5, we introduce a new Cartesian coordinate system $\tilde{x}_1, \ldots, \tilde{x}_N$ in $\mathbb{R}^N$ ($N = 2$ or 3) with origin at the centre of gravity of the face Γ, the coordinate $\tilde{x}_1$ oriented in the direction of the normal $\boldsymbol{n}$ and $\tilde{x}_2, \ldots, \tilde{x}_N$ tangent to Γ. The Euler equations transformed into this coordinate system have the form (3.3.34). Now we neglect the tangential derivatives $\partial/\partial\tilde{x}_s$, $s > 1$, and get the system with one space variable $\tilde{x}_1$ in the form (3.3.36). Further, we linearize this system around the state $\boldsymbol{q}_i^k = \mathbb{Q}(\boldsymbol{n})\boldsymbol{w}_i^k$. As a result we obtain the linear system

$$\frac{\partial \boldsymbol{q}}{\partial t} + \mathbb{A}_1(\boldsymbol{q}_i^k)\frac{\partial \boldsymbol{q}}{\partial \tilde{x}_1} = 0, \tag{3.3.38}$$

which will be considered in the set $(-\infty, 0) \times (0, \infty)$ and equipped with the initial condition

$$\boldsymbol{q}(\tilde{x}_1, 0) = \boldsymbol{q}_i^k, \quad \tilde{x}_1 \in (-\infty, 0), \tag{3.3.39}$$

and the boundary condition

$$\boldsymbol{q}(0, t) = \boldsymbol{q}_j^k, \quad t > 0. \tag{3.3.40}$$

Our goal is to choose the boundary state $\boldsymbol{q}_j^k$ in such a way that the initial-boundary value problem (3.3.38)–(3.3.40) is well-posed, i.e. it has a unique solution. Then we set $\boldsymbol{w}_j^k := \boldsymbol{q}_j^k$. Similarly as in Section 3.2.15, the solution of (3.3.38) can be written in the form

$$\boldsymbol{q}(\tilde{x}_1, t) = \sum_{s=1}^{m} \mu_s(\tilde{x}_1, t)\boldsymbol{r}_s, \tag{3.3.41}$$

where $\boldsymbol{r}_s = \boldsymbol{r}_s(\boldsymbol{q}_i^k)$ are the eigenvectors of the matrix $\mathbb{A}_1(\boldsymbol{q}_i^k)$ corresponding to its eigenvalues $\tilde{\lambda}_s = \tilde{\lambda}_s(\boldsymbol{q}_i^k)$ and creating a basis in $\mathbb{R}^m$. Moreover,

$$\boldsymbol{q}_i^k = \sum_{s=1}^{m} \alpha_s \boldsymbol{r}_s, \quad \boldsymbol{q}_j^k = \sum_{s=1}^{m} \beta_s \boldsymbol{r}_s. \tag{3.3.42}$$

Substituting (3.3.41) into (3.3.38) and using the relation $\mathbb{A}_1(\boldsymbol{q}_i^k)\boldsymbol{r}_s = \tilde{\lambda}_s \boldsymbol{r}_s$, we find that problem (3.3.38)–(3.3.40) is equivalent to m mutually independent linear initial-boundary value scalar problems

$$\begin{aligned} &\frac{\partial \mu_s}{\partial t} + \tilde{\lambda}_s \frac{\partial \mu_s}{\partial \tilde{x}_1} = 0 \quad \text{in } (-\infty, 0) \times (0, \infty), \\ &\mu_s(\tilde{x}_1, 0) = \alpha_s, \quad \tilde{x}_1 \in (-\infty, 0), \\ &\mu_s(0, t) = \beta_s, \quad t \in (0, \infty), \end{aligned} \tag{3.3.43}$$

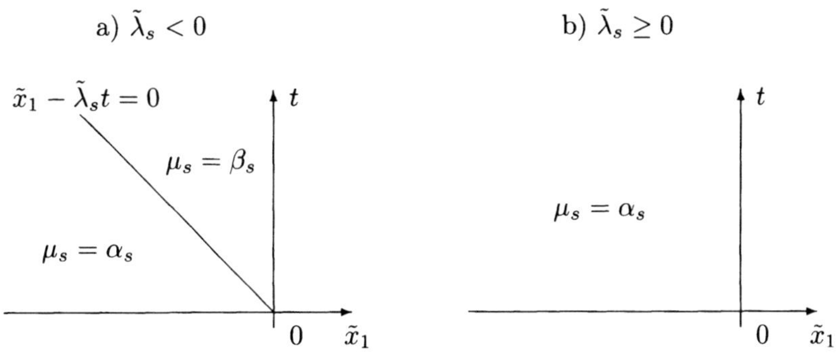

FIG. 3.3.17. Solution of problem (3.3.43)

$s = 1, \dots, m,$

which can be solved by the method of characteristics. The solution is

$$\mu_s(\tilde{x}_1, t) = \begin{cases} \alpha_s, & \tilde{x}_1 - \tilde{\lambda}_s t < 0, \\ \beta_s, & \tilde{x}_1 - \tilde{\lambda}_s t > 0. \end{cases} \tag{3.3.44}$$

The possible situations are shown in Fig. 3.3.17. From this it is clear that

if $\tilde{\lambda}_s > 0$, then $\beta_s = \alpha_s$ (β_s is not prescribed, but it is obtained by the extrapolation of μ_s to the boundary $\tilde{x}_1 = 0$); (3.3.45)

if $\tilde{\lambda}_s = 0$, then β_s is not prescribed (but can again be defined as $\beta_s = \alpha_s$ by the continuous extension of μ_s to the boundary $\tilde{x}_1 = 0$);

if $\tilde{\lambda}_s < 0$, then β_s must be prescribed.

Taking into account the results from Section 3.1, we realize that

$$\tilde{\lambda}_s(\boldsymbol{q}_i^k) = \lambda_s(\boldsymbol{w}_i^k, \boldsymbol{n}), \quad s = 1, \dots, m, \tag{3.3.46}$$

where $\lambda_s(\boldsymbol{w}_i^k, \boldsymbol{n})$ are the eigenvalues of the Jacobi matrix $\mathbb{P}(\boldsymbol{w}_i^k, \boldsymbol{n})$ (see (3.3.5)–(3.3.6)). Hence, on the basis of the above considerations, we come to the following *conclusion*. On $\Gamma = \Gamma_{ij}^\alpha \subset \partial\Omega_h$ (i.e. $i \in J, j \in \gamma(i), \alpha = 1$) with normal $\boldsymbol{n} = \boldsymbol{n}_{ij}^\alpha$, pointing from D_i into D_j, we have to prescribe n_{pr} quantities characterizing the state vector $\boldsymbol{w}$, where n_{pr} is the number of negative eigenvalues of the matrix $\mathbb{P}(\boldsymbol{w}_i^k, \boldsymbol{n})$, whereas we extrapolate n_{ex} quantities to the boundary, where n_{ex} is the number of nonnegative eigenvalues of $\mathbb{P}(\boldsymbol{w}_i^k, \boldsymbol{n})$. The *extrapolation* of a quantity $\boldsymbol{q}$ to the boundary means in this case to set $\boldsymbol{q}_j^k := \boldsymbol{q}_i^k$. On the other hand, if we prescribe the boundary value of $\boldsymbol{q}$, we set $\boldsymbol{q}_j^k := \boldsymbol{q}_{Bj}^k$ with a given value $\boldsymbol{q}_{Bj}^k$, determined by the user on the basis of the physical character of the flow.

It is suitable to use a special treatment if Γ is a part of a *solid impermeable wall*, where $\boldsymbol{v} \cdot \boldsymbol{n} = 0$. Then the flux $\boldsymbol{\mathcal{P}}(\boldsymbol{w}, \boldsymbol{n})$ has the form

$$\begin{aligned} \boldsymbol{\mathcal{P}}(\boldsymbol{w}, \boldsymbol{n}) &= \sum_{s=1}^{N} \boldsymbol{f}_s(\boldsymbol{w}) n_s \qquad (3.3.47) \\ &= (\boldsymbol{v} \cdot \boldsymbol{n})\boldsymbol{w} + p(0, n_1, \ldots, n_N, \boldsymbol{v} \cdot \boldsymbol{n})^{\mathrm{T}} \\ &= p(0, n_1, \ldots, n_N, 0)^{\mathrm{T}}, \end{aligned}$$

which is uniquely determined on Γ by the extrapolated value of the pressure, i.e. by $p_j^k := p_i^k$. Therefore, on the part Γ of the impermeable solid boundary we define the numerical flux

$$\boldsymbol{H}(\boldsymbol{w}_i^k, \boldsymbol{w}_j^k, \boldsymbol{n}) = p_i^k (0, n_1, \ldots, n_N, 0)^{\mathrm{T}}. \qquad (3.3.48)$$

We can see that in view of (3.3.6), on an impermeable boundary, N eigenvalues $\lambda_2, \ldots, \lambda_{m-1}$ of the matrix $\mathbb{P}(\boldsymbol{w}_i^k, \boldsymbol{n})$ are zero, the eigenvalue λ_1 is negative and the eigenvalue λ_m is positive. We prescribe one scalar quantity, namely $\boldsymbol{v} \cdot \boldsymbol{n} = 0$, and extrapolate the pressure p (and possibly the density and tangential components to Γ of the velocity, i.e. we extrapolate $n_{ex} = m - 1$ quantities).

There are several ways to choose what quantities should be prescribed or extrapolated. We present here one possibility, which is often used in practical computations. It is suitable to distinguish several cases given in Table 3.3.1 (for 2D flow, $N = 2,\ m = 4$) and Table 3.3.2 (for 3D flow, $N = 3,\ m = 5$).

Sometimes in practical problems, other quantities than the ones given in Tables 3.3.1 and 3.3.2 are prescribed, e.g. Mach number at the subsonic outlet. Then it is necessary to apply a suitable iterative process for the determination of such outlet boundary pressure which leads to the prescribed outlet boundary Mach number. In some technically relevant problems it is necessary to apply also boundary conditions of other types, such as periodic boundary conditions. See, for example, Section 3.7.

On the basis of the procedure described above, i.e. scheme (3.3.23), equipped with the initial condition (3.3.25) and boundary conditions described in this section, at each time level t_k we obtain an approximate solution of problem (3.3.1)–(3.3.3).

Now we come to *the investigation of qualitative properties* of the finite volume schemes for the solution of the Euler equations. Because of the complexity of this problem, only partial limited results can be achieved in order to support theoretically the applicability of the developed schemes. Usually, the quality of these methods is justified in such a way that their properties are studied and verified on simple model problems and/or under simplified assumptions. Let us mention also the investigation of these schemes applied to scalar or linear problems, or considered on simple structured meshes. A typical example is the investigation of the *stability* of numerical schemes.

Table 3.3.1 *Boundary conditions for 2D flow*

Type of boundary	Character of the flow	The sign of eigenvalues n_{pr} and n_{ex}	Quantities extrapolated	Quantities prescribed
INLET $(\boldsymbol{v} \cdot \boldsymbol{n} < 0)$	supersonic flow $(-\boldsymbol{v} \cdot \boldsymbol{n} > a)$	$\lambda_1 < 0$ $\lambda_2 = \lambda_3 < 0$ $\lambda_4 < 0$ $n_{pr} = 4, \ n_{ex} = 0$	—	ρ, v_1, v_2, p
	subsonic flow $(-\boldsymbol{v} \cdot \boldsymbol{n} \leq a)$	$\lambda_1 < 0$ $\lambda_2 = \lambda_3 < 0$ $\lambda_4 \geq 0$ $n_{pr} = 3, \ n_{ex} = 1$	p	ρ, v_1, v_2
OUTLET $(\boldsymbol{v} \cdot \boldsymbol{n} > 0)$	supersonic flow $(\boldsymbol{v} \cdot \boldsymbol{n} \geq a)$	$\lambda_1 \geq 0$ $\lambda_2 = \lambda_3 > 0$ $\lambda_4 > 0$ $n_{pr} = 0, \ n_{ex} = 4$	ρ, v_1, v_2, p	—
	subsonic flow $(\boldsymbol{v} \cdot \boldsymbol{n} < a)$	$\lambda_1 < 0$ $\lambda_2 = \lambda_3 > 0$ $\lambda_4 > 0$ $n_{pr} = 1, \ n_{ex} = 3$	ρ, v_1, v_2	p
SOLID IMPER-MEABLE BOUNDARY	$\boldsymbol{v} \cdot \boldsymbol{n} = 0$	$\lambda_1 < 0$ $\lambda_2 = \lambda_3 = 0$ $\lambda_4 > 0$ $n_{pr} = 1, \ n_{ex} = 3$	p (ρ, v_t)	$\boldsymbol{v} \cdot \boldsymbol{n} = 0$

3.3.7 *Stability of the finite volume schemes*

Let $\mathbf{w}^k = \{\boldsymbol{w}_i^k\}_{i \in J}$ be an approximate solution on the k-th time level obtained with the aid of the finite volume method (3.3.23). By $\|\mathbf{w}^k\|$ we denote a norm of the approximation $\mathbf{w}^k$. In accordance with Section 3.2.6 we call *scheme* (3.3.23) *stable*, if there exists a constant $c > 0$ independent of τ, h, k such that

$$\|\mathbf{w}^k\| \leq c\|\mathbf{w}^0\|, \quad k = 0, 1, \ldots. \tag{3.3.49}$$

Usually an analogy to the L^p-norm ($p \in [1, \infty]$) is used:

$$\|\mathbf{w}^k\|_\infty = \sup_{i \in J} |\boldsymbol{w}_i^k|, \tag{3.3.50}$$

$$\|\mathbf{w}^k\|_p = \left\{ \sum_{i \in J} |D_i| |\boldsymbol{w}_i^k|^p \right\}^{1/p}, \quad p \in [1, \infty).$$

In what follows we shall be concerned with the stability of the *explicit* FV method (3.3.24). For simplicity we confine our considerations to the 2D case.

Table 3.3.2 *Boundary conditions for 3D flow*

Type of boundary	Character of the flow	The sign of eigenvalues n_{pr} and n_{ex}	Quantities extrapolated	Quantities prescribed
INLET $(\boldsymbol{v} \cdot \boldsymbol{n} < 0)$	supersonic flow $(-\boldsymbol{v} \cdot \boldsymbol{n} > a)$	$\lambda_1 < 0$ $\lambda_2 = \lambda_3 = \lambda_4 < 0$ $\lambda_5 < 0$ $n_{pr} = 5,\ n_{ex} = 0$	—	ρ, v_1, v_2, v_3, p
	subsonic flow $(-\boldsymbol{v} \cdot \boldsymbol{n} \le a)$	$\lambda_1 < 0$ $\lambda_2 = \lambda_3 = \lambda_4 < 0$ $\lambda_5 \ge 0$ $n_{pr} = 4,\ n_{ex} = 1$	p	ρ, v_1, v_2, v_3
OUTLET $(\boldsymbol{v} \cdot \boldsymbol{n} > 0)$	supersonic flow $(\boldsymbol{v} \cdot \boldsymbol{n} \ge a)$	$\lambda_1 \ge 0$ $\lambda_2 = \lambda_3 = \lambda_4 > 0$ $\lambda_5 > 0$ $n_{pr} = 0,\ n_{ex} = 5$	ρ, v_1, v_2, v_3, p	—
	subsonic flow $(\boldsymbol{v} \cdot \boldsymbol{n} < a)$	$\lambda_1 < 0$ $\lambda_2 = \lambda_3 = \lambda_4 > 0$ $\lambda_5 > 0$ $n_{pr} = 1,\ n_{ex} = 4$	ρ, v_1, v_2, v_3	p
SOLID IMPER-MEABLE BOUNDARY	$\boldsymbol{v} \cdot \boldsymbol{n} = 0$	$\lambda_1 < 0$ $\lambda_2 = \lambda_3 = \lambda_4 = 0$ $\lambda_5 > 0$ $n_{pr} = 1,\ n_{ex} = 4$	p (v_{t_1}, v_{t_2}, ρ)	$\boldsymbol{v} \cdot \boldsymbol{n} = 0$

3.3.8 *FV schemes on 2D uniform rectangular meshes*

One possibility is to apply scheme (3.3.24) on a *uniform rectangular mesh* $\mathcal{D}_h = \{D_{\mu\nu}; \mu, \nu \in Z\}$, where $D_{\mu\nu} = \left(\left(\mu - \frac{1}{2}\right) h_1, \left(\mu + \frac{1}{2}\right) h_1\right) \times \left(\left(\nu - \frac{1}{2}\right) h_2, \left(\nu + \frac{1}{2}\right) h_2\right)$ is the rectangle with centre of gravity $P_{\mu\nu} = (\mu h_1, \nu h_2)$, sides of length $h_1 > 0$ parallel to the axis x_1 and sides of length $h_2 > 0$ parallel to the axis x_2. Then scheme (3.3.24) can be written in the form

$$\boldsymbol{w}_{\mu\nu}^{k+1} = \boldsymbol{w}_{\mu\nu}^{k} - \frac{\tau}{h_1 h_2} \left(\Phi^k_{\mu+\frac{1}{2},\nu} h_2 + \Phi^k_{\mu-\frac{1}{2},\nu} h_2 + \Phi^k_{\mu,\nu+\frac{1}{2}} h_1 + \Phi^k_{\mu,\nu-\frac{1}{2}} h_1 \right), \tag{3.3.51}$$

where $\boldsymbol{w}^k_{\mu\nu}$ is the value of the numerical solution on the finite volume $D_{\mu\nu}$ at time t_k and $\Phi^k_{\mu\pm\frac{1}{2},\nu}, \Phi^k_{\mu,\nu\pm\frac{1}{2}}$ are numerical fluxes between $D_{\mu\nu}$ and its neighbours $D_{\mu\pm1,\nu}, D_{\mu,\nu\pm1}$:

$$\Phi^k_{\mu\pm\frac{1}{2},\nu} = \boldsymbol{H}\left(\boldsymbol{w}^k_{\mu\nu}, \boldsymbol{w}^k_{\mu\pm1,\nu}, (\pm1, 0)\right) \tag{3.3.52}$$
$$\Phi^k_{\mu,\nu\pm\frac{1}{2}} = \boldsymbol{H}\left(\boldsymbol{w}^k_{\mu\nu}, \boldsymbol{w}^k_{\mu\pm1,\nu}, (0, \pm1)\right).$$

Here $\boldsymbol{H} = \boldsymbol{H}(\boldsymbol{u}, \boldsymbol{v}, \boldsymbol{n})$ denotes the numerical flux from (3.3.24).

As an example let us consider the 2D *Lax–Friedrichs scheme* (3.3.28) with parameter λ chosen as $2\tau/h_1$ or $2\tau/h_2$ for neighbouring finite volumes with a common face parallel to the axis x_2 or x_1, respectively. Then the Lax–Friedrichs scheme reads

$$\boldsymbol{w}_{\mu\nu}^{k+1} = \frac{1}{4}(\boldsymbol{w}_{\mu+1,\nu}^k + \boldsymbol{w}_{\mu-1,\nu}^k + \boldsymbol{w}_{\mu,\nu+1}^k + \boldsymbol{w}_{\mu,\nu-1}^k) \tag{3.3.53}$$

$$-\tau\Big\{(\boldsymbol{f}_1(\boldsymbol{w}_{\mu+1,\nu}^k) - \boldsymbol{f}_1(\boldsymbol{w}_{\mu-1,\nu}^k))/(2h_1) + (\boldsymbol{f}_2(\boldsymbol{w}_{\mu,\nu+1}^k) - \boldsymbol{f}_2(\boldsymbol{w}_{\mu,\nu-1}^k))/(2h_2)\Big\},$$

$$\mu,\nu \in Z,\ k \in Z^+.$$

3.3.9 *Von Neumann linear stability*

Similarly as in the 1D case, it is possible to investigate the von Neumann linear stability of scheme (3.3.51). We proceed in the following way. First, assuming that we have computed an approximate solution $\boldsymbol{w}_{\mu\nu}^k$ at the time level t_k, we linearize system (3.3.1) (with $N = 2$) at $\boldsymbol{w}_{\mu\nu}^k$:

$$\frac{\partial \boldsymbol{w}}{\partial t} + \sum_{s=1}^{2} \mathbb{A}_s(\boldsymbol{w}_{\mu\nu}^k)\frac{\partial \boldsymbol{w}}{\partial x_s} = 0, \tag{3.3.54}$$

where $\mathbb{A}_s(\boldsymbol{w}) = D\boldsymbol{f}_s(\boldsymbol{w})/D\boldsymbol{w}$. In what follows, we shall simply write $\mathbb{A}$ instead of $\mathbb{A}(\boldsymbol{w}_{\mu\nu}^k)$ and analyse the von Neumann stability for this system, i.e.

$$\frac{\partial \boldsymbol{w}}{\partial t} + \sum_{s=1}^{2} \mathbb{A}_s\frac{\partial \boldsymbol{w}}{\partial x_s} = 0. \tag{3.3.55}$$

The hyperbolicity of system (3.3.1) implies that the matrix $\mathbb{P}(\boldsymbol{n}) = \sum_{s=1}^{2} n_s\mathbb{A}_s$ is diagonalizable and has real eigenvalues for each $\boldsymbol{n} = (n_1, n_2) \in I\!R^2$ with $|\boldsymbol{n}| = 1$. The second step is the application of scheme (3.3.51) to system (3.3.55). We get the formula

$$\boldsymbol{w}_{\mu\nu}^{k+1} = \sum_{(\alpha,\beta)\in M} \mathbb{C}_{\alpha\beta}\boldsymbol{w}_{\mu+\alpha,\nu+\beta}^k, \tag{3.3.56}$$

where $\mathbb{C}_{\alpha\beta}$ are 4×4 matrices and

$$M = \{(0,0),(1,0),(-1,0),(0,1),(0,-1)\}.$$

Let us assume that the approximate solution $\boldsymbol{w}_{\mu\nu}^k$ can be expressed as a *discrete Fourier series in two space variables*:

$$\boldsymbol{w}_{\mu\nu}^k = \sum_{m,n\in Z} \boldsymbol{F}_k(m,n)\exp[i(m\mu h_1 + n\nu h_2)] \quad (i^2 = -1). \tag{3.3.57}$$

Then we say that that *scheme* (3.3.56) *is von Neumann stable*, if there exists a constant $c > 0$, independent of m, n, k, h_1, h_2, τ, such that

$$|\boldsymbol{F}_k(m,n)| \le c|\boldsymbol{F}_0(m,n)| \quad \forall m,n \in Z,\ \forall k \in Z^+,\ \forall h_1,h_2,\tau > 0. \tag{3.3.58}$$

Substituting (3.3.57) into (3.3.56), we find that

$$\sum_{m,n\in Z} \boldsymbol{F}_{k+1}(m,n)\exp[i(m\mu h_1 + n\nu h_2)] \tag{3.3.59}$$
$$= \sum_{m,n\in Z}\sum_{(\alpha,\beta)\in M} \mathbb{C}_{\alpha,\beta}\boldsymbol{F}_k(m,n)\exp[i(m\mu h_1 + n\nu h_2)]\exp[i(m\alpha h_1 + n\beta h_2)],$$
$$\forall \mu,\nu \in Z,\ \forall k \in Z^+,\ \forall h_1,h_2,\tau > 0.$$

The linear independence of the system $\{\exp[i(m\mu h_1+n\nu h_2)]\}$, $m,n \in Z$, implies that

$$\boldsymbol{F}_{k+1}(m,n) = \mathbb{G}\boldsymbol{F}_k(m,n), \tag{3.3.60}$$

where

$$\mathbb{G} = \mathbb{G}(m,n,h_1,h_2) = \sum_{(\alpha,\beta)\in M} \mathbb{C}_{\alpha\beta}\exp[i(m\alpha h_1 + n\beta h_2)] \tag{3.3.61}$$

is called an *amplification matrix*. Obviously, (3.3.60) is obtained using the ansatz

$$\boldsymbol{w}^k_{\mu\nu} = \boldsymbol{F}_k\exp[i(m\mu h_1 + n\nu h_2)]. \tag{3.3.62}$$

Analogously to Theorem 3.33, we can prove the following *stability criterion*:

Theorem 3.56 *Let the matrix* $\mathbb{G} = \mathbb{G}(m,n,h_1,h_2)$ *be diagonalizable,* $\mathbb{G} = \mathbb{T}\mathbb{B}\mathbb{T}^{-1}$*, where* $\mathbb{B}$ *is a diagonal matrix, and let the entries of* $\mathbb{T}$ *and* $\mathbb{T}^{-1}$ *be uniformly bounded independently of* m,n,h_1,h_2*. Then scheme* (3.3.56) *is von Neumann stable if and only if*

$$\sigma(\mathbb{G}) \le 1 \quad \forall m,n \in Z,\ \forall h_1,h_2 > 0, \tag{3.3.63}$$

where $\sigma(\mathbb{G})$ *is the spectral radius of* $\mathbb{G}$.

3.3.10 *Application to the Lax–Friedrichs scheme*

It is easy to see that for problem (3.3.55), scheme (3.3.53) becomes

$$\begin{aligned}\boldsymbol{w}^{k+1}_{\mu\nu} &= \frac{1}{4}\left(\boldsymbol{w}^k_{\mu+1,\nu} + \boldsymbol{w}^k_{\mu-1,\nu} + \boldsymbol{w}^k_{\mu,\nu+1} + \boldsymbol{w}^k_{\mu,\nu-1}\right) \\ &\quad - \frac{\tau}{2h_1}\mathbb{A}_1\left(\boldsymbol{w}^k_{\mu+1,\nu} - \boldsymbol{w}^k_{\mu-1,\nu}\right) - \frac{\tau}{2h_2}\mathbb{A}_2\left(\boldsymbol{w}^k_{\mu,\nu+1} - \boldsymbol{w}^k_{\mu,\nu-1}\right).\end{aligned} \tag{3.3.64}$$

The substitution of ansatz (3.3.62) into (3.3.64) and some calculation leads to relation (3.3.60) with the amplification matrix

$$\mathbb{G} = \frac{1}{2}(\cos(mh_1) + \cos(nh_2))\mathbb{I} - i\left(\frac{\tau}{h_1}\sin(mh_1)\mathbb{A}_1 + \frac{\tau}{h_2}\sin(nh_2)\mathbb{A}_2\right). \tag{3.3.65}$$

Let us set $\nu_1 = \tau \sin(mh_1)/h_1$, $\nu_2 = \tau \sin(nh_2)/h_2$, $\boldsymbol{\nu} = (\nu_1, \nu_2)$. In view of (3.1.60), the matrix $\mathbb{P} = \sum_{s=1}^{2} \nu_s \mathbb{A}_s$ has the eigenvalues

$$\lambda_1 = \boldsymbol{v} \cdot \boldsymbol{\nu} - a|\boldsymbol{\nu}|, \quad \lambda_2 = \lambda_3 = \boldsymbol{v} \cdot \boldsymbol{\nu}, \quad \lambda_4 = \boldsymbol{v} \cdot \boldsymbol{\nu} + a|\boldsymbol{\nu}|, \tag{3.3.66}$$

where $\boldsymbol{v}$ and a is the velocity and speed of sound, respectively, corresponding to the state $\boldsymbol{w}^k_{\mu\nu}$. Moreover, the matrix $\mathbb{P}$ is diagonalizable by the matrices $\mathbb{T}$ and $\mathbb{T}^{-1}$ from Section 4.1. For a fixed value $\boldsymbol{w}^k_{\mu\nu}$, the entries of these matrices are bounded independently of m, n, h_1 and h_2. Obviously, $\mathbb{G}$ is also diagonalizable by the matrices $\mathbb{T}$ and $\mathbb{T}^{-1}$ and has the eigenvalues

$$\mu_j = \frac{1}{2}(\cos(mh_1) + \cos(nh_2)) - i\lambda_j, \quad j = 1, \dots, 4. \tag{3.3.67}$$

Now, using the inequality $2ab \le a^2 + b^2$, we find that

$$\begin{aligned}
\sigma^2(\mathbb{G}) &\le \frac{1}{4}(\cos(mh_1) + \cos(nh_2))^2 + (|\boldsymbol{v}| + a)^2 |\boldsymbol{\nu}|^2 \\
&\le \frac{1}{2}(\cos^2(mh_1) + \cos^2(nh_2)) + \tau^2(|\boldsymbol{v}|+a)^2[\sin^2(mh_1)/h_1^2 + \sin^2(nh_2)/h_2^2] \\
&= \frac{1}{2}\left\{2 - \sin^2(mh_1)\left(1 - \frac{2\tau^2(|\boldsymbol{v}| + a)^2}{h_1^2}\right)\right. \\
&\quad \left. - \sin^2(mh_2)\left(1 - \frac{2\tau^2(|\boldsymbol{v}| + a)^2}{h_2^2}\right)\right\} \le 1,
\end{aligned}$$

provided the *stability condition*

$$\tau \le \frac{\min(h_1, h_2)}{\sqrt{2}(|\boldsymbol{v}| + a)} \mathrm{CFL} \tag{3.3.68}$$

is satisfied with CFL $= 1$. Now, by Theorem 3.56, condition (3.3.68) guarantees the von Neumann linear stability of the Lax–Friedrichs scheme used for the solution of 2D Euler equations. Let us note that in the solution of the Euler equations it is suitable to choose the CFL number less than one (e.g. CFL $\approx$ 0.85), in order to suppress possible instabilities caused by the nonlinearity of the Euler equations.

Exercise 3.57 Extend the above results to 3D flow.

Unfortunately, the von Neumann method cannot be used on irregular unstructured meshes. Some knowledge about the qualitative properties of some numerical method can be obtained, if it is applied to the *scalar Cauchy problem*

$$\frac{\partial w}{\partial t} + \sum_{s=1}^{N} \frac{\partial f_s(w)}{\partial x_s} = 0 \quad \text{in } I\!R^N \times (0, \infty), \tag{3.3.69}$$

$$w(x,0) = w^0(x), \quad x \in I\!R^N.$$

In this case $w : I\!R^N \times (0,\infty) \to I\!R$, $f_s \in C^1(I\!R)$. The explicit FV scheme now has the form

$$w_i^{k+1} = w_i^k - \frac{\tau}{|D_i|} \sum_{j\in S(i)} \sum_{\alpha=1}^{\beta_{ij}} H(w_i^k, w_j^k, \boldsymbol{n}_{ij})|\Gamma_{ij}^\alpha|, \quad i \in J, \tag{3.3.70}$$

$$w_i^0 = \frac{1}{|D_i|}\int_{D_i} w^0(x)\,dx, \quad i \in J$$

(provided $w^0 \in L^1_{\mathrm{loc}}(I\!R^N)$). We assume that the numerical flux $H = H(u,v,\boldsymbol{n}) : I\!R^2 \times \mathcal{S}_1 \to I\!R$ has the properties from Section 3.3.3. Moreover, we shall use a stronger continuity assumption: let H be *locally Lipschitz-continuous*. This means that if $M > 0$, then there exists a constant $c(M) > 0$ such that

$$\begin{aligned} &|H(u,v,\boldsymbol{n}) - H(u^*,v^*,\boldsymbol{n})| \le c(M)(|u-u^*| + |v-v^*|), \\ &u, u^*, v, v^* \in [-M,M], \ \boldsymbol{n} \in \mathcal{S}_1. \end{aligned} \tag{3.3.71}$$

Similarly as in the 1D case (see Section 3.2.21.3), the concept of *monotonicity* plays an important role in the study of stability and convergence of scheme (3.3.70).

Definition 3.58 *Let $M > 0$. We say that the numerical flux H is monotone in the set $[-M,M]$, if the function '$u, v \in [-M,M]$, $\boldsymbol{n} \in \mathcal{S}_1 \to H(u,v,\boldsymbol{n}) \in I\!R$' is nonincreasing with respect to the second variable v. Thus, $H(u,v,\boldsymbol{n}) \le H(u,\tilde{v},\boldsymbol{n})$, provided $u, v, \tilde{v} \in [-M,M], v \ge \tilde{v}, \boldsymbol{n} \in \mathcal{S}_1$.*

It is easy to see that the monotone conservative numerical flux $H = H(u,v,\boldsymbol{n})$ is nondecreasing with respect to the first variable u. Let us recall the notation $\mathrm{w}^k = \{w_i^k\}_{i\in J}$ for the numerical solution at time t_k. We shall show that the monotonicity of the numerical flux implies the stability. Let us use the notation $\mathrm{w}^k = \{w_i^k\}_{i\in J}$ for the numerical solution at time t_k.

Theorem 3.59 *Let $M > 0$ and*

$$\mathrm{w}^0 \in \mathcal{M}_M = \{\mathrm{w} = \{w_j\}_{j\in J}; \|\mathrm{w}\|_\infty := \sup_{j\in J} |w_j| \le M\}. \tag{3.3.72}$$

Let the numerical flux H satisfy the following conditions:

a) (3.3.71) *holds,*

b) H is consistent (i.e. (3.3.26) *holds),*

c) H is monotone in $[-M,M]$,

d) the stability condition

$$\tau c(M)|\partial D_i|/|D_i| \le 1, \quad i \in J \tag{3.3.73}$$

is satisfied.

Then scheme (3.3.70) *is* L^∞*-stable:*

$$\|\mathrm{w}^k\|_\infty \le \|\mathrm{w}^0\|_\infty \quad \forall k \in Z^+. \tag{3.3.74}$$

Proof By induction with respect to k we prove that

$$\text{a)}\ \ \|\mathrm{w}^k\|_\infty \le M, \qquad \text{b)}\ \ \|\mathrm{w}^{k+1}\|_\infty \le \|\mathrm{w}^k\|_\infty, \qquad k \in Z^+. \tag{3.3.75}$$

This already implies (3.3.74). Inequality (3.3.75), a) holds for $k = 0$. Let us assume it is true for some $k \ge 0$. Then we shall establish (3.3.75), b) and, thus, (3.3.75), a) for $k+1$. Using the consistency (3.3.26) and Green's theorem, we find that for each $i \in J$,

$$\sum_{j\in S(i)} \sum_{\alpha=1}^{\beta_{ij}} H(w_i^k, w_i^k, \boldsymbol{n}_{ij}^\alpha)|\Gamma_{ij}^\alpha| = \sum_{j\in S(i)} \sum_{\alpha=1}^{\beta_{ij}} \left(\sum_{s=1}^{N} f_s(w_i^k)(n_{ij}^\alpha)_s \right) |\Gamma_{ij}^\alpha| \tag{3.3.76}$$

$$= \sum_{s=1}^{N} f_s(w_i^k) \left(\sum_{j\in S(i)} \sum_{\alpha=1}^{\beta_{ij}} (n_{ij}^\alpha)_s |\Gamma_{ij}^\alpha| \right) = \sum_{s=1}^{N} f_s(w_i^k) \int_{\partial D_i} n_s \, dS$$

$$= \sum_{s=1}^{N} f_s(w_i^k) \int_{D_i} \frac{\partial 1}{\partial x_s}\, dx = 0.$$

By virtue of (3.3.76), formula (3.3.70) can be rewritten in the form

$$w_i^{k+1} = w_i^k - \frac{\tau}{|D_i|} \sum_{j\in S(i)} \sum_{\alpha=1}^{\beta_{ij}} \left(H(w_i^k, w_j^k, \boldsymbol{n}_{ij}^\alpha) - H(w_i^k, w_i^k, \boldsymbol{n}_{ij}^\alpha) \right) |\Gamma_{ij}^\alpha|$$

$$= w_i^k - \frac{\tau}{|D_i|} \sum_{j\in S(i)} \sum_{\alpha=1}^{\beta_{ij}} \mathcal{H}_{ij}^\alpha |\Gamma_{ij}^\alpha| (w_i^k - w_j^k),$$

where

$$\mathcal{H}_{ij}^\alpha = \begin{cases} \dfrac{H(w_i^k, w_j^k, \boldsymbol{n}_{ij}^\alpha) - H(w_i^k, w_i^k, \boldsymbol{n}_{ij}^\alpha)}{w_i^k - w_j^k}, & \text{if } w_j^k \ne w_i^k, \\ 0, & \text{if } w_j^k = w_i^k. \end{cases} \tag{3.3.77}$$

Hence,

$$w_i^{k+1} = \left(1 - \frac{\tau}{|D_i|} \sum_{j\in S(i)} \sum_{\alpha=1}^{\beta_{ij}} \mathcal{H}_{ij}^\alpha |\Gamma_{ij}^\alpha| \right) w_i^k + \frac{\tau}{|D_i|} \sum_{j\in S(i)} \sum_{\alpha=1}^{\beta_{ij}} \mathcal{H}_{ij}^\alpha |\Gamma_{ij}^\alpha| w_j^k. \tag{3.3.78}$$

From the monotonicity of H it follows that $\mathcal{H}_{ij}^\alpha \ge 0$. Moreover, by (3.3.77) and the Lipschitz-continuity of H in $[-M, M]$, we have $\mathcal{H}_{ij}^\alpha \le c(M)$. From this and the stability condition (3.3.73) we get

$$1 - \frac{\tau}{|D_i|} \sum_{j\in S(i)} \sum_{\alpha=1}^{\beta_{ij}} \mathcal{H}_{ij}^\alpha |\Gamma_{ij}^\alpha| \ge 1 - \frac{\tau c(M)}{|D_i|} \sum_{j\in S(i)} \sum_{\alpha=1}^{\beta_{ij}} |\Gamma_{ij}^\alpha| \tag{3.3.79}$$

$$= 1 - \tau c(M)|\partial D_i|/|D_i| \geq 0.$$

These results now immediately imply that

$$|w_i^{k+1}| \leq \left(1 - \frac{\tau}{|D_i|} \sum_{j\in S(i)} \sum_{\alpha=1}^{\beta_{ij}} \mathcal{H}_{ij}^{\alpha}|\Gamma_{ij}^{\alpha}|\right)|w_i^k|$$

$$+ \frac{\tau}{|D_i|} \sum_{j\in S(i)} \sum_{\alpha=1}^{\beta_{ij}} \mathcal{H}_{ij}^{\alpha}|\Gamma_{ij}^{\alpha}||w_j^k| \leq \|\mathrm{w}^k\|_\infty, \quad i \in J,$$

which we wanted to prove. □

As proven in (Feistauer, 1993), Theorem 7.3.98 or (Kröner, 1997), the assumptions of Theorem 3.59 together with the conservativity property (3.3.27) imply the L^1-stability of scheme (3.3.70):

$$\|w^k\|_1 = \sum_{i\in J} |w_i^k||D_i| \leq \|w^0\|_1, \quad k \in Z^+, \tag{3.3.80}$$

provided $\|w^0\|_1 < \infty$. This and the L^∞-stability yield the L^p-stability for each $p \in (1,\infty)$, since

$$\|w^k\|_p^p \leq \|w^k\|_\infty^{p-1}\|w^k\|_1.$$

Now we come to the question of how to extend the above results to the upwind flux vector splitting schemes of the Godunov type for the solution of the Euler equations. The Vijayasundaram and Steger–Warming schemes applied to the scalar equation (3.3.69) are consistent only in the case that this equation is linear:

$$\frac{\partial w}{\partial t} + \sum_{s=1}^{N} a_s \frac{\partial w}{\partial x_s} = 0, \tag{3.3.81}$$

where $a_s \in I\!R$. Let us denote $\boldsymbol{a} = (a_1, \ldots, a_N)^{\mathrm{T}}$. It is easy to see that the Vijayasundaram, Steger–Warming and Van Leer schemes applied to equation (3.3.81) become identical. The flux of the quantity w has the form

$$\mathcal{P}(w, \boldsymbol{n}) = w \sum_{s=1}^{N} a_s n_s = w(\boldsymbol{a}\cdot\boldsymbol{n}), \quad \boldsymbol{n} = (n_1, \ldots, n_N)^{\mathrm{T}} \in \mathcal{S}_1, \ w \in I\!R,$$

and the corresponding numerical flux becomes

$$H(u, v, \boldsymbol{n}) = (\boldsymbol{a}\cdot\boldsymbol{n})^+ u + (\boldsymbol{a}\cdot\boldsymbol{n})^- v, \quad u, v \in I\!R, \quad \boldsymbol{n} \in \mathcal{S}_1. \tag{3.3.82}$$

We immediately see that this flux is monotone and Lipschitz-continuous with a constant $c(M) = |\boldsymbol{a}|$ for each $M > 0$. Then by virtue of Theorem 3.59, the Vijayasundaram, Steger–Warming and Van Leer schemes applied to equation (3.3.81) are stable under the *stability condition*

$$\tau|\boldsymbol{a}||\partial D_i|/|D_i| \leq 1, \quad i \in J. \tag{3.3.83}$$

Exercise 3.60 Investigate the monotonicity and stability of the Lax–Friedrichs scheme applied to the scalar equation (3.3.69).

3.3.11 *Extension of the stability conditions to the Euler equations*

The approach presented here is heuristic, but its use leads to satisfactory computational results, as we shall see in Section 3.7. We start from the Steger–Warming, Van Leer and Vijayasundaram schemes applied to a linear hyperbolic system (3.3.55) with m equations. The hyperbolicity of (3.3.55) means that the matrix $\mathbb{P}(\boldsymbol{n}) = \sum_{s=1}^{N} n_s \mathbb{A}_s$ has real eigenvalues $\lambda_1(\boldsymbol{n}), \dots, \lambda_m(\boldsymbol{n})$ and is diagonalizable for each $\boldsymbol{n} \in \mathcal{S}_1$: There exists a nonsingular matrix $\mathbb{T} = \mathbb{T}(\boldsymbol{n})$ such that $\mathbb{P}(\boldsymbol{n}) = \mathbb{T}(\boldsymbol{n})\Lambda\!\!\!\Lambda(\boldsymbol{n})\mathbb{T}^{-1}(\boldsymbol{n})$, $\Lambda\!\!\!\Lambda = \Lambda\!\!\!\Lambda(\boldsymbol{n}) = \mathrm{diag}(\lambda_1(\boldsymbol{n}), \dots, \lambda_m(\boldsymbol{n}))$. Using the fact that

$$\frac{1}{|\partial D_i|} \sum_{j \in S(i)} \sum_{\alpha=1}^{\beta_{ij}} \mathbb{T}(\boldsymbol{n}_{ij}^{\alpha})\, \mathbb{I}\, \mathbb{T}^{-1}(\boldsymbol{n}_{ij}^{\alpha})\, |\Gamma_{ij}^{\alpha}| = \mathbb{I},$$

the mentioned schemes can be written in the form

$$\begin{aligned} \boldsymbol{w}_i^{k+1} &= \boldsymbol{w}_i^k - \frac{\tau}{|D_i|} \sum_{j \in S(i)} \sum_{\alpha=1}^{\beta_{ij}} \left(\mathbb{P}^+(\boldsymbol{n}_{ij}^{\alpha})\boldsymbol{w}_i^k + \mathbb{P}^-(\boldsymbol{n}_{ij}^{\alpha})\boldsymbol{w}_j^k)\right) |\Gamma_{ij}^{\alpha}| \qquad (3.3.84) \\ &= \sum_{j \in S(i)} \sum_{\alpha=1}^{\beta_{ij}} \mathbb{T}(\boldsymbol{n}_{ij}^{\alpha}) \left(\frac{1}{|\partial D_i|}\mathbb{I} - \frac{\tau}{|D_i|}\Lambda\!\!\!\Lambda^+(\boldsymbol{n}_{ij}^{\alpha}) \right) |\Gamma_{ij}^{\alpha}| \mathbb{T}^{-1}(\boldsymbol{n}_{ij}^{\alpha})\boldsymbol{w}_i^k \\ &\quad - \frac{\tau}{|D_i|} \sum_{j \in S(i)} \sum_{\alpha=1}^{\beta_{ij}} \mathbb{T}(\boldsymbol{n}_{ij}^{\alpha})\Lambda\!\!\!\Lambda^-(\boldsymbol{n}_{ij}^{\alpha})\mathbb{T}^{-1}(\boldsymbol{n}_{ij}^{\alpha})|\Gamma_{ij}^{\alpha}|\boldsymbol{w}_j^k. \end{aligned}$$

Now we use an analogy with condition (3.3.83) from the previous scalar case and declare scheme (3.3.84) to be '*stable*', if the entries of the matrix

$$\frac{1}{|\partial D_i|}\mathbb{I} - \frac{\tau}{|D_i|}\Lambda\!\!\!\Lambda^+(\boldsymbol{n}_{ij}^{\alpha}), \quad i \in J,\ j \in S(i),\ \alpha = 1, \dots, \beta_{ij},$$

are nonnegative. This is guaranteed by the condition

$$\tau \lambda_{i,\max} |\partial D_i| / |D_i| \le \mathrm{CFL}, \quad i \in J, \qquad (3.3.85)$$

where

$$\lambda_{i,\max} = \max_{\substack{r=1,\dots,m,\, j \in S(i) \\ \alpha=1,\dots,\beta_{ij}}} |\lambda_r(\boldsymbol{n}_{ij}^{\alpha})|$$

and CFL ≤ 1.

In the nonlinear case we replace $\lambda_s(\boldsymbol{n}_{ij}^{\alpha})$ by the eigenvalues of the matrix $\mathbb{P}(\boldsymbol{w}, \boldsymbol{n}_{ij}^{\alpha})$ with $\boldsymbol{w} := (\boldsymbol{w}_i^k + \boldsymbol{w}_j^k)/2$ for the Vijayasundaram and Van Leer schemes and with $\boldsymbol{w} := \boldsymbol{w}_i^k$ for the Steger–Warming scheme and write the *stability condition* again in the form (3.3.85), where we write τ_k instead of τ and $\lambda_{i,\max}^k$ instead of $\lambda_{i,\max}$ with

$$\lambda_{i,\max}^k = \max_{\substack{r=1,\dots,m,j\in S(i)\\ \alpha=1,\dots,\beta_{ij}}} \left|\lambda_r\left(\frac{\boldsymbol{w}_i^k+\boldsymbol{w}_j^k}{2}, \boldsymbol{n}_{ij}^\alpha\right)\right| \tag{3.3.86}$$

for the Vijayasundaram and Van Leer schemes, and

$$\lambda_{i,\max}^k = \max_{\substack{r=1,\dots,m,j\in S(i)\\ \alpha=1,\dots,\beta_{ij}}} |\lambda_r(\boldsymbol{w}_i^k, \boldsymbol{n}_{ij}^\alpha)| \tag{3.3.87}$$

for the Steger–Warming scheme. Usually we choose CFL < 1, e.g. CFL=0.85.

Often an explicit scheme of the form (3.3.24) is used as an iterative process via *time stabilization* as $t_k \to \infty$ for obtaining an approximate *steady-state solution* of the Euler equations with values $\boldsymbol{w}_i$ on D_i, $i \in J$. We expect that $\boldsymbol{w}_i = \lim_{k\to\infty} \boldsymbol{w}_i^k$, where the values $\boldsymbol{w}_i^k$, $i \in J$, $k = 0, 1, \dots$, are computed from (3.3.24) and suitable initial data $\boldsymbol{w}_i^0$, $i \in J$. We also speak of the *time marching method.* The time steps τ_k, $k = 0, 1, \dots$, can be considered as relaxation parameters. In order to speed up the convergence of this process, the *local time stepping* can be used. This means that at any time level t_k, for each $i \in J$, the *local time step* $\tau_k := \tau_{k,i}$ is used in (3.3.24). The local time step is chosen so that it satisfies the condition

$$\tau_{k,i}\lambda_{i,\max}^k|\partial D_i|/|D_i| \le \text{CFL}, \quad i \in J, \tag{3.3.88}$$

with $\lambda_{i,\max}^k$ and CFL defined as above.

3.3.12 *Convergence of the finite volume method*

The convergence of the finite volume method for the numerical solution of the multidimensional Euler equations remains open, similar to the solvability of the continuous problem for this system. The complete analysis of the convergence of approximate FV solutions to an exact entropy weak solution, as $h, \tau \to 0$, was carried out for a scalar conservation law and consistent, conservative, monotone FV methods in (Chen and Liu, 1993), (Champier *et al.*, 1993), (Kröner and Rokyta, 1994), (Cockburn *et al.*, 1994), (Vila, 1994), (Cockburn *et al.*, 1995), (Benharbit *et al.*, 1995), (Kröner *et al.*, 1995), (Noelle, 1995). In (Hou and Le Floch, 1994) the importance of conservativity of schemes for convergence is shown. The analysis of error estimates for numerical schemes for conservation laws is mostly based on the method by N. N. Kuznetsov (Kuznetsov, 1976). (See also (Kröner, 1997), (Godlewski and Raviart, 1991).) Further results can be found in (Cockburn *et al.*, 1994), (Cockburn and Gremaud, 1996*a*), (Cockburn and Gremaud, 1996*b*), (Nessyahu *et al.*, 1994). For systems, a few results are available ((Tveito and Winther, 1993), (Chen and Liu, 1993), (Isaacson and Temple, 1995)).

For a multidimensional hyperbolic system an analogy of the Lax–Wendroff convergence theorem has been established in (Kröner *et al.*, 1996). Let us consider the Cauchy problem for the system of the 2D Euler equations (3.3.1) (i.e. $N = 2, m = 4$) in $Q_T = I\!R^2 \times (0,T)$ with initial condition (3.3.2), where $\boldsymbol{w}^0 \in L^\infty(I\!R^2)^4$. Let $\boldsymbol{w}_i^k$, $i \in J$, $k = 0, \dots$, represent an approximate solution of this problem obtained with the aid of a general explicit FV scheme (3.3.24) on a FV mesh $\mathcal{D}_h = \{D_i\}_{i\in J}$ in $I\!R^2$ and the partition $t_k = k\tau, k = 0, \dots, r$, of the time interval $[0,T]$, where $\tau = T/r$ is a time step. We suppose that the initial values $\boldsymbol{w}_i^0, i \in J$, are defined by

$$\boldsymbol{w}_i^0 = \frac{1}{|D_i|} \int_{D_i} \boldsymbol{w}^0(x)\, dx. \tag{3.3.89}$$

By $\boldsymbol{w}_{h\tau}$ we denote the vector-valued function defined a.e. in Q_T:

$$\boldsymbol{w}_{h\tau}|_{D_i \times (t_k, t_{k+1})} = \boldsymbol{w}_i^k, \quad i \in J,\ k = 0, \dots, r-1. \tag{3.3.90}$$

Let the following *assumptions* be satisfied:

a) $h = \sup_{i\in J} \mathrm{diam}(D_i)$, (3.3.91)

b) $0 < c_1 \le \tau/h \le c_2$,

c) $\sup_{i\in J} \dfrac{h^2}{|D_i|} \le c_3$,

d) $\sum\limits_{D_i\cap\Omega\neq\emptyset} \Big||D_i| - |D_j|\Big| = o(1) \quad \text{as } h \to 0$

for all $j \in \cup\{s(i);\ D_i \cap \Omega \neq 0\}$ and any bounded domain $\Omega \subset I\!R^2$,

e) the numerical flux $H = H(\boldsymbol{u}, \boldsymbol{v}, \boldsymbol{n})$ is locally Lipschitz-continuous, consistent and conservative,

f) $\|\boldsymbol{w}_{h\tau}\|_{L^\infty(Q_T)^4} \le c_4$,

g) $\boldsymbol{w}_{h\tau} \to \boldsymbol{w}$ a.e. in Q_T as $h, \tau \to 0$,

$\boldsymbol{w}_{h\tau}(x,t) \in D$ (= domain of definition of the fluxes $\boldsymbol{f}_s$) for a.a. $(x,t) \in Q_T$,

where $c_1, \dots, c_4$ are constants independent of h and τ. The *FV analogy of the Lax–Wendroff convergence theorem* 3.30 reads:

Theorem 3.61 *If assumptions* (3.3.91), *a)–g) are satisfied, then the limit function* $\boldsymbol{w}$ *is a weak solution of problem* (3.3.1)–(3.3.2).

3.3.13 *Entropy condition*

Our goal is to obtain a numerical solution of a conservation law which approximates an exact, physically admissible *entropy solution.* (See Section 2.3.4.) This means that the approximate solution is expected to converge to the exact entropy solution, provided the mesh size tends to zero. This question was investigated in detail for a scalar conservation law. Let us briefly summarize basic results from (Kröner, 1997), (Godlewski and Raviart, 1991), (Feistauer, 1993).

Let us consider the FV scheme (3.3.70) for the numerical solution of the Cauchy problem (3.3.69). We assume that the numerical flux $H = H(u, v, \boldsymbol{n})$ is consistent, i.e.

$$H(u, u, \boldsymbol{n}) = \sum_{s=1}^{N} f_s(u)\, n_s, \quad u \in I\!R,\ \boldsymbol{n} \in \mathcal{S}_1. \tag{3.3.92}$$

Moreover, let H be conservative and $f_s \in C^1(I\!R)$ for $s = 1, \ldots, N$. Let $\eta \in C^1(I\!R)$ be a convex entropy with fluxes G_s, $s = 1, \ldots, N$. This means that $G_s \in C^1(I\!R)$ and

$$\eta'\, f_s' = G_s', \quad s = 1, \ldots, N. \tag{3.3.93}$$

Now we define a *consistent conservative entropy numerical flux* as a function $\mathcal{G}(u, v, \boldsymbol{n})$, defined in $I\!R^2 \times \mathcal{S}_1$ and satisfying

$$\begin{aligned} \mathcal{G}(u, u, \boldsymbol{n}) &= \sum_{s=1}^{N} G_s(u)\, n_s, \quad u \in I\!R,\ \boldsymbol{n} \in \mathcal{S}_1, \\ \mathcal{G}(u, v, \boldsymbol{n}) &= -\mathcal{G}(v, u, -\boldsymbol{n}), \quad u,\, v \in I\!R,\ \boldsymbol{n} \in \mathcal{S}_1. \end{aligned} \tag{3.3.94}$$

We say that the approximate solution w_i^k, $i \in J$, $k \in Z^+$, of problem (3.3.69) obtained from (3.3.70) satisfies the *entropy condition*, if

$$\eta(w_i^{k+1}) \le \eta(w_i^k) - \frac{\tau}{|D_i|} \sum_{j \in S(i)} \sum_{\alpha=1}^{\beta_{ij}} \mathcal{G}(w_i^k, w_j^k, \boldsymbol{n}_{ij}^\alpha)\, |\Gamma_{ij}^\alpha|, \quad i \in J,\ k \in Z^+. \tag{3.3.95}$$

Given an approximate solution w_i^k, $i \in J$, $k \in Z^+$, we define the piecewise constant function $w_{h\tau}(x, t)$:

$$w_{h\tau}(x, t) = w_i^k \quad \text{for } x \in D_i,\ t \in (t_k, t_{k+1}),\ i \in J,\ k \in Z^+ \tag{3.3.96}$$

(cf. (3.2.24)). The results obtained in (Kröner, 1997), Chapter 3, can be interpreted in the following way.

We introduce the following assumptions:

1) Let H be a differentiable, monotone, conservative, consistent and Lipschitz-continuous numerical flux and let $\mathcal{G}$ be a differentiable, consistent, conservative and Lipschitz-continuous entropy numerical flux for a convex entropy η with fluxes G_s, $s = 1, \ldots, N$. Let the compatibility condition

$$\frac{\partial \mathcal{G}(u, v, \boldsymbol{n})}{\partial v} = \eta'(v)\, \frac{\partial H(u, v, \boldsymbol{n})}{\partial v}, \quad u,\, v \in I\!R,\ \boldsymbol{n} \in \mathcal{S}_1 \tag{3.3.97}$$

be satisfied.

2) Let us consider a sequence of grids $\mathcal{D}_{h_n}$ with $h_n \to 0+$, let $\tau_n \to 0+$ as $n \to \infty$ and let

$$\begin{aligned} &\text{a)} \quad \|w_{h_n\tau_n}\|_{L^\infty(\mathbb{R}^N\times(0,\infty))} \le c, \quad n \in Z^+, &(3.3.98)\\ &\text{b)} \quad w_{h_n\tau_n} \to w \text{ a.e. in } \mathbb{R}^N \times (0,\infty) \quad \text{as } n \to \infty. \end{aligned}$$

Then (3.3.95) is satisfied for $n \in Z^+$ and w is a weak solution of problem (3.3.69) and satisfies the entropy condition

$$\frac{\partial \eta(w)}{\partial t} + \sum_{s=1}^{N} \frac{\partial G_s(w)}{\partial x_s} \le 0 \quad \text{in } \mathbb{R}^N \times (0,\infty \tag{3.3.99}$$

in the sense of distributions.

The *monotonicity* of the numerical flux H plays an important role here. This property is often connected with such concepts as *upwinding* or sufficient *numerical viscosity.* This is the reason that the upwind flux vector splitting schemes of the Godunov type for the numerical solution of the Euler equations usually yield physically admissible entropy solutions.

Further aspects connecting the discrete entropy condition with the so-called *coefficient of numerical viscosity* can be found in (Godlewski and Raviart, 1996) for a scalar conservation law with one space variable.

3.3.14 *Implicit FV methods*

Up to now we have been concerned with explicit FV methods for the numerical solution of the compressible Euler equations. Their advantage is a simple algorithmization without the necessity to solve large systems of algebraic equations in each time step. They are suitable for the solution of nonstationary flow as well as for the simulation of stationary high speed flow by the time stabilization for $t_k \to \infty$. (See Section 3.3.11.) However, for flows with low Mach numbers, the convergence of explicit schemes to a steady-state solution is very slow and the solution of nonstationary flow on a long time interval requires an extremely large number of time steps due to the rather restrictive CFL–stability conditions (see (3.3.85)–(3.3.87)). Therefore, it is suitable to consider implicit methods for the numerical solution of the Euler equations as well. The use of implicit methods contributes to an improvement of the efficiency of numerical schemes for solving the Euler equations in some cases, because implicit methods permit greater time steps and may reduce the computing time in a significant way. Implicit methods for the FV solution of nonlinear conservative laws and the Euler equations were discussed, for example, in (Stouffet, 1983), (Fezoui and Stouffet, 1989), (Fernandez, 1989), (Meister, 1998).

Let us consider the ϑ-scheme (3.3.23) with $\vartheta = 1$:

$$\boldsymbol{w}_i^{k+1} - \boldsymbol{w}_i^k + \frac{\tau_k}{|D_i|} \sum_{j\in S(i)} \sum_{\alpha=1}^{\beta_{ij}} \boldsymbol{H}\left(\boldsymbol{w}_i^{k+1}, \boldsymbol{w}_j^{k+1}, \boldsymbol{n}_{ij}^\alpha\right) |\Gamma_{ij}^\alpha| = 0, \quad i \in J, \tag{3.3.100}$$

where $\boldsymbol{H} = \boldsymbol{H}(\boldsymbol{u}, \boldsymbol{v}, \boldsymbol{n})$ denotes a numerical flux defined for $\boldsymbol{u}$, $\boldsymbol{v} \in D$ (= domain of definition of the fluxes $\boldsymbol{f}_s$) and $\boldsymbol{n} \in \mathcal{S}_1$.

By $\tilde{\mathbf{w}}_h^{k+1}$ we shall denote a vector formed by m-dimensional blocks ($m = N+2$) $\boldsymbol{w}_i^{k+1}$, $i \in J$, of the values of the approximate solution at time t_{k+1} on finite volumes $D_i \in \mathcal{D}_h$ and boundary values $\boldsymbol{w}_j^{k+1}$, $j \in \gamma(i)$, provided $\gamma(i) \neq \emptyset$ for a given $i \in J$. This means that the dimension of $\mathbf{w}_h^{k+1}$ is $n = (\text{card}J + n_B)m$, where card$J$ is the number of the finite volumes $D_i \in \mathcal{D}_h$ and n_B is the number of boundary faces. At each time level t_{k+1} system (3.3.100) is equipped with boundary conditions for $\tilde{\mathbf{w}}_h^{k+1}$ determining the state $\boldsymbol{w}_j^{k+1}$, if $j \in \gamma(i) \neq \emptyset$, as is described in Section 4.1.2.1. The boundary conditions can be written in the form

$$\mathcal{B}_h^{k+1}(\tilde{\mathbf{w}}_h^{k+1}) = 0, \tag{3.3.101}$$

where $\mathcal{B}_h^{k+1}$ is a boundary operator, in general different for different time levels t_{k+1}.

System (3.3.100) together with (3.3.101) can be written in the form

$$\tilde{\mathbf{w}}_h^{k+1} - \tilde{\mathbf{w}}_h^k + \Phi(\tilde{\mathbf{w}}_h^{k+1}) = 0. \tag{3.3.102}$$

For $\mathbf{w} \in I\!R^n$ the vector $\Phi(\mathbf{w})$ consists of m-dimensional blocks $\Phi_i(\mathbf{w})$, $i \in J$, given by

$$\Phi_i(\mathbf{w}) = \frac{\tau_k}{|D_i|} \sum_{j \in S(i)} \sum_{\alpha=1}^{\beta_{ij}} \boldsymbol{H}\left(\boldsymbol{w}_i, \boldsymbol{w}_j, \boldsymbol{n}_{ij}^\alpha\right) |\Gamma_{ij}^\alpha| \tag{3.3.103}$$

and the block given by $\mathcal{B}_h^{k+1}(\mathbf{w})$.

There are several possibilities for the solution of system (3.3.102).

3.3.14.1 *The Newton method* At each time level t_{k+1} the following *iterative process* is applied:

$$\mathbf{w}^{k+1,0} = \tilde{\mathbf{w}}_h^k, \tag{3.3.104}$$

$$\left(\mathbb{I} + \frac{D\Phi(\mathbf{w}^{k+1,r})}{D\mathbf{w}}\right)\left(\mathbf{w}^{k+1,r+1} - \mathbf{w}^{k+1,r}\right) = -\mathbf{w}^{k+1,r} + \tilde{\mathbf{w}}_h^k - \Phi(\mathbf{w}^{k+1,r}),$$

$$r = 0, 1, \ldots .$$

Here $D\Phi(\mathbf{w})/D\mathbf{w}$ is the Jacobian matrix of the mapping Φ. The aim is to obtain the sequence $\{\mathbf{w}^{k+1,r}\}_{r=0}^\infty$ converging to $\tilde{\mathbf{w}}_h^{k+1}$ as $r \to \infty$. Usually we compute a small number of iterations $\mathbf{w}^{k+1,r}$, $r = 0, 1, \ldots, s$, only and set $\tilde{\mathbf{w}}_h^{k+1} = \mathbf{w}^{k+1,s}$. The application of the Newton method is possible, provided the numerical flux $\boldsymbol{H}$ is differentiable. As an example, the Osher–Solomon numerical flux can be used. The Newton method is usually applied with success also in the case of a piecewise smooth Lipschitz-continuous numerical flux, which is the case for the Vijayasundaram, Steger–Warming, Van Leer and Roe schemes.

If we are interested in obtaining a steady-state solution, i.e. a vector $\tilde{\mathbf{w}}_h$ formed by m-dimensional blocks $\boldsymbol{w}_i$, $i \in J$, and $\boldsymbol{w}_j$, $j \in \gamma(i)$, if $\gamma(i) \neq \emptyset$, satisfying the system

$$\Phi(\tilde{\mathbf{w}}_h) = 0, \tag{3.3.105}$$

the Newton method can be applied directly to (3.3.105):

$$\frac{D\Phi(\mathbf{w}^r)}{D\mathbf{w}} \left(\mathbf{w}^{r+1} - \mathbf{w}^r\right) = -\Phi(\mathbf{w}^r), \quad r = 0, 1, \dots, \tag{3.3.106}$$

where $\mathbf{w}^0$ is a suitable initial approximation. The convergence of the sequence $\{\mathbf{w}^r\}_{r=0}^{\infty}$ to a steady-state solution $\tilde{\mathbf{w}}_h$ as $r \to \infty$ depends on the choice of $\mathbf{w}^0$, which should be sufficiently close to $\tilde{\mathbf{w}}_h$.

3.3.14.2 *Linearization of fluxes with the aid of the Taylor expansion* We write

$$\begin{aligned}
&\boldsymbol{H}(\boldsymbol{w}_i^{k+1}, \boldsymbol{w}_j^{k+1}, \boldsymbol{n}_{ij}^{\alpha}) \\
&\approx \boldsymbol{H}(\boldsymbol{w}_i^k, \boldsymbol{w}_j^k, \boldsymbol{n}_{ij}^{\alpha}) + \frac{D\boldsymbol{H}(\boldsymbol{w}_i^k, \boldsymbol{w}_j^k, \boldsymbol{n}_{ij}^{\alpha})}{D\boldsymbol{u}} (\boldsymbol{w}_i^{k+1} - \boldsymbol{w}_i^k) \\
&\quad + \frac{D\boldsymbol{H}(\boldsymbol{w}_i^k, \boldsymbol{w}_j^k, \boldsymbol{w}_{ij}^{\alpha})}{D\boldsymbol{v}} (\boldsymbol{w}_j^{k+1} - \boldsymbol{w}_j^k),
\end{aligned} \tag{3.3.107}$$

where $D\boldsymbol{H}(\boldsymbol{u}, \boldsymbol{v}, \boldsymbol{n})/D\boldsymbol{u}$ and $D\boldsymbol{H}(\boldsymbol{u}, \boldsymbol{v}, \boldsymbol{n})/D\boldsymbol{v}$ are 'partial' Jacobi matrices of the numerical flux $\boldsymbol{H}$ with respect to $\boldsymbol{u}$ and $\boldsymbol{v}$, respectively. Then we get the *linearized implicit scheme* in the form

$$\begin{aligned}
&\boldsymbol{w}_i^{k+1} - \boldsymbol{w}_i^k + \frac{\tau_k}{|D_i|} \sum_{j \in S(i)} \sum_{j=1}^{\beta_{ij}} \Big\{ \boldsymbol{H}(\boldsymbol{w}_i^k, \boldsymbol{w}_j^k, \boldsymbol{n}_{ij}^{\alpha}) \\
&\quad + \frac{D\boldsymbol{H}(\boldsymbol{w}_i^k, \boldsymbol{w}_j^k, \boldsymbol{n}_{ij}^{\alpha})}{D\boldsymbol{u}} (\boldsymbol{w}_i^{k+1} - \boldsymbol{w}_i^k) \\
&\quad + \frac{D\boldsymbol{H}(\boldsymbol{w}_i^k, \boldsymbol{w}_j^k, \boldsymbol{n}_{ij}^{\alpha})}{D\boldsymbol{v}} (\boldsymbol{w}_j^{k+1} - \boldsymbol{w}_j^k) \Big\} |\Gamma_{ij}^{\alpha}| = 0, \quad i \in J,
\end{aligned} \tag{3.3.108}$$

equipped with the boundary equations (3.3.101), for unknown vectors $\boldsymbol{w}_i^{k+1}$, $i \in J$, and boundary states $\boldsymbol{w}_j^{k+1}$, $j \in \gamma(i)$, if $\gamma(i) \neq \emptyset$. This scheme corresponds to the Newton iterative method (3.3.104), where only one iteration is performed. Again, the differentiability of $\boldsymbol{H}$ is required.

3.3.14.3 *Simplified linearization* This employs a special form of individual numerical fluxes. From (3.3.28), (3.3.31), (3.3.32) and (3.3.33) we can conclude that the Lax–Friedrichs, Steger–Warming, Vijayasundaram and Van Leer numerical fluxes can be expressed in the form

$$\boldsymbol{H}(\boldsymbol{u}, \boldsymbol{v}, \boldsymbol{n}) = \mathbb{H}_1(\boldsymbol{u}, \boldsymbol{v}, \boldsymbol{n})\, \boldsymbol{u} + \mathbb{H}_2(\boldsymbol{u}, \boldsymbol{v}, \boldsymbol{n})\, \boldsymbol{v} \tag{3.3.109}$$

with $m \times m$ matrices $\mathbb{H}_1$ and $\mathbb{H}_2$, which inspires us to write down the following *simplified linearized implicit scheme*

$$\boldsymbol{w}_i^{k+1} - \boldsymbol{w}_i^k + \frac{\tau_k}{|D_i|} \sum_{j \in S(i)} \sum_{\alpha=1}^{\beta_{ij}} \Big\{ \mathbb{H}_1(\boldsymbol{w}_i^k, \boldsymbol{w}_j^k, \boldsymbol{n}_{ij}^\alpha)\, \boldsymbol{w}_i^{k+1} + \mathbb{H}_2(\boldsymbol{w}_i^k, \boldsymbol{w}_j^k, \boldsymbol{n}_{ij}^\alpha)\, \boldsymbol{w}_j^{k+1} \Big\} |\Gamma_{ij}^\alpha| = 0, \quad i \in J, \tag{3.3.110}$$

considered together with (3.3.101).

As an example we introduce the *simplified linearized implicit Vijayasundaram scheme*:

$$\boldsymbol{w}_i^{k+1} - \boldsymbol{w}_i^k + \frac{\tau_k}{|D_i|} \sum_{j \in S(i)} \sum_{\alpha=1}^{\beta_{ij}} \left\{ \mathbb{P}^+ \left(\frac{\boldsymbol{w}_i^k + \boldsymbol{w}_j^k}{2}, \boldsymbol{n}_{ij}^\alpha \right) \boldsymbol{w}_i^{k+1} + \mathbb{P}^- \left(\frac{\boldsymbol{w}_i^k + \boldsymbol{w}_j^k}{2}, \boldsymbol{n}_{ij}^\alpha \right) \boldsymbol{w}_j^{k+1} \right\} |\Gamma_{ij}^\alpha| = 0, \quad i \in J. \tag{3.3.111}$$

Further variants of implicit FV schemes can be found in (Stouffet, 1983), (Fernandez, 1989), where also examples of applications to inviscid flow are presented. In order to obtain an efficient method, it is recommended that the stability CFL-condition ((3.3.85)–(3.3.87)) is applied starting from CFL < 1 as in an explicit method and then successively increasing CFL with growing k. In examples from (Fezoui and Stouffet, 1989)), (Stouffet, 1983), CFL varies between 0.8 and 10^3. Similar results are presented in (Fernandez, 1989), where CFL $\in [1, \infty)$.

3.3.14.4 *Linear solvers* In all the discussed implicit schemes an important ingredient is an efficient method for the solution of large systems of linear algebraic equations written in the form

$$\mathbb{A}\boldsymbol{x} = \boldsymbol{b}, \tag{3.3.112}$$

where $\mathbb{A}$ is an $n \times n$ nonsingular matrix, $\boldsymbol{b} \in I\!R^n$ is a given vector and $\boldsymbol{x} \in I\!R^n$ is an unknown vector. The dimension n of system (3.3.112) is large. The matrix $\mathbb{A}$ is sparse, but nonsymmetric with complicated structure. *Iterative solvers* have to be applied for computing the solution $\boldsymbol{x}$.

Originally, the linear systems were solved by the *Gauss–Seidel* or *block Gauss–Seidel* relaxation methods ((Varga, 1962)). However, the speed of convergence of these methods is rather low. Some efficient iterative methods for the solution of symmetric linear systems are described in (Axelsson, 1977). The last decade has seen tremendous progress in the development of *Krylov subspace methods* for arbitrary nonsingular matrices. Let us mention, for example, GMRES, GMRES(ℓ) ((Saad and Schultz, 1986)), BiCG, BiCGSTAB ((van der Vorst, 1992)). See also (Saad, 1996) or (Weiss, 1996). A survey of these methods applied to compressible CFD can be found in (Meister, 1998).

Another efficient technique is the *multigrid method* ((Hackbusch, 1989), (Wesseling, 1992)). It is usually applied to linear systems arising from the discretization of boundary value problems on structured quasiuniform meshes, but it was also adapted to the solution of nonlinear problems. The multigrid methods have been extensively applied to the solution of the stationary Euler equations in (Hemker and Spekreijse, 1986), (Koren and Hemker, 1991), (van der Maarel *et al.*, 1993), (Spekreijse, 1988), (Lallemand *et al.*, 1992), (Dick, 1988), (Dick, 1990), (Riemslagh and Dick, 1994). We refer the reader also to (Mavriplis, 1990), (Mavriplis *et al.*, 1989), (Dick, 1991). The last two works are concerned with the finite volume solution of the compressible Navier–Stokes equations.

Completely unsteady flows can be solved by the schemes explained in this section. Let us mention also the works (Jeltsch and Botta, 1994), (Fey, 1995), (Fey and Jeltsch, 1992*a*), (Fey and Jeltsch, 1992*b*) using the so-called decomposition Euler schemes developed by Fey and Jeltsch.

3.3.14.5 *Low Mach number flows* The convergence of iterative methods or the time marching method with $t \to \infty$ for obtaining a stationary solution of the Euler equations is very slow in the case of *low Mach number flows.* The quality of nonstationary low Mach number flows is often also rather poor. Various techniques are applied for improving the convergence to steady-state solutions or improving the quality of unsteady solutions. One possibility is to use the implicit schemes treated above. Another way is to use explicit time marching methods with a suitable preconditioning ((Turkel, 1987), (Turkel, 1996), (Koren and Hemker, 1991), (Koren and van Leer, 1995), (Koren, 1996), (Sestertenn *et al.*, 1993)). Some works are based on the theory of an incompressible limit of compressible flow, as the Mach number tends to zero ((Klainerman and Majda, 1981), (Schochet, 1986)). Usually the Euler equations are modified in a suitable way in order to approximate well low speed compressible flow. See, for example, (Klein, 1995), (Klein *et al.*, 2000), (Klein and Munz, 1994), (Klein and Munz, 1995), (Bijl and Wesseling, 1998), (Meister, 1999), (Schneider *et al.*, 1999). This approach is also applied to low speed viscous flow (solved by the finite element method), as in (Braack, 2002), (Braack, 2001), (Coré *et al.*, 2002). The choice of a suitable approximation of different dependent variables (velocity, pressure, density, etc.) plays an important role as well. It should be in agreement with their approximations in the case of incompressible flow. See, for example, (Peyret and Taylor, 1983), (Ferziger and Perić, 1996), (Fletcher, 1991), (Roache, 1972), (Roache, 1998*b*).

3.4 Osher–Solomon scheme

One of the most robust numerical schemes of the Godunov type for the solution of the multidimensional Euler equations is the Osher–Solomon method. It was first introduced in (Osher and Solomon, 1982) for 1D flow and in (Spekreijse, 1988) it was extended to 2D problems. Here we shall derive a 3D version of this method according to (Felcman and Šolín, 1998).

In the derivation of the Osher–Solomon scheme for the 3D Euler equations we shall start from the general approach to the multidimensional numerical schemes of the Godunov type treated in Section 3.3.5. Let D_i and D_j be two neighbouring finite volumes with a common interface $\Gamma = \Gamma_{ij}^{\alpha}$ and let $\boldsymbol{n} = \boldsymbol{n}_{ij}^{\alpha}$ be the unit outer normal to ∂D_i on Γ. Using the transformation of the Cartesian coordinates $(x_1, x_2, x_3) \rightarrow (\tilde{x}_1, \tilde{x}_2, \tilde{x}_3)$ given by (3.1.36), (3.1.37), with $\tilde{x}_1$ oriented in the direction $\boldsymbol{n}$ and $\tilde{x}_2, \tilde{x}_3$ tangent to Γ, the Euler equations become the form (3.3.34) for the state vector

$$\boldsymbol{q} = \mathbb{Q}(\boldsymbol{n})\boldsymbol{w}. \tag{3.4.1}$$

Then we use the representation of the numerical flux $\boldsymbol{H}$ in the form (3.3.37), i.e.

$$\boldsymbol{H}(\boldsymbol{w}_i^k, \boldsymbol{w}_j^k, \boldsymbol{n}) = \mathbb{Q}^{-1}(\boldsymbol{n})\boldsymbol{g}_{\mathrm{R}}(\mathbb{Q}(\boldsymbol{n})\boldsymbol{w}_i^k, \mathbb{Q}(\boldsymbol{n})\boldsymbol{w}_j^k). \tag{3.4.2}$$

Here $\boldsymbol{g}_{\mathrm{R}}$ is a numerical flux for the transformed system of the Euler equations with one space variable $\tilde{x}_1$ written in the form (3.3.36):

$$\frac{\partial \boldsymbol{q}}{\partial t} + \frac{\partial \boldsymbol{f}_1(\boldsymbol{q})}{\partial \tilde{x}_1} = 0, \quad \tilde{x}_1 \in I\!R, \ t > 0, \tag{3.4.3}$$

and obtained by neglecting the derivatives with respect to $\tilde{x}_2, \tilde{x}_3$ in (3.3.34). We shall construct $\boldsymbol{g}_{\mathrm{R}}$ as a special approximate Riemann solver to system (3.4.3) – see, Section 3.2.16 for a general framework.

3.4.1 *Approximate Riemann solver*

Let us set

$$\boldsymbol{q}_L = \mathbb{Q}(\boldsymbol{n})\boldsymbol{w}_i^k, \ \boldsymbol{q}_R = \mathbb{Q}(\boldsymbol{n})\boldsymbol{w}_j^k. \tag{3.4.4}$$

Our goal is to determine the value of the approximate Riemann solver $\boldsymbol{g}_{\mathrm{R}}(\boldsymbol{q}_L, \boldsymbol{q}_R)$. We use the same approach as in Section 3.2.16, where $\boldsymbol{f} = \boldsymbol{f}_1, \mathbb{A} = \mathbb{A}_1, \boldsymbol{u} = \boldsymbol{q}_L, \boldsymbol{v} = \boldsymbol{q}_R$. Then, by virtue of (3.2.106), the approximate Riemann solver $\boldsymbol{g}_{\mathrm{R}}$ is heuristically defined as

$$\boldsymbol{g}_{\mathrm{R}}(\boldsymbol{q}_L, \boldsymbol{q}_R) = \boldsymbol{f}_1(\boldsymbol{q}_R) - \int_{\boldsymbol{q}_L}^{\boldsymbol{q}_R} \mathbb{A}_1^+(\boldsymbol{q})\, d\boldsymbol{q}. \tag{3.4.5}$$

Let us recall that the matrix $\mathbb{A}_1$ is diagonalizable with the aid of a nonsingular matrix $\tilde{\mathbb{T}}$ (see (3.1.31) or (3.1.53)):

$$\mathbb{A}_1 = \tilde{\mathbb{T}}\tilde{\mathbb{\Lambda}}\tilde{\mathbb{T}}^{-1}, \tag{3.4.6}$$

where $\tilde{\mathbb{\Lambda}} = \mathrm{diag}(\tilde{\lambda}_1, \ldots, \tilde{\lambda}_5)$. Then we define

$$\begin{gathered}\mathbb{A}_1{}^{\pm} = \tilde{\mathbb{T}}\tilde{\mathbb{\Lambda}}^{\pm}\tilde{\mathbb{T}}^{-1}, \quad \tilde{\mathbb{\Lambda}}^{\pm} = \mathrm{diag}(\tilde{\lambda}_1^{\pm}, \ldots, \tilde{\lambda}_5^{\pm}), \\ \lambda^+ = \max(\lambda, 0), \quad \lambda^- = \min(\lambda, 0).\end{gathered} \tag{3.4.7}$$

In Section 3.2.17, we have derived the Steger–Warming, Van Leer and Vijayasundaram schemes with the aid of suitable quadrature formulae applied to (3.2.106).

The Osher–Solomon method is based on the evaluation of the integral in (3.4.5) along a suitable path in the state space. This method was developed for the 1D Euler equations in (Osher and Solomon, 1982) and in (Spekreijse, 1988) it was derived for 2D flow. The main technique is based on the use of Riemann invariants to the eigenvectors of the matrix $\mathbb{A}_1(\boldsymbol{q})$, defined in Section 2.3.10. The resulting Riemann solver can be expressed as a linear combination of values of the vector function $\boldsymbol{f}_1$ at uniquely determined and analytically expressed points. The results of a number of numerical computations (see Section 3.7) show that the scheme is sufficiently robust and suitable for application on unstructured nonuniform meshes. Here we shall first be concerned with its 3D version.

3.4.2 *The Jacobi matrix of* $\boldsymbol{f}_1$

For simplicity let us summarize some notation and results from Section 3.1. The state vector $\boldsymbol{q} = \boldsymbol{q}(\tilde{x}_1, \tilde{x}_2, \tilde{x}_3, t)$ in (3.4.1) has the form

$$\boldsymbol{q} = (q_1, q_2, q_3, q_4, q_5)^{\mathrm{T}} = (\rho, \rho\tilde{v}_1, \rho\tilde{v}_2, \rho\tilde{v}_3, E)^{\mathrm{T}}, \tag{3.4.8}$$

where

$$\tilde{\boldsymbol{v}} = \begin{pmatrix} \tilde{v}_1 \\ \tilde{v}_2 \\ \tilde{v}_3 \end{pmatrix} = \begin{pmatrix} \cos\alpha\cos\beta & \sin\alpha\cos\beta & \sin\beta \\ -\sin\alpha & \cos\alpha & 0 \\ -\cos\alpha\sin\beta & -\sin\alpha\sin\beta & \cos\beta \end{pmatrix} \begin{pmatrix} v_1 \\ v_2 \\ v_3 \end{pmatrix} \tag{3.4.9}$$

is the transformed velocity. For the sake of simplicity we shall use the notation

$$\begin{aligned} u &= \tilde{v}_1, \\ v &= \tilde{v}_2, \\ w &= \tilde{v}_3. \end{aligned} \tag{3.4.10}$$

Then

$$\boldsymbol{f}_1(\boldsymbol{q}) = \left(\rho u, \rho u^2 + p, \rho uv, \rho uw, (E+p)u\right)^{\mathrm{T}}, \tag{3.4.11}$$

$$\mathbb{A}_1(\boldsymbol{q}) \tag{3.4.12}$$

$$= \begin{pmatrix} 0 & 1 & 0 & 0 & 0 \\ \frac{\gamma_1}{2}|\tilde{\boldsymbol{v}}|^2 - u^2 & (3-\gamma)u & -\gamma_1 v & -\gamma_1 w & \gamma_1 \\ -uv & v & u & 0 & 0 \\ -uw & w & 0 & u & 0 \\ u\left(\gamma_1|\tilde{\boldsymbol{v}}|^2 - \gamma\frac{E}{\rho}\right) & \gamma\frac{E}{\rho} - \gamma_1 u^2 - \frac{\gamma_1}{2}|\tilde{\boldsymbol{v}}|^2 & -\gamma_1 uv & -\gamma_1 uw & \gamma u \end{pmatrix},$$

where $\gamma_1 = \gamma - 1$. The eigenvalues of the matrix $\mathbb{A}_1(\boldsymbol{q})$ have the form

$$\tilde{\lambda}_1(\boldsymbol{q}) = u - a, \quad \tilde{\lambda}_2(\boldsymbol{q}) = \tilde{\lambda}_3(\boldsymbol{q}) = \tilde{\lambda}_4(\boldsymbol{q}) = u \text{ and } \tilde{\lambda}_5(\boldsymbol{q}) = u + a, \tag{3.4.13}$$

where

$$a = \sqrt{\gamma p/\rho} \tag{3.4.14}$$

is the speed of sound. Further, the corresponding eigenvectors read

$$\boldsymbol{r}_1(\boldsymbol{q}) = \left(1, u-a, v, w, \frac{|\tilde{\boldsymbol{v}}|^2}{2} + \frac{a^2}{\gamma-1} - ua\right)^{\mathrm{T}}, \tag{3.4.15}$$

$$\boldsymbol{r}_2(\boldsymbol{q}) = \left(1, u, v, w, \frac{|\tilde{\boldsymbol{v}}|^2}{2}\right)^{\mathrm{T}}, \tag{3.4.16}$$

$$\boldsymbol{r}_3(\boldsymbol{q}) = \left(1, u, v-a, w, \frac{|\tilde{\boldsymbol{v}}|^2}{2} - va\right)^{\mathrm{T}}, \tag{3.4.17}$$

$$\boldsymbol{r}_4(\boldsymbol{q}) = \left(1, u, v, w-a, \frac{|\tilde{\boldsymbol{v}}|^2}{2} - wa\right)^{\mathrm{T}}, \tag{3.4.18}$$

$$\boldsymbol{r}_5(\boldsymbol{q}) = \left(1, u+a, v, w, \frac{|\tilde{\boldsymbol{v}}|^2}{2} + \frac{a^2}{\gamma-1} + ua\right)^{\mathrm{T}}. \tag{3.4.19}$$

They are linearly independent and form a basis in $I\!R^5$.

3.4.3 *Riemann invariants*

Denoting by $\psi^{(\ell)} : D \to I\!R$ a *Riemann invariant* to the eigenvector $\boldsymbol{r}_\ell$, we have $\psi^{(\ell)} \in C^1(D)$ and

$$\nabla\psi^{(\ell)}(\boldsymbol{q}) \cdot \boldsymbol{r}_\ell(\boldsymbol{q}) = 0, \qquad \boldsymbol{q} \in D, \qquad \ell = 1, 2, \ldots, 5, \tag{3.4.20}$$

where $\nabla = \nabla_{\boldsymbol{q}} = (\partial/\partial q_1, \ldots, \partial/\partial q_5)$. For the definition and some properties of the Riemann invariants, see Section 2.3. For every $\boldsymbol{q} \in D$, the condition (3.4.20) can be satisfied by at most four functions $\psi_1{}^{(\ell)}(\boldsymbol{q}), \psi_2^{(\ell)}(\boldsymbol{q}), \psi_3{}^{(\ell)}(\boldsymbol{q}), \psi_4^{(\ell)}(\boldsymbol{q})$, the gradients of which are linearly independent for every $\boldsymbol{q} \in D$. To find a Riemann invariant means to solve the first order partial differential equation (3.4.20). Let us present the Riemann invariants (R. i. for short) we have chosen as the most suitable ones:

R. i. to $\boldsymbol{r}_1(\boldsymbol{q})$: $\psi_1^{(1)}(\boldsymbol{q}) = v$, $\psi_2^{(1)}(\boldsymbol{q}) = w$, $\psi_3^{(1)}(\boldsymbol{q}) = p/\varrho^\gamma$, $\psi_4^{(1)}(\boldsymbol{q}) = u + \frac{2a}{\gamma-1}$,

R. i. to $\boldsymbol{r}_2(\boldsymbol{q})$: $\psi_1^{(2)}(\boldsymbol{q}) = p$, $\psi_2^{(2)}(\boldsymbol{q}) = u$, $\psi_3^{(2)}(\boldsymbol{q}) = v$, $\psi_4^{(2)}(\boldsymbol{q}) = w$,

R. i. to $\boldsymbol{r}_3(\boldsymbol{q})$: $\psi_1^{(3)}(\boldsymbol{q}) = p$, $\psi_2^{(3)}(\boldsymbol{q}) = u$, $\psi_3^{(3)}(\boldsymbol{q}) = v - 2a$, $\psi_4^{(3)}(\boldsymbol{q}) = w$,

R. i. to $\boldsymbol{r}_4(\boldsymbol{q})$: $\psi_1^{(4)}(\boldsymbol{q}) = p$, $\psi_2^{(4)}(\boldsymbol{q}) = u$, $\psi_3^{(4)}(\boldsymbol{q}) = v$, $\psi_4^{(4)}(\boldsymbol{q}) = w - 2a$,

R. i. to $\boldsymbol{r}_5(\boldsymbol{q})$: $\psi_1^{(5)}(\boldsymbol{q}) = v$, $\psi_2^{(5)}(\boldsymbol{q}) = w$, $\psi_3^{(5)}(\boldsymbol{q}) = p/\varrho^\gamma$, $\psi_4^{(5)}(\boldsymbol{q}) = u - \frac{2a}{\gamma-1}$.

Gradients of these Riemann invariants corresponding to an individual eigenvector $\boldsymbol{r}_\ell(\boldsymbol{q})$ are linearly independent for all $\boldsymbol{q} \in D$.

Exercise 3.62 We leave it to the reader to verify that $\psi_i^{(\ell)}$ are Riemann invariants.

3.4.4 *Integration of the eigenvectors* $\boldsymbol{r}_\ell$

Riemann invariants remain constant along every smooth curve in D to which the corresponding eigenvector of $\mathbb{A}_1(\boldsymbol{q})$ is tangential at every point.

Theorem 3.63 *Let for* $\ell \in \{1,\dots,5\}$ $\boldsymbol{\eta}^{(\ell)} : s \in (0,+\infty) \to \boldsymbol{\eta}^{(\ell)}(s) \in D$ *be a smooth curve such that*

$$\frac{d\boldsymbol{\eta}^{(\ell)}(s)}{ds} = \boldsymbol{r}_\ell(\boldsymbol{\eta}^{(\ell)}(s)), \quad s \in (0,+\infty), \tag{3.4.21}$$

where $\boldsymbol{r}_\ell$ *is an eigenvector of the matrix* $\mathbb{A}_1$. *Then the Riemann invariants* $\psi_1^{(\ell)},\dots,\psi_4^{(\ell)}$ *are constant along the curve* $\boldsymbol{\eta}^{(\ell)}$.

Proof We have

$$\frac{d}{ds}(\psi_i^{(\ell)}(\boldsymbol{\eta}^{(\ell)})) = \nabla\psi_i^{(\ell)}(\boldsymbol{\eta}^{(\ell)}) \cdot \frac{d\boldsymbol{\eta}^{(\ell)}}{ds} = \nabla\psi_i^{(\ell)}(\boldsymbol{\eta}^{(\ell)}) \cdot \boldsymbol{r}_\ell(\boldsymbol{\eta}^{(\ell)}) = 0, \quad i = 1,\dots,4.$$

□

These curves $\boldsymbol{\eta}^{(\ell)}$ are very important because the information about physical quantities in the form of the Riemann invariants propagates through them into the domain. We shall see later that the direction of the propagation depends on the sign of the corresponding eigenvalue of the matrix $\mathbb{A}_1$.

Theorem 3.64 *Let* $\tilde{\boldsymbol{q}} = (\tilde{q}_1,\dots,\tilde{q}_5)^{\mathrm{T}} \in D$ *be given. Then for all* $\ell \in \{1,\dots,5\}$ *there exists a unique curve* $\boldsymbol{\eta}^{(\ell)} : (0,+\infty) \to D$, $\boldsymbol{\eta}^{(\ell)} \in C^1(0,+\infty)^5$ *satisfying* (3.4.21) *and the condition*

$$\boldsymbol{\eta}^{(\ell)}(\tilde{q}_1) = \tilde{\boldsymbol{q}}. \tag{3.4.22}$$

Proof The proof is constructive and we can express the curves explicitly. Curve corresponding to $\boldsymbol{r}_1(\boldsymbol{q}) = \left(1, u-a, v, w, \frac{|\boldsymbol{v}|^2}{2} + \frac{a^2}{\gamma-1} - ua\right)^{\mathrm{T}}$:

$$\boldsymbol{\eta}^{(1)}(s) = \begin{pmatrix} s \\ s\psi_4^{(1)} - \frac{2}{\gamma-1}\sqrt{\gamma\psi_3^{(1)}}\, s^{\frac{\gamma+1}{2}} \\ s\psi_1^{(1)} \\ s\psi_2^{(1)} \\ \frac{3\gamma-1}{(\gamma-1)^2}\psi_3^{(1)} s^{\gamma} - \frac{2}{\gamma-1}\psi_4^{(1)}\sqrt{\gamma\psi_3^{(1)}}\, s^{\frac{\gamma+1}{2}} + s\frac{(\psi_4^{(1)})^2+(\psi_1^{(1)})^2+(\psi_2^{(1)})^2}{2} \end{pmatrix}.$$

Curve corresponding to $\boldsymbol{r}_2(\boldsymbol{q}) = \left(1, u, v, w, \frac{|\boldsymbol{v}|^2}{2}\right)^{\mathrm{T}}$:

$$\boldsymbol{\eta}^{(2)}(s) = \begin{pmatrix} s \\ s\psi_2^{(2)} \\ s\psi_3^{(2)} \\ s\psi_4^{(2)} \\ \frac{\psi_1^{(2)}}{\gamma-1} + s\frac{(\psi_2^{(2)})^2+(\psi_3^{(2)})^2+(\psi_4^{(2)})^2}{2} \end{pmatrix}.$$

Curve corresponding to $\boldsymbol{r}_3(\boldsymbol{q}) = \left(1, u, v-a, w, \frac{|\boldsymbol{v}|^2}{2} - va\right)^{\mathrm{T}}$:

$$\boldsymbol{\eta}^{(3)}(s) = \begin{pmatrix} s \\ s\psi_2^{(3)} \\ s\psi_3^{(3)} + 2\sqrt{\gamma\psi_1^{(3)}}\,s^{1/2} \\ s\psi_4^{(3)} \\ \frac{2\gamma^2-2\gamma+1}{\gamma-1}\psi_1^{(3)} + 2\psi_3^{(3)}\sqrt{\gamma\psi_1^{(3)}}\,s^{1/2} + s\frac{(\psi_2^{(3)})^2+(\psi_3^{(3)})^2+(\psi_4^{(3)})^2}{2} \end{pmatrix}.$$

Curve corresponding to $\boldsymbol{r}_4(\boldsymbol{q}) = \left(1, u, v, w-a, \frac{|\boldsymbol{v}|^2}{2} - wa\right)^{\mathrm{T}}$:

$$\boldsymbol{\eta}^{(4)}(s) = \begin{pmatrix} s \\ s\psi_2^{(4)} \\ s\psi_3^{(4)} \\ s\psi_4^{(4)} + 2\sqrt{\gamma\psi_1^{(4)}}\,s^{1/2} \\ \frac{2\gamma^2-2\gamma+1}{\gamma-1}\psi_1^{(4)} + 2\psi_4^{(4)}\sqrt{\gamma\psi_1^{(4)}}\,s^{1/2} + s\frac{(\psi_2^{(4)})^2+(\psi_3^{(4)})^2+(\psi_4^{(4)})^2}{2} \end{pmatrix}.$$

Curve corresponding to $\boldsymbol{r}_5(\boldsymbol{q}) = \left(1, u+a, v, w, \frac{|\boldsymbol{v}|^2}{2} + \frac{a^2}{\gamma-1} + ua\right)^{\mathrm{T}}$:

$$\boldsymbol{\eta}^{(5)}(s) = \begin{pmatrix} s \\ s\psi_4^{(5)} + \frac{2}{\gamma-1}\sqrt{\gamma\psi_3^{(5)}}\,s^{\frac{\gamma+1}{2}} \\ s\psi_1^{(5)} \\ s\psi_2^{(5)} \\ \frac{3\gamma-1}{(\gamma-1)^2}\psi_3^{(5)}s^{\gamma} + \frac{2}{\gamma-1}\psi_4^{(5)}\sqrt{\gamma\psi_3^{(5)}}\,s^{\frac{\gamma+1}{2}} + s\frac{(\psi_4^{(5)})^2+(\psi_1^{(5)})^2+(\psi_2^{(5)})^2}{2} \end{pmatrix}.$$

Here $\psi_i^{(\ell)} = \psi_i^{(\ell)}(\tilde{\boldsymbol{q}})$ for $i = 1, 2, \ldots, 4$ and $\ell = 1, 2, \ldots, 5$. One can see from the form of curves $\boldsymbol{\eta}^{(\ell)}$ that they are parametrized by density, i.e. $s = \rho$. □

Exercise 3.65 Prove that the curves $\boldsymbol{\eta}^{(\ell)}$ are infinitely smooth in $(0, +\infty)$ and for parameter s from $(0, +\infty)$ lie in the admissible state set.
Hint: Show that using the Riemann invariants $\psi_i^{(\ell)}$ from Section 3.4.3 the curve $\boldsymbol{\eta}^{(\ell)}$ has the form $\boldsymbol{\eta}^{(\ell)}(\rho) = (\rho, \rho u, \rho v, \rho w, E)^{\mathrm{T}}$.

3.4.5 *Integration path in the admissible state set*

Let $\boldsymbol{q}_L$, $\boldsymbol{q}_R$ be two different states in D. In this section we are going to construct a continuous path from $\boldsymbol{q}_L$ to $\boldsymbol{q}_R$ in D consisting piecewise of the curves $\boldsymbol{\eta}^{(1)}, \ldots, \boldsymbol{\eta}^{(5)}$. We shall compute four admissible states $\boldsymbol{q}_1$, $\boldsymbol{q}_2$, $\boldsymbol{q}_3$ and $\boldsymbol{q}_4$ in D so that

$$\boldsymbol{\eta}^{(1)}(\rho_L) = \boldsymbol{q}_L,$$

$$\boldsymbol{\eta}^{(1)}(\rho_1) = \boldsymbol{\eta}^{(2)}(\rho_1) = \boldsymbol{q}_1,\ \boldsymbol{\eta}^{(2)}(\rho_2) = \boldsymbol{\eta}^{(3)}(\rho_2) = \boldsymbol{q}_2,\ \boldsymbol{\eta}^{(3)}(\rho_3) = \boldsymbol{\eta}^{(4)}(\rho_3) = \boldsymbol{q}_3,$$
$$\boldsymbol{\eta}^{(4)}(\rho_4) = \boldsymbol{\eta}^{(5)}(\rho_4) = \boldsymbol{q}_4,\ \boldsymbol{\eta}^{(5)}(\rho_R) = \boldsymbol{q}_R,$$

where $\rho_L, \rho_1, \ldots, \rho_4, \rho_R$ is the density corresponding to the state $\boldsymbol{q}_L$, $\boldsymbol{q}_1$, ..., $\boldsymbol{q}_4$, $\boldsymbol{q}_R$, respectively. The situation is shown in Fig. 3.4.1 (here $\boldsymbol{\eta}^{(\ell)}$ are symbolically represented by straight lines). We have five curves $\boldsymbol{\eta}^{(\ell)}$, $\ell = 1, \ldots, 5$, and along

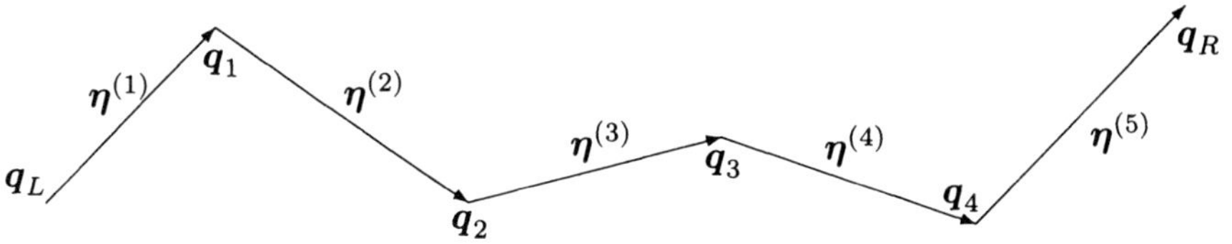

FIG. 3.4.1. Integration path

each of them four Riemann invariants $\psi_i^{(\ell)}$, $i = 1, \ldots, 4$, are constant. Hence, we can write down the following 20 equations for unknown components of $\boldsymbol{q}_1, \ldots, \boldsymbol{q}_4$:

$$\begin{aligned}
\psi_i^{(1)}(\boldsymbol{q}_L) &= \psi_i^{(1)}(\boldsymbol{q}_1), \quad i = 1, \ldots, 4, \\
\psi_i^{(2)}(\boldsymbol{q}_1) &= \psi_i^{(2)}(\boldsymbol{q}_2), \quad i = 1, \ldots, 4, \\
\psi_i^{(3)}(\boldsymbol{q}_2) &= \psi_i^{(3)}(\boldsymbol{q}_3), \quad i = 1, \ldots, 4, \\
\psi_i^{(4)}(\boldsymbol{q}_3) &= \psi_i^{(4)}(\boldsymbol{q}_4), \quad i = 1, \ldots, 4, \\
\psi_i^{(5)}(\boldsymbol{q}_4) &= \psi_i^{(5)}(\boldsymbol{q}_R), \quad i = 1, \ldots, 4.
\end{aligned}$$

In general, it is possible to construct an artificial pair of states $\boldsymbol{q}_L, \boldsymbol{q}_R$ such that this system has no solution, but it is so ridiculous that it would practically never arise. Except that, the solution exists and it is unique.

3.4.6 *Osher–Solomon approximate Riemann solver*

Let $\boldsymbol{q}_L$, $\boldsymbol{q}_R \in D$. We compute the integral in (3.4.5) along the path described in Section 3.4.5. Denoting

$$\boldsymbol{q}_0 = \boldsymbol{q}_L, \quad \boldsymbol{q}_5 = \boldsymbol{q}_R, \tag{3.4.23}$$

for the approximate Riemann solver $\boldsymbol{g}_{\mathrm{R}}$ in (3.4.5) we have

$$\begin{aligned}
\boldsymbol{g}_{\mathrm{R}}(\boldsymbol{q}_L, \boldsymbol{q}_R) &= \boldsymbol{f}_1(\boldsymbol{q}_R) - \sum_{\ell=1}^{5} \int_{\boldsymbol{q}_{\ell-1}}^{\boldsymbol{q}_\ell} \mathbb{A}_1^+(\boldsymbol{q})\, d\boldsymbol{q} \qquad (3.4.24) \\
&= \boldsymbol{f}_1(\boldsymbol{q}_R) - \sum_{\ell=1}^{5} \int_{\rho_{\ell-1}}^{\rho_\ell} \mathbb{A}_1^+\left(\boldsymbol{\eta}^{(\ell)}(s)\right) \boldsymbol{r}_\ell\left(\boldsymbol{\eta}^{(\ell)}(s)\right) ds \\
&= \boldsymbol{f}_1(\boldsymbol{q}_R) - \sum_{\ell=1}^{5} \int_{\rho_{\ell-1}}^{\rho_\ell} \tilde{\lambda}_\ell^+\left(\boldsymbol{\eta}^{(\ell)}(s)\right) \boldsymbol{r}_\ell\left(\boldsymbol{\eta}^{(\ell)}(s)\right) ds
\end{aligned}$$

$$= \boldsymbol{f}_1(\boldsymbol{q}_R) - \sum_{\ell=1}^{5} \int_{\rho_{\ell-1}}^{\rho_\ell} H\left(\tilde{\lambda}_\ell(\boldsymbol{\eta}^{(\ell)}(s))\right) \tilde{\lambda}_\ell\left(\boldsymbol{\eta}^{(\ell)}(s)\right) \boldsymbol{r}_\ell\left(\boldsymbol{\eta}^{(\ell)}(s)\right) ds$$

$$= \boldsymbol{f}_1(\boldsymbol{q}_R) - \sum_{\ell=1}^{5} \int_{\boldsymbol{q}_{\ell-1}}^{\boldsymbol{q}_\ell} H\left(\tilde{\lambda}_\ell(\boldsymbol{q})\right) \mathbb{A}_1(\boldsymbol{q})\, d\boldsymbol{q},$$

where ρ_ℓ is the first component of $\boldsymbol{q}_\ell$, $\ell = 0, \ldots, 5$, and $H : I\!R \to I\!R$ is the *Heaviside function* defined by the relation

$$H(x) = \begin{cases} 0, & x < 0, \\ 1, & x \geq 0. \end{cases}$$

In the following we shall need to know the behaviour of eigenvalues $\tilde{\lambda}_\ell$ of $\mathbb{A}_1$ along the curves $\boldsymbol{\eta}^{(\ell)}$. Expressing $\tilde{\lambda}_\ell$ in terms of Riemann invariants defined in Section 3.4.3, from (3.4.13) with notation (3.4.8), (3.4.10) we get

$$\begin{aligned}
\tilde{\lambda}_1(\boldsymbol{q}) &= u - a = u + \frac{2}{\gamma-1}a - \left(\frac{2}{\gamma-1}a + a\right) \\
&= \left(u + \frac{2}{\gamma-1}a\right) - \frac{\gamma+1}{\gamma-1}\sqrt{\gamma\frac{p}{\rho^\gamma}\rho^{\gamma-1}} \\
&= \psi_4^{(1)}(\boldsymbol{q}) - \frac{\gamma+1}{\gamma-1}\sqrt{\gamma\psi_3^{(1)}(\boldsymbol{q})q_1}^{\frac{\gamma-1}{2}}, \\
\tilde{\lambda}_2(\boldsymbol{q}) &= u = \psi_2^{(2)}(\boldsymbol{q}), \\
\tilde{\lambda}_3(\boldsymbol{q}) &= u = \psi_2^{(3)}(\boldsymbol{q}), \\
\tilde{\lambda}_4(\boldsymbol{q}) &= u = \psi_2^{(4)}(\boldsymbol{q}), \\
\tilde{\lambda}_5(\boldsymbol{q}) &= u + a = \psi_4^{(5)}(\boldsymbol{q}) + \frac{\gamma+1}{\gamma-1}\sqrt{\gamma\psi_3^{(5)}(\boldsymbol{q})q_1}^{\frac{\gamma-1}{2}}.
\end{aligned}$$

Using the fact that the Riemann invariants $\psi_i^{(\ell)}$, $i = 1, \ldots, 4$, are constant along $\boldsymbol{\eta}^{(\ell)}$, $\ell = 1, \ldots, 5$, we see that

$$\tilde{\lambda}_\ell \circ \boldsymbol{\eta}^{(\ell)} = \psi_2^{(\ell)} \circ \boldsymbol{\eta}^{(\ell)} = \text{const}, \quad \ell = 2, 3, 4, \tag{3.4.25}$$

but $\tilde{\lambda}_1, \tilde{\lambda}_5$ can change their sign once along $\boldsymbol{\eta}^{(1)}$ and $\boldsymbol{\eta}^{(5)}$, respectively. If this is the case, we denote by $\boldsymbol{q}_L^*$ and $\boldsymbol{q}_R^*$ the states from the geometrical image of $\boldsymbol{\eta}^{(1)}$ and $\boldsymbol{\eta}^{(5)}$, respectively, satisfying

$$\tilde{\lambda}_1(\boldsymbol{q}_L^*) = 0, \tag{3.4.26}$$

$$\tilde{\lambda}_5(\boldsymbol{q}_R^*) = 0. \tag{3.4.27}$$

We call $\boldsymbol{q}_L^*, \boldsymbol{q}_R^*$ *sonic points*, because the first velocity component equals the speed of sound for them. Taking into account the general case where both $\tilde{\lambda}_1$

and $\tilde{\lambda}_5$ change their sign along $\boldsymbol{\eta}^{(1)}$ and $\boldsymbol{\eta}^{(5)}$, we come to the Osher–Solomon approximate Riemann solver $\boldsymbol{g}_{\mathrm{R}} := \boldsymbol{g}_{\mathrm{OS}}$:

$$
\begin{aligned}
&\boldsymbol{g}_{\mathrm{OS}}(\boldsymbol{q}_L, \boldsymbol{q}_R) \qquad\qquad (3.4.28)\\
&= \boldsymbol{f}_1(\boldsymbol{q}_R) - \left\{ H\left(\tilde{\lambda}_1(\boldsymbol{q}_L)\right) \int_{\boldsymbol{q}_L}^{\boldsymbol{q}_L^*} \mathbb{A}_1(\boldsymbol{q})\, d\boldsymbol{q} + H\left(\tilde{\lambda}_1(\boldsymbol{q}_1)\right) \int_{\boldsymbol{q}_L^*}^{\boldsymbol{q}_1} \mathbb{A}_1(\boldsymbol{q})\, d\boldsymbol{q} \right.\\
&\quad + H\left(\tilde{\lambda}_2(\boldsymbol{q}_1)\right) \sum_{\ell=2}^{4} \int_{\boldsymbol{q}_{\ell-1}}^{\boldsymbol{q}_\ell} \mathbb{A}_1(\boldsymbol{q})\, d\boldsymbol{q}\\
&\quad \left. + H\left(\tilde{\lambda}_5(\boldsymbol{q}_4)\right) \int_{\boldsymbol{q}_4}^{\boldsymbol{q}_R^*} \mathbb{A}_1(\boldsymbol{q})\, d\boldsymbol{q} + H\left(\tilde{\lambda}_5(\boldsymbol{q}_R)\right) \int_{\boldsymbol{q}_R^*}^{\boldsymbol{q}_R} \mathbb{A}_1(\boldsymbol{q})\, d\boldsymbol{q} \right\}.
\end{aligned}
$$

The integrals in (3.4.28) are computed with the aid of the rule

$$
\int_{\boldsymbol{q}_A}^{\boldsymbol{q}_B} \mathbb{A}_1(\boldsymbol{q})\, d\boldsymbol{q} = \boldsymbol{f}_1(\boldsymbol{q}_B) - \boldsymbol{f}_1(\boldsymbol{q}_A). \qquad (3.4.29)
$$

The approximate Riemann solver (3.4.28) can be described by the diagram in Fig. 3.4.2. The following conditions must be tested:

$(\alpha) \ldots \tilde{\lambda}_1(\boldsymbol{q}_L) = u_L - a_L \geq 0,$
$(\beta) \ldots \tilde{\lambda}_1(\boldsymbol{q}_1) = u_1 - a_1 \geq 0,$
$(\gamma) \ldots \tilde{\lambda}_2(\boldsymbol{q}_1) = u_1 \geq 0,$
$(\delta) \ldots \tilde{\lambda}_5(\boldsymbol{q}_4) = u_4 + a_4 = u_1 + a_4 \geq 0,$
$(\omega) \ldots \tilde{\lambda}_5(\boldsymbol{q}_R) = u_R + a_R \geq 0.$

Here the variables u and a with subscripts denote the first velocity component and the speed of sound corresponding to the state $\boldsymbol{q}$ with the same subscript. In the diagram in Fig. 3.4.2, the direction upward means that the condition is satisfied. The numerical flux (3.4.2) defined with the use of the Osher–Solomon Riemann solver (3.4.28) is consistent in the sense of the definition in Section 3.3.3.

3.4.7 *Inlet/outlet boundary conditions*

Let us consider the case when a finite volume $D_i \in \mathcal{D}_h$ has a face $\Gamma = \Gamma_{ij}^1 \subset \partial\Omega_h$ with unit outer normal $\boldsymbol{n} = \boldsymbol{n}_{ij}^1$ (see Section 3.3.6). In this case it is necessary to determine the boundary state $\boldsymbol{w}_j^k$ for the computation of the numerical flux $\boldsymbol{H}(\boldsymbol{w}_i^k, \boldsymbol{w}_j^k, \boldsymbol{n})$ through Γ_{ij}^α using (3.4.2). The state $\boldsymbol{w}_j^k$ is determined in the following way. We set $\boldsymbol{q}_L = \mathbb{Q}(\boldsymbol{n})\boldsymbol{w}_i^k$ and determine $\boldsymbol{q}_B$ so that the linearized system

$$
\begin{aligned}
\frac{\partial \boldsymbol{q}}{\partial t} + \mathbb{A}_1(\boldsymbol{q}_L)\frac{\partial \boldsymbol{q}}{\partial \tilde{x}_1} &= 0 \quad \text{in } (-\infty, 0) \times (0, +\infty), \qquad (3.4.30)\\
\boldsymbol{q}(\tilde{x}_1, 0) = \boldsymbol{q}_L, &\qquad \tilde{x}_1 < 0,
\end{aligned}
$$

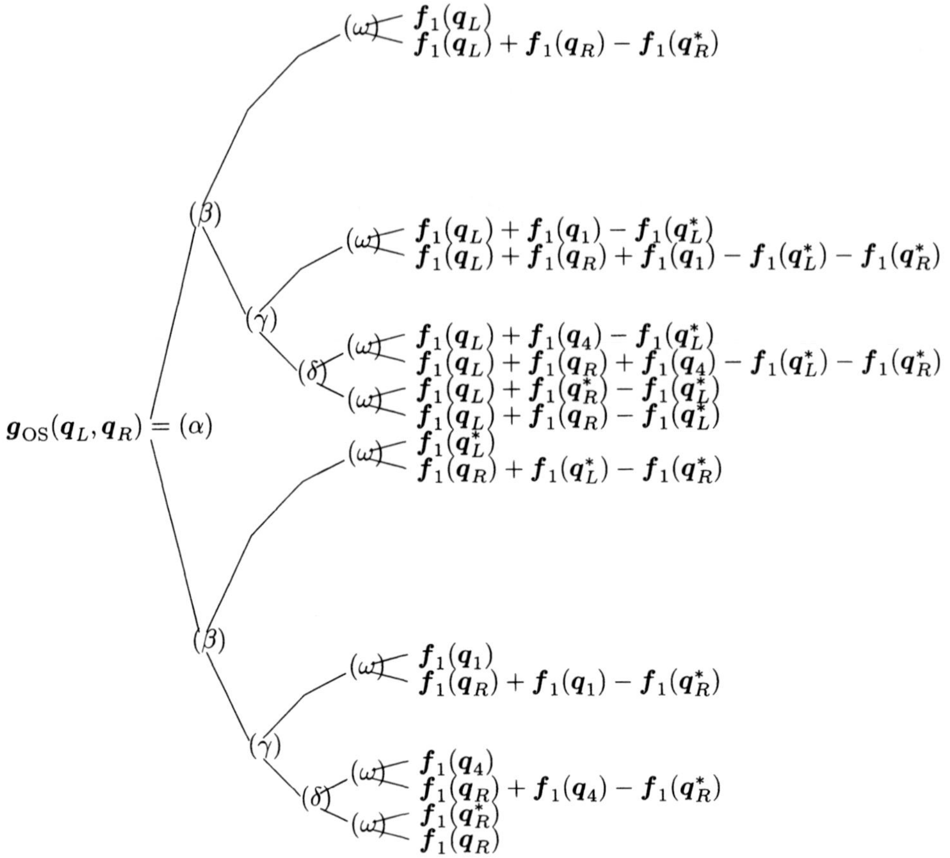

FIG. 3.4.2. Osher–Solomon approximate Riemann solver $\boldsymbol{g}_{\mathrm{OS}}(\boldsymbol{q}_L, \boldsymbol{q}_R)$

$$\boldsymbol{q}(0,t) = \boldsymbol{q}_B, \qquad t > 0,$$

has a unique solution. The state $\boldsymbol{w}_j^k$ is then given by the expression $\boldsymbol{w}_j^k = \mathbb{Q}^{-1}(\boldsymbol{n})\boldsymbol{q}_B$. We seek $\boldsymbol{q}_B$ in the form

$$\boldsymbol{q}_B = \sum_{\ell=1}^{5} \beta_\ell \boldsymbol{r}_\ell(\boldsymbol{q}_L), \tag{3.4.31}$$

where $\boldsymbol{r}_\ell(\boldsymbol{q}_L)$ are the eigenvectors of $\mathbb{A}_1(\boldsymbol{q}_L)$, corresponding to the eigenvalues $\tilde{\lambda}_\ell(\boldsymbol{q}_L)$, $\ell = 1, \ldots, 5$, introduced in Section 3.4.2. If we express

$$\boldsymbol{q}_L = \sum_{\ell=1}^{5} \alpha_\ell \boldsymbol{r}_\ell(\boldsymbol{q}_L), \tag{3.4.32}$$

and

$$\boldsymbol{q}(\tilde{x}_1, t) = \sum_{\ell=1}^{5} \mu_\ell(\tilde{x}_1, t)\boldsymbol{r}_\ell(\boldsymbol{q}_L), \tag{3.4.33}$$

then substituting (3.4.33), (3.4.32) and (3.4.31) into (3.4.30) we get the equivalent system

$$\frac{\partial \mu_\ell}{\partial t} + \tilde{\lambda}_\ell(\boldsymbol{q}_L)\frac{\partial \mu_\ell}{\partial \tilde{x}_1} = 0 \quad \text{in } (-\infty, 0) \times (0, +\infty), \tag{3.4.34}$$

$$\mu_\ell(\tilde{x}_1, 0) = \alpha_\ell, \qquad \tilde{x}_1 < 0, \tag{3.4.35}$$

$$\mu_\ell(0, t) = \beta_\ell, \qquad t > 0. \tag{3.4.36}$$

In Section 3.3.6 we obtained the solution of (3.4.34)–(3.4.36) shown in Fig. 3.3.17:

$$\mu_\ell(\tilde{x}_1, t) = \begin{cases} \alpha_\ell, & \tilde{x}_1 - \tilde{\lambda}_\ell(\boldsymbol{q}_L)t < 0, \\ \beta_\ell, & \tilde{x}_1 - \tilde{\lambda}_\ell(\boldsymbol{q}_L)t > 0. \end{cases} \tag{3.4.37}$$

We see that for $\tilde{\lambda}_\ell(\boldsymbol{q}_L) \geq 0$ the value β_ℓ in (3.4.36) is determined by α_ℓ (see Fig. 3.4.2) and we have to set $\beta_\ell = \alpha_\ell$ in order that the problem is well posed. For $\tilde{\lambda}_\ell(\boldsymbol{q}_L) < 0$ we have to prescribe β_ℓ so that the boundary state $\boldsymbol{q}_B$ in (3.4.31) is admissible. We conclude that we have to prescribe as many components of $\boldsymbol{q}_B$ as the number n_{pr} of negative eigenvalues of $\mathbb{A}_1(\boldsymbol{q}_L)$. If $n_{pr} = 0$ then we set $\boldsymbol{q}_B = \boldsymbol{q}_L$; for $n_{pr} > 0$ we have in (3.4.31)

$$\boldsymbol{q}_B = \sum_{\ell=1}^{n_{pr}} \beta_\ell \boldsymbol{r}_\ell(\boldsymbol{q}_L) + \sum_{\ell=n_{pr}+1}^{5} \alpha_\ell \boldsymbol{r}_\ell(\boldsymbol{q}_L). \tag{3.4.38}$$

Rewriting $\boldsymbol{q}_B$ in the form

$$\boldsymbol{q}_B = \boldsymbol{q}_L + \sum_{\ell=1}^{5} (\beta_\ell - \alpha_\ell)\boldsymbol{r}_\ell(\boldsymbol{q}_L), \tag{3.4.39}$$

we see that $\boldsymbol{q}_L$ and $\boldsymbol{q}_B$ can be connected by n_{pr} linear curves

$$\begin{aligned} \zeta^{(1)} &: s \in [0, s_1] \to \boldsymbol{q}_L + s\boldsymbol{r}_1(\boldsymbol{q}_L), \\ \zeta^{(2)} &: s \in [0, s_2] \to \zeta^{(1)}(s_1) + s\boldsymbol{r}_2(\boldsymbol{q}_L), \\ &\vdots \\ \zeta^{(n_{pr})} &: s \in [0, s_{n_{pr}}] \to \zeta^{(n_{pr}-1)}(s_{n_{pr}-1}) + s\boldsymbol{r}_{n_{pr}}(\boldsymbol{q}_L), \end{aligned}$$

where $s_\ell = \beta_\ell - \alpha_\ell$, $\ell = 1, \ldots, n_{pr}$. We will generalize this idea to the nonlinear system. Let us suppose that analogously to the previous case the states $\boldsymbol{q}_L$ and $\boldsymbol{q}_B$ can be connected by a piecewise smooth curve consisting of parts of the curves $\boldsymbol{\eta}^{(1)}, \boldsymbol{\eta}^{(2)}, \ldots, \boldsymbol{\eta}^{(n_{pr})}$, which we defined in Section 3.4.4. We prescribe n_{pr}

components of $\boldsymbol{q}_B$ and seek n_{pr} states $\boldsymbol{q}_1, \dots, \boldsymbol{q}_{n_{pr}}, \boldsymbol{q}_{n_{pr}} = \boldsymbol{q}_B$ with $5(n_{pr}-1)+(5-n_{pr})$ unknown components. All together we have $4n_{pr}$ unknown components and for these components we have $4n_{pr}$ equations

$$\psi_i^{(\ell)}(\boldsymbol{q}_{\ell-1}) = \psi_i^{(\ell)}(\boldsymbol{q}_\ell), \quad i = 1, \dots, 4, \quad \ell = 1, \dots, m, \tag{3.4.40}$$

where $\psi_i^{(\ell)}$ are the Riemann invariants introduced in Section 3.4.3. In (3.4.40) we put $\boldsymbol{q}_0 = \boldsymbol{q}_L$. The matrix $\mathbb{A}_1$ can have $n_{pr} = 0, 1, 4$ or 5 negative eigenvalues as can be seen from (3.4.13). Therefore we distinguish four cases.

I. If $n_{pr} = 0$, then $\tilde{\lambda}_\ell(\boldsymbol{q}_L) \geq 0$ for all $\ell = 0, \dots, 5$, which corresponds to the sonic or supersonic outlet. We set $\boldsymbol{q}_B = \boldsymbol{q}_L$ and compute

$$\boldsymbol{g}_{\mathrm{OS}}(\boldsymbol{q}_L, \boldsymbol{q}_B) = \boldsymbol{f}_1(\boldsymbol{q}_B) - \int_{\boldsymbol{q}_L}^{\boldsymbol{q}_B} \mathbb{A}_1^+(\boldsymbol{q})\, d\boldsymbol{q} = \boldsymbol{f}_1(\boldsymbol{q}_B). \tag{3.4.41}$$

II. If $n_{pr} = 1$, then we have $\tilde{\lambda}_1(\boldsymbol{q}_L) < 0, \tilde{\lambda}_2(\boldsymbol{q}_L) \geq 0, \dots, \tilde{\lambda}_5(\boldsymbol{q}_L) \geq 0$, i.e. the subsonic outlet. We define one component of $\boldsymbol{q}_B$ in such a way that we prescribe the pressure $p_B > 0$. The prescription of another physical quantity such as ρ_B, u_B requires for the linearized problem (3.4.30) the satisfaction of complementary conditions and moreover v_B and w_B cannot be in the linear case prescribed and are determined by v_L and w_L. The complementary conditions are given by the fact that $\boldsymbol{q}_B$ has to be an admissible state with $\rho_B > 0, p_B > 0$. From the condition that the Riemann invariants $\psi_i^{(1)}$ are constant along $\eta^{(1)}$, we obtain

$$\psi_i^{(1)}(\boldsymbol{q}_B) = \psi_i^{(1)}(\boldsymbol{q}_L), \quad i = 1, \dots, 4 \tag{3.4.42}$$

and, thus,

$$\begin{aligned} v_B &= v_L, \\ w_B &= w_L, \\ \frac{p_B}{\rho_B^\gamma} &= \frac{p_L}{\rho_L^\gamma}, \\ u_B + \frac{2a_B}{\gamma - 1} &= u_L + \frac{2a_L}{\gamma - 1}. \end{aligned}$$

This together with (3.4.14) gives the boundary state

$$\boldsymbol{q}_B = (\rho_B, \rho_B u_B, \rho_B v_B, \rho_B w_B, E_B)^{\mathrm{T}}, \tag{3.4.43}$$

where p_B is prescribed and

$$\begin{aligned} \rho_B &= \left(\frac{p_B}{p_L}\right)^{\frac{1}{\gamma}} \rho_L, \\ u_B &= u_L + \frac{2}{\gamma - 1}(a_L - a_B), \quad a_B = \sqrt{\gamma p_B/\rho_B}, \end{aligned} \tag{3.4.44}$$

$$
\begin{aligned}
v_B &= v_L, \\
w_B &= w_L, \\
E_B &= \frac{p_B}{\gamma - 1} - \frac{1}{2}\rho_B|\tilde{\boldsymbol{v}}_B^2|.
\end{aligned}
$$

Using the results from Section 3.4.6, we can express

$$
\boldsymbol{g}_{\mathrm{OS}}(\boldsymbol{q}_L, \boldsymbol{q}_B) = \boldsymbol{f}_1(\boldsymbol{q}_B) - \int_{\boldsymbol{q}_L}^{\boldsymbol{q}_B} H\left(\tilde{\lambda}_1(\boldsymbol{q})\right) \mathbb{A}_1(\boldsymbol{q})\, d\boldsymbol{q}. \tag{3.4.45}
$$

If $\tilde{\lambda}_1(\boldsymbol{q}_B) = u_B - a_B \geq 0$, then there exists the sonic point $\boldsymbol{q}_L^*$ satisfying the condition (3.4.26) in which $\tilde{\lambda}_1$ changes its sign. In (3.4.45) we have

$$
\boldsymbol{g}_{\mathrm{OS}}(\boldsymbol{q}_L, \boldsymbol{q}_B) = \left\{ \begin{array}{ll} \boldsymbol{f}_1(\boldsymbol{q}_B), & \tilde{\lambda}_1(\boldsymbol{q}_B) < 0 \\ \boldsymbol{f}_1(\boldsymbol{q}_B) - \int_{\boldsymbol{q}_L^*}^{\boldsymbol{q}_B} \mathbb{A}_1(\boldsymbol{q})\, d\boldsymbol{q} = \boldsymbol{f}_1(\boldsymbol{q}_L^*), & \tilde{\lambda}_1(\boldsymbol{q}_B) \geq 0 \end{array} \right\}.
$$

III. For $n_{pr} = 4$ we have $\tilde{\lambda}_1(\boldsymbol{q}_L), \ldots, \tilde{\lambda}_4(\boldsymbol{q}_L) < 0, \tilde{\lambda}_5(\boldsymbol{q}_L) \geq 0$, which corresponds to the subsonic or sonic inlet. There are three intersection states $\boldsymbol{q}_1, \boldsymbol{q}_2, \boldsymbol{q}_3$, between the states $\boldsymbol{q}_L$ and $\boldsymbol{q}_B$. The eigenvalues $\tilde{\lambda}_2 = \tilde{\lambda}_3 = \tilde{\lambda}_4$ satisfy (3.4.25). Hence, $\tilde{\lambda}_2(\boldsymbol{q}_1) = \tilde{\lambda}_2(\boldsymbol{q}_2) = \tilde{\lambda}_2(\boldsymbol{q}_3) = \tilde{\lambda}_2(\boldsymbol{q}_B)$ and the Osher–Solomon approximate Riemann solver can be expressed as

$$
\begin{aligned}
\boldsymbol{g}_{\mathrm{OS}}(\boldsymbol{q}_L, \boldsymbol{q}_B) &= \boldsymbol{f}_1(\boldsymbol{q}_B) - \int_{\boldsymbol{q}_L}^{\boldsymbol{q}_1} H\left(\tilde{\lambda}_1(\boldsymbol{q})\right) \mathbb{A}_1(\boldsymbol{q})\, d\boldsymbol{q} \\
&\quad - \int_{\boldsymbol{q}_1}^{\boldsymbol{q}_2} H\left(\tilde{\lambda}_2(\boldsymbol{q})\right) \mathbb{A}_1(\boldsymbol{q})\, d\boldsymbol{q} - \int_{\boldsymbol{q}_2}^{\boldsymbol{q}_3} H\left(\tilde{\lambda}_3(\boldsymbol{q})\right) \mathbb{A}_1(\boldsymbol{q})\, d\boldsymbol{q} \\
&\quad - \int_{\boldsymbol{q}_3}^{\boldsymbol{q}_B} H\left(\tilde{\lambda}_4(\boldsymbol{q})\right) \mathbb{A}_1(\boldsymbol{q})\, d\boldsymbol{q} \\
&= \boldsymbol{f}_1(\boldsymbol{q}_B) - \int_{\boldsymbol{q}_L}^{\boldsymbol{q}_1} H\left(\tilde{\lambda}_1(\boldsymbol{q})\right) \mathbb{A}_1(\boldsymbol{q})\, d\boldsymbol{q} \\
&\quad - H\left(\tilde{\lambda}_2(\boldsymbol{q}_B)\right) \left\{ \int_{\boldsymbol{q}_1}^{\boldsymbol{q}_2} \mathbb{A}_1(\boldsymbol{q})\, d\boldsymbol{q} + \int_{\boldsymbol{q}_2}^{\boldsymbol{q}_3} \mathbb{A}_1(\boldsymbol{q})\, d\boldsymbol{q} + \int_{\boldsymbol{q}_3}^{\boldsymbol{q}_B} \mathbb{A}_1(\boldsymbol{q})\, d\boldsymbol{q} \right\} \\
&= \boldsymbol{f}_1(\boldsymbol{q}_B) - \int_{\boldsymbol{q}_L}^{\boldsymbol{q}_1} H\left(\tilde{\lambda}_1(\boldsymbol{q})\right) \mathbb{A}_1(\boldsymbol{q})\, d\boldsymbol{q} \qquad (3.4.46) \\
&\quad - H\left(\tilde{\lambda}_2(\boldsymbol{q}_B)\right) \left(\boldsymbol{f}_1(\boldsymbol{q}_B) - \boldsymbol{f}_1(\boldsymbol{q}_1)\right).
\end{aligned}
$$

(See also (3.4.24).) Let us note that in (3.4.46) only the states $\boldsymbol{q}_L, \boldsymbol{q}_1, \boldsymbol{q}_B$ are involved. The state $\boldsymbol{q}_L$ is known. Since $n_{pr} = 4$ we prescribe four variables, namely ρ_B, u_B, v_B and w_B. We have the following relations for the determination of the state $\boldsymbol{q}_1$:

$$v_1 = \psi_1^{(1)}(\boldsymbol{q}_1) = \psi_1^{(1)}(\boldsymbol{q}_L) = v_L, \tag{3.4.47}$$

$$w_1 = \psi_2^{(1)}(\boldsymbol{q}_1) = \psi_2^{(1)}(\boldsymbol{q}_L) = w_L, \tag{3.4.48}$$

$$\begin{aligned} u_1 &= \psi_2^{(2)}(\boldsymbol{q}_1) = \psi_2^{(2)}(\boldsymbol{q}_2) = \psi_2^{(3)}(\boldsymbol{q}_2) \\ &= \psi_2^{(3)}(\boldsymbol{q}_3) = \psi_2^{(4)}(\boldsymbol{q}_3) = \psi_2^{(4)}(\boldsymbol{q}_B) = u_B, \end{aligned} \tag{3.4.49}$$

$$u_1 + \frac{2a_1}{\gamma - 1} = \psi_4^{(1)}(\boldsymbol{q}_1) = \psi_4^{(1)}(\boldsymbol{q}_L) = u_L + \frac{2a_L}{\gamma - 1},$$

$$a_1 = a_L + \frac{\gamma - 1}{2}(u_L - u_1), \tag{3.4.50}$$

$$\gamma \frac{p_1}{\rho_1} \frac{1}{\rho_1^{\gamma-1}} = \psi_3^{(1)}(\boldsymbol{q}_1) = \psi_3^{(1)}(\boldsymbol{q}_L) = \gamma \frac{p_L}{\rho_L} \frac{1}{\rho_L^{\gamma-1}},$$

$$\rho_1 = \left(\frac{a_1^2}{a_L^2} \right)^{\frac{1}{\gamma-1}} \rho_L, \tag{3.4.51}$$

$$p_1 = \frac{a_1^2 \rho_1}{\gamma}. \tag{3.4.52}$$

From relations (3.4.47)–(3.4.52) we successively get $v_1, w_1, u_1, a_1, \rho_1, p_1$ which determine the state $\boldsymbol{q}_1$. The state $\boldsymbol{q}_B$ is determined by the prescribed values ρ_B, u_B, v_B, w_B and by the pressure p_B satisfying

$$\begin{aligned} p_1 &= \psi_1^{(2)}(\boldsymbol{q}_1) = \psi_1^{(2)}(\boldsymbol{q}_2) = \psi_1^{(3)}(\boldsymbol{q}_2) = \psi_1^{(3)}(\boldsymbol{q}_3) \\ &= \psi_1^{(4)}(\boldsymbol{q}_3) = \psi_1^{(4)}(\boldsymbol{q}_B) = p_B. \end{aligned} \tag{3.4.53}$$

IV. If $n_{pr} = 5$, we have a supersonic inlet and prescribe $\boldsymbol{q}_B \in D$ and for the computation of $\boldsymbol{g}_{\mathrm{OS}}(\boldsymbol{q}_L, \boldsymbol{q}_B)$ we use the scheme shown in Fig. 3.4.2.

3.4.8 *Solid wall boundary conditions*

We use here the slip condition on the solid wall:

$$\boldsymbol{v} \cdot \boldsymbol{n} = 0, \tag{3.4.54}$$

which is (in view of (3.4.1), (3.4.8), (3.4.9), (3.4.10)) equivalent to the condition

$$u = 0. \tag{3.4.55}$$

It follows from (3.1.40) and (3.4.11) that under the condition (3.4.55),

$$\begin{aligned} \mathcal{P}(\boldsymbol{w}, \boldsymbol{n}) &= \mathbb{Q}^{-1}(\boldsymbol{n}) \left(\rho u, \rho u^2 + p, \rho uv, \rho uw, (E + p)u \right)^{\mathrm{T}} \\ &= \mathbb{Q}^{-1}(\boldsymbol{n}) \left(0, p, 0, 0, 0 \right)^{\mathrm{T}}. \end{aligned} \tag{3.4.56}$$

Approximation (3.3.21) of the flux by the numerical flux leads to

$$\boldsymbol{g}_{\mathrm{OS}}(\boldsymbol{q}_L, \boldsymbol{q}_B) := (0, p_L, 0, 0, 0)^{\mathrm{T}} \tag{3.4.57}$$

without being forced to specify the boundary state $\boldsymbol{q}_B$.

Table 3.4.1 *Two-dimensional Osher–Solomon scheme*

	$u_R \geq -a_R$, $u_L \leq a_L$	$u_R \geq -a_R$, $u_L > a_L$	$u_R < -a_R$, $u_L \leq a_L$	$u_R < -a_R$, $u_L > a_L$
$a_1 \leq u_1$	$\boldsymbol{f}_1(\boldsymbol{q}_L^*)$	$\boldsymbol{f}_1(\boldsymbol{q}_L)$	$\boldsymbol{f}_1(\boldsymbol{q}_R)-\boldsymbol{f}_1(\boldsymbol{q}_R^*)+\boldsymbol{f}_1(\boldsymbol{q}_L^*)$	$\boldsymbol{f}_1(\boldsymbol{q}_L)-\boldsymbol{f}_1(\boldsymbol{q}_R^*)+\boldsymbol{f}_1(\boldsymbol{q}_R)$
$0 < u_1$, $u_1 < a_1$	$\boldsymbol{f}_1(\boldsymbol{q}_1)$	$\boldsymbol{f}_1(\boldsymbol{q}_L)-\boldsymbol{f}_1(\boldsymbol{q}_L^*)+\boldsymbol{f}_1(\boldsymbol{q}_1)$	$\boldsymbol{f}_1(\boldsymbol{q}_R)-\boldsymbol{f}_1(\boldsymbol{q}_R^*)+\boldsymbol{f}_1(\boldsymbol{q}_1)$	$\boldsymbol{f}_1(\boldsymbol{q}_L)-\boldsymbol{f}_1(\boldsymbol{q}_R^*)+\boldsymbol{f}_1(\boldsymbol{q}_R)-\boldsymbol{f}_1(\boldsymbol{q}_L^*)+\boldsymbol{f}_1(\boldsymbol{q}_1)$
$-a_3 \leq u_1$, $u_1 \leq 0$	$\boldsymbol{f}_1(\boldsymbol{q}_3)$	$\boldsymbol{f}_1(\boldsymbol{q}_L)-\boldsymbol{f}_1(\boldsymbol{q}_L^*)+\boldsymbol{f}_1(\boldsymbol{q}_3)$	$\boldsymbol{f}_1(\boldsymbol{q}_R)-\boldsymbol{f}_1(\boldsymbol{q}_R^*)+\boldsymbol{f}_1(\boldsymbol{q}_3)$	$\boldsymbol{f}_1(\boldsymbol{q}_L)-\boldsymbol{f}_1(\boldsymbol{q}_R^*)+\boldsymbol{f}_1(\boldsymbol{q}_R)-\boldsymbol{f}_1(\boldsymbol{q}_L^*)+\boldsymbol{f}_1(\boldsymbol{q}_3)$
$u_1 < -a_3$	$\boldsymbol{f}_1(\boldsymbol{q}_R^*)$	$\boldsymbol{f}_1(\boldsymbol{q}_L)-\boldsymbol{f}_1(\boldsymbol{q}_L^*)+\boldsymbol{f}_1(\boldsymbol{q}_R^*)$	$\boldsymbol{f}_1(\boldsymbol{q}_R)$	$\boldsymbol{f}_1(\boldsymbol{q}_L)+\boldsymbol{f}_1(\boldsymbol{q}_R)-\boldsymbol{f}_1(\boldsymbol{q}_L^*)$

3.4.9 *Osher–Solomon scheme for the 2D Euler equations*

A 2D version of the Osher–Solomon numerical flux $\boldsymbol{g}_{\rm OS}(\boldsymbol{q}_L, \boldsymbol{q}_R)$ is defined in Table 3.4.9. The states

$$\begin{aligned} \boldsymbol{q}_1 &= (\rho_1, \rho_1 u_1, \rho_1 v_1, E_1), & \boldsymbol{q}_3 &= (\rho_3, \rho_3 u_3, \rho_3 v_3, E_3), \\ \boldsymbol{q}_L^* &= (\rho_L^*, \rho_L^* u_L^*, \rho_L^* v_L^*, E_L^*), & \boldsymbol{q}_R^* &= (\rho_R^*, \rho_R^* u_R^*, \rho_R^* v_R^*, E_R^*) \end{aligned} \tag{3.4.58}$$

are defined in the following way. We set

$$\begin{aligned} a_L &= \left(\tfrac{\gamma p_L}{\rho_L}\right)^{1/2}, \qquad s_L = \tfrac{p_L}{\rho_L^\gamma}, \\ a_R &= \left(\tfrac{\gamma p_R}{\rho_R}\right)^{1/2}, \qquad s_R = \tfrac{p_R}{\rho_R^\gamma}, \qquad \alpha = \left(\tfrac{s_R}{s_L}\right)^{(1/2\gamma)}, \\ z_L &= \tfrac{1}{2}(\gamma-1)u_L + a_L, \; z_R = \tfrac{1}{2}(\gamma-1)u_R - a_R \end{aligned}$$

and furthermore

$$\begin{aligned} a_1 &= \tfrac{z_L - z_R}{1+\alpha}, \quad \rho_1 = \left(\tfrac{a_1}{a_L}\right)^{\frac{2}{\gamma-1}} \rho_L, \quad u_1 = 2\tfrac{z_L - a_1}{\gamma-1}, \quad v_1 = v_L, \quad p_1 = \tfrac{a_1^2 \rho_1}{\gamma}, \\ a_3 &= \alpha a_1, \qquad \rho_3 = \tfrac{\rho_1}{\alpha^2}, \qquad u_3 = u_1, \qquad v_3 = v_R, \quad p_3 = \tfrac{a_3^2 \rho_3}{\gamma}, \\ a_L^* &= 2\tfrac{z_L}{\gamma+1}, \quad \rho_L^* = \left(\tfrac{a_L^*}{a_L}\right)^{\frac{2}{\gamma-1}} \rho_L, \quad u_L^* = a_L^*, \qquad v_L^* = v_L, \quad p_L^* = \tfrac{(a_L^*)^2 \rho_L^*}{\gamma}, \\ a_R^* &= -2\tfrac{z_R}{\gamma+1}, \; \rho_R^* = \left(\tfrac{a_R^*}{a_R}\right)^{\frac{2}{\gamma-1}} \rho_R, \; u_R^* = -a_R^*, \quad v_R^* = v_R, \; p_R^* = \tfrac{(a_R^*)^2 \rho_R^*}{\gamma}. \end{aligned}$$

The mentioned construction is possible under the condition

$$a_L + a_R + (\gamma - 1)(u_L - u_R)/2 > \max\{0, (v_L - v_R)/2\}. \tag{3.4.59}$$

Exercise 3.66 Show that the 2D Osher–Solomon numerical flux $\boldsymbol{g}_{\mathrm{OS}}$ through the boundary is defined by the following formulae (cf. Section 3.4.7):

I. *Supersonic outlet*: We simply have

$$\boldsymbol{g}_{\mathrm{OS}}(\boldsymbol{q}_L, \boldsymbol{q}_B) = \boldsymbol{f}_1(\boldsymbol{q}_L).$$

II. *Subsonic outlet*: We prescribe p_B and use the relations

$$\rho_B = \rho_L (p_B/p_L)^{1/\gamma}, \quad u_B = u_L + \frac{2}{\gamma - 1}(a_L - \sqrt{\gamma p_B/\rho_B}), \quad v_B = v_L.$$

Moreover we define the state $\boldsymbol{q}_L^*$ by

$$a_L^* = \frac{\gamma - 1}{\gamma + 1} u_L + \frac{2a_L}{\gamma + 1}, \; \rho_L^* = \left(\frac{a_L^*}{a_L}\right)^{\frac{2}{\gamma - 1}} \rho_L, \; u_L^* = a_L^*, v_L^* = v_L,$$

$$p_L^* = \frac{\rho_L^* (a_L^*)^2}{\gamma}.$$

Then we distinguish the two following subcases:

a) $u_B < a_B$

$$\boldsymbol{g}_{\mathrm{OS}}(\boldsymbol{q}_L, \boldsymbol{q}_B) = \boldsymbol{f}_1(\boldsymbol{q}_B),$$

b) $u_B \geq a_B$

$$\boldsymbol{g}_{\mathrm{OS}}(\boldsymbol{q}_L, \boldsymbol{q}_B) = \boldsymbol{f}_1(\boldsymbol{q}_L^*).$$

III. *Subsonic inlet*: We prescribe ρ_B, u_B, v_B and compute p_B using a state $\boldsymbol{q}_1$:

$$a_1 = a_L + \frac{\gamma - 1}{2}(u_L - u_B), \rho_1 = (a_1^2 \rho_L / \gamma p_L)^{1/(\gamma - 1)} \rho_L, u_1 = u_B, v_1 = v_L,$$

$$p_B = \rho_1 a_1^2 / \gamma.$$

Furthermore, we define the state $\boldsymbol{q}_L^*$ by

$$a_L^* = \frac{\gamma - 1}{\gamma + 1} u_L + \frac{2a_L}{\gamma + 1}, \rho_L^* = \left(\frac{a_L^*}{a_L}\right)^{\frac{2}{\gamma - 1}} \rho_L, u_L^* = a_L^*, v_L^* = v_L,$$

$$p_L^* = \frac{\rho_L^* (a_L^*)^2}{\gamma}.$$

Then we distinguish the four following subcases:

a) $u_B < 0$

(i) $u_B < a_1$

$$\boldsymbol{g}_{\mathrm{OS}}(\boldsymbol{q}_L, \boldsymbol{q}_B) = \boldsymbol{f}_1(\boldsymbol{q}_B),$$

(ii) $u_B \geq a_1$

$$\boldsymbol{g}_{\mathrm{OS}}(\boldsymbol{q}_L, \boldsymbol{q}_B) = \boldsymbol{f}_1(\boldsymbol{q}_B) + \boldsymbol{f}_1(\boldsymbol{q}_L^*) - \boldsymbol{f}_1(\boldsymbol{q}_1),$$

b) $u_B \geq 0$

(i) $u_B < a_1$

$$g_{OS}(q_L, q_B) = f_1(q_1),$$

(ii) $u_B \geq a_1$

$$g_{OS}(q_L, q_B) = f_1(q_L^*).$$

IV. *Supersonic inlet*: We prescribe ρ_B, u_B, v_B, p_B and

$$g_{OS}(q_L, q_B) = f_1(q_B).$$

V. *Impermeable wall*:

$$u_B = 0,\ a_B = a_L + (\gamma - 1)u_L/2,\ \rho_B = (a_B^2 \rho_L/\gamma p_L)^{1/(\gamma-1)}\rho_L,$$

$$p_B = \rho_B a_B^2/\gamma.$$

and we have $g_{OS}(q_L, q_B) = (0, p_B, 0, 0)^{\mathrm{T}}$.

3.5 Higher order finite volume schemes

The finite volume schemes derived in Section 3.3.2 are formally of first order of accuracy. This causes that discontinuities in a solution are often smeared and some details are not resolved sufficiently accurately. To avoid this deficiency, two techniques can be used:

a) the use of the so-called higher order finite volume schemes,
b) adaptive mesh refinement in the vicinity of discontinuities.

In this section we shall be concerned with the construction of 'second order' (in space) MUSCL-type (monotone upwind schemes for conservation laws) finite volume schemes. We proceed in analogy with the 1D case discussed in Section 3.2.21. This approach, introduced by (Van Leer, 1979), is now widely used. It is based on a piecewise linear reconstruction of a piecewise constant finite volume solution. Similarly to the 1D case a suitable limitation procedure is used in order to avoid spurious oscillations. (See, for example, (Durlofsky *et al.*, 1992), (Stoufflet *et al.*, 1987), (Cournede *et al.*, 1998), (Fürst and Kozel, 2002*a*).) The increase of the accuracy in time can be achieved with the aid of the Runge–Kutta methods (cf. 3.2.2).

3.5.1 *General form of a 'second order' MUSCL-type FV scheme*

We start from the explicit finite volume scheme derived in 3.3.2 and written in the form (3.3.24). The higher order modification of this scheme reads

$$w_i^{k+1} = w_i^k - \frac{\tau_k}{|D_i|} \sum_{j \in S(i)} \sum_{\alpha=1}^{\beta_{ij}} H\left(w_{ij}^{k,\alpha}, w_{ji}^{k,\alpha}, n_{ij}^{\alpha}\right) |\Gamma_{ij}^{\alpha}|. \tag{3.5.1}$$

By $w_{ij}^{k,\alpha}, w_{ji}^{k,\alpha}$ we denote the 'second order' approximations of the solution on the face Γ_{ij}^{α}, computed from the side of D_i and D_j, respectively. They are obtained in the following way:

1. We compute an approximation of the gradient of dependent variables on each finite volume D_i. That is, for a variable u_ℓ ($\ell = 1, \dots, m$) we denote the approximation of its gradient on D_i by $(\nabla u_\ell)_i$. (We omit the superscript k.)
2. Linear extrapolation on each edge Γ_{ij}^α of quantities u_ℓ is carried out.
3. A limitation procedure is applied.

3.5.2 *Computation of the approximate gradient*

There are several strategies for evaluating the approximate gradient $(\nabla u)_i$ of a variable u on a finite volume D_i. Let us mention some of them. We consider a piecewise constant function u with values u_i on D_i, $i \in J$.

Let P_i be the centre of gravity of D_i or an element of the set σ_h, with which the dual finite volume D_i is associated, or an element of the set $\mathcal{Q}_h$, with which the barycentric finite volume D_i is associated. See Sections 3.3.1.1 or 3.3.1.2. Let Q_{ij}^α denote the centre of gravity of the face Γ_{ij}^α.

It is possible to choose one of the following strategies:

(i) *The use of Green's theorem (GREEN)* We write

$$|D_i|(\nabla u)_i \approx \int_{D_i} \nabla u \, dx = \int_{\partial D_i} u \boldsymbol{n} \, dS \tag{3.5.2}$$
$$\approx \sum_{j \in S(i)} \sum_{\alpha=1}^{\beta_{ij}} \tilde{u}_{ij} \cdot \boldsymbol{n}_{ij}^\alpha,$$

where we set

$$\tilde{u}_{ij} = (u_i + u_j)/2. \tag{3.5.3}$$

A more accurate approximation uses the weighted average

$$\tilde{u}_{ij} = \frac{u_i|D_j| + u_j|D_i|}{|D_i| + |D_j|}. \tag{3.5.4}$$

(ii) *Least squares evaluation of* $(\nabla u)_i$ *(LS∇)* We approximate the derivatives of u on D_i in the directions $P_j - P_i$, $j \in S(i)$:

$$\nabla u(P_i) \cdot (P_j - P_i) \approx \frac{u_j - u_i}{|P_j - P_j|} \quad \text{for } j \in S(i). \tag{3.5.5}$$

From this we obtain the conditions

$$(\nabla u)_i \cdot (P_j - P_i) = \frac{u_j - u_i}{|P_j - P_i|}, \quad j \in S(i), \tag{3.5.6}$$

for $(\nabla u)_i$. If $j \in \gamma(i)$, which means that the finite volume D_i is adjacent to the boundary of Ω_h, we use Q_{ij}^α as the point P_j and boundary conditions for the determination of the value u_j. Conditions (3.5.6) form a linear system for the approximate partial derivatives $(\partial u/\partial x_s)_i$, $s = 1, \dots, N$, on D_i, which may be in general overdetermined. Therefore, the least squares method is used for the evaluation of $(\nabla u)_i$.

(iii) *Least squares linear interpolant approach (LSLI)* We construct a linear function L_i which minimizes the functional

$$\Phi(L_i) = (L_i(P_i) - u_i)^2 + \sum_{j \in S(i)} (L_i(P_j) - u_j)^2 . \tag{3.5.7}$$

Then we set

$$(\nabla u)_i = \nabla L_i. \tag{3.5.8}$$

(iv) *The use of linear interpolants over a triangular mesh (DEO)* This approach was proposed by (Durlofsky *et al.*, 1992). In this case we have $\Gamma_{ij}^{\alpha} = \Gamma_{ij}^{1} = \Gamma_{ij}$ for any finite volume $D_i \in \mathcal{D}_h$. By $Q_{ij}^{\alpha} = Q_{ij}^{1} = Q_{ij}$ we denote the centre of gravity of the face Γ_{ij}. Further, by D_j, D_k, D_ℓ we denote neighbours of D_i and construct linear functions L_{ijk}, $L_{ij\ell}$, $L_{ik\ell}$ such that

$$\begin{aligned} L_{ijk}(P_i) &= L_{ij\ell}(P_i) = L_{ik\ell}(P_i) = u_i, \\ L_{ijk}(P_j) &= L_{ij\ell}(P_j) = u_j, \\ L_{ijk}(P_k) &= L_{ik\ell}(P_k) = u_k, \\ L_{ij\ell}(P_\ell) &= L_{ik\ell}(P_\ell) = u_\ell. \end{aligned} \tag{3.5.9}$$

Then we get three approximations of $(\nabla u)_i$, namely

$$\nabla L_{ijk}, \quad \nabla L_{ij\ell}, \quad \nabla L_{ik\ell}, \tag{3.5.10}$$

from which we choose a suitable candidate on the basis of one of the following strategies leading to the determination of $(\nabla u)_i$.

We redenote the functions L_{ijk}, $L_{ij\ell}$ and $L_{ik\ell}$ as L_1, L_2, L_3 so that

$$|\nabla L_1| \geq |\nabla L_2| \geq |\nabla L_3|. \tag{3.5.11}$$

Now we distinguish the following strategies:

(DEO1) a) We set $L := L_1$ and verify the following conditions:

$$L(Q_{i\beta}) \text{ is between } u_i \text{ and } u_\beta \text{ for } \beta = j, k, \ell. \tag{3.5.12}$$

If (3.5.12) is satisfied, we set

$$(\nabla u)_i := \nabla L, \tag{3.5.13}$$

otherwise we continue to the next step.

b) We set $L := L_2$ and again test conditions (3.5.12). If they are satisfied, then we define $(\nabla u)_i$ by (3.5.13), otherwise we set $L := L_3$ and use (3.5.13) for the definition of $(\nabla u)_i$.

(DEO2) This method proceeds in the same way as (DEO1), but conditions (3.5.12) are also tested for $L = L_3$. If they are satisfied, then $(\nabla u)_i$ is defined by (3.5.13), otherwise we set $(\nabla u)_i = 0$.

(DEO3) In this case we set directly

$$(\nabla u)_i = \nabla L_3, \tag{3.5.14}$$

i.e. $(\nabla u)_i$ is defined as the gradient of the linear interpolant whose gradient has minimal magnitude.

3.5.3 *Linear extrapolation*

In this step the prediction $\overline{u}_{ij}^{\alpha}$ of the value u_{ij}^{α} is computed:

$$\overline{u}_{ij}^{\alpha} = u_i + (\nabla u)_i \cdot \left(Q_{ij}^{\alpha} - P_i\right). \tag{3.5.15}$$

Numerical experiments show that the definition $u_{ij}^{\alpha} := \overline{u}_{ij}^{\alpha}$ leads to approximate solutions suffering from spurious oscillations. Therefore, it is necessary to apply the next step.

3.5.4 *Limitation procedure*

The limited value u_{ij}^{α} on Γ_{ij}^{α} is written in the form

$$u_{ij}^{\alpha} = u_i + \phi_i (\nabla u)_i \cdot \left(Q_{ij}^{\alpha} - P_i\right), \tag{3.5.16}$$

where the choice of the parameter ϕ_i, called a *slope limiter*, is based on the idea that the value u_{ij}^{α} should belong to the interval between u_i and u_j. If u_i is a local extremum relative to its neighbours, then we usually set $\phi_i = 0$.

There are a number of limitation strategies (see (Godlewski and Raviart, 1991), Chapter IV, Section 5). Numerical experiments show that not all proposed approaches lead to the improvement of accuracy of the original FV method from Section 3.3.2, particularly if unstructured meshes are used. Here we describe the method proposed in (Barth and Jespersen, 1989).

We define $\overline{u}_{ij}^{\alpha}$ ($j \in S(i)$, $\alpha = 1, \dots, \beta_{ij}$) by (3.5.15) and successively define

$$u_i^{\min} = \min\left(u_i, \min_{j \in S(i)} u_j\right), \tag{3.5.17}$$

$$u_i^{\max} = \max\left(u_i, \max_{j \in S(i)} u_j\right),$$

$$\phi_{ij}^{\alpha} = \begin{cases} 1, & \overline{u}_{ij}^{\alpha} - u_i = 0, \\ \min\left(1, \frac{u_i^{\max} - u_i}{u_{ij}^{\alpha} - u_i}\right), & \overline{u}_{ij}^{\alpha} - u_i > 0, \\ \min\left(\frac{u_i^{\min} - u_i}{\overline{u}_{ij}^{\alpha} - u_i}\right), & \overline{u}_{ij}^{\alpha} - u_i < 0, \end{cases} \tag{3.5.18}$$

$$\phi_i = \min_{\substack{j \in S(i) \\ \alpha = 1, \dots, \beta_{ij}}} \phi_{ij}^{\alpha}. \tag{3.5.19}$$

Then u_{ij}^{α} is computed from (3.5.16).

Of course, if the mesh $\mathcal{D}_h$ is triangular, then the superscript $\alpha = 1$ can be omitted. This is, for example, the case for the methods (DEO1)–(DEO3).

3.5.5 *The choice of variables u_ℓ*

There are a number of possibilities of how to choose the dependent variables u_ℓ, $\ell = 1, \dots, m$. This is motivated by stability considerations, maximum principles, positivity of pressure or temperature, etc. (see, for example, (Mehlman, 1991), (Larrouturou, 1991)), (Perthame and Qiu, 1994), (Mulder and van Leer, 1985)), (Arminjon *et al.*, 1989), (Cournede *et al.*, 1998)). We mention the basic variants of the choice of variables u_ℓ:

a) $u_\ell = w_\ell$ *are conservative variables*: $\rho, \rho v_i, E$.
b) u_ℓ are *primitive variables*: ρ, v_i, p.
c) u_ℓ are *physical variables*: ρ, v_i, θ.
d) u_ℓ are *characteristic variables*, i.e. the components of the vector $\boldsymbol{W} = \mathbb{T}^{-1}(\boldsymbol{w}, \boldsymbol{n})\boldsymbol{w}$, where $\mathbb{T}$ and $\mathbb{T}^{-1}$ diagonalize the matrix $\mathbb{P}(\boldsymbol{w}, \boldsymbol{n})$ and $\boldsymbol{n} = \boldsymbol{n}_{ij}^\alpha$.
e) u_ℓ are *entropy variables*: since the function $\eta = -\rho S$, where S is the physical entropy, is strictly convex with respect to the state vector $\boldsymbol{w}$, then $\nabla_w \eta$ is a one-to-one mapping. Then the entropy variables form the vector

$$\boldsymbol{u} = \nabla_w \eta(\boldsymbol{w}). \tag{3.5.20}$$

Assuming that we have already computed the state vector $\boldsymbol{w}$ on the finite volumes $D_i \in \mathcal{D}_h$ at time t_k, i.e. we have the vectors $\boldsymbol{w}_i^k \in \mathbb{R}^m$, $i \in J$, with components $(w_\ell)_i^k$, $\ell = 1, \dots, m$, we define various types of individual variables u in the following way:

a) conservative variables:

$$\rho_i^k, (\rho v_1)_i^k, \dots, (\rho v_N)_i^k, E_i^k;$$

b) primitive variables:

$$\rho_i^k, (v_s)_i^k = (\rho v_s)_i^k / \rho_i^k, p_i^k = (\gamma - 1)\left(E_i^k - \rho_i^k \sum_{s=1}^N (v_s^2)_i^k / 2\right);$$

c) physical variables:

$$\rho_i^k, (v_s)_i^k = (\rho v_s)_i^k / \rho_i^k, \theta_i^k = \frac{1}{c_v}\left(E_i^k / \rho_i^k - \sum_{s=1}^n (v_s^2)_i^k / 2\right).$$

We proceed similarly in the case of characteristic and entropy variables. After the computation of all limited individual variables u_{ij}^α, we express the corresponding state vectors $\boldsymbol{w}_{ij}^\alpha$, substitute into scheme (3.5.1) and compute the values $\boldsymbol{w}_i^{k+1}$ of the approximate solution at the new time level t_{k+1}.

3.5.6 *Test of the accuracy of MUSCL-type schemes*

In order to determine the order of accuracy of the described schemes, we solve the shock-tube problem (see Remark 3.21 and Section 3.7.1) considered in the domain $\Omega = (-3,3) \times (0, 0.1)$ using the 2D FV method with the Osher–Solomon numerical flux from Section 3.4.9. The 2D Euler equations are equipped with the initial conditions

$$\begin{aligned} \boldsymbol{v}(x,0) &= 0 \\ \rho(x,0) &= \begin{cases} \rho_L, & x < x_0, \\ \rho_R, & x > x_0, \end{cases} \\ p(x,0) &= \begin{cases} p_L, & x < x_0, \\ p_R, & x > x_0, \end{cases} \end{aligned} \tag{3.5.21}$$

where

$$\begin{aligned} &x_0 = 0, \quad \rho_L = 1, \quad p_L = 2, \\ &\rho_R = 1.101463, \quad p_R = 1. \end{aligned} \tag{3.5.22}$$

The boundary conditions are the same as the initial conditions. The analytical solution (obtained on the basis of results from Section 3.1.6) and the numerical solution are compared at the time level $t = 1$. The solution depends on time t and the space variable x_1 only and is independent of x_2, but it can be, of course, considered as a function of two space variables x_1 and x_2. Moreover, if we define the velocity component in the direction x_2 as $v_2 = 0$, then we get a solution of the 2D Euler equations. The graph of the density of this solution depending on x_1 and t is shown in Fig. 3.7.1 below.

The numerical schemes were applied on three uniform triangulations containing 75×4, 150×8 and 300×16 vertices. For each mesh $\mathcal{D}_h$, the error in the density was measured at time $t = 1$ in the L^1-norm:

$$\begin{aligned} \|e_h\| &= \|\rho(\cdot,1) - \rho_h(\cdot,1)\|_{L^1(\tilde{\Omega})} \\ &= \int_{\tilde{\Omega}} |\rho(x,1) - \rho_h(x,1)|\, dx. \end{aligned} \tag{3.5.23}$$

The time step $\tau_k = t_{k+1} - t_k$ was determined on the basis of the stability condition (3.3.85). However, in order to suppress the discretization errors in time we use the 'overkill' in time multiplying the resulting τ from (3.3.85) by 10^{-2}.

Two cases were distinguished in (3.5.23):

1) $\tilde{\Omega} = \Omega$,
2) $\tilde{\Omega} = \Omega - K$, where K is formed by the union of narrow strips along discontinuities (shock wave and contact discontinuity).

Assuming that

$$\|e_h\| \approx Ch^q \tag{3.5.24}$$

and applying the logarithm to (3.5.24), we get

$$\log \|e_h\| \approx \log C + q \log h. \tag{3.5.25}$$

We denote

$$y = \log \|e_h\|, \quad z = \log h, \quad \beta = \log C, \tag{3.5.26}$$

and use the *least squares method* for the determination of the parameters C and q on the basis of results obtained over n $(= 3)$ meshes with size h_i and yielding the values $z_i = \log h_i$, $y_i = \log e_{h_i}$, $i = 1, 2, \ldots, n$. Then q and β are obtained as the minimum point of the function

$$\Phi(q, \beta) = \sum_{i=1}^{n} (y_i - \beta - q z_i)^2. \tag{3.5.27}$$

The solution reads

$$q = \frac{n \sum_{i=1}^{n} y_i z_i - \sum_{i=1}^{n} y_i \sum_{i=1}^{n} z_i}{n \sum_{i=1}^{n} z_i^2 - \left(\sum_{i=1}^{n} z_i\right)^2}, \tag{3.5.28}$$

$$\beta = \frac{1}{n} \left(\sum_{i=1}^{n} y_i - q \sum_{i=1}^{n} z_i \right)$$

The quality of the obtained results can be measured with the aid of the *correlation coefficient* r:

$$r = \frac{n \sum_{i=1}^{n} y_i z_i - \sum_{i=1}^{n} y_i \sum_{i=1}^{n} z_i}{\left(n \sum_{i=1}^{n} z_i^2 - \left(\sum_{i=1}^{n} z_i\right)^2 \right)^{1/2} \left(n \sum_{i=1}^{n} y_i^2 - \left(\sum_{i=1}^{n} y_i\right)^2 \right)^{1/2}}. \tag{3.5.29}$$

We have $r \in [-1, 1]$. If $r \approx 1$, then the relation (3.5.25) (and thus (3.5.24)) is satisfied very precisely (for $r = 1$ it is valid exactly) and if $r \ll 1$, then this relation is valid inaccurately. The number q is called the *experimental order of convergence* (EOC).

The numerical experiments were carried out for the following methods:

- first order scheme,
- MUSCL schemes: (GREEN) – see 4.5.2 (i),
  (LSLI) – see 4.5.2 (iii),
  (DEO1)–(DEO3) – see 4.5.2 (iv).

In all cases the extrapolation and limiting procedures described in 4.5.3 and 4.5.4 are applied.

In Table 3.5.1, the experimental order of accuracy q, the constant C (see (3.5.24)) and the correlation r (see (3.5.29)) are given for the mentioned methods and the domains $\tilde{\Omega} = \Omega$ (whole domain including discontinuities in the solution) and $\tilde{\Omega} = \Omega - K$ (domain where the solution is regular).

Table 3.5.1 *Order of accuracy*

	Domain Ω			Domain $\Omega - K$		
Method	q	C	r	q	C	r
1st order	0.627	0.065	0.999	1.062	0.106	0.099
GREEN	0.599	0.061	0.999	1.022	0.097	0.999
LSLI	0.594	0.044	0.998	1.270	0.113	0.999
DEO1	0.210	0.021	0.981	0.074	0.009	0.546
DEO2	0.203	0.016	0.630	0.010	0.004	0.019
DEO3	0.848	0.063	0.991	1.399	0.094	0.987

As we can see, the most accurate results are obtained with the aid of the simplest DEO strategy, i.e. (DEO3). All other methods are worse in the whole domain Ω as well as in the regularity region than the original first order scheme. Also the method (LSLI) is comparably accurate with (DEO3) in the regularity region. None of these methods is of second order in the area where the solution is regular. This indicates that *one must be careful concerning the statement that 'some FV method is of second (or higher) order'.* According to our experience, as well as the experience of other authors ((Pike, 1987), (Jeng and Chen, 1992), (Durlofsky *et al.*, 1992), a second order method is reduced to a first order method on irregular meshes.

For further numerical experiments we refer the reader to Sections 3.5 and 5.3 from (Kröner, 1997), where computational results and the experimental order of convergence for a scalar conservation law and the shock-tube problem, respectively, are presented.

On the basis of numerical experiments we can conclude that a sufficiently accurate numerical solution of the Euler equations is not in general guaranteed by the use of 'higher order' MUSCL schemes. Usually an efficient and accurate solution requires a suitable mesh refinement, particularly in the vicinity of discontinuities. This will be the subject of the next section.

Another approach to the construction of higher order finite volume schemes is based on the extension of the ENO or WENO strategies, explained in Section 3.2.22 for problems with one space variable, to problems with more space dimensions. See, for instance, (Abgrall, 1994), (Fürst and Kozel, 1999), (Angot *et al.*, 1997), (Fürst and Kozel, 2001), (Fürst and Kozel, 2002*b*), (Sonar, 1997*a*), (Sonar, 1997*b*). A nice explanation of multidimensional ENO and WENO schemes can be found in (Shu, 1999).

In Chapter 4 we shall be concerned with finite element techniques which allow us to develop theoretically based higher order schemes in a natural way with the aid of a weak formulation.

3.6 Adaptive methods

In practical computations we try to achieve sufficiently accurate solutions with precise resolutions of important characteristic flow features, e.g. boundary layers or shock waves and their interactions. For complex compressible flow the global mesh refinement leads to a huge increase of degrees of freedom. Consequently, the CPU time (typically hours) of the procedure exceeds the user limits (which are very severe for the weather-forecast-related flow problems, for example) and the computer memory becomes insufficient for running the program. To be able to solve complex engineering problems not just on supercomputers, local adaptive remeshing has to be used.

Very often our goal is to obtain a steady-state solution of system (3.3.1)–(3.3.3) by the time marching process $t_k \to \infty$. The computation is stopped if

$$\left\| \frac{1}{\tau} \left(\mathbf{w}_h^{k+1} - \mathbf{w}_h^k \right) \right\| \leq \varepsilon, \tag{3.6.1}$$

where $\| \cdot \|$ is a suitable norm and ε is a user-prescribed accuracy. The density ρ can be chosen as a significant flow quantity to control the stationarity of the approximate solution, i.e. the first component of $\mathbf{w}_h^k$ is used in (3.6.1).

In what follows, we shall deal with the adaptation techniques for computation of the steady-state solution of the Euler equations. We shall present mesh adaptation techniques for triangular and tetrahedral computational meshes because of their flexibility and applicability and because they reflect well some important characteristics of flow problems, e.g. curved boundaries, fast changes of shape of computational domain or periodic boundary conditions. If the barycentric or dual finite volumes are used in the FV algorithm, the triangular (in 2D) or tetrahedral (in 3D) mesh, over which they are constructed, is adapted. Then the computation continues on the barycentric or dual FV mesh constructed over the adapted triangular (in 2D) or tetrahedral (in 3D) mesh.

3.6.1 *Geometrical data structure*

For the sake of simplicity, we suppose that $\Omega \subset I\!R^N$ is a polygonal domain, if $N = 2$, and a polyhedral domain, if $N = 3$. (We do not deal with the 1D adaptation in this section.)

Let $\tilde{P}_1, \ldots, \tilde{P}_{N+1}$ be different points in $I\!R^N$, $1 \leq N \leq 3$, which are not colinear for $N = 2$ and not coplanar for $N = 3$. For $N = 1$ or $N = 2$ or $N = 3$ we denote by N-simplex the straight segment or the triangle or the tetrahedron with vertices $\tilde{P}_1, \ldots, \tilde{P}_{N+1}$, respectively. The $(N-1)$-simplexes forming the boundary of an N-simplex are called faces. The straight segment connecting the vertices $\tilde{P}_i$, $\tilde{P}_j$, $i, j \in \{1, \ldots, N+1\}$, $i \neq j$, of an N-simplex is called an edge. If e is an edge with end points $\tilde{P}_i$ and $\tilde{P}_j$, we denote by $\boldsymbol{e}$ a vector $\tilde{P}_j - \tilde{P}_i$.

Let $\mathcal{T}_h = \{K_i\}_{i \in I}$ be an N-simplicial grid in $\Omega \subset I\!R^N$, $N = 2$ or $N = 3$, with the usual properties used in the finite element method (see (3.3.15) and (3.3.16)).

Let $\mathcal{D}_h = \{D_i\}_{i\in J}$ be an N-simplicial, dual or barycentric finite volume mesh, constructed over the N-simplicial grid $\mathcal{T}_h$. Its construction is described in Sections 3.3.1.1 and 3.3.1.2. If the finite volume mesh is formed by N-simplexes, then we set $\mathcal{D}_h = \mathcal{T}_h$. The finite volume solution of problem (3.3.1)–(3.3.3) is sought in the space of discontinuous piecewise constant functions on $\mathcal{D}_h$

$$\boldsymbol{\mathcal{C}}_{\mathcal{D}_h} = \mathcal{C}^m_{\mathcal{D}_h} = \{v \in L^\infty(\Omega); v|_D \text{ is constant } \forall\, D \in \mathcal{D}_h\}^m . \tag{3.6.2}$$

Analogously to (3.6.2) we define the space of piecewise constant functions on $\mathcal{T}_h$

$$\boldsymbol{\mathcal{C}}_{\mathcal{T}_h} = \mathcal{C}^m_{\mathcal{T}_h} = \{v \in L^\infty(\Omega); v|_K \text{ is constant } \forall\, K \in \mathcal{T}_h\}^m . \tag{3.6.3}$$

Let us further introduce the space of piecewise linear functions on $\mathcal{T}_h$

$$\boldsymbol{\mathcal{L}}_{\mathcal{T}_h} = \mathcal{L}^m_{\mathcal{T}_h} = \left\{v \in C(\overline{\Omega}); v|_K \text{ is linear } \forall\, K \in \mathcal{T}_h\right\}^m . \tag{3.6.4}$$

3.6.2 *Adaptation algorithm*

Let $\mathbf{w}_h^k$ be the finite volume solution of problem (3.3.1)–(3.3.3) at time level t_k (see Definition 3.54). In further considerations the time level t_k will be fixed and therefore we omit the upper index k in the notation of the finite volume solution $\mathbf{w}_h^k$. In the numerical process, the fixed time level t_k is the instant when we stop the computation of non-stationary flow or when we reach the steady-state in the time marching procedure (3.3.23) for the solution of the stationary flow. The flowchart of the adaptive algorithm for the numerical solution $\boldsymbol{w}_h$ of the Euler equations (3.3.1)–(3.3.3) consists of following steps:

Step 1 – Process initialization
Step 2 – Computation of $\mathbf{w}_h$
Step 3 – Accuracy test
Step 4 – Adaptation
Step 5 – Post processing (e.g. computation of physical quantities such as Mach number, lift, drag, distribution of quantities along boundaries, isolines, graphical output, video, etc.)

In Step 1 the finite volume mesh $\mathcal{D}_h$ over the N-simplicial grid $\mathcal{T}_h$ is constructed and the process is initialized by defining the initial finite volume approximation $\mathbf{w}_h^0 \in \boldsymbol{\mathcal{C}}_{\mathcal{D}_h}$ and boundary conditions.

The initial grid $\mathcal{T}_h$ is constructed using a suitable mesh generator. Among many, the mesh generator NETGEN written by Joachim Schöberl can be mentioned. It is sufficiently robust and covers a wide class of 2D and 3D geometries. It can be freely downloaded from

`http://www.sfb013.uni-linz.ac.at/~joachim/netgen`

The geometrical data structure for the construction of the finite volume mesh $\mathcal{D}_h$ is described in Section 3.6.1. The initial approximation $\mathbf{w}_h^0 \in \boldsymbol{\mathcal{C}}_{\mathcal{D}_h}$ is constructed in such a way that

$$\mathbf{w}_h^0|_{D_i} = \boldsymbol{w}_i^0, \quad D_i \in \mathcal{D}_h, \tag{3.6.5}$$

where $\boldsymbol{w}_i^0$ is defined in (3.3.25).

Having defined the initial approximation $\mathbf{w}_h^0$ together with the boundary conditions, the finite volume scheme (3.3.23) is applied on the mesh $\mathcal{D}_h$ in Step 2.

Step 3 represents the stopping criterion in the adaptive procedure and it is up to the user to define it. An example might be the computation of the steady-state solution via the time marching process. The approximation $\mathbf{w}_h \in \mathcal{C}_{\mathcal{D}_h}$ satisfying condition (3.6.1) is computed on the mesh $\mathcal{D}_h$. Then the mesh $\mathcal{D}_h$ is adapted as described in Step 4 and the computation continues on the modified mesh. The process is repeated in several steps until the desired accuracy of the solution and sharpness of discontinuities is obtained.

In Step 4 the recovery $\mathbf{w}_h^* \in \mathcal{C}_{\mathcal{T}_h}$ or $\mathbf{w}_h^* \in \mathcal{L}_{\mathcal{T}_h}$, respectively, of the finite volume solution $\mathbf{w}_h \in \mathcal{C}_{\mathcal{D}_h}$ is constructed. Then using $\mathbf{w}_h^*$, the adaptation of the grid $\mathcal{T}_h$ is performed and the new N-simplicial grid $\mathcal{T}_h^{\text{new}}$ is constructed. Over the grid $\mathcal{T}_h^{\text{new}}$ the mesh $\mathcal{D}_h^{\text{new}}$ is constructed and the recovery $\mathbf{w}_h^{\text{new}} \in \mathcal{C}_{\mathcal{D}_h^{\text{new}}}$ of $\mathbf{w}_h \in \mathcal{C}_{\mathcal{D}_h}$ is defined. The reinitialization

$$\begin{aligned} \mathcal{T}_h &:= \mathcal{T}_h^{\text{new}}, \\ \mathcal{D}_h &:= \mathcal{D}_h^{\text{new}}, \\ \mathbf{w}_h^0 &:= \mathbf{w}_h^{\text{new}} \end{aligned} \tag{3.6.6}$$

is made and the process returns to Step 2.

3.6.3 *Mesh refinement*

One useful way of mesh adaptation is the construction of nested meshes using mesh refinement. Let $\mathcal{T}_h = \{K_i\}_{i\in I}$ be an N-simplicial mesh and

$$g_{\text{ind}} : i \in I \to g_{\text{ind}}(i) \in [0, \infty)$$

a given function called a refinement indicator. The relative indicator $g : i \in I \to g(i) \in [0, 1]$

$$g(i) = g_{\text{ind}}(i) / \max_{j\in I} g_{\text{ind}}(j), \quad i \in I \tag{3.6.7}$$

is calculated. If for some $K_i \in \mathcal{T}_h$

$$g(i) \geq TOL, \tag{3.6.8}$$

where $TOL \in [0, 1]$ is a given tolerance, then the N-simplex K_i is refined and the resulting partition of Ω is modified so that the conforming N-simplicial mesh (see (3.3.15) and (3.3.16)) is obtained. The following 2D and 3D examples illustrate the geometrical aspects of this technique. The construction of the refinement indicators is presented in Sections 3.6.4 and 3.6.5 .

3.6.3.1 *Red–green refinement in 2D* If for some $K_i \in \mathcal{T}_h$ (3.6.8) holds, then the 'mother' triangle K_i is divided into four equal 'daughter' subtriangles (red refinement), see Fig. 3.6.1 left. The resulting partition of Ω is modified so that the conforming triangulation of Ω is obtained (green refinement), see Fig. 3.6.1 right. In order to reduce the number of obtuse angles in the triangulation, red* refinement could be used, see Fig. 3.6.1 bottom.

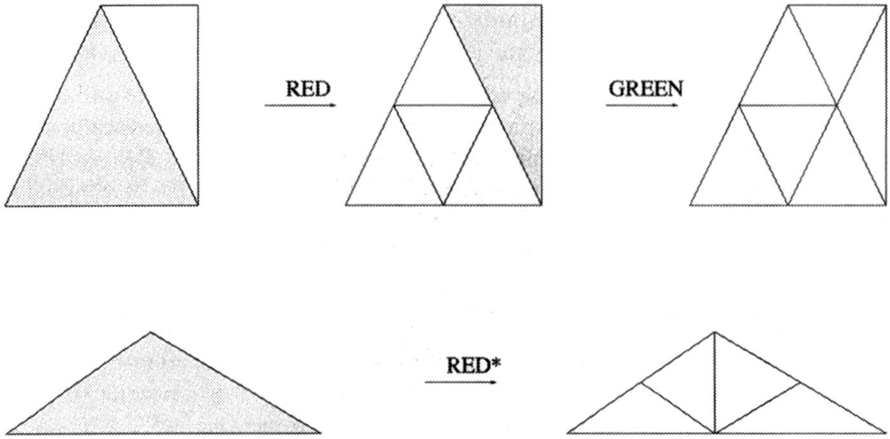

FIG. 3.6.1. Red–green refinement

3.6.3.2 *Bisection refinement in 3D* The basic idea of the mesh refinement in 3D is to divide all tetrahedra $K \in \mathcal{T}_h$ marked for the refinement into two tetrahedra K_i^1, K_i^2 by a plane cut passing through the midpoint of the longest edge of K_i (see Fig. 3.6.2). We call such a cut the *longest edge bisection* (see Fig. 3.6.3).

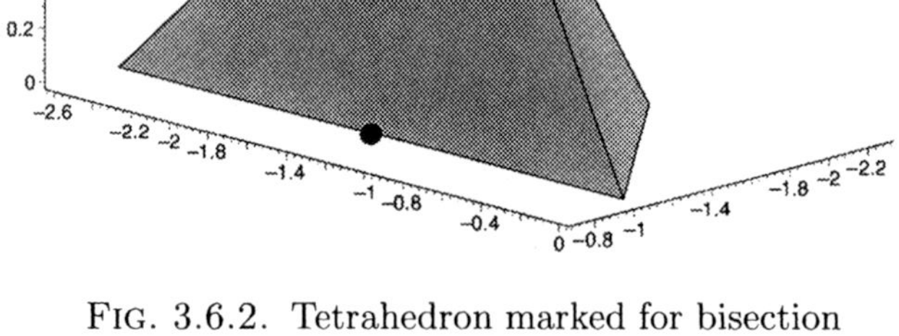

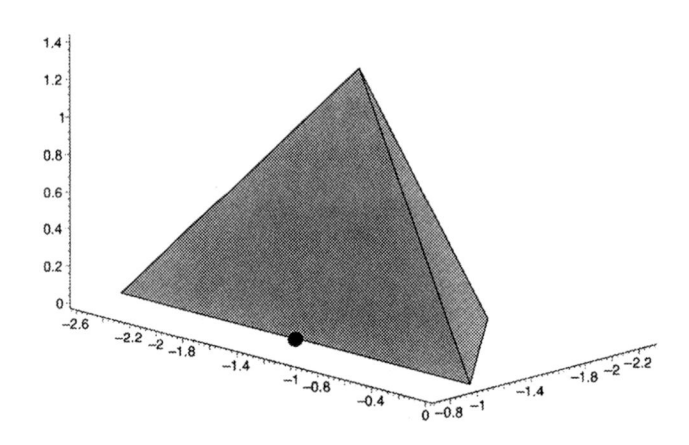

FIG. 3.6.2. Tetrahedron marked for bisection

Let $g : I \to I\!R$ be the refinement indicator. By $REF = \{i \in I; g(i) \geq TOL\}$ we denote the index set of tetrahedra to be refined. The algorithm reads (see (Rivara, 1996))

$LEB(\mathcal{T}_h, I, REF)\{$

  while $(REF \neq \emptyset)\{$

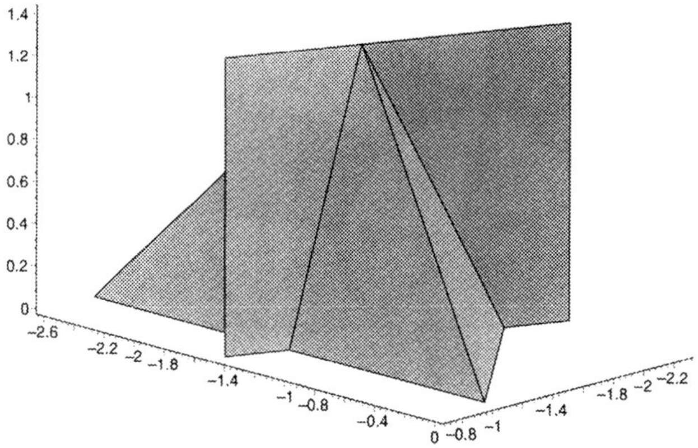

FIG. 3.6.3. Bisection of a tetrahedron

forall ($i \in REF$)

Longest Edge Bisect (K_i)

$REF := \{i \in I; i$ has a hanging node on an edge$\}$

}

},

where the procedure Longest Edge Bisect ( ) is defined as follows:

Longest Edge Bisect (T) {

Bisect T into two tetrahedra T^1, T^2 by the longest edge cut

Remove T from $\mathcal{T}_h$

Include T^1, T^2 in $\mathcal{T}_h$

Update I

}.

The algorithm run is illustrated for simplicity in Figs 3.6.4–3.6.7 for the 2D case. Having marked the elements with hanging nodes (Fig. 3.6.5 right), the algorithm continues with their bisection (Fig. 3.6.7 right).

It has been proven in 2D (see (Rivara, 1996)) that the algorithm is finite and produces non-degenerate elements. The 3D case remains open.

Moreover, the algorithm is very sensitive to the determination of the *longest edge* (of the equilateral tetrahedron, for example) and due to the computer arithmetic can produce a nonconforming mesh. This drawback can be avoided by the following modification of the algorithm using the recursive procedure Longest Edge Bisect Rec ( ) (see (Persiano *et al.*, 1993)):

$LEBm(\mathcal{T}_h, I, REF)\{$

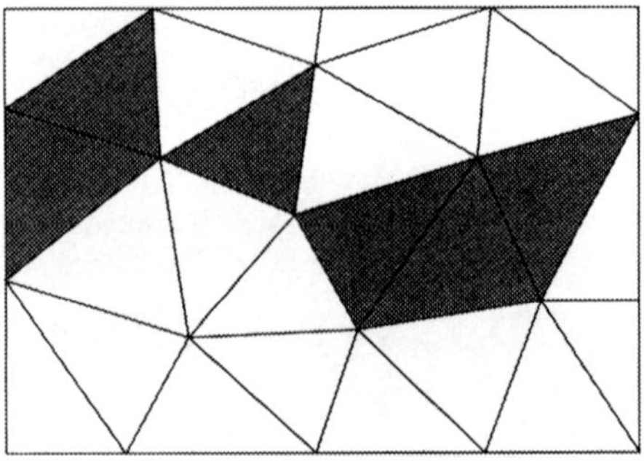

FIG. 3.6.4. Initial grid with marked elements

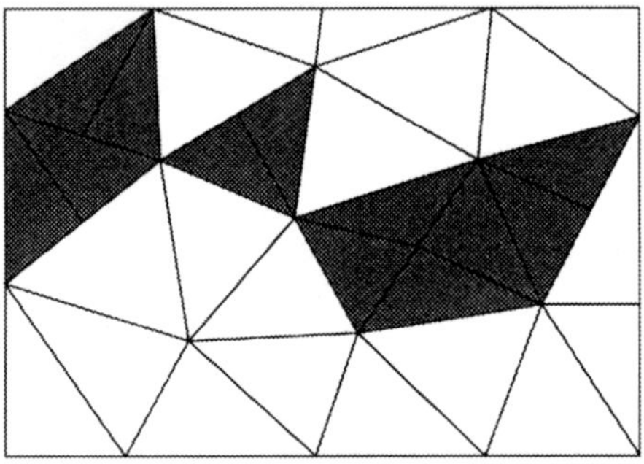

FIG. 3.6.5. Bisection of marked elements

while ($\exists\, i \in REF$)
  Longest Edge Bisect Rec(K_i)
},

where the procedure Longest Edge Bisect Rec() is defined as follows:

Longest Edge Bisect Rec(T){
  $e :=$ the longest edge of T
  while ($\exists\, i \in I; e \in K_i$ and e is not the longest edge of K_i)
    Longest Edge Bisect Rec (K_i)
  $REF := REF \setminus \{i; e \in K_i\}$
  Edge Bisect (e)
  Update I
  }.

Here the procedure Edge Bisect () puts the new vertex in the centre of the edge e and bisects all tetrahedra having the edge e in common (the so-called edge bisection). It can be shown that this procedure produces the conforming mesh.

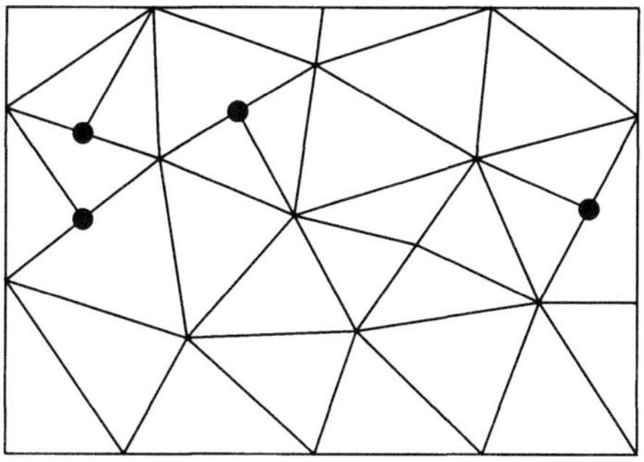

FIG. 3.6.6. Hanging nodes

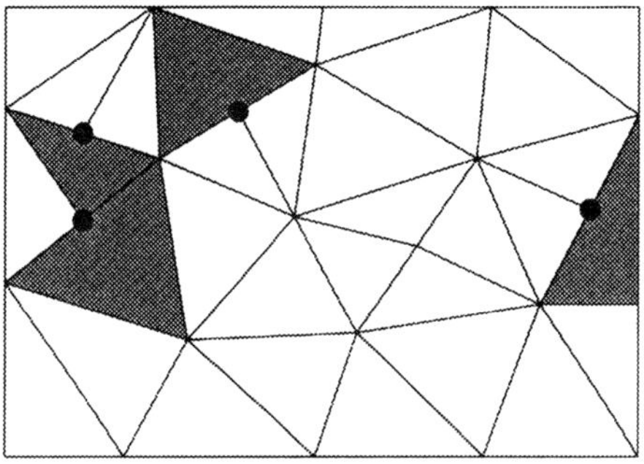

FIG. 3.6.7. Marked elements with hanging nodes

However, the procedure suffers from inaccuracy in the computer arithmetic – in the longest edge search recursions can become infinite. The finiteness of the algorithm is proven under the assumption that there are no different edges with the same length. Another algorithm of bisection was presented by D. N. Arnold, A. Mukherjee and L. Pouly in

`ftp://ftp.math.psu.edu/pub/dna/papers/bistet.ps`.

Since we adapt only N-simplicial meshes, the piecewise constant approximate solution $\mathbf{w}_h^k$ on $\mathcal{D}_h$ is recovered by a suitable function $\mathbf{w}_h^*$ defined on $\mathcal{T}_h$ and for this function the mesh adaptation algorithm is applied. The mesh adaptation techniques and criteria can be classified according to the type of recovery of the finite volume solution $\mathbf{w}_h \in \boldsymbol{\mathcal{C}}_{\mathcal{D}_h}$.

3.6.4 *Adaptation techniques based on constant recovery*

Let $\mathbf{w}_h \in \boldsymbol{\mathcal{C}}_{\mathcal{D}_h}$ be a finite volume solution. Depending on the finite volume mesh $\mathcal{D}_h$ used, we distinguish the following cases for the construction of the constant recovery $\mathbf{w}_h^*$:

a) *N-simplicial mesh $\mathcal{D}_h$* As in this case $\mathcal{D}_h = \mathcal{T}_h$, the recovery $\mathbf{w}_h^* \in \boldsymbol{\mathcal{C}}_{\mathcal{T}_h}$ is defined

$$\mathbf{w}_h^* \equiv \mathbf{w}_h. \tag{3.6.9}$$

b) *Dual finite volume mesh $\mathcal{D}_h$ over the grid $\mathcal{T}_h$* Let $K_i \in \mathcal{T}_h$ be an N-simplex with vertices $P_1^{K_i}, \ldots, P_{N+1}^{K_i}$. We define the recovery $\mathbf{w}_h^* \in \boldsymbol{C}_{\mathcal{T}_h}$ such that

$$\mathbf{w}_h^*|_{K_i} = \frac{1}{N+1} \sum_{s=1}^{N+1} \mathbf{w}_h(P_s^{K_i}). \tag{3.6.10}$$

(Note that $\mathbf{w}_h(P_s^{K_i}) = \mathbf{w}_h|_{D_j}$, where D_j is the finite volume associated with the vertex $P_s^{K_i}$.)

c) *Barycentric finite volume mesh $\mathcal{D}_h$ over the grid $\mathcal{T}_h$* Let $K_i \in \mathcal{T}_h$ be an N-simplex with centres of gravity of faces $Q_1^{K_i}, \ldots, Q_{N+1}^{K_i}$. We define the recovery $\mathbf{w}_h^* \in \boldsymbol{C}_{\mathcal{T}_h}$ such that

$$\mathbf{w}_h^*|_{K_i} = \frac{1}{N+1} \sum_{s=1}^{N+1} \mathbf{w}_h(Q_s^{K_i}). \tag{3.6.11}$$

(Note that $\mathbf{w}_h(Q_s^{K_i}) = \mathbf{w}_h|_{D_j}$, where D_j is the finite volume associated with the centre of gravity $Q_s^{K_i}$.)

The recovery $\mathbf{w}_h^*$ of the finite volume solution $\mathbf{w}_h$ is piecewise constant. We shall denote the constant value $\mathbf{w}_h^*|_{K_i}$ by

$$\boldsymbol{w}_i^* = (w_{i1}^*, w_{i2}^*, \ldots, w_{im}^*)^{\mathrm{T}}. \tag{3.6.12}$$

Now we shall concentrate on the various refinement indicators.

3.6.4.1 *Shock indicator* Let Γ be a shock wave. (Note that its position is not a priori known.) We use here the fact that the density jumps up passing the shock wave in the direction of the flow. This leads us to the shock indicator applied on each N-simplex $K_i \in \mathcal{T}_h$:

$$g_{\text{shock}}(i) = \max_{j \in s(i)} \left((\rho_j^* - \rho_i^*)\, \boldsymbol{v}_i^* \cdot \boldsymbol{n}_{ij} \right)^+ \Big/ h_{ij}, \quad i \in I, \tag{3.6.13}$$

where h_{ij} is the distance of the centres of gravity of K_i and $K_j \in \mathcal{T}_h$ and ρ_i^* and $\boldsymbol{v}_i^*$ are the density and the velocity vector on K_i, respectively.

For the density ρ_i^* and velocity $\boldsymbol{v}_i^*$ we have from (3.1.8)

$$\begin{aligned} \rho_i^* &= w_{i1}^* \\ \boldsymbol{v}_i^* &= \frac{1}{\rho_i^*} \left(w_{i2}^*, \ldots, w_{i(N+1)}^* \right)^{\mathrm{T}}. \end{aligned} \tag{3.6.14}$$

The set $s(i)$ in (3.6.13) denotes the neigbouring N-simplexes to K_i and $\boldsymbol{n}_{ij}$ denotes the unit outer normal to ∂K_i in the sense of the finite volume geometry described in Section 3.3.1. The symbol q^+ denotes the positive part of the quantity q, i.e. $q^+ = \max(q, 0)$. The jumps in the density across faces of K_i in the flow direction are computed and only those that are jumps up are taken into account in (3.6.13). Indeed, for unphysical jumps down, the expression inside the brackets in (3.6.13) is negative and so its positive part is zero. Now we set

$$g_{\text{ind}} := g_{\text{shock}}$$

in (3.6.7) and apply the refinement technique from Sections 3.6.3.1 or 3.6.3.2.

3.6.4.2 *Residual error indicator* The ultimate goal of any numerical method is to compute the approximate solution with the prescribed tolerance TOL

$$\|\boldsymbol{w} - \mathbf{w}_h\|_X \leq \text{TOL}, \tag{3.6.15}$$

where $\boldsymbol{w}$ and $\mathbf{w}_h$ are the exact and approximate solutions of the problem and $\|\cdot\|_X$ is a suitable norm. The error estimation theory deals with the derivation of an *error estimator* η, depending on the approximate solution $\mathbf{w}_h$ and bounding the error from below and above

$$C_1\, \eta(\mathbf{w}_h) \leq \|\boldsymbol{w} - \mathbf{w}_h\|_X \leq C_2\, \eta(\mathbf{w}_h), \tag{3.6.16}$$

with constants C_1, $C_2 > 0$. The error estimator is expressed in terms of *error indicators* $\bar{\eta}(i)$, $i \in I$, defined for $K_i \in \mathcal{T}_h$:

$$\eta(\mathbf{w}_h) = \sqrt{\sum_{K_i \in \mathcal{T}_h} \bar{\eta}^2(i)}. \tag{3.6.17}$$

$\bar{\eta}$ can be used as refinement indicator g in the mesh refinement algorithm described in Section 3.6.3. See also the derivation of error indicators for elliptic problems in the framework of the finite element method in Section 4.1.12.1. The error estimation problem for hyperbolic systems has not yet been satisfactorily solved. Therefore, at least the norm of the residual is approximated in terms of error indicators which are then used in the mesh refinement algorithm. The relation of the norm of the *residual* to the norm of the *error* can be simply explained for the linear operator equation with bounded inverse operator. Let $\boldsymbol{w}$ be the exact solution of an abstract problem

$$\boldsymbol{\mathcal{A}}(\boldsymbol{w}) = 0 \tag{3.6.18}$$

and let $\mathbf{w}_h$ be its approximate solution. Let the form of (3.6.18) allow us to substitute $\mathbf{w}_h$ into (3.6.18) and to define the residual

$$\boldsymbol{r}_h = \boldsymbol{\mathcal{A}}(\mathbf{w}_h). \tag{3.6.19}$$

Let $\|\cdot\|_X$ be a suitable norm in which the residual is measured and let $\|\cdot\|_Y$ be the norm in which the inverse operator $\mathcal{A}^{-1}$ is measured. Under the above assumptions we have

$$\|\mathbf{w}_h - \boldsymbol{w}\|_X \leq \|\boldsymbol{\mathcal{A}}^{-1}\|_Y\, \|\boldsymbol{r}_h\|_X, \tag{3.6.20}$$

which demonstrates, at least for a linear problem, the role of the residual in the error estimation. The proposed residual error indicator is based on the approximation of the norm of the residual. To this end, the concept of the residual for the Euler equations has to be defined.

We shall confine our considerations to the *stationary case*. This means that $\partial/\partial t \equiv 0$ and the Euler equations take the form

$$\sum_{s=1}^{N} \frac{\partial \boldsymbol{f}_s(\boldsymbol{w})}{\partial x_s} = 0. \tag{3.6.21}$$

It makes no sense to substitute a numerical solution containing discontinuities into (3.6.21) and to compute some norm of the obtained residual as an error indicator. Therefore, one has to introduce a weak formulation.

Let $\boldsymbol{w}$ be a classical solution of problem (3.6.21). Multiplying (3.6.21) by any $\boldsymbol{\varphi} = (\varphi^1, \varphi^2, \dots, \varphi^m)^{\mathrm{T}} \in \boldsymbol{H}_0^1(\Omega) = H_0^1(\Omega)^m$, integrating over Ω and using Green's theorem, we obtain the identity

$$0 = \int_\Omega \sum_{s=1}^{N} \frac{\partial \boldsymbol{f}_s(\boldsymbol{w})}{\partial x_s} \boldsymbol{\varphi}\, dx = -\int_\Omega \sum_{s=1}^{N} \boldsymbol{f}_s(\boldsymbol{w}) \frac{\partial \boldsymbol{\varphi}}{\partial x_s}\, dx. \tag{3.6.22}$$

This integral identity leads us to the following definition.

Definition 3.67 *A vector function* $\boldsymbol{w} \in \boldsymbol{L}^\infty(\Omega) = L^\infty(\Omega)^m$ *is called a weak solution of equation* (3.6.21), *if*

$$\boldsymbol{a}(\boldsymbol{w}, \boldsymbol{\varphi}) = 0 \quad \forall \boldsymbol{\varphi} \in \boldsymbol{H}_0^1(\Omega), \tag{3.6.23}$$

where

$$\begin{aligned} \boldsymbol{a}(\boldsymbol{w}, \boldsymbol{\varphi}) &= (a^1(\boldsymbol{w}, \boldsymbol{\varphi}), a^2(\boldsymbol{w}, \boldsymbol{\varphi}), \dots, a^m(\boldsymbol{\varphi}, \boldsymbol{w}))^{\mathrm{T}}, \\ a^\ell(\boldsymbol{w}, \boldsymbol{\varphi}) &= \int_\Omega \sum_{s=1}^{m} f_{s\ell}(\boldsymbol{w}) \frac{\partial \varphi^\ell}{\partial x_s}\, dx, \quad \ell = 1, \dots, m, \\ \boldsymbol{f}_s(\boldsymbol{w}) &= (f_{s1}(\boldsymbol{w}), f_{s2}(\boldsymbol{w}), \dots, f_{sm}(\boldsymbol{w}))^{\mathrm{T}} \quad s = 1, \dots, N. \end{aligned} \tag{3.6.24}$$

The mapping

$$\varphi \in H_0^1(\Omega) \to \sum_{s=1}^{m} f_{s\ell}(\boldsymbol{w}) \frac{\partial \varphi}{\partial x_S}\, dx \tag{3.6.25}$$

is for a fixed integer $\ell \in \{1, \dots, m\}$ and $\boldsymbol{w} \in \boldsymbol{L}^\infty(\Omega)$ a continuous linear functional on $H_0^1(\Omega)$. By virtue of the Riesz representation theorem, there exists an $\mathcal{A}^\ell(\boldsymbol{w}) \in H_0^1(\Omega)$ such that

$$a^\ell(\boldsymbol{w}, \boldsymbol{\varphi}) = \left(\mathcal{A}^\ell(\boldsymbol{w}), \varphi^\ell\right), \tag{3.6.26}$$

where $(\cdot,\cdot)$ denotes the scalar product in $H_0^1(\Omega)$. Hence, if we write

$$\boldsymbol{\mathcal{A}}(\boldsymbol{w}) = \left(\mathcal{A}^1(\boldsymbol{w}), \mathcal{A}^2(\boldsymbol{w}), \dots, \mathcal{A}^m(\boldsymbol{w})\right)^{\mathrm{T}},$$

we see that (3.6.23) can be written in the form

$$\mathcal{A}(\boldsymbol{w}) = 0. \tag{3.6.27}$$

Now let us approximate the exact solution $\boldsymbol{w}$ by the recovery $\mathbf{w}_h{}^* \in \mathcal{C}_{\mathcal{T}_h}$ of the approximate solution of $\mathbf{w}_h$ and substitute $\mathbf{w}_h{}^*$ in (3.6.27) instead of $\boldsymbol{w}$. We obtain the residual $\boldsymbol{r}_h$

$$\boldsymbol{r}_h = \mathcal{A}(\mathbf{w}_h^*) \tag{3.6.28}$$

(in the sense of distributions). Based on this residual we compute a residual error indicator. We use the $\boldsymbol{H}^{-1}$-norm of the residual

$$\|\boldsymbol{r}_h\|_{\boldsymbol{H}^{-1}(\Omega)} = \left(\sum_{\ell=1}^{m} \|r_h^\ell\|_{H^{-1}(\Omega)}^2 \right)^{1/2}, \tag{3.6.29}$$

where

$$\|r_h^\ell\|_{H^{-1}(\Omega)} = \sup_{\substack{\varphi \neq 0 \\ \varphi \in H_0^1(\Omega)}} \frac{|(\mathcal{A}^\ell(\mathbf{w}_h^*), \varphi)|}{\|\varphi\|_{H_0^1(\Omega)}}. \tag{3.6.30}$$

The use of the quantities (3.6.28) and (3.6.30) is motivated by the 2D error estimates of U. Göhner and G. Warnecke in (Göhner and Warnecke, 1995) for compressible potential flows as well as by the analysis of J. Mackenzie, E. Süli and G. Warnecke in (Mackenzie *et al.*, 1994). Related discrete approximations of these dual norms were considered in (Sonar, 1993; Sonar and Warnecke, 1996).

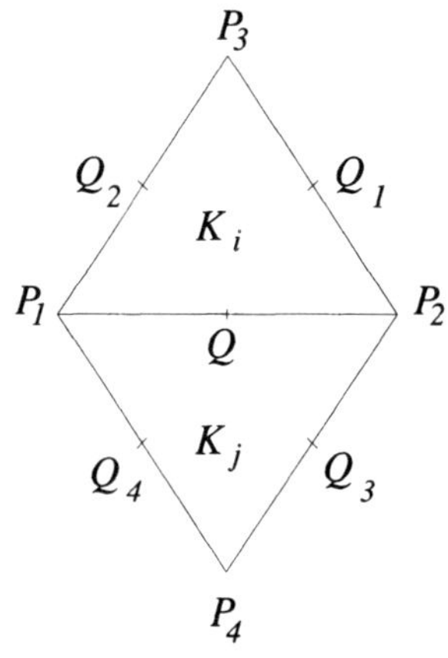

FIG. 3.6.8. Local notation of vertices and centres of gravity of faces in 2D

The supremum in (3.6.30) will be approximated by the maximum over a finite set of piecewise linear functions $\mathcal{E}_{\mathcal{T}_h}$. For its definition some geometrical preliminaries are needed.

Let K_i, $K_j \in \mathcal{T}_h$, $i < j$, be neighbours with a common face Γ_{ij}. Let $\tilde{P}_1, \ldots, \tilde{P}_{N+2}$ be the vertices of K_i, K_j in the local notation from Figs 3.6.8 and 3.6.9.

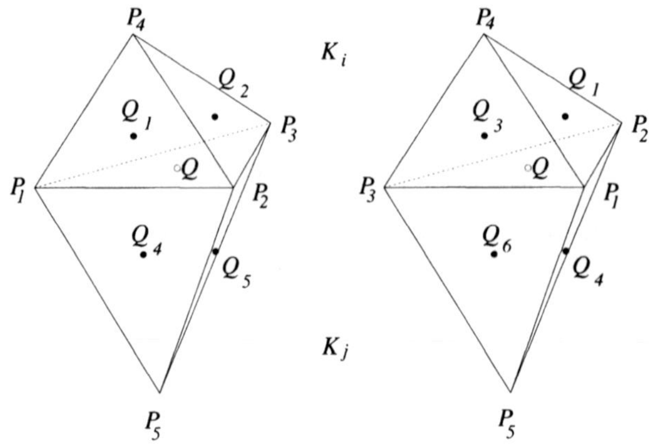

FIG. 3.6.9. Local notation of vertices and centres of gravity of faces in 3D

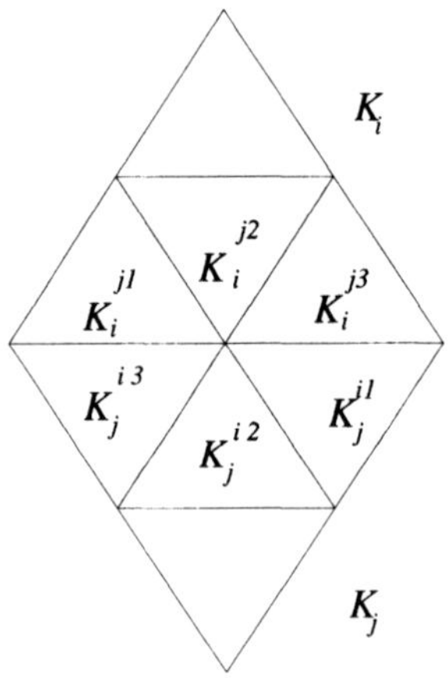

FIG. 3.6.10. The set $\mathcal{K}_{ij}$ in 2D (shaded)

Let $\tilde{Q}$ be the centre of gravity of Γ_{ij} and let $\tilde{Q}_1, \dots, \tilde{Q}_{2N}$ be the other centres of gravity of faces of K_i, K_j in the local notation from Figs 3.6.8 and 3.6.9. Let K_i^{jk}, K_j^{ik}, $k = 1, \dots, 2^N - 1$, be the N-simplexes from Figs 3.6.10–3.6.12. The vertices of K_i, K_j and the centres of gravity of faces of K_i, K_j are used for the construction K_i^{jk}, K_j^{ik}, $k = 1, \dots, 2^N - 1$. Since K_i^{jk}, K_j^{ik} have the centre of gravity $\tilde{Q}$ of Γ_{ij} in common, their construction is clearly seen from Figs 3.6.10–3.6.12. For example, the N-simplex K_i^{j7} from Fig. 3.6.11 has vertices $\tilde{Q}$, $\tilde{Q}_1$, $\tilde{Q}_2$, $\tilde{Q}_3$. The N-simplexes forming the shaded set from Figs 3.6.10–3.6.12 are used in the definition of the set $\mathcal{K}_{ij}$:

$$\mathcal{K}_{ij} = \bigcup_{k=1}^{2^N - 1} (K_i^{jk} \cup K_j^{ik}). \tag{3.6.31}$$

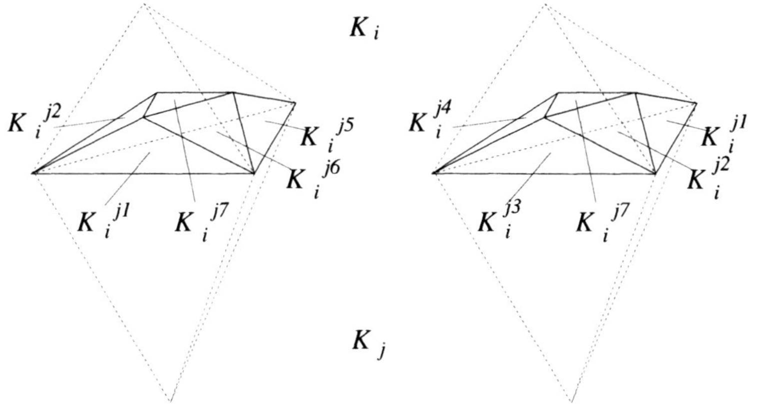

FIG. 3.6.11. The set $\mathcal{K}_{ij} \cap K_i$ in 3D (shaded).

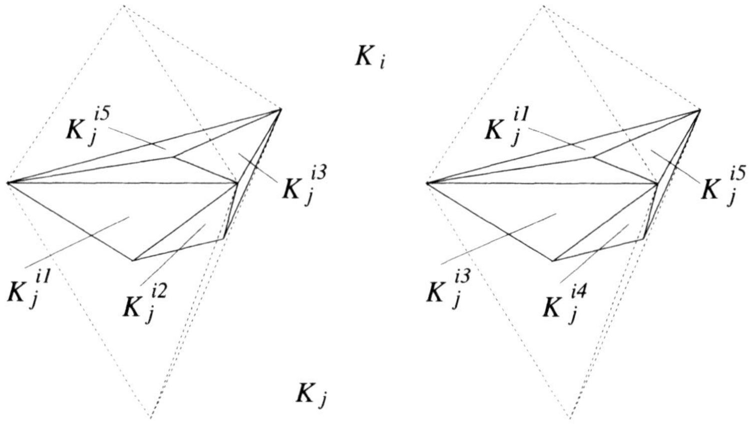

FIG. 3.6.12. The set $\mathcal{K}_{ij} \cap K_j$ in 3D (shaded)

We define the set $\mathcal{E}_{\mathcal{T}_h}$

$$\begin{aligned}\mathcal{E}_{\mathcal{T}_h} = \Big\{ & \varphi_{ij} \in C(\overline{\Omega});\ \varphi_{ij} \text{ is linear on } K_i^{jk},\ K_j^{ik},\ k = 1, \ldots, 2^N - 1, \\ & \varphi_{ij} = 1 \text{ at the centre of gravity of } \Gamma_{ij},\ \varphi_{ij} = 0 \text{ on } \partial\mathcal{K}_{ij}, \\ & \operatorname{supp} \varphi_{ij} = \mathcal{K}_{ij},\ K_i,\ K_j \in \mathcal{T}_h,\ i < j, \text{ are neighbours} \Big\}.\end{aligned}$$

The norm of the residual $\|r_h^\ell\|_{H^{-1}(\Omega)}$ in (3.6.30) is approximated by taking the maximum over the set $\mathcal{E}_{\mathcal{T}_h}$:

$$\|r_h^\ell\|_{H^{-1}(\Omega)} \approx \max_{\varphi_{ij} \in \mathcal{E}_{\mathcal{T}_h}} \frac{\left|(\mathcal{A}^\ell(\mathbf{w}_h^*), \varphi_{ij})\right|}{|\varphi_{ij}|_{H_0^1(\Omega)}}, \tag{3.6.32}$$

where $|\cdot|_{H_0^1(\Omega)}$ denotes the seminorm involving only the derivatives, which is equivalent to the norm $\|\cdot\|_{H_0^1(\Omega)}$ in $H_0^1(\Omega)$ due to the Friedrichs inequality. Let us denote the index set of neighbouring N-simplexes to K_i by $s(i)$ (cf. the notation in Section 3.3.1) and put

$$g^\ell(i) = \max_{j\in s(i)} \frac{|(\mathcal{A}^\ell(\mathbf{w}_h^*), \varphi_{ij})|}{|\varphi_{ij}|_{H_0^1(\Omega)}}, \quad i \in I. \tag{3.6.33}$$

Then

$$\max_{\varphi_{ij}\in\mathcal{E}_{\mathcal{T}_h}} \frac{|(\mathcal{A}^\ell(\mathbf{w}_h^*), \varphi_{ij})|}{|\varphi_{ij}|_{H_0^1(\Omega)}} = \max_{K_i\in\mathcal{T}_h} g^\ell(i) \le \left(\sum_{K_i\in\mathcal{T}_h} (g^\ell(i))^2\right)^{1/2}.$$

From this, (3.6.29) and (3.6.32) we obtain the approximation

$$\|\boldsymbol{r}_h\|^2_{\boldsymbol{H}^{-1}(\Omega)} \approx \sum_{K_i\in\mathcal{T}_h} \left[\sum_{\ell=1}^m (g^\ell(i))^2\right]. \tag{3.6.34}$$

On the basis of (3.6.34) we define the *residual error indicator*

$$g_{\mathrm{res}}(i) = \left(\sum_{\ell=1}^m (g^\ell(i))^2\right)^{1/2}. \tag{3.6.35}$$

The approximation

$$\|\boldsymbol{r}_h\|_{\boldsymbol{H}^1(\Omega)} \approx \left(\sum_{K_i\in\mathcal{T}_h} g^2_{\mathrm{res}}(i)\right)^{1/2}$$

replaces the relations (3.6.16)–(3.6.17) and is used instead of them for the adaptive mesh refinement.

The evaluation of $g^\ell(i)$ in (3.6.35) is accomplished in the following way. The norm $|\varphi_{ij}|_{H_0^1(\Omega)}$ in (3.6.33) can be computed exactly, because φ_{ij} is linear on N-simplexes K_i^{jk} and K_j^{ik} for neighbours K_i, $K_j \in \mathcal{T}_h$, $k = 1, \dots, 2^N - 1$:

$$\begin{aligned} |\varphi_{ij}|^2_{H_0^1(\Omega)} &= \int_\Omega \sum_{s=1}^N \left(\frac{\partial\varphi_{ij}}{\partial x_s}\right)^2 dx = \int_{K_{ij}} \sum_{s=1}^N \left(\frac{\partial\varphi_{ij}}{\partial x_s}\right)^2 dx \\ &= \sum_{k=1}^{2^N-1} \int_{K_i^{jk}} \sum_{s=1}^N \left(\frac{\partial\varphi_{ij}}{\partial x_s}\right)^2 dx + \sum_{k=1}^{2^N-1} \int_{K_j^{ik}} \sum_{s=1}^N \left(\frac{\partial\varphi_{ij}}{\partial x_s}\right)^2 dx. \end{aligned} \tag{3.6.36}$$

From (3.6.36) it follows that

$$|\varphi_{ij}|_{H_0^1(\Omega)} = \left[\sum_{k=1}^{2^N-1}\sum_{s=1}^N \left(|K_i^{jk}| \left(\frac{\partial\varphi_{ij}}{\partial x_s}\Big|_{K_i^{jk}}\right)^2 + |K_j^{ik}| \left(\frac{\partial\varphi_{ij}}{\partial x_s}\Big|_{K_j^{ik}}\right)^2\right)\right]^{1/2}, \tag{3.6.37}$$

where $|K_i^{jk}|$ and $|K_j^{ik}|$ are N-dimensional measures of K_i^{jk} and K_j^{ik}, respectively.

Using (3.6.26), (3.6.24) and setting $\boldsymbol{\varphi}_{ij} = (\varphi_{ij}, \varphi_{ij}, \ldots, \varphi_{ij})^{\mathrm{T}}$, we have for the numerator in (3.6.33)

$$
\begin{aligned}
\left|(\mathcal{A}^\ell(\mathbf{w}_h^*), \boldsymbol{\varphi}_{ij})\right| &= |a^\ell(\mathbf{w}_h^*, \boldsymbol{\varphi}_{ij})| = \left| \int_\Omega \sum_{s=1}^N f_{s\ell}(\mathbf{w}_h^*) \frac{\partial \varphi_{ij}}{\partial x_s} \, dx \right| \\
&= \left| \sum_{s=1}^N f_{s\ell}(\boldsymbol{w}_i^*) \sum_{k=1}^{2^N-1} \int_{K_i^{jk}} \frac{\partial \varphi_{ij}}{\partial x_s} \, dx + \sum_{s=1}^N f_{s\ell}(\boldsymbol{w}_j^*) \sum_{k=1}^{2^N-1} \int_{K_j^{ik}} \frac{\partial \varphi_{ij}}{\partial x_s} \, dx \right| \\
&= \left| \sum_{s=1}^N f_{s\ell}(\boldsymbol{w}_i^*) \, (n_{ij})_s \int_{\Gamma_{ij}} \varphi_{ij} \, dS + \sum_{s=1}^N f_{s\ell}(\boldsymbol{w}_j^*) \, (-n_{ij})_s \int_{\Gamma_{ij}} \varphi_{ij} \, dS \right| \\
&= \left| \sum_{s=1}^N f_{s\ell}(\boldsymbol{w}_i^*) \, (n_{ij})_s \frac{1}{N} \, |\Gamma_{ij}| + \sum_{s=1}^N f_{s\ell}(\boldsymbol{w}_j^*) \, (-n_{ij})_s \frac{1}{N} \, |\Gamma_{ij}| \right| \\
&= \left| \frac{|\Gamma_{ij}|}{N} \sum_{s=1}^N (n_{ij})_s \, (f_{s\ell}(\boldsymbol{w}_i^*) - f_{s\ell}(\boldsymbol{w}_j^*)) \right| .
\end{aligned} \tag{3.6.38}
$$

In (3.6.38) notation (3.1.9) and (3.3.11) is used. If we substitute (3.6.38) into (3.6.33), we get

$$
g^\ell(i) = \max_{j \in s(i)} \frac{|\Gamma_{ij}|}{N |\varphi_{ij}|_{H_0^1(\Omega)}} \left| \sum_{s=1}^N (n_{ij})_s \, (f_{s\ell})(\boldsymbol{w}_i^*) - f_{s\ell}(\boldsymbol{w}_j^*) \right| , \tag{3.6.39}
$$

with $|\varphi_{ij}|_{H_0^1(\Omega)}$ given by relation (3.6.37). Then the computation of $g_{\mathrm{res}}(i)$ in (3.6.35) is straightforward. Finally, we set

$$
g_{\mathrm{ind}} := g_{\mathrm{res}}
$$

in (3.6.7) and apply the refinement technique from Sections 3.6.3.1 or 3.6.3.2.

3.6.5 *Adaptation techniques based on linear recovery*

In this section we construct a piecewise linear recovery of a piecewise constant approximate solution and use it in the adaptation algorithm.

Let $\mathbf{w}_h \in \boldsymbol{\mathcal{C}}_{\mathcal{D}_h}$ be the piecewise constant finite volume solution (see Definition 3.54). Let us recall that σ_h denotes the set of all vertices of all elements $K_i \in \mathcal{T}_h$ and $\mathcal{Q}_h$ denotes the set of all centres of gravity of all faces of all elements $K_i \in \mathcal{T}_h$.

a) *N-simplicial mesh $\mathcal{D}_h$* In this case $\mathcal{D}_h = \mathcal{T}_h$ and the recovery $\mathbf{w}_h^* \in \boldsymbol{\mathcal{L}}_{\mathcal{T}_h}$ is determined uniquely by its values calculated at each vertex $P_j \in \sigma_h$ of the grid $\mathcal{T}_h$. These values are weighted averages of the values $\boldsymbol{w}_i$ given on N-simplexes which have the corresponding vertex in common. The areas of these N-simplexes are used as weights . Hence, we set

$$\mathbf{w}_h^*(P_j) = \frac{\sum_{K_i \in \mathcal{K}_{P_j}} |K_i| \, \boldsymbol{w}_i}{\sum_{K_i \in \mathcal{K}_{P_j}} |K_i|}, \tag{3.6.40}$$

where

$$\mathcal{K}_{P_j} = \{K_i \in \mathcal{T}_h; K_i \ni P_j\}, \quad P_j \in \sigma_h. \tag{3.6.41}$$

b) *Dual finite volume mesh $\mathcal{D}_h$ over the grid $\mathcal{T}_h$* The recovery $\mathbf{w}_h^* \in \boldsymbol{\mathcal{L}}_{\mathcal{T}_h}$ is determined uniquely by its values calculated at each vertex $P_j \in \sigma_h$ of the grid $\mathcal{T}_h$. We set

$$\mathbf{w}_h^*(P_j) = \boldsymbol{w}_j. \tag{3.6.42}$$

c) *Barycentric finite volume mesh $\mathcal{D}_h$ over the grid $\mathcal{T}_h$* The recovery $\mathbf{w}_h^* \in \boldsymbol{\mathcal{L}}_{\mathcal{T}_h}$ is determined uniquely by its values at each vertex $P \in \sigma_h$ of the grid $\mathcal{T}_h$. We define the average A_K on the N-simplex $K \in \mathcal{T}_h$ computed from values of $\mathbf{w}_h$ at centres of gravity of faces $Q \in \mathcal{Q}_h$ of K

$$A_K = \frac{1}{N+1} \sum_{Q \in K \cap \mathcal{Q}_h} \mathbf{w}_h(Q) \tag{3.6.43}$$

and we set

$$\mathbf{w}_h^*(P) = \frac{\sum_{K \in \mathcal{K}_P} |K| \, A_K}{\sum_{K \in \mathcal{K}_P} |K|} \tag{3.6.44}$$

where the summation is made over the N-simplexes $K \in \mathcal{T}_h$ having the vertex $P \in \sigma_h$ in common:

$$\mathcal{K}_P = \{K \in \mathcal{T}_h; K \ni P\}, \quad P \in \sigma_h.$$

3.6.5.1 *Superconvergence error indicator* The idea of the superconvergence indication comes from the superconvergence properties of the approximate solution to elliptic problems. It was proposed by Zienkiewicz and Zhou and analysed in (Verfürth, 1996). In the framework of the finite volume method, the superconvergence error indicator is applied on N-simplicial finite volume meshes $\mathcal{D}_h$. We proceed in the following way.

From the finite volume solution $\mathbf{w}_h$ (see Definition 3.54) and its linear recovery $\mathbf{w}_h^*$ determined uniquely by its values at vertices in (3.6.40), (3.6.42) and (3.6.44) we construct a function

$$\tilde{\boldsymbol{w}}_h := \mathbf{w}_h^* - \mathbf{w}_h, \tag{3.6.45}$$

which is piecewise linear and discontinuous in the domain Ω. The array of values

$$g_{\sup}(i) = \|\tilde{\boldsymbol{w}}_h\|_{L^2(K_i)}, \quad i \in I, \tag{3.6.46}$$

is called the *superconvergence error indicator*. We set

$$g_{\text{ind}} := g_{\text{sup}}$$

in (3.6.7) and apply the refinement technique from Sections 3.6.3.1 or 3.6.3.2.

3.6.6 *Data reinitialization for the mesh refinement*

The adaptation of the N-simplicial grid $\mathcal{T}_h$ leading to the new grid $\mathcal{T}_h^{\text{new}}$ is followed by the construction of the finite volume mesh $\mathcal{D}_h^{\text{new}}$ by means of Sections 3.3.1.1 and 3.3.1.2, respectively. To run the finite volume algorithm (3.3.23) on the mesh $\mathcal{D}_h^{\text{new}}$, the initial condition (3.3.25) has to be defined. To this end, the finite volume solution $\mathbf{w}_h$ on the mesh $\mathcal{D}_h$ is recomputed on the mesh $\mathcal{D}_h^{\text{new}}$.

a) *N-simplicial mesh* $\mathcal{D}_h^{\text{new}}$ The initial approximation on the new mesh is defined by taking the same values on daughter N-simplexes as on the mother N-simplex. Since we have $\mathcal{D}_h^{\text{new}} = \mathcal{T}_h^{\text{new}}$, we set

$$\mathbf{w}_h^{\text{new}}\Big|_D = \mathbf{w}_h^*\Big|_D, \quad D \in \mathcal{D}_h^{\text{new}}, \tag{3.6.47}$$

where $\mathbf{w}_h^* \in \boldsymbol{C}_{\mathcal{T}_h}$ is the recovery from (3.6.9).

b) *Dual finite volume mesh* $\mathcal{D}_h^{\text{new}}$ In this case the mesh $\mathcal{D}_h^{\text{new}}$ is constructed over the grid $\mathcal{T}_h^{\text{new}}$ with vertices that form the set σ_h^{new}. We can write $\mathcal{D}_h^{\text{new}} = \{D_P; P \in \sigma_h^{\text{new}}\}$, where D_P is the dual finite volume associated with vertex P. Starting from the piecewise constant recovery $\mathbf{w}_h^* \in \boldsymbol{C}_{\mathcal{T}_h}$ of the approximate solution, we set

$$\mathbf{w}_h^{\text{new}}\Big|_{D_P} = \frac{\sum\limits_{\substack{K \in \mathcal{T}_h^{\text{new}} \\ K \ni P}} \mathbf{w}_h^*\Big|_K |K|}{\sum\limits_{\substack{K \in \mathcal{T}_h^{\text{new}} \\ K \ni P}} |K|}, \quad P \in \sigma_h^{\text{new}}, \tag{3.6.48}$$

where σ_h^{new} is a set of vertices of $\mathcal{T}_h^{\text{new}}$.

c) *Barycentric finite volume mesh* $\mathcal{D}_h^{\text{new}}$ In this case $\mathcal{D}_h^{\text{new}} = \{D_Q; Q \in \mathcal{Q}_h^{\text{new}}\}$, where D_Q denotes the barycentric finite volume associated with $Q \in \mathcal{Q}_h^{\text{new}}$. Using the piecewise constant recovery $\mathbf{w}_h^* \in \boldsymbol{C}_{\mathcal{T}_h}$, we define

$$\mathbf{w}_h^{\text{new}}\Big|_{D_Q} = \frac{\sum\limits_{\substack{K \in \mathcal{T}_h^{\text{new}} \\ K \ni Q}} \mathbf{w}_h^*\Big|_K |K|}{\sum\limits_{\substack{K \in \mathcal{T}_h^{\text{new}} \\ K \ni Q}} |K|}, \quad Q \in \mathcal{Q}_h^{\text{new}}. \tag{3.6.49}$$

3.6.7 *Anisotropic mesh adaptation*

In this section we present the so-called anisotropic mesh adaptation (AMA) algorithm, which can be successfully applied to numerical solutions of a wide range of problems in physics and engineering described by partial differential equations. The motivation can be found in (Fortin *et al.*, 1996), (Habashi *et al.*, 1996), (Castro-Díaz *et al.*, 1996). A review of the anisotropic mesh generation research can be found in (Simpson, 1994) and references therein. For the application of the AMA technique to the numerical solution of compressible flow see (Dolejší and Angot, 1996; Dolejší, 1998*b*; Dolejší, 1998*a*; Dolejší, 2001; Dolejší, 2000).

In what follows, we formulate the necessary condition for the properties of the N-simplicial mesh, on which the discretization error is below the prescribed tolerance, and control this necessary condition by the interpolation error. For a sufficiently smooth function, we apply the AMA technique. We apply this strategy to the smoothed approximate solution and demonstrate the practical aspects of the adaptation procedure: the Riemann norm in which the length of edges is measured, the notion of optimal mesh, the control of the ratio between the longest and the shortest edge of the mesh and the control of the transition of the coarsest part of the mesh to the finest one. We can put this technique in the framework of mesh generation methods for the construction of an almost equilateral mesh. A problem with isotropic properties in the Riemann norm should produce an almost equilateral mesh in the Riemann norm.

If there are regions where the approximate solution changes very rapidly, i.e. in flows with shocks and/or boundary layers, a directionally refined mesh is needed. Such a refinement is called the grid alignment. AMA takes this feature into account by a refinement in the direction where the interpolation error is high. For an application see the numerical examples in Section 3.7.

a) *Discretization error* To measure how close the approximate solution $\mathbf{w}_h^*$ is to the exact solution $\boldsymbol{w}(\cdot, t_k)$, we choose a significant scalar flow quantity $u(\cdot, t_k)$ in which we measure the discretization error. Since the time instant t_k is fixed, we identify the function $u(\cdot, t_k)$ with the function $u^k(x) = u(x, t_k)$ and omit the superscript k in the sequel. The function $u(\cdot, t_k)$ is supposed to be an element of the space of trial functions W, which has to be specified. Let $u_h^* \in \mathcal{L}_{\mathcal{T}_h}$ be the corresponding physical quantity computed from the linear recovery $\mathbf{w}_h^*$ of the approximate solution $\mathbf{w}_h$. We define the discretization error (at the time level t_k) by

$$e_h = \|u - u_h^*\|_X, \tag{3.6.50}$$

where $\|\cdot\|_X$ is a suitable norm.

We say that the approximate solution is computed with a tolerance TOL, if

$$e_h \leq \text{TOL}. \tag{3.6.51}$$

This is the ultimate goal of any numerical method: to compute the approximate solution with the prescribed tolerance. In what follows, we present the necessary condition for the relation (3.6.51) and show how it can be used for the construction of adaptive meshes including the grid alignment.

b) *Necessary condition* Let $\pi_h : W \to \mathcal{L}_{\mathcal{T}_h}$ be an operator of the best approximation such that

$$\|u - \pi_h u\|_X = \min_{v \in \mathcal{L}_{\mathcal{T}_h}} \|u - v\|_X. \tag{3.6.52}$$

The operator π_h may not be uniquely determined; its uniqueness depends on the choice of $\|\cdot\|_X$.

Since $u_h^* \in \mathcal{L}_{\mathcal{T}_h}$, it is evident from (3.6.52) that

$$\|u - \pi_h u\|_X \leq e_h. \tag{3.6.53}$$

This is the crucial point of the proposed strategy: the discretization error (3.6.50) is bounded from below by $\|u - \pi_h u\|_X$. The *necessary condition* to fulfil (3.6.51) is

$$\|u - \pi_h u\|_X \leq \mathrm{TOL}. \tag{3.6.54}$$

Remark 3.68 Since condition (3.6.54) is only necessary, it says nothing more than, if it is not satisfied, then the numerical method for the solution of (3.3.1)–(3.3.3) cannot compute the approximate solution with the error satisfying (3.6.51). To give the numerical method the chance to satisfy (3.6.51), the mesh on which (3.6.54) holds is needed. It is evident that the magnitude of $\|u - \pi_h u\|_X$ strongly depends on the choice of $\mathcal{T}_h$ and it is completely independent of the numerical solution u_h^* itself. We shall modify the given mesh $\mathcal{T}_h$ in order to satisfy the necessary condition while keeping the number of N-simplexes as small as possible.

Let us illustrate the idea of the mesh adaptation process on a 1D example. Let the tolerance $\mathrm{TOL} = \omega_1$ be prescribed as in Fig. 3.6.13 right and let $\mathcal{T}_h$ be the initial grid from Fig. 3.6.13 left. If the exact solution is as in Fig. 3.6.13 left, that on the mesh $\mathcal{T}_h$, the necessary condition (3.6.54), where we set $X = C([A, B])$, is not satisfied. The computational mesh satisfying condition (3.6.54) is shown in Fig. 3.6.13 right. It is anisotropic and reflects the steep gradient in the solution. In the numerical algorithm for the construction of the adapted mesh the smoothed approximate solution computed on the starting mesh is used instead of the exact solution u.

c) *Interpolation operator* In order to control the necessary condition (3.6.54), we introduce the interpolation operator and apply the AMA technique. For more details about AMA we refer to (Bank *et al.*, 1983; Buscaglia and Dari, 1977; D'Azevedo and Simpson, 1989; D'Azevedo and Simpson, 1991; Castro-Díaz *et al.*, 1996; Fortin *et al.*, 1996; Habashi *et al.*, 1996; Apel *et al.*, 2001; Simpson, 1994).

The assumptions of the sufficient smoothness of the exact solution are required for AMA. Let $x_K \in K$ be a given point of an element $K \in \mathcal{T}_h$ and let

$$u \in C^1(\overline{\Omega}). \tag{3.6.55}$$

Since we are mainly interested in the non-smooth generalized solutions of compressible flow, we shall show later how to overcome the assumption (3.6.55) using a suitable smoothing of the approximate solution u_h^*. We define the *interpolation operator* $r_h : C^1(\overline{\Omega}) \to \mathcal{L}_{\mathcal{T}_h}$ such that

$$r_h\, u(x_K) = u(x_K) \quad \forall\, K \in \mathcal{T}_h$$

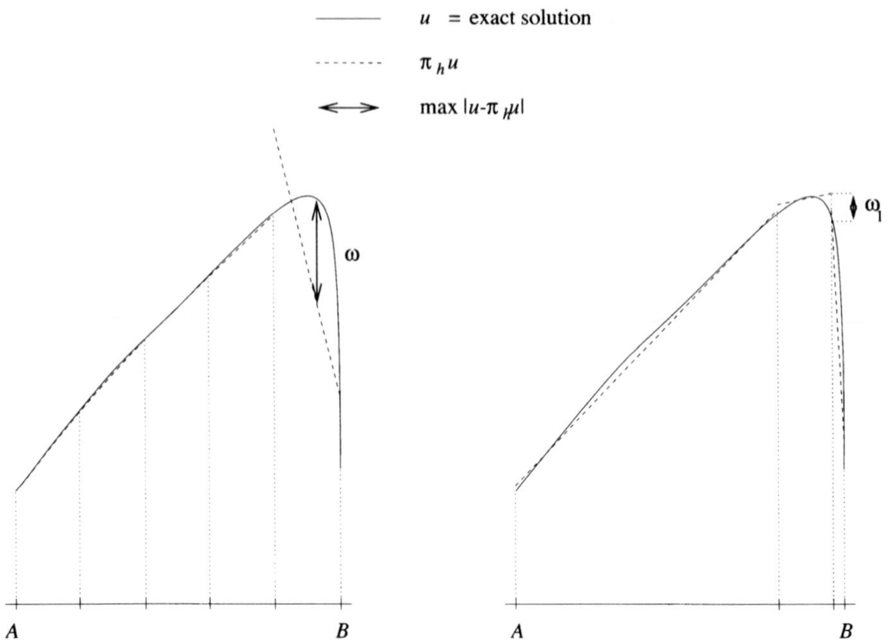

FIG. 3.6.13. Exact solution u and its best approximation $\pi_h u$ on the given mesh.

$$\nabla r_h\, u(x_K) = \nabla u(x_K) \quad \forall K \in \mathcal{T}_h. \tag{3.6.56}$$

The definition of r_h is unique for a given set of points $\{x_K; K \in \mathcal{T}_h\}$ and $r_h u \in \mathcal{L}_{\mathcal{T}_h}$ can easily be constructed for a given $u \in C^1(\overline{\Omega})$. Since $r_h u \in \mathcal{L}_{\mathcal{T}_h}$, we have from (3.6.52)

$$\|u - \pi_h u\|_X \le \|u - r_h u\|_X. \tag{3.6.57}$$

Clearly, the necessary condition (3.6.54) is satisfied, if

$$\|u - r_h u\|_X \le \mathrm{TOL}. \tag{3.6.58}$$

We shall guarantee the condition (3.6.54) by condition (3.6.58). Under assumption (3.6.55) we can set $X = C(\overline{\Omega})$ in (3.6.57).

d) *Riemann norm* The main tool in the interpolation error control is the adaptation of the mesh in such a way that the interpolation error function

$$I_u(x) \equiv |u(x) - r_h u(x)| \tag{3.6.59}$$

is equidistributed over the computational domain Ω, i.e.

$$I_u(x) \approx C \quad \forall x \in \Omega, \tag{3.6.60}$$

where $C > 0$ is a constant. If $u \in C^2(\overline{\Omega})$, then using the Taylor expansion at the point x_K we have

$$u(x) - r_h u(x) = \frac{1}{2}(x - x_K)^{\mathrm{T}}\, \mathbb{H}(x_K)\,(x - x_K) + o\left(|x - x_K|^2\right), \qquad (3.6.61)$$

where

$$\mathbb{H}(x) \equiv \begin{pmatrix} \frac{\partial^2 u}{\partial x_1^2} & \frac{\partial^2 u}{\partial x_1 \partial x_2} & \cdots & \frac{\partial^2 u}{\partial x_1 \partial x_N} \\ \frac{\partial^2 u}{\partial x_2 \partial x_1} & \frac{\partial^2 u}{\partial x_2^2} & \cdots & \frac{\partial^2 u}{\partial x_2 \partial x_N} \\ \vdots & \vdots & & \vdots \\ \frac{\partial^2 u}{\partial x_N \partial x_1} & \frac{\partial^2 u}{\partial x_N \partial x_2} & \cdots & \frac{\partial^2 u}{\partial x_N^2} \end{pmatrix}(x) \qquad (3.6.62)$$

is the Hesse matrix of the function u. Up to now we have considered the interpolation error function over N-simplexes. Further, we shall consider the interpolation error function I_u over edges. Let e be an edge of the N-simplex K and x_e its midpoint. By $|e|$ we denote the length of the edge e and by $\boldsymbol{e}$ the vector forming the edge e. We approximate $I_u|_e$ by the mean value of I_u over e. Then, substituting $x_K := x_e$ and omitting terms of higher order in (3.6.61), we can approximate

$$\begin{aligned} I_u|_e &\approx \frac{1}{|e|}\int_e |u(x) - r_h\, u(x)|\, dS \\ &\approx \frac{1}{2}\frac{1}{|e|}\int_e |(x - x_e)^{\mathrm{T}}\, \mathbb{H}(x_e)(x - x_e)| dS \\ &= \frac{1}{24}|\boldsymbol{e}^{\mathrm{T}}\, \mathbb{H}(x_e)\, \boldsymbol{e}|, \end{aligned} \qquad (3.6.63)$$

where the integral is evaluated using the substitution

$$x - x_e = \xi\left(\frac{b - a}{2}\right), \qquad \xi \in [-1, 1],$$

and a, b are the end points of the edge e. We define the solution-dependent Riemann norm of the edge e corresponding to the matrix $\mathbb{H}(x_e)$ by

$$\|e\|_u = \left(\boldsymbol{e}^{\mathrm{T}}\, \mathbb{H}(x_e)\, \boldsymbol{e}\right)^{1/2}, \qquad (3.6.64)$$

provided $\mathbb{H}(x_e)$ is positive definite. It follows from (3.6.63)–(3.6.64) that the interpolation error function is uniformly distributed over Ω, if

$$I_u|_e \approx \frac{1}{24}\, \|e\|_u^2 \approx c \qquad (3.6.65)$$

for all edges e of the grid $\mathcal{T}_h$, where c is a constant.

On the basis of (3.6.58) we specify the constant c in (3.6.65),

$$c := \tilde{c}_N = \begin{cases} \frac{1}{4}\mathrm{TOL}, & N = 2, \\ \frac{2}{9}\,\mathrm{TOL}, & N = 3, \end{cases} \tag{3.6.66}$$

where TOL is a user-defined tolerance. (For a detailed analysis see (Dolejší, 2003).) This means that the interpolation error function is uniformly distributed over Ω, if

$$\|e\|_u \approx c_N := \begin{cases} \sqrt{3}\,\sqrt{2\,\mathrm{TOL}}, & N = 2, \\ \sqrt{\frac{8}{3}}\,\sqrt{2\,\mathrm{TOL}}, & N = 3. \end{cases} \tag{3.6.67}$$

This leads us to the following concept:

Definition 3.69 *We say that the N-simplicial grid $\mathcal{T}_h$ is optimal, if*

$$\|e\|_u = c_N \tag{3.6.68}$$

for all edges e of the grid $\mathcal{T}_h$, where c_N is a constant defined in (3.6.67).

To find a mesh $\mathcal{T}_h$ satisfying (3.6.68) is rather difficult. Therefore, we reformulate this condition in the least squares sense. We introduce the so-called quality parameter $\mathrm{Q}_{\mathcal{T}_h}$ of the mesh $\mathcal{T}_h$

$$\mathrm{Q}_{\mathcal{T}_h} = \frac{1}{\#\mathcal{T}_h} \sum_{K \in \mathcal{T}_h} \mathrm{dist}\,(K) \sum_{e = \,\mathrm{edge\,of}\, K} (\|e\|_u - c_N)^2 \tag{3.6.69}$$

where c_N is defined in (3.6.67), $\#\mathcal{T}_h$ denotes the number of elements of $\mathcal{T}_h$ and $\mathrm{dist}\,(K)$ is the so-called parameter of distortion, which in 3D characterizes the distortion of a tetrahedron. For $N = 2$ we set $\mathrm{dist}\,(K) \equiv 1$ and for $N = 3$,

$$\mathrm{dist}\,(K) = \frac{2 \sum_{\triangle \subset K} |\triangle|^2}{27\sqrt{3}}\,|K|^2. \tag{3.6.70}$$

In (3.6.70) the sum is taken over all faces $\triangle$ forming the boundary of the tetrahedron K and $|\triangle|$ and $|K|$ denote the 2D and 3D measure of $\triangle$ and K, respectively. The parameter of distortion (3.6.70) was introduced in the 3D case in order to avoid highly distorted tetrahedra in the mesh. It takes its values from the interval $[1, \infty)$; the value 1 is taken, if the tetrahedron is equilateral with respect to the Euclidean norm of edges.

e) *Geometrical interpretation of the Riemann norm* Let $x_0 \in \Omega$ be a given point. We define the global interpolation operator $\tilde{r}_h : C^1(\overline{\Omega}) \to P^1(\Omega)$ such that

$$\tilde{r}_h\, u(x_0) = u(x_0), \tag{3.6.71}$$

$$\nabla \tilde{r}_h u(x_0) = \nabla u(x_0), \quad u \in C^1(\overline{\Omega}).$$

(Compare with (3.6.56).) Let $u \in C^2(\overline{\Omega})$. Analogously as in (3.6.59) we define the global interpolation error function

$$\tilde{I}_u(x) = |u(x) - \tilde{r}_h u(x)|. \tag{3.6.72}$$

Let us investigate the set of points

$$\left\{x \in I\!R^N; \tilde{I}_u(x) \le \mathrm{TOL}\right\}. \tag{3.6.73}$$

Writing the Taylor expansion at the point x_0 in the form

$$u(x) = \tilde{r}_h u(x) + \frac{1}{2}(x - x_0)^{\mathrm{T}} \mathbb{H}(x_0)(x - x_0) + o\left(|x - x_0|^2\right) \tag{3.6.74}$$

and neglecting terms of higher order, the set (3.6.73) can be approximated by the set

$$\epsilon_{x_0} = \left\{x \in I\!R^N; |(x - x_0)^{\mathrm{T}} \mathbb{H}(x_0)(x - x_0)| \le 2\,\mathrm{TOL}\right\}. \tag{3.6.75}$$

The set (3.6.75) is the ellipse or the ellipsoid for $N = 2$ or $N = 3$, respectively, provided the Hesse matrix $\mathbb{H}(x_0)$ is symmetric and positive definite. Let K_{x_0} be such an N-simplex that $K_{x_0} \subset \epsilon_{x_0}$ and $|K_{x_0}| \ge |K|$ for any N-simplex $K \subset \epsilon_{x_0}$. It can be proven that

$$\left(\boldsymbol{e}^{\mathrm{T}} \mathbb{H}(x_0)\, \boldsymbol{e}\right)^{1/2} = \sqrt{3}\,\sqrt{2\,\mathrm{TOL}}\ \forall e,\ e \text{ is the edge of } K_{x_0},\ N = 2, \tag{3.6.76}$$

$$\left(\boldsymbol{e}^{\mathrm{T}} \mathbb{H}(x_0)\, \boldsymbol{e}\right)^{1/2} = \sqrt{\frac{8}{3}}\,\sqrt{2\,\mathrm{TOL}}\ \forall e,\ e \text{ is the edge of } K_{x_0},\ N = 3. \tag{3.6.77}$$

In this case the N-simplex with the maximal N-dimensional measure inscribed into ϵ_{x_0} is equilateral with the Riemann norm of edges induced by the matrix $\mathbb{H}(x_0)$, equal to the constant c_N from (3.6.67), which corresponds to Definition 3.69. For $\mathbb{H}(x_0) = \mathbb{I}$ (the identity matrix) the proof is evident. For a general case see (Dolejší, 1998*b*) and (Dolejší, 2003).

f) *Iterative process* The almost equilateral N-simplicial grid with respect to the Riemann norm $\|\cdot\|_u$ is constructed by minimizing the quality parameter $\mathrm{Q}_{\mathcal{T}_h}$ in order to approach the optimal number of tetrahedra in the sense of Definition 3.69.

For a given $\mathcal{T}_h$ the quality parameter $\mathrm{Q}_{\mathcal{T}_h}$ is a computable quantity provided the Hesse matrices (3.6.62) of u are evaluated at midpoints of edges. (See item n) of this section for its construction.) We adapt the grid $\mathcal{T}_h$ in order to decrease $\mathrm{Q}_{\mathcal{T}_h}$ and we want to find a new grid $\mathcal{T}_h^{\mathrm{new}}$ such that the quality parameter $\mathrm{Q}_{\mathcal{T}_h^{\mathrm{new}}}$ of $\mathcal{T}_h^{\mathrm{new}}$ is smaller. To this end, the iterative process including the face swappings

($\mathcal{F}$), the edge swappings ($\mathcal{E}$), the edge bisections ($\mathcal{B}$), the removal of edges ($\mathcal{R}$) and the moving of vertices ($\mathcal{M}$) is used. The iterative process reads

$$m \times [\mathcal{L} + n \times [\mathcal{B} + \mathcal{L} + \mathcal{R} + \mathcal{L}] + r \times \mathcal{M}], \tag{3.6.78}$$

where $\mathcal{L} = o \times [\mathcal{F} + \mathcal{E}]$. It has been found by numerical experiments (see (Dolejší, 2003)) that suitable repetition parameters are $o = m = n = 10$, $r = 3$.

The above-mentioned local operations are performed as long as the quality parameter $\mathrm{Q}_{\mathcal{T}_h}$ decreases. The process is stopped if no local operation leads to a decrease of the quality parameter $\mathrm{Q}_{\mathcal{T}_h}$.

g) *Face swappping $\mathcal{F}$ in the 2D geometry* Let ABC and BDC be two adjacent triangles with the common edge BC. Then we can replace the diagonal BC of the quadrilateral $ABCD$ by the diagonal AD and we obtain the triangles ABD and ADC (see Fig. 3.6.14).

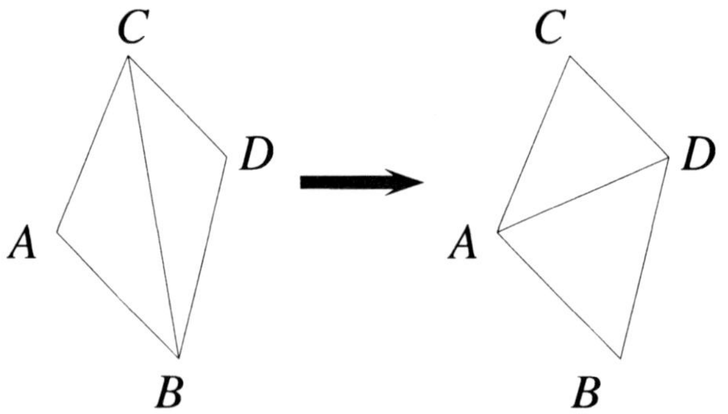

FIG. 3.6.14. Swapping the diagonal of the quadrilateral formed by a pair of adjacent triangles

h) *Face swapping $\mathcal{F}$ in the 3D geometry* The face swapping is performed on two tetrahedra with common face ABC. The $(2 \to 2)$ swapping and the $(2 \to 3)$ swapping (the creation of two and three new tetrahedra, respectively) are demonstrated in Figs 3.6.15 and 3.6.16.

i) *Edge swapping $\mathcal{E}$* In the 2D geometry the edge swapping coincides with that of the face. In the 3D geometry the edge swapping operation is performed on the set of M tetrahedra, $M \geq 3$, having the edge AB in common. Numerical experiments show that it is sufficient to consider $M \leq 7$. By the edge swapping, $2M - 4$ new tetrahedra from M tetrahedra are created, which is denoted as $(M \to 2M - 4)$ swapping. The example of $(5 \to 6)$ swapping is shown in Fig. 3.6.17.

j) *Edge bisection $\mathcal{B}$* The new vertex is put at the midpoint of the edge e and all N-simplexes having the edge e in common are bisected, if the Riemann norms of the new created edges will be closer to c_N than the original edge. This can be formulated as $|\|e\|_u - c_N| > \left|\frac{1}{2}\|e\|_u - c_N\right|$, where c_N is defined in (3.6.67). Equivalently, the edge e is bisected if

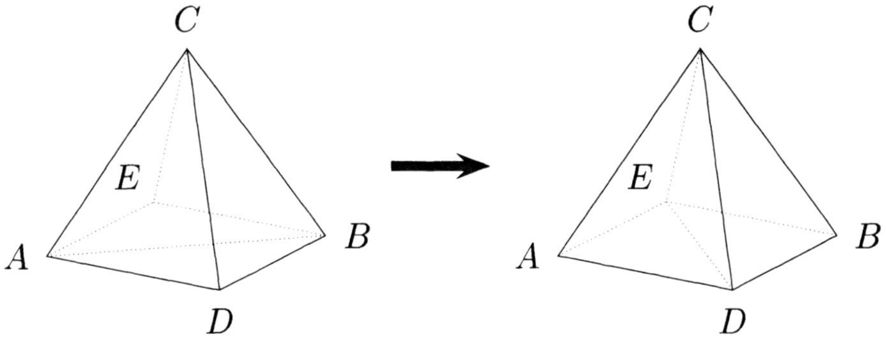

FIG. 3.6.15. $(2 \to 2)$ swapping

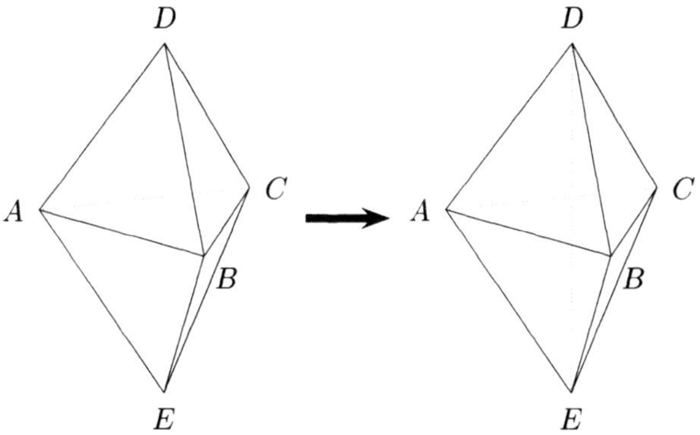

FIG. 3.6.16. $(2 \to 3)$ swapping

$$\|e\|_u > L_{\max}, \tag{3.6.79}$$

where $L_{\max} = \frac{4}{3}\, c_N$ is the positive solution of the equation

$$|L - c_N| = \left|\frac{1}{2}L - c_N\right| .$$

k) *Removing the edge* $\mathcal{R}$ Contrary to case j), removing the edge e is performed if

$$\|e\|_u < L_{\min}, \tag{3.6.80}$$

where $L_{\min} = \frac{2}{3}\, c_N$. This means that a coarsening of the mesh can be done without increasing the quality parameter $\mathrm{Q}_{\mathcal{T}_h}$. The value of $L_{\min}$ follows from a similar consideration as in case j) and is based heuristically on the implication

$$|\|e\|_u - c_N| > |2\|e\|_u - c_N| \Rightarrow \|e\|_u < \frac{2}{3}\, c_N .$$

l) *Moving a vertex* $\mathcal{M}$ Let $x \in \sigma_h$ be a vertex of the N-simplicial grid $\mathcal{T}_h$. The quality parameter $\mathrm{Q}_{\mathcal{T}_h}$ is considered as a function of the coordinates $x_1, \ldots, x_N$

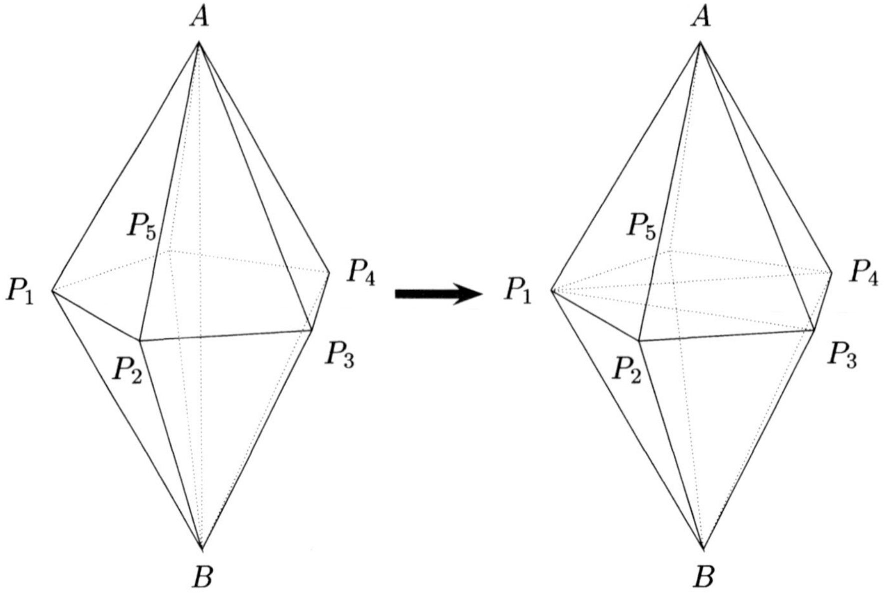

FIG. 3.6.17. (5 → 6) swapping

of the vertex x, i.e. $\mathrm{Q}_{\mathcal{T}_h} = \overline{\mathrm{Q}}_{\mathcal{T}_h}(x)$. The minimization of the function $\overline{\mathrm{Q}}_{\mathcal{T}_h}(x)$ with respect to $x_1, \ldots, x_N$ is used for the determination of the new position of the vertex x, with the aim of decreasing the quality parameter $\mathrm{Q}_{\mathcal{T}_h}$. The steepest descent method is applied and the loop over all vertices $P_i \in \sigma_h$ of the N-simplicial grid $\mathcal{T}_h$ sorted by the descent of the quality parameter $\overline{\mathrm{Q}}_{\mathcal{T}_h}(P_i)$ is performed. Another strategy can be found in (Dolejší, 1998*b*) for the 2D case.

m) *Minimal angle condition* Each of the local operations described above is performed, if the magnitude of angles of all N-simplexes is bounded from below. This can be formulated as the condition

$$\sum_{e=\text{edge of } K} |e|^N \leq \alpha|K|, \quad K \in \mathcal{T}_h, \tag{3.6.81}$$

with a user-defined constant α. The value of $\alpha = 0.002$ or $\alpha = 0.0035$ is a reasonable choice in 2D or 3D computations, respectively.

n) *Hesse matrix of* u_h^* It has been described in item f) of this section, how to construct an N-simplicial grid $\mathcal{T}_h$, suitable for obtaining a numerical approximation of the exact, sufficiently smooth solution u. In practical examples, of course, we do not know the exact solution u and therefore we must use a numerical approximation u_h^* on the mesh $\mathcal{T}_h$. We apply a suitable postprocessing of the function u_h^*, defined by its values at vertices, for the evaluation of the Hesse matrix of u_h^* at the midpoints of edges.

Let $u_h^* \in \mathcal{L}_{\mathcal{T}_h}$ be given by its values at vertices $P_i \in \sigma_h$ (obtained from (3.6.40) or (3.6.41) or (3.6.44) for the N-simplicial or dual or barycentric finite volume method, respectively). We compute the elements of the Hesse matrix $\mathbb{H}$ using the evaluation of the second order derivatives of u_h^* in the following sense. We approximate $\partial^2 u_h^*/\partial x_1^2$ by the function $u_h^{*x_1x_1} \in \mathcal{L}_{\mathcal{T}_h}$ such that

$$\left(\frac{\partial^2 u_h^*}{\partial x_1^2}, \varphi\right) \approx (u_h^{*x_1x_1}, \varphi) = \int_{\partial\Omega} \frac{\partial u_h^*}{\partial x_1} \varphi n_1 \, dS - \left(\frac{\partial u_h^*}{\partial x_1}, \frac{\partial \varphi}{\partial x_1}\right) \quad \forall \varphi \in \mathcal{L}_{\mathcal{T}_h}, \tag{3.6.82}$$

where $(\cdot,\cdot)$ means the scalar product in $L^2(\Omega)$. The right-hand side of (3.6.82) makes sense because u_h^*, $\varphi \in \mathcal{L}_{\mathcal{T}_h}$ and, therefore, the first order derivatives exist and are piecewise constant on $K \in \mathcal{T}_h$. The choice of the test functions

$$\{\varphi_i; \varphi_i \in \mathcal{L}_{\mathcal{T}_h}, \varphi_i(P_j) = \delta_{ij}, \, i, \, j \in I\},$$

where $P_j \in \sigma_h$ are vertices of $\mathcal{T}_h$ and δ_{ij} is the Kronecker delta, leads to the explicit formula for $u_h^{*x_1x_1}(P)$, $P \in \sigma_h$. We proceed in the following way:

$$\begin{aligned} (u_h^{*x_1x_1}, \varphi_i) &= \sum_{K\in\mathcal{T}_h} \int_K u_h^{*x_1x_1} \varphi_i \, dx \\ &\approx \sum_{K\in\mathcal{T}_h} \frac{|K|}{N+1} \sum_{P_j \in K\cap\sigma_h} u_h^{*x_1x_1}(P_j)\, \varphi_i(P_j) \\ &= \left(\sum_{K\cap\sigma_h \ni P_i} \frac{|K|}{N+1}\right) u_h^{*x_1x_1}(P_i), \quad i \in I. \end{aligned} \tag{3.6.83}$$

In (3.6.83) the 'mass lumping' numerical quadrature is used (see also Section 4.1.8). For the integral on the right-hand side of (3.6.82) we get

$$\begin{aligned} \int_{\partial\Omega} \frac{\partial u_h^*}{\partial x_1} \varphi_i n_1 \, dS &= \sum_{j\in I_b} \left.\frac{\partial u_h^*}{\partial x_1}\right|_{S_j} n_1^j \int_{S_j} \varphi_i \, dS \\ &= \frac{1}{N} \sum_{S_j \ni P_i} |S_j| \left.\frac{\partial u_h^*}{\partial x_1}\right|_{S_j} n_1^j, \end{aligned} \tag{3.6.84}$$

where the notation (3.3.12) is adopted for the N-simplexes K_i instead of finite volumes D_i and for I, I_b instead of J, J_b. The summation in the last term of (3.6.84) is taken over all boundary faces S_j having the vertex P_i in common. The unit outer normal to the face S_j is denoted (n_1^j, n_2^j).

The scalar product on the right-hand side of (3.6.82) gives

$$\left(\frac{\partial u_h^*}{\partial x_1}, \frac{\partial \varphi_i}{\partial x_1}\right) = \sum_{K\ni P_i} |K| \left.\frac{\partial u_h^*}{\partial x_1}\right|_K \left.\frac{\partial \varphi_i}{\partial x_1}\right|_K, \quad i \in I. \tag{3.6.85}$$

If we substitute relations (3.6.83), (3.6.84) and (3.6.85) into (3.6.82), we get an explicit formula for $u_h^{*x_1x_1}(P_i)$ for each $P_i \in \sigma_h$. The other elements of the Hesse matrix $\mathbb{H}(P_i)$ are computed analogously. It can be proven that the matrix

$$\mathbb{H}(P_i) = \left(u_h^{*x_i x_j}\right)_{i,j=1}^{N} \tag{3.6.86}$$

is the symmetric matrix.

If e is an edge from the grid $\mathcal{T}_h$ with the end points P_1 and P_2, we put

$$\mathbb{H}\left(\frac{P_1 + P_2}{2}\right) \approx \left(\mathbb{H}(P_1) + \mathbb{H}(P_2)\right)/2. \tag{3.6.87}$$

Remark 3.70 The matrix $\mathbb{H}$ given by (3.6.86) is symmetric but not positive definite as required in d). Moreover, if u_h^* is linear in Ω then $\mathbb{H} \equiv 0$ in Ω and consequently the iterative process from (3.6.78) cannot be applied for minimization of $Q_{\mathcal{T}_h}$. Since for any N-simplicial mesh the zero interpolation error is obtained in this case, only the number of tetrahedra is to be controlled by the user. For these reasons the matrix $\mathbb{H}$ is replaced its modification $\mathbb{M}$.

o) *Modification of the Hesse matrix* Let $\mathbb{H}(P)$ be a Hesse matrix (3.6.86) evaluated at the vertex $P \in \sigma_h$ of the N-simplicial mesh $\mathcal{T}_h$. Since $\mathbb{H}(P)$ is symmetric, we decompose it in the following way:

$$\mathbb{H}(P) = \mathbb{R}\,\mathrm{diag}(\lambda_1, \ldots, \lambda_N)\,\mathbb{R}^{-1}, \tag{3.6.88}$$

where $\mathbb{R}$ are orthogonal matrices. We construct the symmetric positive definite matrix

$$\bar{\mathbb{H}}(P) = \mathbb{H}(P) = \mathbb{R}\,\mathrm{diag}(|\lambda_1|, \ldots, |\lambda_N|)\,\mathbb{R}^{-1}. \tag{3.6.89}$$

To avoid the problems mentioned in Remark 3.70 we put

$$\mathbb{M}(P) = \bar{c}\left[\mathbb{I} + \bar{\alpha}\left(\|\bar{\mathbb{H}}(P)\|\right)\bar{\mathbb{H}}(P)\right)\right], \tag{3.6.90}$$

where $\mathbb{I}$ is a unit $N \times N$ matrix, $\bar{c} > 0$ is a constant and $\bar{\alpha} : (0, \infty) \to (0, \infty)$ is a function. In (3.6.90) we put $\|\bar{\mathbb{H}}\| = \max_{i,j=1,\ldots,N} |\bar{h}_{ij}|$, where $\bar{h}_{ij},\ i,j = 1, \ldots, N$, are the elements of $\bar{\mathbb{H}}$. The first term in the square brackets of (3.6.90) guarantees that the matrices $\mathbb{M}(P)$ are always regular. By the parameter $\bar{c}$ we control the number of tetrahedra, and by the parameter $\bar{\alpha}$ the ratio between the longest and shortest edge in the N-simplicial mesh, and the transition of the coarsest part of the mesh to the finest one, which is very important mainly in the viscous flow computation, where the aspect ratio of tetrahedra near the fixed boundary should be proportional to the square root of the Reynolds number.

p) *Setting the parameter* $\bar{c}$ If $\bar{\mathbb{H}}(P) = 0$ for all vertices P of the N-simplicial mesh $\mathcal{T}_h$ then all eigenvalues of $\mathbb{M}(P)$ are equal to $\bar{c}$ and if the equilateral (with respect to the Riemann norm given by the matrices $\mathbb{M}$) N-simplicial mesh of $\mathcal{T}_h^{\mathrm{new}}$ exists, then it is formed by the tetrahedra with length of edges (in the

Euclidean norm) equal to $c_N/\sqrt{\bar{c}}$, where c_N is given by (3.6.67). Let us consider the case when $N = 3$. For $N = 3$, the volume of the N-simplex $K \in \mathcal{T}_h{}^{\text{new}}$ is

$$|K| = \frac{c_3^3\sqrt{2}}{12\bar{c}^{3/2}}. \tag{3.6.91}$$

Let $|\Omega|$ be the volume of the computational domain. It holds that

$$\#\mathcal{T}_h{}^{\text{new}} \cdot |K| = |\Omega| \tag{3.6.92}$$

and for the parameter $\bar{c}$ we have

$$\bar{c} = \left(\frac{c_3^3\sqrt{2}\,\#\mathcal{T}_h{}^{\text{new}}}{12|\Omega|} \right)^{2/3}. \tag{3.6.93}$$

Having prescribed the parameter $\bar{c}$, the number of N-simplexes of the mesh $\mathcal{T}_h{}^{\text{new}}$ could be estimated from the formula (3.6.93). Setting the parameter $\bar{c}$ in the 2D case is analogous.

q) *Setting the function* $\bar{\alpha}$ The function $\bar{\alpha}(\|H\|)$ is defined (see (Dolejší and Felcman, 2001)) as

$$\bar{\alpha}(\|\bar{H}(P)\|) = \frac{\varepsilon_1}{\varepsilon_2 + \|\bar{H}(P)\|}, \tag{3.6.94}$$

where ε_1 and ε_2 are constants. This choice of $\bar{\alpha}$ ensures that the Riemann norms of edges in the N-simplicial mesh are bounded from below, which is a consequence of the fact that

$$\lim_{\|\bar{H}(P)\| \to \infty} \|\mathbb{M}(P)\| = \bar{c}(1 + \varepsilon_1). \tag{3.6.95}$$

In what follows we shall specify the role of constants ε_1 and ε_2.

Let $\mathbb{H}(P) = 0$ for all vertices P. If the equilateral N-simplicial mesh exists, then the length of edges (in the Euclidean norm) is

$$l_0 = \frac{c_N}{\sqrt{\bar{c}}}. \tag{3.6.96}$$

Let $\beta \in I\!R, \beta \boldsymbol{g} 1$ and $\mathbb{H}(P) = \beta \mathbb{I}$ for all vertices P. If the equilateral N-simplicial mesh exists, then for $\beta \to \infty$ the length of edges (in the Euclidean norm) tends to

$$l_1 = \frac{c_N}{\sqrt{\bar{c}(1 + \varepsilon_1)}}. \tag{3.6.97}$$

It could be shown that l_0 and l_1 are the minimal and maximal length of edges in the N-simplicial mesh, respectively. Let us define the ratio of the longest and shortest edge we allow in the mesh generation:

$$\text{ratio} = \frac{l_0}{l_1}. \tag{3.6.98}$$

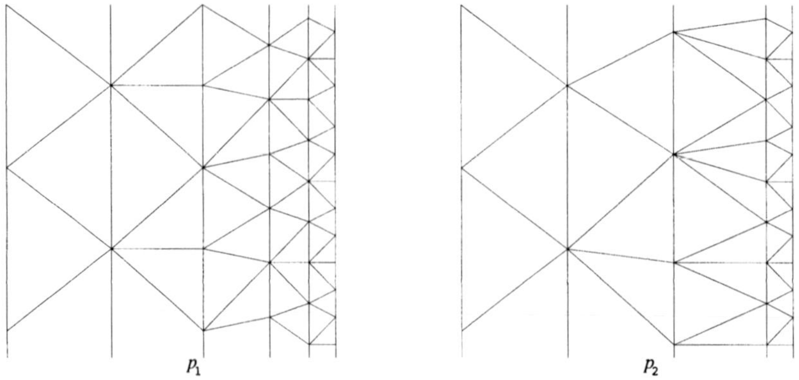

FIG. 3.6.18. Transition of the coarsest part to the finest one for two different $p_1 > p_2$

It follows from (3.6.96) and (3.6.97) that

$$\text{ratio} = \sqrt{1+\varepsilon_1} \tag{3.6.99}$$

which clearly demonstrate how the ratio can be controlled by the parameter ε_1 by setting

$$\varepsilon_1 = \text{ratio}^2 - 1 \tag{3.6.100}$$

for the prescribed parameter ratio.

In the practical computation we use the following relation for ε_2:

$$\varepsilon_2 = \frac{\varepsilon_1}{p} \tag{3.6.101}$$

where p is some positive number. The number p controls the transition of the coarsest part of the mesh (with edge length l_0 given by (3.6.96)) to the finest one (with edge length l_1 given by (3.6.97)) as demonstrated in 2D in Fig. 3.6.18.

3.6.8 *Data reinitialization for anisotropic mesh refinement*

The mesh adaptation technique described in Section 3.6.7 includes the mesh refinement, coarsening and moving of vertices. The auxiliary recovery function $\tilde{\mathbf{w}}_h \in \boldsymbol{C}_{\mathcal{T}_h^{\text{new}}}$ is defined as

$$\tilde{\boldsymbol{w}}_h\Big|_K = \frac{1}{N+1}\sum_{i=1}^{N+1} \mathbf{w}_h^*(P_i^K), \quad K \in \mathcal{T}_h^{\text{new}}, \tag{3.6.102}$$

where $P_1^K, \ldots, P_{N+1}^K$ are vertices of the N-simplex $K \in \mathcal{T}_h^{\text{new}}$ and $\mathbf{w}_h^* \in \boldsymbol{\mathcal{L}}_{\mathcal{T}_h}$ is the linear recovery of the approximate solution $\mathbf{w}_h \in \boldsymbol{C}_{\mathcal{D}_h}$ (see Section 3.6.5). The relations (3.6.47), (3.6.48) and (3.6.49) with $\mathbf{w}_h^* := \tilde{\boldsymbol{w}}_h$ are used for the construction of $\mathbf{w}_h^{\text{new}}\Big|_D$ for N-simplicial, dual and barycentric finite volumes, respectively.

3.7 Examples of finite volume simulations

The qualitative properties (e.g. accuracy, stability, efficiency, etc.) of numerical methods for the solution of compressible flow, established theoretically under simplified assumptions, are usually confirmed with the aid of various test problems. For example, computational methods for inviscid compressible flow can be tested on the so-called Ringleb flow. In this case the exact solution constructed by the hodograph method is available. See (Ringleb, 1940), (Chiocchia, 1985). This test problem allows a quantitative comparison of the computed approximate solution with the exact solution. In the literature it is possible to find a number of computational results obtained with the aid of various methods. The comparison with these results also indicates the quality of the numerical method under consideration.

In this section we shall present some applications of the finite volume method for the Euler equations to several problems, namely to channel flow, flow around an isolated profile and cascade flow. We shall use the explicit finite volume scheme (3.3.24) for the computation of values $\boldsymbol{w}_i^k$ of the approximate solution on the finite volume $D_i \subset \overline{\Omega}_h, i \in J$, at time instant t_k, where J is an index set.

To run the scheme (3.3.24) we need the initial approximation $\boldsymbol{w}_i^0$, which is given by relation (3.3.25) for $i \in J$, and the data for boundary conditions in order to evaluate the numerical flux $\boldsymbol{H}(\boldsymbol{w}_i^k, \boldsymbol{w}_j^k, \boldsymbol{n}_{ij}^\alpha)$ for $i \in J$ such that $j \in \gamma(i) \neq \emptyset$. (i.e. $\Gamma_{ij}^\alpha \subset \partial\Omega_h, \alpha = 1$). See (3.3.10). The use of Tables 3.3.1 and 3.3.2 for 2D and 3D flow, respectively, leads to the evaluation of the boundary state $\boldsymbol{w}_j^k$, $j \in \gamma(i)$, using the primitive variables ρ_j^k, $\boldsymbol{v}_j^k$ and p_j^k, which are computed from $\boldsymbol{w}_i^k$ by extrapolation and a user-defined state $\boldsymbol{w}_{Bj}^k$. See Table 3.7.1. On a solid impermeable boundary we compute the numerical flux from relation (3.3.48).

By (3.1.8) and (3.1.5),

$$\boldsymbol{w}_j^k = \left(\rho_j^k, \rho_j^k v_{1j}^k, \ldots, \rho_j^k v_{Nj}^k, E_j^k\right)^{\mathrm{T}}, \tag{3.7.1}$$

$$E_j^k = \frac{p_j^k}{\gamma - 1} + \frac{1}{2}\rho_j^k |\boldsymbol{v}_j^k|^2. \tag{3.7.2}$$

In all examples the Poisson adiabatic constant γ is set as

$$\gamma = 1.4, \tag{3.7.3}$$

i.e. the flow of air is considered. In connection with the Osher–Solomon approximate Riemann solver, see Section 3.4, the inlet/outlet boundary conditions from Section (3.4.7) are used.

As scheme (3.3.24) is explicit, a suitable stability condition has to be used (see Section 3.3.11). The FV solution of all problems treated in the sequel is realized by explicit schemes. We concentrate on the following questions:

- description of the geometry of the computational domain,
- initial conditions,
- boundary conditions,

Table 3.7.1 *Evaluation of* $\boldsymbol{w}_j^k$, $j \in \gamma(i)$, $i \in J$, *using the primitive variables*

Type of boundary	Character of the flow	Quantities extrapolated	Quantities prescribed
INLET $(\boldsymbol{v}_i^k \cdot \boldsymbol{n}_{ij}^\alpha < 0)$	supersonic flow $(-\boldsymbol{v}_i^k \cdot \boldsymbol{n}_{ij}^\alpha > a_i^k)$	–	$\rho_j^k := \rho_{Bj}^k$ $\boldsymbol{v}_j^k := \boldsymbol{v}_{Bj}^k$ $p_j^k := p_{Bj}^k$
	subsonic flow $(-\boldsymbol{v}_i^k \cdot \boldsymbol{n}_{ij}^\alpha \leq a_i^k)$	$p_j^k := p_i^k$	$\rho_j^k := \rho_{Bj}^k$ $\boldsymbol{v}_j^k := \boldsymbol{v}_{Bj}^k$
OUTLET $(\boldsymbol{v}_i^k \cdot \boldsymbol{n}_{ij}^\alpha > 0)$	supersonic flow $(\boldsymbol{v}_i^k \cdot \boldsymbol{n}_{ij}^\alpha \geq a_k^k)$	$\rho_j^k := \rho_i^k$ $\boldsymbol{v}_j^k := \boldsymbol{v}_i^k$ $p_j^k := p_i^k$	
	subsonic flow $(\boldsymbol{v}_i^k \cdot \boldsymbol{n}_{ij}^\alpha < 0)$	$\rho_j^k := \rho_i^k$ $\boldsymbol{v}_j^k := \boldsymbol{v}_i^k$	$p_j^k := p_{Bj}^k$

- choice of numerical flux,
- stability condition, choice of CFL number,
- adaptation technique,
- stopping criterion for the steady-state computation.

Some of the following results are presented:

- computational (adapted) mesh,
- convergence history to the steady-state,
- Mach number, density or entropy isolines,
- boundary distribution of computed quantities.

In order to obtain isolines or the boundary distribution, the computed piecewise constant numerical solution is postprocessed with the aid of a linear recovery from Section 3.6.5, leading to a continuous piecewise linear approximation given by its values at vertices. If the aim is to obtain the steady-state solution, we use a time marching process for $t_k \to \infty$. This means that scheme (3.3.24) is applied as an iterative process. The density ρ is chosen as the significant quantity for which we measure the convergence. The algorithm is stopped when the L^1-norm or L^∞-norm of the relative error e_k in density is below the prescribed tolerance ε:

$$e_k := \sum_{i \in J} |D_i| \left| \frac{\rho_i^{k+1} - \rho_i^k}{\tau_k \rho_i^k} \right| < \varepsilon, \tag{3.7.4}$$

$$e_k := \max_{i \in J} \left| \frac{\rho_i^{k+1} - \rho_i^k}{\tau_k \rho_i^k} \right| < \varepsilon. \tag{3.7.5}$$

Usually we set $\varepsilon := 10^{-7}$.

3.7.1 *Shock-tube problem*

A simple example for which we know an exact solution is the shock-tube problem. The physical set-up is an infinite tube filled with gas, initially divided by a membrane into two sections. The gas has a higher density and pressure in one half of the tube than in the other half with zero velocity everywhere. At time $t = 0$, the membrane is suddenly removed or broken, and the gas allowed to flow. We expect motion in the direction of low pressure. Assuming the flow is uniform across the tube, there is variation in only one direction and the 1D Euler equations apply. This physical problem is reasonably well approximated by solving the Riemann problem introduced in Section 2.3.8 and studied in detail in Section 3.1.6 with the aid of analytical tools. Given the density ρ_L, the pressure p_L in the left half of the tube and the density ρ_R and the pressure p_R in the right half of the tube and zero velocity on both sides, the initial condition (3.1.69) is given by the constant states $\boldsymbol{w}_L$, $\boldsymbol{w}_R$, where

$$\boldsymbol{w}_L = (\rho_L, 0, E_L)^{\mathrm{T}} \tag{3.7.6}$$

$$\boldsymbol{w}_R = (\rho_R, 0, E_R)^{\mathrm{T}} \tag{3.7.7}$$

and E_L and E_R are given by relation (3.1.5), taking into account the vanishing initial velocity:

$$E_i = p_i/(\gamma - 1), \quad i = L,\, R. \tag{3.7.8}$$

The following initial values were chosen for the numerical test:

$$\begin{aligned} \rho_L &= 1, \quad \rho_R = 1.101463, \\ p_L &= 2, \quad p_R = 1. \end{aligned} \tag{3.7.9}$$

The entropy weak solution $\boldsymbol{w} : Q_T \to I\!R^3$ of system (2.3.49) equipped with the initial condition (2.3.50) with $\boldsymbol{w}_L$, $\boldsymbol{w}_R$ given by (3.7.6)–(3.7.9) is described in Exercise 3.23.

For the purpose of comparison of analytical and numerical solutions, the density ρ was chosen as the significant flow quantity. The density distribution for the above initial data is shown in Fig. 3.7.1. The density distribution at time $t = 1$, i.e. the graph of function $\rho(x, 1)$, $x \in [-3, 3]$, is shown in Fig. 3.7.2.

The above-described shock-tube problem can be, of course, modelled by 2D Euler equations. The jump in the initial data only occurs in the x_1-direction. Thus, the solution should not depend on the x_2-direction. Therefore, the shock-tube problem is useful in assessing the performance of the finite volume method and checking the correctness of the computer program. We solve the 2D Euler equations by the finite volume method in the time-space cylinder

$$Q_T = \Omega \times [0, T), \quad T > 1, \tag{3.7.10}$$

where

$$\Omega = (-3, 3) \times (0, 0.1). \tag{3.7.11}$$

(Because of the numerical solution, the infinitely long tube is truncated to a bounded domain.) The explicit finite volume scheme (3.3.24) is applied on the

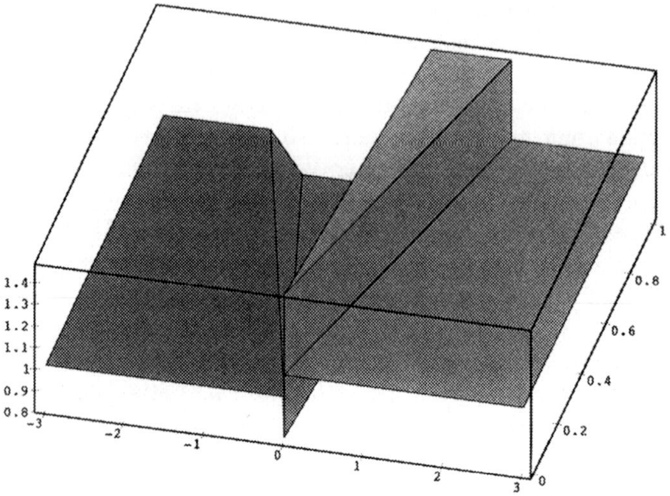

FIG. 3.7.1. Density distribution in the (x, t)-plane

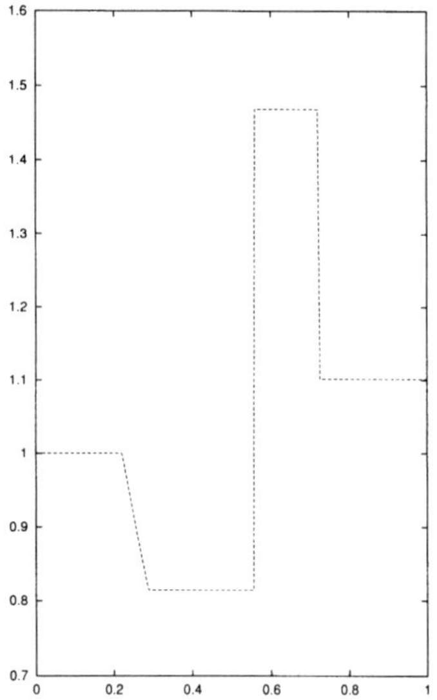

FIG. 3.7.2. Density distribution at time $t = 1$

triangular mesh of the computational domain $\Omega_h = \Omega$. The Osher–Solomon approximate Riemann solver from Fig. 3.4.2 is used in the construction of the numerical flux $\boldsymbol{H}$ in (3.4.2). The initial condition (3.3.25) is given according to (3.7.6)–(3.7.9):

$$\boldsymbol{w}^0(x) = \begin{cases} (1, 0, 0, 5)^{\mathrm{T}}, & x < 0, \\ (1.101463, 0, 0, 2.5)^{\mathrm{T}}, & x > 0. \end{cases} \tag{3.7.12}$$

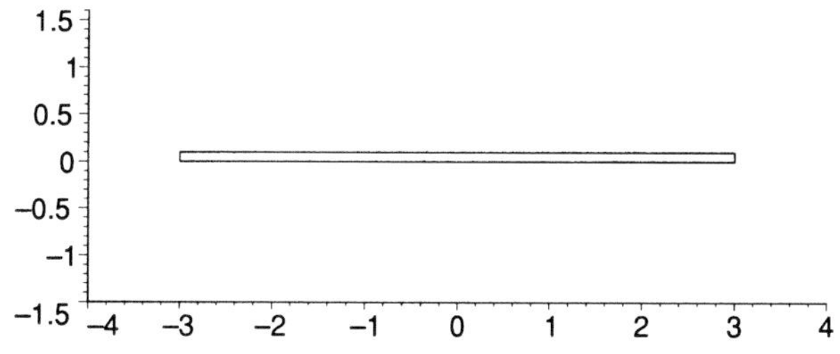

FIG. 3.7.3. Computational domain

The inlet/outlet boundary conditions from Section 3.4.7 and the solid wall boundary conditions from Section 3.4.8 are applied on $\partial\Omega_h = \Gamma_I \cup \Gamma_O \cup \Gamma_W^1 \cup \Gamma_W^2$, where

$$\begin{aligned} \text{inlet:}\ \Gamma_I &= \left\{x \in \mathbb{R}^2; x_1 = -3,\ x_2 \in [0, 0.1]\right\}, \qquad (3.7.13)\\ \text{outlet:}\ \Gamma_O &= \left\{x \in \mathbb{R}^2; x_1 = 3,\ x_2 \in [0, 0.1]\right\},\\ \text{solid wall:}\ \Gamma_W^1 &= \left\{x \in \mathbb{R}^2; x_1 \in [-3, 3],\ x_2 = 0\right\},\\ \text{solid wall:}\ \Gamma_W^2 &= \left\{x \in \mathbb{R}^2; x_1 \in [-3, 3],\ x_2 = 0.1\right\}, \end{aligned}$$

see Fig. 3.7.3.

At time $t = 1$ we compare the density distribution in the cut $x_2 = 0$ computed by the finite volume method with the analytical solution described in Exercise 3.23 and drawn in Fig. 3.7.2. Figures 3.7.4–3.7.6 show the comparison of the analytical solution (dashed line), the numerical solution obtained by the scheme (3.3.24) (dotted line) and the numerical solution obtained by the higher order scheme (3.5.1) (full line). The DEO3 strategy for the computation of the approximate gradient in (3.5.15) was used.

3.7.2 *GAMM channel*

The 2D channel from Fig. 3.7.7 (10 % circular arc bump on the lower wall Γ_W^1) was proposed by the Gesellschaft für Angewandte Mathematik und Mechanik (GAMM) as a test channel for the transonic compressible flow simulation. The size of the channel is as follows: $x_1 \in [-1, 1]$, $x_2 \in [0, 1]$, the circular arc bump is a part of the circle with centre at the point $C = (0, -1.2)$ and radius $r = 1.3$. Since the height of the bump is 0.1, we speak of a 10 % circular bump with respect to the height of the channel. We denote

$$\begin{aligned} \text{inlet:}\ \Gamma_I &= \left\{x \in \mathbb{R}^2; x_1 = -1,\ x_2 \in (0, 1)\right\}, \qquad (3.7.14)\\ \text{outlet:}\ \Gamma_O &= \left\{x \in \mathbb{R}^2; x_1 = 1,\ x_2 \in (0, 1)\right\},\\ \text{lower wall:}\ \Gamma_W^1 &= \left\{x \in \mathbb{R}^2; x_1 \in [-1, -0.5],\ x_2 = 0\right\}\\ &\quad \cup \left\{x \in \mathbb{R}^2; x_1 \in [-0.5, 0.5],\ x_2 = \sqrt{1.69 - x_1^2} - 1.2\right\}\\ &\quad \cup \left\{x \in \mathbb{R}^2; x_1 \in [0.5, 1],\ x_2 = 0\right\}, \end{aligned}$$

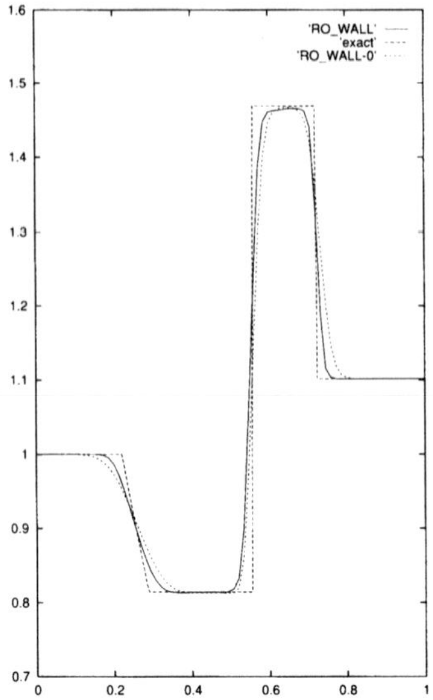

FIG. 3.7.4. Distribution of density on mesh with 75×4 nodes

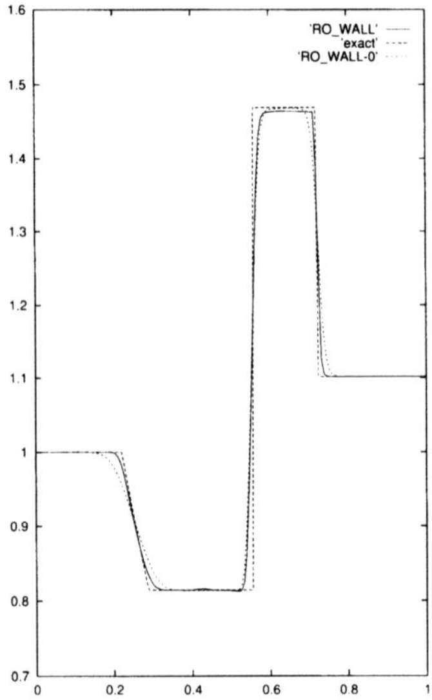

FIG. 3.7.5. Distribution of density on mesh with 150×8 nodes

upper wall: $\Gamma_W^2 = \left\{x \in I\!R^2; x_1 \in [-1,1],\, x_2 = 1\right\}.$

The boundary of the GAMM channel Ω is

$$\partial\Omega = \Gamma_I \cup \Gamma_O \cup \Gamma_W^1 \cup \Gamma_W^2. \tag{3.7.15}$$

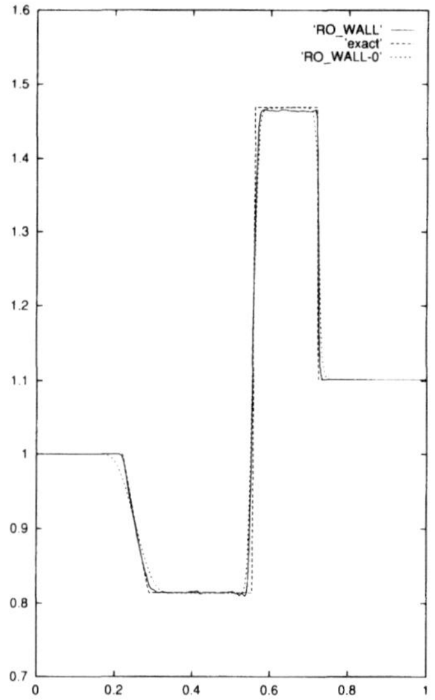

FIG. 3.7.6. Distribution of density on mesh with 300×16 nodes

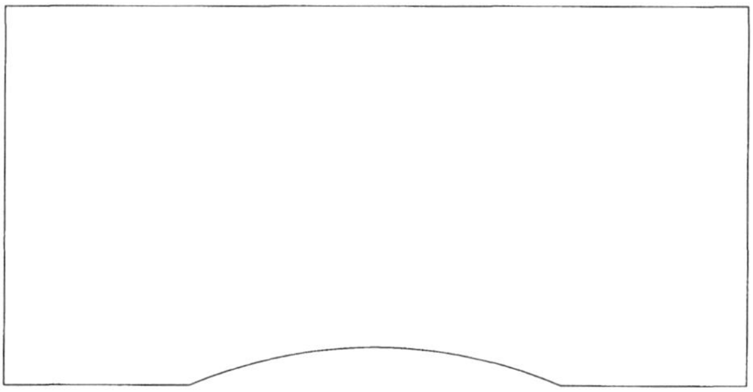

FIG. 3.7.7. The GAMM channel

The direction of the flow is from left to right with the inlet velocity angle $\alpha = 0$. The numerical solution of the inviscid compressible GAMM channel flow is very sensitive to the prescribed inlet Mach number. For the GAMM channel benchmark problem the inlet Mach number is set to $M = 0.67$.

The initial condition is supposed to be constant in the whole domain Ω. Expressed in the primitive variables it reads

$$\begin{aligned} \rho^0 &= 1.5\,\mathrm{kg\,m^{-3}}, \\ v_1^0 &= 205.709277\,\mathrm{m\,s^{-1}}, \\ v_2^0 &= 0\,\mathrm{m\,s^{-1}}, \\ p^0 &= 101000\,\mathrm{Pa}. \end{aligned} \tag{3.7.16}$$

In the flow through the GAMM channel, under the above data, an isolated shock

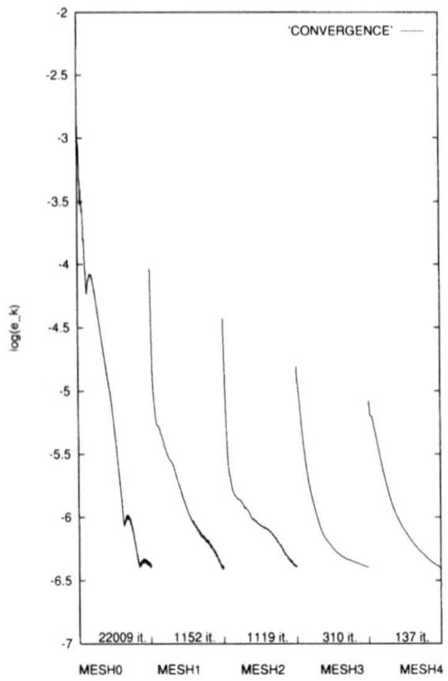

FIG. 3.7.8. Convergence behaviour on the basic mesh and four refinements MESH1–MESH4

wave is generated near the lower wall. A good resolution of this shock wave requires the application of a suitable mesh refinement strategy. As the position of the shock wave is not a priori known, the mesh refinement has to be applied during the computational process. Here we present some results characterizing the various aspects of the finite volume method such as adaptivity or the choice of finite volumes.

The explicit finite volume scheme (3.3.24) is applied on a triangular, dual or barycentric finite volume mesh. The following numerical fluxes $\boldsymbol{H}$ were tested:

- Osher–Solomon numerical flux given by (3.4.2) and scheme from Fig. 3.4.2,
- Vijayasundaram numerical flux (3.3.32),
- numerical flux given by (3.4.2) with the exact Riemann solver from Section 3.1.6.4.

We shall sort the results according to various mesh adaptation strategies.

3.7.2.1 *Shock indicator* The adaptive mesh refinement using the shock indicator described in Section 3.6.4.1 was applied during the computational process. The convergence behaviour of the quantity $\log(e_k)$ with e_k defined in (3.7.4), depending on k, is given in Fig. 3.7.8. The Osher–Solomon numerical flux was used. Four successive red–green refinements were performed during the iterative process. The resulting finite volume mesh is drawn in Fig. 3.7.9. A detail of the refined mesh is shown in Fig. 3.7.10. The isolines of the Mach number in Fig. 3.7.11 clearly show the position of the shock wave. Its intensity is well seen on the Mach number distribution along the lower wall, which is shown in Fig. 3.7.12

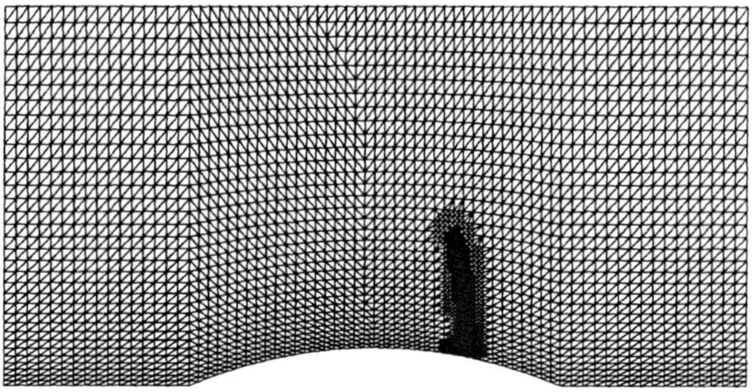

FIG. 3.7.9. Shock indicator (fourth refinement) – MESH4

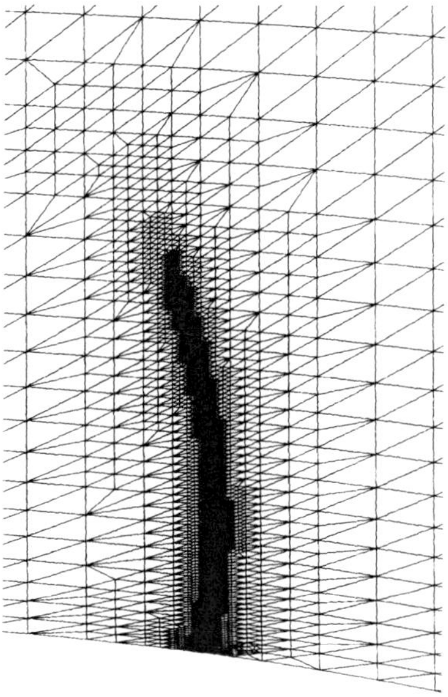

FIG. 3.7.10. Detail of MESH4

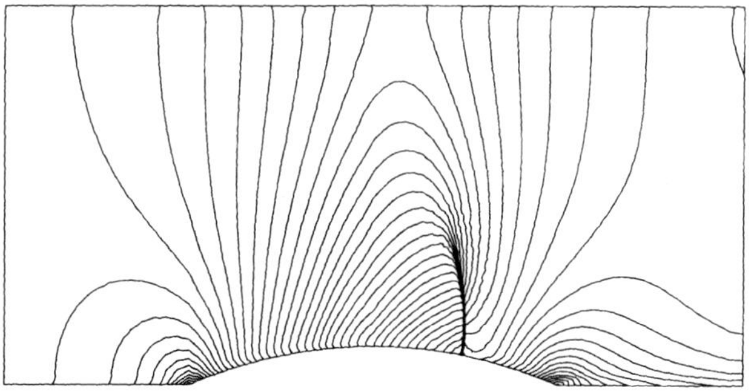

FIG. 3.7.11. Isolines of the Mach number on MESH4

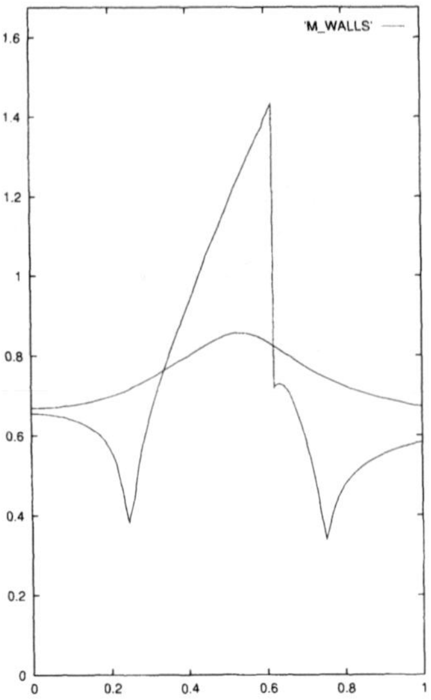

FIG. 3.7.12. Mach number distribution along the walls

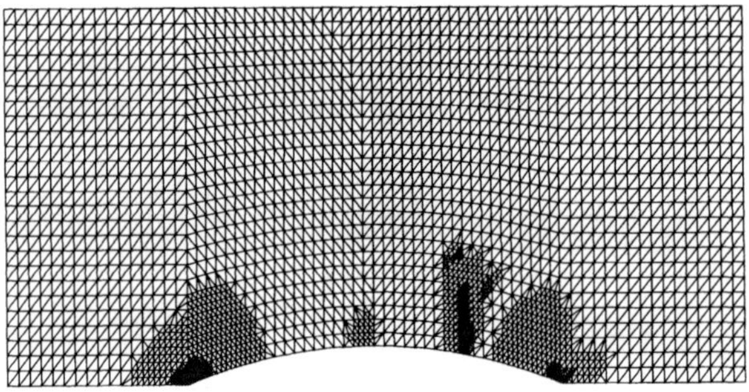

FIG. 3.7.13. Residual error indicator – mesh refinement

together with the Mach number distribution along the upper wall . The distribution is drawn with respect to the relative lengths of walls, i.e. on the interval $[0,1]$. The local minimum behind the shock wave in the Mach number distribution along the lower wall located near the point 0.6 on the x_1-axis is called the *Zierep singularity*. A good numerical method should be able to capture this singularity.

3.7.2.2 *Residual error indicator* In Fig. 3.7.13 the mesh refined with the aid of the residual error indicator is plotted. The regions where the approximation of the norm of the residual is greater than the prescribed tolerance are refined. A small neighbourhood of the shock wave is also included in the refined region. The Mach number isolines and the Mach number distribution along the walls are shown in Figs 3.7.14 and 3.7.15. The Osher–Solomon numerical flux was used.

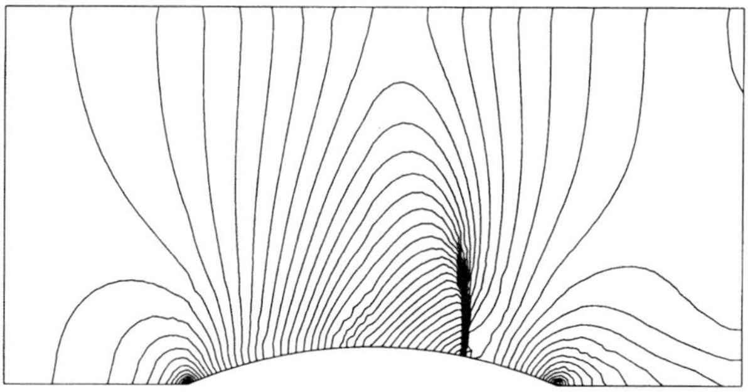

FIG. 3.7.14. Mach number isolines

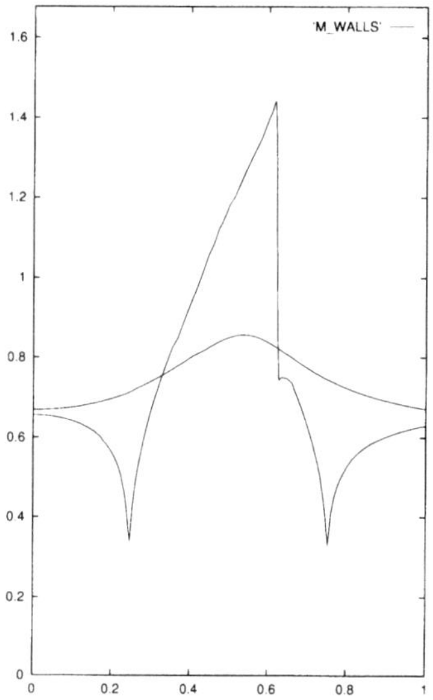

FIG. 3.7.15. Mach number distribution along the walls

3.7.2.3 *Superconvergence error indicator* The use of the superconvergence error indicator described in Section 3.6.5.1 is demonstrated in Figs 3.7.16–3.7.18. The Osher–Solomon numerical flux was used.

3.7.2.4 *User-defined refinement* This was used on the basis of a priori knowledge of the flow character in the case when dual finite volumes over a triangular grid were applied. In Figs 3.7.19 and 3.7.20, the triangulation of the domain and the corresponding dual FV mesh are shown. The Vijayasundaram numerical flux was used in the finite volume method. Figure 3.7.21 shows the convergence history to the steady-state measured in the L^1-norm; $\log e_k$, where e_k is defined in (3.7.4), is drawn against the number of iterations k. In Figs 3.7.22 and 3.7.23 we see the Mach number isolines and the Mach number distribution along the walls, respectively. To demonstrate the production of the entropy (see Section 2.3.5)

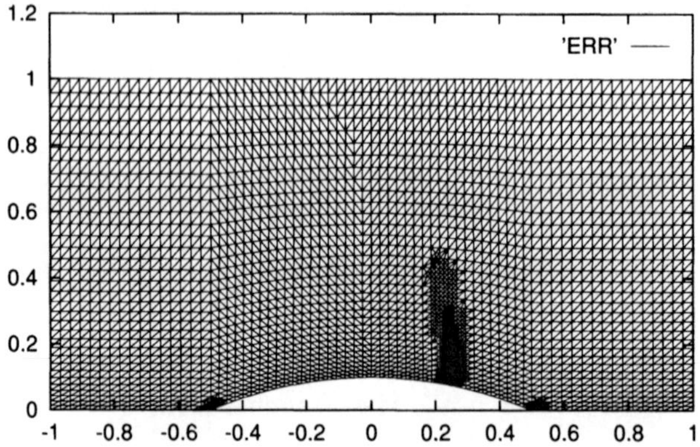

FIG. 3.7.16. Superconvergence error indicator – mesh refinement

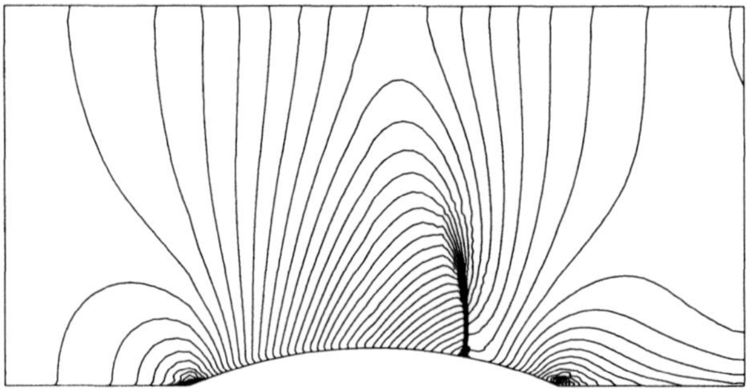

FIG. 3.7.17. Detail of the mesh refinement

on the shock wave, the entropy isolines are shown in Fig. 3.7.24.

3.7.3 *The 3D channel – 10 % cylindrical bump*

A useful 3D test case is a 3D version of the GAMM channel introduced in Section 3.7.2. It allows us to compare the computed flow quantities such as

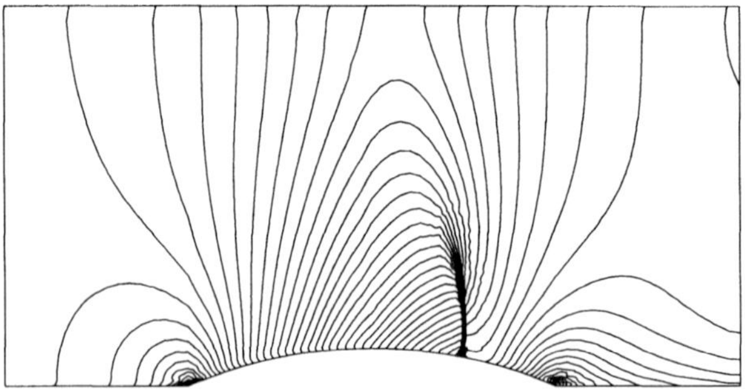

FIG. 3.7.18. Mach number isolines

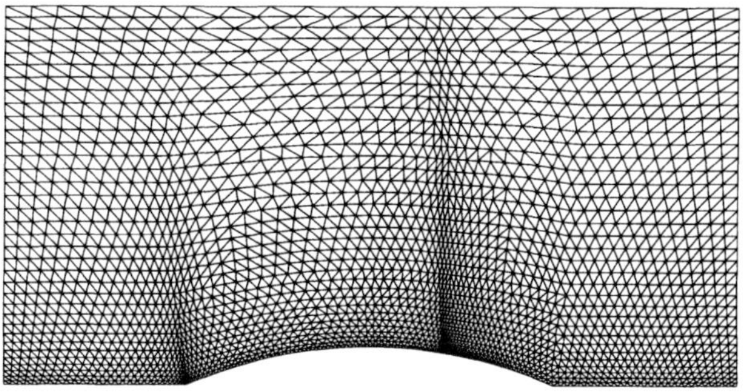

FIG. 3.7.19. User-defined mesh refinement

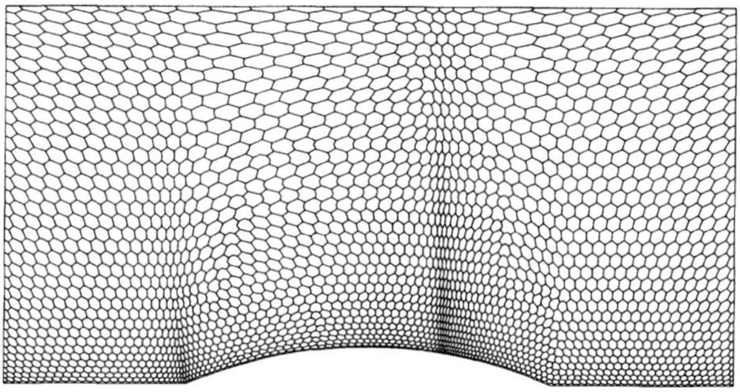

FIG. 3.7.20. Dual finite volume mesh

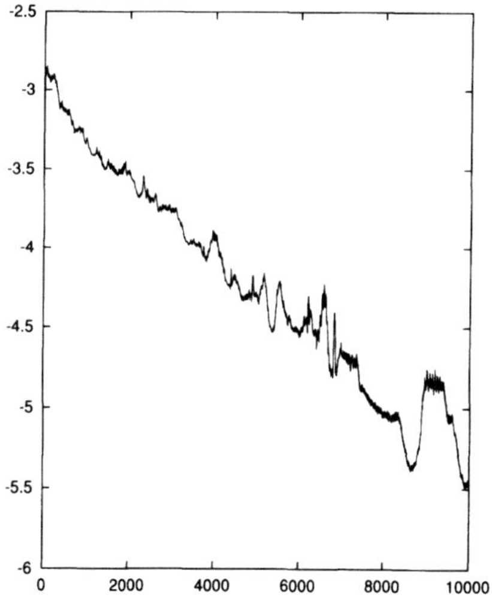

FIG. 3.7.21. Convergence history measured in the L^1-norm

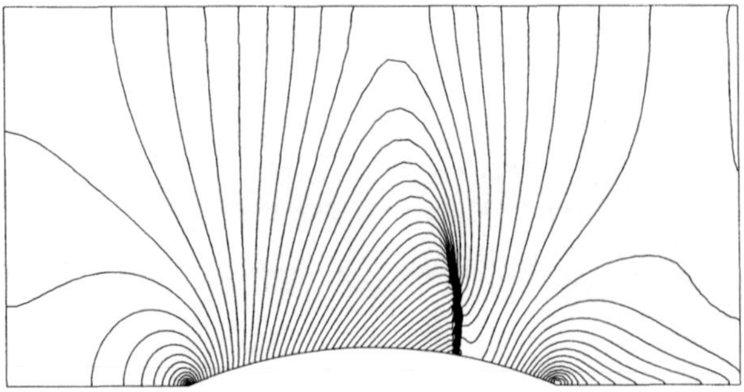

FIG. 3.7.22. Mach number isolines

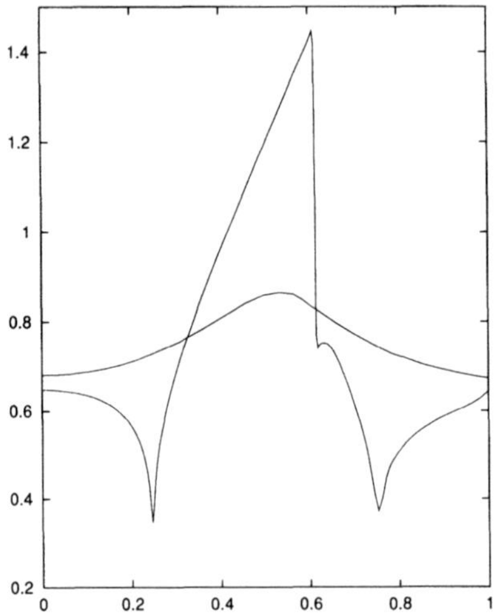

FIG. 3.7.23. Mach number distribution along the walls

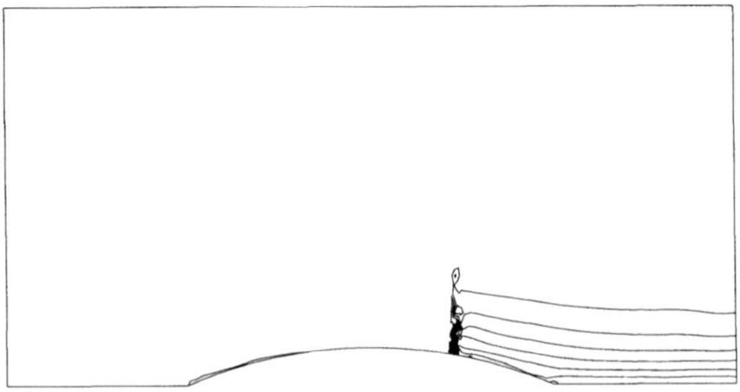

FIG. 3.7.24. Entropy isolines

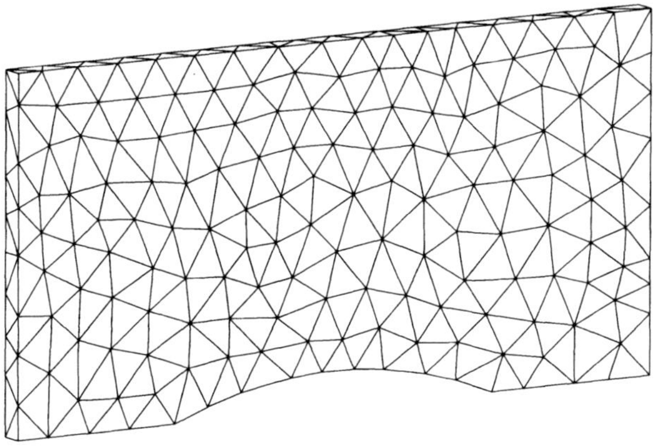

FIG. 3.7.25. 10 % cylindrical bump

Mach number, density, etc., in suitable plane cuts with 2D results and so to test the correctness of a 3D code. The natural extension of the GAMM channel flow to 3D is the flow past a cylindrical bump in the channel shown in Fig. 3.7.25. The 3D Euler equations are solved in the domain

$$\Omega_3 = \left\{x \in I\!R^3; (x_1, x_2) \in \Omega,\ x_3 \in (0, 0.07)\right\}, \tag{3.7.17}$$

where Ω is the domain from Fig. 3.7.7. We denote

$$\begin{aligned} \text{inlet: } \Gamma_I &= \left\{x \in I\!R^3; x_1 = -1,\ x_2 \in (0,1),\ x_3 \in (0, 0.07)\right\} \\ \text{outlet: } \Gamma_O &= \left\{x \in I\!R^3; x_1 = 1,\ x_2 \in (0,1),\ x_3 \in (0, 0.07)\right\} \\ \text{solid walls: } \Gamma_W &= \partial\Omega_3 \setminus (\Gamma_I \cup \Gamma_O). \end{aligned} \tag{3.7.18}$$

The direction of the flow is from left to right. At the inlet, the constant velocity parallel with the axis x_1 is assumed. The flow with the same inlet Mach number as for the GAMM channel flow, i.e. $M = 0.67$, was simulated. The initial condition $\boldsymbol{w}^0$ (see (3.3.25)) is supposed to be constant in the domain Ω_3. Expressed in primitive variables, the data in (3.7.16) are completed by the initial condition for the velocity in the x_3-direction

$$v_3^0 = 0\,\mathrm{m\,s^{-1}}. \tag{3.7.19}$$

The steady-state solution was computed with the aid of the Godunov flux defined by (3.2.71) and results from Section 3.1.7 – see Fig. 3.1.15 and Remark 3.28. In Fig. 3.7.25 the initial tetrahedral mesh is plotted. In Figs 3.7.26 and 3.7.27 the Mach number isolines on the front wall and the Mach number distribution along the lower wall, respectively, are drawn. The obtained results are further improved using the 3D adaptation strategy described in Section 3.6.

3.7.3.1 *Bisection refinement using the shock indicator* The bisection refinement from Section 3.6.3.2 was applied in the region where the relative shock

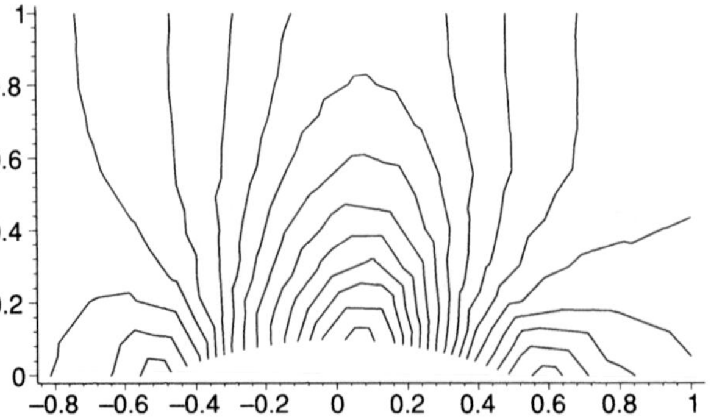

FIG. 3.7.26. Mach number isolines on the front wall

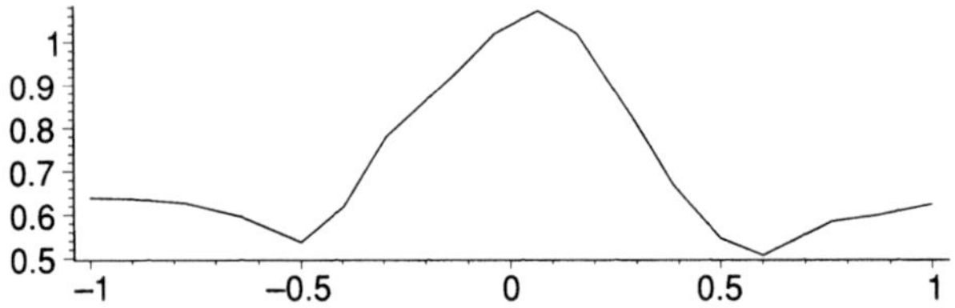

FIG. 3.7.27. Mach number distribution along the lower wall

indicator defined in Section 3.6.4.1 was greater than the prescribed tolerance. Six successive mesh refinements were made during the computation process. The results are shown in Figs 3.7.28–3.7.30.

3.7.3.2 *Bisection refinement using the residual error indicator* Figures 3.7.31–3.7.33 demonstrate the computational results of the bisection refinement from Section 3.6.3.2 based on the use of residual error indicator from Section 3.6.4.2. The results are plotted after six refinements.

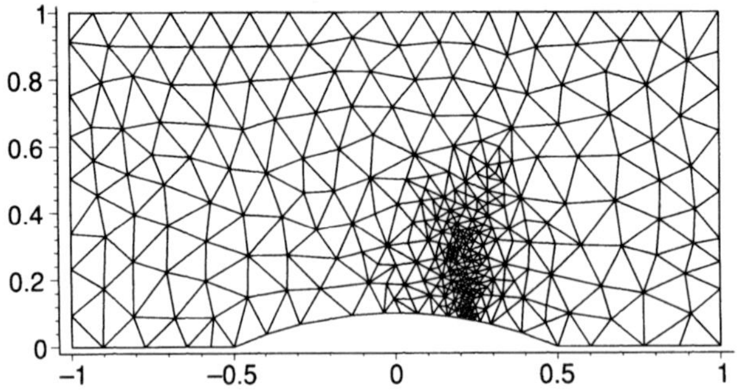

FIG. 3.7.28. Front view of final computational tetrahedral mesh – shock indicator refinement

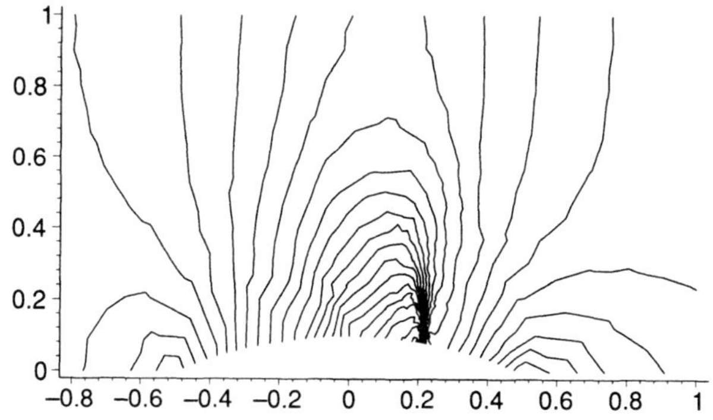

FIG. 3.7.29. Mach number isolines on the front wall

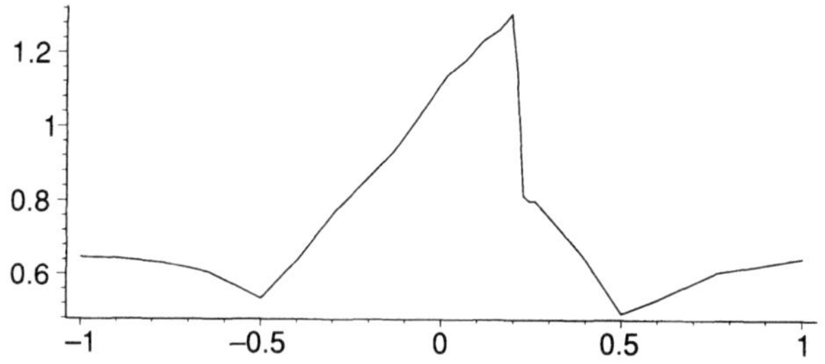

FIG. 3.7.30. Mach number distribution along the lower wall

3.7.3.3 *Anisotropic mesh adaptation* The mesh adaptation from Section 3.6.7 was applied five times during the solution process. The results concerning the mesh adaptation and Mach number distribution are plotted in Figs 3.7.34–3.7.36.

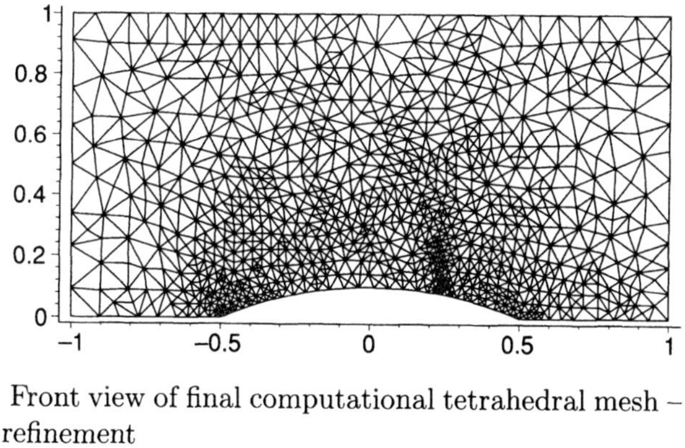

FIG. 3.7.31. Front view of final computational tetrahedral mesh – residual error indicator refinement

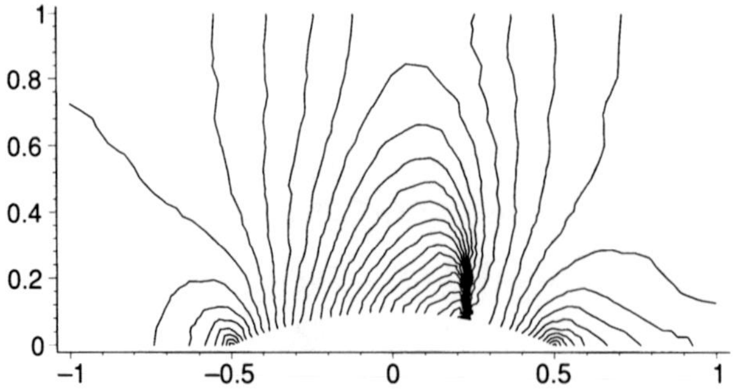

FIG. 3.7.32. Mach number isolines on the front wall

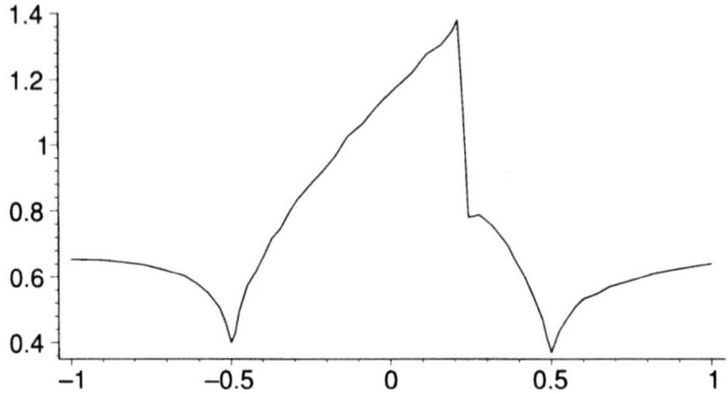

FIG. 3.7.33. Mach number distribution along the lower wall

3.7.4 *The 3D channel with 25 % spherical bump*

Another 3D analogy of the GAMM channel, in which the flow has a more 3D character than in the 10 % cylindrical bump, is drawn in Fig. 3.7.37. The three-dimensional Euler equations are solved in the domain Ω:

$$\Omega = \left\{x \in I\!R^3; x_1 \in (0,2),\, x_2 \in (0,1.5),\, x_3 \in (0,1)\right\} \setminus \Omega_{sph},$$

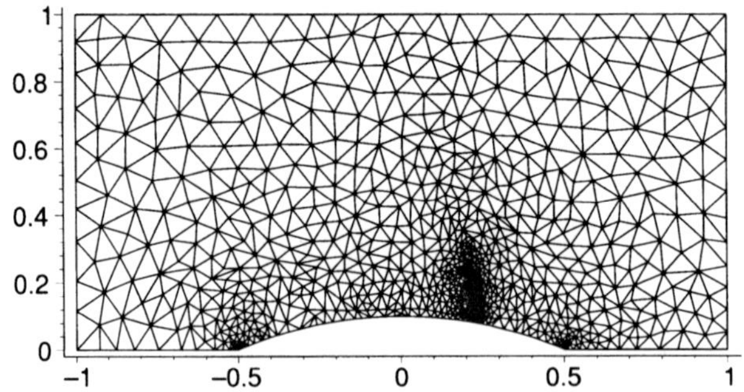

FIG. 3.7.34. Front view of final computational tetrahedral mesh – anisotropic mesh adaptation

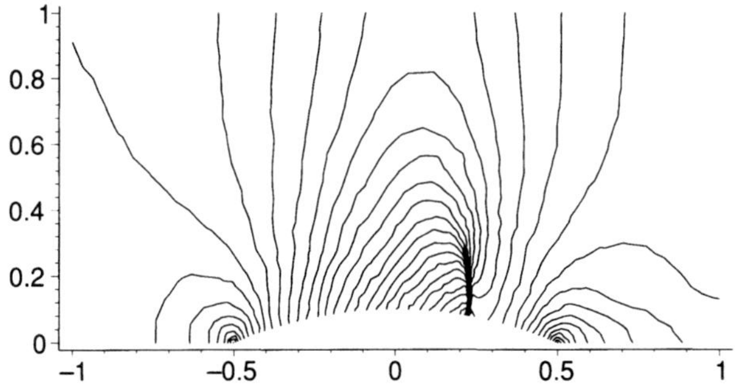

FIG. 3.7.35. Mach number isolines on the front wall

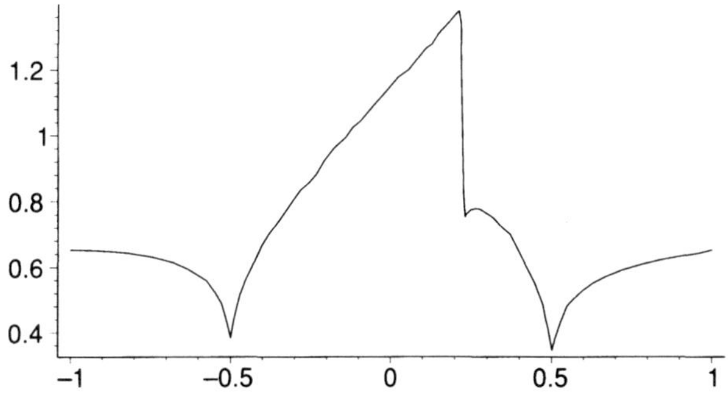

FIG. 3.7.36. Mach number distribution along the lower wall

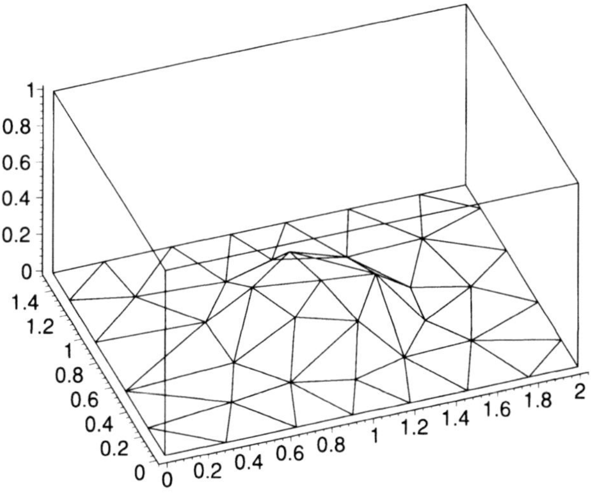

FIG. 3.7.37. 25 % spherical bump

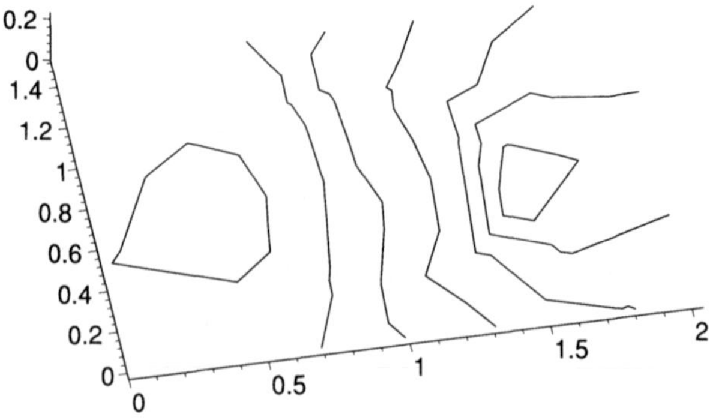

FIG. 3.7.38. Mach number isolines on the lower wall

where

$$\Omega_{sph} = \{x \in \mathbb{R}^3; (x_1 - 1)^2 + (x_2 - 0.75)^2 + (x_3 + 0.375)^2 \leq 0.390625\}.$$

We denote

$$\begin{aligned} \text{inlet:} \quad & \Gamma_I = \{x \in \mathbb{R}^3; x_1 = 0,\ x_2 \in (0, 1.5),\ x_3 \in (0,1)\}, \qquad (3.7.20) \\ \text{outlet:} \quad & \Gamma_O = \{x \in \mathbb{R}^3; x_1 = 2,\ x_2 \in (0, 1.5),\ x_3 \in (0,1)\}, \\ \text{solid walls:} \quad & \Gamma_W = \partial\Omega \setminus (\Gamma_I \cup \Gamma_O). \end{aligned}$$

The initial condition, the boundary conditions and the numerical flux used are the same as in Section 3.7.3. The steady-state solution was computed with three different adaptation strategies. In each case five bisection refinements were performed. To visualize the computational results, we present some data related to the mesh on the lower wall. The aim of the computation is an accurate capturing of the shock wave indicated in the Mach number distribution on the lower wall. In Fig. 3.7.37 the surface mesh of the initial unstructured tetrahedral mesh is plotted. It consists of 637 tetrahedra and was created by the software NETGEN, see

`http://www.sfb013.uni-linz.ac.at/~joachim/netgen`

In Fig. 3.7.38 the Mach number isolines on the lower wall are drawn. The results obtained are further improved using the 3D adaptation strategy described in Section 3.6.

3.7.4.1 *Bisection refinement using the shock indicator* The relative shock indicator from Section 3.6.4.1 was computed. In Fig. 3.7.39 the final surface mesh on the lower wall is shown. The Mach number isolines and the Mach number distribution along the lower wall are depicted in Figs 3.7.40–3.7.41. The sharpness of the computed shock wave is demonstrated with the aid of the cut by the plane $x_2 = 0.75$.

The natural question is: what amount of the total computational time is necessary for the mesh adaptation? A comparison of the CPU time of the numerical

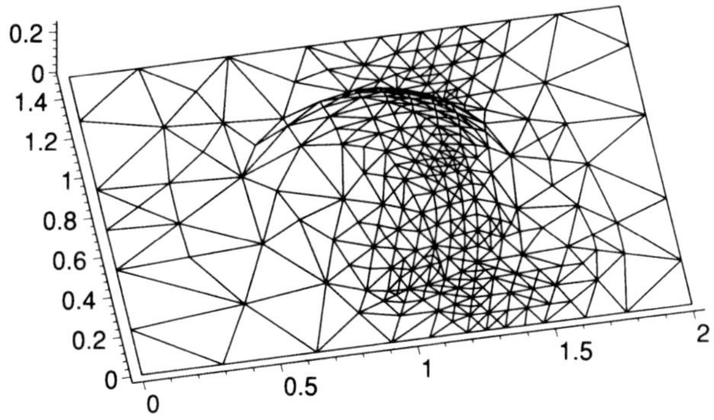

FIG. 3.7.39. Shock indicator mesh refinement on the lower wall

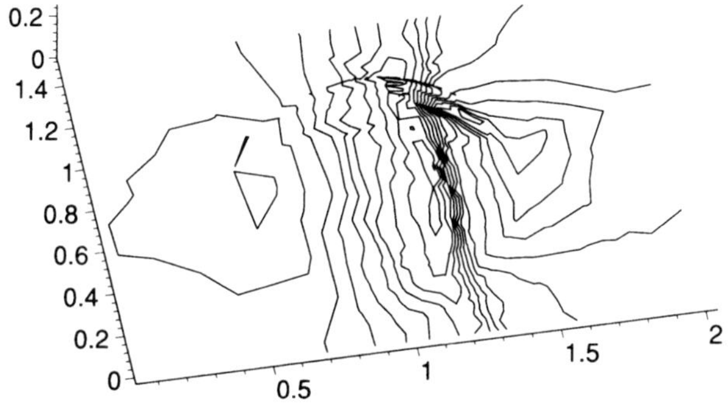

FIG. 3.7.40. Mach number isolines on the lower wall

solution on the mesh $\mathcal{T}_h$ (denoted CPU-Comp($\mathcal{T}_h$)) with the CPU time for the adaptation of the mesh $\mathcal{T}_h$ (denoted CPU-Ref($\mathcal{T}_h \rightarrow \mathcal{T}_h^{\text{new}}$)) is presented in Table 3.7.2. The number of tetrahedra in $\mathcal{T}_h$ is denoted by $\#\mathcal{T}_h$. From the last row it follows that the refinement/computation time rate is 0.03 %. The results were obtained on a DEC Alpha EV56/500 MHz workstation.

Table 3.7.2 *Comparison of the computational and adaptation time – shock indicator*

$\mathcal{T}_h \rightarrow \mathcal{T}_h^{\text{new}}$	$\#\mathcal{T}_h$	CPU-Comp($\mathcal{T}_h$) [s]	CPU-Ref($\mathcal{T}_h \rightarrow \mathcal{T}_h^{\text{new}}$) [s]
1	637	1.44	0.01
2	1142	6.51	0.05
3	1919	17.73	0.08
4	4071	93.75	0.20
5	8387	260.41	0.48
6	18677	2586.59	
total		2966.59	0.82

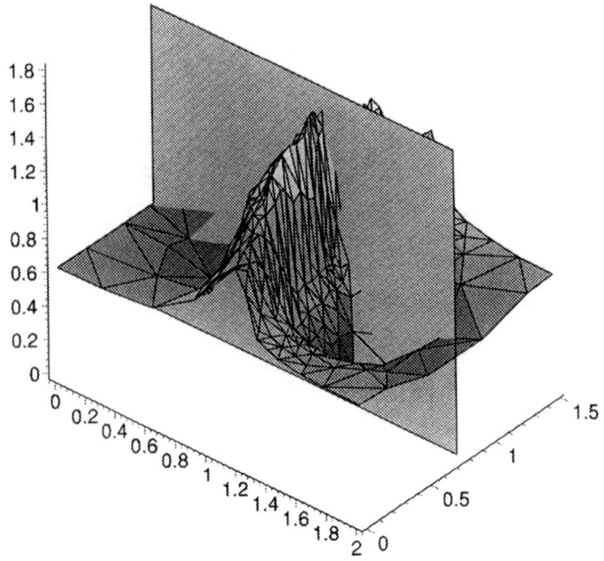

FIG. 3.7.41. Mach number distribution along the lower wall with a plane cut

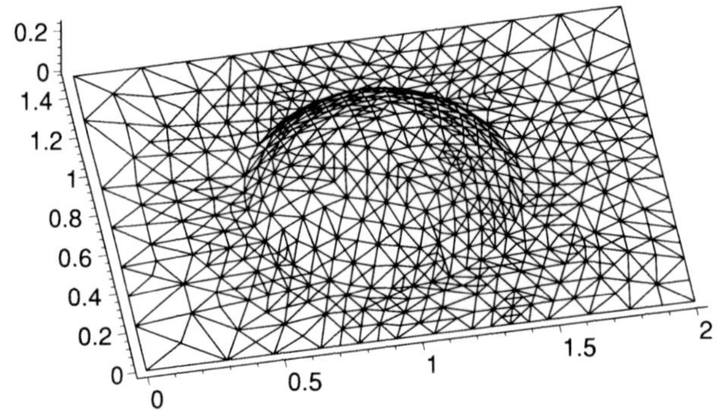

FIG. 3.7.42. Residual error indicator mesh refinement on the lower wall

3.7.4.2 *Bisection refinement using the residual error indicator* Analogously to Section 3.7.4.1 we present the computational results in Figs 3.7.42 and 3.7.43. The comparison of the CPU time (DEC Alpha EV56/500 MHz) of the numerical solution on the mesh $\mathcal{T}_h$ (denoted CPU-Comp($\mathcal{T}_h$)) with the CPU time of the adaptation of the mesh $\mathcal{T}_h$ (denoted CPU-Ref($\mathcal{T}_h - \mathcal{T}_h^{\text{new}}$)) is presented in Table 3.7.3. The refinement/computation time rate is 0.01 %.

3.7.4.3 *Anisotropic mesh adaptation* Analogously to the two previous sections, the computational results are presented in Figs 3.7.44 and 3.7.45. The comparison of the CPU time (DEC Alpha EV56/500 MHz) of the numerical solution on the mesh $\mathcal{T}_h$ (denoted CPU-Comp($\mathcal{T}_h$)) with the CPU time for the adaptation of

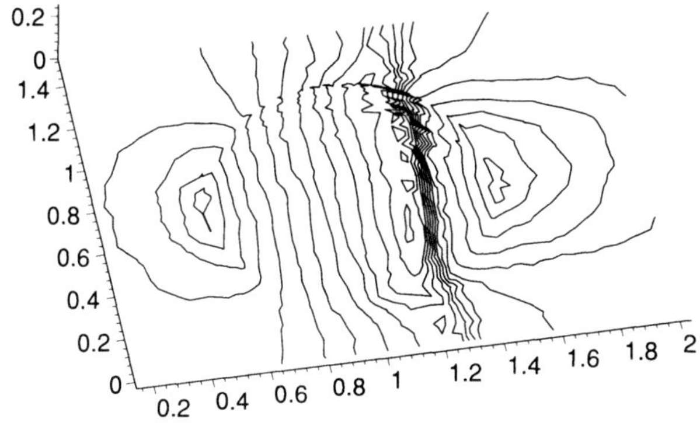

FIG. 3.7.43. Mach number isolines on the lower wall

Table 3.7.3 *Comparison of the computational and adaptation time – residual error indicator*

$\mathcal{T}_h \to \mathcal{T}_h^{\text{new}}$	$\#\mathcal{T}_h$	CPU-Comp($\mathcal{T}_h$) [s]	CPU-Ref($\mathcal{T}_h \to \mathcal{T}_h^{\text{new}}$) [s]
1	637	1.46	0.06
2	2072	9.09	0.18
3	5624	67.50	0.46
4	14235	312.66	0.93
5	28521	1431.29	1.51
6	46887	22924.81	
total		24746.81	3.14

the mesh $\mathcal{T}_h$ (denoted CPU-Ref($\mathcal{T}_h \to \mathcal{T}_h^{\text{new}}$)) is presented in Table 3.7.4. The adaptation/computation time rate is 25.2 %.

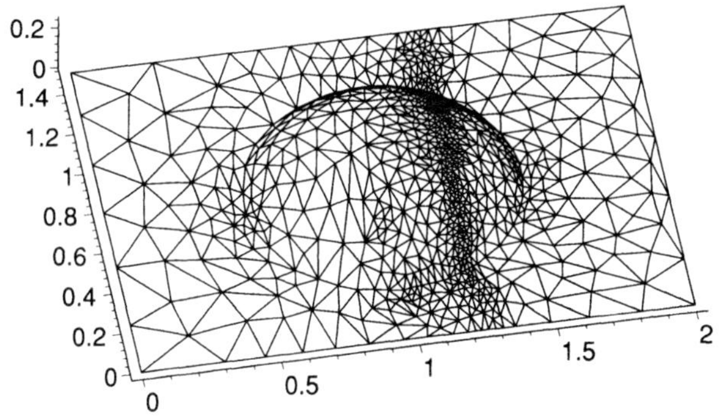

FIG. 3.7.44. Anisotropic mesh adaptation on the lower wall

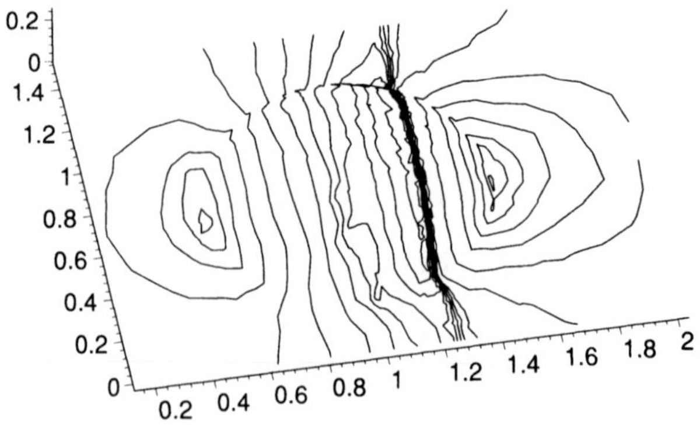

FIG. 3.7.45. Mach number isolines on the lower wall

Table 3.7.4 *Comparison of the computational and adaptation time – anisotropic mesh adaptation*

$\mathcal{T}_h \to \mathcal{T}_h^{\text{new}}$	$\#\mathcal{T}_h$	CPU-Comp($\mathcal{T}_h$) [s]	CPU-Ref($\mathcal{T}_h \to \mathcal{T}_h^{\text{new}}$) [s]
1	637	1.48	191.40
2	3096	29.55	409.18
3	6434	160.00	647.53
4	11053	741.08	1345.08
5	17823	1743.58	2282.41
6	23721	16652.93	
total		19328.62	4875.60

3.7.5 *Flow past NACA 0012 airfoil*

The Euler equations are used extensively in aerodynamics for example, in the modelling of flow past aircraft or other vehicles. These are typically 3D problems, although 2D or even 1D models are sometimes of interest. A typical 2D problem is the flow of air past an airfoil, which is simply the cross section of a wing. Small changes in the shape of an airfoil can lead to very different flow patterns, and so the ability to experiment by performing calculations with a wide variety of shapes is required. Of particular interest to the aerodynamic engineer is the pressure distribution along the airfoil surface. From this the engineer can calculate the lift and drag (the vertical and horizontal components of the force acting on the wing) which are crucial in evaluating performance. Here we present some results for flow past the NACA 0012 airfoil, which is drawn in Fig. 3.7.46. NACA 0012 is a symmetric airfoil with an upper surface described by the function $f : [0,1] \to I\!R$,

$$\begin{aligned} f(\xi) &= \frac{0.6}{\xi_0}\Big(0.2969\sqrt{\xi\xi_0} - 0.126\,\xi\xi_0 \\ &\quad - 0.3516\,(\xi\xi_0)^2 + 0.2843\,(\xi\xi_0)^3 - 0.1015\,(\xi\xi_0)^4\Big), \\ \xi_0 &= 1.008930411365. \end{aligned} \tag{3.7.21}$$

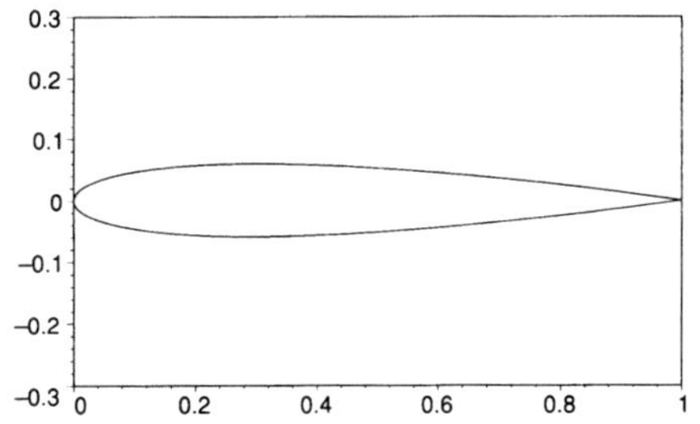

FIG. 3.7.46. NACA 0012 airfoil

The region occupied by the gas is the exterior of the profile, which is an unbounded domain. It is approximated by a bounded but sufficiently large domain, as, for example,

$$\begin{aligned}\Omega = &\left\{x \in I\!R^2; (x_1 - 0.5)^2 + x_2^2 < 36\right\} \\ &\setminus \left\{x \in I\!R^2; x_1 \in [0,1],\, x_2 \in [-f(x_1), f(x_1)]\right\},\end{aligned} \tag{3.7.22}$$

where f is defined in (3.7.21). The boundary $\partial\Omega$ of the domain Ω consists of two components,

$$\partial\Omega = \Gamma_{I/O} \cup \Gamma_{\text{prof}}, \tag{3.7.23}$$

where

$$\begin{aligned}\Gamma_{I/O} &= \left\{x \in I\!R^2 + (x_1 - 0.5)^2 + x_2^2 = 36\right\}, \\ \Gamma_{\text{prof}} &= \left\{x \in I\!R^2; x_2 = \pm f(x_1),\, x_1 \in [0,1]\right\}.\end{aligned} \tag{3.7.24}$$

On $\Gamma_{I/O}$ the inlet/outlet boundary conditions are applied, and the airfoil Γ_{prof} is considered as a solid impermeable wall. The steady-state solution is computed for a particular case when the angle of attack (i.e. the far-field velocity angle measured from the positive part of the x_1-axis in a counterclockwise direction) $\alpha = 1.25°$ and the far-field Mach number $M_\infty = 0.8$. The initial condition corresponds to these data. Expressed in the primitive variables, we set

$$\begin{aligned}\rho^0 &= 1.5\,\text{kg}\,\text{m}^{-3}, \\ v_1^0 &= 245.564\,\text{m}\,\text{s}^{-1}, \\ v_2^0 &= 5.358\,\text{m}\,\text{s}^{-1}, \\ p^0 &= 101000\,\text{Pa}.\end{aligned} \tag{3.7.25}$$

The problem formulated above was solved by the FV method on a triangular grid which was successively refined with the aid of the mesh adaptation techniques described in Section 3.6. In what follows we are concerned with results obtained with the aid of the Osher–Solomon numerical flux.

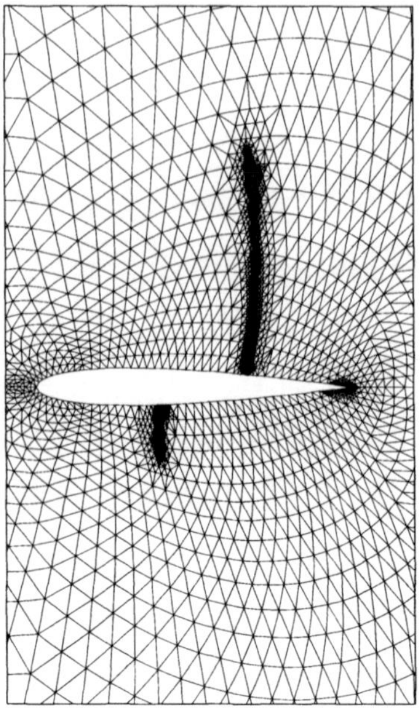

FIG. 3.7.47. Shock indicator mesh refinement

3.7.5.1 *Application of the shock indicator* Figure 3.7.47 shows a detail of the fourth refinement of the initial mesh using the shock indicator from Section 3.7.2.1. In Figs 3.7.48 and 3.7.49 Mach number isolines and the Mach number distribution are shown. Two shock waves are visible here, a strong one on the upper surface and a weak one on the lower surface of the airfoil.

3.7.5.2 *Anisotropic mesh adaptation – triangular finite volumes* The mesh adaptation from Section 3.6.7 was applied. A detail of the adapted mesh, Mach number isolines and the Mach number distribution along the lower and upper surface of the airfoil are shown in Figs 3.7.50–3.7.52 for triangular finite volumes.

3.7.5.3 *Anisotropic mesh adaptation – barycentric finite volumes* The barycentric finite volumes and the mesh adaptation from Section 3.6.7 are applied. A detail of the adapted mesh, Mach number isolines and the Mach number distribution along the airfoil are shown in Figs 3.7.53–3.7.55.

3.7.5.4 *Anisotropic mesh adaptation – triangular finite volumes at Mach number $M = 2$* In Fig. 3.7.56 a detail of triangulation produced by AMA and the corresponding distribution of Mach number around the NACA 0012 profile for the case $M_\infty = 2$ and inlet angle of attack $\alpha = 0°$ are presented. The dimensionless form of the Euler equations was used for the computation (see Section 1.2.23). The dimensionless initial conditions were set as

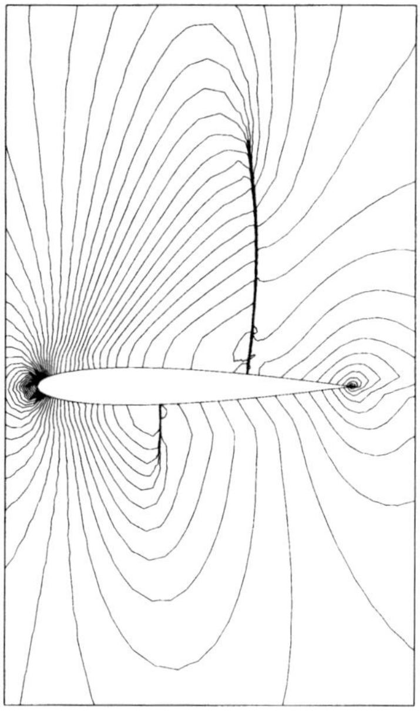

FIG. 3.7.48. Mach number isolines

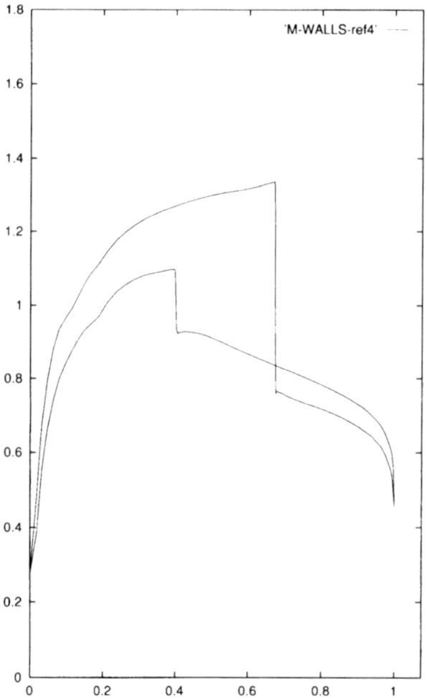

FIG. 3.7.49. Mach number distribution along the lower and upper surface of the airfoil

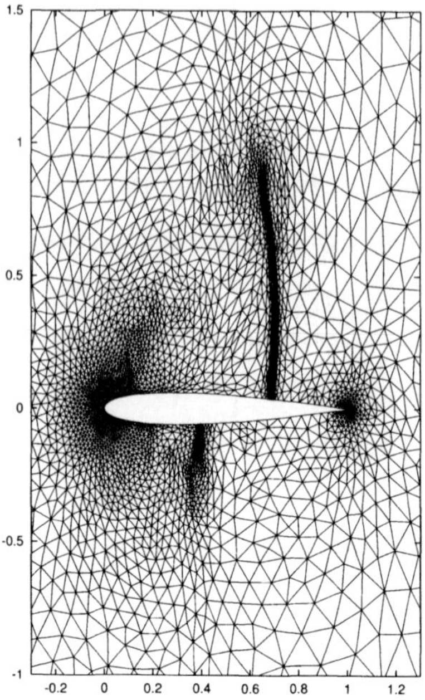

FIG. 3.7.50. Adaptive triangular mesh – anisotropic mesh adaptation

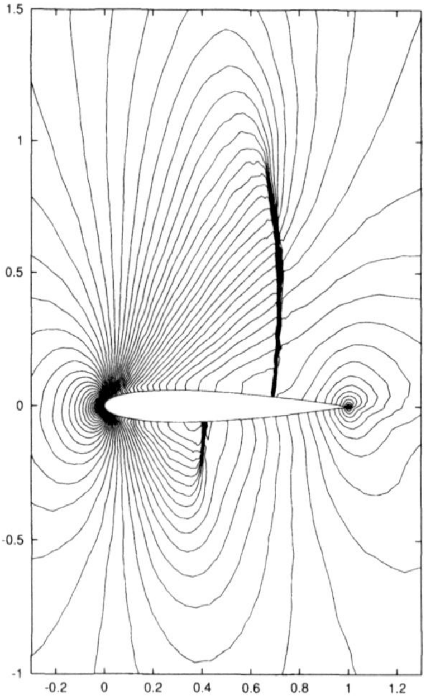

FIG. 3.7.51. Mach number isolines

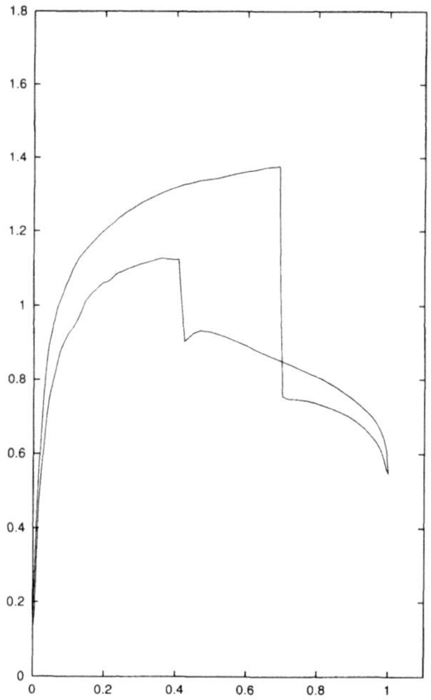

FIG. 3.7.52. Mach number distribution along lower and upper surface of the airfoil

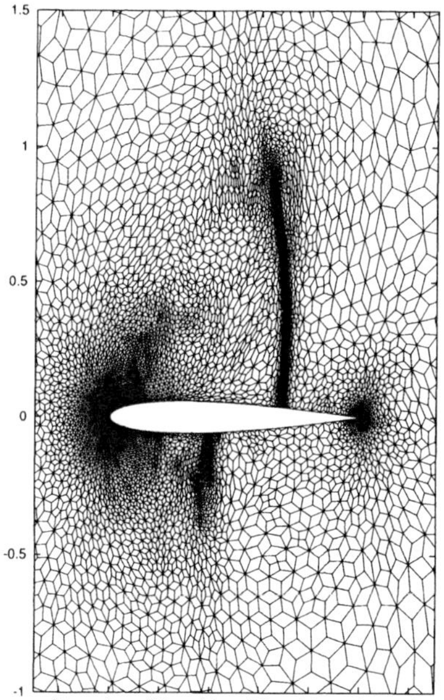

FIG. 3.7.53. Adaptive barycentric mesh – anisotropic mesh adaptation

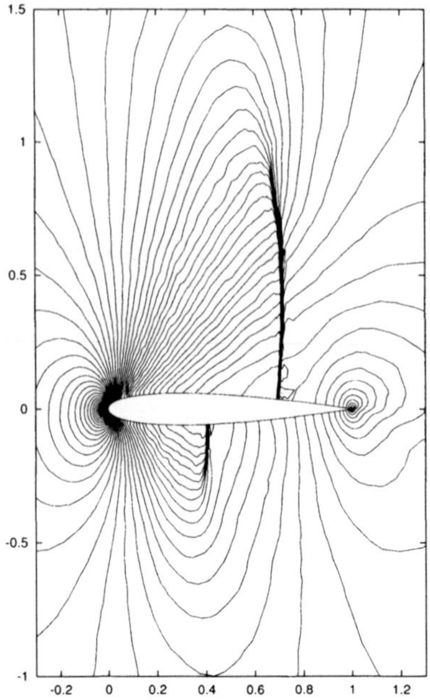

FIG. 3.7.54. Mach number isolines

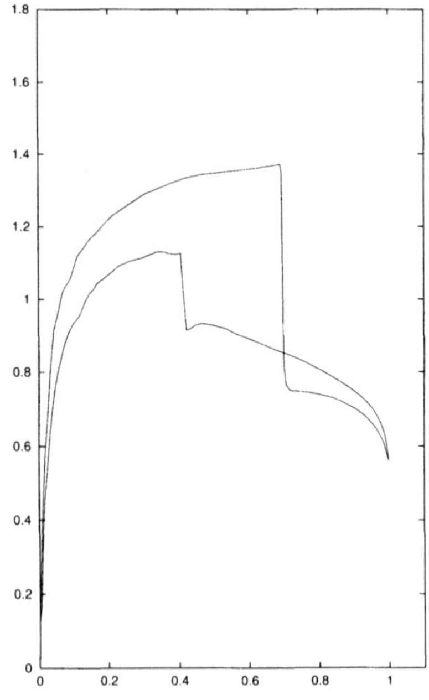

FIG. 3.7.55. Mach number distribution along lower and upper surface of the airfoil

$$\begin{aligned} \rho'^0 &= 1, \\ v_1'^0 &= 1, \\ v_2'^0 &= 0, \\ p'^0 &= \frac{1}{\gamma M_\infty^2}. \end{aligned} \tag{3.7.26}$$

The constant γ is given in (3.7.3).

3.7.6 *Flow past a cascade of profiles*

In this section we present the numerical solution of a technically relevant problem arising in the investigation of the aerodynamical properties of blade machines.

A widely used model for the simulation of flow through blade rows of turbines and compressors is the flow past a plane cascade of profiles. In this case the region occupied by the fluid is represented by a plane, infinitely connected domain $\tilde{\Omega}$, bounded in one space direction, say x_1, and unbounded but periodic in the direction x_2. The domain $\tilde{\Omega}$ is the exterior of a *cascade of profiles* formed by an infinite number of disjoint profiles, periodically spaced in the direction x_2 with period $\tilde{\tau} > 0$ (called a pitch in turbine terminology), as shown in Fig. 3.7.57 below. Assuming also the periodicity of the flow field, the computational domain Ω can be chosen in the form of one period of the original domain $\tilde{\Omega}$.

The boundary of Ω has the form

$$\partial\Omega = \Gamma_W \cup \Gamma_I \cup \Gamma_O \cup \Gamma^- \cup \Gamma^+, \tag{3.7.27}$$

where Γ_W, Γ_I and Γ_O represent the profile, inlet and outlet, respectively, and Γ^-, Γ^+ are two artificial cuts represented by piecewise linear arcs with initial points on Γ_I and terminal points on Γ_O, satisfying the condition

$$\Gamma^+ = \left\{(x_1, x_2 + \tilde{\tau}); (x_1, x_2) \in \Gamma^-\right\}. \tag{3.7.28}$$

The profile Γ_W is considered as a fixed impermeable wall.

The inviscid gas flow past the cascade of profiles is described by the system (3.3.1) of the Euler equations in $Q_T = \Omega \times (0, T)$, equipped with initial conditions (3.3.2) and suitable boundary conditions (3.3.3). In the FV numerical solution of the problem, the inlet, outlet and impermeable wall boundary conditions on Γ_I, Γ_O and Γ_W, respectively, are treated according to Section 3.3.7 – see also Table 3.7.1. On Γ^- and Γ^+ the periodicity condition is used:

$$\boldsymbol{w}(x_1, x_2 + \tilde{\tau}) = \boldsymbol{w}(x_1, x_2), \quad (x_1, x_2) \in \Gamma^-. \tag{3.7.29}$$

3.7.6.1 *FV discretization* This is carried out in a standard way as described in Section 3.3. The domain Ω is approximated by a polygonal domain Ω_h with boundary $\partial\Omega_h = \Gamma_I \cup \Gamma_O \cup \Gamma^- \cup \Gamma^+ \cup \Gamma_{Wh}$, where Γ_{Wh} is a simple, closed, piecewise linear curve approximating Γ_W. By $\mathcal{D}_h = \{D_i\}_{i \in J}$ we denote an FV mesh in Ω_h with properties from Section 3.3. Moreover, in order to apply the

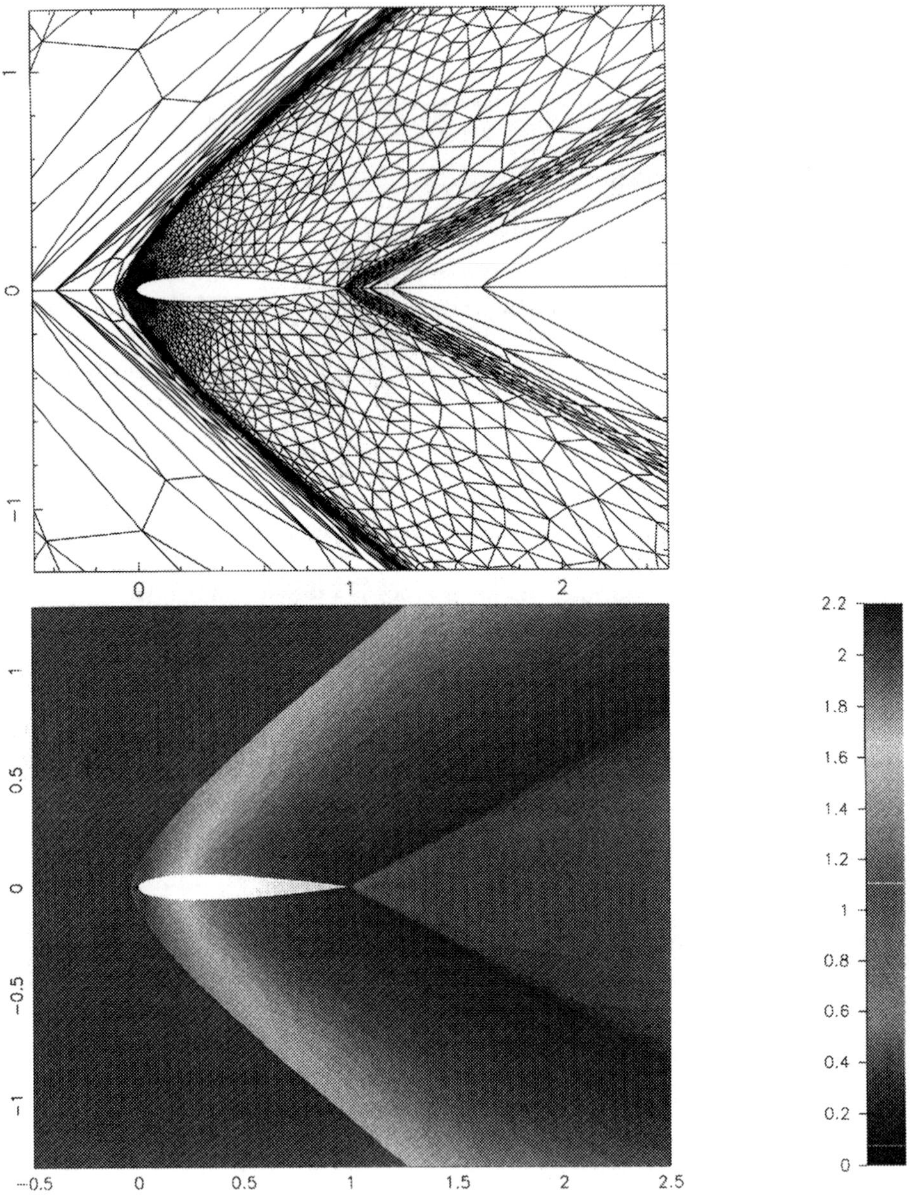

FIG. 3.7.56. Inviscid flow around NACA 0012 profile with $M_\infty = 2$; triangulation produced by AMA and the corresponding distribution of Mach number

periodicity condition in the numerical scheme (3.3.24), we assume that the mesh $\mathcal{D}_h$ possesses the *periodicity property*: $\emptyset \neq S_i^- = \partial D_i \cap \Gamma^-$ for some $D_i \in \mathcal{D}_h$ if and only if there exists $D_j \in \mathcal{D}_h$ such that

$$S_i^+ = \{(x_1, x_2 + \tilde{\tau}); (x_1, x_2) \in S_i^-\} = \partial D_j \cap \Gamma^+. \tag{3.7.30}$$

In this case we put $\Gamma_{ij}^1 = \Gamma_{ij} = S_i^-$ and $\Gamma_{ji}^1 = \Gamma_{ji} = S_i^+$. If the mesh is triangular or barycentric, we set

$$\begin{aligned} S(i) &= s(i) \cup \{j\}, \\ S(j) &= s(j) \cup \{i\}. \end{aligned} \tag{3.7.31}$$

However, for a *dual mesh* we introduce new finite volumes $D_i' = D_j' = D_i \cup D_j$ and put

$$S(i) = S(j) = s(i) \cup s(j). \tag{3.7.32}$$

For simplicity of notation we omit the prime. Then all the notation in Section 3.3.1 remains without change, and the periodicity boundary condition given by (3.7.29) need not be written explicitly for the FV scheme (3.3.24).

Here we present the computation of 2D flow past a cascade of turbine profiles. The goal is to obtain a *steady-state* solution comparable with wind tunnel experiments carried out at the Institute of Thermodynamics of the Academy of Sciences of the Czech Republic in Prague. (See (Šťastný and Šafařík, 1990).)

The cascade of profiles is shown in Fig. 3.7.57 below. The profile Γ_W represents a test case of the blade row SE 1050 of the ŠKODA Pilsen Turbine Company. The geometry of Γ_W is described in (Šťastný and Šafařík, 1990). The experiment as well as the computation were performed for the following data: angle of attack $\alpha = 19^\circ\, 18'$, inlet Mach number $M_I = 0.32$, outlet Mach number $M_O = 1.18$. Figure 3.7.58 shows an interferogram of density resulting from the experiment. A detail of the interferogram corresponding to the rectangle in Fig. 3.7.57 is shown in Fig. 3.7.59. We see here a number of interesting details (as a system of shock waves), which are quite difficult to resolve. The geometry of the computational domain is shown in Fig. 3.7.60. The inlet and outlet is given as

$$\begin{aligned} \Gamma_I &= \{x \in I\!R^2; x_1 = -0.06,\, x_2 \in (-0.046716, 0.0084)\}, \\ \Gamma_O &= \{x \in I\!R^2; x_1 = 0.14,\, x_2 \in (-0.177617, -0.1225)\}. \end{aligned} \tag{3.7.33}$$

The artificial cut Γ^- is given pointwise with initial point at $(-0.06, -0.046716)$ and terminal point at $(0.14, -0.177617)$. The artificial cut Γ^+ is given by relation (3.7.28) and the period of the cascade $\tilde{\tau} = 0.05511679$. We use the dimensionless form of the Euler equations (see Section 1.2.23). The initial conditions and the inlet and outlet boundary conditions are defined in such a way that they respect the change of the inlet and outlet angle of the velocity. The initial condition $\mathbf{w}_h^0$ in (3.3.25) is supposed to be given by two different constant states $\mathbf{w}_L^0$ and $\mathbf{w}_R^0$ in the left and right parts of the domain Ω. In the dimensionless primitive variables (see Section 1.2.23) we set

$$\rho_L'^0 = \rho_R'^0 = 1, \tag{3.7.34}$$

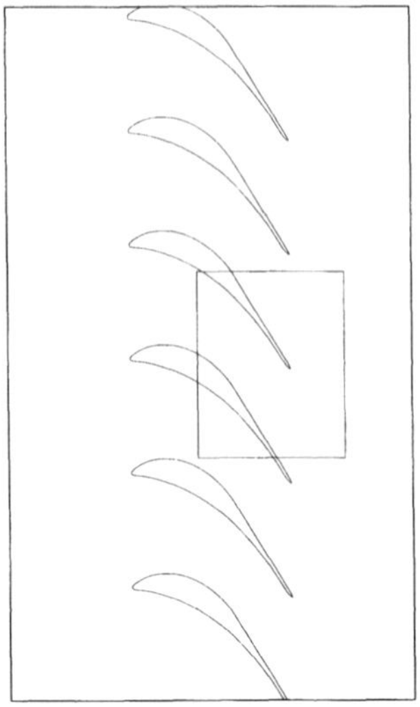

FIG. 3.7.57. Zooming rectangle

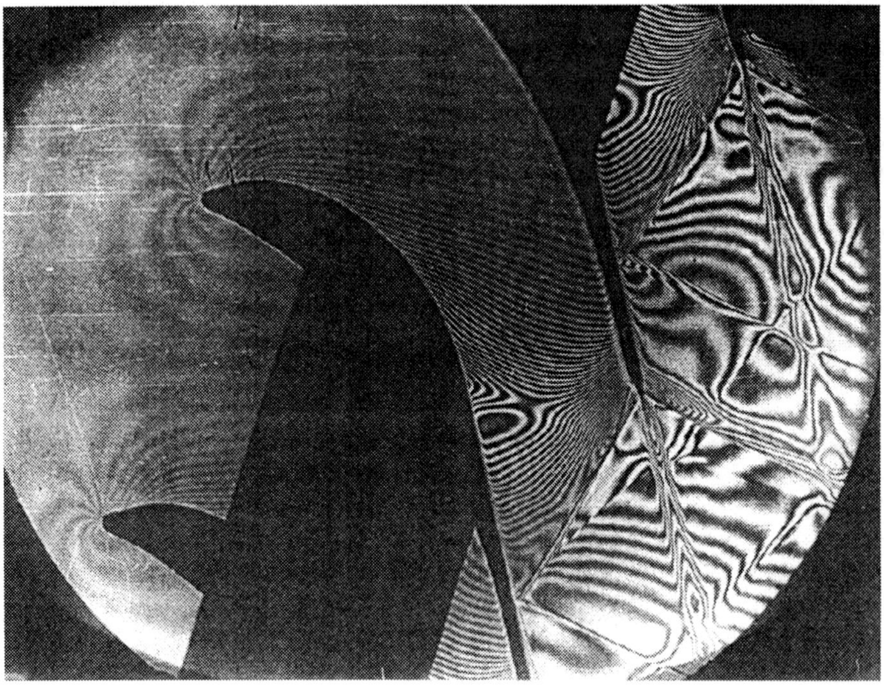

FIG. 3.7.58. Interferogram of the density

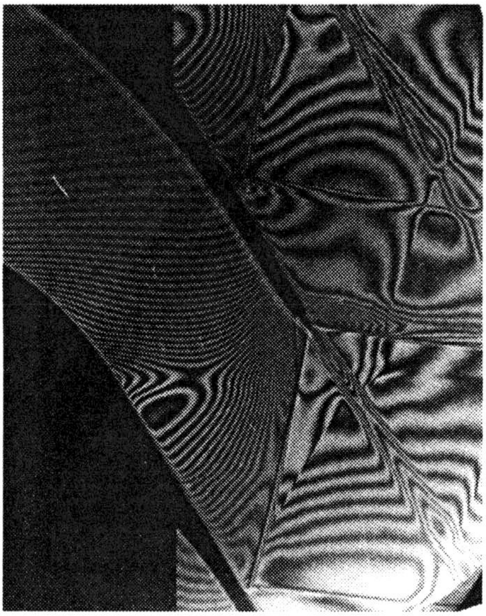

FIG. 3.7.59. Interferogram of the density – detail

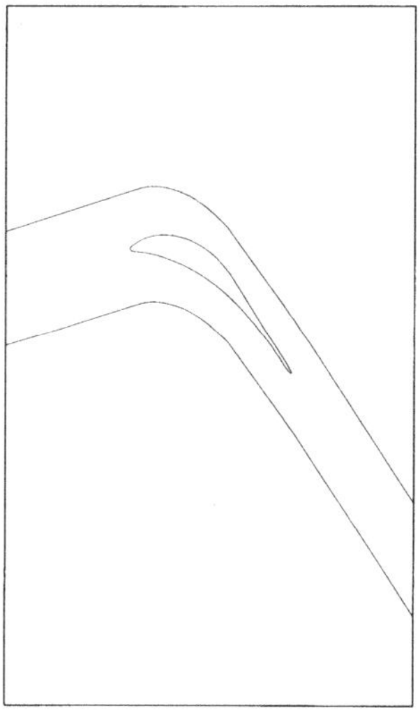

FIG. 3.7.60. Computational domain Ω

$$|\boldsymbol{v}_L'^0| = |\boldsymbol{v}_R'^0| = 1, \tag{3.7.35}$$

$$\alpha_i = \begin{cases} 19^\circ\, 18', & i = L, \\ 57^\circ, & i = R, \end{cases}$$

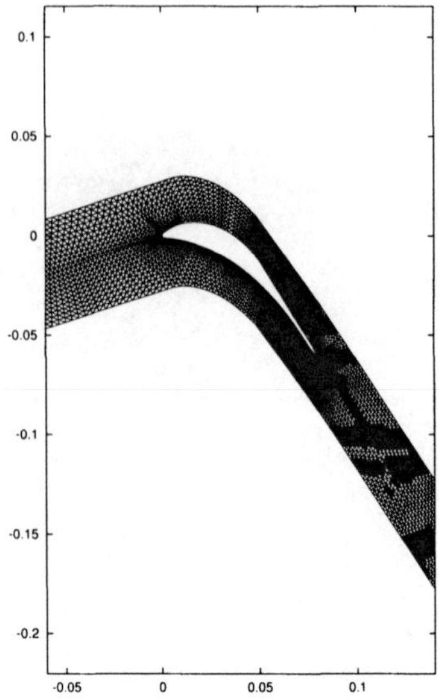

FIG. 3.7.61. Mesh refinement using the shock indicator

$$p_i'^0 = \begin{cases} \frac{1}{\gamma M_I^2}, & i = L, \\ \frac{1}{\gamma M_\infty^2} & i = R, \end{cases}$$

where α_i, $i = L, R$, is the angle of dimensionless velocity $\boldsymbol{v}'$. So the initial approximation of the velocity has the outlet angle which is supposed for the operational regime of the turbine cascade.

Due to the complexity of the flow past a cascade of profiles, a suitable mesh having the periodicity property is of great importance. It is shown in (Felcman *et al.*, 1994) that the structured quadrilateral FV meshes give poor results. Here we present several examples of adaptive mesh refinement and the corresponding numerical results. In the numerical solution procedure, the Osher–Solomon numerical flux was applied.

3.7.6.2 *Adaptive mesh refinement using the shock indicator* In Fig. 3.7.61 the mesh refinement using the shock indicator from Section 3.6.4.1 is plotted. Shown in Fig. 3.7.62 are the Mach number isolines.

3.7.6.3 *Adaptive mesh refinement using the superconvergence error indicator* The use of the superconvergence error indicator from Section 3.6.5.1 is demonstrated in Figs 3.7.63–3.7.64.

3.7.6.4 *Anisotropic mesh adaptation* A detail of the mesh adapted by the algorithm described in Section 3.6.7 is presented in Fig. 3.7.65. Figure 3.7.66 shows the density isolines. The coincidence of experimental and computational results is relatively satisfactory, although the real viscous gas flow is modelled with the

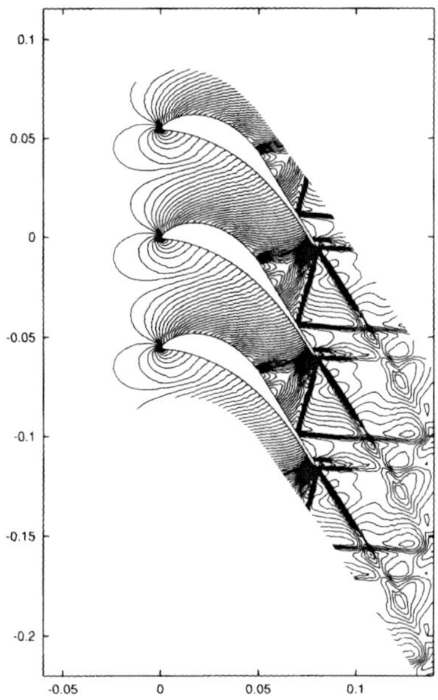

FIG. 3.7.62. Mach number isolines

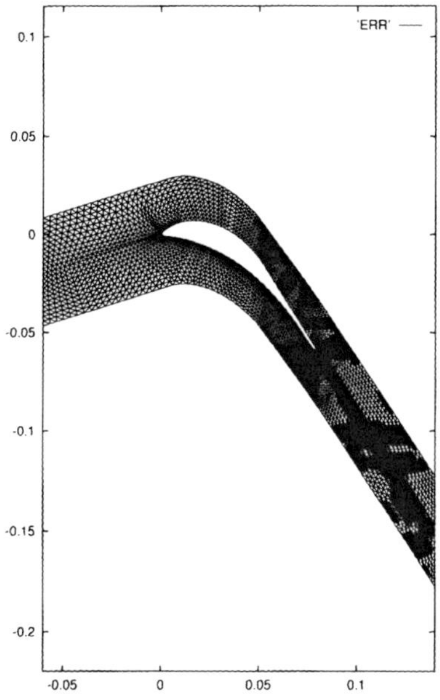

FIG. 3.7.63. Mesh refinement using the superconvergence error indicator

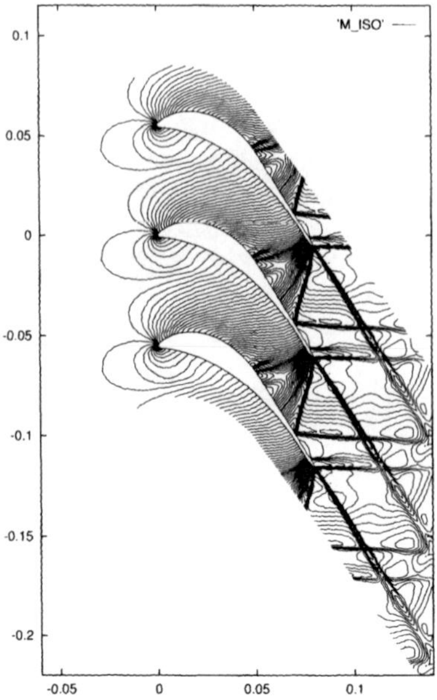

FIG. 3.7.64. Mach number isolines

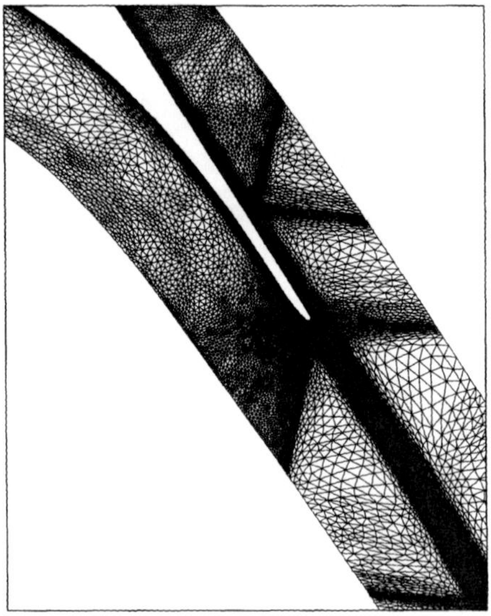

FIG. 3.7.65. Anisotropic mesh adaptation

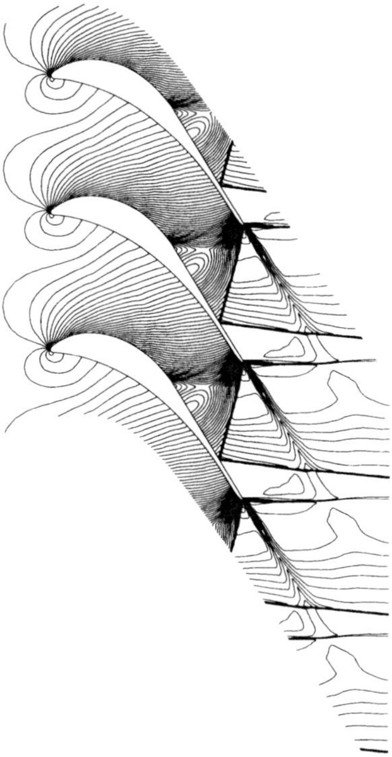

FIG. 3.7.66. Density isolines

aid of the inviscid Euler system. Figure 3.7.67 shows the computed (−) and measured (◇) dimensionless pressure distributions along the lower and upper part of the profile, drawn from the leading edge to the trailing edge depending on relative arc length. The mesh adaptation, density isolines and pressure distribution for barycentric finite volumes are shown in Figs 3.7.68–3.7.70.

3.7.7 *Scramjet*

The supersonic flow problem with a complicated structure of shock waves was introduced in (Díaz *et al.*, 1995). The flow is considered in a channel with inserted obstacles. The numerical solution is performed in the domain Ω shown in Fig. 3.7.71. The inlet/outlet boundary conditions from Section 3.4.7 and solid wall boundary conditions from Section 3.4.8 are applied on $\partial\Omega = \Gamma_I \cup \Gamma_O \cup \bigcup_{i=1}^{4} \Gamma_W^i$, where

$$\begin{aligned} \text{inlet:}\quad \Gamma_I &= \left\{x \in I\!R^2; x_1 = 0,\ x_2 \in (-3.5, 3.5)\right\}, \\ \text{outlet:}\quad \Gamma_O &= \left\{x \in I\!R^2; x_1 = 16.9,\ x_2 \in (-1.7, 1.7)\right\}. \end{aligned} \tag{3.7.36}$$

The lower wall, upper wall and obstacles are given as piecewise linear curves passing through the given set of points:

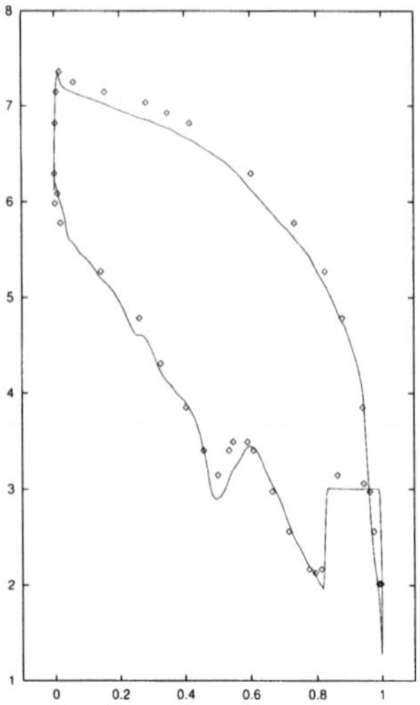

FIG. 3.7.67. Dimensionless pressure distribution along the profile – triangular finite volumes

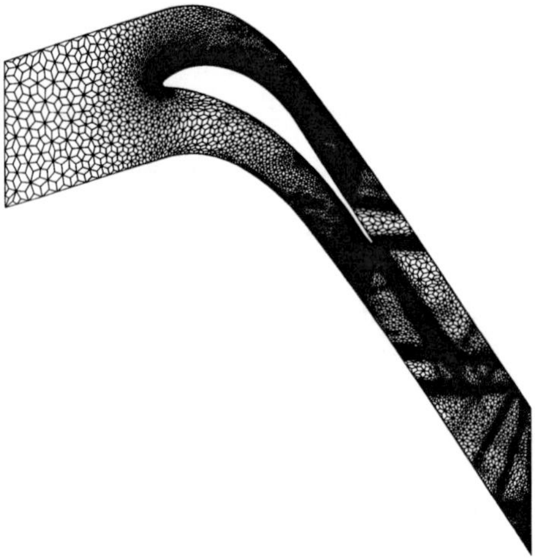

FIG. 3.7.68. Anisotropic mesh adaptation – barycentric mesh

$$\begin{aligned} \text{lower wall } \Gamma_W^1 : \ & \{(0,-3.5),(0.4,-3.5),(4.9,-2.9), \\ & (12.6,-2.12),(14.25,-1.92),(16.9,-1.7)\}, \\ \text{upper wall } \Gamma_W^2 : \ & \Gamma_W^2 = \left\{x \in I\!R^2; x = (\tilde{x}_1,-\tilde{x}_2),\ \tilde{x} \in \Gamma_W^1\right\}, \\ \text{obstacle } \Gamma_W^3 : \ & \{(4.9,-1.4),(8.9,-0.5),(9.4,-0.5), \\ & (14.25,-1.2),(12.6,-1.4)\}, \\ \text{obstacle } \Gamma_W^4 : \ & \Gamma_W^4 = \left\{x \in I\!R^2; x \in (\tilde{x}_1,-\tilde{x}_2),\ \tilde{x} \in \Gamma_W^3\right\}. \end{aligned} \tag{3.7.37}$$

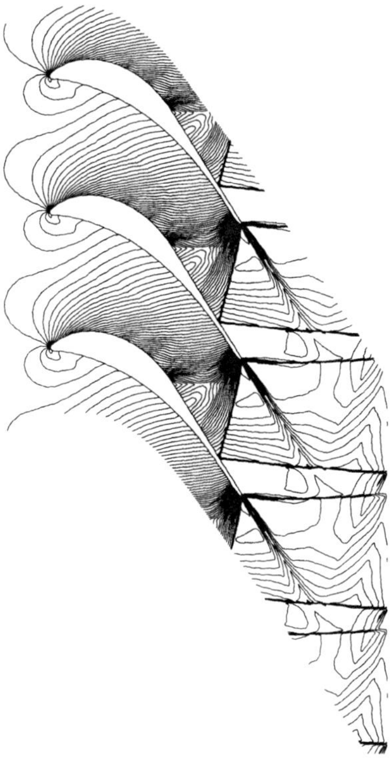

FIG. 3.7.69. Density isolines

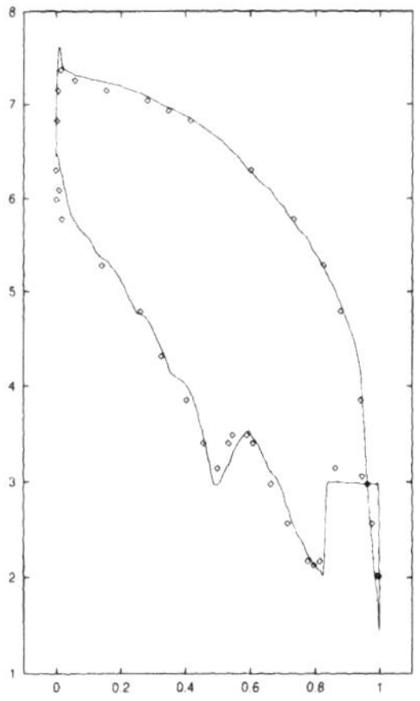

FIG. 3.7.70. Dimensionless pressure distribution along the profile – barycentric finite volumes

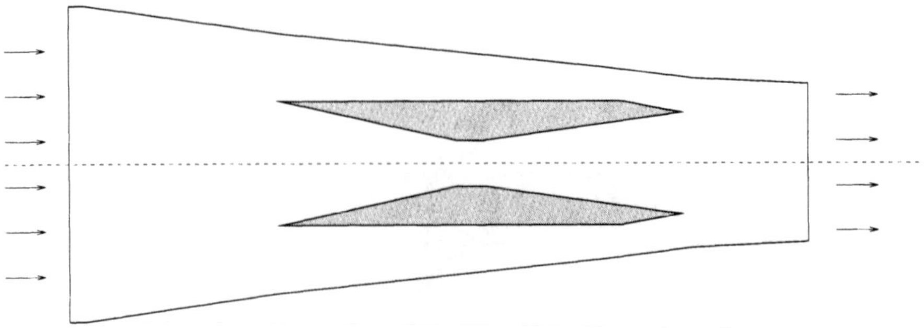

FIG. 3.7.71. Scramjet

The direction of the flow is from left to right with the inlet velocity angle $\alpha = 0°$ and inlet Mach number $M_I = 3$. The initial condition $\mathbf{w}_h^0$ (see (3.3.25)) is supposed to be constant in the whole domain Ω in agreement with given α and M_I. Expressed in dimensionless primitive variables the initial data are

$$\begin{aligned} \rho'^0 &= 1, \\ v_1'^0 &= 1, \\ v_2'^0 &= 0, \\ p'^0 &= \frac{1}{\gamma M_I^2}. \end{aligned} \tag{3.7.38}$$

The constant γ is given in (3.7.3). The explicit finite volume scheme (3.3.24) with the Osher–Solomon numerical flux was applied. The computational mesh and the Mach number isolines are shown in Figs 3.7.72–3.7.73. The anisotropic mesh adaptation was used.

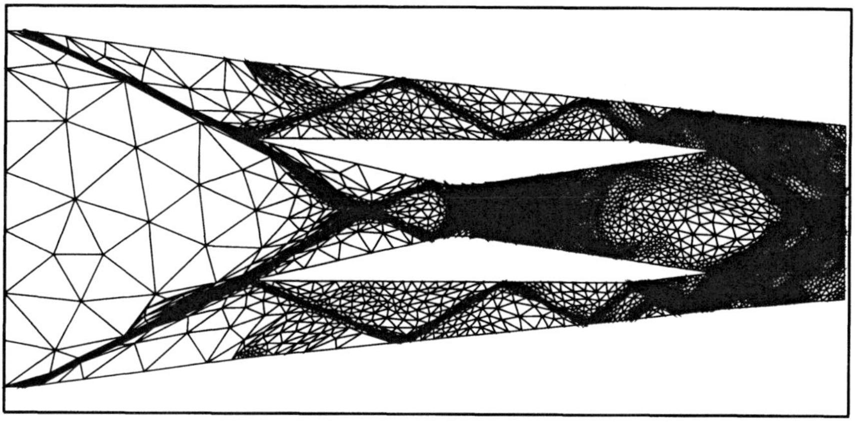

FIG. 3.7.72. Anisotropic mesh adaptation

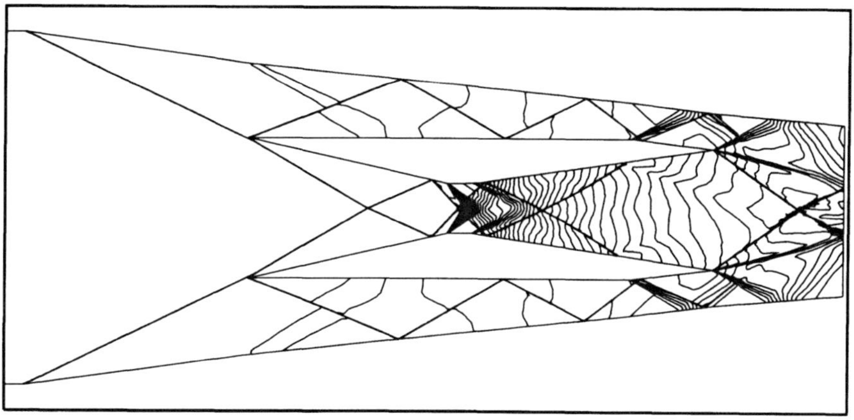

FIG. 3.7.73. Mach number isolines

4

FINITE ELEMENT SOLUTION OF COMPRESSIBLE FLOW

In Chapter 3, our attention was on the numerical solution of inviscid flow described by the Euler equations. We discussed the finite difference method applied to the problem with one space variable and the finite volume schemes for the numerical solution of multidimensional problems. The inviscid model gives physically acceptable results in some cases of internal and external aerodynamics. However, if the bodies immersed in the gas have a complicated structure or the angles of attack of flow past airfoils are large, then the inviscid model can no longer be applied. In order to obtain a correct physical solution of the flow of a real fluid, in many cases of practical interest one must use models taking the viscosity effects into account.

In this chapter we shall be mainly concerned with the numerical simulation of viscous compressible flow. This belongs to the most difficult areas of computational fluid dynamics (CFD) since all the difficulties appearing in CFD occur here: strong nonlinearities and mixed hyperbolic–parabolic (or hyperbolic–elliptic) types of governing equations, the treatment of convection dominating diffusion, shock waves, boundary layers, wakes and their interaction.

The starting point for the numerical solution of viscous gas flow is usually a numerical scheme for the solution of inviscid flow, which is completed by the discretization of viscous terms. In recent years, rapid progress has been made in this area. The finite volume method (FVM) appears to be the most popular in the engineering community. However, from the point of view of higher order schemes, the importance of the finite element method (FEM) and its combinations with finite volume techniques increases rapidly.

In this chapter we shall discuss the following methods for the numerical solution of viscous compressible flow: the streamline diffusion conforming finite element method for barotropic flow as well as for flow of a real heat-conductive gas, combined finite volume–finite element (FV–FE) methods and the discontinuous Galerkin finite element method (DGFEM).

From the point of view of the fact that the viscosity and heat conduction of gases are rather small, the goal is to develop robust schemes which would work also in the limit case of vanishing viscosity and heat conduction. Therefore, some sections will also be devoted to the application of the FEM to the solution of conservation laws and the inviscid Euler equations.

Because of a better understanding of the FE approximations of compressible flow problems, we start with an introductory section explaining the basic principles of the FEM applied to scalar linear elliptic, parabolic and hyperbolic equations. We pay attention to the derivation of FE schemes as well as to the analysis of their stability, convergence and error estimates. Then we approach the extension and applications of the FE techniques to complicated problems of gas dynamics.

4.1 Finite element method – elementary treatment

The finite element method (FEM) is a modern and efficient technique for the numerical solution of partial differential equations. It is based on the so-called variational formulation of the problem under consideration, represented by an integral identity satisfied for suitable test functions, and on the piecewise polynomial approximation of the sought solution. A detailed theoretical treatment and numerous applications of the FEM are the subject of thousands of papers and a number of books. From this extensive literature let us mention a few monographs, such as (Babuška and Strouboulis, 2001), (Brenner and Scott, 1994), (Ciarlet, 1979), (Girault and Raviart, 1979), (Girault and Raviart, 1986), (Glowinski, 1984), (Hinton and Owen, 1977), (Johnson, 1987), (Křížek and Neittaanmäki, 1990), (Pironneau, 1989), (Quarteroni and Valli, 1997), (Schwab, 1998), (Strang and Fix, 1973), (Szabo and Babuška, 1991), (Thomée, 1997), (Ženíšek, 1990), (Zienkiewicz and Morgan, 1983).

There is a widely held opinion that finite element techniques should be mainly used for the solution of problems with large diffusion and solid mechanics problems, whereas the finite volume method is more suitable for problems with small or vanishing diffusion and fluid dynamics problems. However, the FEM scores a considerable success also in CFD.

Concerning the FE solution of incompressible viscous flow (incompressible Navier–Stokes equations), there are the well-known monographs (Temam, 1977), (Girault and Raviart, 1979), (Girault and Raviart, 1986), (Gresho and Sani, 2000), (Turek, 1999). Most specialists prefer to use the so-called conforming finite elements which yield approximate solutions continuous in the whole computational domain. In the numerical solution of incompressible Navier–Stokes equations, sometimes nonconforming finite elements are used. In this case the requirement of the interelement continuity is relaxed to some points on the interelement interfaces.

There have also been attempts to apply the FEM to hyperbolic problems, nonlinear conservation laws and the compressible Euler and Navier–Stokes equations. However, the use of standard conforming finite element techniques, useful for elliptic or parabolic problems with dominating diffusion, leads to the Gibbs phenomenon in the numerical solution, i.e. nonphysical spurious oscillations, undershoots or overshoots (cf. Section 3.2.21.2). In order to avoid this phenomenon, it is necessary to apply special techniques. One possibility is to use the so-called streamline diffusion method based on introducing suitable stabilization terms in the original finite element scheme. Another approach uses completely discontinuous piecewise polynomial approximations and the concept of the numerical flux, which is an important ingredient in the finite volume method. It is called the discontinuous Galerkin method. Both the streamline diffusion method and the discontinuous Galerkin method are suitable for the solution of first order hyperbolic problems or singularly perturbed problems with dominating convection and for the approximation of solutions with discontinuities or steep gradients.

In this section we present the simple explanation of basic principles of the FEM which will be useful for understanding the treatment of the FEM applied to the solution of compressible flow.

4.1.1 *Elliptic problems*

Let us start with a simple elliptic problem to find $u : \overline{\Omega} \to I\!R$ such that

$$\begin{aligned} &\text{a) } -\Delta u = f \quad \text{in } \Omega, \\ &\text{b) } u\Big|_{\Gamma_\mathrm{D}} = u_\mathrm{D}, \\ &\text{c) } \frac{\partial u}{\partial n}\Big|_{\Gamma_\mathrm{N}} = \varphi_\mathrm{N}. \end{aligned} \tag{4.1.1}$$

Here $\Omega \subset I\!R^N$ is a bounded domain with a Lipschitz-continuous boundary $\partial\Omega = \overline{\Gamma}_\mathrm{D} \cup \overline{\Gamma}_\mathrm{N}$, where the disjoint subsets $\Gamma_\mathrm{D}, \Gamma_\mathrm{N} \subset \partial\Omega$ are formed by a finite number of connected sets, open with respect to the topology on $\partial\Omega$, and $\mathrm{meas}_{N-1}(\Gamma_\mathrm{D}) > 0$. (For example, if $N = 2$, then Γ_D and Γ_N are formed by a finite number of open arcs.) The symbol $\partial/\partial n$ here stands for the derivative with respect to the unit outer normal to $\partial\Omega$ and f, u_D and φ_N are given functions. Equation (4.1.1), b) is the *Dirichlet boundary condition*, also called *essential boundary condition*; c) is called the *Neumann* or *natural* boundary condition.

The FEM for the numerical solution of problem (4.1.1) is based on the concepts of a variational formulation and a weak solution.

4.1.1.1 *Weak solution* We shall work here with Sobolev spaces introduced in Section 1.3.3, where the necessary notation is also introduced.

Let us put

$$\mathcal{V} = \left\{ v \in C^\infty(\overline{\Omega});\ \mathrm{supp}\, v \subset \Omega \cup \Gamma_N \right\} \tag{4.1.2}$$

and define the space V as the closure of the set $\mathcal{V}$ in the topology of the space $H^1(\Omega)$:

$$V = \overline{\mathcal{V}}^{H^1(\Omega)}. \tag{4.1.3}$$

Obviously, $H^1_0(\Omega) \subset V \subset H^1(\Omega)$. It can be proven that

$$V = \left\{ v \in H^1(\Omega);\ v|_{\Gamma_\mathrm{D}} = 0 \text{ (in the sense of traces)} \right\}. \tag{4.1.4}$$

It is easy to see that V is a Hilbert space.

Let $u \in C^2(\overline{\Omega})$ be the classical solution of problem (4.1.1). Multiplying equation (4.1.1), a) by any $v \in \mathcal{V}$, integrating over Ω and using Green's theorem, we obtain

$$\int_\Omega f v \, dx = -\int_\Omega (\Delta u) v \, dx \tag{4.1.5}$$

$$= \int_\Omega \nabla u \cdot \nabla v \, dx - \int_{\partial\Omega} \nabla u \cdot \boldsymbol{n} v \, dS.$$

Taking into account that $v|_{\Gamma_D} = 0$, $\nabla u \cdot \boldsymbol{n} = \partial u/\partial n$ and using condition (4.1.1), c), we arrive at the *integral identity*

$$\int_\Omega \nabla u \cdot \nabla v \, dx = \int_\Omega f v \, dx + \int_{\Gamma_N} \varphi_N v \, dS, \quad v \in \mathcal{V}, \tag{4.1.6}$$

which is the basis for introducing the *variational formulation* of problem (4.1.1).

Let us define the forms $a: H^1(\Omega) \times H^1(\Omega) \to I\!R$, L^Ω, L^Γ and $L: H^1(\Omega) \to I\!R$:

$$a(u, v) = \int_\Omega \nabla u \cdot \nabla v \, dx, \tag{4.1.7}$$
$$L^\Omega(v) = \int_\Omega f v \, dx,$$
$$L^\Gamma(v) = \int_{\Gamma_N} \varphi_N v \, dS,$$
$$L(v) = L^\Omega(v) + L^\Gamma(v) \quad \text{for } u, v \in H^1(\Omega).$$

We can show with the aid of the Cauchy inequality that the form a is well defined and that the forms L^Ω and L^Γ are well defined, if, for example,

$$f \in L^2(\Omega) \quad \text{and} \quad \varphi_N \in L^2(\Gamma_N). \tag{4.1.8}$$

Then L^Ω and L are continuous linear functionals on the space $H^1(\Omega)$ and a is a continuous bilinear form on $H^1(\Omega)$. Let us assume that there exists $u^* \in H^1(\Omega)$ such that

$$u^*|_{\Gamma_D} = u_D. \tag{4.1.9}$$

Definition 4.1 *A function u is called a weak solution of problem* (4.1.1), *if the following conditions are satisfied:*

$$a)\ u \in H^1(\Omega), \qquad b)\ u - u^* \in V, \tag{4.1.10}$$
$$c)\ a(u, v) = L(v) \quad \forall v \in V.$$

We speak of the weak (or variational) formulation of problem (4.1.1).

In applications in mechanics, the functions from the space $H^1(\Omega)$, where we seek a week solution, are called *trial functions* and the elements $v \in V$ are called *test functions*.

There is a close relation between the classical and weak solutions:

Theorem 4.2 *Any classical solution $u \in C^2(\overline{\Omega})$ of problem* (4.1.1) *is a weak solution. Conversely, if u is a weak solution of problem* (4.1.1) *satisfying the additional condition $u \in C^2(\overline{\Omega})$, then u is a classical solution.*

Proof The proof can be carried out on the basis of (4.1.2)–(4.1.7) and properties of the forms a, L^Ω and L^Γ. □

Theorem 4.2 establishes the *formal equivalence* of the classical and weak formulations.

Now we turn our attention to the existence and uniqueness of a weak solution:

Theorem 4.3 *Problem* (4.1.10) *has a unique solution. This solution is independent of the choice of the function* $u^* \in H^1(\Omega)$ *satisfying condition* (4.1.9).

Proof We use the well-known *Lax–Milgram lemma* cited in Section 1.4.6. In view of (4.1.10), b), we seek the solution in the form $u = u^* + z$, where $z \in V$. Then (4.1.10) is equivalent to the condition

$$a(z, v) = \langle \varphi, v \rangle, \quad \forall v \in V, \tag{4.1.11}$$

where

$$\varphi \in V^*, \qquad \langle \varphi, v \rangle = L(v) - a(u^*, v), \quad v \in V. \tag{4.1.12}$$

Let us verify the conditions of the Lax–Milgram lemma 1.4.6. We work with the Hilbert space $H = V$, equipped with the norm $\|\cdot\|_{1,\Omega}$. It is obvious that the form $a(z, v)$ is a continuous bilinear form on V. Further, from the assumption that $\text{meas}_{N-1}(\Gamma_D) > 0$, we can use the Friedrichs inequality. Hence, we have

$$a(z, z) = \int_\Omega |\nabla z|^2 \, dx = |z|^2_{1,\Omega} \geq C_F \|z\|^2_{1,\Omega}, \quad z \in V, \tag{4.1.13}$$

where the constant $C_F > 0$ is independent of z. This means that the form $a(z, v)$ is V-elliptic. The right-hand side of (4.1.11) is clearly a continuous linear functional. Hence, the assumptions of the Lax–Milgram lemma are satisfied and problem (4.1.11) as well as (4.1.10) have a unique solution.

Further, let us consider two functions $u_1^*, u_2^* \in H^1(\Omega)$ satisfying (4.1.9) and denote by u_1 and u_2 the corresponding solutions of (4.1.10). Then

$$a(u_i, v) = L(v) \quad \forall v \in V, \ i = 1, 2,$$

and $u_1 - u_2 \in V$. Subtracting the above equations, substituting there $v := u_1 - u_2$ and using (4.1.13), we obtain

$$0 = a(u_1 - u_2, u_1 - u_2) \geq C_F \|u_1 - u_2\|^2_{1,\Omega} \geq 0$$

and thus $u_1 = u_2$ in Ω. □

4.1.2 *Finite element discretization of the elliptic problem*

Now we shall be concerned with the FE approximation of problem (4.1.1). For simplicity we confine our considerations to the 2D case, i.e. $N = 2$. We start from the weak formulation (4.1.10) and proceed in several steps:

I. The bounded domain $\Omega \subset \mathbb{R}^2$ is approximated by a *polygonal* Lipschitz domain Ω_h, i.e. a domain with a boundary $\partial\Omega_h$ consisting of a finite number of simple, closed, piecewise linear curves. (h is a discretization parameter whose meaning will be explained later.) In $\overline{\Omega}_h$ a partition $\mathcal{T}_h$ is constructed consisting of a finite number of closed triangles K with the following properties:

$$\overline{\Omega}_h = \cup_{K\in\mathcal{T}_h} K \tag{4.1.14}$$

and

$$\begin{aligned} &\text{if } K_1, K_2 \in \mathcal{T}_h, K_1 \neq K_2, \quad \text{then either} \quad K_1 \cap K_2 = \emptyset, \\ &\text{or } K_1 \cap K_2 \text{ is a common vertex of } K_1 \text{ and } K_2, \\ &\text{or } K_1 \cap K_2 \ \text{ is a common side of } K_1 \text{ and } K_2. \end{aligned} \tag{4.1.15}$$

Such a partition of $\overline{\Omega}_h$ is called the *triangulation* of Ω_h. For each $K \in \mathcal{T}_h$ we denote by h_K the length of the maximum side of K. Hence, we can write $h_K = \operatorname{diam} K$. The subscript h usually represents the length of the maximum side of all triangles $K \in \mathcal{T}_h$. This means that

$$h = \max_{K\in\mathcal{T}_h} h_K. \tag{4.1.16}$$

Further, we denote by $\sigma_h = \{P_1, \dots, P_{N_h}\}$ the set of all vertices of all $K \in \mathcal{T}_h$. (Cf. Section 3.3.1.1.) We usually suppose that $\sigma_h \cap \partial\Omega_h \subset \partial\Omega, \sigma_h \subset \overline{\Omega}$, $\overline{\Gamma}_D \cap \overline{\Gamma}_N \subset \sigma_h$. (Cf. Section 3.3.)

II. For an integer $p \geq 0$ we denote by P^p the set of all polynomials of degree $\leq p$ depending on the variables x_1 and x_2 and put $P^p(K) = \{\varphi|_K;\ p \in P^p\}$. An approximate solution of the variational problem (4.1.10) will be sought in the finite dimensional space

$$X_h := X_h^{(p)} = \left\{v_h \in C(\overline{\Omega}_h);\ v_h|_K \in P^p(K)\ \forall K \in \mathcal{T}_h\right\}, \tag{4.1.17}$$

called the *finite element space* or the space of *trial functions*. It can be shown that $X_h \subset H^1(\Omega_h)$. (See (Ciarlet, 1979), Theorem 2.1.1.)

Any function $v_h \in X_h$ is determined by its values and the values of some of its derivatives at certain points of triangles $K \in \mathcal{T}_h$. These points and the corresponding values are called *nodes* and *nodal parameters* (or *degrees of freedom*), respectively. More generally, the function v_h is uniquely determined on each $K \in \mathcal{T}_h$ by a set Σ_K of linear functionals. This means that the degrees of freedom are expressed as the values $\sigma(v_h)$ for $\sigma \in \Sigma_K$. We call triangles $K \in \mathcal{T}_h$ *elements* and the triples (K, P^p, Σ_K) are called *finite elements*. Often, also the functions $v_h \in X_h$ are called finite elements.

Example 4.4 a) If $p = 1$, then we speak of *linear finite elements*. It is clear that in this case any function $v_h \in X_h$ is uniquely determined by its values at the vertices of the triangulation. This means that each $K \in \mathcal{T}_h$ contains three nodes

P_i, P_j, P_k, which are its vertices, and the function v_h has three nodal parameters $v_h(P_i), v_h(P_j), v_h(P_k)$ on K.

b) *Quadratic elements* ($p = 2$) are determined by their values at the vertices and the midpoints of sides of all triangles $K \in \mathcal{T}_h$.

c) *Cubic elements* ($p = 3$). The vertices and the centre of gravity of all triangles are chosen for the nodes. The nodal parameters are the values at all nodes and the values of the first order derivatives with respect to x_1 and x_2 at the vertices. Another possibility is to use in each element K with vertices P_i, P_j, P_k as nodes these vertices, the centre of gravity of K and points $P_{\alpha,\beta} = (2P_\alpha + P_\beta)/3, \alpha \neq \beta, \alpha, \beta = 1, 2, 3$. The nodal parameters are values at the nodes.

Further examples can be found in the literature mentioned above. The finite elements, whose nodal parameters are only their values at nodes, are called *Lagrange elements*. If the values of derivatives are also used as nodal parameters, we speak of *Hermite elements*.

Exercise 4.5 Consider the space X_h of piecewise linear finite elements from Example 4.4, a). Show that any $\psi \in X_h$ can be expressed on a triangle $K \in \mathcal{T}_h$ with vertices $P_i = (x_1^i, x_2^i)$, $P_j = (x_1^j, x_2^j)$, $P_k = (x_1^k, x_2^k)$ in the form

$$\psi(x) = \alpha_0 + \alpha_1 x_1 + \alpha_2 x_2, \quad x = (x_1\, x_2) \in K, \tag{4.1.18}$$

where

$$\alpha_0 = \frac{1}{D}\det\begin{pmatrix} x_1^i, & x_2^i, & \psi(P_i) \\ x_1^j, & x_2^j, & \psi(P_j) \\ x_1^k, & x_2^k, & \psi(P_k) \end{pmatrix}, \quad \alpha_1 = -\frac{1}{D}\det\begin{pmatrix} x_2^i, & \psi(P_i), & 1 \\ x_2^j, & \psi(P_j), & 1 \\ x_2^k, & \psi(P_k), & 1 \end{pmatrix}, \tag{4.1.19}$$

$$\alpha_2 = \frac{1}{D}\det\begin{pmatrix} x_1^i, & \psi(P_i), & 1 \\ x_1^j, & \psi(P_j), & 1 \\ x_1^k, & \psi(P_k), & 1 \end{pmatrix}, \quad D = \det\begin{pmatrix} x_1^i, & x_2^i, & 1 \\ x_1^j, & x_2^j, & 1 \\ x_1^k, & x_2^k, & 1 \end{pmatrix} \quad (= 2|K| \neq 0).$$

Hint: Write (4.1.18) for $x := P_i, P_j, P_k$. From this system of equations express the coefficients $\alpha_0, \alpha_1, \alpha_2$ with the aid of the Cramér rule.

Let us consider a finite element space X_h defined by (4.1.17). Let $K \in \mathcal{T}_h$ and $u : K \to \mathbb{R}$ be a function for which the nodal parameters on K make sense. Then we define the *P^p-interpolant* $\pi_K u$ of u as the element of $P^p(K)$ which has the same values of nodal parameters as the function u. If $u : \Omega_h \to \mathbb{R}$ is a function for which all nodal parameters make sense, then we define the *X_h-interpolant* $r_h u \in X_h$ of u by

$$(r_h u)|_K = \pi_K(u|_K), K \in \mathcal{T}_h. \tag{4.1.20}$$

The mapping r_h is called the *X_h-interpolation operator* (or, shortly, interpolation).

For example, if $u \in C(\overline{\Omega}_h)$ and $p = 1$, then $r_h u$ is such an element of the space X_h that $(r_h u)(P_i) = u(P_i)$ for all vertices $P_i \in \sigma_h$. In this case we speak of the *Lagrange piecewise linear interpolation*.

III. Now let us denote the parts of $\partial\Omega_h$ approximating $\overline{\Gamma}_D$ and $\overline{\Gamma}_{\mathrm{N}}$ by $\overline{\Gamma}_{\mathrm{D}h}$ and $\overline{\Gamma}_{\mathrm{N}h}$, respectively (again, $\Gamma_{\mathrm{D}h}$ and $\Gamma_{\mathrm{N}h}$ are open in $\partial\Omega_h$), and put

$$V_h = \{v_h \in X_h;\ v_h|_{\Gamma_{\mathrm{D}h}} = 0\}. \tag{4.1.21}$$

Let u^* be a function satisfying (4.1.9) and let the X_h-interpolation $r_h u^*$ make sense. Then we set $u_h^* = r_h u^*$. Further, we approximate the function $\varphi_{\mathrm{N}} : \overline{\Gamma}_{\mathrm{N}} \to \mathbb{R}$ from the Neumann boundary condition by a function $\varphi_{\mathrm{N}h} : \overline{\Gamma}_{\mathrm{N}h} \to \mathbb{R}$ in a suitable way, assume that $f \in L^2(\Omega \cup \Omega_h)$ and introduce the forms

$$\begin{aligned}
a_h(u_h, v_h) &= \int_{\Omega_h} \nabla u_h \cdot \nabla v_h \, dx, \\
L_h^\Omega(v_h) &= \int_{\Omega_h} f v_h \, dx, \\
L_h^\Gamma(v_h) &= \int_{\Gamma_{\mathrm{N}h}} \varphi_{\mathrm{N}h} v_h \, dS, \\
L_h(v_h) &= L_h^\Omega(v_h) + L_h^\Gamma(v_h), \\
u_h, v_h &\in X_h.
\end{aligned} \tag{4.1.22}$$

Obviously, L_h^Ω, L_h^Γ and L_h are linear forms and a_h is a symmetric bilinear form.

Now we can already introduce a *discrete problem* to (4.1.10).

Definition 4.6 *A function* $u_h : \overline{\Omega}_h \to \mathbb{R}$ *is called the approximate solution of problem* (4.1.10), *if*

$$\begin{aligned}
&a) \quad u_h \in X_h, \qquad b) \quad u_h - u_h^* \in V_h, \\
&c) \ a_h(u_h, v_h) = L_h(v_h), \quad \forall v_h \in V_h.
\end{aligned} \tag{4.1.23}$$

4.1.2.1 *Algebraic system equivalent to the discrete problem* Let $n = n_h$ denote the dimension of the space V_h and let $\{w_i^*\}_{i=1}^n$ be a basis in V_h. Then each function $z_h \in V_h$ can be uniquely expressed as the linear combination

$$z_h = \sum_{j=1}^n z_j w_j^* \tag{4.1.24}$$

with coefficients $z_j \in \mathbb{R}$ and the solution of problem (4.1.23) can be written in the form

$$u_h = u_h^* + \sum_{j=1}^n z_j w_j^*. \tag{4.1.25}$$

Substituting (4.1.25) into (4.1.23), c) and taking into account the obvious fact that this identity is satisfied if and only if it holds for all test functions v_h equal to the basis functions w_i^*, $i = 1, \ldots, n$, we obtain the system of linear algebraic equations

$$\sum_{j=1}^{n} a_{ij} z_j = b_i, \quad i = 1, \ldots, n, \tag{4.1.26}$$

where

$$\begin{aligned} a_{ij} &= a_h(w_j^*, w_i^*), \\ b_i &= L_h(w_i^*) - a_h(u_h^*, w_i^*). \end{aligned} \tag{4.1.27}$$

The symmetry of the bilinear form a_h implies that $a_{ij} = a_{ji}$ for all $i, j = 1, \ldots, n$, which means that the matrix $\mathbb{A} = (a_{ij})_{i,j=1}^n$ is symmetric. Moreover, let us show that $\mathbb{A}$ is positive definite. If $0 \neq \boldsymbol{z} = (z^1, \ldots, z^n)^{\mathrm{T}} \in I\!R^n$, then

$$\boldsymbol{z}^{\mathrm{T}} \mathbb{A} \boldsymbol{z} = \sum_{i,j=1}^{n} a_h(w_j^*, w_i^*) z_i z_j = a_h(z_h, z_h), \tag{4.1.28}$$

where $z_h \in V_h$ is the function defined by (4.1.24). Obviously, $z_h \not\equiv 0$ in $\overline{\Omega}_h$, which together with (4.1.28), (4.1.22) and the Friedrichs inequality (valid for functions $z_h \in V_h$ due to (4.1.21) and the fact that $\mathrm{meas}_1(\Gamma_{\mathrm{D}h}) > 0$) imply that

$$\boldsymbol{z}^{\mathrm{T}} \mathbb{A} \boldsymbol{z} = \int_{\Omega_h} |\nabla z_h|^2 \, dx \geq c \|z_h\|_{1,\Omega_h}^2 > 0. \tag{4.1.29}$$

Here $c > 0$ is a constant independent of z_h. (Under some assumptions on the mesh $\mathcal{T}_h$ it is even possible to show that c is independent of h.)

Hence, the matrix $\mathbb{A}$ is nonsingular and *system* (4.1.26) *possesses a unique solution.* Consequently, the discrete problem (4.1.23) has a unique solution. (This is also a consequence of the Lax–Milgram lemma.)

It obviously follows from the definitions of the spaces X_h and V_h that it is possible to construct bases consisting of functions with small supports. Then the matrix $\mathbb{A}$ of system (4.1.26) has only $O(n)$ nonzero elements. We say that it is *sparse.* If a suitable ordering of nodes is used, then $\mathbb{A}$ is a *band matrix*, which means that there exists a positive integer $d < n$ such that $a_{ij} = 0$, if $|i - j| > d$.

Example 4.7 In the case of linear elements ($p = 1$) it is suitable to use the basis of the space X_h formed by the functions $w_i \in X_h$, $i = 1, \ldots, N_h$ ($= \mathrm{card}\, \sigma_h =$ number of the vertices of $\mathcal{T}_h$), such that

$$w_i(P_j) = \delta_{ij}, \quad i, j = 1, \ldots, N_h. \tag{4.1.30}$$

If the vertices $P_i \in \sigma_h$ are ordered in such a way that $P_1, \ldots, P_{n_h} \in \Omega_h \cup \Gamma_{\mathrm{N}h}$ and $P_{n_h+1}, \ldots, P_{N_h} \in \overline{\Gamma}_{\mathrm{D}h}$, then the functions $w_i^* = w_i$, $i = 1, \ldots, n_h$, form a basis in V_h. Clearly, supp w_i consists of only those triangles $K \in \mathcal{T}_h$ which have a common vertex P_i.

4.1.3 *Convergence of the FEM*

There is an important question: how large is the error of the approximation of the exact solution u by the solution u_h of the discrete problem? Moreover, we are interested in the behaviour of the error, provided the triangulation is consequently refined so that the parameter h tends to zero. For simplicity, we confine our considerations to the case when the *domain* Ω *is polygonal* and we can assume that $\Omega_h = \Omega$, $\Gamma_q Dh = \Gamma_D$, $\Gamma_{\mathrm{N}h} = \Gamma_{\mathrm{N}}$ and $\varphi_{\mathrm{N}h} = \varphi_{\mathrm{N}}$. Further, let $u_{\mathrm{D}} = 0$, so that $u^* = u_h^* = 0$. Obviously, $a_h = a$ and $L_h = L$. Now the continuous problem and the discrete problem become

$$u \in V, \qquad a(u,v) = L(v) \quad \forall v \in V, \tag{4.1.31}$$

and

$$u_h \in V_h, \qquad a(u_h, v_h) = L(v_h) \quad \forall v_h \in V_h. \tag{4.1.32}$$

The process of approximating problem (4.1.31) by the discrete problem (4.1.32) with a finite dimensional subspace $V_h \subset V$ is called the *Galerkin* or *Ritz–Galerkin method.* Since $X_h \subset H^1(\Omega)$ and $V_h \subset V$, we speak of *conforming finite elements.*

In what follows, we shall use the symbol c to denote a positive constant independent of h, K, v_h and u_h, attaining, in general, different values at different places.

We have the following *abstract error estimate* for the Galerkin method.

Theorem 4.8 *Let a space* $V \subset H^1(\Omega)$ *and a form* $a(u,v)$ *satisfy the assumptions of the Lax–Milgram lemma* 1.4.6. *Moreover, let* $V_h \subset V$ *be a finite dimensional subspace and* $L \in V^*$. *Then there exists a constant* $c > 0$ *independent of* u, u_h, L *and* V_h *such that*

$$\|u - u_h\|_{1,\Omega} \le c \inf_{v_h \in V_h} \|u - v_h\|_{1,\Omega}, \tag{4.1.33}$$

where u *and* u_h *are solutions of* (4.1.31) *and* (4.1.32), *respectively.*

Proof Let $v_h \in V_h$. Then $u_h - v_h \in V_h$. By virtue of (4.1.31), (4.1.32) and the inclusion $V_h \subset V$,

$$a(u - u_h, u_h - v_h) = 0.$$

The V-ellipticity and continuity of the form a imply that

$$\begin{aligned}
&\alpha \|u - u_h\|_{1,\Omega}^2 \le a(u - u_h, u - u_h) \\
&= a(u - u_h, u - v_h) \le M \|u - u_h\|_{1,\Omega} \|u - v_h\|_{1,\Omega},
\end{aligned}$$

so that

$$\|u - u_h\|_{1,\Omega} \le \frac{M}{\alpha} \|u - v_h\|_{1,\Omega} \quad \forall v_h \in V_h.$$

This immediately yields (4.1.33). □

The above theorem converts the problem of the error estimate to the investigation of *approximation properties of finite element spaces.* Let us deal with this question in more detail in the case of finite elements from Example 4.4.

By h_K, ρ_K and ϑ_K we denote the length of the largest side of a triangle K, the radius of the largest circle inscribed in K and the magnitude of the smallest angle of K, respectively. Let us set

$$\vartheta_h = \min_{K \in \mathcal{T}_h} \vartheta_K. \tag{4.1.34}$$

The system $\{\mathcal{T}_h\}_{h \in (0,h_0)}$ ($h_0 > 0$) of triangulations is called *regular*, if there exists a constant $c > 0$ such that

$$h_K/\rho_K \le c \quad \forall K \in \mathcal{T}_h, \ \forall h \in (0, h_0). \tag{4.1.35}$$

This condition can be used in general N-dimensional problems. If $N = 2$, then it is equivalent to the following *minimum angle condition*: there exists a constant $\vartheta_0 > 0$ independent of h such that

$$\vartheta_h \ge \vartheta_0 \quad \forall h \in (0, h_0). \tag{4.1.36}$$

If $u \in V$ is sufficiently regular, then $r_h u \in V_h$ and, therefore,

$$\inf_{v_h \in V_h} \|u - v_h\|_{1,\Omega} \le \|u - r_h u\|_{1,\Omega}. \tag{4.1.37}$$

Moreover, for $v \in H^k(\Omega)$ we have

$$\|v\|^2_{k,\Omega} = \sum_{K \in \mathcal{T}_h} \|v\|^2_{k,K}, \tag{4.1.38}$$

which follows from the definition of the norm $\|\cdot\|_{k,\Omega}$ and the properties of the triangulation $\mathcal{T}_h$. These results imply that it is sufficient to find a bound of $\|u - r_h u\|_{1,K}$ for any $K \in \mathcal{T}_h$. To this end, we introduce the *reference triangle* $\hat{K}$ with vertices (0,0), (1,0) and (0,1) and consider a linear (i.e. affine from the point of view of functional analysis) one-to-one mapping

$$F_K(\hat{x}) = \mathbb{B}_K \hat{x} + \mathbf{b}_K \tag{4.1.39}$$

of $\hat{K}$ onto $K \in \mathcal{T}_h$. Here $\mathbb{B}_K$ is a 2×2 nonsingular matrix, $\mathbf{b}_K \in I\!R^2$ and $\hat{x}$ denotes a point of $\hat{K}$. It can be proven that there exist constants $c_1, c_2, c_3 > 0$ independent of $K \in \mathcal{T}_h$ and $h \in (0, h_0)$ such that

$$\begin{aligned} &\|\mathbb{B}_K\| \le c_1 h_K, \qquad \|\mathbb{B}_K^{-1}\| \le c_2/\rho_K, \\ &\pi \rho_K^2 \le \text{meas}(K) \le \frac{\sqrt{3}}{4} h_K^2, \\ &|\text{det}\mathbb{B}_K| = \text{meas}(K)/\text{meas}(\hat{K}) = 2\,\text{meas}(K), \end{aligned} \tag{4.1.40}$$

where $\|\mathbb{B}_K\|$ is the norm of the matrix $\mathbb{B}_K$. (Cf., for example, (Ciarlet, 1979), Section 3.1.)

For $v: K \to \mathbb{R}$ we define the function $\hat{v}: \hat{K} \to \mathbb{R}$ as $\hat{v}(\hat{x}) = v(F(\hat{x}))$, $\hat{x} \in \hat{K}$. By (Ciarlet, 1979), Theorem 3.1.2, $v \in H^k(K)$ if and only if $\hat{v} \in H^k(\hat{K})$ ($k \geq 0$ is an integer) and there exists a constant $\tilde{c} > 0$ such that

$$\begin{aligned} &\text{a)}\ |\hat{v}|_{k,\hat{K}} \leq \tilde{c}\|\mathbb{B}_K\|^k |\det\mathbb{B}_K|^{-1/2} |v|_{k,K}, \\ &\text{b)}\ |v|_{k,K} \leq \tilde{c}\|\mathbb{B}_K^{-1}\|^k |\det\mathbb{B}_K|^{1/2} |\hat{v}|_{k,\hat{K}} \end{aligned} \tag{4.1.41}$$

for every $v \in H^k(K)$. We can find that $v \in P^p(K)$ if and only if $\hat{v} \in P^p(\hat{K})$. In general, if

$$X_h = \{v; v|_K = \hat{v} \circ F_K^{-1}, \hat{v} \in \hat{P}, K \in \mathcal{T}_h\}, \tag{4.1.42}$$

where $\hat{P}$ is a space of functions defined on the reference element $\hat{K}$, we speak of *affine equivalent finite elements*. We require that $P^p(\hat{K}) \subset \hat{P}$, but usually we choose $\hat{P} = P^p(\hat{K})$.

The analysis of error estimates in the FEM is based on the following:

Theorem 4.9 (Bramble–Hilbert lemma) *Let $k, m \geq 0$ be integers such that $k + 1 \geq m$ and let $\hat{\pi}$ be a continuous linear mapping of $H^{k+1}(\hat{K})$ into $H^m(\hat{K})$ satisfying the condition*

$$\hat{\pi}\hat{\varphi} = \hat{\varphi} \quad \textit{for any } \hat{\varphi} \in P^k(\hat{K}). \tag{4.1.43}$$

Then there exists a constant $\hat{c} > 0$ such that

$$|\hat{v} - \hat{\pi}\hat{v}|_{m,\hat{K}} \leq \hat{c}|\hat{v}|_{k+1,\hat{K}}, \quad \hat{v} \in H^{k+1}(\hat{K}). \tag{4.1.44}$$

Proof This is carried out in (Ciarlet, 1979), Theorem 3.1.4, for a more general situation. □

We say that an operator $\hat{\pi}$ is *polynomial preserving*, if it has property (4.1.43). Theorem 4.9 and relations (4.1.40)–(4.1.41) yield the following:

Lemma 4.10 *Let the assumptions of Theorem 4.9 be satisfied and let*

$$\hat{\pi}\hat{v} = \widehat{\pi_K v} \quad \textit{for any } v \in H^{k+1}(K). \tag{4.1.45}$$

Then there exists a constant $c > 0$ such that

$$|v - \pi_K v|_{m,K} \leq c\frac{h_K^{k+1}}{\rho_K^m}|v|_{k+1,K}, \quad v \in H^{k+1}(K). \tag{4.1.46}$$

Let us consider such finite elements that the corresponding P^p-interpolation operator π_K is defined on the space $H^{k+1}(K)$. This is true, for example, for the finite elements from Example 4.4, provided $k = 1$ and $N \leq 3$. The operator π_K is polynomial preserving:

$$\pi_K\varphi = \varphi \quad \text{for any } \varphi \in P^p(K). \tag{4.1.47}$$

Let us define the operator $\hat{\pi}$ by relation (4.1.45). In view of (4.1.47) and the fact that F is a linear mapping, $\hat{\pi}$ is also polynomial preserving, i.e. $\hat{\pi}$ satisfies

(4.1.43). Hence, for $p \ge k \ge 1$ the assumptions of Lemma 4.10 are satisfied. This means, for example, that estimate (4.1.46) is valid with $p = 1$, $p = 2$ and $p = 3$ for the finite elements from Example 4.4, a), b) and c), respectively. As an important consequence we obtain:

Lemma 4.11 *Let $\{\mathcal{T}_h\}_{h\in(0,h_0)}$ be a regular system of triangulations of the domain Ω and let $p \ge k \ge 1$. Then there exists a constant $c > 0$ such that the X_h-interpolation operator r_h defined by* (4.1.20) *satisfies*

$$
\begin{aligned}
|v - r_h v|_{1,K} &\le c h_K^k |v|_{k+1,K}, \\
\|v - r_h v\|_{0,K} &\le c h_K^{k+1} |v|_{k+1,K}, \\
v \in H^{k+1}(K),\ & K \in \mathcal{T}_h,\ h \in (0,h_0).
\end{aligned}
\tag{4.1.48}
$$

Proof This follows from estimate (4.1.46) with $m = 0$ or $m = 1$ and assumption (4.1.35). □

Corollary 4.12 *Under the same assumptions as in Lemma 4.11, we have*

$$
\begin{aligned}
|v - r_h v|_{1,\Omega} &\le c h^k |v|_{k+1,\Omega}, \\
\|v - r_h v\|_{0,\Omega} &\le c h^{k+1} |v|_{k+1,\Omega}, \quad v \in H^{k+1}(\Omega),\ h \in (0,h_0).
\end{aligned}
\tag{4.1.49}
$$

Proof The proof is a consequence of (4.1.48) and (4.1.38). □

Remark 4.13 The above approximation results can be extended to more general Sobolev spaces $W^{k,\alpha}$. That is, if $\{\mathcal{T}_h\}_{h\in(0,h_0)}$ is a regular system of triangulations, $p \ge k \ge 1, 0 \le m \le k+1, \alpha \in [1,\infty]$, then there exists a constant $c > 0$ such that

$$
|v - r_h v|_{m,\alpha,K} \le c \frac{h_K^{k+1}}{\rho_K^m} |v|_{k+1,\alpha,K}, \quad v \in W^{k+1,\alpha}(K), \quad K \in \mathcal{T}_h, \quad h \in (0,h_0).
\tag{4.1.50}
$$

This implies that

$$
|v - r_h v|_{m,\alpha,\Omega} \le c h^{k+1-m} |v|_{k+1,\alpha,\Omega}, \quad v \in W^{k+1,\alpha}(\Omega), \quad h \in (0,h_0).
\tag{4.1.51}
$$

See (Ciarlet, 1979), Theorem 3.1.5.

Theorem 4.8 and Corollary 4.12 imply the following fundamental result:

Theorem 4.14 (Error estimate of the FEM) *Let $\{\mathcal{T}_h\}_{h\in(0,h_0)}$ be a regular system of triangulations of the domain Ω and $p \ge k \ge 1$. Let $u \in H^{k+1}(\Omega)$ and $u_h \in X_h$ be the solutions of problems* (4.1.31) *and* (4.1.32), *respectively. Then there exists a constant $c > 0$ such that*

$$
\|u - u_h\|_{1,\Omega} \le c h^k |u|_{k+1,\Omega}, \quad h \in (0,h_0).
\tag{4.1.52}
$$

Proof Since $r_h u \in V_h$ for $u \in H^{k+1}(\Omega) \cap V$, we find from (4.1.33) that

$$
\|u - u_h\|_{1,\Omega} \le c \|u - r_h u\|_{1,\Omega}.
$$

Now it is sufficient to use (4.1.49). □

The verification of the assumption that $u \in H^{k+1}(\Omega)$ belongs to the *theory of regularity* of solutions of partial differential equations. This assumption is unrealistic in a number of problems and the only information available is the fact that $u \in H^1(\Omega)$. Then we obtain:

Theorem 4.15 (On the convergence of the FEM) *If* $\{\mathcal{T}_h\}_{h\in(0,h_0)}$ *is a regular system of triangulations of the domain* Ω, *then*

$$\lim_{h\to 0} u_h = u \quad \text{in } H^1(\Omega).$$

Proof Let $\varepsilon > 0$. Since $u \in V$ and the set $\mathcal{V}$ (defined by (4.1.2)) is *dense* in V, there exists $v \in \mathcal{V}$ such that

$$\|u - v\|_{1,\Omega} < \frac{\varepsilon}{2c}, \tag{4.1.53}$$

where $c > 0$ is the constant from (4.1.33). Obviously, the infinitely differentiable function v satisfies estimate (4.1.49) for $k = 1$, which implies the existence of $h_\varepsilon \in (0, h_0)$ such that

$$\|v - r_h v\|_{1,\Omega} < \frac{\varepsilon}{2c} \quad \forall h \in (0, h_\varepsilon). \tag{4.1.54}$$

Since $r_h v \in V_h$, (4.1.33), (4.1.53) and (4.1.54) imply the inequalities

$$\begin{aligned} &\|u - u_h\|_{1,\Omega} \le c\|u - r_h v\|_{1,\Omega} \\ &\le c\,(\|u - v\|_{1,\Omega} + \|v - r_h v\|_{1,\Omega}) < \varepsilon \quad \forall h \in (0, h_\varepsilon), \end{aligned}$$

which proves the desired result. □

Remark 4.16 Under the same assumptions as in Theorem 4.14 and a sufficient regularity of problem (4.1.31), with the duality technique by Aubin–Nitsche, it is possible to prove the optimal error estimate of the FEM in the $L^2(\Omega)$-norm:

$$\|u - u_h\|_{0,\Omega} \le ch^{k+1}|u|_{k+1,\Omega}. \tag{4.1.55}$$

See (Ciarlet, 1979), Section 3.2. In (Ciarlet, 1979), Section 3.3, the reader will find the error estimates in the L^∞-norm:

$$\begin{aligned} \|u - u_h\|_{0,\infty,\Omega} &\le ch^2|\ln h||u|_{2,\infty,\Omega}, \\ |u - u_h|_{1,\infty,\Omega} &\le ch|\ln h||u|_{2,\infty,\Omega}, \quad h \in (0, h_0). \end{aligned} \tag{4.1.56}$$

We see from the above results that the accuracy of the finite element method depends on the degree p of the polynomial approximation and the regularity of the exact solution represented by the assumption that $u \in H^{k+1}(\Omega)$.

4.1.4 *Several additional remarks*

a) According to G. Strang ((Strang, 1972)), so-called finite element *variational crimes* are committed in the following cases:

- a domain Ω with a *curved boundary is approximated* by a polygonal one (as was described at the beginning of Section 4.1.2),
- integrals appearing in the definitions of the forms a_h, L_h^Ω and L_h^Γ are approximated with the use of *numerical integration*,
- we admit that $X_h \not\subset H^1(\Omega)$ and/or $V_h \not\subset V$. We speak of an *external approximation* of Sobolev spaces. This variational crime may be committed due to the approximation of a general domain by a polygonal one. Another possibility is the application of *nonconforming finite elements*, for which the continuity requirement in $\overline{\Omega}_h$ is relaxed and thus $X_h \not\subset H^1(\Omega_h)$.

To increase the accuracy of the FEM in the case of nonpolygonal domains, *curved isoparametric finite elements* are often used. These elements are based on the approximation of curved parts of the boundary by the quadratic or cubic interpolation and on the use of a suitable modification of basis functions on elements adjacent to the boundary. Details are explained, for example, in (Ciarlet, 1979) or (Zienkiewicz and Morgan, 1983).

b) Sometimes *quadrilateral* (quadrangular) finite element meshes (with possibly *curved isoparametric* quadrilateral elements near the boundary) are used. (See, for example, (Ciarlet, 1979), (Zienkiewicz and Morgan, 1983).) It is relatively rare but quite correct to combine triangles with quadrilaterals.

c) Further, let us mention the fact, important from a practical point of view, that the *minimum angle condition* (4.1.36) is not necessary for the convergence of the FEM. For example, in (Křížek, 1991) estimate (4.1.49) with $k = 1$ for the piecewise linear finite elements was proven under the *maximum angle condition*: there exists a constant $\vartheta_{\max} < \pi$ such that for every $h \in (0, h_0)$ all angles of all triangles $K \in \mathcal{T}_h$ are bounded from above by $\vartheta_{\max}$. This allows us to use very 'thin' triangular elements with one angle arbitrarily small, which is necessary for the construction of suitable finite element meshes in domains having very 'narrow' parts, such as a slot between close profiles or boundary layers.

d) Modern computers allow us to apply the FEM also to the numerical solution of *three-dimensional problems*. The approximation is carried out similarly as described above with the use of a partition of the domain into *tetrahedra*. In this case the domain Ω is approximated by a polyhedral domain Ω_h, whose closure is written as the union of tetrahedral elements $K \in \mathcal{T}_h$ satisfying the following properties:

$$
\begin{aligned}
&\text{if } K_i, K_j \in \mathcal{T}_h, K_i \neq K_j, \text{ then either } K_i \cap K_j = \emptyset, \\
&\text{or } K_i \cap K_j \text{ is a common vertex of } K_i \text{ and } K_j, \\
&\text{or } K_i \cap K_j \text{ is a common edge of } K_i \text{ and } K_j,
\end{aligned}
\tag{4.1.57}
$$

or $K_i \cap K_j$ is a common face of K_i and K_j.

In this case piecewise linear finite elements ($p = 1$) are determined by their values at the vertices of tetrahedra $K \in \mathcal{T}_h$. The piecewise quadratic finite elements ($p = 2$) are given by their values at the vertices and midpoints of edges. Often hexahedral elements are used. In order to improve the accuracy, elements with curved sides are used with isoparametric approximations.

e) From the results presented in this section we see that the *refinement of the finite element mesh* increases the accuracy of the approximate solution. It turns out, however, that it is not necessary to carry out the refinement of the whole computational domain, but it is sufficient to do it in some parts of the domain only. This idea is used in *adaptive methods*, based on the automatic refinement of the grid in the parts of the domain where the accuracy should be increased.

In Section 3.6 we have discussed several mesh adaptation techniques based on various refinement indicators. Most of these methods were derived in a more or less heuristic way (often on the basis of a physical ground). Therefore, we speak of *ad hoc criteria*. They indicate regions for the mesh refinement usually in a reliable way, can be applied within the finite volume as well as finite element methods, but do not give qualitative information about their impact on the accuracy of the approximate solution. This can be achieved on the basis of the so-called *a posteriori error estimates*, which will be briefly discussed in Section 4.1.12.

f) In this section we have been concerned with the FE strategy for the solution of elliptic problems using a piecewise polynomial approximation with a fixed degree p. The increases of accuracy and the convergence are achieved by diminishing the mesh size $h \to 0+$. We speak of the *h-version* of the FEM. Another possibility is to use a fixed mesh $\mathcal{T}_h$ and to increase the polynomial degree p. This approach is called the *p-version* of the FEM. If the exact solution is sufficiently regular, then the p-version is more efficient than the h-version. That is, for analytical u, the p-version gives approximate solutions converging to u *exponentially* with respect to the number of degrees of freedom, as $p \to +\infty$. The combination of both h- and p-versions gives the *hp-version*. This can be extremely efficient for problems with solutions having singularities at boundary points (e.g., corners and edges). It is applied in such a way that in the vicinity of singular boundary points the mesh is refined according to the law of geometric progression towards the boundary (keeping the polynomial degree fixed), whereas in the area of high regularity of u the degree p is increased. In this way, the exponential convergence can again be achieved even for solutions irregular at boundary points. For details we refer the reader to the monograph (Schwab, 1998). In CFD the p- and hp-versions of the FEM have not yet domesticated.

4.1.5 *Parabolic problems*

Now we shall address the finite element approximation of a simple parabolic heat conduction problem. Let $\Omega \subset \mathbb{R}^2$ be a bounded domain with Lipschitz-continuous boundary. For simplicity let us assume that Ω is polygonal. Let $T > 0$ and $Q_T = \Omega \times (0, T)$. Our goal is to find $u : Q_T \to I\!R$ such that

$$\begin{aligned} &\text{a)} \quad \frac{\partial u}{\partial t} = \nu \Delta u + f \quad \text{in } Q_T, \\ &\text{b)} \quad u|_{\partial\Omega \times (0,T)} = 0, \\ &\text{c)} \quad u(x, 0) = u^0(x), \quad x \in \Omega. \end{aligned} \tag{4.1.58}$$

Here $u = u(x,t)$, $f = f(x,t)$, $x \in \Omega$, $t \in (0,T)$ and we assume that $\nu > 0$ is constant. It is called a *diffusion coefficient* and represents, for example, the heat conduction coefficient.

Similarly as in the FE approximation of an elliptic problem, the application of the FEM to problem (4.1.58) is based on a suitable integral identity. We use the notation $V = H_0^1(\Omega), \mathcal{V} := C_0^\infty(\Omega)$. The space $\mathcal{V}$ is dense in V. Let $u \in C^2(\overline{Q}_T)$ be a classical solution. Multiplying equation (4.1.58), a) by any $v \in \mathcal{V}$, integrating over Ω and using Green's theorem, we obtain the relation

$$\int_\Omega \frac{\partial u(x,t)}{\partial t} v(x)\, dx + \int_\Omega \nu \nabla u(x,t) \cdot \nabla v(x)\, dx = \int_\Omega f(x,t) v(x)\, dx. \tag{4.1.59}$$

Under the notation

$$\begin{aligned} (u, v) &= \int_\Omega uv\, dx, \\ a(u, v) &= \nu \int_\Omega \nabla u \cdot \nabla v\, dx, \end{aligned} \tag{4.1.60}$$

(4.1.59) can be rewritten as

$$\left(\frac{\partial u(t)}{\partial t}, v \right) + a(u(t), v) = (f(t), v), \quad v \in \mathcal{V}. \tag{4.1.61}$$

(Note that $u(t) : \Omega \to I\!R$, $(u(t))(x) = u(x,t)$ for $x \in \Omega$.) Interchanging the integration over Ω with the differentiation with respect to t, we obtain from (4.1.61)

$$\frac{d}{dt}(u(t), v) + a(u(t), v) = (f(t), v), \quad v \in \mathcal{V}, \tag{4.1.62}$$

which is the basis of the definition of a weak solution to problem (4.1.58).

Definition 4.17 *Let $f \in L^2(Q_T)$, $u^0 \in L^2(\Omega)$. We call a function $u = u(x,t)$ a weak solution of problem* (4.1.58), *if it satisfies the following conditions:*

$$\begin{aligned} &a) \quad u \in L^2(0, T; V), \\ &b) \quad \frac{d}{dt}(u(t), v) + a(u(t), v) = (f(t), v) \quad \forall v \in V \\ &\qquad \textit{in the sense of distributions on } (0, T), \\ &c) \quad u(0) = u^0. \end{aligned} \tag{4.1.63}$$

Condition (4.1.63), b) means that

$$-\int_0^T (u(t), v)\vartheta'(t)\,dt + \int_0^T a(u(t), v)\vartheta(t)\,dt$$
$$= \int_0^T (f(t), v)\vartheta(t)\,dt \quad \forall v \in V,\ \forall \vartheta \in C_0^\infty(0, T).$$

It is possible to prove the existence and uniqueness of a solution u to problem (4.1.63). Moreover, one can show that this solution possesses the generalized time derivative $\partial u/\partial t \in L^2(0, T; V^*)$ and that $u \in C([0, T]; L^2(\Omega))$. This indicates that the initial condition (4.1.63), c) makes sense. (Cf., for example, (Lions, 1969), (Rektorys, 1982).) Moreover, if $\partial u/\partial t \in L^2(Q_T)$, then (4.1.63), b) can be written in the form (4.1.61), which is satisfied for all $v \in V$ and a.a. $t \in (0, T)$. (For the definitions of the spaces used above, see Section 1.3.4.)

4.1.6 *Finite element discretization of the parabolic problem*

Let $\mathcal{T}_h$ be a triangulation of Ω introduced in Section 4.1.2 and let X_h be defined by (4.1.17), where we set $\Omega_h := \Omega$. By V_h we denote the finite dimensional subspace of V:

$$V_h = \{v_h \in X_h; v_h|_{\partial\Omega} = 0\}. \tag{4.1.64}$$

Now we can introduce the *discrete problem.*

Definition 4.18 *We define the finite element approximate solution of the continuous problem* (4.1.58) *as a function* $u_h : \overline{Q}_T \to \mathbb{R}$ *satisfying the following conditions:*

$$\begin{aligned} &a)\ u_h \in C^1([0, T]; V_h), \\ &b)\ \left(\frac{\partial u_h(t)}{\partial t}, v_h\right) + a(u_h(t), v_h) = (f(t), v_h) \\ &\quad \forall v_h \in V_h,\ \forall t \in (0, T), \\ &c)\ u_h(0) = u_h^0, \end{aligned} \tag{4.1.65}$$

where u_h^0 *is a* V_h*-approximation of* u^0*.*

The *initial condition* u_h^0 from the discrete problem can be defined, for example, as the L^2*-projection* of u^0 on V_h:

$$u_h^0 \in V_h, \quad (u_h^0, v_h) = (u^0, v_h) \quad \forall v_h \in V_h. \tag{4.1.66}$$

If u^0 is sufficiently regular, u_h^0 can also be defined as an X_h-interpolation of u^0.

Let us notice that the discrete problem (4.1.65) has been obtained by the use of the discretization with respect to the space variable $x \in \Omega$. We speak of the *semidiscretization in space*, also called the *method of lines* (cf. Section 3.2.2). The full space-time discretization will be introduced later.

4.1.6.1 *System of ordinary differential equations equivalent to the discrete problem* Similarly as in Section 4.1.2.1 we denote by $\{w_i^*\}_{i=1}^n$ the basis of the space V_h, where $n = n_h$ is the dimension of V_h. Then we have

$$u_h(t) = \sum_{j=1}^{n} \xi_j(t) w_j^*, \quad t \in [0,T], \tag{4.1.67}$$
$$u_h^0 = \sum_{j=1}^{n} \xi_j^0 w_j^*.$$

The substitution of (4.1.67) into (4.1.65), b), where we set $v_h := w_i^*$ for $i = 1, \dots, n$, yields the following initial value problem for a system of ordinary differential equations written in the form

$$\mathbb{B}\dot{\xi} + \mathbb{A}\xi = g \quad \text{in } (0,T), \tag{4.1.68}$$
$$\xi(0) = \xi^0,$$

equivalent to (4.1.65). We use the following *notation:* $\xi = \xi(t) : [0,T] \to I\!R^n$, $\xi = (\xi_1, \dots, \xi_n)^{\mathrm{T}}$, $\dot{\xi} = (\dot{\xi}_1, \dots, \dot{\xi}_n)^{\mathrm{T}}$, $\dot{\xi}_i = d\xi_i/dt$, $\mathbb{B} = (b_{ij})_{i,j=1}^n$, $b_{ij} = (w_j^*, w_i^*)$, $\mathbb{A} = (a_{ij})_{i,j=1}^n$, $a_{ij} = a(w_j^*, w_i^*)$, $g = g(t) : (0,T) \to I\!R^n$, $g_i(t) = (f(t), w_i^*)$. If

$$f \in C([0,T]; L^2(\Omega)), \tag{4.1.69}$$

then $g_i \in C([0,T])$ for $i = 1, \dots, n$.

Exercise 4.19 Prove that $\mathbb{A}$ and $\mathbb{B}$ are symmetric positive definite matrices and that under assumption (4.1.69), problem (4.1.68) has a unique solution in the interval $[0,T]$.

4.1.6.2 *Full space-time discretization* In practical computations, the system (4.1.68), which can be rewritten in the form (3.2.18), is solved by a suitable method for the solution of ordinary differential equations. We can use, for instance, the *Runge–Kutta methods* from Section 3.2.2.

Another possibility is to use the so-called ϑ-scheme. Let us consider a partition $0 = t_0 < t_1 < t_2 < \dots$ of the time interval and put $\tau_k = t_{k+1} - t_k$ (time step). Then, using in (4.1.65), b) the approximations $u_h^k \approx u_h(t_k)$ and

$$\frac{\partial u_h}{\partial t}(t_k) \approx \frac{u_h^{k+1} - u_h^k}{\tau_k}, \tag{4.1.70}$$
$$\frac{\partial u_h}{\partial t}(t_{k+1}) \approx \frac{u_h^{k+1} - u_h^k}{\tau_k},$$

we obtain

$$\frac{1}{\tau_k}(u_h^{k+1} - u_h^k, v_h) + \vartheta a(u_h^{k+1}, v_h) + (1-\vartheta) a(u_h^k, v_h) \tag{4.1.71}$$

$$= \vartheta(f(t_{k+1}), v_h) + (1 - \vartheta)(f(t_k), v_h), \quad v_h \in V_h, \ k = 0, 1, \dots .$$

Usually one of the following cases is used:

$$\vartheta = 0\text{: Euler forward scheme,}$$
$$\vartheta = 1\text{: Euler backward scheme,}$$
$$\vartheta = \frac{1}{2}\text{: Crank–Nicolson scheme.}$$

4.1.7 *Stability and convergence*

Now we shall pay attention to the derivation of error estimates and investigate the stability of the FE schemes for the numerical solution of parabolic problems.

4.1.7.1 *Convergence of the method of lines* First we shall be concerned with error estimates for the discrete problem (4.1.65) in the special case $p = 1$ (see (4.1.17)), i.e. for piecewise linear finite elements. To this end, we introduce some assumptions and auxiliary results.

Let $\{\mathcal{T}_h\}_{h\in(0,h_0)}$ be a regular system of triangulations of the domain Ω. By Π we denote a suitable V_h-interpolation defined on $H^{k+1}(\Omega) \cap V$. If $k = 1$, then we can put

$$\Pi = r_h = \text{the Lagrange interpolation defined by (4.1.20).}$$

The Lagrange interpolation has the approximation properties specified in Corollary 4.12. That is, for $p = k = 1$ we have

$$\begin{aligned} &\|v - r_h v\|_{0,\Omega} \le ch^2 |v|_{2,\Omega}, \\ &|v - r_h v|_{1,\Omega} \le ch|v|_{2,\Omega}, \\ &\qquad v \in H^2(\Omega), \ h \in (0, h_0). \end{aligned} \tag{4.1.72}$$

If $k = 0$, then we can use the operator $\Pi = \Pi_C : V \to V_h$, called *Clément's interpolation*, defined in detail in (Ciarlet, 1979), Exercise 3.2.3. Its approximation properties read

$$\begin{aligned} &\|v - \Pi_C v\|_{0,\Omega} \le ch^m |v|_{m,\Omega}, \quad v \in H^m(\Omega) \cap V, \quad m = 0, 1, 2, \\ &|v - \Pi_C v|_{1,\Omega} \le ch|v|_{2,\Omega}, \quad v \in H^2(\Omega) \cap V, \\ &\qquad h \in (0, h_0). \end{aligned} \tag{4.1.73}$$

Let us assume that the weak solution of problem (4.1.58) is sufficiently regular:

$$u \in L^\infty(0, T; H^2(\Omega)), \quad \frac{\partial u}{\partial t} \in L^2(0, T; H^1(\Omega)). \tag{4.1.74}$$

This implies that

$$u \in C([0, T]; H^1(\Omega)) \tag{4.1.75}$$

and u satisfies identity (4.1.61) for all $v \in V$.

We set

$$e_h = u_h - u, \qquad (4.1.76)$$
$$\eta = \Pi u - u \in V, \qquad \xi = u_h - \Pi u \in V_h,$$

where Π is a V_h-interpolation operator (either r_h or Π_C). Then the error e_h can be written as $e_h = \xi + \eta$. For all $t \in [0, T]$ we have $\eta(t) \in V$, $\xi(t) \in V_h$. Subtracting (4.1.61) from (4.1.65), b) and using the test function $v = v_h = \xi(t)$ for any fixed $t \in (0, T)$, we obtain the relation

$$\left(\frac{\partial \xi}{\partial t}, \xi\right) + a(\xi, \xi) = -\left(\frac{\partial \eta}{\partial t}, \xi\right) - a(\eta, \xi) \text{ a.e. in } (0, T). \qquad (4.1.77)$$

(For simplicity we do not write the argument t here.) Using the relations

$$\left(\frac{\partial \xi(t)}{\partial t}, \xi(t)\right) = \frac{1}{2}\frac{d}{dt}\|\xi(t)\|_{0,\Omega}^2 \qquad (4.1.78)$$

and

$$a(\xi, \xi) = \nu|\xi|_{1,\Omega}^2, \qquad (4.1.79)$$

with the aid of the Cauchy inequality and the Friedrichs inequality, from (4.1.77) we obtain

$$\frac{1}{2}\frac{d}{dt}\|\xi\|_{0,\Omega}^2 + \nu|\xi|_{1,\Omega}^2 \le \left\|\frac{\partial \eta}{\partial t}\right\|_{0,\Omega} \|\xi\|_{0,\Omega} + \nu|\eta|_{1,\Omega}|\xi|_{1,\Omega}$$
$$\le \left(C_F \left\|\frac{\partial \eta}{\partial t}\right\|_{0,\Omega} + \nu|\eta|_{1,\Omega}\right)|\xi|_{1,\Omega}. \qquad (4.1.80)$$

Now the application of the Young inequality written in the form $2ab \le \nu a^2/2 + 2b^2/\nu$ and $2\nu ab \le \nu a^2/2 + 2\nu b^2$ yields the estimate

$$\frac{d}{dt}\|\xi(\vartheta)\|_{0,\Omega}^2 + \nu|\xi(\vartheta)|_{1,\Omega}^2 \le 2\left(\nu^{-1}C_F^2 \left\|\frac{\partial \eta(\vartheta)}{\partial t}\right\|_{0,\Omega}^2 + \nu|\eta(\vartheta)|_{1,\Omega}^2\right), \qquad (4.1.81)$$

for a.a. $\vartheta \in (0, T)$.

Finally, the integration of (4.1.81) with respect to ϑ from 0 to any $t \in (0, T]$ gives

$$\max_{t\in[0,T]} \|\xi(t)\|_{0,\Omega}^2 + \nu \int_0^T |\xi(\vartheta)|_{1,\Omega}^2 d\vartheta$$
$$\le 4\left(\nu^{-1}C_F^2 \int_0^T \left\|\frac{\partial \eta(\vartheta)}{\partial t}\right\|_{0,\Omega}^2 d\vartheta + \nu\int_0^T |\eta(\vartheta)|_{1,\Omega}^2 d\vartheta\right) + 2\|\xi(0)\|_{0,\Omega}^2. \qquad (4.1.82)$$

In what follows we shall set $\Pi := \Pi_C$. Then, in view of (4.1.74), (4.1.76) and (4.1.73),

$$\left\|\frac{\partial \eta}{\partial t}(\vartheta)\right\|_{0,\Omega}^2 \le ch^2 \left|\frac{\partial u}{\partial t}(\vartheta)\right|_{1,\Omega}^2, \qquad |\eta(\vartheta)|_{1,\Omega}^2 \le ch^2|u(\vartheta)|_{2,\Omega}^2$$

$$\text{for a.a. } \vartheta \in (0,T), \qquad (4.1.83)$$
$$\|\eta(\vartheta)\|_{0,\Omega}^2 \le ch^2 |u(\vartheta)|_{1,\Omega}^2 \quad \text{for all } \vartheta \in [0,T].$$

By integration we get

$$\int_0^T \left\| \frac{\partial \eta}{\partial t}(\vartheta) \right\|_{0,\Omega}^2 d\vartheta \le ch^2 \left\| \frac{\partial u}{\partial t} \right\|_{L^2(0,T;H^1(\Omega))}^2, \qquad (4.1.84)$$
$$\int_0^T |\eta(\vartheta)|_{1,\Omega}^2 d\vartheta \le ch^2 \|u\|_{L^2(0,T;H^2(\Omega))}^2.$$

Moreover,

$$\max_{t\in[0,T]} \|\eta(t)\|_{0,\Omega}^2 \le ch^2 \|u\|_{C([0,T];H^1(\Omega))}^2.$$

For simplicity let us set $u_h^0 := \Pi_C u^0$. Then $\xi(0) = 0$.

Now, since (4.1.76) implies that

$$\|e_h\|_{0,\Omega}^2 \le 2\left(\|\xi\|_{0,\Omega}^2 + \|\eta\|_{0,\Omega}^2\right),$$
$$|e_h|_{1,\Omega}^2 \le 2\left(|\xi|_{1,\Omega}^2 + |\eta|_{1,\Omega}^2\right),$$

from (4.1.82), (4.1.83) and (4.1.84) we find that

$$\max_{t\in[0,T]} \|e_h(t)\|_{0,\Omega}^2 + \nu \int_0^T |e_h(\vartheta)|_{1,\Omega}^2 d\vartheta \qquad (4.1.85)$$
$$\le ch^2 \left(\nu \|u\|_{L^2(0,T;H^2(\Omega))}^2 + \nu^{-1} \left\| \frac{\partial u}{\partial t} \right\|_{L^2(0,T;H^1(\Omega))}^2 + \|u\|_{C([0,T];H^1(\Omega))}^2 \right).$$

If we take the square root of (4.1.85), we see that we obtain the error estimate in the norms of the spaces $L^\infty(0,T;L^2(\Omega))$ and $L^2(0,T;H^1(\Omega))$.

Theorem 4.20 *Let $\{\mathcal{T}_h\}_{h\in(0,h_0)}$ be a regular system of triangulations of the domain Ω, let the exact solution u of problem* (4.1.63) *satisfy the regularity assumptions* (4.1.74) *and let $u_h^0 = \Pi_C u^0$. Then the error $e_h = u_h - u$ of the method of lines* (4.1.65) *with piecewise linear finite elements satisfies estimate* (4.1.85).

Remark 4.21 Note that estimate (4.1.85) is not optimal with respect to the norm $\|\cdot\|_{L^\infty(0,T;L^2(\Omega))}$. Using the duality technique by Aubin–Nitsche (see, for example, (Thomée, 1997), Chapter 1), if $\partial u/\partial t \in L^2(0,T;H^2(\Omega))$, it is possible to obtain the estimate

$$\|e_h\|_{L^\infty(0,T;L^2(\Omega))} \le ch^2. \qquad (4.1.86)$$

In practical computations one must use, of course, a fully discrete problem. For the discrete problems from Section 4.1.6.2 it is possible to prove discrete versions of estimate (4.1.85). We shall not deal with this topic, but we shall pay attention to the problem of stability of ϑ-schemes (4.1.71) with $\vartheta = 0$ and $\vartheta = 1$.

Exercise 4.22 Adapt the error estimates to the case when we set $\Pi := r_h$.
Hint: Use (4.1.72) and consider a stronger regularity of u:

$$u \in L^\infty(0,T;H^2(\Omega)),\ \frac{\partial u}{\partial t} \in L^2(0,T;H^2(\Omega)).$$

4.1.7.2 *Stability of the Euler backward scheme* Similarly as in Section 3.2.6, under the stability of a numerical scheme we understand the boundedness of the approximate solution, provided the data are bounded.

First let us consider the Euler backward method (4.1.71) with $\vartheta = 1$. For simplicity we use a constant time step τ, i.e. $\tau_k = \tau$ for all $k \geq 0$. This scheme can be written in the form

$$(u_h^{k+1} - u_h^k, v_h) + \tau a(u_h^{k+1}, v_h) = \tau(f(t_{k+1}), v_h), \qquad \forall v_h \in V_h,\ k = 0, 1, \dots. \tag{4.1.87}$$

Substituting here $v_h := u_h^{k+1}$ and using the relation

$$(u - v, u) = \frac{1}{2}\left(\|u\|_{0,\Omega}^2 - \|v\|_{0,\Omega}^2 + \|u - v\|_{0,\Omega}^2\right), \qquad u, v \in L^2(\Omega), \tag{4.1.88}$$

we obtain

$$\begin{aligned} &\|u_h^{k+1}\|_{0,\Omega}^2 - \|u_h^k\|_{0,\Omega}^2 + \|u_h^{k+1} - u_h^k\|_{0,\Omega}^2 + 2\tau a(u_h^{k+1}, u_h^{k+1}) \\ &= 2\tau(f(t_{k+1}), u_h^{k+1}), \\ &\qquad k = 0, 1, \dots. \end{aligned} \tag{4.1.89}$$

This, the Cauchy, Friedrichs and Young inequalities imply that

$$\begin{aligned} &\|u_h^{k+1}\|_{0,\Omega}^2 - \|u_h^k\|_{0,\Omega}^2 + \|u_h^{k+1} - u_h^k\|_{0,\Omega}^2 + 2\tau\nu|u_h^{k+1}|_{1,\Omega}^2 \\ &\leq C_F^2 \tau\nu^{-1}\|f(t_{k+1})\|_{0,\Omega}^2 + \tau\nu|u_h^{k+1}|_{1,\Omega}^2 \end{aligned}$$

and, hence, the summation over $k = 0, \dots, m-1$ yields

$$\begin{aligned} &\|u_h^m\|_{0,\Omega}^2 + \sum_{k=0}^{m-1} \|u_h^{k+1} - u_h^k\|_{0,\Omega}^2 + \tau\nu \sum_{k=0}^{m-1} |u_h^{k+1}|_{1,\Omega}^2 \\ &\leq C_F^2 \tau\nu^{-1} \sum_{k=0}^{m-1} \|f(t_{k+1})\|_{0,\Omega}^2 + \|u_h^0\|_{0,\Omega}^2, \quad t_m \in [0,T]. \end{aligned} \tag{4.1.90}$$

Let us define the function $u_{h\tau} : [0,T] \to V_h$:

$$u_{h\tau}(t) = u_h^{k+1} \quad \text{for } t_k < t \leq t_{k+1}, \quad k = 0, 1, \dots, \tag{4.1.91}$$

$$u_{h\tau}(0) = u_h^0,$$

and assume, for example, that $f \in C([0,T]; L^2(\Omega))$. Then from (4.1.90) we obtain the estimate

$$\max_{t\in[0,T]} \|u_{h\tau}(t)\|_{0,\Omega}^2 + \nu \int_0^T |u_{h\tau}(t)|_{1,\Omega}^2 \, dt \tag{4.1.92}$$
$$\leq 2C_F^2 T \nu^{-1} \|f\|_{C([0,T];L^2(\Omega))}^2 + 2\|u_h^0\|_{0,\Omega}^2.$$

Hence, if $\|u_h^0\|_{0,\Omega}$ is bounded (which is true, if $u^0 \in H^1(\Omega)$ and $u_h^0 = \Pi_C u^0$ – see (4.1.73) with $k = 1$), we obtain the boundedness of the approximate solution in the norms $\|\cdot\|_{L^\infty(0,T;L^2(\Omega))}$ and $\|\cdot\|_{L^2(0,T;H^1(\Omega))}$ independently of h and τ. We conclude that the Euler backward *scheme is unconditionally stable.* (Cf. Remark 3.34.)

4.1.7.3 *Stability of the Euler forward scheme* In this case the investigation of the stability is more complicated. As a special tool we need here the so-called inverse inequality. Let us assume that the regular system of triangulations $\{\mathcal{T}_h\}_{h\in(0,h_0)}$ satisfies the *inverse assumption*

$$h \leq c_{\text{inv}} h_K \quad \forall K \in \mathcal{T}_h, \ \forall h \in (0, h_0) \tag{4.1.93}$$

with a constant c_{inv} independent of K and h. Then the *inverse inequality* holds: there exists a constant C_{inv} such that

$$|v_h|_{1,\Omega} \leq C_{\text{inv}} h^{-1} \|v_h\|_{0,\Omega} \quad \forall v_h \in X_h, \ \forall h \in (0, h_0). \tag{4.1.94}$$

(For a proof, see, (Ciarlet, 1979), Section 3.2.)

We start from the substitution $v_h := u_h^k$ in the forward Euler scheme, i.e. (4.1.71) with $\vartheta = 0$:

$$(u_h^{k+1} - u_h^k, u_h^k) + \tau\nu |u_h^k|_{1,\Omega}^2 = \tau(f(t_k), u_h^k). \tag{4.1.95}$$

Again, we consider constant time steps $\tau_k = \tau, k \geq 0$. Using the relation

$$(u - v, v) = \frac{1}{2}\left(\|u\|_{0,\Omega}^2 - \|v\|_{0,\Omega}^2 - \|u - v\|_{0,\Omega}^2\right), \quad u, v \in L^2(\Omega), \tag{4.1.96}$$

and the Cauchy, Friedrichs and Young inequalities, we get

$$\|u_h^{k+1}\|_{0,\Omega}^2 - \|u_h^k\|_{0,\Omega}^2 - \|u_h^{k+1} - u_h^k\|_{0,\Omega}^2 + 2\tau\nu |u_h^k|_{1,\Omega}^2$$
$$\leq 2\tau \|f(t_k)\|_{0,\Omega} \|u_h^k\|_{0,\Omega}$$
$$\leq 2\tau C_F \|f(t_k)\|_{0,\Omega} |u_h^k|_{1,\Omega}$$

$$\leq \frac{2\tau C_F^2}{\nu}\|f(t_k)\|_{0,\Omega}^2 + \frac{\tau\nu}{2}|u_h^k|_{1,\Omega}^2.$$

Hence,

$$\|u_h^{k+1}\|_{0,\Omega}^2 - \|u_h^k\|_{0,\Omega}^2 + \frac{3\tau\nu}{2}|u_h^k|_{1,\Omega}^2 \leq \|u_h^{k+1} - u_h^k\|_{0,\Omega}^2 + \frac{2\tau C_F^2}{\nu}\|f(t_k)\|_{0,\Omega}^2. \quad (4.1.97)$$

Now let us set $v_h := u_h^{k+1} - u_h^k$ in (4.1.71). We get

$$\|u_h^{k+1} - u_h^k\|_{0,\Omega}^2 + \tau a(u_h^k, u_h^{k+1} - u_h^k) = \tau(f(t_k), u_h^{k+1} - u_h^k).$$

This and the Cauchy inequality yield

$$\|u_h^{k+1} - u_h^k\|_{0,\Omega}^2 \leq \tau\|f(t_k)\|_{0,\Omega}\|u_h^{k+1} - u_h^k\|_{0,\Omega} + \tau\nu|u_h^k|_{1,\Omega}|u_h^{k+1} - u_h^k|_{1,\Omega}.$$

Applying the inverse inequality (4.1.94) and dividing by $\|u_h^{k+1} - u_h^k\|_{0,\Omega}$, we obtain

$$\|u_h^{k+1} - u_h^k\|_{0,\Omega} \leq \tau\|f(t_k)\|_{0,\Omega} + \tau\nu C_{\mathrm{inv}} h^{-1}|u_h^k|_{1,\Omega}.$$

Raising this inequality to the second power and assuming that the condition

$$0 < \tau \leq \frac{h^2}{2\nu C_{\mathrm{inv}}^2} \quad (4.1.98)$$

is satisfied, we arrive at the inequality

$$\|u_h^{k+1} - u_h^k\|_{0,\Omega}^2 \leq 2(\tau^2\|f(t_k)\|_{0,\Omega}^2 + \tau\nu|u_h^k|_{1,\Omega}^2). \quad (4.1.99)$$

Substituting (4.1.99) into (4.1.97), we conclude that

$$\|u_h^{k+1}\|_{0,\Omega}^2 - \|u_h^k\|_{0,\Omega}^2 + \frac{\tau\nu}{2}|u_h^k|_{1,\Omega}^2 \quad (4.1.100)$$
$$\leq 2\tau(\tau + C_F^2/\nu)\|f(t_k)\|_{0,\Omega}^2.$$

Now we proceed in the same way as in Section 4.1.7.2. We sum (4.1.100) over $k = 0, \ldots, m-1$ and get the estimate

$$\|u_h^m\|_{0,\Omega}^2 + \frac{\tau\nu}{2}\sum_{k=0}^{m-1}|u_h^k|_{1,\Omega}^2 \quad (4.1.101)$$
$$\leq c\tau\sum_{k=0}^{m-1}\|f(t_k)\|_{0,\Omega}^2 + \|u_h^0\|_{0,\Omega}^2, \quad t_m \in [0,T].$$

From this we see that the stability of the method is guaranteed by conditions (4.1.93) and (4.1.98). The first one represents a restriction on the space mesh $\mathcal{T}_h$. The second of these conditions means that $\tau = O(h^2)$. This represents a strong restriction on the time step similarly as in explicit finite difference schemes for the numerical solution of parabolic problems. Therefore, we say that the Euler forward scheme (4.1.71) with $\vartheta = 0$ is *conditionally stable.* (Cf. Section 3.2.9.)

Exercise 4.23 Prove relations (4.1.88), (4.1.96) and a similar relation for the form $a(u - v, u)$.

4.1.8 *Mass lumping*

The finite element Euler forward scheme (4.1.71) with $\vartheta = 0$ cannot be considered fully explicit in contrast to corresponding finite difference or finite volume methods. In this scheme, it is necessary to solve a linear system for u_h^{k+1} with the matrix $\mathbb{B}$, called the *mass matrix.* As already mentioned in Section 4.1.4, the integrals appearing in the finite element formulation of the discrete problem are often evaluated by *numerical integration.* If we apply a quadrature formula to the computation of the integrals in the definition of the elements

$$b_{ij} = (w_i^*, w_j^*) = \int_\Omega w_i^* w_j^* \, dx \tag{4.1.102}$$

of the matrix $\mathbb{B}$, we obtain an approximation $\tilde{\mathbb{B}}$ of $\mathbb{B}$.

In the case of piecewise linear triangular finite elements ($p = 1$) a suitable quadrature formula gives a *diagonal* approximation $\tilde{\mathbb{B}}$ of the mass matrix and the Euler forward scheme (4.1.71) with $\vartheta = 0$ becomes *fully explicit.* Actually, let us associate each vertex $P_i \in \Omega \cap \sigma_h$ with the basis function w_i^* from Example 4.7 and compute integrals of a function $f \in C(\overline{\Omega})$ over the domain Ω approximately by the following formulae:

$$\int_\Omega f \, dx = \sum_{K \in \mathcal{T}_h} \int_K f \, dx, \tag{4.1.103}$$

$$\int_K f \, dx \approx \frac{1}{3} |K| \sum_{s=1}^{3} f(P_s^K),$$

where $|K|$ is the measure of K and P_s^K, $s = 1, 2, 3$, are the vertices of the triangle K. If we apply this procedure to the integrals from (4.1.102), we get

$$\begin{aligned} b_{ij} \approx \tilde{b}_{ij} &= \frac{1}{3} \sum_{K \in \mathcal{T}_h} |K| \sum_{s=1}^{3} w_i^*(P_s^K) w_j^*(P_s^K) \\ &= \frac{1}{3} \sum_{K \subset \mathrm{supp} w_i^*} |K| \delta_{ij} \quad \text{for } P_i, P_j \in \sigma_h \cap \Omega. \end{aligned} \tag{4.1.104}$$

Hence, the approximate mass matrix $\tilde{\mathbb{B}} = (\tilde{b}_{ij})_{i,j=1}^n$ ($n = \mathrm{card}(\sigma_h \cap \Omega)$) is diagonal. The described approximation of the mass matrix is called *mass lumping.*

4.1.9 *Singularly perturbed and hyperbolic problems*

Initial-boundary value problems of fluid dynamics are usually much more complicated than the standard elliptic and parabolic problems (4.1.1) and (4.1.58). In fluid dynamics, typically convection–diffusion problems with small or even vanishing diffusion occur. This means that these problems are either singularly perturbed (elliptic or parabolic) or hyperbolic. The standard application of the finite element method explained above leads to numerical schemes which frequently do not give reasonable results.

A simple prototype of a parabolic *convection–diffusion equation* can be written in the form

$$\frac{\partial u}{\partial t} + \operatorname{div}(u\boldsymbol{v}) + \sigma u - \varepsilon \Delta u = f \quad \text{in } Q_T = \Omega \times (0,T). \tag{4.1.105}$$

Here $\Omega \subset I\!R^N$ is a bounded domain, $(0,T)$ is a time interval, u is a scalar unknown representing a variable transported in a fluid with velocity $\boldsymbol{v} \in I\!R^N$, σ is an absorption coefficient and $\varepsilon > 0$ is a diffusion coefficient. The ratio between convection and diffusion is characterized by the number $|\boldsymbol{v}|/\varepsilon$. If $1 \ll |\boldsymbol{v}|/\varepsilon$, we have *dominating convection* and speak of a *singularly perturbed convection–diffusion equation.* On the other hand, for $|\boldsymbol{v}|/\varepsilon \ll 1$ diffusion dominates. If $\varepsilon = 0$, then equation (4.1.105) becomes hyperbolic.

In dynamics of viscous gases, the Navier–Stokes equations and the energy equation have a parabolic character, but the continuity equation is of hyperbolic type. We call such a system *incompletely parabolic.* For vanishing viscosity, the whole system degenerates to the hyperbolic system of the Euler equations.

Equation (4.1.105) can be written in the form

$$\frac{\partial u}{\partial t} + \boldsymbol{v} \cdot \nabla u + \gamma u - \varepsilon \Delta u = f \quad \text{in } Q_T, \tag{4.1.106}$$

where $\gamma = \sigma + \operatorname{div} \boldsymbol{v}$. In view of Section 1.2, the expressions $\mathrm{D}u/\mathrm{D}t = \partial u/\partial t + \boldsymbol{v} \cdot \nabla u$ and $u_{\boldsymbol{v}} = \boldsymbol{v} \cdot \nabla u$ are the derivatives along trajectories and along streamlines, respectively. In the stationary case we get the elliptic equations

$$\operatorname{div}(u\boldsymbol{v}) + \sigma u - \varepsilon \Delta u = f \quad \text{or} \quad \boldsymbol{v} \cdot \nabla u + \gamma u - \varepsilon \Delta u = f \quad \text{in } \Omega. \tag{4.1.107}$$

For a singularly perturbed equation the standard application of the Galerkin FEM gives rise to the *Gibbs phenomenon*, manifested by *spurious oscillations* in the numerical solution. The same appears in the case when $\varepsilon = 0$ and the data as well as the exact solution are not regular. (Cf. Section 3.2.21.2.)

Example 4.24 Let us consider the problem

$$\begin{aligned} vu' - \varepsilon u'' &= 1 \quad \text{in } [0,1], \\ u(0) = u(1) &= 0, \end{aligned} \tag{4.1.108}$$

where $v \in I\!R$ and $\varepsilon > 0$ are constants. Let us assume that $v \neq 0$. The solution has the form

$$u(x) = \frac{1}{v}\left\{x - \frac{\exp(vx/\varepsilon) - 1}{\exp(v/\varepsilon) - 1}\right\}, \quad x \in [0,1]. \tag{4.1.109}$$

See Fig. 4.1.1, where this function is shown for $\varepsilon = 10^{-2}$ and $v = 1$. If $\varepsilon \to 0$ and $v > 0$, then $u(x) \to x/v$ for $x \in [0,1)$. The limit function is the solution of the problem

$$vu' = 1, \quad u(0) = 0, \tag{4.1.110}$$

obtained as the limit of equation (4.1.108) for $\varepsilon \to 0$.

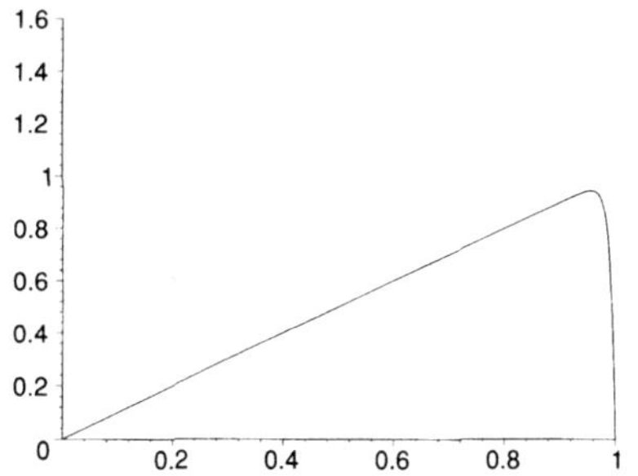

FIG. 4.1.1. Exact solution of problem (4.1.108)

Let us apply the Galerkin FEM with conforming linear elements on a uniform mesh $\mathcal{T}_h = \{K_i\}_{i=0}^n$, where $K_i = [x_i, x_{i+1}]$, $x_i = ih$, $h = 1/(n+1)$, to problem (4.1.108). The approximate solution u_h as well as test functions φ_h are elements of the space

$$V_h = \{\varphi_h \in C([0,1]); \varphi_h|_{K_i} \in P^1(K_i), i = 0, \ldots, n, \quad \varphi_h(0) = \varphi_h(1) = 0\}. \tag{4.1.111}$$

Using the identity valid for the exact solution

$$\int_0^1 (\varepsilon u'\varphi' + vu'\varphi)\, dx = \int_0^1 \varphi\, dx \quad \forall \varphi \in C^1([0,1]), \quad \varphi(0) = \varphi(1) = 0,$$

we define the approximate solution as a function u_h such that

$$\begin{aligned} &\text{a)}\ u_h \in V_h, \\ &\text{b)}\ \int_0^1 (\varepsilon u_h'\varphi_h' + vu_h'\varphi_h)\, dx = \int_0^1 \varphi_h\, dx \quad \forall \varphi_h \in V_h. \end{aligned} \tag{4.1.112}$$

This is equivalent to the linear system of the form

$$\begin{aligned} &-\frac{\varepsilon}{h^2}(u_h(x_{i+1}) - 2u_h(x_i) + u_h(x_{i-1})) \\ &+ \frac{v}{2h}(u_h(x_{i+1}) - u_h(x_{i-1})) = 1, \\ &i = 1, \ldots, n, \ u_h(x_0) = u_h(x_{n+1}) = 0, \end{aligned} \tag{4.1.113}$$

for the values $u_h(x_i)$, $i = 0, \ldots, n+1$, determining uniquely the approximate solution. We notice that (4.1.113) can also be obtained by the finite difference method with central difference approximation $v(u_h(x_{i+1}) - u_h(x_{i-1}))/2h$ for the convective term vu'.

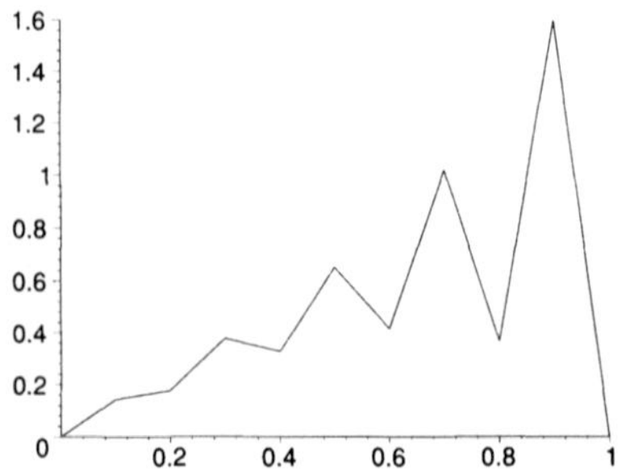

FIG. 4.1.2. Central difference solution of problem (4.1.108)

The matrix of system (4.1.113) is nonsymmetric, but for $h < 2\varepsilon/|v|$ it is *irreducibly diagonally dominant*,[3] which guarantees good properties of the approximate solution. However, if $h \geq 2\varepsilon/|v|$, the approximate solution need not make physical sense because of spurious oscillations shown in Fig. 4.1.2 (where we use $h = 1/10$). This means that the Gibbs phenomenon arises here. We see that the *mesh Péclet number* defined as

$$Pe = \frac{h|v|}{2\varepsilon} \tag{4.1.114}$$

must satisfy the condition

$$Pe < 1 \tag{4.1.115}$$

in order to avoid the Gibbs phenomenon.

In the finite difference method, the Gibbs phenomenon can be avoided by *upwinding* in the approximation of the derivative $u'(x_i)$. Let us assume that $v > 0$. We use the idea that the information propagates in the fluid in the direction of the velocity v and, therefore, we approximate the derivative $u'(x_i)$ by a one-sided difference oriented against the velocity direction:

$$u'(x_i) \approx \frac{u_h(x_i) - u_h(x_{i-1})}{h}. \tag{4.1.116}$$

Then we obtain the linear system

$$\begin{aligned} &-\frac{\varepsilon}{h^2}\left(u_h(x_{i+1}) - 2u_h(x_i) + u_h(x_{i-1})\right) \\ &+\frac{v}{h}\left(u_h(x_i) - u_h(x_{i-1})\right) = 1, \quad i = 1, \ldots, n, \\ &u_h(x_0) = u_h(x_{n+1}) = 0, \end{aligned} \tag{4.1.117}$$

with a diagonally dominant matrix for all $h > 0$ and the physically reasonable solution shown in Fig. 4.1.3 (with $h = 1/10$).

[3]The matrix $\mathbb{A} = (a_{ij})_{i,j=1}^n$ is diagonally dominant, if $\sum_{j=1, j\neq i}^n |a_{ij}| \leq |a_{ii}|$ for all $i = 1, \ldots, n$ with strict inequality for at least one i. If $\mathbb{A}$ is also irreducible (i.e. the system $\mathbb{A}\boldsymbol{y} = \boldsymbol{b}$ does not consist of subsystems that are independent of each other), then the matrix $\mathbb{A}$ is called irreducibly diagonally dominant.

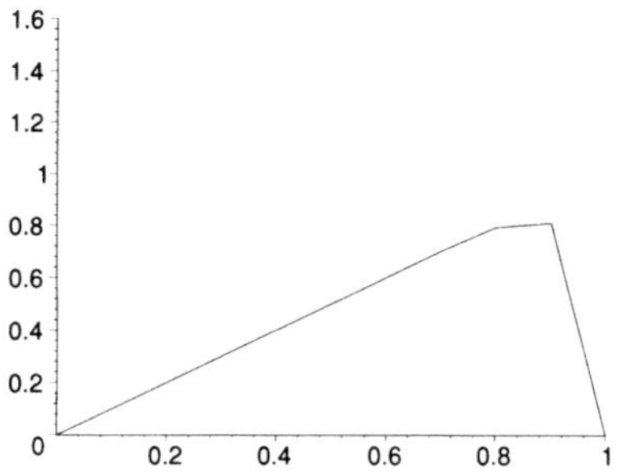

FIG. 4.1.3. Upwinding solution of problem (4.1.108)

Let us assume that the exact solution $u \in C^4([0,1])$. Then, with the aid of the Taylor formula, we find that (4.1.117) is equivalent to the second order discretization (4.1.113) with ε replaced by $\varepsilon + \frac{vh}{2}$. Hence, the diffusion coefficient ε is enlarged by an additional *numerical (artificial) diffusion* $vh/2$, also called *numerical (artificial) viscosity.* (Cf. Remark 3.36.)

Exercise 4.25 a) Prove that the function u from (4.1.109) is a solution of problem (4.1.108).
b) Derive system (4.1.113) equivalent to the discrete problem (4.1.112).
c) Analyse in detail the truncation error of schemes (4.1.113) and (4.1.117) provided $u \in C^4([0,1]))$. (Use the Taylor formula.)
d) Prove that the values $u_h(x_i)$ of the approximate solution obtained from scheme (4.1.113) can be expressed for $v = 1$ as

$$u_h(x_i) = x_i - \frac{\left(\frac{\varepsilon+h/2}{\varepsilon-h/2}\right)^i - 1}{\left(\frac{\varepsilon+h/2}{\varepsilon-h/2}\right)^{n+1} - 1}, \quad i = 0, \dots, n+1,$$

and the values $u_h(x_h)$ obtained from (4.1.117) read

$$u_h(x_i) = x_i - \frac{\left(\frac{\varepsilon+h}{\varepsilon}\right)^i - 1}{\left(\frac{\varepsilon+h}{\varepsilon}\right)^{n+1} - 1}, \quad i = 0, \dots, n+1.$$

Hint: Use the ansatz

$$u_h(x_i) = x_i + A\alpha^i + B, \quad i = 0, \dots, n+1,$$

with unknown constants A, B, α and substitute it in (4.1.113) and (4.1.117).

Remark 4.26 It is also possible to use the approximation

$$u'(x_i) \approx (1-\xi)\frac{u(x_{i+1}) - u(x_{i-1})}{2h} + \xi\frac{u(x_i) - u(x_{i-1})}{h}, \tag{4.1.118}$$

where $\xi \in [0,1]$ is a parameter to be chosen. For $\xi = 0$ and $\xi = 1$ we get the scheme (4.1.113) and (4.1.117), respectively. The optimal choice is

$$\xi = \xi_{\text{opt}} = \coth Pe - \frac{1}{Pe} \tag{4.1.119}$$

for which $u(x_i) = u_h(x_i)$, $i = 0, \ldots, n+1$. We say that the obtained scheme is nodally exact. (See (Iljin, 1969).)

We can note that (4.1.118) leads to scheme (4.1.113) for equation (4.1.108) with ε enlarged by artificial viscosity $\xi hv/2$. This corresponds to the discrete FE problem to find $u_h \in V_h$ such that

$$\int_0^1 \left\{ \left(\varepsilon + \frac{\xi hv}{2} \right) u_h' \varphi_h' + v u_h' \varphi_h \right\} dx = \int_0^1 \varphi_h \, dx, \quad \forall \varphi_h \in V_h,$$

or

$$\int_0^1 (\varepsilon u_h' \varphi_h' + v u_h' (\varphi_h + \delta v \varphi_h')) \, dx = \int_0^1 \varphi_h \, dx, \quad \forall \varphi_h \in V_h, \tag{4.1.120}$$

with $\delta = \frac{\xi h}{2|v|}$. These results lead us to the concept of the streamline diffusion method treated in the next section.

4.1.10 *Streamline diffusion method*

Example 4.24 shows us how to handle the difficulties arising when the standard FEM is applied to the solution of singularly perturbed equations (4.1.105)–(4.1.107) with $\varepsilon < |v|h/2$. The simplest way is to use the *artificial diffusion* (or artificial viscosity) method. This means that to the physical diffusion $-\varepsilon \Delta u$ one adds the artificial diffusion term $-\delta \Delta u$ with $\delta = O(h)$, which yields a new equation with diffusion $-(\varepsilon + \delta)\Delta u$. If the artificial diffusion is appropriately chosen, then this method produces non-oscillating solutions, but introduces a considerable amount of additional diffusion. In particular, it introduces diffusion acting in the direction orthogonal to streamlines (i.e. the so-called crosswind diffusion) and causing that sharp fronts, steep gradients or jumps are smeared. Moreover, this method is at most first order accurate.

Taking into account that the convective term $\boldsymbol{v} \cdot \nabla u$ represents the derivative along streamlines, the analogy with Example 4.24 leads us to the idea to add the term $-\delta u_{\boldsymbol{v}\boldsymbol{v}}$, i.e. a diffusion term acting only in the direction of the streamlines. In this way we obtain the so-called *streamline diffusion finite element method* (SDFEM).

4.1.10.1 *Streamline diffusion method for equation* (4.1.107) *with* $\varepsilon = 0$ We shall demonstrate the use of the streamline diffusion method on an example of a boundary value problem for equation (4.1.107) with $\varepsilon = 0$ and $\gamma = \sigma + \operatorname{div} \boldsymbol{v} = 1$ written in the form

$$\boldsymbol{v} \cdot \nabla u + u = f \quad \text{in } \Omega, \tag{4.1.121}$$

i.e.

$$u_{\boldsymbol{v}} + u = f \quad \text{in } \Omega. \tag{4.1.122}$$

In order to equip this equation with a boundary condition, we split the boundary $\Gamma = \partial\Omega$ into two parts:

$$\begin{aligned} \Gamma^- &= \{x \in \Gamma; \boldsymbol{v}(x)\cdot\boldsymbol{n}(x) < 0\} \text{ (inlet)}, \\ \Gamma^+ &= \Gamma - \Gamma^- \qquad \text{(outlet or impermeable wall)}, \end{aligned} \tag{4.1.123}$$

where $\boldsymbol{n}$ denotes the unit outer normal to Γ. Now we introduce the boundary condition

$$u|_{\Gamma^-} = g \tag{4.1.124}$$

and consider problem (4.1.121), (4.1.124). The boundary condition (4.1.124) prescribes the quantity u on the inlet Γ^-, from where it is transported by the fluid, moving with velocity $\boldsymbol{v}$, into the domain Ω.

In the sequel, let us assume that $N = 2, \boldsymbol{v} \in I\!R^2$ is constant and the domain $\Omega \subset I\!R^2$ is polygonal with a Lipschitz-continuous boundary $\Gamma = \partial\Omega$, split as in (4.1.123). Further, let $\{\mathcal{T}_h\}_{h\in(0,h_0)}$ be a regular system of triangulations of Ω (see (4.1.35)) and let the points between Γ^- and Γ^+ be vertices of triangles $K \in \mathcal{T}_h$ (i.e. elements of σ_h). We define the space X_h by (4.1.17) with an integer $p \geq 1$.

Problem (4.1.121), (4.1.124) makes sense for $u \in H^1(\Omega), f \in L^2(\Omega)$ and $g \in L^2(\Gamma)$. It can be shown that it is equivalent to the conditions

$$\begin{aligned} &\text{a)}\ \ u \in H^1(\Omega), \\ &\text{b)}\ \ \int_\Omega (\boldsymbol{v}\cdot\nabla u + u)\varphi\,dx - (1+\delta)\int_{\Gamma^-} u\varphi\boldsymbol{v}\cdot\boldsymbol{n}\,dS \\ &\qquad = \int_\Omega f\varphi\,dx - (1+\delta)\int_{\Gamma^-} g\varphi\boldsymbol{v}\cdot\boldsymbol{n}\,dS \quad \forall\varphi \in H^1(\Omega), \end{aligned} \tag{4.1.125}$$

where $\delta \geq 0$ is a parameter introduced here for our further convenience. We see that the Dirichlet boundary condition (4.1.124) is formulated in a 'weak sense'.

Problem (4.1.125) has a discrete analogy obtained by replacing $H^1(\Omega)$ by X_h and u, φ by u_h, $\varphi_h \in X_h$. In this way we would obtain a classical Galerkin finite element formulation with bad properties. In order to obtain a more convenient scheme, we shall use the streamline diffusion method which also starts from identity (4.1.125), but in the integrals over Ω, the test function φ is replaced by $\varphi + \delta\boldsymbol{v}\cdot\nabla\varphi = \varphi + \delta\varphi_{\boldsymbol{v}}$, where $\delta > 0$ is a suitable *stabilization parameter*. This leads to the definition of the forms

$$\begin{aligned} B(u,\varphi) &= \int_\Omega (\boldsymbol{v}\cdot\nabla u + u)(\varphi + \delta\boldsymbol{v}\cdot\nabla\varphi)dx \\ &\quad - (1+\delta)\int_{\Gamma^-} u\varphi\boldsymbol{v}\cdot\boldsymbol{n}\,dS \\ &= \int_\Omega (u_{\boldsymbol{v}} + u)(\varphi + \delta\varphi_{\boldsymbol{v}})dx - (1+\delta)\int_{\Gamma^-} u\varphi\boldsymbol{v}\cdot\boldsymbol{n}\,dS, \\ L(\varphi) &= \int_\Omega f(\varphi + \delta\boldsymbol{v}\cdot\nabla\varphi)\,dx - (1+\delta)\int_{\Gamma^-} g\varphi\boldsymbol{v}\cdot\boldsymbol{n}\,dS \end{aligned} \tag{4.1.126}$$

$$= \int_\Omega f(\varphi + \delta\varphi_{\boldsymbol{v}})\,dx - (1+\delta)\int_{\Gamma^-} g\varphi\boldsymbol{v}\cdot\boldsymbol{n}\,dS.$$

Obviously, the exact solution $u \in H^1(\Omega)$ of problem (4.1.121), (4.1.124) satisfies the identity

$$B(u,\varphi) = L(\varphi) \quad \forall\varphi \in H^1(\Omega). \tag{4.1.127}$$

If we compare (4.1.125) with (4.1.127), we notice the presence of the additional term $\delta\int_\Omega u_{\boldsymbol{v}}\varphi_{\boldsymbol{v}}\,dx$ representing the streamline diffusion. Actually, integration by parts of this term yields an expression containing

$$-\delta\int_\Omega u_{\boldsymbol{v}\boldsymbol{v}}\varphi\,dx,$$

which represents artificial diffusion applied in the streamline direction.

Now we come to the *discrete problem* to find an approximate solution u_h such that

$$\begin{aligned} &\text{a)}\quad u_h \in X_h,\\ &\text{b)}\quad B(u_h,\varphi_h) = L(\varphi_h) \quad \forall\varphi_h \in X_h. \end{aligned} \tag{4.1.128}$$

Since $X_h \subset H^1(\Omega)$, $\varphi_h \in X_h$ can be used as test functions in (4.1.127). This implies that the error $e_h = u_h - u$ satisfies the identity

$$B(e_h,\varphi_h) = 0 \quad \forall\varphi_h \in X_h. \tag{4.1.129}$$

We shall establish an error estimate in the *streamline diffusion norm*

$$\|\varphi\|_{\mathrm{SD}} = \left(\delta\|\boldsymbol{v}\cdot\nabla\varphi\|^2_{0,\Omega} + \|\varphi\|^2_{0,\Omega} + \frac{1+\delta}{2}\int_\Gamma \varphi^2|\boldsymbol{v}\cdot\boldsymbol{n}|\,dS\right)^{1/2}. \tag{4.1.130}$$

First we prove the following

Lemma 4.27 *We have*

$$B(\varphi,\varphi) = \|\varphi\|^2_{\mathrm{SD}}, \quad \varphi \in H^1(\Omega). \tag{4.1.131}$$

Proof By Green's theorem,

$$\int_\Omega (\boldsymbol{v}\cdot\nabla\varphi)\varphi\,dx = \int_\Omega \boldsymbol{v}\cdot\boldsymbol{n}\varphi^2\,dS - \int_\Omega \varphi(\boldsymbol{v}\cdot\nabla\varphi)\,dx$$

and, thus,

$$\int_\Omega (\boldsymbol{v}\cdot\nabla\varphi)\varphi\,dx = \frac{1}{2}\int_\Gamma \boldsymbol{v}\cdot\boldsymbol{n}\varphi^2\,dS.$$

This, (4.1.126), the relation $\int_\Gamma = \int_{\Gamma^+} + \int_{\Gamma^-}$ and the inequalities $\boldsymbol{v}\cdot\boldsymbol{n} < 0$ on Γ^- and $\boldsymbol{v}\cdot\boldsymbol{n} \geq 0$ on Γ^+ yield

$$B(\varphi,\varphi) = \delta\|\varphi_{\boldsymbol{v}}\|^2_{0,\Omega} + \|\varphi\|^2_{0,\Omega} + \frac{1+\delta}{2}\int_\Gamma \varphi^2\boldsymbol{v}\cdot\boldsymbol{n}\,dS$$

$$
\begin{aligned}
&- (1+\delta) \int_{\Gamma^-} \varphi^2 \boldsymbol{v} \cdot \boldsymbol{n} \, dS \\
&= \delta \|\varphi_{\boldsymbol{v}}\|_{0,\Omega}^2 + \|\varphi\|_{0,\Omega}^2 + \frac{1+\delta}{2} \left(\int_{\Gamma^+} \varphi^2 \boldsymbol{v} \cdot \boldsymbol{n} \, dS - \int_{\Gamma^-} \varphi^2 \boldsymbol{v} \cdot \boldsymbol{n} \, dS \right) \\
&= \delta \|\varphi_{\boldsymbol{v}}\|_{0,\Omega}^2 + \|\varphi\|_{0,\Omega}^2 + \frac{1+\delta}{2} \int_{\Gamma} \varphi^2 |\boldsymbol{v} \cdot \boldsymbol{n}| \, dS \\
&= \|\varphi\|_{\mathrm{SD}}^2,
\end{aligned}
$$

which we wanted to prove. □

Now we come to the proof of the *error estimate* for the streamline diffusion method (4.1.128).

Theorem 4.28 *Let $\{\mathcal{T}_h\}_{h\in(0,h_0)}$ be a regular system of triangulations of the domain Ω and let $u \in H^{p+1}(\Omega)$. Then, provided $\delta = h$, there exists a constant $c > 0$ such that*

$$\|u - u_h\|_{\mathrm{SD}} \le c h^{p+1/2} |u|_{p+1,\Omega}, \quad h \in (0, h_0), \tag{4.1.132}$$

where u and u_h satisfy (4.1.127) *and* (4.1.128), *respectively.*

Proof Let us set $\eta = \Pi u - u$, $\xi = u_h - \Pi u$, where Π is an X_h-interpolation operator. Then $e_h = u_h - u = \eta + \xi$, $\xi \in X_h$, $\eta \in H^1(\Omega)$. By virtue of (4.1.131), (4.1.129) and (4.1.126), if $\delta = h$,

$$
\begin{aligned}
\|e_h\|_{\mathrm{SD}}^2 &= B(e_h, e_h) = B(e_h, \eta) + B(e_h, \xi) = B(e_h, \eta) \\
&= \int_{\Omega} (e_{h\boldsymbol{v}} + e_h)(\eta + h\eta_{\boldsymbol{v}}) dx - (1+h) \int_{\Gamma^-} e_h \eta \boldsymbol{v} \cdot \boldsymbol{n} \, dS.
\end{aligned}
$$

This and the Cauchy inequality imply that

$$
\begin{aligned}
\|e_h\|_{\mathrm{SD}}^2 \le\ & \|e_{h\boldsymbol{v}}\|_{0,\Omega} \|\eta\|_{0,\Omega} + h \|e_{h\boldsymbol{v}}\|_{0,\Omega} \|\eta_{\boldsymbol{v}}\|_{0,\Omega} \\
&+ \|e_h\|_{0,\Omega} \|\eta\|_{0,\Omega} + \|e_h\|_{0,\Omega} \left(h \|\eta_{\boldsymbol{v}}\|_{0,\Omega} \right) \\
&+ (1+h) \|e_h |\boldsymbol{v} \cdot \boldsymbol{n}|^{1/2}\|_{0,\Gamma} \|\eta |\boldsymbol{v} \cdot \boldsymbol{n}|^{1/2}\|_{0,\Gamma}.
\end{aligned}
$$

($\|\cdot\|_{0,\Gamma}$ denotes the $L^2(\Gamma)$-norm.)

Further, using (4.1.130) and the Young inequality in the form $ab \le \vartheta a^2/4 + \vartheta^{-1} b^2$ with $\vartheta = h$ or $\vartheta = 1$, we obtain

$$
\begin{aligned}
\|e_h\|_{\mathrm{SD}}^2 &= h \|e_{h\boldsymbol{v}}\|_{0,\Omega}^2 + \|e_h\|_{0,\Omega}^2 + \frac{1+h}{2} \|e_h |\boldsymbol{v} \cdot \boldsymbol{n}|^{1/2}\|_{0,\Gamma}^2 \\
&\le \frac{h}{4} \|e_{h\boldsymbol{v}}\|_{0,\Omega}^2 + h^{-1} \|\eta\|_{0,\Omega}^2 + \frac{h}{4} \|e_{h\boldsymbol{v}}\|_{0,\Omega}^2 + h \|\eta_{\boldsymbol{v}}\|_{0,\Omega}^2 \\
&\quad + \frac{1}{4} \|e_h\|_{0,\Omega}^2 + \|\eta\|_{0,\Omega}^2 + \frac{1}{4} \|e_h\|_{0,\Omega}^2 + h^2 \|\eta_{\boldsymbol{v}}\|_{0,\Omega}^2
\end{aligned}
$$

$$+\frac{1+h}{4}\|e_h|\boldsymbol{v}\cdot\boldsymbol{n}|^{1/2}\|^2_{0,\Gamma}+(1+h)\|\eta|\boldsymbol{v}\cdot\boldsymbol{n}|^{1/2}\|^2_{0,\Gamma}.$$

Thus,

$$\|e_h\|^2_{\text{SD}}\le 2(1+h^{-1})\|\eta\|^2_{0,\Omega}+2(1+h)h\|\eta_{\boldsymbol{v}}\|^2_{0,\Omega} \qquad (4.1.133)$$
$$+2(1+h)\|\eta|\boldsymbol{v}\cdot\boldsymbol{n}|^{1/2}\|^2_{0,\Gamma}.$$

If $\Pi = r_h = X_h$-interpolation from Section 4.1.2, then, by virtue of Corollary 4.12,

$$\|\eta\|^2_{0,\Omega}\le ch^{2p+2}|u|^2_{p+1,\Omega}, \qquad (4.1.134)$$

$$h\|\eta_{\boldsymbol{v}}\|^2_{0,\Omega}=h\int_\Omega|\boldsymbol{v}\cdot\nabla\eta|^2\,dx\le h|\boldsymbol{v}|^2|\eta|^2_{1,\Omega}$$
$$\le ch^{2p+1}|u|^2_{p+1,\Omega}.$$

Further, the Cauchy inequality and Lemma 4.30 (see later) imply that

$$\|\eta|\boldsymbol{v}\cdot\boldsymbol{n}|^{1/2}\|^2_{0,\Gamma}\le|\boldsymbol{v}|\|\eta\|^2_{0,\Gamma} \qquad (4.1.135)$$
$$\le ch^{2p+1}|u|^2_{p+1,\Omega}.$$

From (4.1.133)–(4.1.135) we conclude that

$$\|e_h\|^2_{\text{SD}}\le ch^{2p+1}|u|^2_{p+1,\Omega},$$

which gives (4.1.132). □

Remark 4.29 From the error estimate (4.1.132) and the definition of the norm $\|\cdot\|_{\text{SD}}$ we see that

$$\|u-u_h\|_{0,\Omega}\le ch^{p+1/2}|u|_{p+1,\Omega}, \qquad (4.1.136)$$
$$\|u_{\boldsymbol{v}}-u_{h\boldsymbol{v}}\|_{0,\Omega}\le ch^p|u|_{p+1,\Omega}. \qquad (4.1.137)$$

The comparison with (4.1.55) shows that estimate (4.1.136) for the error in the L^2-norm is suboptimal, whereas the L^2-error estimate (4.1.137) of the derivative in the streamline direction is optimal, i.e. it cannot be improved.

Lemma 4.30 *Under the assumptions of Theorem 4.28, for $u\in H^{p+1}(\Omega)$,*

$$\|\eta\|_{0,\Gamma}=\|u-r_hu\|_{0,\Gamma}\le ch^{p+1/2}|u|_{p+1,\Omega}, \qquad (4.1.138)$$

where r_h is the X_h-interpolation.

Proof a) Write $\|\eta\|^2_{0,\Gamma}=\sum_E\|\eta\|^2_{0,E}$, where we sum over all edges $E\subset\Gamma$ of elements $K_E\in\mathcal{T}_h$ adjacent to Γ.

b) Transform K_E with edge $E \subset \Gamma$ on the reference triangle $\hat{K}$ (see Section 4.1.3) so that E is transformed on the edge $\hat{E} = \{(\hat{x}_1, \hat{x}_2); \hat{x}_1 \in [0,1], \hat{x}_2 = 0\}$ of $\hat{K}$. Then, using the notation for $\hat{\eta}$ from Section 4.1.3, we can write

$$\|\eta\|^2_{0,E} = |E| \|\hat{\eta}\|^2_{0,\hat{E}}, \quad |E| \le h_{K_E},$$

where $|E|$ is the length of E.

c) Use the theorem on traces in $H^1(\hat{K})$:

$$\|\hat{\eta}\|^2_{0,\hat{E}} \le \hat{c} \|\hat{\eta}\|^2_{1,\hat{K}} = \hat{c}\left(\|\hat{\eta}\|^2_{0,\hat{K}} + |\hat{\eta}|^2_{1,\hat{K}}\right).$$

Hence,

$$\|\eta\|^2_{0,E} \le \hat{c} h_{K_E}\left(\|\hat{\eta}\|^2_{0,\hat{K}} + |\hat{\eta}|^2_{1,\hat{K}}\right).$$

d) Apply the backward transformation. Using (4.1.35), (4.1.40) and (4.1.41), we find that

$$\|\eta\|^2_{0,E} \le c\left(h^{-1}_{K_E} \|\eta\|^2_{0,K_E} + h_{K_E} |\eta|^2_{1,K_E}\right). \tag{4.1.139}$$

e) Further, (4.1.48) (with $k := p$) yields

$$\|\eta\|^2_{0,E} \le c h^{2p+1}_{K_E} |u|^2_{p+1,K_E}.$$

The summation over all edges $E \subset \Gamma$ together with the inequalities $h_{K_E} \le h$ and $\sum_{E \subset \Gamma} |u|^2_{p+1,K_E} \le |u|^2_{p+1,\Omega}$ yield (4.1.138). □

4.1.10.2 *SDFEM for equation* (4.1.107) *with* $\varepsilon > 0$ Let us consider the convection–diffusion equation (4.1.107) with $\varepsilon, \gamma > 0$, $\varepsilon, \gamma, \boldsymbol{v}$ constant, equipped with zero boundary condition on $\partial\Omega$. The *weak formulation* is to find $u \in V = H^1_0(\Omega)$ such that

$$a(u,v) := \varepsilon(\nabla u, \nabla\varphi) + (\boldsymbol{v} \cdot \nabla u + \gamma u, \varphi) = (f, \varphi) \quad \forall \varphi \in V. \tag{4.1.140}$$

(By $(\cdot,\cdot)$ we again denote the $L^2(\Omega)$-scalar product and assume that $f \in L^2(\Omega)$.)

Let Ω be polygonal. Consider a triangulation $\mathcal{T}_h$ of Ω and set

$$V_h = \left\{\varphi_h \in H^1_0(\Omega); \varphi_h|_K \in P^p(K)\ \forall K \in \mathcal{T}_h\right\} \tag{4.1.141}$$

with an integer $p \ge 1$. Similarly as above, the test functions $\varphi_h \in V_h$ in the streamline diffusion approximation of (4.1.140) are augmented by $\delta\varphi_{\boldsymbol{v}}$, but we shall assume that δ depends on K. Multiplying equation (4.1.107) by the test function $\varphi_h + \delta\varphi_{h\boldsymbol{v}}$, with $\delta : \Omega \to I\!R$ such that $\delta > 0$ and $\delta|_K = \delta_K = \text{const}$, integrating over Ω and applying Green's theorem, we arrive at the identity

$$a_h(u, \varphi_h) = L_h(\varphi_h) \quad \forall \varphi_h \in V_h, \tag{4.1.142}$$

where

$$a_h(u, \varphi_h) = \varepsilon(\nabla u, \nabla\varphi_h) + (\boldsymbol{v} \cdot \nabla u + \gamma u, \varphi_h) \tag{4.1.143}$$

$$+ \sum_{K\in\mathcal{T}_h} \delta_K(-\varepsilon\Delta u + \boldsymbol{v}\cdot\nabla u + \gamma u, \boldsymbol{v}\cdot\nabla\varphi_h)_K,$$

$$L_h(\varphi_h) = (f,\varphi_h) + \sum_{K\in\mathcal{T}_h} \delta_K(f,\boldsymbol{v}\cdot\nabla\varphi_h)_K$$

and where $(\cdot,\cdot)_K$ is the $L^2(K)$-scalar product.

Now the SDFEM is defined as follows: find u_h such that

$$\begin{aligned} &\text{a)}\ \ u_h \in V_h, \\ &\text{b)}\ \ a_h(u_h,\varphi_h) = L_h(\varphi_h) \quad \forall\varphi_h \in V_h. \end{aligned} \tag{4.1.144}$$

Let us summarize the main results of the analysis of this method. We assume that

$$\begin{aligned} &\text{a)}\ \ \{\mathcal{T}_h\}_{h\in(0,h_0)} \text{ is a regular system of triangulations of } \Omega, \\ &\text{b)}\ \ u \in H^{p+1}(\Omega), \\ &\text{c)}\ \ 0 < \delta_K \le \frac{1}{2}\min\left(\frac{1}{\gamma}, \frac{h_K^2}{\varepsilon\tilde{C}_{\rm inv}^2}\right), \end{aligned} \tag{4.1.145}$$

where $\tilde{C}_{\rm inv}$ is the constant from the inverse inequality

$$\|\Delta\varphi_h\|_{0,K} \le \tilde{C}_{\rm inv} h_K^{-1}|\varphi_h|_{1,K} \quad \forall\varphi_h \in V_h,\ \forall h \in (0,h_0). \tag{4.1.146}$$

Then, defining the norm

$$\|\varphi\|_{\rm SD} = \left(\varepsilon|\varphi|_{1,\Omega}^2 + \gamma\|\varphi\|_{0,\Omega}^2 + \sum_{K\in\mathcal{T}_h}\delta_K\|\varphi\boldsymbol{v}\|_{0,K}^2\right)^{1/2}, \tag{4.1.147}$$

we get the estimates

$$a_h(\varphi_h,\varphi_h) \ge \frac{1}{2}\|\varphi_h\|_{\rm SD}^2 \quad \forall\varphi_h \in V_h \tag{4.1.148}$$

(V_h – ellipticity of the form a_h) and

$$\|u_h\|_{\rm SD} \le 2\max\{1,\gamma^{1/2}\}\left(\|f\|_{0,\Omega}^2 + \sum_{K\in\mathcal{T}_h}\delta_K\|f\|_{0,K}^2\right)^{1/2} \tag{4.1.149}$$

(a priori estimate of u_h). Further, if we denote by $r_h u$ the interpolant of u and use estimates (4.1.48), we get

$$\|u - r_h u\|_{\rm SD} \le ch^p\Big(\sum_{K\in\mathcal{T}_h}(\varepsilon + \delta_K + h_K^2)|u|_{p+1,K}^2\Big)^{1/2}|u|_{p+1,\Omega}. \tag{4.1.150}$$

Moreover, with the aid of (4.1.148), we find that

$$\|r_h u - u_h\|_{\mathrm{SD}} \le ch^p \left(\sum_{K\in\mathcal{T}_h} (\varepsilon + \delta_K + \delta_K^{-1}h_K^2 + h_K^2)|u|_{p+1,K}^2 \right)^{1/2}. \tag{4.1.151}$$

Now, in order to get the best possible convergence rate, we have to balance the terms $\varepsilon, \delta_K, \delta_K^{-1}h_K^2$ under condition (4.1.146). To this end, we set

$$\delta_K = \begin{cases} \delta_0 h_K, & \text{if } Pe_K > 1 \text{ (convection-dominated case)}, \\ \delta_1 h_K^2/\varepsilon, & \text{if } Pe_K \le 1 \text{ (diffusion-dominated case)}, \end{cases} \tag{4.1.152}$$

with suitable positive constants δ_0, δ_1. Here Pe_K is the *mesh Péclet number*:

$$Pe_K = \frac{|\boldsymbol{v}|h_K}{2\varepsilon} \tag{4.1.153}$$

– see also (4.1.114). Under the above estimates and the choice (4.1.152) of δ_K, we obtain the *global error estimate*

$$\|u - u_h\|_{\mathrm{SD}} \le C(\varepsilon^{1/2} + h^{1/2})h^p|u|_{p+1,\Omega}. \tag{4.1.154}$$

All proofs and further comments can be found in (Roos *et al.*, 1996), Section 3.2.1.

4.1.10.3 *Galerkin least squares FEM (GLSFEM)* The streamline diffusion approach can be further generalized. Let us again consider the problem for equation (4.1.107), namely

$$\mathcal{L}u := -\varepsilon\Delta u + \boldsymbol{v}\cdot\nabla u + \gamma u = f \quad \text{in } \Omega, \qquad u|_{\partial\Omega} = 0, \tag{4.1.155}$$

under the same assumptions on $\varepsilon, \boldsymbol{v}, \gamma, f$ and Ω as in the preceding section. The original idea of the GLSFEM is to find an approximate solution of problem (4.1.155) as $u_h \in W_h$ such that

$$\|\mathcal{L}u_h - f\|_{0,\Omega}^2 = \min_{\varphi_h\in W_h} \|\mathcal{L}\varphi_h - f\|_{0,\Omega}^2,$$

where $W_h \subset V = H_0^1(\Omega)$ is a suitable finite dimensional space. This is equivalent to the condition

$$(\mathcal{L}u_h - f, \mathcal{L}\varphi_h) = 0 \quad \forall\varphi_h \in W_h.$$

This formulation makes sense, provided the inclusion $W_h \subset H^2(\Omega)$ holds, which is not true for V_h from (4.1.141). In order to avoid this disadvantage, we shall start from the standard Galerkin FE formulation of the discrete problem and add there the residual treated elementwise in the form

$$\sum_{K\in\mathcal{T}_h} \delta_K(\mathcal{L}u_h - f, \mathcal{L}\varphi_h)_K, \tag{4.1.156}$$

which makes sense for $u_h, \varphi_h \in V_h$ defined in (4.1.141). (When we write $(\mathcal{L}u_h - f, \mathcal{L}\varphi_h)_K$, we have in mind the expression $(\mathcal{L}(u_h|_K) - f, \mathcal{L}(\varphi_h|_K))_K$, which is well defined for $u_h, \varphi_h \in V_h$.)

Thus, we come to the GLSFE *discrete problem* to find u_h such that

$$\begin{aligned} &\text{a) } u_h \in V_h, \\ &\text{b) } a_h(u_h, \varphi_h) := \varepsilon(\nabla u_h, \nabla\varphi_h) + (\boldsymbol{v}\cdot\nabla u_h + \gamma u_h, \varphi_h) \\ &\qquad + \sum_{K\in\mathcal{T}_h} \delta_K(\mathcal{L}u_h, \mathcal{L}\varphi_h)_K = L_h(\varphi_h) \\ &\qquad := (f, \varphi_h) + \sum_{K\in\mathcal{T}_h} \delta_K(f, \mathcal{L}\varphi_h)_K \quad \forall\varphi_h \in V_h. \end{aligned} \tag{4.1.157}$$

The GLSFEM can be generalized in such a way that the expression (4.1.156) is replaced by

$$\sum_{K\in\mathcal{T}_h} (\mathcal{L}u_h - f, \psi_h(\varphi_h))_K, \tag{4.1.158}$$

where ψ_h is some user-chosen operator. If $\psi_h(\varphi_h)|_K = \delta_K\mathcal{L}(\varphi_h)|_K$, we get the GLSFEM. For $\psi_h(\varphi_h)|_K = \delta_K\boldsymbol{v}\cdot\nabla(\varphi_h)|_K = \delta_K\varphi_{\boldsymbol{v}}|_K$, the SDFEM is obtained.

It is easy to show that provided $\delta_K > 0$ for all $K \in \mathcal{T}_h$ the form a_h is V_h-elliptic in the mesh-dependent norm

$$\|\varphi_h\|_{\mathrm{GLS}} = \left(\varepsilon|\varphi_h|_{1,\Omega}^2 + \gamma\|\varphi_h\|_{0,\Omega}^2 + \sum_{K\in\mathcal{T}_h} \delta_K\|\mathcal{L}\varphi_h\|_{0,K}^2\right)^{1/2}. \tag{4.1.159}$$

That is, we have

$$a_h(\varphi_h, \varphi_h) = \|\varphi_h\|_{\mathrm{GLS}}^2 \quad \forall\varphi_h \in V_h. \tag{4.1.160}$$

Moreover, using the Cauchy inequality, we find that

$$|L_h(\varphi_h)| \le c\left(\|f\|_{0,\Omega} + \left(\sum_{K\in\mathcal{T}_h} \delta_K\|f\|_{0,K}^2\right)^{1/2}\right)\|\varphi_h\|_{\mathrm{GLS}} \quad \forall\varphi_h \in V_h.$$

From (4.1.157)–(4.1.160), the a priori estimate of the approximate solution u_h follows:

$$\|u_h\|_{\mathrm{GLS}} \le c\left\{\|f\|_{0,\Omega} + \left(\sum_{K\in\mathcal{T}_h} \delta_K\|f\|_{0,K}^2\right)^{1/2}\right\}. \tag{4.1.161}$$

One can easily see that the exact solution $u \in H^2(\Omega)$ satisfies the condition

$$a_h(u, \varphi_h) = L_h(\varphi_h) \quad \forall \varphi_h \in V_h. \tag{4.1.162}$$

If $\{\mathcal{T}_h\}_{h \in (0,h_0)}$ is a regular system of triangulations of the domain Ω, all parameters δ_K are positive and the exact solution $u \in H^{p+1}(\Omega)$, then from the above results it is possible to derive the error estimate

$$\|u - u_h\|_{\text{GLS}} \leq Ch^p \Big\{ \sum_{K \in \mathcal{T}_h} (\varepsilon + \varepsilon^2 \delta_K h_K^{-2} + \delta_K + h_K^2 + \delta_K^{-1} h_K^2) |u|^2_{p+1,K} \Big\}^{1/2}. \tag{4.1.163}$$

In order to minimize the expression $\varepsilon^2 \delta_K h_K^{-2} + \delta_K + \delta_K^{-1} h_K^2$, we set

$$\delta_K = \delta_0 \frac{h_K}{\sqrt{1 + (\varepsilon/h_K)^2}}, \tag{4.1.164}$$

where $\delta_0 > 0$ is some user-chosen parameter. Then we get

$$\|u - u_h\|_{\text{GLS}} \leq C(\varepsilon^{1/2} + h^{1/2}) h^p |u|_{p+1,\Omega}, \tag{4.1.165}$$

which is the same as the estimate (4.1.153) obtained for the SDFEM.

The optimization of estimate (4.1.163) can be carried out for one space variable ($N = 1$), $\gamma = 0$ (i.e. for problem (4.1.108)) and a uniform grid with step size h. Then, similarly as for the generalized upwind finite difference method, the optimal choice is

$$\delta_K = \frac{h}{2|v|} \left(\coth Pe - \frac{1}{Pe} \right) \quad \text{with} \quad Pe = \frac{|v| h}{2\varepsilon}. \tag{4.1.166}$$

(See also Example 4.24.)

The detailed analysis of the above results can be found in (Roos *et al.*, 1996), Section 3.2.2.

Remark 4.31 Let us mention that the streamline diffusion FEM was introduced by Hughes and Brooks ((Hughes and Brooks, 1979)). Sometimes it is known as the *streamline upwind Petrov–Galerkin method* (SUPG method). The concept of the Petrov–Galerkin FEM is used for a finite element technique when test functions are different from trial functions for the representation of an approximate solution. In the standard FEM (see Sections 4.1.2 and 4.1.6) the test and trial functions are the same (modulo boundary conditions). In the case of the SD-FEM, $u_h \in X_h$, but we use test functions $\varphi_h + \delta \varphi_{hv} \notin X_h$ (SUPG method). The finite element method for linear hyperbolic problems was analysed in (Johnson *et al.*, 1984).

The SDFEM was applied to the solution of compressible flow by several authors, e.g. (Hughes and Tezduyar, 1984), (Hughes, Franca and Mallet, 1986),

(Rachowicz, 1997). The Galerkin least squares finite element method was proposed in (Hughes and Shakib, 1988) and (Hughes *et al.*, 1989). These methods applied to the solution of compressible flow will be discussed in Sections 4.2 and 4.3.

Theoretically, the SDFEM and GLSFEM (applied to scalar problems) combine good stability properties with high accuracy. However, the quality of the numerical solution is strongly influenced by the choice of stabilization parameters. In the case of problem (4.1.121), (4.1.124), the choice of the stabilization parameter was quite simple: $\delta := h$. For more complicated problems (with $\varepsilon > 0$ and/or nonconstant coefficients), the stabilization parameter should depend on $K \in \mathcal{T}_h$. We shall not deal with this topic here. For a detailed analysis, see (Roos *et al.*, 1996), Chapter III, (Apel and Lube, 1996), (Apel and Lube, 1998), (Lube and Weiss, 1995).

4.1.11 *Discontinuous Galerkin FEM for a linear hyperbolic problem*

For complex problems of compressible flow the choice of optimal stabilization parameters becomes quite sophisticated. This is the reason that a number of authors have tried to develop methods which would not need to tune various parameters on one hand, and would allow one to obtain a good resolution of important details, such as steep gradients, discontinuities, boundary layers and wakes, not suffering from spurious oscillations and nonphysical entropy production, on the other hand. One possibility is to get rid of the requirement of the conformity (i.e. continuity) of the FE solutions. This idea leads us to the so-called discontinuous Galerkin finite element method (DGFEM).

Here we shall derive and analyse the discontinuous Galerkin finite element method for a simple problem (4.1.121), (4.1.124) assuming again that $\Omega \subset I\!R^2$ is a polygonal domain with a Lipschitz boundary $\Gamma = \partial\Omega$, $f \in L^2(\Omega)$ and that $\boldsymbol{v} \in I\!R^2$ is constant. Let $\mathcal{T}_h$ be a triangulation of Ω with standard properties from Section 4.1.2. For each $K \in \mathcal{T}_h$ we introduce the notation

$$\begin{aligned} \partial K^- &= \{x \in \partial K; \boldsymbol{v} \cdot \boldsymbol{n}(x) < 0\}, \\ \partial K^+ &= \{x \in \partial K; \boldsymbol{v} \cdot \boldsymbol{n}(x) \geq 0\}. \end{aligned} \tag{4.1.167}$$

By $H^k(\Omega, \mathcal{T}_h)$ we denote the so-called *broken Sobolev space*:

$$H^k(\Omega, \mathcal{T}_h) = \left\{v \in L^2(\Omega); v\big|_K \in H^k(K)\ \forall K \in \mathcal{T}_h\right\}. \tag{4.1.168}$$

For $u \in H^1(\Omega, \mathcal{T}_h)$ we set

$$u_K^+ = \text{trace of } u|_K \text{ on } \partial K \tag{4.1.169}$$

(i.e. the interior trace of u on ∂K). For each edge $E \subset \partial K \setminus \Gamma$ of K, there exists $K' \neq K$, $K' \in \mathcal{T}_h$, adjacent to E from the opposite side than K. Then we put

$$u_K^- = \text{trace of } u|_{K'} \text{ on } E. \tag{4.1.170}$$

In this way we obtain the exterior trace u_K^- of u on $\partial K \setminus \Gamma$ and define the jump of u on $\partial K \setminus \Gamma$:

$$[u]_K = u_K^+ - u_K^-. \tag{4.1.171}$$

4.1.11.1 *Derivation of the DGFEM* Let $u \in H^1(\Omega)$ be a solution of problem (4.1.121), (4.1.124). Then u satisfies the identity

$$\int_K f\varphi\, dx = \int_K u\varphi\, dx + \int_K (\boldsymbol{v}\cdot\nabla u)\varphi\, dx, \quad \varphi \in H^1(\Omega, \mathcal{T}),\ K \in \mathcal{T}_h. \tag{4.1.172}$$

The application of Green's theorem to the last term in (4.1.172) gives

$$\begin{aligned}\int_K (\boldsymbol{v}\cdot\nabla u)\varphi\, dx &= \int_{\partial K} \boldsymbol{v}\cdot\boldsymbol{n} u_K^+\varphi_K^+\, dS - \int_K \operatorname{div}(\boldsymbol{v}\varphi)u\, dx \qquad (4.1.173)\\ &= \int_{\partial K^-} \boldsymbol{v}\cdot\boldsymbol{n} u_K^+\varphi_K^+\, dS + \int_{\partial K^+} \boldsymbol{v}\cdot\boldsymbol{n} u_K^+\varphi_K^+ dS - \int_K \operatorname{div}(\boldsymbol{v}\varphi)u\, dx.\end{aligned}$$

(We also use the above notation here for φ.) Of course, as $u \in H^1(\Omega)$, we have $u_K^- = u_K^+$. Moreover, $u_K^-|_{\partial K^-\cap\Gamma^-} := u|_{\partial K^-\cap\Gamma^-} = g$. Then we can write

$$\begin{aligned}&\int_K (\boldsymbol{v}\cdot\nabla u)\varphi\, dx \qquad (4.1.174)\\ &= \int_{\partial K^-} \boldsymbol{v}\cdot\boldsymbol{n} u_K^-\varphi_K^+\, dS + \int_{\partial K^+} \boldsymbol{v}\cdot\boldsymbol{n} u_K^+\varphi_K^+\, dS - \int_K \operatorname{div}(\boldsymbol{v}\varphi)u\, dx.\end{aligned}$$

Applying Green's theorem again to the last term, we arrive at the identity

$$\int_K f\varphi\, dx = \int_K (\boldsymbol{v}\cdot\nabla u + u)\varphi\, dx - \int_{\partial K^-} \boldsymbol{v}\cdot\boldsymbol{n}(u_K^+ - u_K^-)\varphi_K^+\, dS$$

or, equivalently,

$$\begin{aligned}&\int_K (\boldsymbol{v}\cdot\nabla u + u)\varphi\, dx - \int_{\partial K^-\setminus\Gamma} \boldsymbol{v}\cdot\boldsymbol{n}[u]_K\varphi_K^+\, dS \qquad (4.1.175)\\ &- \int_{\partial K^-\cap\Gamma} \boldsymbol{v}\cdot\boldsymbol{n} u_K^+\varphi_K^+\, dS = \int_K f\varphi\, dx - \int_{\partial K^-\cap\Gamma} \boldsymbol{v}\cdot\boldsymbol{n} g\varphi_K^+\, dS,\\ &\qquad \varphi \in H^1(K),\ K \in \mathcal{T}_h.\end{aligned}$$

Setting

$$\begin{aligned}a_K(u,\varphi) = &\int_K (\boldsymbol{v}\cdot\nabla u + u)\varphi dx - \int_{\partial K^-\setminus\Gamma} \boldsymbol{v}\cdot\boldsymbol{n}[u]_K\varphi_K^+\, dS \qquad (4.1.176)\\ &- \int_{\partial K^-\cap\Gamma} \boldsymbol{v}\cdot\boldsymbol{n} u_K^+\varphi_K^+\, dS,\end{aligned}$$

$$L_K(\varphi) = \int_K f\varphi\,dx - \int_{\partial K^- \cap \Gamma} \boldsymbol{v}\cdot\boldsymbol{n} g \varphi_K^+\,dS,$$

we can rewrite equation (4.1.175) as

$$a_K(u,\varphi) = L_K(\varphi), \quad \varphi \in H^1(K),\ K \in \mathcal{T}_h. \tag{4.1.177}$$

This identity makes sense also for $u \in H^1(\Omega, \mathcal{T}_h)$. In this case, we can note that in the first term of the second line of (4.1.173), on ∂K^- (= the inlet of K with respect to the velocity $\boldsymbol{v}$) we replace the value u_K^+ (the interior trace of u) by u_K^-. This means that *upwinding* is used here, because the value of the trace of u on ∂K^- is taken from the side of ∂K^- against the velocity direction.

Now, on the basis of (4.1.177) we come to the definition of the *discrete problem*. The *approximate solution* is a function u_h satisfying the conditions

$$\begin{aligned} \text{a)}\ & u_h \in S_h = S^{p,-1}(\Omega, \mathcal{T}_h) \\ & := \left\{\varphi \in L^2(\Omega); \varphi|_K \in P^p(K)\ \forall K \in \mathcal{T}_h\right\}, \\ \text{b)}\ & a_K(u_h, \varphi_h) = L_K(\varphi_h) \quad \forall \varphi_h \in S_h,\ \forall K \in \mathcal{T}_h. \end{aligned} \tag{4.1.178}$$

Here, $p \geq 0$ is an integer. The approximate solution and test functions are piecewise polynomial functions without any continuity requirement on interfaces between neighbouring elements. The continuity requirement is replaced here by the jump term $\int_{\partial K^-} \boldsymbol{v}\cdot\boldsymbol{n}[u]_K \varphi_K^+\,dS$.

4.1.11.2 *Error estimate for the DGFEM* If we define the forms

$$\begin{aligned} a_h(u,\varphi) &= \sum_{K\in\mathcal{T}_h} a_K(u,\varphi), \quad u,\varphi \in H^1(\Omega,\mathcal{T}_h), \\ L_h(\varphi) &= \sum_{K\in\mathcal{T}_h} L_K(\varphi), \quad \varphi \in H^1(\Omega,\mathcal{T}_h), \end{aligned} \tag{4.1.179}$$

problem (4.1.178) can be written equivalently as

$$\begin{aligned} \text{a)}\ & u_h \in S_h, \\ \text{b)}\ & a_h(u_h,\varphi_h) = L_h(\varphi_h) \quad \forall \varphi_h \in S_h. \end{aligned} \tag{4.1.180}$$

As follows from (4.1.177), the exact solution $u \in H^1(\Omega)$ satisfies the relation

$$a_h(u,\varphi_h) = L_h(\varphi_h) \quad \forall \varphi_h \in S_h. \tag{4.1.181}$$

Hence, for the error $e_h = u_h - u$ we have

$$a_h(e_h,\varphi_h) = 0 \quad \forall \varphi_h \in S_h. \tag{4.1.182}$$

Our goal now is to establish the error estimate in the following mesh-dependent norm $\|\cdot\|_{\mathrm{DG}}$ defined by

$$\|\varphi\|_{\mathrm{DG}}^2 = \|\varphi\|_{0,\Omega}^2 + \frac{1}{2}\sum_{K\in\mathcal{T}_h}\left\{\int_{\partial K^- \cap \Gamma^-} |\boldsymbol{v}\cdot\boldsymbol{n}|(\varphi_K^+)^2\,dS\right. \tag{4.1.183}$$

$$+\int_{\partial K^-\setminus\Gamma} |\boldsymbol{v}\cdot\boldsymbol{n}|(\varphi_K^+ - \varphi_k^-)^2\, dS + \int_{\partial K^+\cap\Gamma} |\boldsymbol{v}\cdot\boldsymbol{n}|(\varphi_K^+)^2\, dS\Big\}.$$

Theorem 4.32 *Let $\{\mathcal{T}_h\}_{h\in(0,h_0)}$ be a regular system of triangulations of the domain Ω and let $u\in H^{p+1}(\Omega)$, $p\ge 0$. Then there exists a constant $c>0$ such that*

$$\|u-u_h\|_{\mathrm{DG}} \le ch^{p+1/2}|u|_{p+1,\Omega}, \tag{4.1.184}$$

where u is the exact solution of problem (4.1.121), (4.1.124) *and u_h is the approximate DG solution obtained from* (4.1.180).

Proof We proceed in several steps:

1. We set $\xi = u_h - \Pi u$, $\eta = \Pi u - u$, where Π is the L^2-projection of u on $S_h = S^{p,-1}(\Omega,\mathcal{T}_h)$:

$$\text{a)}\ \ \Pi u \in S_h, \tag{4.1.185}$$

$$\text{b)}\ \ \int_\Omega (\Pi u - u)\varphi_h\, dx = 0 \quad \forall \varphi_h\in S_h.$$

We can see that (4.1.185) is equivalent to the condition

$$\int_K (\Pi u - u)\varphi dx = 0 \quad \forall\varphi\in P^p(K), \tag{4.1.186}$$

and, thus, $\Pi u|_K$ is the L^2-projection of $u|_K$ on $P^p(K)$. Obviously, Π is $P^p(K)$-polynomial preserving: $\Pi v = v$ for $v\in P^p(K)$. Using the linear transformation F_K of the reference triangle $\hat K$ onto K from Section 4.1.3 and defining the operator $\hat\Pi : H^{k+1}(\hat K)\to P^p(\hat K)$ by (4.1.45), we find that $\hat\Pi\hat v = \hat v$ for $\hat v\in P^p(\hat K)$. Hence, we can use Theorem 4.9 and Lemma 4.10 (with $k := p$ and $m = 0,1$). For $\eta = \Pi u - u$ we get the estimates

$$\begin{aligned}\|\eta\|_{0,K} &\le ch_K^{p+1}|u|_{p+1,K}, \\ |\eta|_{1,K} &\le ch_K^{p}|u|_{p+1,K}, \\ & K\in\mathcal{T}_h,\ h\in(0,h_0).\end{aligned} \tag{4.1.187}$$

Further, by virtue of (4.1.139),

$$\|\eta\|_{0,\partial K} \le ch_K^{p+1/2}|u|_{p+1,K}, \quad K\in\mathcal{T}_h,\ h\in(0,h_0). \tag{4.1.188}$$

2. Now, the application of Green's theorem and a suitable rearrangement yield

$$\begin{aligned} a_h(\xi,\xi) = \sum_{K\in\mathcal{T}_h}\Big\{&\int_K (\boldsymbol{v}\cdot\nabla\xi + \xi)\xi\, dx \\ &-\int_{\partial K^-\setminus\Gamma} \boldsymbol{v}\cdot\boldsymbol{n}[\xi]_K\xi_K^+\, dS - \int_{\partial K^-\cap\Gamma}\boldsymbol{v}\cdot\boldsymbol{n}(\xi_K^+)^2 dS\Big\}\end{aligned}$$

$$
\begin{aligned}
&= \sum_{K\in\mathcal{T}_h} \Big\{ \frac{1}{2}\int_{\partial K} \boldsymbol{v}\cdot\boldsymbol{n}(\xi_K^+)^2\,dS + \int_K \xi^2\,dx \\
&\quad - \int_{\partial K^-\setminus\Gamma} \boldsymbol{v}\cdot\boldsymbol{n}(\xi_K^+ - \xi_K^-)\xi_K^+\,dS - \int_{\partial K^-\cap\Gamma} \boldsymbol{v}\cdot\boldsymbol{n}(\xi_K^+)^2 dS \Big\} \\
&= \|\xi\|_{0,\Omega}^2 + \frac{1}{2}\sum_{K\in\mathcal{T}_h} \Big\{ \int_{\partial K^-\cap\Gamma^-} |\boldsymbol{v}\cdot\boldsymbol{n}|(\xi_K^+)^2\,dS \\
&\quad + \int_{\partial K^-\setminus\Gamma} |\boldsymbol{v}\cdot\boldsymbol{n}|(\xi_K^+ - \xi_K^-)^2\,dS + \int_{\partial K^+\cap\Gamma} |\boldsymbol{v}\cdot\boldsymbol{n}|(\xi_K^+)^2\,dS \Big\} \\
&= \|\xi\|_{\mathrm{DG}}^2.
\end{aligned}
$$

(We have used the relation $a^2 - ab = \frac{1}{2}(a^2 - b^2 + (a-b)^2)$ and (4.1.167).)

3. Obviously, $e_h = \xi + \eta$, $\xi \in S_h$, and, in view of (4.1.182),

$$\|\xi\|_{\mathrm{DG}}^2 = a_h(\xi,\xi) = -a_h(\eta,\xi). \tag{4.1.189}$$

Moreover, using Green's theorem, after an easy but lengthy calculation, we find that

$$
\begin{aligned}
&a_h(\eta,\xi) \qquad\qquad (4.1.190)\\
&= \sum_{K\in\mathcal{T}_h} \Big\{ \int_K \eta\xi\,dx - \int_K (\boldsymbol{v}\cdot\nabla\xi)\eta\,dx + \int_{\partial K^+} \boldsymbol{v}\cdot\boldsymbol{n}\eta_K^+\xi_K^+\,dS \\
&\quad + \int_{\partial K^-\setminus\Gamma} \boldsymbol{v}\cdot\boldsymbol{n}\eta_K^+\xi_K^+\,dS + \int_{\partial K^-\cap\Gamma} \boldsymbol{v}\cdot\boldsymbol{n}\eta_K^+\xi_K^+\,dS \\
&\quad - \int_{\partial K^-\setminus\Gamma} \boldsymbol{v}\cdot\boldsymbol{n}(\eta_K^+ - \eta_K^-)\xi_K^+\,dS - \int_{\partial K^-\cap\Gamma} \boldsymbol{v}\cdot\boldsymbol{n}\eta_K^+\xi_K^+\,dS \Big\} \\
&= \sum_{K\in\mathcal{T}_h} \Big\{ \int_K \eta\xi\,dx - \int_K (\boldsymbol{v}\cdot\nabla\xi)\eta\,dx + \int_{\partial K^+\cap\Gamma} \boldsymbol{v}\cdot\boldsymbol{n}\eta_K^+\xi_K^+\,dS \\
&\quad + \int_{\partial K^+\setminus\Gamma} \boldsymbol{v}\cdot\boldsymbol{n}\eta_K^+\xi_K^+\,dS + \int_{\partial K^-\setminus\Gamma} \boldsymbol{v}\cdot\boldsymbol{n}\eta_K^-\xi_K^+\,dS \Big\}.
\end{aligned}
$$

Now we can notice that $\boldsymbol{v}\cdot\nabla\xi|_K \in P^{p-1}(K)$ for each $K \in \mathcal{T}_h$ and, hence,

$$\int_K (\boldsymbol{v}\cdot\nabla\xi)\eta\,dx = 0, \quad K \in \mathcal{T}_h, \tag{4.1.191}$$

as follows from (4.1.186). Taking into account that for any edge E common to two adjacent triangles K and K' we have $\xi_K^- = \xi_{K'}^+$, $\eta_K^- = \eta_{K'}^+$ and $(\boldsymbol{v}\cdot\boldsymbol{n})|_{\partial K} = -(\boldsymbol{v}\cdot\boldsymbol{n})|_{\partial K'}$, we obtain from (4.1.190) and (4.1.191):

$$
\begin{aligned}
a_h(\eta,\xi) &= \sum_{K\in\mathcal{T}_h} \Big\{ \int_K \eta\xi\,dx + \int_{\partial K^+\cap\Gamma} \boldsymbol{v}\cdot\boldsymbol{n}\xi_K^+\eta_K^+\,dS \qquad (4.1.192)\\
&\quad + \int_{\partial K^-\setminus\Gamma} \boldsymbol{v}\cdot\boldsymbol{n}(\xi_K^+ - \xi_K^-)\eta_K^-\,dS \Big\}.
\end{aligned}
$$

4. The use of the Young inequality gives

$$|a_h(\eta,\xi)| \le \frac{1}{2}\|\xi\|_{\mathrm{DG}}^2 + \frac{1}{2}\|\eta\|_{0,\Omega}^2 + \sum_{K\in\mathcal{T}_h} \Big\{ \int_{\partial K^- \setminus \Gamma} |\boldsymbol{v}\cdot\boldsymbol{n}|(\eta_K^-)^2\, dS + \int_{\partial K^+ \cap \Gamma} |\boldsymbol{v}\cdot\boldsymbol{n}|(\eta_K^+)^2\, dS \Big\}.$$

This and (4.1.189) lead to the estimate

$$\|\xi\|_{\mathrm{DG}}^2 \le \|\eta\|_{0,\Omega}^2 + 2|\boldsymbol{v}| \sum_{K\in\mathcal{T}_h} \Big\{ \int_{\partial K^- \setminus \Gamma} (\eta_K^-)^2\, dS + \int_{\partial K^+ \cap \Gamma} (\eta_K^+)^2\, dS \Big\}.$$

5. Finally, using estimates (4.1.187) and (4.1.188), we arrive at

$$\|\xi\|_{\mathrm{DG}}^2 \le c\left(h^{2p+2} + h^{2p+1}\right)|u|_{p+1,\Omega}^2,$$

which we wanted to prove.

□

Remark 4.33 In contrast to (4.1.49), the obtained error bound (4.1.184) yields a suboptimal estimate in the L^2-norm. Similar results in more complex situations were obtained in (Johnson and Pitkäranta, 1986) for the h-version of DGFEM and in (Houston *et al.*, 2002) for the hp-version of DGFEM. Nevertheless, these estimates represent an improvement on the result from (Le Saint and Raviart, 1974), where only $O(h^p)$ rate of convergence was proven. It is important that DGFEM, in contrast to the streamline diffusion method, does not require any stabilization parameters. As we saw in Section 4.1.11.1, *upwinding is included here in the framework of a higher order scheme* in a natural way. From this point of view the DGFEM seems to be a very promising method for the solution of singularly perturbed and hyperbolic problems.

Remark 4.34 The original DGFEM was introduced by Reed and Hill ((Reed and Hill, 1973)) for the solution of the neutron transport equation. The first analysis of the DGFEM for a linear problem was carried out by Le Saint and Raviart ((Le Saint and Raviart, 1974)), and later an improved estimate was achieved by Johnson and Pitkäranta ((Johnson and Pitkäranta, 1986)). The hp-version of the DGFEM was analysed by Houston, Süli and Schwab in (Houston *et al.*, 2002). For a survey of DGFE techniques and applications, see (Cockburn, 1999). In Section 4.6 we shall pay attention to the use of the DGFEM for the numerical solution of conservation laws and compressible flow.

4.1.12 *Adaptive mesh refinement and a posteriori error estimates*

In CFD we are interested in the computation of sufficiently accurate solutions to various flow problems. The goal of the design of any computational method is *reliability* and *efficiency*. Reliability means that the computational error is controlled at a given tolerance level. The efficiency means that the computational work to compute a solution within a given tolerance is as small as possible. These two requirements are usually achieved with the aid of mesh refinement techniques. In Section 3.6, several *ad hoc* methods have been discussed. They indicate well the regions for the mesh refinement, but do not give a picture of the behaviour of the error. Here we shall explain methods based on *a posteriori error estimates* that can be obtained on the basis of the computed approximate solution and the data of the problem. The goal is to achieve reliability either in the sense that the numerical solution approximates the exact solution in a given norm to within a given tolerance, or in the sense that some physically relevant quantities (e. g., flux through a part of a boundary, drag, lift) are computed to within a given tolerance. There is an extensive literature devoted to a posteriori error estimates (particularly for elliptic and parabolic problems). We refer the reader to the monographs (Babuška and Strouboulis, 2001), (Verfürth, 1996), (Apel, 1999) and references therein.

4.1.12.1 *Simple example* For example, let us consider the linear finite element approximation of problem (4.1.1) in a polygonal domain Ω with a solution $u \in H^2(\Omega)$. Then, in view of Theorem 4.8 and Lemma 4.11,

$$\|u - u_h\|_{1,\Omega} \leq c \left(\sum_{K \in \mathcal{T}_h} h_K^2 |u|_{2,K}^2 \right)^{1/2} . \tag{4.1.193}$$

It follows from (4.1.193) that the error could be reduced by a suitable *mesh refinement* so that all the contributions $h_K^2 \, |u|_{2,K}^2$ on the right-hand side of (4.1.193) are sufficiently small, e.g.

$$h_K^2 |u|_{2,K}^2 < \frac{\delta}{M_h}, \tag{4.1.194}$$

where $\delta > 0$ is a prescribed *tolerance* and $M_h = \#\mathcal{T}_h$ is the number of elements of $\mathcal{T}_h$. The expression

$$\left(\sum_{K \in \mathcal{T}_h} h_K^2 |u|_{2,K}^2 \right)^{1/2}$$

is called the *error estimator* and the terms $h_K |u|_{2,K}$ are called *error indicators*.

The adaptive mesh refinement can be carried out in the following steps:

(i) The approximate solution u_h is computed on a given triangulation $\mathcal{T}_h$.

(ii) A suitable postprocessing $\tilde{u}_h$ of u_h is constructed so that it can be considered as a more accurate approximation of the exact solution u. It is used instead of u for evaluating the error indicators.

(iii) Each triangle $K \in \mathcal{T}_h$, for which

$$h_K^2 |\tilde{u}_h|_{2,K}^2 \geq \frac{\delta}{M_h}, \tag{4.1.195}$$

is divided into four congruent subtriangles by midlines.

(iv) The resulting partition of the domain Ω is modified in such a way that a new triangulation $\mathcal{T}_h^{\text{new}}$ of Ω satisfying condition (4.1.15) is obtained.

The main problem is the determination of the constant c in (4.1.193). Moreover, the reliability of this approach is based on the choice of the function $\tilde{u}_h$.

In what follows we shall describe some more sophisticated techniques which are used in scientific computing and are more and more extensively applied in CFD.

4.1.12.2 *A posteriori error estimates à la Johnson* For complicated linear and nonlinear problems one can employ a posteriori error estimates proposed by Johnson ((Johnson and Hansbo, 1992), (Johnson, 1994)).

Let us suppose that V is a Hilbert space with scalar product $(\cdot,\cdot)$ and norm $\|\cdot\|$, $\mathcal{L} : V \to T$ is a linear operator and $f \in V$. We consider the problem to find $u \in V$ such that

$$\mathcal{L}u = f, \tag{4.1.196}$$

which can be written as

$$(\mathcal{L}u, v) = (f, v) \quad \forall\, v \in V. \tag{4.1.197}$$

The *Galerkin method* uses a subspace $V_h \subset V$ and an approximate solution is defined as $u_h \in V_h$ such that

$$(\mathcal{L}u_h, v_h) = (f, v_h) \quad \forall\, v_h \in V_h. \tag{4.1.198}$$

In order to obtain a computable bound on the error $e_h = u - u_h$, we shall define the *residual*

$$r_h = f - \mathcal{L}u_h \tag{4.1.199}$$

and use the *Galerkin orthogonality property*

$$(r_h, v_h) = 0 \quad \forall\, v_h \in V_h, \tag{4.1.200}$$

which follows from (4.1.197)–(4.1.199). Further, by $\mathcal{L}^*$ we denote the adjoint of $\mathcal{L}$ (see Section 1.4.5.6) and consider an auxiliary *dual problem*: find $z \in V$ such that

$$\mathcal{L}^* z = e_h, \tag{4.1.201}$$

i.e.

$$(v, \mathcal{L}^* z) = (e_h, v) \quad \forall\, v \in V. \tag{4.1.202}$$

Then we have

$$\begin{aligned} \|e_h\|^2 &= (e_h, e_h) = (e_h, \mathcal{L}^* z) \\ &= (\mathcal{L}e_h, z) = (\mathcal{L}u - \mathcal{L}u_h, z) \end{aligned} \tag{4.1.203}$$

$$= (f - \mathcal{L}u_h, z) = (r_h, z).$$

Now, using the Galerkin orthogonality property (4.1.200), for any $z_h \in V_h$ we get

$$\|e_h\|^2 = (r_h, z - z_h) = (h^s r_h, h^{-s}(z - z_h)), \tag{4.1.204}$$

where $s \geq 0$ is a suitable number and h is a suitable positive function. The application of the Cauchy inequality to (4.1.204) yields the estimate

$$\|u - u_h\|^2 \leq \|h^s r_h\| \, \|h^{-s}(z - z_h)\|, \quad z_h \in V_h. \tag{4.1.205}$$

Let us consider the above procedure in the framework of the finite element method over a mesh $\mathcal{T}_h$. Then $\mathcal{L}$ and $\mathcal{L}^*$ can be defined by $(\mathcal{L}u, v) = a(u, v) = (u, \mathcal{L}^*v)$ for all $u, v \in V$, where a is a continuous bilinear form on $V \times V$. By h we denote here a function defined (a.e.) in Ω so that $h|_{\mathring{K}} = h_K$, where $\mathring{K}$ is the interior of $K \in \mathcal{T}_h$. We now use two important ingredients:

a) *Approximation property* There exists a constant C_{appr} (independent of h) such that for each $z \in V$ sufficiently regular, i.e. $z \in V \cap H^s(\Omega)$, there exists $z_h \in V_h$ satisfying the condition

$$\|h^{-s}(z - z_h)\| \leq C_{\text{appr}} \|z\|_{H^s(\Omega)}, \tag{4.1.206}$$

cf. Corollary 4.12.

b) *Stability property of $\mathcal{L}^*$* There exists a constant $C_{\text{stab}} > 0$ (independent of h, z, e_h) such that

$$\|z\|_{H^s(\Omega)} \leq C_{\text{stab}} \|e_h\|, \tag{4.1.207}$$

where z is the solution of problem (4.1.202).

Then, from (4.1.205)–(4.1.207) we get the a posteriori error estimate in the form

$$\|u - u_h\| \leq C_{\text{appr}} C_{\text{stab}} \|h^s r_h\|. \tag{4.1.208}$$

It is clear that the applicability of this estimate depends on properties of the constants C_{appr} and C_{stab}. It is desirable that they are computable, or at least can be estimated. Moreover, the magnitude of these constants plays an important role. Our aim is to obtain an efficient error estimate in the form

$$\|u - u_h\| \leq \eta_h, \tag{4.1.209}$$

where η_h is a computable quantity called the *error estimator*. In the FEM, η_h is often expressed in the form

$$\eta_h = \left(\sum_{K \in \mathcal{T}_h} \eta_{hK}^2 \right)^{1/2}, \tag{4.1.210}$$

where η_{hK} are the so-called *error indicators*, which serve as an indication of mesh refinement. (Cf. Section 4.1.12.1.) We require that estimate (4.1.209) is reliable, i.e. not underestimated, and efficient, i.e. not overestimated. To this end, Babuška introduced the concept of the *effectivity index*

$$I_{\text{eff}} = \eta_h / \|u - u_h\|, \tag{4.1.211}$$

which should be close to one.

The above method can also be applied to obtain a posteriori error estimates for functionals depending on the computed approximate solution. Let us consider the problem to find $u \in V$ such that

$$a(u, v) = (f, v) \quad \forall\, v \in V, \tag{4.1.212}$$

with a continuous bilinear form a on $V \times V$. We are interested in the approximation of the value $M(u)$ of a continuous linear functional M defined on V. If we obtain an approximate solution $u_h \in V_h$ satisfying

$$a(u_h, v_h) = (f, v_h) \quad \forall\, v_h \in V_h, \tag{4.1.213}$$

we can compute the approximation $M(u_h)$ of $M(u)$. Our goal is to derive an a posteriori bound of the error $M(u) - M(u_h)$. Obviously, for the error $e_h = u - u_h$ we have the Galerkin orthogonality property

$$a(e_h, v_h) = 0 \quad \forall\, v_h \in V_h. \tag{4.1.214}$$

Introducing the *dual problem* to find $z \in V$ such that

$$a(w, z) = M(w) \quad \forall\, w \in V, \tag{4.1.215}$$

and using (4.1.214), we get

$$\begin{aligned} M(u) - M(u_h) &= M(e_h) = a(e_h, z - z_h) \\ &= (f, z - z_h) - a(u_h, z - z_h) \end{aligned} \tag{4.1.216}$$

for any $z_h \in V_h$. Here $(f, \cdot) - a(u_h, \cdot)$ is a continuous linear functional r_h defined on V, i.e. $r_h \in V^*$, representing the residual. Hence, we can write

$$M(u) - M(u_h) = \langle r_h, z - z_h \rangle, \qquad z_h \in V_h, \tag{4.1.217}$$

where $\langle \cdot, \cdot \rangle$ is the duality between V^* and V (cf. Section 1.4.3.2). Using properties analogous to (4.1.206) and (4.1.207) from (4.1.217) we can derive an a posteriori error estimate for the functional M:

$$|M(u) - M(u_h)| \le C_{\text{appr}}\, C_{\text{stab}} \|h^s r_h\|_{V^*}. \tag{4.1.218}$$

Of course, there is again the same problem as above concerning the magnitude of the constants C_{appr} and C_{stab} and the possibility of computing or estimating them.

Various aspects of the method à la Johnson, including applications to nonlinear conservation laws and a number of references, can be found in (Süli, 1999).

4.1.12.3 *A posteriori error estimates à la Rannacher* In order to avoid the problems with the constants C_{appr} and C_{stab} appearing in a posteriori error estimates à la Johnson, Rannacher introduced a modification of this method. His idea is to start from formula (4.1.204), i.e.

$$\|e_h\|^2 = (r_h, z - z_h)$$

and to write it as the sum

$$(r_h, z - z_h) = \sum_{K \in \mathcal{T}_h} \eta_K. \qquad (4.1.219)$$

For example, in the case of problem (3.1.1) with a homogeneous Dirichlet boundary condition prescribed on the whole boundary $\partial\Omega$, we get

$$\eta_K = \int_K (f + \Delta u_h)(z - z_h)\, dx - \frac{1}{2} \int_{\partial K} (\boldsymbol{n} \cdot [\nabla u_h])(z - z_h)\, dS, \qquad (4.1.220)$$

where $\boldsymbol{n}$ is the unit outer normal to ∂K and $[\nabla u_h]$ the jump of ∇u_h on ∂K computed in the direction of $\boldsymbol{n}$. The term η_K is estimated by

$$|\eta_K| \le \rho_K \omega_K, \qquad (4.1.221)$$

where ρ_K represents the local residual

$$\rho_K = h_K \|f + \Delta u_h\|_{0,K} + \frac{1}{2} h_K^{1/2} \|\boldsymbol{n} \cdot [\nabla u_h]\|_{0,\partial K} \qquad (4.1.222)$$

and ω_K is the weight

$$\omega_K = \max\left\{ h_K^{-1} \|z - z_h\|_{0,K},\, h_K^{-1/2} \|z - z_h\|_{0,\partial K} \right\}. \qquad (4.1.223)$$

The local residuals are computable on the basis of the obtained approximate solution u_h, and the weights are approximated in a suitable way, e.g.

$$\omega_K \approx \tilde{\omega}_K = h_K^2 |\nabla_h^2 \tilde{z}_h(x_K)|,$$

where $\tilde{z}_h \in V_h$ is the approximate solution of the dual problem (4.1.215):

$$\tilde{z}_h \in V_h, \quad a(w_h, \tilde{z}_h) = M(w_h) \quad \forall w_h \in V_h, \qquad (4.1.224)$$

computed on the current mesh $\mathcal{T}_h$. The symbol ∇_h^2 means a second order difference and x_K is the midpoint of K. The expression $\nabla_h^2 \tilde{z}_h|_K$ can be evaluated similarly as in the algorithm from point n) of Section 3.6.7 for the approximation of second order derivatives of a piecewise polynomial function.

Numerical experiments show the reliability and efficiency of this approach. See (Becker and Rannacher, 1996) for details. This method can also be adapted to nonlinear problems. For example, in (Hartmann and Houston, 2001*b*) and (Hartmann and Houston, 2001*a*) it was applied to the numerical solution of the stationary Euler equations.

4.2 Finite element solution of viscous barotropic flow

There are several models describing viscous compressible flow. The most general is the model of a real heat-conductive gas which has to be used for the simulation of high speed flow. If the flow speed is not too high (i.e. the flow is subsonic, shock waves and contact discontinuities are not present and the entropy production can be neglected) and the viscosity coefficients are constant, it is possible to use the *model of barotropic flow* formulated in Section 1.2.18. In this case the energy equation is not considered, because it is replaced by a given relation between the pressure and the density.

As one can see from Chapter 2, there exists an extensive qualitative mathematical theory of viscous barotropic flow problems. In this section we shall be concerned with the application of the conforming finite element method to viscous compressible barotropic flow. Following the results from (Kellog and Liu, 1996), (Kellog and Liu, 1997), (Liu, 2000), we shall describe the finite element discretization of the initial-boundary value problem for barotropic flow with the streamline diffusion stabilization in the continuity equation and analyse the solvability of the discrete problem.

4.2.1 *Continuous problem*

The model of nonstationary viscous compressible barotropic flow is described by the continuity equation, the Navier–Stokes equations and the equation of state, written in the form

$$\rho_t + \boldsymbol{u} \cdot \nabla \rho + \rho \operatorname{div} \boldsymbol{u} = 0, \tag{4.2.1}$$

$$\rho(\boldsymbol{u}_t + (\boldsymbol{u} \cdot \nabla)\boldsymbol{u}) = \rho \boldsymbol{f} - \nabla p + \mu \Delta \boldsymbol{u} + \eta \nabla(\operatorname{div} \boldsymbol{u}), \tag{4.2.2}$$

$$p = p(\rho), \tag{4.2.3}$$

considered in the space-time cylinder $Q_T = \Omega \times (0, T)$ and equipped with the boundary condition

$$\boldsymbol{u} = 0 \quad \text{on } \partial\Omega \times (0, T) \tag{4.2.4}$$

and the initial conditions

$$\boldsymbol{u}(x, 0) = \boldsymbol{u}^0(x), \quad \rho(x, 0) = \rho^0(x) > 0, \quad x \in \Omega. \tag{4.2.5}$$

Here $\Omega \subset I\!R^N$ ($N = 2$ or 3) is a bounded domain with a Lipschitz-continuous boundary $\Gamma = \partial\Omega$, $T \in (0, \infty)$ – the length of the time interval, ρ – density, p – pressure, $\boldsymbol{u} = (u_1, \ldots, u_N)$ – velocity vector, $\boldsymbol{f}$ – density of outer volume force, μ – dynamical viscosity and η – bulk viscosity. We assume that $\mu, \eta > 0$ and $\mu, \eta =$ const. See Section 1.2.10. We have $\boldsymbol{u} = \boldsymbol{u}(x, t)$, $\rho = \rho(x, t)$ for $x \in \Omega$, $t \in (0, T)$, and assume that $\boldsymbol{f} = \boldsymbol{f}(x)$, $x \in \Omega$. By (1.2.94),

$$p \in C^1(0, +\infty), \quad p,\ p' > 0, \tag{4.2.6}$$

and $p = p(\rho(x, t))$. Hence, $\nabla p = \frac{dp}{d\rho} \nabla \rho$. As an example, homoentropic flow with the equation of state in the form

$$p(\rho) = C\rho^{\gamma} \tag{4.2.7}$$

with constants $C > 0$ and $\gamma > 1$ can be used. (See Section 1.2.18.)

The continuity equation (4.2.1) is obtained from (1.2.26) by differentiation and the Navier–Stokes equations (4.2.2) represent (1.2.53), rewritten in the nonconservative form. (We use here the notation $\boldsymbol{u}$ for the velocity instead of $\boldsymbol{v}$.) System (4.2.1)–(4.2.3) is usually simply called the *compressible barotropic Navier–Stokes equations.*

The assumption of the barotropic flow and the nonconservative form of the governing equations cause that this model is not suitable for the simulation of high speed flow with strong shock waves. The boundary condition (4.2.4) is rather simplified assuming that the whole boundary $\partial\Omega$ is impermeable. In practical problems it is also necessary to consider inflow and outflow boundaries. In this case, we prescribe, for example, the velocity, which is not identically zero on $\partial\Omega \times (0, T)$:

$$\boldsymbol{u} = \boldsymbol{u}_D \quad \text{on } \partial\Omega \times (0, T). \tag{4.2.8}$$

Further, we define the inlet

$$\Gamma_I(t) = \{x \in \Gamma; \boldsymbol{u}_D(x,t) \cdot \boldsymbol{n}(x) < 0\}, \quad t \in (0,T), \tag{4.2.9}$$

where $\boldsymbol{n}(x)$ is the unit outer normal to Γ, and consider there the boundary condition for the density

$$\rho(x,t) = \rho_D(x,t), \quad x \in \Gamma_I(t),\ t \in (0,T). \tag{4.2.10}$$

The starting point of the application of the FEM to problem (4.2.1)–(4.2.5) is its *weak formulation.* (For simplicity we consider the problem with the homogeneous boundary condition (4.2.4). The discrete problem with nonhomogeneous boundary conditions (4.2.8), (4.2.10) is briefly discussed in Section 4.2.4.) We proceed in a standard way. We multiply equations (4.2.1) and (4.2.2) by test functions

$$q \in Q = L^2(\Omega) \quad \text{and} \quad \boldsymbol{v} \in \boldsymbol{V} = H_0^1(\Omega)^N \tag{4.2.11}$$

respectively, integrate over Ω and apply Green's theorem in viscous terms. As a result we get the identities

$$\begin{aligned} &(\rho_t, q) + (\boldsymbol{u} \cdot \nabla\rho, q) + (\rho \operatorname{div} \boldsymbol{u}, q) = 0 \quad \forall q \in Q, \\ &(\rho\boldsymbol{u}_t, \boldsymbol{v}) + (\rho(\boldsymbol{u} \cdot \nabla)\boldsymbol{u}, \boldsymbol{v}) + \mu(\nabla\boldsymbol{u}, \nabla\boldsymbol{v}) \\ &\qquad + \eta(\operatorname{div} \boldsymbol{u}, \operatorname{div} \boldsymbol{v}) = (\rho\boldsymbol{f}, \boldsymbol{v}) + (p, \operatorname{div} \boldsymbol{v}) \quad \forall \boldsymbol{v} \in \boldsymbol{V}. \end{aligned} \tag{4.2.12}$$

By $(\cdot,\cdot)$ we denote the scalar products in $L^2(\Omega)$, $L^2(\Omega)^N$ and $L^2(\Omega)^{N\times N}$ according to the situation. In view of the boundary condition (4.2.4), we can assume that for $t \in [0,T]$, the function $u(t) = u(\cdot,t) \in \boldsymbol{V}$. Moreover, $\rho(t), p(t) \in H^1(\Omega)$. For simplicity we define the following forms:

$$a(\boldsymbol{u}, \boldsymbol{v}) = \mu(\nabla\boldsymbol{u}, \nabla\boldsymbol{v}) + \eta(\operatorname{div} \boldsymbol{u}, \operatorname{div} \boldsymbol{v}), \tag{4.2.13}$$

$$b(\boldsymbol{u}, q) = (\operatorname{div} \boldsymbol{u}, q),$$
$$\alpha(\boldsymbol{u}, \rho, q) = (\boldsymbol{u} \cdot \nabla\rho, q),$$
$$d(\rho, \boldsymbol{w}, \boldsymbol{u}, \boldsymbol{v}) = (\rho(\boldsymbol{w} \cdot \nabla)\boldsymbol{u}, \boldsymbol{v}),$$
$$e(\rho, \boldsymbol{u}, q) = (\rho \operatorname{div} \boldsymbol{u}, q).$$

Then (4.2.12) can be rewritten in the form

$$(\rho_t, q) + \alpha(\boldsymbol{u}, \rho, q) + e(\rho, \boldsymbol{u}, q) = 0 \quad \forall q \in Q,\ t \in (0, T), \tag{4.2.14}$$
$$(\rho \boldsymbol{u}_t, \boldsymbol{v}) + a(\boldsymbol{u}, \boldsymbol{v}) + d(\rho, \boldsymbol{u}, \boldsymbol{u}, \boldsymbol{v}) = b(\boldsymbol{v}, p) + (\rho \boldsymbol{f}, \boldsymbol{v}) \tag{4.2.15}$$
$$\forall \boldsymbol{v} \in \boldsymbol{V},\ t \in (0, T),$$

where p satisfies (4.2.3): $p = p(\rho)$.

Let us recall the *notation* of norms in various spaces: $\|\cdot\|_{k,\alpha,\Omega} = \|\cdot\|_{W^{k,\alpha}(\Omega)}$, ($k \geq 0$ integer, $\alpha \in [1, \infty]$), $\|\cdot\|_{k,\Omega} = \|\cdot\|_{H^k(\Omega)} = \|\cdot\|_{W^{k,2}(\Omega)}$, $\|\cdot\|_\infty = \|\cdot\|_{L^\infty(Q_T)}$. The symbol $|\cdot|_{k,\Omega}$ denotes the seminorm in $H^k(\Omega)$. (See Section 1.3.)

The existence of a weak solution to problem (4.2.1), (4.2.2), (4.2.3), (4.2.5), (4.2.8) and (4.2.10) was discussed in Chapter 2.

4.2.2 *Discrete problem*

For simplicity we shall assume that the domain Ω is polygonal in the case $N = 2$ and polyhedral if $N = 3$. (If this assumption is not satisfied, then Ω is approximated by a polygonal (polyhedral) domain Ω_h.)

Let $\{\mathcal{T}_h\}_{h \in (0,h_0)}$ be a regular system of triangulations of the domain Ω (formed by closed triangles or tetrahedra for $N = 2$ or 3, respectively) with standard properties from Section 4.1. In the time interval $[0, T]$ we construct a partition $t_n = n\tau$, $n = 0, \ldots, r$, with *time step* $\tau = T/r$, where $r > 0$ is an integer. For a function w defined in $\Omega \times \{t_0, t_1, \ldots\}$ we set

$$d_t w^n = (w(t_n) - w(t_{n-1}))/\tau. \tag{4.2.16}$$

The approximate solution and test functions will be sought at each time level t_n in finite dimensional spaces of *conforming finite elements* $\boldsymbol{V}_h \subset \boldsymbol{V}$ and $Q_h \subset Q$. We set

$$Q_h = X_h^{(m)}, \quad \boldsymbol{V}_h = \left\{\boldsymbol{v}_h \in [X_h^{(k)}]^N; \boldsymbol{v}_h\big|_{\partial\Omega} = 0\right\}, \tag{4.2.17}$$

where $X_h^{(p)}$ is defined by (4.1.17). We shall consider the *inverse assumption*

$$h \leq c h_K \quad \forall K \in \mathcal{T}_h,\ \forall h \in (0, h_0) \tag{4.2.18}$$

with a positive constant c. As proven in (Ciarlet, 1979), Theorem 3.2.6, assumption (4.2.18) implies that the *inverse inequalities* hold:

$$\|\boldsymbol{w}_h\|_{0,\infty,\Omega} \leq c h^{-N/2} \|\boldsymbol{w}_h\|_{0,\Omega}, \quad \boldsymbol{w}_h \in \boldsymbol{V}_h, \tag{4.2.19}$$
$$\|q_h\|_{0,\infty,\Omega} \leq c h^{-N/2} \|q_h\|_{0,\Omega}, \quad q_h \in Q_h,$$
$$h \in (0, h_0),$$

where the constant c is independent of $\boldsymbol{u}_h, q_h$ and h. (Cf. Section 4.1.7.3.)

In the discretization of the compressible Navier–Stokes equations, which would allow us to compute a physically reasonable approximate solution, it is necessary to overcome two main obstacles: a) nonlinearity and b) the hyperbolic character of the continuity equation with respect to the density ρ. We shall use a suitable *linearization* leading to the solution of a linear algebraic system at each time level $t = t_n$, and apply *stabilization* avoiding spurious oscillations in the density and pressure. On the basis of results obtained in Section 4.1.10.1 for a scalar linear hyperbolic problem, we use the *streamline diffusion technique* in the weak form (4.2.14) of the continuity equation.

The discretization is carried out in the following way: we start from the weak formulation (4.2.14)–(4.2.15), where we approximate the spaces $\boldsymbol{V}$ and Q by $\boldsymbol{V}_h$ and Q_h, respectively, and use the approximations $\boldsymbol{u}(t_n) \approx \boldsymbol{u}_h^n \in \boldsymbol{V}_h$, $\rho(t_n) \approx \rho_h^n \in Q_h$, $\boldsymbol{u}_t(t_n) \approx d_t u_h^n = (u_h^n - u_h^{n-1})/\tau$, $\rho_t(t_n) \approx d_t \rho_h^n = (\rho_h^n - \rho_h^{n-1})/\tau$. Moreover, we carry out a linearization of the nonlinear system and in the terms with space derivatives of the continuity equation, the streamline diffusion test function

$$q_h + \delta q_{h\beta} \quad \text{with } q_{h\beta} = \delta \boldsymbol{u}_h^{n-1} \cdot \nabla q_h \tag{4.2.20}$$

and a suitable constant $\delta > 0$ will be used instead of q_h. In this way we arrive at the *formulation of the discrete problem:*

Definition 4.35 *We define an approximate solution of the compressible barotropic Navier–Stokes problem as functions $\boldsymbol{u}_h^n, \rho_h^n$, $n = 0, \dots, r$, satisfying the following conditions:*

$$\begin{aligned}
&a)\ \boldsymbol{u}_h^n \in \boldsymbol{V}_h, \quad \rho_h^n \in Q_h, \quad n = 0, \dots, r, \\
&b)\ (d_t \rho_h^n, q_h) + a(u_h^{n-1}, \rho_h^n, q_h + \delta q_{h\beta}) \\
&\qquad + e(\rho_h^{n-1}, \boldsymbol{u}_h^n, q_h + \delta q_{h\beta}) = 0 \quad \forall q_h \in Q_h, \\
&c)\ (\rho_h^{n-1} d_t \boldsymbol{u}_h^n, \boldsymbol{v}_h) + a(\boldsymbol{u}_h^n, \boldsymbol{v}_h) + d(\rho_h^{n-1}, \boldsymbol{u}_h^{n-1}, \boldsymbol{u}_h^n, \boldsymbol{v}_h) \\
&\qquad = b(\boldsymbol{v}_h, p_h^{n-1}) + (\rho_h^{n-1} \boldsymbol{f}, \boldsymbol{v}_h) \quad \forall \boldsymbol{v}_h \in \boldsymbol{V}_h.
\end{aligned} \tag{4.2.21}$$

The functions $\boldsymbol{u}_h^0$ and ρ_h^0 represent $\boldsymbol{V}_h$- and Q_h-approximations of the initial data $\boldsymbol{u}^0$ and ρ^0, respectively.

Provided the functions $\boldsymbol{u}_h^{n-1}, \rho_h^{n-1}, p_h^{n-1}$ are known, from (4.2.21) the approximate solution $\boldsymbol{u}_h^n, \rho_h^n$ should be computed. Then the corresponding discrete pressure at time t_n is evaluated from (4.2.3):

$$p_h^n = p(\rho_h^n). \tag{4.2.22}$$

System (4.2.21), c) with respect to the unknown vector function $\boldsymbol{u}_h^n$ has a parabolic character, whereas (4.2.21), b) represents a streamline diffusion discrete form of a linear first order hyperbolic equation for ρ_h^n.

4.2.3 *Existence and uniqueness of the approximate solution*

Our goal is to establish the unique solvability of the discrete problem (4.2.21).

Theorem 4.36 *Let us assume that we have computed the approximate solution* $\boldsymbol{u}_h^{n-1}, \rho_h^{n-1}$ *at time level* t_{n-1} *such that* $\rho_h^{n-1} \geq \rho_0$ *with a constant* $\rho_0 > 0$ *and let*

$$K_{n-1} = max\{\|\boldsymbol{u}_h^{n-1}\|_{0,\infty,\Omega}, \|\rho_h^{n-1}\|_{0,\infty,\Omega}\}. \tag{4.2.23}$$

If

$$\tau \leq \frac{\mu\rho_0}{2K_{n-1}^4}, \quad \frac{3}{2}\tau \leq \delta \leq \frac{\mu}{4NK_{n-1}^2}, \tag{4.2.24}$$

then there exists a unique solution $\boldsymbol{u}_h^n, \rho_h^n$ *of problem* (4.2.21) *at time level* t_n.

Proof If we introduce the simplified notation $\psi = \rho_h^{n-1}/\tau$, $\varphi = \rho_h^{n-1}$, $\boldsymbol{U} = \boldsymbol{u}_h^{n-1}$, $g = 1/\tau, K = K_{n-1}$ and the forms

$$\begin{aligned}\tilde{a}(\boldsymbol{u}, \rho, \boldsymbol{v}, q) &= (\psi\boldsymbol{u}, \boldsymbol{v}) + \mu(\nabla\boldsymbol{u}, \nabla\boldsymbol{v}) \\ &\quad + \eta(\operatorname{div}\boldsymbol{u}, \operatorname{div}\boldsymbol{v}) + ((\varphi\boldsymbol{U}\cdot\nabla)\boldsymbol{u}, \boldsymbol{v}) + (g\rho, q) \\ &\quad + (\boldsymbol{U}\cdot\nabla\rho, q + \delta q_\beta) + (\varphi\operatorname{div}\boldsymbol{u}, q + \delta q_\beta) \\ F(\boldsymbol{v}, q) &= b(\boldsymbol{v}, p(\varphi)) + (\varphi\boldsymbol{f}, \boldsymbol{v}) + (\psi\boldsymbol{U}, \boldsymbol{v}) + (g\varphi, q),\end{aligned} \tag{4.2.25}$$

the problem (4.2.21) for unknowns $\boldsymbol{u} = \boldsymbol{u}_h^n \in \boldsymbol{V}_h$, $\rho = \rho_h^n \in Q_h$ reads

$$\tilde{a}(\boldsymbol{u}, \rho, \boldsymbol{v}, q) = F(\boldsymbol{v}, q) \quad \forall\boldsymbol{v} \in \boldsymbol{V}_h, \ \forall q \in Q_h. \tag{4.2.26}$$

It is enough to prove the positivity of the bilinear form $\tilde{a}$.

Using the Cauchy inequality and the Young inequality in the form $\alpha\beta \leq \varepsilon\alpha^2 + \beta^2/(4\varepsilon)$ and (4.2.23), we find that for arbitrary $\varepsilon_1, \ldots, \varepsilon_4 > 0$

$$\begin{aligned}((\varphi\boldsymbol{U}\cdot\nabla)\boldsymbol{u}, \boldsymbol{u}) &\leq \|\varphi\boldsymbol{U}\|_{0,\infty,\Omega}\|\nabla\boldsymbol{u}\|_{0,\Omega}\|\boldsymbol{u}\|_{0,\Omega} \\ &\leq \varepsilon_1\|\nabla\boldsymbol{u}\|_{0,\Omega}^2 + \frac{K^4}{4\varepsilon_1}\|\boldsymbol{u}\|_{0,\Omega}^2, \\ (\psi\boldsymbol{u}, \boldsymbol{u}) &\geq \frac{\rho_0}{\tau}\|\boldsymbol{u}\|_{0,\Omega}^2, \\ (\boldsymbol{U}\cdot\nabla\rho, \rho) &\leq \varepsilon_2\|\rho_\beta\|_{0,\Omega}^2 + \frac{1}{4\varepsilon_2}\|\rho\|_{0,\Omega}^2, \\ (\varphi\operatorname{div}\boldsymbol{u}, \rho + \delta\rho_\beta) &\leq \varepsilon_3\|\nabla\boldsymbol{u}\|_{0,\Omega}^2 + \frac{NK^2}{4\varepsilon_3}\|\rho\|_{0,\Omega}^2 \\ &\quad + \varepsilon_4\|\nabla\boldsymbol{u}\|_{0,\Omega}^2 + \frac{NK^2\delta^2}{4\varepsilon_4}\|\rho_\beta\|_{0,\Omega}^2.\end{aligned} \tag{4.2.27}$$

From (4.2.25) and (4.2.27) it follows that

$$\begin{aligned}\tilde{a}(\boldsymbol{u}, \rho, \boldsymbol{u}, \rho) &\geq \left(\frac{\rho_0}{\tau} - \frac{K^4}{4\varepsilon_1}\right)\|\boldsymbol{u}\|_{0,\Omega}^2 \\ &\quad + (\mu - \varepsilon_1 - \varepsilon_3 - \varepsilon_4)\|\nabla\boldsymbol{u}\|_{0,\Omega}^2 + \eta\,\|\operatorname{div}\boldsymbol{u}\|_{0,\Omega}^2\end{aligned} \tag{4.2.28}$$

$$+ \left(\delta - \varepsilon_2 - \frac{N\delta^2 K^2}{4\varepsilon_4} \right) \|\rho_\beta\|_{0,\Omega}^2 + \left(\frac{1}{\tau} - \frac{1}{4\varepsilon_2} - \frac{NK^2}{4\varepsilon_3} \right) \|\rho\|_{0,\Omega}^2.$$

If we choose $\varepsilon_i = \mu/4$ for $i = 1, 3, 4$, $\varepsilon_2 = \delta/2$ and assume that

$$\tau \le \frac{\mu \rho_0}{2K^4}, \qquad (4.2.29)$$

$$\frac{3}{2}\tau \le \delta \le \frac{\mu}{4NK^2},$$

from (4.2.28) we get

$$\tilde{a}(\boldsymbol{u}, \rho, \boldsymbol{u}, \rho) \ge \frac{\rho_0}{2\tau} \|\boldsymbol{u}\|_{0,\Omega}^2 + \frac{\mu}{4} \|\nabla \boldsymbol{u}\|_{0,\Omega}^2 + \eta \|\operatorname{div} \boldsymbol{u}\|_{0,\Omega}^2 \qquad (4.2.30)$$

$$+ \frac{\delta}{4} \|\rho_\beta\|_{0,\Omega}^2 + \frac{1}{2\tau} \|\rho\|_{0,\Omega}^2.$$

Hence, the bilinear form $\tilde{a}$ is positive and problem (4.2.26) has a unique solution. □

Remark 4.37 We can note that problem (4.2.21) has a convenient structure. First, from (4.2.21), c) the approximate velocity $\boldsymbol{u}_h^n$ can be computed. Then, having substituted $\boldsymbol{u}_h^n$ in (4.2.21), b), we obtain ρ_h^n. Thus, the solution of problem (4.2.21) is split into two partially independent linear algebraic systems. Their matrices are obviously nonsymmetric. They can be solved by some iterative process suitable for nonsymmetric systems, as Krylov subspace methods, mentioned in Section 3.3.14.

An important and interesting question is the relation between the approximate and exact solution, the error estimate and the convergence of the approximate solution to the exact one as $h, \tau \to 0$ in a suitable way. Because of the complexity of the continuous as well as discrete problems, there exist very few theoretical results concerning the error analysis. Usually some convergence results and/or error estimates are obtained for methods of the numerical solution of compressible flow when they are applied to strongly simplified (e.g. scalar) problems, although these methods work in practice very well and give good results for complicated, technically relevant problems. The error estimates for the streamline diffusion finite element method (4.2.21) were analysed in the work (Liu, 2000). Unfortunately, the proof contains an error and we do not know of a corrected version. Therefore, we do not cite the error estimates from this paper.

4.2.4 *Discrete problem with nonhomogeneous boundary conditions*

Now let us pay attention to the FE solution of viscous compressible barotropic flow with nonhomogeneous boundary conditions, which is of great interest from the point of view of practical applications.

That is, we shall be concerned with the problem to solve system (4.2.1)–(4.2.3) equipped with initial conditions (4.2.5) and boundary conditions (4.2.8) and (4.2.10). For simplicity, we shall assume that the inlet $\Gamma_I \subset \Gamma = \partial\Omega$ is independent of t. This means that $\{x \in \Gamma; \boldsymbol{u}_D(x,t) \cdot \boldsymbol{n}(x) < 0\} = \Gamma_I$ for all $t \in [0,T]$.

In the discretization of this problem we proceed similarly as in Section 4.2.2. We start from the weak formulation of the continuous problem. The boundary condition (4.2.8) will be treated similarly as in Section 4.1.1. Let us assume that there exists $\boldsymbol{u}^*\ (0,T) \to \boldsymbol{H}^1(\Omega) = H^1(\Omega)^N$ such that

$$\boldsymbol{u}^*(x,t) = \boldsymbol{u}_D(x,t), \quad x \in \Gamma,\ t \in (0,T) \tag{4.2.31}$$

(in the sense of traces on Γ). Then the *weak formulation* of the Navier–Stokes equations reads

$$\text{a)}\ \boldsymbol{u}(t) - \boldsymbol{u}^*(t) \in \boldsymbol{V}, \quad t \in (0,T), \tag{4.2.32}$$

$$\begin{aligned}\text{b)}\ &(\rho\boldsymbol{u}_t, \boldsymbol{v}) + a(\boldsymbol{u},\boldsymbol{v}) + d(\rho,\boldsymbol{u},\boldsymbol{u},\boldsymbol{v}) \\ &= b(\boldsymbol{v},p) + (\rho\boldsymbol{f},\boldsymbol{v}) \quad \forall \boldsymbol{v} \in \boldsymbol{V},\ t \in (0,T).\end{aligned} \tag{4.2.33}$$

Thus, the prescribed Dirichlet boundary condition (4.2.8) is satisfied in the sense of traces on Γ for $t \in (0,T)$.

Concerning the boundary condition (4.2.10) for the density ρ prescribed on the inlet Γ_I, we shall use an analogy with Section 4.1.10.1, where the boundary condition was formulated in a 'weak integral sense' on Γ_I. That is, the weak form of the continuity equation including condition (4.2.10) reads

$$\begin{aligned}&(\rho_t, q) + \alpha(\boldsymbol{u},\rho,q) + e(\rho,\boldsymbol{u},q) - \lambda\int_{\Gamma_I} \rho\boldsymbol{u}_D \cdot \boldsymbol{n} q\, dS \\ &= -\lambda\int_{\Gamma_I} \rho_D\boldsymbol{u}_D \cdot \boldsymbol{n} q\, dS \quad \forall q \in Q,\ t \in (0,T),\end{aligned} \tag{4.2.34}$$

where $\lambda \neq 0$ is a suitable parameter.

Now we derive the discrete problem on the basis of (4.2.31), (4.2.34), completed by (4.2.3). We use the same notation as in Section 4.2.2 and set $\boldsymbol{X}_h = [X_h^{(k)}]^N$, where we shall seek the approximate velocity. The approximation of the density will be sought in Q_h. By $\boldsymbol{u}_h^* : [0,T] \to \boldsymbol{X}_h$ we denote an approximation of $\boldsymbol{u}^*$. If $\boldsymbol{u}^*$ is sufficiently regular, we can use Lagrangian finite elements and define $\boldsymbol{u}_h^*$ in such a way that

$$\begin{aligned}&\boldsymbol{u}_h^*(P_i,t) = \boldsymbol{u}_D(P_i,t) \quad \text{for all nodes } P_i \in \Gamma \text{ and } t \in [0,T], \\ &\boldsymbol{u}_h^*(P_i,t) = 0 \quad \text{for all nodes } P_i \in \Omega.\end{aligned} \tag{4.2.35}$$

We set $\boldsymbol{u}_h^{*n} = \boldsymbol{u}_h^*(t_n)$, $\rho_D^n = \rho_D(t_n)$, $\boldsymbol{u}_D^n = \boldsymbol{u}_D(t_n)$.

The *discrete problem* with nonhomogeneous boundary conditions is formulated as follows:

Definition 4.38 *We define an approximate solution of the compressible barotropic Navier–Stokes problem with nonhomogeneous boundary conditions as functions* $\boldsymbol{u}_h^n, \rho_h^n$, $n = 0, \dots, r$, *satisfying the conditions*

$$a)\ \boldsymbol{u}_h^n \in \boldsymbol{X}_h,\ \boldsymbol{u}_h^n - \boldsymbol{u}_h^{*n} \in \boldsymbol{V}_h,\ \rho_h^n \in Q_h, \tag{4.2.36}$$

$$\begin{aligned} b)\ & (d_t\rho_h^n, q_h) + \alpha(\boldsymbol{u}_h^{n-1}, \rho_h^n, q_h + \delta q_{h\beta}) \\ & + e(\rho_h^{n-1}, \boldsymbol{u}_h^n, q_h + \delta q_{h\beta}) - \lambda \int_{\Gamma_I} \rho_h^n \boldsymbol{u}_D^n \cdot \boldsymbol{n} q_h \, dS \\ & = -\lambda \int_{\Gamma_I} \rho_D^n \boldsymbol{u}_D^n \cdot \boldsymbol{n} q_h \, dS \quad \forall q_h \in Q_h, \end{aligned}$$

$$\begin{aligned} c)\ & (\rho_h^{n-1} d_t \boldsymbol{u}_h^n, \boldsymbol{v}_h) + a(\boldsymbol{u}_h^n, \boldsymbol{v}_h) \\ & + d(\rho_h^{n-1}, \boldsymbol{u}_h^{n-1}, \boldsymbol{u}_h^n, \boldsymbol{v}_h) = b(\boldsymbol{v}, p_h^{n-1}) + (\rho_h^{n-1}\boldsymbol{f}, v_h) \quad \forall \boldsymbol{v}_h \in \boldsymbol{V}_h, \end{aligned}$$

$$d)\ p_h^n = p(\rho_h^n).$$

As we see, there is a small difference between this problem and problem (4.2.21) with homogeneous boundary conditions. We can write

$$\boldsymbol{u}_h^n = \boldsymbol{u}_h^{*n} + \boldsymbol{z}_h^n, \quad \text{with} \quad \boldsymbol{z}_h^n \in \boldsymbol{V}_h. \tag{4.2.37}$$

Assuming that $\boldsymbol{u}_h^{n-1}, \rho_h^{n-1}, p_h^{n-1}$ are known, and substituting (4.2.37) in (4.2.36), we get a linear system for parameters determining the unknown functions $\boldsymbol{z}_h^n$ and ρ_h^n.

Exercise 4.39 Prove that the discrete problem (4.2.36) has a unique solution, provided $\lambda > 0$ and condition (4.2.23) is satisfied.
Hint: Write $\boldsymbol{u}_h^n$ in the form (4.2.37). Under the notation from the proof of Theorem 4.36 and $\hat{\boldsymbol{U}} = \boldsymbol{u}_h^{*n}$, the discrete problem is equivalent to finding $\boldsymbol{z}_h^n \in \boldsymbol{V}_h$ such that

$$\hat{a}(\boldsymbol{z}_h^n, \rho_h^n, \boldsymbol{v}_h, q_h) = \hat{F}(\boldsymbol{v}_h, q_h) \quad \forall \boldsymbol{v}_h \in \boldsymbol{V}_h,\ \forall q_h \in Q_h, \tag{4.2.38}$$

where

$$\hat{a}(\boldsymbol{z}, \rho, \boldsymbol{v}, q) = \tilde{a}(\boldsymbol{z}, \rho, \boldsymbol{v}, q) - \lambda \int_{\Gamma_I} \rho \boldsymbol{u}_D^n \cdot \boldsymbol{n} q \, dS, \tag{4.2.39}$$

$$\begin{aligned} \hat{F}(\boldsymbol{v}, q) = {} & F(\boldsymbol{v}, q) - (\psi\hat{\boldsymbol{U}}, \boldsymbol{v}) - \mu(\nabla\hat{\boldsymbol{U}}, \nabla\boldsymbol{v}) \\ & - \eta(\operatorname{div}\hat{\boldsymbol{U}}, \operatorname{div}\boldsymbol{v}) - ((\varphi\boldsymbol{U}\cdot\nabla)\hat{\boldsymbol{U}}, \boldsymbol{v}) \\ & - (\varphi \operatorname{div}\hat{\boldsymbol{U}}, q + \delta q_\beta) - \lambda \int_{\Gamma_I} \rho_D^n \boldsymbol{u}_D^n \cdot \boldsymbol{n} q \, dS. \end{aligned}$$

Here $\tilde{a}$ and F are defined by (4.2.25). Then

$$\hat{a}(\boldsymbol{z}, \rho, \boldsymbol{z}, \rho) = \tilde{a}(\boldsymbol{z}, \rho, \boldsymbol{z}, \rho) - \lambda \int_{\Gamma_I} \rho^2 \boldsymbol{u}_D^n \cdot \boldsymbol{n} \, dS.$$

Taking into account that $\boldsymbol{u}_D^n \cdot \boldsymbol{n} < 0$ on Γ_I and using (4.2.30), we see that the form $\hat{a}$ is positive, under the conditions $\lambda > 0$ and (4.2.24).

Exercise 4.40 Construct a linear system for parameters defining $\boldsymbol{z}_h^n$ and ρ_h^n, equivalent to (4.2.38), in the case when $k = m = 2$ and Lagrange elements from Section 4.1.2 are used.

Remark 4.41 In many cases, the Dirichlet boundary condition (4.2.31) prescribing the velocity on the whole boundary Γ of the domain Ω occupied by the gas appears rather strict on the outlet, through which the gas should leave Ω. Therefore, on the outlet one often uses a 'softer' boundary condition such as the prescribed normal component of the stress tensor. We leave the formulation of such a problem to the reader as an exercise. More details concerning the so-called natural soft outlet boundary conditions can be found in the next sections.

4.3 Finite element solution of a heat-conductive gas flow

For the description of gas flow with the Mach number exceeding possibly one, it is necessary to use a model that allows a good resolution of all phenomena appearing in the flow field. We have in mind mainly boundary layers, wakes, shock waves and shear layers (i.e. contact discontinuities smeared due to viscosity). The model of compressible barotropic (homoentropic) flow treated in Section 4.2 does not offer a sufficiently accurate description of such a flow, when strong shock waves and nonnegligible entropy production are present. For this purpose it is necessary to use the full system of conservation laws describing the flow of a real heat-conductive gas. This system consists of the continuity equation, the Navier–Stokes equations, the energy equation and thermodynamical relations and is equipped with appropriate initial and boundary conditions.

In this section, the discretization of the initial-boundary value problem describing the flow of a heat-conductive gas will be carried out by the conforming FEM. The boundary and internal layers cause that the conforming finite element solution suffers from the Gibbs phenomenon manifested by spurious oscillations. To cure this disease, it is necessary to apply a suitable stabilization. It appears that the use of the simple streamline diffusion method applied in Section 4.1.9 and Section 4.2 or the Galerkin least squares method from Section 4.1.9 are not sufficient to avoid the Gibbs phenomenon completely and it is necessary to introduce more sophisticated stabilization techniques. This section will be devoted to the derivation and explanation of various stabilized schemes proposed by Hughes, Mallet, Franca, Tezduyar, Johnson and Hansbo which can be applied with success to the numerical simulation of real gas flow with high speed and large Reynolds numbers. As numerical experiments prove, these methods can be adopted even for the case of vanishing viscosity leading to inviscid flow described by the Euler equations.

In contrast to Section 4.1, where a rigorous convergence analysis for the FEM applied to linear model problems was carried out, for the real heat-conductive gas flow no comparable results exist. Here one must rely on theoretical results obtained for simplified problems – see Section 4.1.9 on their heuristic extension to the complex problems and on their verification with the aid of suitable test problems.

4.3.1 *Continuous problem*

The complete system describing viscous compressible flow in a domain $\Omega \subset I\!R^N$ with Lipschitz-continuous boundary $\Gamma = \partial\Omega$ and in a time interval $(0, T)$ can be written in the form

$$\frac{\partial \boldsymbol{w}}{\partial t} + \sum_{i=1}^{N} \frac{\partial \boldsymbol{f}_i(\boldsymbol{w})}{\partial x_i} = \sum_{i=1}^{N} \frac{\partial \boldsymbol{R}_i(\boldsymbol{w}, \nabla \boldsymbol{w})}{\partial x_i} + \boldsymbol{F}(\boldsymbol{w}) \quad \text{in } Q_T, \tag{4.3.1}$$

where $Q_T = \Omega \times (0, T)$ and

$$\begin{aligned}
\boldsymbol{w} &= (\rho, \rho v_1, \dots, \rho v_N, E)^{\mathrm{T}} \in I\!R^m, \qquad (4.3.2)\\
m &= N + 2, \ \boldsymbol{w} = \boldsymbol{w}(x, t), \ x \in \Omega, \ t \in (0, T),\\
\boldsymbol{f}_i(\boldsymbol{w}) &= (f_{i1}, \dots, f_{im})^{\mathrm{T}}\\
&= (\rho v_i, \rho v_1 v_i + \delta_{1i} p, \dots, \rho v_N v_i + \delta_{Ni} p, (E + p) v_i)^{\mathrm{T}}\\
\boldsymbol{R}_i(\boldsymbol{w}, \nabla \boldsymbol{w}) &= (R_{i1}, \dots, R_{im})^{\mathrm{T}}\\
&= (0, \tau_{i1}, \dots, \tau_{iN}, \tau_{i1} v_1 + \cdots + \tau_{iN} v_N + k \partial\theta / \partial x_i)^{\mathrm{T}},\\
\tau_{ij} &= \lambda \operatorname{div} \boldsymbol{v} \delta_{ij} + 2\mu d_{ij}(\boldsymbol{v}), \ d_{ij}(\boldsymbol{v}) = \frac{1}{2} \left(\frac{\partial v_i}{\partial x_j} + \frac{\partial v_j}{\partial x_i} \right),\\
\boldsymbol{F}(\boldsymbol{w}) &= \rho(0, f_1, \dots, f_N, q)^{\mathrm{T}}.
\end{aligned}$$

To system (4.3.1) we add the thermodynamical relations

$$p = (\gamma - 1)(E - \rho |\boldsymbol{v}|^2 / 2), \quad \theta = \left(\frac{E}{\rho} - \frac{1}{2} |\boldsymbol{v}|^2 \right) / c_v. \tag{4.3.3}$$

As usual, we use the following *notation:* $\boldsymbol{v} = (v_1, \dots, v_N)^{\mathrm{T}}$ – velocity vector, ρ – density, p – pressure, θ – absolute temperature, E – total energy, γ – Poisson adiabatic constant, c_v – specific heat at constant volume, μ, λ – viscosity coefficients, k – heat conduction coefficient. We assume $\mu, k > 0$, $2\mu + 3\lambda \geq 0$. Usually we set $\lambda = -2\mu/3$. For details see Section 1.2. By τ_{ij} we denote here the of the viscous part of the stress tensor. For the definition of the symbol $\nabla \boldsymbol{w}$, see (1.1.13).

Under notation (4.3.2), system (4.3.1) represents the continuity equation (1.2.26), the compressible Navier–Stokes equations (1.2.45) and the energy equation (1.2.61). Relations (4.3.3) are obtained from the equation of state (1.2.66) and relations (1.2.56), b) and (1.2.69).

Similarly as in Chapter 3 devoted to the numerical solution of the Euler equations, the conservative variables forming the state vector $\boldsymbol{w}$ are used. The terms $\boldsymbol{f}_i$, $i = 1, \dots, N$, are inviscid Euler fluxes and $\boldsymbol{R}_i$, $i = 1, \dots, N$, represent viscous (diffusion) terms also called viscous fluxes. For the representation of the fluxes $\boldsymbol{f}_i$ as functions of the components w_i, $i = 1, \dots, m$, of the state vector $\boldsymbol{w}$, see (3.1.9). The explicit dependence of $\boldsymbol{R}_i$ on $\boldsymbol{w}$ and $\nabla \boldsymbol{w}$ will be given in Section 4.3.1.2.

Remark 4.42 The vector $\boldsymbol{F}$ represents the outer volume force and heat sources. Since gases are very light, the outer volume force is usually neglected. Hence, $\boldsymbol{f} \equiv 0$. Moreover, in a number of applications, heat sources are not considered and $q \equiv 0$. Then the term $\boldsymbol{F} \equiv 0$ on the right-hand side of system (4.3.1).

Using the notation from Section 1.2.23, the dimensionless form of system (4.3.1) can be written as follows (see the dimensionless form of gas dynamics equations (1.2.132)–(1.2.134) written in physical variables)

$$\frac{\partial \boldsymbol{w}'}{\partial t'} + \sum_{i=1}^{N} \frac{\partial \boldsymbol{f}_i(\boldsymbol{w}')}{\partial x_i'} = \sum_{i=1}^{N} \frac{\partial \boldsymbol{R}_i'(\boldsymbol{w}', \nabla \boldsymbol{w}')}{\partial x_i'} + \boldsymbol{F}'(\boldsymbol{w}') \quad \text{in } Q_{T'}, \tag{4.3.4}$$

where $Q_{T'} = \Omega' \times (0, T')$, $\Omega' = \frac{1}{L^*}\Omega$, $T' = TU^*/L^*$ and

$$\begin{aligned}
\boldsymbol{w}' &= (\rho', \rho' v_1', \dots, \rho' v_N', E')^{\mathrm{T}} \in I\!R^m, \\
m &= N + 2, \ \boldsymbol{w}' = \boldsymbol{w}'(x', t'), \ x' \in \Omega', \ t' \in (0, T'), \\
\boldsymbol{f}_i(\boldsymbol{w}') &= (f_{i1}, \dots, f_{im})^{\mathrm{T}} \\
&= (\rho' v_i', \rho' v_1' v_i' + \delta_{1i} p', \dots, \rho' v_N' v_i' + \delta_{Ni} p', (E' + p') v_i')^{\mathrm{T}} \\
\boldsymbol{R}_i'(\boldsymbol{w}', \nabla \boldsymbol{w}') &= (R_{i1}', \dots, R_{im}')^{\mathrm{T}} \\
&= \left(0, \tau_{i1}', \dots, \tau_{iN}', \tau_{i1}' v_1' + \dots + \tau_{iN}' v_N' + \frac{\gamma k'}{Re\, Pr} \partial \theta / \partial x_i\right)^{\mathrm{T}}, \\
\tau_{ij}' &= \frac{1}{Re} \{\lambda' \operatorname{div} \boldsymbol{v}' \delta_{ij} + 2\mu' d_{ij}(\boldsymbol{v}')\}, \ d_{ij}(\boldsymbol{v}') = \frac{1}{2}\left(\frac{\partial v_i'}{\partial x_j'} + \frac{\partial v_j'}{\partial x_i'}\right), \\
\boldsymbol{F}'(\boldsymbol{w}') &= \rho' \left(0, \frac{1}{Fr^2} f_1, \dots, \frac{1}{Fr^2} f_N, q'\right)^{\mathrm{T}}.
\end{aligned} \tag{4.3.5}$$

In (4.3.5) the partial derivatives in operators div and ∇ are with respect to x'. To system (4.3.4) we add the thermodynamical relations (4.3.3) in dimensionless form, i.e.

$$p' = (\gamma - 1)(E' - \rho' |\boldsymbol{v}'|^2 / 2), \quad \theta' = \frac{E'}{\rho'} - \frac{1}{2} |\boldsymbol{v}'|^2. \tag{4.3.6}$$

Formally, system (4.3.4) has the same form and properties as system (4.3.1) and therefore the primes are omitted if we consider the equations in dimensionless form.

System (4.2.1) is equipped with *initial conditions* written in the form

$$\boldsymbol{w}(x,0) = \boldsymbol{w}^0(x), \quad x \in \Omega, \tag{4.3.7}$$

where $\boldsymbol{w}^0(x)$ is a given vector-valued function defined in Ω.

4.3.1.1 *Boundary conditions* The choice of appropriate boundary conditions represents an important problem in CFD. Boundary conditions have to reflect physical behaviour of the flow on the boundary of the domain occupied by the fluid on one hand, and should be in agreement with the character of partial differential equations on the other hand. There are several approaches to the formulation of the boundary conditions, depending on the problem and the geometry of the domain Ω. Some basic ideas are explained in Section 1.2.22.

In what follows, let us assume that Ω is a bounded domain. (In the flow past profiles their exterior is replaced by a bounded, sufficiently large domain Ω with boundary formed by the profiles and an artificial exterior component.) We write $\partial\Omega = \Gamma_I \cup \Gamma_O \cup \Gamma_W$, where Γ_I represents the inlet through which the gas enters the domain Ω, Γ_O is the outlet through which the gas should leave Ω and Γ_W represents impermeable fixed walls.

On Γ_I one can prescribe the conditions

$$\begin{aligned} &\text{a)}\ \rho\big|_{\Gamma_I\times(0,T)} = \rho_D, \quad \text{b)}\ \boldsymbol{v}\big|_{\Gamma_I\times(0,T)} = \boldsymbol{v}_D = (v_{D1},\dots,v_{DN})^{\mathrm{T}}, \\ &\text{c)}\ \theta\big|_{\Gamma_I\times(0,T)} = \theta_D \end{aligned} \tag{4.3.8}$$

with given functions $\rho_D, \boldsymbol{v}_D, \theta_D$. The inlet Γ_I is characterized, of course, by the condition $\boldsymbol{v}_D \cdot \boldsymbol{n} < 0$ on Γ_I, where $\boldsymbol{n}$ is the unit outer normal to $\partial\Omega$. Sometimes condition (4.3.8), c) is replaced by a 'softer' *natural boundary condition*

$$\text{c}^*)\quad \sum_{j=1}^{N}\left(\sum_{i=1}^{N}\tau_{ij}n_i\right)v_j + k\frac{\partial\theta}{\partial n} = g_N \quad \text{on } \Gamma_I \times (0,T),$$

with a given function g_N. Often we set $g_N = 0$.

On Γ_W we use the no-slip boundary conditions (1.2.118). Moreover, either the temperature θ is prescribed or the *condition of adiabatic wall* with zero heat flux is applied. Hence, we have

$$\begin{aligned} &\text{a)}\ \boldsymbol{v}\big|_{\Gamma_W\times(0,T)} = 0 \quad \text{and either} \\ &\text{b)}\ \theta\big|_{\Gamma_W\times(0,T)} = \theta_D \quad \text{or} \quad \text{b}^*)\ \frac{\partial\theta}{\partial n}\Big|_{\Gamma_W\times(0,T)} = 0. \end{aligned} \tag{4.3.9}$$

The choice of suitable conditions on the outlet Γ_O is the most difficult one. If we prescribe $\boldsymbol{v}$ and θ here, we can obtain a solution which exhibits unrealistic behaviour from the physical point of view. Therefore, we try to find 'soft' *natural* outlet boundary conditions. One possibility is to prescribe, for instance, the normal component of the viscous part of the stress tensor and the heat flux:

$$\sum_{i=1}^{N} \tau_{ij} n_i = 0, \quad j = 1, \dots, N, \tag{4.3.10}$$
$$\frac{\partial \theta}{\partial n} = 0 \quad \text{on } \Gamma_O \times (0, T).$$

Note that in practical computations, the boundary conditions (4.3.8), a), b), c*), (4.3.9), a), b*) and (4.3.10) are sometimes completed by a prescribed pressure on Γ_O, applied in the inviscid terms.

4.3.1.2 *Representation of viscous fluxes in terms of conservative variables* The viscous terms $\boldsymbol{R}_i(\boldsymbol{w}, \nabla \boldsymbol{w})$ can be expressed in the form

$$\boldsymbol{R}_i(\boldsymbol{w}, \nabla \boldsymbol{w}) = \sum_{j=1}^{N} \mathbb{K}_{ij}(\boldsymbol{w}) \boldsymbol{w}_{x_j}, \quad i = 1, \dots, N, \tag{4.3.11}$$

where $\mathbb{K}_{ij}$ are $m \times m$ matrices dependent on $\boldsymbol{w}$ and $\boldsymbol{w}_{x_j} = \partial \boldsymbol{w}/\partial x_j$.

Lemma 4.43 *In the case of 2D flow ($N = 2$), the matrices $\mathbb{K}_{ij}$ have the following form:*

$$\mathbb{K}_{11} = \tag{4.3.12}$$
$$\begin{pmatrix} 0, & 0, & 0, & 0 \\ -(2\mu+\lambda)\frac{w_2}{w_1^2}, & (2\mu+\lambda)\frac{1}{w_1}, & 0, & 0 \\ -\mu\frac{w_3}{w_1^2}, & 0, & \frac{\mu}{w_1}, & 0 \\ -(2\mu+\lambda)\frac{w_2^2}{w_1^3} - \mu\frac{w_3^2}{w_1^3} + \frac{k}{c_v}\left(-\frac{w_4}{w_1^2} + \frac{w_2^2+w_3^2}{w_1^3}\right), & \left(2\mu+\lambda-\frac{k}{c_v}\right)\frac{w_2}{w_1^2}, & \left(\mu-\frac{k}{c_v}\right)\frac{w_3}{w_1^2}, & \frac{k}{c_v w_1} \end{pmatrix},$$

$$\mathbb{K}_{12} = \begin{pmatrix} 0, & 0, & 0, & 0 \\ -\lambda\frac{w_3}{w_1^2}, & 0, & \frac{\lambda}{w_1}, & 0 \\ -\mu\frac{w_2}{w_1^2}, & \frac{\mu}{w_1}, & 0, & 0 \\ -(\lambda+\mu)\frac{w_2 w_3}{w_1^3}, & \mu\frac{w_3}{w_1^2}, & \lambda\frac{w_2}{w_1^2}, & 0 \end{pmatrix},$$

$$\mathbb{K}_{21} = \begin{pmatrix} 0, & 0, & 0, & 0 \\ -\mu\frac{w_3}{w_1^2}, & 0, & \frac{\mu}{w_1}, & 0 \\ -\lambda\frac{w_2}{w_1^2}, & \frac{\lambda}{w_1}, & 0, & 0 \\ -(\lambda+\mu)\frac{w_2 w_3}{w_1^3}, & \lambda\frac{w_3}{w_1^2}, & \mu\frac{w_2}{w_1^2}, & 0 \end{pmatrix},$$

$$\mathbb{K}_{22} =$$
$$\begin{pmatrix} 0, & 0, & 0, & 0 \\ -\mu\frac{w_2}{w_1^2}, & \frac{\mu}{w_1}, & 0, & 0 \\ -(2\mu+\lambda)\frac{w_3}{w_1^2}, & 0, & (2\mu+\lambda)\frac{1}{w_1}, & 0 \\ -\mu\frac{w_2^2}{w_1^3} - (2\mu+\lambda)\frac{w_3^2}{w_1^3} + \frac{k}{c_v}\left(-\frac{w_4}{w_1^2} + \frac{w_2^2+w_3^2}{w_1^3}\right), & \left(\mu-\frac{k}{c_v}\right)\frac{w_2}{w_1^2}, & \left(2\mu+\lambda-\frac{k}{c_v}\right)\frac{w_3}{w_1^2}, & \frac{k}{c_v w_1} \end{pmatrix}.$$

Proof The proof can be carried out by the substitution of (4.3.12) into (4.3.11) and the use of (4.3.2) (where $N = 2$, $m = 4$). □

Exercise 4.44 By analogy with (4.3.12), derive the matrices $\mathbb{K}_{ij}$ from (4.3.11) for $N = 3$.

4.3.1.3 *Formulation of a compressible flow problem* In what follows, we shall pay attention to the following initial-boundary value problem describing viscous compressible flow of a heat-conductive gas:

Problem (CFP) Find a function $\boldsymbol{w} = \boldsymbol{w}(x,t)$ satisfying the Navier–Stokes system (4.3.1), the initial condition (4.3.7) and the boundary conditions (4.3.8), a), b), c*), (4.3.9), a), b*) and (4.3.10).

We define a *classical solution* of problem (CFP) as a vector-valued function $\boldsymbol{w}$ sufficiently regular in $\overline{Q}_T$ satisfying pointwise system (4.3.1) and the considered initial and boundary conditions.

4.3.1.4 *Weak formulation of problem (CFP)* As usual, the FEM for the solution of real viscous gas flow is based on the concept of a weak formulation.

The Dirichlet boundary conditions can be expressed in terms of the conservative variables in the form

$$
\begin{aligned}
w_1 = \rho_D,\ (w_2,\dots,w_{m-1})^{\mathrm{T}} &= \rho_D \boldsymbol{v}_D \quad \text{on } \Gamma_I \times (0,T), \\
w_2 = \dots = w_{m-1} &= 0 \quad \text{on } \Gamma_W \times (0,T).
\end{aligned}
\tag{4.3.13}
$$

This is reflected in the definition of the *space of test functions*

$$
\begin{aligned}
\boldsymbol{V} = \{ &\boldsymbol{\varphi} = (\varphi_1,\dots,\varphi_m)^{\mathrm{T}}; \varphi_i \in H^1(\Omega),\ i = 1,\dots,m, \\
&\varphi_1,\varphi_2,\dots,\varphi_{m-1} = 0 \text{ on } \Gamma_I,\ \varphi_2,\dots,\varphi_{m-1} = 0 \text{ on } \Gamma_W \}.
\end{aligned}
\tag{4.3.14}
$$

The boundary conditions (4.3.13) can be formulated similarly as in Section 4.1.1. Let $\boldsymbol{w}^* : [0,T] \to \boldsymbol{H}^1(\Omega) = H^1(\Omega)^m$ be a function satisfying conditions (4.3.13). Then the fact that a solution $\boldsymbol{w}$ satisfies the Dirichlet boundary conditions (4.3.13) can be expressed as the condition

$$
\boldsymbol{w}(t) - \boldsymbol{w}^*(t) \in \boldsymbol{V}, \quad t \in (0,T), \tag{4.3.15}
$$

or

$$
\boldsymbol{w}(t) \in \boldsymbol{w}^*(t) + \boldsymbol{V} := \{\boldsymbol{w}^*(t) + \boldsymbol{\varphi}; \boldsymbol{\varphi} \in \boldsymbol{V}\}, \quad t \in (0,T). \tag{4.3.16}
$$

Now, assuming that $\boldsymbol{w}$ is a classical solution of problem (CFP), we multiply equation (4.3.1) by any $\boldsymbol{\varphi} \in \boldsymbol{V}$, integrate over Ω and apply Green's theorem to viscous terms. We obtain the identity

$$
\int_\Omega \frac{\partial \boldsymbol{w}}{\partial t} \cdot \boldsymbol{\varphi}\, dx + \int_\Omega \sum_{i=1}^N \frac{\partial \boldsymbol{f}_i(\boldsymbol{w})}{\partial x_i} \cdot \boldsymbol{\varphi}\, dx \tag{4.3.17}
$$

$$+\int_{\Omega}\sum_{i=1}^{N}\boldsymbol{R}_i(\boldsymbol{w},\nabla\boldsymbol{w})\cdot\frac{\partial\boldsymbol{\varphi}}{\partial x_i}\,dx$$
$$-\int_{\partial\Omega}\sum_{i=1}^{N}n_i\boldsymbol{R}_i(\boldsymbol{w},\nabla\boldsymbol{w})\cdot\boldsymbol{\varphi}\,dS=\int_{\Omega}\boldsymbol{F}(\boldsymbol{w})\cdot\boldsymbol{\varphi}\,dx.$$

From the representation of $\boldsymbol{R}_i$ in (4.3.2) we find that

$$\int_{\partial\Omega}\sum_{i=1}^{N}n_i\boldsymbol{R}_i(\boldsymbol{w},\nabla\boldsymbol{w})\cdot\boldsymbol{\varphi}\,dS$$
$$=\int_{\partial\Omega}\sum_{i=1}^{N}\sum_{j=1}^{N}\tau_{ij}n_i\varphi_{j+1}\,dS+\int_{\partial\Omega}\left\{\sum_{j=1}^{N}\left(\sum_{i=1}^{N}\tau_{ij}v_jn_i\right)+k\frac{\partial\theta}{\partial n}\right\}\varphi_m\,dS.$$

Using the definition of the space $\boldsymbol{V}$ and boundary conditions (4.3.8), c*), (4.3.9), a), b*) and (4.3.10), we obtain

$$\int_{\partial\Omega}\sum_{i=1}^{N}n_i\boldsymbol{R}_i(\boldsymbol{w},\nabla\boldsymbol{w})\cdot\boldsymbol{\varphi}\,dS=\int_{\partial\Omega}\boldsymbol{B}\cdot\boldsymbol{\varphi}\,dS, \tag{4.3.18}$$

where $\boldsymbol{B}$ represents a boundary term corresponding to the boundary conditions (4.3.8), c*), (4.3.9), a), b*) and (4.3.10): $\boldsymbol{B}=(0,\ldots,0,g_N)^{\mathrm{T}}$ on $\Gamma_I\times(0,T)$ and $\boldsymbol{B}=0$ on $(\partial\Omega\setminus\Gamma_I)\times(0,T)$. Let us introduce the notation

$$(\boldsymbol{w},\boldsymbol{\varphi})=\int_{\Omega}\boldsymbol{w}\cdot\boldsymbol{\varphi}\,dx, \tag{4.3.19}$$
$$a(\boldsymbol{w},\boldsymbol{\varphi})=\int_{\Omega}\sum_{i=1}^{N}\boldsymbol{R}_i(\boldsymbol{w},\nabla\boldsymbol{w})\cdot\frac{\partial\boldsymbol{\varphi}}{\partial x_i}\,dx,$$
$$b(\boldsymbol{w},\boldsymbol{\varphi})=\int_{\Omega}\sum_{i=1}^{N}\frac{\partial\boldsymbol{f}_i(\boldsymbol{w})}{\partial x_i}\cdot\boldsymbol{\varphi}\,dx,$$
$$\beta(\boldsymbol{w},\boldsymbol{\varphi})=\int_{\Omega}\boldsymbol{F}(\boldsymbol{w})\cdot\boldsymbol{\varphi}dx+\int_{\partial\Omega}\boldsymbol{B}\cdot\boldsymbol{\varphi}\,dS.$$

With respect to (4.3.2), the form a can be expressed as

$$a(\boldsymbol{w},\boldsymbol{\varphi})=\sum_{i=1}^{m}a^i(\boldsymbol{w},\boldsymbol{\varphi}), \tag{4.3.20}$$

where

$$a^1(\boldsymbol{w},\boldsymbol{\varphi})\equiv 0, \tag{4.3.21}$$

$$a^{j+1}(\boldsymbol{w},\boldsymbol{\varphi}) = \int_\Omega \sum_{i=1}^{N} \left\{ \lambda \operatorname{div} \boldsymbol{v} \delta_{ij} + \mu \left(\frac{\partial v_i}{\partial x_j} + \frac{\partial v_j}{\partial x_i} \right) \right\} \frac{\partial \varphi_{j+1}}{\partial x_i}\, dx, \quad j = 1, \dots, N,$$

$$a^{m}(\boldsymbol{w},\boldsymbol{\varphi}) = \int_\Omega \sum_{j=1}^{N} \left\{ \sum_{i=1}^{N} \left(\lambda \operatorname{div} \boldsymbol{v} \delta_{ij} + \mu \left(\frac{\partial v_i}{\partial x_j} + \frac{\partial v_j}{\partial x_i} \right) \right) v_i + k \frac{\partial \theta}{\partial x_j} \right\} \frac{\partial \varphi_m}{\partial x_j}\, dx.$$

Obviously, the forms given in (4.3.19)–(4.3.21) are linear with respect to $\boldsymbol{\varphi}$ and make sense for functions $\boldsymbol{w}$ with weaker regularity than that of the classical solution. We shall not specify it here. From the point of view of the FE solution, it is sufficient to write the *weak formulation* of problem (CFP) as the conditions

$$\begin{aligned} &\text{a)}\ \ \boldsymbol{w}(t) - \boldsymbol{w}^*(t) \in \boldsymbol{V}, \quad t \in (0,T), \\ &\text{b)}\ \ \left(\frac{\partial \boldsymbol{w}(t)}{\partial t}, \boldsymbol{\varphi} \right) + a(\boldsymbol{w}(t),\boldsymbol{\varphi}) + b(\boldsymbol{w}(t),\boldsymbol{\varphi}) \\ &\qquad - \beta(\boldsymbol{w}(t),\boldsymbol{\varphi}) = 0 \quad \forall \boldsymbol{\varphi} \in \boldsymbol{V},\ t \in (0,T), \\ &\text{c)}\ \ \boldsymbol{w}(0) = \boldsymbol{w}^0. \end{aligned} \tag{4.3.22}$$

(Let us recall that $\boldsymbol{w}(t)$ is such a function that $\boldsymbol{w}(t)(x) = \boldsymbol{w}(x,t)$ for $x \in \Omega$.) A function $\boldsymbol{w}$ for which the individual terms in (4.3.22), b) make sense, satisfying conditions (4.3.22), a)-c) is called a *weak solution* of the compressible flow problem (CFP).

Identity (4.3.22), b) can also be rewritten as

$$\begin{aligned} &\left(\partial \boldsymbol{w}/\partial t + \sum_{i=1}^{N} \partial \boldsymbol{f}_i(\boldsymbol{w})/\partial x_i - \boldsymbol{F}(\boldsymbol{w}), \boldsymbol{\varphi} \right) \\ &+ \sum_{i=1}^{N} (\boldsymbol{R}_i(\boldsymbol{w}, \nabla \boldsymbol{w}), \partial \boldsymbol{\varphi}/\partial x_i) = (\boldsymbol{B}, \boldsymbol{\varphi})_{\partial\Omega} \quad \forall \boldsymbol{\varphi} \in \boldsymbol{V}. \end{aligned} \tag{4.3.23}$$

Here $(\cdot,\cdot)_{\partial\Omega}$ is the $L^2(\partial\Omega)$-scalar product. (For simplicity we omit the variable t in $\boldsymbol{w}$.) Using the representation (4.3.11) of the viscous fluxes $\boldsymbol{R}_i$ and writing the equation for $\boldsymbol{w}$ in the nonconservative form

$$\frac{\partial \boldsymbol{w}}{\partial t} + \sum_{i=1}^{N} \mathbb{A}_i(\boldsymbol{w}) \frac{\partial \boldsymbol{w}}{\partial x_i} = \sum_{i,j=1}^{N} \frac{\partial}{\partial x_i} \left(\mathbb{K}_{ij}(\boldsymbol{w}) \frac{\partial \boldsymbol{w}}{\partial x_j} \right) + \boldsymbol{F}(\boldsymbol{w}), \tag{4.3.24}$$

where $\mathbb{A}_i = D\boldsymbol{f}_i/D\boldsymbol{w}$, the weak formulation reads

$$\begin{aligned} &\text{a)}\ \ \boldsymbol{w}(t) - \boldsymbol{w}^*(t) \in \boldsymbol{V}, \quad t \in (0,T), \\ &\text{b)}\ \ \left(\frac{\partial \boldsymbol{w}(t)}{\partial t} + \sum_{i=1}^{N} \mathbb{A}_i(\boldsymbol{w}(t)) \frac{\partial \boldsymbol{w}(t)}{\partial x_i} - \boldsymbol{F}(\boldsymbol{w}(t)), \boldsymbol{\varphi} \right) \\ &\qquad + \sum_{i,j=1}^{N} \left(\mathbb{K}_{ij}(\boldsymbol{w}(t)) \frac{\partial \boldsymbol{w}(t)}{\partial x_j}, \frac{\partial \boldsymbol{\varphi}}{\partial x_i} \right) = (\boldsymbol{B}, \boldsymbol{\varphi})_{\partial\Omega} \end{aligned} \tag{4.3.25}$$

$$\forall \boldsymbol{\varphi} \in \boldsymbol{V}, \ t \in (0,T),$$
$$\text{c)} \ \ \boldsymbol{w}(0) = \boldsymbol{w}^0.$$

This formulation will be used in the following part of this section.

Exercise 4.45 a) Express the forms a^i, $i = 1, \dots, m$, with the aid of matrices $\mathbb{K}_{ij}$ from (4.3.11).

b) Derive the weak formulation of the compressible viscous flow provided conditions (4.3.8), c) and/or (4.3.9), b) are used instead of (4.3.8), c*) and/or (4.3.9), b*), respectively.

Hint: It is enough to change the definition of the space $\boldsymbol{V}$, assuming now that $\varphi_m = 0$ on Γ_I and/or Γ_W.

4.3.2 *Symmetrization of the Euler and Navier–Stokes equations*

In the thermodynamics of gases treated in Section 1.2.13 the entropy

$$S = \ln \frac{p}{\rho^\gamma} \tag{4.3.26}$$

was introduced (for simplicity here we omit the constant factor c_v) and in Section 2.3.5 a generalized (mathematical) entropy function

$$\eta = -\rho S \tag{4.3.27}$$

was defined. As mentioned in Section 2.3.5, η is a strictly convex function of the state vector $\boldsymbol{w}$ and, thus, the Hesse matrix

$$\mathbb{H}(\boldsymbol{w}) = \eta_{ww} = \left(\partial^2 \eta / \partial w_i \partial w_j\right)_{i,j=1}^{m} \tag{4.3.28}$$

is positive definite. This implies that the mapping

$$\boldsymbol{w} \to \boldsymbol{u} = \nabla_w \eta(\boldsymbol{w}) \tag{4.3.29}$$

is one-to-one and allows us to change the variables. The components u_i of $\boldsymbol{u}$ are called *entropy variables*. (Cf. 3.5.5.) For $N = 3$ the explicit form of (4.3.29) reads

$$\boldsymbol{u} = \frac{1}{\rho e} \begin{pmatrix} -w_5 + \rho e(\gamma + 1 - S) \\ w_2 \\ w_3 \\ w_4 \\ -w_1 \end{pmatrix}, \tag{4.3.30}$$

where

$$S = \ln\left((\gamma - 1)\rho e / w_1^\gamma\right), \tag{4.3.31}$$

$$\rho e = w_5 - (w_2^2 + w_3^2 + w_4^2)/(2w_1).$$

The inverse mapping $\boldsymbol{u} \to \boldsymbol{w}$ is given by

$$\boldsymbol{w} = \rho e \begin{pmatrix} -u_5 \\ u_2 \\ u_3 \\ u_4 \\ 1 - (u_2^2 + u_3^2 + u_4^2)/(2u_5) \end{pmatrix}, \tag{4.3.32}$$

where

$$\begin{aligned} \rho e &= ((\gamma - 1)/(-u_5)^\gamma)^{1/(\gamma-1)} \exp(-S/(\gamma - 1)), \\ S &= \gamma - u_1 + (u_2^2 + u_3^2 + u_4^2)/(2u_5). \end{aligned} \tag{4.3.33}$$

Using the change of variables (4.3.29), system (4.3.24) is transformed to

$$\tilde{\mathbb{A}}_0(\boldsymbol{u})\frac{\partial \boldsymbol{u}}{\partial t} + \sum_{i=1}^{N} \tilde{\mathbb{A}}_i(\boldsymbol{u})\frac{\partial \boldsymbol{u}}{\partial x_i} - \sum_{i,j=1}^{N} \frac{\partial}{\partial x_i}\left(\tilde{\mathbb{K}}_{ij}(\boldsymbol{u})\frac{\partial \boldsymbol{u}}{\partial x_j}\right) = \tilde{\boldsymbol{F}}(\boldsymbol{u}), \tag{4.3.34}$$

where

$$\begin{aligned} \tilde{\mathbb{A}}_0(\boldsymbol{u}) &= \mathbb{H}(\boldsymbol{w}(\boldsymbol{u}))^{-1}, \\ \tilde{\mathbb{A}}_i(\boldsymbol{u}) &= \mathbb{A}_i(\boldsymbol{w}(\boldsymbol{u}))\tilde{\mathbb{A}}_0(\boldsymbol{u}), \\ \tilde{\mathbb{K}}_{ij}(\boldsymbol{u}) &= \mathbb{K}_{ij}(\boldsymbol{w}(\boldsymbol{u}))\tilde{\mathbb{A}}_0(\boldsymbol{u}), \\ \tilde{\boldsymbol{F}}(\boldsymbol{u}) &= \boldsymbol{F}(\boldsymbol{w}(\boldsymbol{u})). \end{aligned} \tag{4.3.35}$$

The expression of the matrices $\tilde{\mathbb{A}}_i$ and $\tilde{\mathbb{K}}_{ij}$ in terms of $\boldsymbol{u}$-variables can be found in (Shakib *et al.*, 1991). Due to the structure of η, these matrices have the following properties:

Lemma 4.46 *(i) $\tilde{\mathbb{A}}_0$ is a symmetric positive definite $m \times m$ matrix,*
(ii) $\tilde{\mathbb{A}}_i$, $i = 1, \ldots, N$, are symmetric $m \times m$ matrices,
(iii) if $\mu \geq 0$, $2\mu + 3\lambda \geq 0$, then $\tilde{\mathbb{K}} = (\tilde{\mathbb{K}}_{ij})_{i,j=1}^N$ is a symmetric positive semi-definite $Nm \times Nm$ matrix.

For the proof, see (Harten, 1983).

Hence, the system

$$\tilde{\mathbb{A}}_0(\boldsymbol{u})\frac{\partial \boldsymbol{u}}{\partial t} + \sum_{i=1}^{N} \tilde{\mathbb{A}}_i(\boldsymbol{u})\frac{\partial \boldsymbol{u}}{\partial x_i} = 0 \tag{4.3.36}$$

is a *symmetric hyperbolic quasilinear system.* (Symmetric hyperbolic systems and the notion of entropy are linked together, as was shown in (Harten, 1983).)

The *weak formulation* of the compressible flow problem in terms of the entropy variable $\boldsymbol{u}$ can be written in the form

$$
\begin{aligned}
&\text{a)}\ \boldsymbol{u}(t) - \boldsymbol{u}^*(t) \in \boldsymbol{V}, \quad t \in (0,T),\\
&\text{b)}\ \Big(\tilde{\mathbb{A}}_0(\boldsymbol{u}(t))\frac{\partial \boldsymbol{u}(t)}{\partial t} + \sum_{i=1}^{N} \tilde{\mathbb{A}}_i(\boldsymbol{u}(t))\frac{\partial \boldsymbol{u}(t)}{\partial x_i} - \tilde{\boldsymbol{F}}(\boldsymbol{u}(t)), \boldsymbol{\varphi}\Big)\\
&\qquad + \sum_{i,j=1}^{N} \Big(\tilde{\mathbb{K}}_{ij}(\boldsymbol{u})\frac{\partial \boldsymbol{u}(t)}{\partial x_j}, \frac{\partial \boldsymbol{\varphi}}{\partial x_i}\Big) = (\tilde{\boldsymbol{B}}, \boldsymbol{\varphi})_{\partial\Omega}\\
&\qquad \forall \boldsymbol{\varphi} \in \boldsymbol{V},\ t \in (0,T),\\
&\text{c)}\ \boldsymbol{u}(0) = \boldsymbol{u}^0,
\end{aligned}
$$

where $\boldsymbol{u}^*$ is a function satisfying the prescribed Dirichlet boundary conditions.

4.3.3 *Galerkin finite element space semidiscretization and its stabilization*

Let us assume that the domain Ω is polygonal ($N = 2$) or polyhedral ($N = 3$) (otherwise, Ω is approximated by a domain Ω_h with this property) and let $\mathcal{T}_h$ be a triangulation of Ω formed by closed triangles or tetrahedra K. We assume that $\mathcal{T}_h$ has the standard properties from Section 4.1. It is formed by a finite number of closed triangles ($N = 2$) or tetrahedra ($N = 3$) K covering the closure of Ω:

$$\overline{\Omega}_h = \cup_{K \in \mathcal{T}_h} K, \tag{4.3.37}$$

and satisfying (4.1.15) ($N = 2$) or (4.1.56) ($N = 3$). By σ_h we denote the set of all vertices of all elements $K \in \mathcal{T}_h$.

The approximate solution will be sought in the finite dimensional space

$$\boldsymbol{X}_h = \left\{\boldsymbol{\varphi}_h \in C(\overline{\Omega})^m; \boldsymbol{\varphi}_h|_K \in P^p(K)^m\, \forall K \in \mathcal{T}_h\right\}, \tag{4.3.38}$$

where p is a positive integer. The space $\boldsymbol{V}$ is approximated by its subspace

$$
\begin{aligned}
\boldsymbol{V}_h &= \{\boldsymbol{\varphi}_h \in \boldsymbol{V}; \boldsymbol{\varphi}_h|_K \in P^p(K)^m\, \forall K \in \mathcal{T}_h\} \\
&= \boldsymbol{X}_h \cap \boldsymbol{V}.
\end{aligned} \tag{4.3.39}
$$

Similarly as in Section 4.1, by $P^p(K)$ we denote the space of all polynomial functions on K of degree at most p. Nevertheless, in many problems of CFD the simplest possibility $p = 1$ is considered. By $\boldsymbol{w}_h^*$ and $\boldsymbol{u}_h^*$ we denote an $\boldsymbol{X}_h$-approximation of the functions $\boldsymbol{w}^*$ and $\boldsymbol{u}^*$ representing the Dirichlet boundary conditions in (4.3.25) and (4.3.37), respectively. These functions are usually defined as elements of $\boldsymbol{X}_h$ with components satisfying the Dirichlet conditions at nodes $P_i \in \partial\Omega \cap \sigma_h$, where these conditions are prescribed. (For a more detailed formulation, see Section 4.4.3.) Further, we need $\boldsymbol{X}_h$-approximations $\boldsymbol{w}_h^0$ and $\boldsymbol{u}_h^0$ of the initial data $\boldsymbol{w}^0$ and $\boldsymbol{u}^0$, respectively. They can be defined as L^2-projections of $\boldsymbol{w}^0$ and $\boldsymbol{u}^0$ onto $\boldsymbol{X}_h$, or $\boldsymbol{X}_h$-interpolants, provided $\boldsymbol{w}^0$ and $\boldsymbol{u}^0$ are sufficiently regular. (Cf. Section 4.1.6.)

Let us apply the standard *Galerkin space semidiscretization* to problem (4.3.25) formulated with the aid of conservative variables. We define its *approximate solution* as a function $\boldsymbol{w}_h$ satisfying the conditions

$$
\begin{aligned}
&\text{a)}\ \boldsymbol{w}_h \in C^1([0,T];\boldsymbol{X}_h), && (4.3.40)\\
&\text{b)}\ \boldsymbol{w}_h(t) - \boldsymbol{w}_h^*(t) \in \boldsymbol{V}_h \quad \forall t \in (0,T),\\
&\text{c)}\ \left(\frac{\partial \boldsymbol{w}_h(t)}{\partial t} + \sum_{i=1}^{N} \mathbb{A}_i(\boldsymbol{w}_h(t))\frac{\partial \boldsymbol{w}_h(t)}{\partial x_i} - \boldsymbol{F}(\boldsymbol{w}_h(t)), \boldsymbol{\varphi}_h\right)\\
&\qquad + \sum_{i,j=1}^{N}\left(\mathbb{K}_{ij}(\boldsymbol{w}_h(t))\frac{\partial \boldsymbol{w}_h(t)}{\partial x_j}, \frac{\partial \boldsymbol{\varphi}_h}{\partial x_i}\right) = (\boldsymbol{B}, \boldsymbol{\varphi}_h)_{\partial\Omega}\\
&\qquad\qquad \forall \boldsymbol{\varphi}_h \in \boldsymbol{V}_h,\ \forall t \in (0,T),\\
&\text{d)}\ \boldsymbol{w}_h(0) = \boldsymbol{w}_h^0.
\end{aligned}
$$

The *discrete problem* to (4.3.37) in terms of the entropy variables can be formulated similarly: find $\boldsymbol{u}_h$ such that

$$
\begin{aligned}
&\text{a)}\ \boldsymbol{u}_h \in C^1([0,T];\boldsymbol{X}_h), && (4.3.41)\\
&\text{b)}\ \boldsymbol{u}_h(t) - \boldsymbol{u}_h^*(t) \in \boldsymbol{V}_h \quad \forall t \in (0,T),\\
&\text{c)}\ \left(\tilde{\mathbb{A}}_0(\boldsymbol{u}_h(t))\frac{\partial \boldsymbol{u}_h(t)}{\partial t} + \sum_{i=1}^{N} \tilde{\mathbb{A}}_i(\boldsymbol{u}_h(t))\frac{\partial \boldsymbol{u}_h(t)}{\partial x_i} - \tilde{\boldsymbol{F}}(\boldsymbol{u}_h(t)), \boldsymbol{\varphi}_h\right)\\
&\qquad + \sum_{i,j=1}^{N}\left(\tilde{\mathbb{K}}_{ij}(\boldsymbol{u}_h(t))\frac{\partial \boldsymbol{u}_h(t)}{\partial x_j}, \frac{\partial \boldsymbol{\varphi}_h}{\partial x_i}\right) = \left(\tilde{\boldsymbol{B}}, \boldsymbol{\varphi}_h\right)_{\partial\Omega}\\
&\qquad\qquad \forall \boldsymbol{\varphi}_h \in \boldsymbol{V}_h\ \forall t \in (0,T),\\
&\text{d)}\ \boldsymbol{u}_h(0) = \boldsymbol{u}^0.
\end{aligned}
$$

For our further considerations we define the operators $\mathcal{L}$ and $\tilde{\mathcal{L}}$ on $C^1([0,T];\boldsymbol{X}_h)$ in such a way that for $\boldsymbol{w}_h \in C^1([0,T];\boldsymbol{X}_h)$, we set

$$
\begin{aligned}
(\mathcal{L}\boldsymbol{w}_h)(x,t) = {} & \frac{\partial \boldsymbol{w}_h(x,t)}{\partial t} + \sum_{i=1}^{N} \mathbb{A}_i(\boldsymbol{w}_h(x,t))\frac{\partial \boldsymbol{w}_h(x,t)}{\partial x_i} && (4.3.42)\\
& - \sum_{i,j=1}^{N} \frac{\partial}{\partial x_i}\left(\mathbb{K}_{ij}(\boldsymbol{w}_h(x,t))\frac{\partial \boldsymbol{w}_h(x,t)}{\partial x_j}\right),\\
& x \in \overset{\circ}{K},\ t \in (0,T),\ K \in \mathcal{T}_h,
\end{aligned}
$$

where $\overset{\circ}{K}$ denotes the interior of K, i.e. $\overset{\circ}{K} = K \setminus \partial K$. If $\boldsymbol{u}_h \in C^1([0,T];\boldsymbol{X}_h)$, we define

$$
(\tilde{\mathcal{L}}\boldsymbol{u}_h)(x,t) = \tilde{\mathbb{A}}_0(\boldsymbol{u}_h(x,t))\frac{\partial \boldsymbol{u}_h(x,t)}{\partial t} \tag{4.3.43}
$$

$$+\sum_{i=1}^{N} \tilde{\mathbb{A}}_i(\boldsymbol{u}_h(x,t)) \frac{\partial \boldsymbol{u}_h(x,t)}{\partial x_i}$$
$$-\sum_{i,j=1}^{N} \frac{\partial}{\partial x_i} \left(\tilde{\mathbb{K}}_{ij}(\boldsymbol{u}_h(x,t)) \frac{\partial \boldsymbol{u}_h(x,t)}{\partial x_j} \right),$$
$$x \in \overset{\circ}{K}, \ t \in (0,T), \ K \in \mathcal{T}_h.$$

It has been observed that the Galerkin FEM gives rise to the *Gibbs phenomenon* (cf. 3.2.21.2) manifested by spurious oscillations, undershoots and overshoots caused by unresolved internal and boundary layers. In order to avoid the Gibbs phenomenon, suitable stabilization is used. It is based on the application of the streamline diffusion finite element method (SDFEM), also called the streamline upwind Petrov–Galerkin (SUPG) method, or its extension, called the Galerkin least squares finite element method (GLSFEM). In Section 4.1.10, we were concerned with the analysis of these methods applied to scalar linear problems and showed that they combine good stability properties with high accuracy. These methods have been developed and applied to complex problems of CFD starting from the 1980s, in particular by Hughes, Mallet, Franca, Tezduyar, Johnson and Hansbo in a series of papers ((Hughes and Brooks, 1979), (Hughes, Franca and Mizukami, 1986), (Hughes, Franca and Mallet, 1986), (Hughes and Mallet, 1986*a*), (Hughes and Mallet, 1986*b*), (Hughes *et al.*, 1987), (Hughes *et al.*, 1989), (Hansbo and Johnson, 1991), (Hansbo, 1993), (Aliabadi *et al.*, 1993), (Le Beau *et al.*, 1993)). Here we explain briefly some basic ideas and refer the reader to the mentioned literature for a deeper insight into these techniques.

We start by introducing a general form of the stabilized Galerkin scheme to problem (4.3.41). It will be obtained analogously as in Section 4.1.10.3. Under the notation (4.3.42), to identity (4.3.40), c) we add the stabilization term

$$\sum_{K \in \mathcal{T}_h} (\mathcal{L}\boldsymbol{w}_h - \boldsymbol{F}(\boldsymbol{w}_h), \boldsymbol{\psi}_h(\boldsymbol{\varphi}_h))_K, \tag{4.3.44}$$

where $(\cdot,\cdot)_K$ is the $L^2(K)$-scalar product and $\boldsymbol{\psi}_h(\boldsymbol{\varphi}_h)$ is a *stabilization perturbation* of the test function $\boldsymbol{\varphi}_h \in \boldsymbol{V}_h$. The mapping $\boldsymbol{\psi}_h$ defined on the space $\boldsymbol{V}_h$ is called the *streamline diffusion stabilization operator.* As follows from numerical experiments (see, for example, (Hughes *et al.*, 1986)), the numerical solution obtained with the aid of the SDFEM captures discontinuities or steep gradients of the exact solution in a thin numerical layer. However, within this layer the approximate solution may exhibit overshoots or undershoots. Therefore, beside the expression (4.3.44), additional term is added in the form

$$\sum_{K \in \mathcal{T}_h} (\mathcal{L}\boldsymbol{w}_h - \boldsymbol{F}(\boldsymbol{w}_h), \boldsymbol{d}_h(\boldsymbol{\varphi}_h))_K, \tag{4.3.45}$$

where $\boldsymbol{d}_h$ is the so-called *discontinuity capturing operator*. Both operators $\boldsymbol{\psi}_h$ and $\boldsymbol{d}_h$ depend linearly on the test functions $\boldsymbol{\varphi}_h$, but, in general, also on the sought approximate solution $\boldsymbol{w}_h$. Hence,

$$\begin{aligned} \boldsymbol{\psi}_h &= \boldsymbol{\psi}_h(\boldsymbol{w}_h, \boldsymbol{\varphi}_h), \\ \boldsymbol{d}_h &= \boldsymbol{d}_h(\boldsymbol{w}_h, \boldsymbol{\varphi}_h). \end{aligned} \tag{4.3.46}$$

For simplicity, we do not emphasize the dependence on $\boldsymbol{w}_h$ by notation. Therefore, we usually write $\boldsymbol{\psi}_h = \boldsymbol{\psi}_h(\boldsymbol{\varphi}_h)$, $\boldsymbol{d}_h = \boldsymbol{d}_h(\boldsymbol{\varphi}_h)$.

Using the stabilization terms (4.3.44) and (4.3.45), we replace (4.3.40), c) by the identity

$$\begin{aligned} &\left(\frac{\partial \boldsymbol{w}_h}{\partial t} + \sum_{i=1}^{N} \mathbb{A}_i(\boldsymbol{w}_h)\frac{\partial \boldsymbol{w}_h}{\partial x_i} - \boldsymbol{F}(\boldsymbol{w}_h), \boldsymbol{\varphi}_h\right) \\ &+ \sum_{i,j=1}^{N} \left(\mathbb{K}_{ij}(\boldsymbol{w}_h)\frac{\partial \boldsymbol{w}_h}{\partial x_j}, \frac{\partial \boldsymbol{\varphi}_h}{\partial x_i}\right) \\ &+ \sum_{K\in\mathcal{T}_h} (\mathcal{L}\boldsymbol{w}_h - \boldsymbol{F}(\boldsymbol{w}_h), \boldsymbol{\psi}_h(\boldsymbol{\varphi}_h))_K \\ &+ \sum_{K\in\mathcal{T}_h} (\mathcal{L}\boldsymbol{w}_h - \boldsymbol{F}(\boldsymbol{w}_h), \boldsymbol{d}_h(\boldsymbol{\varphi}_h))_K = (\boldsymbol{B}, \boldsymbol{\varphi}_h)_{\partial\Omega}, \\ &\boldsymbol{\varphi}_h \in \boldsymbol{V}_h. \end{aligned} \tag{4.3.47}$$

For simplicity we do not write here the argument t in $\boldsymbol{w}_h$.

Similarly we can write the stabilized formulation of (4.3.41), c):

$$\begin{aligned} &\left(\tilde{\mathbb{A}}_0(\boldsymbol{u}_h)\frac{\partial \boldsymbol{u}_h}{\partial t} + \sum_{i=1}^{N} \tilde{\mathbb{A}}_i(\boldsymbol{u}_h)\frac{\partial \boldsymbol{u}_h}{\partial x_i} - \tilde{\boldsymbol{F}}(\boldsymbol{u}_h), \boldsymbol{\varphi}_h\right) \\ &+ \sum_{i,j=1}^{N} \left(\tilde{\mathbb{K}}_{ij}(\boldsymbol{u}_h)\frac{\partial \boldsymbol{u}_h}{\partial x_j}, \frac{\partial \boldsymbol{\varphi}_h}{\partial x_i}\right) \\ &+ \sum_{K\in\mathcal{T}_h} \left(\tilde{\mathcal{L}}\boldsymbol{u}_h - \tilde{\boldsymbol{F}}(\boldsymbol{u}_h), \tilde{\boldsymbol{\psi}}_h(\boldsymbol{\varphi}_h)\right)_K \\ &+ \sum_{K\in\mathcal{T}_h} \left(\tilde{\mathcal{L}}\boldsymbol{u}_h - \tilde{\boldsymbol{F}}(\boldsymbol{u}_h), \tilde{\boldsymbol{d}}_h(\boldsymbol{\varphi}_h)\right)_K = \left(\tilde{\boldsymbol{B}}, \boldsymbol{\varphi}_h\right)_{\partial\Omega}, \\ &\boldsymbol{\varphi}_h \in \boldsymbol{V}_h. \end{aligned} \tag{4.3.48}$$

Now the main problem is to choose the stabilization perturbations $\boldsymbol{\psi}_h(\boldsymbol{\varphi}_h)$, $\tilde{\boldsymbol{\psi}}_h(\boldsymbol{\varphi}_h)$ of $\boldsymbol{\varphi}_h \in \boldsymbol{V}_h$ and the discontinuity capturing operators $\boldsymbol{d}_h$ and $\tilde{\boldsymbol{d}}_h$.

4.3.4 *Analysis of a linear model system with one space variable*

Let us consider the problem to find $\boldsymbol{u} : (0,1) \times (0,T) \to \mathbb{R}^m$ satisfying the equation

$$\tilde{\mathcal{L}}\boldsymbol{u} := \tilde{\mathbb{A}}_0 \frac{\partial \boldsymbol{u}}{\partial t} + \tilde{\mathbb{A}} \frac{\partial \boldsymbol{u}}{\partial x} - \tilde{\mathbb{K}} \frac{\partial^2 \boldsymbol{u}}{\partial x^2} = \tilde{\boldsymbol{F}}, \quad \text{in } Q_T = (0,1) \times (0,T), \tag{4.3.49}$$

equipped with zero Dirichlet boundary conditions $\boldsymbol{u}(0,t) = \boldsymbol{u}(1,t) = 0$, $t \in (0,T)$, and initial condition $\boldsymbol{u}(x,0) = \boldsymbol{u}^0(x)$, $x \in (0,1)$. We assume that $\tilde{\mathbb{A}}_0$, $\tilde{\mathbb{K}}$ and $\tilde{\mathbb{A}}$ are constant $m \times m$ matrices, $\tilde{\mathbb{A}}_0$ and $\tilde{\mathbb{K}}$ are symmetric positive definite, $\tilde{\mathbb{A}}$ is symmetric and $\tilde{\boldsymbol{F}} \in I\!R^m$ is a constant vector.

Let us construct the Cholesky decomposition of the matrix $\tilde{\mathbb{A}}_0$,

$$\tilde{\mathbb{A}}_0 = \mathbb{L}\mathbb{L}^{\mathrm{T}}, \tag{4.3.50}$$

where $\mathbb{L}$ is a nonsingular lower triangular matrix, and put

$$\begin{aligned} \hat{\mathbb{A}} &= \mathbb{L}^{-1}\tilde{\mathbb{A}}\mathbb{L}^{-\mathrm{T}}, \\ \hat{\mathbb{K}} &= \mathbb{L}^{-1}\tilde{\mathbb{K}}\mathbb{L}^{-\mathrm{T}}, \\ \hat{\boldsymbol{u}} &= \mathbb{L}^{\mathrm{T}}\boldsymbol{u}, \\ \hat{\boldsymbol{F}} &= \mathbb{L}^{-1}\tilde{\boldsymbol{F}}. \end{aligned} \tag{4.3.51}$$

The matrices $\hat{\mathbb{A}}$ and $\hat{\mathbb{K}}$ are symmetric and, therefore, diagonalizable. Let us assume that they both are diagonalizable with the aid of a same nonsingular matrix $\mathbb{T}$:

$$\begin{aligned} \hat{\mathbb{A}} &= \mathbb{T}\Lambda\mathbb{T}^{-1}, \quad \Lambda = \operatorname{diag}(\lambda_1, \ldots, \lambda_m), \\ \hat{\mathbb{K}} &= \mathbb{T}\mathcal{E}\mathbb{T}^{-1}, \quad \mathcal{E} = \operatorname{diag}(\varepsilon_1, \ldots, \varepsilon_m), \quad \varepsilon_1, \ldots, \varepsilon_m > 0. \end{aligned} \tag{4.3.52}$$

It is possible to choose $\mathbb{T}$ so that $\mathbb{T}^{-1} = \mathbb{T}^{\mathrm{T}}$. Defining the vectors

$$\boldsymbol{\chi} = \mathbb{T}^{-1}\hat{\boldsymbol{u}}, \quad \boldsymbol{f} = \mathbb{T}^{-1}\hat{\boldsymbol{F}}, \tag{4.3.53}$$

we get from (4.3.49) the equations for $\hat{\boldsymbol{u}}$ and $\boldsymbol{\chi}$ in the form

$$\hat{\mathcal{L}}\hat{\boldsymbol{u}} := \frac{\partial \hat{\boldsymbol{u}}}{\partial t} + \hat{\mathbb{A}} \frac{\partial \hat{\boldsymbol{u}}}{\partial x} - \hat{\mathbb{K}} \frac{\partial^2 \hat{\boldsymbol{u}}}{\partial x^2} = \hat{\boldsymbol{F}} \tag{4.3.54}$$

and

$$\mathcal{L}\boldsymbol{\chi} := \frac{\partial \boldsymbol{\chi}}{\partial t} + \Lambda \frac{\partial \boldsymbol{\chi}}{\partial x} - \mathcal{E} \frac{\partial^2 \boldsymbol{\chi}}{\partial x^2} = \boldsymbol{f}, \tag{4.3.55}$$

respectively. Both $\hat{\boldsymbol{u}}$ and $\boldsymbol{\chi}$ satisfy homogeneous boundary conditions and appropriate initial conditions. System (4.3.55) is in fact split into m independent equations for components χ_i, $i = 1, \ldots, m$, of $\boldsymbol{\chi}$:

$$\mathcal{L}_i \chi_i := \frac{\partial \chi_i}{\partial t} + \lambda_i \frac{\partial \chi_i}{\partial x} - \varepsilon_i \frac{\partial^2 \chi_i}{\partial x^2} = f_i, \quad i = 1, \ldots, m. \tag{4.3.56}$$

(Cf. Section 2.2.4.)

System (4.3.54) can be considered as a viscous perturbation of the diagonally *hyperbolic* system

$$\hat{\boldsymbol{u}}_t + \hat{\mathbb{A}}\hat{\boldsymbol{u}}_x = 0.$$

For simplicity, let us assume that in this section the matrices $\hat{\mathbb{A}}$ and $\tilde{\mathbb{A}}$ are nonsingular and, hence, $\lambda_i \neq 0$ for all $i = 1, \ldots, m$.

4.3.4.1 *Stabilized FE discrete problems* Let us consider a partition $\mathcal{T}_h$ of the interval $[0,1]$ formed by intervals $K_i = [x_i, x_{i+1}]$, $i = 0, \ldots, n$, where $x_i = ih$, $i = 0, \ldots, n+1$, and $h = 1/(n+1)$. We define the FE spaces

$$V_h = \left\{\varphi_h \in C([0,1]); \varphi_h|_{K_i} \in P^1(K_i),\ i = 0, \ldots, n, \varphi_h(0) = \varphi_h(1)\right\}, \quad (4.3.57)$$
$$\boldsymbol{V}_h = V_h^m.$$

Then the stabilized FE scheme to the initial-boundary value problems for unknown functions $\boldsymbol{u}, \hat{\boldsymbol{u}}$ and $\boldsymbol{\chi}$ read:

a) Find $\boldsymbol{u}_h \in C^1([0,T]; \boldsymbol{V}_h)$ satisfying the initial condition and

$$\begin{aligned} &\int_0^1 \left(\tilde{\mathbb{A}} \frac{\partial \boldsymbol{u}_h}{\partial x} \cdot \boldsymbol{\varphi}_h + \tilde{\mathbb{K}} \frac{\partial \boldsymbol{u}_h}{\partial x} \cdot \frac{\partial \boldsymbol{\varphi}_h}{\partial x}\right) dx \\ &\quad + \sum_{i=0}^{n} \int_{K_i} (\tilde{\mathcal{L}} \boldsymbol{u}_h - \tilde{\boldsymbol{F}}) \cdot \tilde{\boldsymbol{\psi}}_h(\boldsymbol{\varphi}_h)\, dx \\ &= \int_0^1 \left(\tilde{\boldsymbol{F}} - \tilde{\mathbb{A}}_0 \frac{\partial \boldsymbol{u}_h}{\partial t}\right) \cdot \boldsymbol{\varphi}_h\, dx \quad \forall \boldsymbol{\varphi}_h \in \boldsymbol{V}_h. \end{aligned} \tag{4.3.58}$$

b) Find $\hat{\boldsymbol{u}}_h \in C^1([0,T]; \boldsymbol{V}_h)$ satisfying the initial condition and

$$\begin{aligned} &\int_0^1 \left(\hat{\mathbb{A}} \frac{\partial \hat{\boldsymbol{u}}_h}{\partial x} \cdot \hat{\boldsymbol{\varphi}}_h + \tilde{\mathbb{K}} \frac{\partial \hat{\boldsymbol{u}}_h}{\partial x} \cdot \frac{\partial \hat{\boldsymbol{\varphi}}_h}{\partial x}\right) dx \\ &\quad + \sum_{i=0}^{n} \int_{K_i} (\hat{\mathcal{L}} \hat{\boldsymbol{u}}_h - \hat{\boldsymbol{F}}) \cdot \hat{\boldsymbol{\psi}}_h(\hat{\boldsymbol{\varphi}}_h)\, dx \\ &= \int_0^1 \left(\hat{\boldsymbol{F}} - \frac{\partial \hat{\boldsymbol{u}}_h}{\partial t}\right) \cdot \hat{\boldsymbol{\varphi}}_h\, dx \quad \forall \hat{\boldsymbol{\varphi}}_h \in \boldsymbol{V}_h. \end{aligned} \tag{4.3.59}$$

c) Find $\chi_{hi} \in V_h$, $i = 1, \ldots, m$, satisfying the initial conditions and

$$\begin{aligned} &\int_0^1 \left(\lambda_i \frac{\partial \chi_{hi}}{\partial x} \varphi_h + \varepsilon_i \frac{\partial \chi_{hi}}{\partial x} \frac{\partial \varphi_h}{\partial x}\right) \\ &\quad + \sum_{i=0}^{n} \int_{K_i} (\mathcal{L}_i \chi_{hi} - f_i) \psi_{hi}(\varphi_h)\, dx \\ &= \int_0^1 \left(f_i - \frac{\partial \chi_{hi}}{\partial t}\right) \varphi_h\, dx \quad \forall \varphi_h \in V_h,\ i = 1, \ldots, m. \end{aligned} \tag{4.3.60}$$

There is the same relation between test functions $\boldsymbol{\varphi}_h$ and $\hat{\boldsymbol{\varphi}}_h$ as between $\boldsymbol{u}_h$ and $\hat{\boldsymbol{u}}_h$: $\hat{\boldsymbol{\varphi}}_h = \mathbb{L}^{\mathrm{T}} \boldsymbol{\varphi}_h$.

Our goal is to determine the stabilization perturbations $\tilde{\boldsymbol{\psi}}_h$, $\hat{\boldsymbol{\psi}}_h$, and ψ_{hi}. Let us start from problems (4.3.60), $i = 1, \ldots, m$. If we consider a steady case and assume that $f_i = \text{const}$, we obtain the situation investigated in Example 4.24

with $v = \lambda_i$ and in Section 4.1.10. The results obtained there lead us to the idea to use the *streamline diffusion stabilization*

$$\psi_{hi}(\varphi_h) = \delta\lambda_i \partial\varphi_h/\partial x. \tag{4.3.61}$$

The optimal case (i.e. nodally exact) for the i-th component χ_i is obtained, if

$$\delta = \delta_i := \frac{h\tilde{\xi}(\alpha_i)}{2|\lambda_i|}, \tag{4.3.62}$$

where

$$\alpha_i = \frac{|\lambda_i|h}{2\varepsilon_i} \tag{4.3.63}$$

and

$$\tilde{\xi}(\alpha) = \coth\alpha - \alpha^{-1}. \tag{4.3.64}$$

Here α_i is the element Péclet number for the i-th component χ_i. It is possible to find that

$$\tilde{\xi}(\alpha) \to 1 \quad \text{as} \quad \alpha \to +\infty. \tag{4.3.65}$$

Hence, if convection strongly dominates diffusion, then $\tilde{\xi}(\alpha) \approx 1$ and we can set

$$\delta_i = \frac{1}{2}h/|\lambda_i|. \tag{4.3.66}$$

This parameter is also used in the limit case when $\varepsilon_i = 0$ and, thus, $\alpha_i = +\infty$. This allows us to cover even the case when $\mathbb{K}$ and $\hat{\mathbb{K}}$ are positive semidefinite.

We see that, in general, it is necessary to choose different streamline diffusion parameters for different components, because for δ too small for a particular component, spurious oscillations will result in that component, and for δ too large, rather diffuse results will be obtained.

Now let us concentrate on problem (4.3.59), but assume that

$$\hat{\mathbb{K}} = \operatorname{diag}(\varepsilon_1, \dots, \varepsilon_m)$$

is diagonal with $\varepsilon_1, \dots, \varepsilon_m > 0$. Using the notation $\hat{\mathbb{A}} = (\hat{a}_{ij})_{i,j=1}^m$ and taking into account that in the i-th component of the convective term $\hat{\mathbb{A}}\partial\boldsymbol{u}/\partial x$, the j-th component $\partial u_j/\partial x$ is multiplied by the 'velocity' $\hat{a}_{ij}$, we come to the idea that a suitable streamline diffusion perturbation $\hat{\psi}_h(\varphi_h)$ might be written in the form

$$\hat{\psi}_h(\varphi_h) = \delta\hat{\mathbb{A}}\frac{\partial\varphi_h}{\partial x}. \tag{4.3.67}$$

However, because of the above reasons, it is not convenient to use only one parameter δ, so we set

$$\hat{\psi}_h(\varphi_h) = \hat{\underline{\underline{\delta}}}\hat{\mathbb{A}}\frac{\partial\varphi_h}{\partial x}, \tag{4.3.68}$$

where $\hat{\underline{\underline{\delta}}}$ is a *matrix* defined by

$$\hat{\underline{\underline{\delta}}} = \mathbb{T}\mathrm{diag}(\delta_1, \dots, \delta_m)\mathbb{T}^{-1} \tag{4.3.69}$$

with parameters δ_i given by (4.3.62) and (4.3.63). In the case of unequal elements K with length h_K, δ_i needs to be replaced on K by

$$\delta_i^K = \frac{h_K \tilde{\xi}(\alpha_i^K)}{2|\lambda_i|}, \tag{4.3.70}$$

where

$$\alpha_i^K = \frac{|\lambda_i| h_K}{2\varepsilon_i}. \tag{4.3.71}$$

In analogy with (4.3.64)–(4.3.66), for strongly dominating convection on $K \in \mathcal{T}_h$ we can set

$$\hat{\underline{\underline{\delta}}} = \frac{1}{2} h_K \mathbb{T}\mathrm{diag}\left(|\lambda_1|^{-1}, \dots, |\lambda_m|^{-1}\right)\mathbb{T}^{-1}. \tag{4.3.72}$$

Now, if we substitute the representation (4.3.68) of the stabilization perturbation $\hat{\boldsymbol{\psi}}_h$ into system (4.3.59), and take into account the form (4.3.54) of the operator $\hat{\mathcal{L}}$, we obtain there the terms

$$\left(\hat{\mathbb{A}}\frac{\partial \hat{\boldsymbol{u}}_h}{\partial x}, \hat{\underline{\underline{\delta}}}\hat{\mathbb{A}}\frac{\partial \hat{\boldsymbol{\varphi}}_h}{\partial x}\right)_K = \int_K \left(\frac{\partial \hat{\boldsymbol{u}}_h}{\partial x}\right)^{\mathrm{T}} \hat{\mathbb{A}}^{\mathrm{T}}\hat{\underline{\underline{\delta}}}\hat{\mathbb{A}}\frac{\partial \hat{\boldsymbol{\varphi}}_h}{\partial x}\, dx. \tag{4.3.73}$$

The expression

$$\hat{\mathbb{K}}_v = \hat{\mathbb{A}}^{\mathrm{T}}\hat{\underline{\underline{\delta}}}\hat{\mathbb{A}} \tag{4.3.74}$$

represents an *artificial viscosity (diffusivity) matrix*. It is symmetric and positive semidefinite (even in the case of a nonsymmetric matrix $\hat{\mathbb{A}}$).

The operator $\tilde{\boldsymbol{\psi}}_h$ in (4.3.58) will be obtained with the aid of the transformation of problem (4.3.59) to (4.3.58) on the basis of relations (4.3.51). The transformation of the artificial viscosity term (4.3.73) yields the expression

$$\begin{aligned} &\int_K \mathbb{L}^{-1}\tilde{\mathbb{A}}\mathbb{L}^{-\mathrm{T}}\mathbb{L}^{\mathrm{T}}\frac{\partial \boldsymbol{u}_h}{\partial x}\,\hat{\underline{\underline{\delta}}}\,\mathbb{L}^{-1}\tilde{\mathbb{A}}\mathbb{L}^{-\mathrm{T}}\mathbb{L}^{\mathrm{T}}\frac{\partial \boldsymbol{\varphi}_h}{\partial x}\, dx \\ &\qquad = \int_K \left(\frac{\partial \boldsymbol{u}_h}{\partial x}\right)^{\mathrm{T}} \tilde{\mathbb{A}}^{\mathrm{T}}\mathbb{L}^{-\mathrm{T}}\,\hat{\underline{\underline{\delta}}}\,\mathbb{L}^{-1}\tilde{\mathbb{A}}\frac{\partial \boldsymbol{\varphi}_h}{\partial x}\, dx. \end{aligned} \tag{4.3.75}$$

This leads us to the definition of the matrix $\tilde{\underline{\underline{\delta}}}$ in the form

$$\tilde{\underline{\underline{\delta}}} = \mathbb{L}^{-\mathrm{T}}\,\hat{\underline{\underline{\delta}}}\,\mathbb{L}^{-1} = \mathbb{L}^{-\mathrm{T}}\mathbb{T}\,\mathrm{diag}(\delta_1, \dots, \delta_m)\mathbb{T}^{-1}\mathbb{L}^{-1}, \tag{4.3.76}$$

where δ_i are given by (4.3.70) on each $K \in \mathcal{T}_h$. On the basis of these considerations we define

$$\tilde{\boldsymbol{\psi}}_h(\boldsymbol{\varphi}_h) = \tilde{\underline{\underline{\delta}}}\,\tilde{\mathbb{A}}\frac{\partial \boldsymbol{\varphi}_h}{\partial x}, \tag{4.3.77}$$

which induces the artificial viscosity matrix in the form

$$\tilde{\mathbb{K}}_v = \tilde{\mathbb{A}}^{\mathrm{T}} \underline{\underline{\tilde{\delta}}} \tilde{\mathbb{A}}. \tag{4.3.78}$$

In the case when a general diffusion matrix $\hat{\mathbb{K}}$ appears in (4.3.54) instead of a diagonal matrix $\hat{\mathbb{K}} = \mathcal{E} = \mathrm{diag}(\varepsilon_1, \dots, \varepsilon_m)$ the coefficient ε is replaced in (4.3.71) by

$$\varepsilon_i := \hat{\boldsymbol{r}}_i^{\mathrm{T}} \hat{\mathbb{K}} \hat{\boldsymbol{r}}_i, \tag{4.3.79}$$

where $\hat{\boldsymbol{r}}_i$ is the eigenvector of the matrix $\hat{\mathbb{A}}$ associated with the eigenvalue λ_i:

$$(\hat{\mathbb{A}} - \lambda_i \mathbb{I})\hat{\boldsymbol{r}}_i = 0. \tag{4.3.80}$$

Obviously, the vectors $\hat{\boldsymbol{r}}_i$, $i = 1, \dots, m$, are the columns of the matrix $\mathbb{T}$. In the case of system (4.3.49) we write $\hat{\boldsymbol{r}}_i = \mathbb{L}^{\mathrm{T}} \tilde{\boldsymbol{r}}_i$ and ε_i is expressed in the form

$$\varepsilon_i = (\mathbb{L}^{\mathrm{T}} \tilde{\boldsymbol{r}}_i)^{\mathrm{T}} \mathbb{L}^{-1} \tilde{\mathbb{K}} \mathbb{L}^{-\mathrm{T}} (\mathbb{L}^{\mathrm{T}} \tilde{\boldsymbol{r}}_i) = \tilde{\boldsymbol{r}}_i^{\mathrm{T}} \tilde{\mathbb{K}} \tilde{\boldsymbol{r}}_i, \tag{4.3.81}$$

as follows from (4.3.50) and (4.3.51). Here $\tilde{\boldsymbol{r}}_i$ are the eigenvectors corresponding to the generalized eigenvalue problem

$$(\tilde{\mathbb{A}} - \lambda_i \tilde{\mathbb{A}}_0)\tilde{\boldsymbol{r}}_i = 0. \tag{4.3.82}$$

4.3.5 *Multidimensional problems*

The results obtained in the previous section are generalized to problems with more space dimensions. Since the mesh $\mathcal{T}_h$ is, in general, unstructured and/or anisotropic, the geometry of elements $K \in \mathcal{T}_h$ is taken into account with the aid of the mapping (4.1.39), i.e. $x = F_K(\xi) = \mathbb{B}_K \xi + \boldsymbol{b}_K$ of the reference element $\hat{K}$ onto K. (We write ξ here instead of $\hat{x}$.) Hence, $\xi = F_K^{-1}(x)$. By $\partial \xi_i / \partial x_j$ we denote the derivative of the i-th component of the inverse F_K^{-1} with respect to the variable x_j.

In the sequel, we introduce some examples of the streamline diffusion and shock capturing stabilization operators, following (Hughes and Mallet, 1986*a*), (Hughes and Mallet, 1986*b*), (Le Beau *et al.*, 1993) and (Rachowicz, 1997).

4.3.5.1 *Compact form of governing equations* Let us introduce the notation

$$\boldsymbol{w} := (w_1, \dots, w_m)^{\mathrm{T}}, \tag{4.3.83}$$

$$\boldsymbol{\nabla}^{\mathrm{T}} := (\mathbb{I}\partial_1, \dots, \mathbb{I}\partial_N), \tag{4.3.84}$$

$$(\boldsymbol{\nabla} \boldsymbol{w})^{\mathrm{T}} = (\partial_1 \boldsymbol{w}^{\mathrm{T}}, \dots, \partial_N \boldsymbol{w}^{\mathrm{T}}), \tag{4.3.85}$$

$$\mathbb{A}^{\mathrm{T}} := (\mathbb{A}_1, \dots, \mathbb{A}_N), \tag{4.3.86}$$

$$\mathbb{A} \cdot \boldsymbol{\nabla} \boldsymbol{w} := \mathbb{A}^{\mathrm{T}} \boldsymbol{\nabla} \boldsymbol{w} = \sum_{i=1}^{N} \mathbb{A}_i \frac{\partial \boldsymbol{w}}{\partial x_i}, \tag{4.3.87}$$

$$\mathbb{K} := \begin{pmatrix} \mathbb{K}_{11}, & \ldots, & \mathbb{K}_{1N} \\ \vdots & & \vdots \\ \mathbb{K}_{N1}, & \ldots, & \mathbb{K}_{NN} \end{pmatrix}, \tag{4.3.88}$$

$$\boldsymbol{\nabla} \cdot \mathbb{K}\boldsymbol{\nabla}\boldsymbol{w} := \boldsymbol{\nabla}^{\mathrm{T}}(\mathbb{K}\boldsymbol{\nabla}\boldsymbol{w}) = \sum_{i,j=1}^{N} \left(\mathbb{K}_{ij}\boldsymbol{w}_{x_j}\right)_{x_i}. \tag{4.3.89}$$

Here $\mathbb{I}$ is the $m \times m$ identity matrix, $\mathbb{A}_i$ and $\mathbb{K}_{ij}$ are $m \times m$ matrices, $\boldsymbol{\nabla}$ is an $(Nm) \times m$ differential operator matrix, $\mathbb{A}$ is an $(Nm) \times m$ matrix, $\boldsymbol{\nabla}\boldsymbol{w}$ is an Nm-dimensional column vector and $\mathbb{K}$ is a symmetric and positive semidefinite $(Nm) \times (Nm)$ matrix.

Under the above notation, system (4.3.24) can be written in a *compact form*

$$\mathcal{L}\boldsymbol{w} := \boldsymbol{w}_t + \mathbb{A} \cdot \boldsymbol{\nabla}\boldsymbol{w} - \boldsymbol{\nabla} \cdot \mathbb{K}\boldsymbol{\nabla}\boldsymbol{w}. \tag{4.3.90}$$

Then identity (4.3.44) reads

$$\begin{aligned} &\int_\Omega (\boldsymbol{\varphi}_h \cdot (\mathbb{A} \cdot \boldsymbol{\nabla}\boldsymbol{w}_h) + \boldsymbol{\nabla}\boldsymbol{\varphi}_h \cdot (\mathbb{K}\boldsymbol{\nabla}\boldsymbol{w}_h))\, dx \\ &\quad + \sum_{K \in \mathcal{T}_h} \int_K (\mathcal{L}\boldsymbol{w}_h - \boldsymbol{F}) \cdot (\boldsymbol{\psi}_h(\boldsymbol{\varphi}_h) + \boldsymbol{d}_h(\boldsymbol{\varphi}_h))\, dx \\ &= \int_\Omega \boldsymbol{\varphi}_h \cdot (\boldsymbol{F} - \partial\boldsymbol{w}_h/\partial t)\, dx + \int_{\partial\Omega} \boldsymbol{B} \cdot \boldsymbol{\varphi}_h\, dS \\ &\qquad \forall \boldsymbol{\varphi}_h \in \boldsymbol{V}_h. \end{aligned} \tag{4.3.91}$$

(For simplicity we omit the argument $\boldsymbol{w}_h$ in $\mathbb{A}, \mathbb{K}, \boldsymbol{F}$ and $\boldsymbol{B}$ and t in $\boldsymbol{w}_h$).

Similarly we rewrite system (4.3.34) for the entropy variables:

$$\tilde{\mathcal{L}}\boldsymbol{u} := \tilde{\mathbb{A}}_0\boldsymbol{u}_t + \tilde{\mathbb{A}} \cdot \boldsymbol{\nabla}\boldsymbol{u} - \boldsymbol{\nabla} \cdot \tilde{\mathbb{K}}\boldsymbol{\nabla}\boldsymbol{u}. \tag{4.3.92}$$

Then (4.3.48) becomes

$$\begin{aligned} &\int_\Omega \left(\boldsymbol{\varphi}_h \cdot (\tilde{\mathbb{A}} \cdot \boldsymbol{\nabla}\boldsymbol{u}_h) + \boldsymbol{\nabla}\boldsymbol{\varphi}_h \cdot (\tilde{\mathbb{K}}\boldsymbol{\nabla}\boldsymbol{u}_h)\right) dx \\ &\quad + \sum_{K \in \mathcal{T}_h} \int_K (\tilde{\mathcal{L}}\boldsymbol{u}_h - \tilde{\boldsymbol{F}}) \cdot \left(\tilde{\boldsymbol{\psi}}_h(\boldsymbol{\varphi}_h) + \tilde{\boldsymbol{d}}_h(\boldsymbol{\varphi}_h)\right) dx \\ &= \int_\Omega \boldsymbol{\varphi}_h \cdot (\tilde{\boldsymbol{F}} - \tilde{\mathbb{A}}_0\partial\boldsymbol{u}_h/\partial t)\, dx + \int_{\partial\Omega} \tilde{\boldsymbol{B}} \cdot \boldsymbol{\varphi}_h\, dS \\ &\qquad \forall \boldsymbol{\varphi}_h \in \boldsymbol{V}_h. \end{aligned} \tag{4.3.93}$$

Let us assume that system (4.3.92) is linear and that the matrices $\tilde{\mathbb{A}}_0, \tilde{\mathbb{A}}_i, \tilde{\mathbb{K}}_{ij}$ and the vectors $\tilde{\boldsymbol{F}}$ and $\tilde{\boldsymbol{B}}$ are constant. Hence, also $\tilde{\mathbb{A}}$ and $\tilde{\mathbb{K}}$ are constant. Moreover, we assume that $\tilde{\mathbb{A}}_0$ is symmetric positive definite, $\tilde{\mathbb{A}}_i$ are symmetric and

$\tilde{\mathbb{K}}$ is symmetric positive semidefinite. Now we apply transformations similar to (4.3.50), (4.3.51):

$$\begin{aligned} \tilde{\mathbb{A}}_0 &= \mathbb{L}\mathbb{L}^{\mathrm{T}}, \\ \hat{\boldsymbol{u}} &= \mathbb{L}^{\mathrm{T}}\boldsymbol{u},\ \hat{\boldsymbol{\varphi}} = \mathbb{L}^{\mathrm{T}}\boldsymbol{\varphi}, \\ \hat{\mathbb{A}}_i &= \mathbb{L}^{-1}\tilde{\mathbb{A}}_i\mathbb{L}^{-\mathrm{T}}, \\ \hat{\mathbb{K}}_{ij} &= \mathbb{L}^{-1}\tilde{\mathbb{K}}_{ij}\mathbb{L}^{-\mathrm{T}}, \\ \hat{\boldsymbol{F}} &= \mathbb{L}^{-1}\tilde{\boldsymbol{F}},\ \hat{\boldsymbol{B}} = \mathbb{L}^{-1}\tilde{\boldsymbol{B}}. \end{aligned} \tag{4.3.94}$$

Then $\hat{\mathbb{A}}_i$ are symmetric $m \times m$ matrices, $\hat{\mathbb{K}}$ is symmetric positive semidefinite and (4.3.92) is equivalent to

$$\hat{\mathcal{L}}\hat{\boldsymbol{u}} := \hat{\mathbb{A}}_0\hat{\boldsymbol{u}}_t + \hat{\mathbb{A}} \cdot \boldsymbol{\nabla}\hat{\boldsymbol{u}} - \boldsymbol{\nabla} \cdot \hat{\mathbb{K}}\boldsymbol{\nabla}\hat{\boldsymbol{u}} = \hat{\boldsymbol{F}}. \tag{4.3.95}$$

The corresponding discrete stabilized problem reads

$$\begin{aligned} &\int_\Omega \left(\hat{\boldsymbol{\varphi}}_h \cdot (\hat{\mathbb{A}} \cdot \boldsymbol{\nabla}\hat{\boldsymbol{u}}_h) + \boldsymbol{\nabla}\hat{\boldsymbol{\varphi}}_h \cdot (\hat{\mathbb{K}}\boldsymbol{\nabla}\hat{\boldsymbol{u}}_h)\right) dx \\ &\quad + \sum_{K\in\mathcal{T}_h} \int_K (\hat{\mathcal{L}}\hat{\boldsymbol{u}}_h - \hat{\boldsymbol{F}}) \cdot \left(\hat{\boldsymbol{\psi}}_h(\hat{\boldsymbol{\varphi}}_h) + \hat{\boldsymbol{d}}_h(\hat{\boldsymbol{\varphi}}_h)\right) dx \\ &= \int_\Omega \hat{\boldsymbol{\varphi}}_h \cdot (\hat{\boldsymbol{F}} - \hat{\mathbb{A}}_0\partial\hat{\boldsymbol{u}}_h/\partial t)\, dx + \int_{\partial\Omega} \hat{\boldsymbol{B}} \cdot \hat{\boldsymbol{\varphi}}_h\, dS \\ &\qquad \forall \hat{\boldsymbol{\varphi}}_h \in \boldsymbol{V}_h. \end{aligned} \tag{4.3.96}$$

4.3.5.2 *Matrix functions* Let f be a real function and let $\mathbb{G}$ be a diagonalizable $m \times m$ matrix with real eigenvalues $\mu_1, \ldots, \mu_m$. Then there exists a nonsingular matrix $\mathbb{T}$ such that

$$\mathbb{G} = \mathbb{T}\,\mathrm{diag}(\mu_1, \ldots, \mu_m)\mathbb{T}^{-1}. \tag{4.3.97}$$

The columns $\boldsymbol{v}_i$, $i = 1, \ldots, m$, of $\mathbb{T}$ are eigenvectors of $\mathbb{G}$ associated with its eigenvalues μ_i. Now we define

$$f(\mathbb{G}) = \mathbb{T}\,\mathrm{diag}\left(f(\mu_1), \ldots, f(\mu_m)\right)\mathbb{T}^{-1}, \tag{4.3.98}$$

provided the values $f(\mu_i)$, $i = 1, \ldots, m$, make sense.

This allows us to define, for example, the matrix

$$|\mathbb{G}|^p = \mathbb{T}\,\mathrm{diag}\left(|\mu_1|^p, \ldots, |\mu_m|^p\right)\mathbb{T}^{-1} \tag{4.3.99}$$

for $p \geq 0$.

If $\mu_i \neq 0$ for all $i = 1, \ldots, m$, which means that $\mathbb{G}$ is nonsingular, we have

$$\mathbb{G}^{-1} = \mathbb{T}\,\mathrm{diag}\left(\mu_1^{-1}, \ldots, \mu_m^{-1}\right)\mathbb{T}^{-1}. \tag{4.3.100}$$

If $\mathbb{G}$ is symmetric, then $\boldsymbol{v}_i$ can be chosen mutually orthonormal, which means that $\mathbb{T}^{-1} = \mathbb{T}^{\mathrm{T}}$, $\boldsymbol{v}_i^{\mathrm{T}}\boldsymbol{v}_j = \delta_{ij}$ and

$$\mathbb{G}^{-1} = \sum_{i=1}^{m} \mu_i^{-1} \boldsymbol{v}_i \boldsymbol{v}_i^{\mathrm{T}}. \tag{4.3.101}$$

In the case of a *symmetric singular* matrix $\mathbb{G}$, when $\mu_i \neq 0$ for $i = 1, \dots, \ell < m$ and $\mu_i = 0$ for $i = \ell + 1, \dots, m$, we define the *generalized inverse* $\mathbb{G}^{-1}$ of the matrix $\mathbb{G}$ by

$$\mathbb{G}^{-1} = \sum_{i=1}^{\ell} \mu_i^{-1} \boldsymbol{v}_i \boldsymbol{v}_i^{\mathrm{T}}. \tag{4.3.102}$$

In other words, $\mathbb{G}^{-1}$ is the inverse of the operator $\mathbb{G} : I\!R^m \to I\!R^m$ restricted on its nondegenerate subspace spanned by all eigenvectors $\boldsymbol{v}_i$ corresponding to nonzero eigenvalues μ_i. Setting

$$\begin{aligned} \overline{\mathbb{T}} &= (\boldsymbol{v}_1, \dots, \boldsymbol{v}_\ell) \quad (m \times \ell \,\text{matrix}), \\ \overline{\mathbb{M}} &= \operatorname{diag}\left(\mu_1^{-1}, \dots, \mu_\ell^{-1}\right) \quad (\ell \times \ell \,\text{matrix}), \end{aligned} \tag{4.3.103}$$

we can write $\mathbb{G}^{-1}$ in the matrix form

$$\mathbb{G}^{-1} = \overline{\mathbb{T}}\,\overline{\mathbb{M}}\,\overline{\mathbb{T}}^{\mathrm{T}}. \tag{4.3.104}$$

4.3.5.3 *Determination of the streamline diffusion operator* $\hat{\psi}_h$ For simplicity, in the sequel we shall sometimes use the Einstein summation convention over repeated indices.

By analogy with representation (4.3.68) of the streamline diffusion operator for a system with one space variable, we set

$$\hat{\psi}_h(\hat{\varphi}_h) = \underline{\underline{\hat{\delta}}}\, \hat{\mathbb{A}} \cdot \boldsymbol{\nabla} \hat{\varphi}_h = \underline{\underline{\hat{\delta}}} \sum_{i=1}^{N} \hat{\mathbb{A}}_i \frac{\partial \hat{\varphi}_h}{\partial x_i}. \tag{4.3.105}$$

We assume that $\underline{\underline{\hat{\delta}}}$ is a symmetric positive semidefinite $m \times m$ matrix. This yields the artificial viscosity term with the artificial viscosity $(Nm) \times (Nm)$ matrix

$$\hat{\mathbb{K}}_v = \hat{\mathbb{A}}\, \underline{\underline{\hat{\delta}}}\, \hat{\mathbb{A}}^{\mathrm{T}}. \tag{4.3.106}$$

Now we shall be concerned with the determination of the matrix $\underline{\underline{\hat{\delta}}}$. Let us set

$$\begin{aligned} \hat{\mathbb{B}}_i &= 2 \frac{\partial \xi_i}{\partial x_j} \hat{\mathbb{A}}_j \quad (\text{summation over}\, j), \\ \hat{\mathbb{B}}^{\mathrm{T}} &= \left(\hat{\mathbb{B}}_1^{\mathrm{T}}, \dots, \hat{\mathbb{B}}_N^{\mathrm{T}}\right). \end{aligned} \tag{4.3.107}$$

Observe that for a uniform triangulation $\mathcal{T}_h$ with elements having the same size h in all directions x_j, we get

$$\hat{\mathbb{B}}_i = \frac{2}{h} \hat{\mathbb{A}}_i. \tag{4.3.108}$$

For $p \in [1, \infty)$ we introduce the $m \times m$ *matrix-valued p-norm* of the matrix $\hat{\mathbb{B}}$:

$$|\hat{\mathbb{B}}|_p = \left(\sum_{i=1}^{N} |\hat{\mathbb{B}}_i|^p \right)^{1/p} . \tag{4.3.109}$$

See (4.3.99). Then we define the matrix $\underline{\underline{\hat{\delta}}}$ as

$$\underline{\underline{\hat{\delta}}} = \vartheta |\hat{\mathbb{B}}|_p^{-1} \tag{4.3.110}$$

with the generalized inverse of the symmetric matrix $|\hat{\mathbb{B}}|_p$. Here $\vartheta > 0$ is a suitable factor taking into account the relation between the magnitude of convection and diffusion. For strongly dominating convection we set $\vartheta = 1$. Denoting by $\boldsymbol{v}_1, \ldots, \boldsymbol{v}_m$ orthonormal eigenvectors of $|\hat{\mathbb{B}}|_p$ associated with eigenvalues $\mu_1, \ldots, \mu_m$ such that $\mu_i \neq 0$ for $i = 1, \ldots, \ell$ and $\mu_i = 0$ for $i = \ell + 1, \ldots, m$, and using (4.3.102)–(4.3.104), we get

$$\underline{\underline{\hat{\delta}}} = \vartheta \sum_{i=1}^{\ell} \mu_i^{-1} \boldsymbol{v}_i \boldsymbol{v}_i^{\mathrm{T}} = \vartheta \overline{\mathbb{T}}\, \overline{\mathbb{M}}\, \overline{\mathbb{T}}^{\mathrm{T}}. \tag{4.3.111}$$

It can be verified that this choice is in agreement with results obtained in Section 4.3.4 for a system with one space variable as well as with results obtained in Section 4.1.10 for a multidimensional scalar stationary equation. Actually, if $N = 1$ and $\Omega = (0, 1)$, then $\hat{\mathbb{A}} = \hat{\mathbb{A}}_1$ and $\hat{\mathbb{B}} = \hat{\mathbb{B}}_1$ are symmetric $m \times m$ matrices. Assuming that $\hat{\mathbb{A}}$ is nonsingular, all its eigenvalues $\lambda_i \neq 0$, $i = 1, \ldots, m$. Moreover, $\hat{\mathbb{A}}$ is diagonalizable in the form

$$\hat{\mathbb{A}} = \mathbb{T} \operatorname{diag}(\lambda_1, \ldots, \lambda_m) \mathbb{T}^{\mathrm{T}}.$$

If the mesh $\mathcal{T}_h$ in $(0, 1)$ is uniform with size h, then $\hat{\mathbb{B}} = \frac{2}{h} \hat{\mathbb{A}}$ has the same eigenvectors $\boldsymbol{v}_i$, $i = 1, \ldots, m$, as $\hat{\mathbb{A}}$ and eigenvalues $\mu_i = 2\lambda_i / h$. For $p = 1$ and $\vartheta = 1$ we get

$$\underline{\underline{\hat{\delta}}} = \frac{1}{2} h |\hat{\mathbb{A}}|^{-1} = \frac{1}{2} h \mathbb{T} \operatorname{diag}\left(|\lambda_1|^{-1}, \ldots, |\lambda_m|^{-1}\right) \mathbb{T}^{\mathrm{T}},$$

which is in agreement with (4.3.72) derived in Section 4.3.4 for dominating convection.

Further, let us consider equation (4.1.107) discretized on a uniform mesh $\mathcal{T}_h$ and choose $p = 2$. Then $\partial \xi_i / \partial x_j = h^{-1} \delta_{ij}$, $\hat{\mathbb{B}}_i = 2 v_i / h$ and

$$\hat{\psi}_h(\hat{\varphi}_h) = \hat{\delta} \sum_{i=1}^{N} v_i \frac{\partial \hat{\varphi}_h}{\partial x_i}$$

and

$$\hat{\delta} = \frac{h}{2|v|},$$

which is obtained from (4.1.166) for $Pe \to \infty$.

It is possible to introduce a further *generalization* of the definition of the matrix $\underline{\underline{\hat{\delta}}}$, combining the above results with the choice (4.3.69)–(4.3.71) and (4.3.79) of the matrix $\underline{\underline{\hat{\delta}}}$ in the case with one space variable. We set

$$\underline{\underline{\hat{\delta}}} = \overline{\mathbb{T}}\,\mathrm{diag}(\delta_1, \dots, \delta_\ell)\overline{\mathbb{T}}^{\mathrm{T}}, \tag{4.3.112}$$

where the parameters δ_i are determined by the following relations:

$$\begin{aligned} \delta_i &= \tilde{\xi}(\alpha_i)/\mu_i, \\ \alpha_i &= \mu_i/\sigma_i, \\ \sigma_i &= \frac{1}{N}\sum_{j,k=1}^{N} \boldsymbol{v}_i^{\mathrm{T}}(\hat{\mathbb{K}}_\xi)_{jk}\boldsymbol{v}_i, \\ (\hat{\mathbb{K}}_\xi)_{jk} &= 4\sum_{r,s}^{N} \frac{\partial \xi_j}{\partial x_r}\frac{\partial \xi_k}{\partial x_s}\hat{\mathbb{K}}_{rs}, \end{aligned} \tag{4.3.113}$$

where $\boldsymbol{v}_i$ are eigenvectors of the matrix $|\mathbb{B}|_p$ associated with eigenvalues μ_i. We easily find that for $N = 1$ and $p = 1$ we get the stabilization matrix (4.3.69) with parameters δ_i and ε_i given by (4.3.70) and (4.3.79), respectively.

The choice $p = 1$ and $p = 2$ is of particular interest. This yields

$$\begin{aligned} |\hat{\mathbb{B}}|_1 &= \sum_{i=1}^{N} |\hat{\mathbb{B}}_i|, \\ |\hat{\mathbb{B}}|_2 &= \Big(\sum_{i=1}^{N} \hat{\mathbb{B}}_i^2\Big)^{1/2} = (\hat{\mathbb{B}}^{\mathrm{T}}\hat{\mathbb{B}})^{1/2}. \end{aligned} \tag{4.3.114}$$

If we set $\hat{\mathbb{B}}_i = 2\hat{\mathbb{A}}_i/h_K$ (on each element $K \in \mathcal{T}_h$), then the choice $p = 2$ leads to the representation

$$\begin{aligned} \underline{\underline{\hat{\delta}}} &= \frac{1}{2}h_K\vartheta\,(\hat{\mathbb{A}}^{\mathrm{T}}\hat{\mathbb{A}})^{-1/2} \\ &= \frac{1}{2}h_K\vartheta\,\Big(\sum_{i=1}^{N} \hat{\mathbb{A}}_i^{\mathrm{T}}\hat{\mathbb{A}}_i\Big)^{-1/2} \end{aligned} \tag{4.3.115}$$

often used in applications – see Section 4.3.5.6.

4.3.5.4 *Stabilization operator* $\tilde{\boldsymbol{\psi}}_h$ The stabilization streamline diffusion operator $\tilde{\boldsymbol{\psi}}_h$ in the discrete formulation (4.3.93) for the entropy variables is obtained from the situation discussed in the previous section with the aid of the transformations (4.3.94). We obtain the formulae

$$\tilde{\boldsymbol{\psi}}_h(\boldsymbol{\varphi}_h) = \underline{\underline{\tilde{\delta}}}\,\tilde{\mathbb{A}} \cdot \nabla\boldsymbol{\varphi}_h = \underline{\underline{\tilde{\delta}}}\sum_{i=1}^{N} \tilde{\mathbb{A}}_i\frac{\partial \boldsymbol{\varphi}_h}{\partial x_i}, \tag{4.3.116}$$

$$\begin{aligned}
\tilde{\underline{\underline{\delta}}} &= \mathbb{L}^{-\mathrm{T}}\,\hat{\underline{\underline{\delta}}}\,\mathbb{L}^{-1} = \mathbb{L}^{-\mathrm{T}}\,\overline{\mathbb{T}}\,\mathrm{diag}(\delta_1,\ldots,\delta_\ell)\,\overline{\mathbb{T}}^{\mathrm{T}}\,\mathbb{L}^{-1}\\
&= \mathbb{R}\,\mathrm{diag}(\delta_1,\ldots,\delta_\ell)\,\mathbb{R}^{\mathrm{T}},\\
\mathbb{R} &= \mathbb{L}^{-\mathrm{T}}\,\overline{\mathbb{T}} = (\boldsymbol{r}_1,\ldots,\boldsymbol{r}_\ell),\\
\boldsymbol{r}_i &= \mathbb{L}^{-\mathrm{T}}\boldsymbol{v}_i,\\
\delta_i &= \tilde{\xi}(\alpha_i)/\mu_i,\\
\alpha_i &= \mu_i/\sigma_i,\\
\sigma_i &= \frac{4}{N}\sum_{j,k=1}^{\ell}\frac{\partial\xi_j}{\partial x_r}\frac{\partial\xi_k}{\partial x_s}\tilde{\mathbb{K}}_{rs}.
\end{aligned}$$

4.3.5.5 *Determination of the discontinuity capturing operators* As already mentioned, even when the streamline diffusion stabilization is applied, the numerical solution may suffer from undershoots and overshoots appearing in a thin layer near a discontinuity of the exact solution. The goal is to avoid this effect by introducing a discontinuity capturing term as a vector parallel to the gradient of the solution into the formulation of the discrete problem. Since the discontinuity capturing term is a function of the approximate solution gradient, the numerical method becomes nonlinear even when the original equation is linear. Here we shall briefly describe the derivation of the discontinuity capturing operators carried out in (Hughes and Mallet, 1986*b*) for the entropy variables formulation.

In (Hughes and Mallet, 1986*b*), the operator $\tilde{\boldsymbol{d}}_h$ is defined in the form

$$\tilde{\boldsymbol{d}}_h(\boldsymbol{\varphi}_h) = \tilde{\underline{\underline{\delta}}}_d\tilde{\mathbb{A}}_\pi\cdot\boldsymbol{\nabla}\boldsymbol{\varphi}_h = \tilde{\underline{\underline{\delta}}}\,\tilde{\mathbb{A}}_\pi^{\mathrm{T}}\boldsymbol{\nabla}\boldsymbol{\varphi}_h, \tag{4.3.117}$$

where $\tilde{\underline{\underline{\delta}}}_d$ is the sought *discontinuity capturing* $m\times m$ *matrix* and $\tilde{\mathbb{A}}_\pi$ is an $(Nm)\times m$ matrix defined in the following way. First we introduce an $(Nm)\times(Nm)$ matrix $\hat{\Pi}$ defined by

$$\hat{\Pi} = \boldsymbol{\nabla}\hat{\boldsymbol{u}}_h(\boldsymbol{\nabla}\hat{\boldsymbol{u}}_h)^{\mathrm{T}}/[(\boldsymbol{\nabla}\hat{\boldsymbol{u}}_h)^{\mathrm{T}}\boldsymbol{\nabla}\hat{\boldsymbol{u}}_h], \tag{4.3.118}$$

and set

$$\hat{\mathbb{A}}_\pi^{\mathrm{T}} = \hat{\mathbb{A}}^{\mathrm{T}}\hat{\Pi}. \tag{4.3.119}$$

It is possible to prove the following properties of $\hat{\Pi}$:

(i) $\mathrm{Range}(\hat{\Pi}) = \mathrm{span}(\boldsymbol{\nabla}\hat{\boldsymbol{u}}_h) =$ space spanned by the vector $\boldsymbol{\nabla}\hat{\boldsymbol{u}}$;

(ii) $\hat{\Pi}^2 = \hat{\Pi}$;

(iii) $\hat{\Pi} = \hat{\Pi}^{\mathrm{T}}$;

(iv) for all (Nm)-dimensional vectors $\boldsymbol{\varphi}$ which are orthogonal to $\boldsymbol{\nabla}\hat{\boldsymbol{u}}_h$ (i.e. $\boldsymbol{\varphi}\cdot\boldsymbol{\nabla}\hat{\boldsymbol{u}}_h = 0$), we have $\hat{\mathbb{A}}_\pi\boldsymbol{\varphi} = 0$;

(v) $\hat{\mathbb{A}}_\pi\cdot\boldsymbol{\nabla}\hat{\boldsymbol{u}}_h = \hat{\mathbb{A}}\cdot\boldsymbol{\nabla}\hat{\boldsymbol{u}}_h$.

Properties (i)–(iii) imply that $\hat{\Pi} : I\!R^{Nm} \to I\!R^{Nm}$ is the orthogonal projection onto $\mathrm{span}(\boldsymbol{\nabla}\hat{\boldsymbol{u}}_h)$. By properties (iv) and (v) we see that the null space of $\hat{\mathbb{A}}_\pi$ is the orthogonal complement $(\mathrm{span}(\boldsymbol{\nabla}\hat{\boldsymbol{u}}_h))^\perp$ of the space $\mathrm{span}(\boldsymbol{\nabla}\hat{\boldsymbol{u}}_h)$ and that the action of $\hat{\mathbb{A}}_\pi$ on $\mathrm{span}(\boldsymbol{\nabla}\hat{\boldsymbol{u}}_h)$ is identical to that of $\hat{\mathbb{A}}$.

Now, transformation (4.3.94) is applied. As a result we get

$$\tilde{\mathbb{A}}_\pi^{\mathrm{T}} = \tilde{\mathbb{A}}^{\mathrm{T}}\tilde{\Pi}, \qquad (4.3.120)$$

where

$$\tilde{\Pi} = \boldsymbol{\nabla}\boldsymbol{u}_h(\boldsymbol{\nabla}\boldsymbol{u}_h)^{\mathrm{T}}\,\mathrm{diag}(\tilde{\mathbb{A}}_0,\ldots,\tilde{\mathbb{A}}_0)\,/[(\boldsymbol{\nabla}\boldsymbol{u}_h)^{\mathrm{T}}\mathrm{diag}(\tilde{\mathbb{A}}_0,\ldots,\tilde{\mathbb{A}}_0)\boldsymbol{\nabla}\boldsymbol{u}_h]. \qquad (4.3.121)$$

Here $\mathrm{diag}(\tilde{\mathbb{A}}_0,\ldots,\tilde{\mathbb{A}}_0)$ is the $(Nm)\times(Nm)$ matrix formed by N diagonal blocks $\tilde{\mathbb{A}}_0$. The matrix $\underline{\underline{\delta}}$ may be given by

$$\tilde{\underline{\underline{\delta}}}_d = \boldsymbol{\varphi}_\pi\delta_\pi\boldsymbol{\varphi}_\pi^{\mathrm{T}}, \qquad (4.3.122)$$

where the vector $\boldsymbol{\varphi}_\pi$ and the scalar δ_π are defined by the following formulae:

$$\begin{aligned}
\boldsymbol{\varphi}_\pi &= \left(\tilde{\mathbb{A}}_0^{-1}\tilde{\mathbb{A}}\cdot\boldsymbol{\nabla}\boldsymbol{u}_h\right)\left((\tilde{\mathbb{A}}\cdot\boldsymbol{\nabla}\boldsymbol{u}_h)^{\mathrm{T}}\tilde{\mathbb{A}}_0^{-1}(\tilde{\mathbb{A}}\cdot\boldsymbol{\nabla}\boldsymbol{u}_h)\right)^{-1/2}, \qquad (4.3.123)\\
\mu_\pi^2 &= \frac{((\tilde{\mathbb{A}}\cdot\boldsymbol{\nabla}\boldsymbol{u}_h)^{\mathrm{T}}\tilde{\mathbb{A}}_0^{-1}\cdot\boldsymbol{\nabla}\boldsymbol{u}_h)((\mathbb{D}\boldsymbol{u}_h)^{\mathrm{T}}\mathrm{diag}(\tilde{\mathbb{A}}_0,\ldots\tilde{\mathbb{A}}_0)\mathbb{D}\boldsymbol{u}_h)}{((\boldsymbol{\nabla}\boldsymbol{u}_h)^{\mathrm{T}}\mathrm{diag}(\tilde{\mathbb{A}}_0,\ldots,\tilde{\mathbb{A}}_0)\boldsymbol{\nabla}\boldsymbol{u}_h)^2},\\
\delta_\pi &= \tilde{\xi}(\alpha_\pi)/\mu_\pi,\\
\alpha_\pi &= \mu_\pi/\sigma_\pi,\\
\sigma_\pi &= \frac{1}{N}\sum_{j,k=1}^{N}\boldsymbol{\varphi}_\pi^{\mathrm{T}}(\tilde{\mathbb{K}}_\xi)_{jk}\boldsymbol{\varphi}_\pi,
\end{aligned}$$

where

$$\mathbb{D}\boldsymbol{u}_h = \begin{pmatrix} \boldsymbol{u}_{hx_j}\frac{\partial\xi_1}{\partial x_j}\\ \vdots \\ \boldsymbol{u}_{hx_j}\frac{\partial\xi_N}{\partial x_j}\end{pmatrix} \quad \text{(summation over } j\text{)},$$

$\tilde{\xi}$ is the function from (4.3.64) and $(\tilde{\mathbb{K}}_\xi)_{jk}$ are matrices defined in (4.3.113). However, as described in (Hughes, Franca and Mizukami, 1986) devoted to the solution of a scalar equation, this choice may cause a doubling effect under certain circumstances and numerical results appear to be too diffusive. This deficiency is reduced by an alternative using the scalar

$$\delta = \frac{(\tilde{\mathbb{A}}_0\boldsymbol{\varphi}_\pi)^{\mathrm{T}}\tilde{\underline{\underline{\delta}}}\,\tilde{\mathbb{A}}_0\boldsymbol{\varphi}_\pi}{(\boldsymbol{\varphi}_\pi\cdot\tilde{\mathbb{A}}_0\boldsymbol{\varphi}_\pi)^2} \qquad (4.3.124)$$

and defining finally $\tilde{\underline{\underline{\delta}}}_d$ as

$$\tilde{\underline{\underline{\delta}}}_d = \boldsymbol{\varphi}_\pi\max(0,\delta_\pi-\delta)\boldsymbol{\varphi}_\pi^{\mathrm{T}}. \qquad (4.3.125)$$

Exercise 4.47 Verify the properties (i)–(v) of the matrix $\hat{\Pi}$ for special cases when $N\geq 1, m=1$ (scalar equation with N independent space variables) and $N=1, m\geq 1$ (system of equations with one space variable).

4.3.5.6 *Some special representations of stabilization operators* The above derivation of the stabilization operators $\tilde{\boldsymbol{\psi}}_h$ and $\tilde{\boldsymbol{d}}_h$ was realized for the linear system (4.3.92) under the assumption that the coefficient matrices $\tilde{\mathbb{A}}_0, \tilde{\mathbb{A}}$ and $\tilde{\mathbb{K}}$ are constant and the matrices $\tilde{\mathbb{A}}_i$ are symmetric. The derived formulae for the representation of $\tilde{\boldsymbol{\psi}}_h$ and $\tilde{\boldsymbol{d}}_h$ are also applied in the case of nonlinear problems for the Euler and Navier–Stokes equations formulated in terms of the entropy variables $\boldsymbol{u}$. In this case the matrices $\tilde{\mathbb{A}}_i$ and $\tilde{\mathbb{K}}_{ij}$ are, of course, functions of the sought approximate solution $\boldsymbol{u}_h$.

The computation of the stabilization operators introduced in the above sections is rather complicated. In practice, several simplifications of the stabilization operators are used. For example, in (Le Beau *et al.*, 1993) also the following *stabilization* for the Euler equations *in entropy variables* is proposed:

$$\tilde{\boldsymbol{\psi}}_h(\boldsymbol{\varphi}_h) = \underline{\underline{\tilde{\delta}}} \sum_{i=1}^{N} \tilde{\mathbb{A}}_i \frac{\partial \boldsymbol{\varphi}_h}{\partial x_i}, \tag{4.3.126}$$

where

$$\underline{\underline{\tilde{\delta}}} = \left(\frac{\partial \xi_i}{\partial x_j} \tilde{\mathbb{A}}_j \tilde{\mathbb{A}}_0^{-1} \frac{\partial \xi_i}{\partial x_k} \tilde{\mathbb{A}}_k \right)^{-1/2}. \tag{4.3.127}$$

(We use the summation convention.) By the expression $(.)^{-1/2}$ we understand the square root of the generalized inverse of a symmetric matrix defined in Section 4.3.5.2. In this paper the discontinuity capturing operator recommended for the Euler equations is written in the form

$$\tilde{\boldsymbol{d}}_h(\boldsymbol{\varphi}_h) = \sum_{K \in \mathcal{T}_h} \int_K \tilde{\delta}_d \sum_{i=1}^{N} \frac{\partial \boldsymbol{u}_h}{\partial x_i} \frac{\partial \boldsymbol{\varphi}_h}{\partial x_i}\, dx, \tag{4.3.128}$$

with scalar $\tilde{\delta}_d$ given by

$$\tilde{\delta}_d = \max(\tilde{\nu}_d - \tilde{\nu}, 0). \tag{4.3.129}$$

Here the scalar parameters $\tilde{\nu}_d$ and $\tilde{\nu}$ are given by formulae

$$\tilde{\nu} = \frac{(\tilde{\mathbb{A}}_i \boldsymbol{u}_{hx_i})^{\mathrm{T}} \tilde{\mathbb{A}}_0^{-1} (\tilde{\mathbb{A}}_j \boldsymbol{u}_{hx_j})}{\boldsymbol{u}_{hx_j}^{\mathrm{T}} \tilde{\mathbb{A}}_0^{-1} \boldsymbol{u}_{hx_j}} \tag{4.3.130}$$

and

$$\tilde{\nu}_d = \left(\frac{(\tilde{\mathbb{A}}_i \boldsymbol{u}_{hx_i})^{\mathrm{T}} \tilde{\mathbb{A}}_0^{-1} (\tilde{\mathbb{A}}_j \boldsymbol{u}_{hx_j})}{\frac{\partial \xi_\ell}{\partial x_i} \boldsymbol{u}_{hx_i}^{\mathrm{T}} \tilde{\mathbb{A}}_0^{-1} \frac{\partial \xi_\ell}{\partial x_j} \boldsymbol{u}_{hx_j}} \right)^{1/2}. \tag{4.3.131}$$

The representation of the operators $\tilde{\boldsymbol{\psi}}_h$ and $\tilde{\boldsymbol{d}}_h$ is also adapted to the discrete problem (4.3.47) formulated with the aid of the conservative variables $\boldsymbol{w}$, when the matrices $\mathbb{A}_i = \mathbb{A}_i(\boldsymbol{w}_h)$ are not symmetric. The streamline diffusion operator $\boldsymbol{\psi}_h$ is often expressed in the form analogous to (4.3.105), i.e.

$$\boldsymbol{\psi}_h(\boldsymbol{\varphi}_h) = \underline{\underline{\delta}} \sum_{i=1}^{N} \mathbb{A}_i \frac{\partial \boldsymbol{\varphi}_h}{\partial x_i}. \tag{4.3.132}$$

In (Stiller, 1996), the matrix $\underline{\underline{\delta}}$ is computed on each element $K \in \mathcal{T}_h$ similarly as in (4.3.115):

$$\underline{\underline{\delta}} = \frac{1}{2} h_K \vartheta \Big(\sum_{i=1}^{N} \mathbb{A}_i^{\mathrm{T}} \mathbb{A}_i \Big)^{-1/2}. \tag{4.3.133}$$

The matrix $(\,\cdot\,)^{-1/2}$ is again computed in accordance with Section 4.3.5.2. The scaling parameter $\vartheta > 0$ is defined as $\vartheta = 1$ for the case of strongly dominating convection (including the Euler equations). Another possibility is to set $\vartheta = \tilde{\xi}(\alpha)$ with $\alpha = \frac{1}{2} h_K (a + |\boldsymbol{v}|)/\max(\mu, k)$, suitable for the Navier–Stokes equations. Here $\boldsymbol{v}$ is the velocity vector and a denotes the speed of sound.

In (Rachowicz, 1997), the operator $\boldsymbol{\psi}_h$ is written in the form

$$\boldsymbol{\psi}_h(\boldsymbol{\varphi}_h) = \delta \sum_{i=1}^{N} \mathbb{A}_i \frac{\partial \boldsymbol{\varphi}_h}{\partial x_i}, \tag{4.3.134}$$

where δ is a scalar parameter given by

$$\delta = \frac{\max_{i=1,\dots,N}(h_i |\beta_i|)}{2(a + |\boldsymbol{v} \cdot \boldsymbol{\beta}|)}. \tag{4.3.135}$$

The following notation is used here: h_i – the length of a particular element K (on which we evaluate the value $\boldsymbol{\psi}_h(\boldsymbol{\varphi}_h)$) in the direction x_i,

$$\boldsymbol{\beta} = (\beta_1, \dots, \beta_N)^{\mathrm{T}} = \nabla(\boldsymbol{w}_h^{\mathrm{T}} \tilde{\mathbb{A}}_0^{-1} \boldsymbol{w}_h) \Big/ \left| \nabla(\boldsymbol{w}_h^{\mathrm{T}} \tilde{\mathbb{A}}_0^{-1} \boldsymbol{w}_h) \right|, \tag{4.3.136}$$

a – speed of sound, $\boldsymbol{v}$ – velocity vector. We can note that the number $a + |\boldsymbol{v} \cdot \boldsymbol{\beta}|$ is the spectral radius of the matrix $\sum_{i=1}^{N} \mathbb{A}_i \beta_i = \mathbb{P}(\boldsymbol{w}_h, \boldsymbol{\beta})$ (cf. (3.1.14)).

The shock capturing operator appears in the literature in several variants. In (Rachowicz, 1997) and (Stiller, 1996) the shock capturing term is rather simplified. Instead of its representation (4.3.45) the reduced form

$$\sum_{K \in \mathcal{T}_h} \int_K \nu_d \sum_{i=1}^{N} \frac{\partial \boldsymbol{w}_h}{\partial x_i} \cdot \frac{\partial \boldsymbol{\varphi}_h}{\partial x_i}\, dx \tag{4.3.137}$$

is used. The scalar ν_d is defined in various ways. In (Stiller, 1996), on each element $K \in \mathcal{T}_h$

$$\nu_d = \frac{1}{2} h_K \|(\mathcal{L}\boldsymbol{w}_h - \boldsymbol{F})|_K\|_{\underline{\underline{\delta}}} \Big/ \|\boldsymbol{\nabla} \boldsymbol{w}_h|_K\|_{\tilde{\mathbb{A}}_0} \tag{4.3.138}$$

with norms defined according to (Shakib *et al.*, 1991)

$$\|\boldsymbol{a}\|_{\underline{\underline{\delta}}}^2 = \boldsymbol{a}^{\mathrm{T}} \underline{\underline{\delta}}\, \boldsymbol{a}, \tag{4.3.139}$$

$$\|\nabla \boldsymbol{w}_h\|^2_{\tilde{\mathbb{A}}_0} = (\nabla \boldsymbol{w}_h)^{\mathrm{T}} \mathrm{diag}(\tilde{\mathbb{A}}_0, \ldots, \tilde{\mathbb{A}}_0)(\nabla \boldsymbol{w}_h),$$

where $\underline{\underline{\delta}}$ is the streamline diffusion matrix. The matrix $\tilde{\mathbb{A}}_0$ is expressed as a function depending on $\boldsymbol{w}_h$: $\tilde{\mathbb{A}}_0(\boldsymbol{w}_h) = \mathbb{H}(\boldsymbol{w}_h)$ – see Section 4.3.2. In (Le Beau *et al.*, 1993) and (Rachowicz, 1997)

$$\nu_d = \left(\frac{(\mathbb{A}_i \boldsymbol{w}_{hx_i})^{\mathrm{T}} \tilde{\mathbb{A}}_0^{-1} (\mathbb{A}_j \boldsymbol{w}_{hx_j})}{\frac{\partial \xi_\ell}{\partial x_i} \boldsymbol{w}^{\mathrm{T}}_{hx_i} \tilde{\mathbb{A}}_0^{-1} \frac{\partial \xi_\ell}{\partial x_j} \boldsymbol{w}_{hx_j}} \right)^{1/2} . \tag{4.3.140}$$

Remark 4.48 If we solve the Euler or Navier–Stokes equations, we have to realize that the matrices $\mathbb{A}_j, \mathbb{K}_{ij}, \tilde{\mathbb{A}}_j$ and $\tilde{\mathbb{K}}_{ij}$ depend on the sought solution and the stabilized discrete problem becomes strongly nonlinear.

Remark 4.49 Another class of stabilized FE schemes for the numerical simulation of compressible flow is based on the generalization of the Galerkin least squares finite element method (GLSFEM) explained in Section 4.1.10.3 for a linear scalar equation. The Galerkin least squares stabilization of the finite method for the compressible Euler and Navier–Stokes equations can be found in (Shakib *et al.*, 1991). From a large number of other works on the FE simulation of compressible flow let us also quote the works by Löhner, Morgan, Zienkiewicz et al. (Löhner *et al.*, 1984), (Löhner *et al.*, 1985), (Morgan *et al.*, 1987), (Morgan *et al.*, 1997). The *hp*-version of the FEM was used, for example, in (Lomtev *et al.*, 1998).

4.3.6 *Time discretization*

Up to now, in this section we have been concerned with the space semidiscretization of the compressible Euler and Navier–Stokes equations (i.e. the method of lines), leading to a large system of ordinary differential equations. In practical computations, it is necessary to apply a suitable time discretization. To this end, we construct a partition of the time interval $(0, T)$ formed by time instants $0 = t_0 < t_1 < \ldots < t_{n+1} = T$. By $\tau_k = t_{k+1} - t_k$ we denote the time step, set $\mathcal{I}_k = (t_k, t_{k+1})$, $\boldsymbol{w}_h^{*k} = \boldsymbol{w}_h^*(t_k)$ (representation of the Dirichlet boundary conditions – see Section 4.3.3) and use the approximation $\boldsymbol{w}_h^k \approx \boldsymbol{w}_h(t_k)$. The function $\boldsymbol{w}_h^k \in \boldsymbol{X}_h$ of the variable x is called the *approximate solution at time* t_k. There are several possibilities of the time discretization used in practice, as follows.

4.3.6.1 *The Runge–Kutta methods* The semidiscrete problem can be written as a system of ordinary differential equations of the form (3.2.18). Then the Runge–Kutta methods from Section 3.2.2 are used for the time discretization. In general, the realization of the Runge–Kutta schemes requires the solution of linear systems of equations for parameters defining the approximate solution $\boldsymbol{w}_h^{k+1}$ at each time level with a positive definite mass matrix. Of course, in practical computations the integrals are evaluated with the aid of numerical integration. Using the *mass lumping* quadrature formula (see Section 4.1.8), we obtain the *fully explicit scheme*. Its algorithmization is quite simple. On the other hand, the Runge–Kutta schemes are conditionally stable and it is necessary to use a suitable CFL stability condition, strongly limiting the time step (cf., e.g. Section 4.1.7.3).

4.3.6.2 *The Euler forward scheme* The simplest and rather popular version of the Runge–Kutta methods is the Euler forward scheme (3.2.23). For example, in the case of problem (4.3.47), using the approximation $\partial \boldsymbol{w}_h(t_k)/\partial t \approx (\boldsymbol{w}_h^{k+1} - \boldsymbol{w}_h^k)/\tau_k$, we get the following *fully discrete scheme*: starting from the initial condition $\boldsymbol{w}_h^0$, for $k = 0, 1, \ldots, n$ we compute $\boldsymbol{w}_h^{k+1}$ satisfying the conditions

$$\text{a)} \quad \boldsymbol{w}_h^{k+1} - \boldsymbol{w}_h^{*k+1} \in \boldsymbol{V}_h, \tag{4.3.141}$$

$$\begin{aligned}
\text{b)} \quad (\boldsymbol{w}_h^{k+1}, \boldsymbol{\varphi}_h) = (\boldsymbol{w}_h^k, \boldsymbol{\varphi}_h) - \tau_k \Big\{ \Big(\sum_{i=1}^{N} \mathbb{A}_i(\boldsymbol{w}_h^k) \frac{\partial \boldsymbol{w}_h^k}{\partial x_i} \\
- \boldsymbol{F}(\boldsymbol{w}_h^k), \boldsymbol{\varphi}_h \Big) + \sum_{i,j=1}^{N} \Big(\mathbb{K}_{ij}(\boldsymbol{w}_h^k) \frac{\partial \boldsymbol{w}_h^k}{\partial x_j}, \frac{\partial \boldsymbol{\varphi}_h}{\partial x_i} \Big) \\
+ \sum_{K \in \mathcal{T}_h} \big(\mathcal{L}\boldsymbol{w}_h^k - \boldsymbol{F}(\boldsymbol{w}_h^k), \boldsymbol{\psi}_h(\boldsymbol{w}_h^k, \boldsymbol{\varphi}_h) \big)_K \\
+ \sum_{K \in \mathcal{T}_h} \big(\mathcal{L}\boldsymbol{w}_h^k - \boldsymbol{F}(\boldsymbol{w}_h^k), \boldsymbol{d}_h(\boldsymbol{w}_h^k, \boldsymbol{\varphi}_h) \big)_K \\
- \big(\boldsymbol{B}, \boldsymbol{\varphi}_h \big)_{\partial\Omega} \Big\} \quad \forall \boldsymbol{\varphi}_h \in \boldsymbol{V}_h.
\end{aligned}$$

It is necessary to specify the approximation of $\partial \boldsymbol{w}_h/\partial t$ at time t_k in the stabilization terms containing the operator $\mathcal{L}$. There are several possibilities:

$$\frac{\partial \boldsymbol{w}_h}{\partial t}(t_k) \approx \frac{\boldsymbol{w}_h^{k+1} - \boldsymbol{w}_h^k}{\tau_k} \quad \text{forward form,} \tag{4.3.142}$$

or

$$\frac{\partial \boldsymbol{w}_h}{\partial t}(t_k) \approx \frac{\boldsymbol{w}_h^{k} - \boldsymbol{w}_h^{k-1}}{\tau_k} \quad \text{backward form.} \tag{4.3.143}$$

Sometimes the time derivative $\partial \boldsymbol{w}_h/\partial t$ in $\mathcal{L}$ is simply omitted in the stabilization terms. See, for example, (Rachowicz, 1997). This is used particularly in the case, when the time marching procedure $t_k \to \infty$ is applied to obtain a steady-state solution.

If the mass lumping is used for the computation of the mass matrix, then the Euler forward scheme becomes fully explicit.

4.3.6.3 *Implicit linearized scheme* In order to diminish the limitation of the time step by the CFL stability condition, one can apply a *linearized Euler backward scheme.* We keep the terms depending on $\boldsymbol{w}_h$ and the stabilization terms at time t_k, whereas the space derivatives of $\boldsymbol{w}_h$ are considered at t_{k+1}.

Starting from an initial condition $\boldsymbol{w}_h^0$, for $k = 0, 1, \ldots, n$, we compute $\boldsymbol{w}_h^{k+1}$ such that

$$\text{a)}\quad \boldsymbol{w}_h^{k+1} - \boldsymbol{w}_h^{*k+1} \in \boldsymbol{V}_h, \tag{4.3.144}$$

$$\begin{aligned}
\text{b)}\quad & \left(\boldsymbol{w}_h^{k+1} + \tau_k \sum_{i=1}^{N} \mathbb{A}_i(\boldsymbol{w}_h^k)\frac{\partial \boldsymbol{w}_h^{k+1}}{\partial x_i} - \tau_k \boldsymbol{F}(\boldsymbol{w}_h^k), \boldsymbol{\varphi}_h\right) \\
& + \tau_k \sum_{i,j=1}^{N} \left(\mathbb{K}_{ij}(\boldsymbol{w}_h^k)\frac{\partial \boldsymbol{w}_h^{k+1}}{\partial x_j}, \frac{\partial \boldsymbol{\varphi}_h}{\partial x_i}\right) = (\boldsymbol{w}_h^k, \boldsymbol{\varphi}_h) \\
& - \tau_k \Big\{ \sum_{K \in \mathcal{T}_h} \big(\mathcal{L}\boldsymbol{w}_h^k - \boldsymbol{F}(\boldsymbol{w}_h^k), \boldsymbol{\psi}_h(\boldsymbol{w}_h^k, \boldsymbol{\varphi}_h)\big)_K \\
& + \sum_{K \in \mathcal{T}_h} \big(\mathcal{L}\boldsymbol{w}_h^k - \boldsymbol{F}(\boldsymbol{w}_h^k), \boldsymbol{d}_h(\boldsymbol{w}_h^k, \boldsymbol{\varphi}_h)\big)_K \Big\} + \tau_k(\boldsymbol{B}, \boldsymbol{\varphi}_h)_{\partial\Omega} \quad \forall \boldsymbol{\varphi}_h \in \boldsymbol{V}_h.
\end{aligned}$$

In this scheme, at each time level one has to solve a linear system for parameters defining $\boldsymbol{w}_h^{k+1}$. There are also other variants of this scheme, treating implicitly also some expressions with derivatives of $\boldsymbol{w}_h$ in the stabilization terms. The time derivative $\partial \boldsymbol{w}_h/\partial t$ at time t_k in the stabilization terms is again expressed as described above.

4.3.6.4 *Discontinuous Galerkin time discretization* The increase of accuracy in the time discretization can be achieved, for example, with the aid of higher order Runge–Kutta methods. Another possibility is to use the DGFEM applied to the time discretization.

Let us formulate our semidiscrete problem in the following way: find $\boldsymbol{w}_h \in C^1([0,T]; \boldsymbol{X}_h)$ such that $\boldsymbol{w}_h(t) \in \boldsymbol{w}_h^* + \boldsymbol{V}_h$ for $t \in (0,T)$, $\boldsymbol{w}_h(0) = \boldsymbol{w}_h^0$, and

$$\left(\frac{\partial \boldsymbol{w}_h(t)}{\partial t}, \boldsymbol{\varphi}_h\right) + A(\boldsymbol{w}_h(t), \boldsymbol{\varphi}_h) = L(\boldsymbol{\varphi}_h), \quad \boldsymbol{\varphi}_h \in \boldsymbol{V}_h, t \in (0,T). \tag{4.3.145}$$

Here the form $A : \boldsymbol{X}_h \times \boldsymbol{X}_h \to \mathbb{R}$ is linear in $\boldsymbol{\varphi}_h$ and, in general, nonlinear with respect to $\boldsymbol{w}_h$ and L is a linear functional on $\boldsymbol{V}_h$. For simplicity we assume that the boundary conditions represented by the function $\boldsymbol{w}_h^*$ are independent of time. To formulate the DG time discretization over a partition $\{\mathcal{I}_k; k = 0, 1, \ldots, n\}$ of the time interval $(0,T)$ with $\mathcal{I}_k = (t_k, t_{k+1})$, we introduce the spaces

$$\begin{aligned}
\boldsymbol{X}_{h\tau} &= \{\boldsymbol{\varphi} : [0,T] \to \boldsymbol{X}_h; \boldsymbol{\varphi}|_{\mathcal{I}_k} \in \mathcal{P}^q(\mathcal{I}_k)^m, k = 0, \ldots, n\}, \\
\boldsymbol{V}_{h\tau} &= \{\boldsymbol{\varphi} \in \boldsymbol{X}_{h\tau}; \boldsymbol{\varphi}(t) \in \boldsymbol{V}_h \,\text{for all}\, t \in [0,T]\},
\end{aligned} \tag{4.3.146}$$

where

$$\mathcal{P}^q(\mathcal{I}_k) = \left\{\boldsymbol{\varphi} : \mathcal{I}_k \to \boldsymbol{X}_h; \boldsymbol{\varphi}(t) = \sum_{i=0}^{q} t^i \boldsymbol{\varphi}_i \text{ with } \boldsymbol{\varphi}_i \in \boldsymbol{X}_h\right\} \tag{4.3.147}$$

and $q \geq 0$ is a given integer. This means that $\boldsymbol{X}_{h\tau}$ is the space of functions on $[0,T]$ with values in $\boldsymbol{X}_h$ that on each time interval $\mathcal{I}_k$ vary as polynomials depending on t of degree at most q. These functions may be discontinuous in time at the discrete time instants t_k. Therefore, we introduce the notation

$$\varphi_{\pm}^{k} = \lim_{s\to 0+} \varphi(t_k \pm s), \tag{4.3.148}$$
$$[\varphi^k] = \varphi_{+}^{k} - \varphi_{-}^{k}.$$

Now the DG time discretization of (4.3.145) is formulated as follows: find $\boldsymbol{W} : [0,T] \to \boldsymbol{X}_h$ such that

$$\text{a)}\ \ \boldsymbol{W} - \boldsymbol{w}_h^* \in \boldsymbol{V}_{h\tau}, \tag{4.3.149}$$

$$\text{b)}\ \ \sum_{k=0}^{n} \int_{\mathcal{I}_k} \left\{ \left(\frac{\partial \boldsymbol{W}(t)}{\partial t}, \varphi(t) \right) + A(\boldsymbol{W}(t), \varphi(t)) \right\} dt$$
$$+ \sum_{k=1}^{n} \left([\boldsymbol{W}^k], \varphi_{+}^{k} \right) + (\boldsymbol{W}_{+}^{0}, \varphi_{+}^{0})$$
$$= \int_0^T L(\varphi(t))\, dt + (\boldsymbol{w}_h^0, \varphi_{+}^{0}) \quad \forall \varphi \in \boldsymbol{V}_{h\tau}.$$

Since $\varphi \in \boldsymbol{V}_{h\tau}$ varies independently on each subinterval $\mathcal{I}_k$, it is possible to reformulate (4.3.149) in the following way: for $k = 0, \ldots, n$, given $\boldsymbol{W}_{-}^{k}$, find $\boldsymbol{W} \equiv \boldsymbol{W}|_{\mathcal{I}_k} \in \mathcal{P}^q(\mathcal{I}_k)^m$ such that $(\boldsymbol{W} - \boldsymbol{w}_h^*)|_{\partial\Omega \times \mathcal{I}_k} = 0$ and

$$\int_{\mathcal{I}_k} \left(\left(\frac{\partial \boldsymbol{W}(t)}{\partial t}, \varphi(t) \right) + A(\boldsymbol{W}(t), \varphi(t)) \right) dt + \left(\boldsymbol{W}_{+}^{k}, \varphi_{+}^{k} \right) \tag{4.3.150}$$
$$= \int_{\mathcal{I}_k} L(\varphi(t))\, dt + \left(\boldsymbol{W}_{-}^{k}, \varphi_{+}^{k} \right) \quad \forall \varphi \in \mathcal{P}^q(\mathcal{I}_k),$$

where $\boldsymbol{W}_{-}^{0} = \boldsymbol{w}_h^0$.

For $q = 0$, we obtain the Euler backward scheme of the following form: for $j = 0, \ldots, n+1$ find $\boldsymbol{W}^j \in \boldsymbol{V}_h$ such that

$$\left(\boldsymbol{W}^{k+1} - \boldsymbol{W}^{k}, \varphi \right) + \tau_k A(\boldsymbol{W}^{k+1}, \varphi) = \tau_k L(\varphi) \quad \forall \varphi \in \boldsymbol{V}_h,\ k = 0, 1, \ldots, n, \tag{4.3.151}$$

where we use the notation $\boldsymbol{W}^k = \boldsymbol{W}_{-}^{k} = \boldsymbol{W}_{+}^{k-1}$ and $\boldsymbol{W}^0 = \boldsymbol{w}_h^0$.

The DGFE time discretization was applied to the solution of the compressible Euler and Navier–Stokes equations in (Shakib *et al.*, 1991) and (Hansbo and Johnson, 1991). We shall not give further details here.

Remark 4.50 In all time discretization techniques, the goal is to transform the solution of the nonlinear discrete problem to (a sequence of) linear algebraic systems. They are large, usually sparse, but nonsymmetric. In most cases, they are solved iteratively by the Krylov subspace methods mentioned in Section 3.3.14.

The computational results obtained with the aid of various stabilized finite element methods for the compressible Euler or Navier–Stokes equations can be found, for example, in (Shakib *et al.*, 1991), (Le Beau *et al.*, 1993), (Stiller, 1996), (Rachowicz, 1997), (Dinkler *et al.*, 1997). Some further aspects are contained in (Aliabadi *et al.*, 1993) and (Almeida and Galeao, 1996). In (Demkowicz *et al.*, 1990), an operator splitting FEM for the simulation of viscous compressible flow is proposed.

Theory of the streamline diffusion method has been developed mainly for linear problems (see citations in Section 4.1.10). The streamline diffusion and shock capturing methods applied to scalar nonlinear conservation laws are analysed in (Johnson and Szepessy, 1987) and (Johnson *et al.*, 1990).

The computational results show that the stabilized finite element methods give good results in various situations. The advantage of these techniques is the fact that it is not necessary to use the concept of an approximate Riemann solver. On the other hand, the stabilizations for obtaining sufficiently accurate and physically admissible results are quite sophisticated and complicated, as seen in the above sections, and require tuning of various stabilization parameters.

4.4 Combined finite volume–finite element method for viscous compressible flow

As we saw in Chapter 3, the finite volume method (FVM) represents an efficient and robust method for the solution of inviscid compressible flow. The FVM, expressing the balance of fluxes of conserved quantities through boundaries of control volumes, combined with approximate Riemann solvers, is widely used for the numerical solution of conservation laws. On the other hand, the finite element method (FEM), based on the concept of a weak solution defined with the aid of suitable test functions is quite natural for the solution of elliptic and parabolic problems. However, it is not mandatory to adhere to these paths of discretization in their respective regimes of common use. The finite (control) volume is known as the box method for elliptic problems (see (Eymard *et al.*, 2000), (Heinrich, 1987)), the finite element method is, on the other hand, applied to convection problems, as was shown in Section 4.1.9 and Section 4.3. Often the control volume approach is used in the framework of the FE methods for obtaining upwinding ((Angermann, 1991), (Angermann, 1993), (Angermann, 1995), (Ohmori and Ushijima, 1984), (Schieweck and Tobiska, 1989)).

In the solution of convection–diffusion problems, including viscous compressible flow, it is quite natural to try to employ the advantages of both FV and FE methods in such a way that the FVM is used for the discretization of inviscid Euler fluxes, whereas the FEM is applied to the approximation of viscous terms. This idea leads us to the *combined finite volume–finite element method* (FV–FE method) proposed in (Feistauer *et al.*, 1995). (Sometimes it is also called the mixed FV–FE method.) Further versions and analysis of this method were investigated in (Feistauer *et al.*, 1997), (Feistauer *et al.*, 1999*a*), (Feistauer *et al.*, 1999*b*), (Angot *et al.*, 1998) (Dolejší *et al.*, 2002). The numerical computations for the system of compressible viscous flow ((Feistauer *et al.*, 1996), (Feistauer and Felcman, 1997), (Dolejší *et al.*, 2002), (Dolejší, 1998*c*), (Kliková, 2000) demonstrate that the combined FV–FE method is feasible and produces good numerical results for technically relevant problems. The idea of using a combination of the FV and FE methods appears also in (Arminjon and Madrane, 1999*a*)), (Ghidaglia and Pascal, 1999) and (Pascal and Ghidaglia, 2001).

In this section we shall explain a new unified treatment of the combined FV–FE methods for nonlinear convection–diffusion problems and compressible viscous flow. Similarly as in Section 4.3, we shall be concerned with problem (CFP) to solve system (4.3.1) equipped with conditions (4.3.7), (4.3.8), a), b), c*), (4.3.9), a), b*) and (4.3.10). (See Section 4.3.1.3.) As formulated in (4.3.22), find $\boldsymbol{w}$ such that

$$\begin{aligned} &\text{a)}\ \ \boldsymbol{w}(t) - \boldsymbol{w}^*(t) \in \boldsymbol{V}, \quad t \in (0,T), \\ &\text{b)}\ \ \left(\frac{\partial \boldsymbol{w}(t)}{\partial t}, \boldsymbol{\varphi}\right) + a(\boldsymbol{w}(t), \boldsymbol{\varphi}) + b(\boldsymbol{w}(t), \boldsymbol{\varphi}) - \beta(\boldsymbol{w}(t), \boldsymbol{\varphi}) = 0 \\ &\qquad \forall \boldsymbol{\varphi} \in \boldsymbol{V},\ t \in (0,T), \\ &\text{c)}\ \ \boldsymbol{w}(0) = \boldsymbol{w}^0. \end{aligned} \tag{4.4.1}$$

Here $\boldsymbol{w}^* : [0,T] \to H^1(\Omega)^m$ is a function satisfying the Dirichlet boundary conditions (4.3.13), i.e.

$$\begin{aligned} &w_1^* = \rho_D, (w_2^*, \ldots, w_{m-1}^*) = \rho_D \boldsymbol{v}_D \quad \text{on } \Gamma_I \times (0,T), \\ &w_2^* = \ldots = w_{m-1}^* = 0 \quad \text{on } \Gamma_W \times (0,T). \end{aligned} \tag{4.4.2}$$

The forms $(\cdot,\cdot)$ a, b, β are defined by (4.3.19) and the space $\boldsymbol{V}$ is given in (4.3.14).

4.4.1 *Computational grids*

By Ω_h we denote a polygonal ($N = 2$) or polyhedral ($N = 3$) approximation of the domain Ω. In the combined FV–FE method we work with two meshes constructed in the domain Ω_h: a finite element mesh $\mathcal{T}_h = \{K_i\}_{i \in I}$ and a finite volume mesh $\mathcal{D}_h = \{D_j\}_{j \in J}$. Here, I and $J \subset Z^+$ are suitable index sets.

The FE mesh $\mathcal{T}_h$ satisfies the standard properties from the FEM treated in Section 4.1. It is formed by a finite number of closed triangles ($N = 2$) or tetrahedra ($N = 3$) $K = K_i$ covering the closure of Ω_h,

$$\overline{\Omega}_h = \bigcup_{K \in \mathcal{T}_h} K, \tag{4.4.3}$$

and satisfying (4.1.15) ($N = 2$) or (4.1.56) ($N = 3$). By σ_h we denote the set of all vertices of all elements $K \in \mathcal{T}_h$ and assume that $\sigma_h \cap \partial\Omega_h \subset \partial\Omega$. The symbol $\mathcal{Q}_h$ will denote the set of all midpoints of sides of all elements $K \in \mathcal{T}$. By $|K|$

we denote the N-dimensional measure of $K \in \mathcal{T}_h$ (i.e. $|K|$ is the area of K, if $N = 2$, and $|K|$ is the volume of K, if $N = 3$), $h_K = \text{diam}(K)$ and ρ_K is the radius of the largest ball inscribed in K. We set $h = \max_{K \in \mathcal{T}_h} h_K$.

We shall also work with an FV mesh $\mathcal{D}_h$ in Ω_h, formed by a finite number of closed polygons ($N = 2$) or polyhedra ($N = 3$) such that

$$\overline{\Omega}_h = \bigcup_{D \in \mathcal{D}_h} D. \tag{4.4.4}$$

Various types of FV meshes were introduced in Section 3.3.

We use the same notation as in Section 3.3.1. The boundary ∂D_i of each finite volume $D_i \in \mathcal{D}_h$ can be expressed as

$$\partial D_i = \bigcup_{j \in S(i)} \bigcup_{\alpha=1}^{\beta_{ij}} \Gamma_{ij}^{\alpha}, \tag{4.4.5}$$

where Γ_{ij}^{α} are straight segments ($N = 2$) or plane manifolds ($N = 3$), called faces of D_i, $\Gamma_{ij}^{\alpha} = \Gamma_{ji}^{\alpha}$, which either form the common boundary of neighbouring finite volumes D_i and D_j or are part of $\partial\Omega_h$. We denote by $|D_i|$ the N-dimensional measure of D_i, $|\Gamma_{ij}^{\alpha}|$– the $(N-1)$-dimensional measure of Γ_{ij}^{α}, $\boldsymbol{n}_{ij}^{\alpha}$ – the unit outer normal to ∂D_i on Γ_{ij}^{α}. Clearly, $\boldsymbol{n}_{ij}^{\alpha} = -\boldsymbol{n}_{ji}^{\alpha}$. $S(i)$ is a suitable index set written in the form

$$S(i) = s(i) \cup \gamma(i), \tag{4.4.6}$$

where $s(i)$ contains indexes of neighbours D_j of D_i and $\gamma(i)$ is formed by indexes j of $\Gamma_{ij}^1 \subset \partial\Omega_h$ (in this case we set $\beta_{ij} = 1$). For details see Section 3.3.1.

4.4.2 *FV and FE spaces*

The FE approximate solution will be sought in a finite dimensional space

$$\boldsymbol{X}_h = X_h^m, \tag{4.4.7}$$

called a *finite element space*. We shall consider two cases of the definition of X_h:

$$X_h = \left\{\varphi_h \in C(\overline{\Omega}_h); \varphi_h|_K \in P^1(K)\ \forall K \in \mathcal{T}_h\right\} \tag{4.4.8}$$

(conforming piecewise linear elements) and

$$X_h = \left\{\varphi_h \in L^2(\Omega); \varphi_h|_K \in P^1(K),\ \varphi_h \text{ are continuous at midpoints } Q_j \in \mathcal{Q} \text{ of all faces of all } K \in \mathcal{T}_h\right\} \tag{4.4.9}$$

(nonconforming Crouzeix–Raviart piecewise linear elements – they were originally proposed for the approximation of the velocity of incompressible flow, see (Crouzeix and Raviart, 1973), (Feistauer, 1993)).

The finite volume approximation is an element of the finite volume space

$$\boldsymbol{Z}_h = Z_h^m, \tag{4.4.10}$$

where

$$Z_h = \left\{\varphi_h \in L^2(\Omega); \varphi_h|_D = \text{const } \forall D \in \mathcal{D}_h\right\}. \tag{4.4.11}$$

One of the most important concepts is a relation between the spaces $\boldsymbol{X}_h$ and $\boldsymbol{Z}_h$. We assume the existence of a one-to-one mapping $L_h : \boldsymbol{X}_h \to \boldsymbol{Z}_h$, called a *lumping operator.*

Example 4.51 In practical computations the following combinations of the FV and FE spaces are used (see, for example, (Feistauer and Felcman, 1997), (Feistauer *et al.*, 1995), (Feistauer *et al.*, 1996), (Dolejší *et al.*, 2002)).

a) *Conforming finite elements combined with dual finite volumes* In this case the FE space $\boldsymbol{X}_h$ is defined by (4.4.7) – (4.4.8). The mesh $\mathcal{D}_h$ is formed by dual FVs D_i constructed over the mesh $\mathcal{T}_h$, associated with vertices $P_i \in \sigma_h = \{P_i\}_{i\in J}$, defined in Sections 3.3.1.1, c) for $N = 2$ and 3.3.1.2, c) for $N = 3$. In this case, the lumping operator is defined as such a mapping $L_h : \boldsymbol{X}_h \to \boldsymbol{Z}_h$ that for each $\varphi_h \in \boldsymbol{X}_h$

$$L_h\varphi_h \in \boldsymbol{Z}_h, \quad L_h\varphi_h|_{D_i} = \varphi_h(P_i) \quad \forall i \in J. \tag{4.4.12}$$

Obviously, L_h is a one-to-one mapping of $\boldsymbol{X}_h$ onto $\boldsymbol{Z}_h$.

b) *Nonconforming finite elements combined with barycentric finite volumes* Now let $\mathcal{Q}_h = \{Q_i; i \in J\}$ denote centres of faces of all $K \in \mathcal{T}_h$. Then $\mathcal{D}_h = \{D_i\}_{i\in J}$ is the mesh formed by barycentric FVs constructed over $\mathcal{T}_h$, associated with Q_i, $i \in J$, as described in Sections 3.3.1.1, d) for $N = 2$ and 3.3.1.2, d) for $N = 3$. The space $\boldsymbol{X}_h$ is given in (4.4.7) and (4.4.9) and L_h is defined by

$$L_h\varphi_h \in \boldsymbol{Z}_h, \quad L_h\varphi_h|_{D_i} = \varphi_h(Q_i), \quad i \in J, \tag{4.4.13}$$

for any $\varphi_h \in \boldsymbol{X}_h$. Again, L_h is a one-to-one mapping of $\boldsymbol{X}_h$ onto $\boldsymbol{Z}_h$.

c) *Combination of conforming finite elements with triangular finite volumes* We assume that $N = 2$ and start from a triangular FV mesh $\mathcal{D}_h = \{D_i\}_{i\in J}$ with properties (3.3.15), considered as a *primary FV mesh.* Then we construct an *adjoint FE mesh* $\mathcal{T}_h$ in such a way that the vertices of triangles $K \in \mathcal{T}_h$ are barycentres of $D_i \in \mathcal{D}_h$ or vertices of $D_i \in \mathcal{D}_h$ lying on $\partial\Omega_h$. In this case the space $\boldsymbol{X}_h$ is defined by (4.4.7) – (4.4.8). The lumping operator is defined by

$$\begin{aligned} L_h\varphi_h|_{D_i} = \varphi_h(P_i),\ P_i &= \text{centre of gravity of } D_i \\ &\text{and simultaneously a vertex of some } K \in \mathcal{T}_h. \end{aligned} \tag{4.4.14}$$

(See (Feistauer and Felcman, 1997), (Feistauer *et al.*, 1996), (Feistauer and Kliková, 2001).) In this case, L_h is not one-to-one.

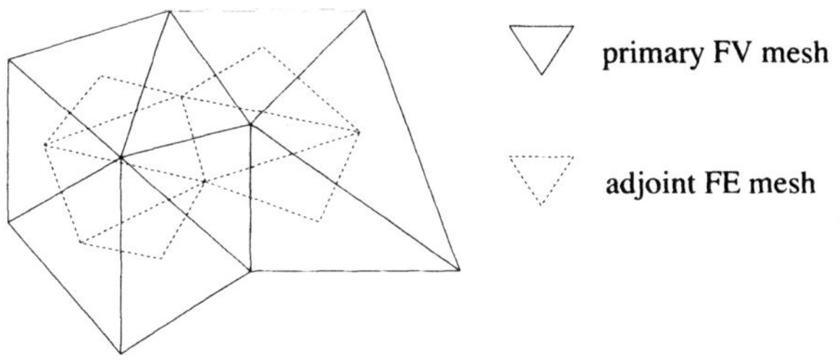

FIG. 4.4.1. Adjoint FEs $K \in \mathcal{T}_h$ to the primary FV mesh $\mathcal{D}_h$

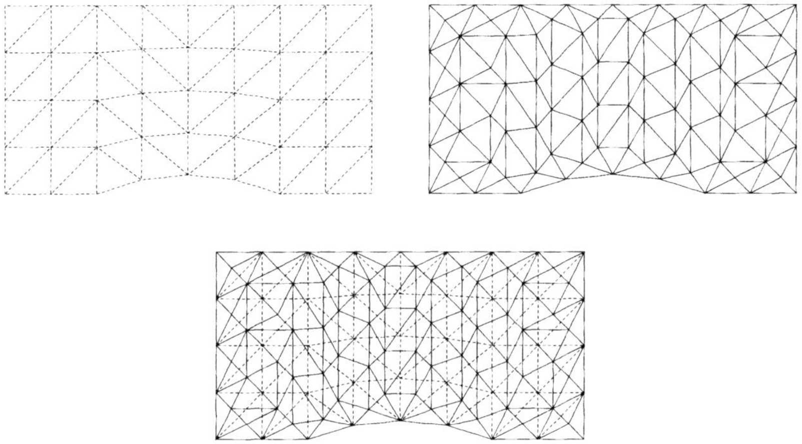

FIG. 4.4.2. Primary triangular FV and adjoint FE mesh in a channel, separately and together

Another possibility (but more complicated) of how to define the lumping operator in the all above examples is to use an *averaging procedure*: for $\boldsymbol{\varphi}_n \in \boldsymbol{X}_h$ we set

$$L_h \boldsymbol{\varphi}_h|_{D_i} = \frac{1}{|D_i|} \int_{D_i} \boldsymbol{\varphi}_h dx, \quad i \in J. \tag{4.4.15}$$

An adjoint FE mesh $\mathcal{T}_h$ to a primary triangular FV mesh $\mathcal{D}_h$ can be constructed in the following way. For each vertex $\tilde{P}_i$ of the mesh $\mathcal{D}_h$, consider system Σ_i of all triangles having the vertex $\tilde{P}_i$ in common. Connecting the barycentres of neighbouring triangles K and $K' \in \Sigma_i$, we obtain a contour which forms the boundary of a polygon Π_i containing $\tilde{P}_i$. Then this polygon is divided into triangles with vertices equal to the corners Π_i. At the boundary, one also uses vertices of $\mathcal{D}_h$ lying on $\partial\Omega_h$. As a result one obtains the mesh $\mathcal{T}_h$. Since the partition of Σ_i is not unique, the goal is to divide Π_i into triangles with angles as small as possible. One possibility is to construct a *Delaunay triangulation* over the set of barycentres of triangles $D_i \in \mathcal{D}_h$ and vertices of $D_i \in \mathcal{D}_h$ lying on $\partial\Omega_h$ (see (Delaunay, 1934), (Frey and George, 2000), (Bern, 1999)).

In Fig. 3.3.2 from Chapter 3, examples of a dual FV mesh and a barycentric FV mesh constructed over a triangular FE grid are shown. In Fig. 4.4.1, adjoint FEs constructed on a primary triangular mesh are presented. Figure 4.4.2 shows a primary triangular FV mesh and an adjoint triangular FE mesh, separately and together.

4.4.3 *Space semidiscretization of the problem*

We use the following approximations: $\Omega \approx \Omega_h$, $\Gamma_I \approx \Gamma_{Ih} \subset \partial\Omega_h$, $\Gamma_W \approx \Gamma_{Wh} \subset \partial\Omega_h$, $\Gamma_O \approx \Gamma_{Oh} \subset \partial\Omega_h$, $\boldsymbol{w}(t) \approx \boldsymbol{w}_h(t) \in \boldsymbol{X}_h$, $\varphi \approx \varphi_h \in \boldsymbol{V}_h \approx \boldsymbol{V}$, where

$$\boldsymbol{V}_h = \Big\{\varphi_h = (\varphi_{h1}, \dots, \varphi_{hm}) \in \boldsymbol{X}_h; \varphi_{hn}(P_i) = 0 \text{ for } n = 1, \dots, m-1 \quad \text{at } P_i \in \Gamma_{Ih}, \varphi_{hn}(P_i) = 0 \text{ for } n = 2, \dots, m-1 \text{ at } P_i \in \Gamma_{Wh}\Big\}. \tag{4.4.16}$$

Here P_i denote nodes, i.e. vertices $P_i \in \sigma_h$ or midpoints of faces $P_i \in \mathcal{Q}_h$ in the case of conforming or nonconforming finite elements, respectively.

The forms $a(\boldsymbol{w}, \varphi)$ and $\beta(\boldsymbol{w}, \varphi)$ defined in (4.3.19) are approximated by

$$\tilde{a}(\boldsymbol{w}_h, \varphi_h) = \sum_{K \in \mathcal{T}_h} \int_K \sum_{s=1}^{N} \boldsymbol{R}_s(\boldsymbol{w}_h, \nabla \boldsymbol{w}_h) \cdot \frac{\partial \varphi_h}{\partial x_s}\, dx, \tag{4.4.17}$$

$$\tilde{\beta}(\boldsymbol{w}_h, \varphi_h) = \int_{\Omega_h} \boldsymbol{F}(\boldsymbol{w}_h) \cdot \varphi_h dx + \int_{\partial\Omega} \boldsymbol{B} \cdot \varphi_h\, dS, \quad \boldsymbol{w}_h, \varphi_h \in \boldsymbol{X}_h.$$

In order to approximate the nonlinear convective terms containing inviscid fluxes $\boldsymbol{f}_s$, we start from the analogy with the form b from (4.3.19) written as $\int_\Omega \sum_{s=1}^N (\partial \boldsymbol{f}_s(\boldsymbol{w})/\partial x_s) \cdot \varphi\, dx$, where we use the approximation $\varphi \approx L_h \varphi_h$. Then Green's theorem is applied and the flux $\sum_{s=1}^N f_s(\boldsymbol{w}) n_s$ is approximated with the aid of a numerical flux $\boldsymbol{H}(\boldsymbol{w}, \boldsymbol{w}', \boldsymbol{n})$ from the FVM treated in Section 3.3:

$$\begin{aligned}
\int_\Omega \sum_{s=1}^N \frac{\partial \boldsymbol{f}_s(\boldsymbol{w})}{\partial x_s} \cdot \varphi dx &\approx \sum_{i \in J} \int_{D_i} \sum_{s=1}^N \frac{\partial \boldsymbol{f}_s(\boldsymbol{w})}{\partial x_s} \cdot L_h \varphi_h\, dx \\
&= \sum_{i \in J} L_h \varphi_h|_{D_i} \cdot \int_{D_i} \sum_{s=1}^N \frac{\partial \boldsymbol{f}_s(\boldsymbol{w})}{\partial x_s}\, dx \\
&= \sum_{i \in J} L_h \varphi_h|_{D_i} \cdot \int_{\partial D_i} \sum_{s=1}^N \boldsymbol{f}_s(\boldsymbol{w}) n_s\, dS \\
&= \sum_{i \in J} L_h \varphi_h|_{D_i} \cdot \sum_{j \in S(i)} \sum_{\alpha=1}^{\beta_{ij}} \int_{\Gamma_{ij}^\alpha} \sum_{s=1}^N \boldsymbol{f}_s(\boldsymbol{w}) n_s\, dS \\
&\approx \sum_{i \in J} L_h \varphi_h|_{D_i} \cdot \sum_{j \in S(i)} \sum_{\alpha=1}^{\beta_{ij}} \boldsymbol{H}(L_h \boldsymbol{w}_h|_{D_i}, L_h \boldsymbol{w}_h|D_j, \boldsymbol{n}_{ij}^\alpha)|\Gamma_{ij}^\alpha|.
\end{aligned} \tag{4.4.18}$$

Hence, we set

$$b_h(\boldsymbol{w}_h, \boldsymbol{\varphi}_h) = \sum_{i \in J} L_h \boldsymbol{\varphi}_h|_{D_i} \cdot \sum_{j \in S(i)} \sum_{\alpha=1}^{\beta_{ij}} \boldsymbol{H}(L_h \boldsymbol{w}_h|_{D_i}, L_h \boldsymbol{w}_h|_{D_j}, \boldsymbol{n}_{ij}^{\alpha})|\Gamma_{ij}^{\alpha}|. \quad (4.4.19)$$

If $\Gamma_{ij}^{\alpha} \subset \partial\Omega_h$, it is necessary to give an interpretation of $L_h w_h|_{D_j}$ – see Section 4.4.5.

In practical computations, the integrals are evaluated approximately with the aid of numerical quadratures. It is suitable to use the *mass lumping* from Section 4.1.8 leading to diagonal mass matrices. This means that in the case of conforming finite elements from Example 4.51, a) and c), the vertices $P_i \in \sigma_h$ are used as integration points:

$$\int_K F dx \approx \frac{1}{N+1}|K| \sum_{i=1}^{N+1} F(P_i^K), \quad (4.4.20)$$

for $F \in C(K)$ and element $K \in \mathcal{T}_h$ with vertices P_i^K, $i = 1, \ldots, N+1$. Then the $L^2(\Omega)$-scalar product is approximated by the form

$$(\boldsymbol{w}_h, \boldsymbol{\varphi}_h)_h = \frac{1}{N+1} \sum_{K \in \mathcal{T}_h} |K| \sum_{i=1}^{N+1} \boldsymbol{w}_h(P_i^K) \cdot \boldsymbol{\varphi}_h(P_i^K), \quad \boldsymbol{w}_h, \boldsymbol{\varphi}_h \in \boldsymbol{X}_h. \quad (4.4.21)$$

If nonconforming piecewise linear finite elements from Example 4.51, b) are used, the integration points are chosen as midpoints $Q_i^K \in \mathcal{Q}_h$, $i = 1, \ldots, N+1$, of faces of $K \in \mathcal{T}_h$. Then we write

$$\int_K F dx \approx \frac{1}{N+1}|K| \sum_{i=1}^{N+1} F(Q_i^K) \quad (4.4.22)$$

for $F \in C(K)$ and set

$$(\boldsymbol{w}_h, \boldsymbol{\varphi}_h)_h = \frac{1}{N+1} \sum_{K \in \mathcal{T}_h} |K| \sum_{i=1}^{N+1} \boldsymbol{w}_h(Q_i^K) \cdot \boldsymbol{\varphi}_h(Q_i^K), \quad \boldsymbol{w}_h, \boldsymbol{\varphi}_h \in \boldsymbol{X}_h. \quad (4.4.23)$$

Formula (4.4.20) is exact for all polynomials of degree 1, whereas formula (4.4.22) is exact for all polynomials of degree 2. (See (Ciarlet, 1979), Section 4.1.) Therefore, in the case of nonconforming finite elements, for the approximation (4.4.21) of the L^2-scalar product we have $(.,.)_h = (.,.)$ on the space $\boldsymbol{X}_h$ defined by (4.4.7) and (4.4.9).

Now, we define a_h and β_h as the forms approximating $\tilde{a}_h$ and $\tilde{\beta}_h$, respectively, with aid of the above numerical integration. Moreover, we take into account that

$$\nabla \boldsymbol{w}_h|_K = \text{const}, \quad \nabla \boldsymbol{\varphi}_h|_K = \text{const} \quad (4.4.24)$$

for each $K \in \mathcal{T}_h$, because $\boldsymbol{w}_h|_K$ and $\boldsymbol{\varphi}_h|_K$ are linear vector-valued functions.

Exercise 4.52 a) Verify that in the case of conforming finite elements, the form a_h can be expressed as

$$a_h(\boldsymbol{w}_h, \boldsymbol{\varphi}_h) = \sum_{i=1}^{m} a_h^i(\boldsymbol{w}_h, \boldsymbol{\varphi}_h), \tag{4.4.25}$$

where

$$a_h^1(\boldsymbol{w}_h, \boldsymbol{\varphi}_h) \equiv 0, \tag{4.4.26}$$

$$a_h^{j+1}(\boldsymbol{w}_h, \boldsymbol{\varphi}_h) = \sum_{K\in\mathcal{T}_h} |K| \sum_{s=1}^{N} \Big\{ \lambda \operatorname{div} \boldsymbol{v}_h|_K \delta_{sj}$$
$$+ \mu \Big(\frac{\partial v_{hs}}{\partial x_j}\Big|_K + \frac{\partial v_{hj}}{\partial x_s}\Big|_K \Big) \Big\} \frac{\partial \varphi_{h(j+1)}}{\partial x_s}\Big|_K, \quad j = 1, \dots, N,$$

$$a_h^{m}(\boldsymbol{w}_h, \boldsymbol{\varphi}_h) = \sum_{K\in\mathcal{T}_h} |K| \sum_{s=1}^{N} \Big\{ \sum_{i=1}^{N} \Big[\lambda \operatorname{div} \boldsymbol{v}_h|_K \delta_{si}$$
$$+ \mu \Big(\frac{\partial v_{hs}}{\partial x_i}\Big|_K + \frac{\partial v_{hi}}{\partial x_s}\Big|_K \Big) \frac{1}{N+1} \sum_{r=1}^{N+1} v_{hi}(P_r^K) \Big] + k \frac{\partial \theta_h}{\partial x_s} \Big\} \frac{\partial \varphi_{hm}}{\partial x_s}.$$

Here a_h^i approximate the forms a^i from (4.3.21). By $\boldsymbol{v}_h = (v_{h1}, \dots, v_{hN})^{\mathrm{T}} \in \boldsymbol{X}_h$ and $\theta_h \in X_h$ we denote the approximations of the velocity $\boldsymbol{v}$ and the temperature θ defined by their values at vertices $P_i \in \sigma_h$:

$$v_{hj}(P_i) = w_{h(j+1)}(P_i)/w_{h1}(P_i), \quad j = 1, \dots, N, \tag{4.4.27}$$
$$\theta_h(P_i) = \frac{1}{c_v} \left(\frac{w_{hm}(P_i)}{w_{h1}(P_i)} - \frac{1}{2} \sum_{j=1}^{N} v_{hj}^2(P_i) \right).$$

b) Express the form β_h in a similar way.
c) Derive similar representations for nonconforming elements.

In order to realize the Dirichlet boundary conditions (4.4.2), we introduce an auxiliary function $\boldsymbol{w}_h^* = (w_{h1}^*, \dots, w_{hm}^*)^{\mathrm{T}} \in \boldsymbol{X}_h$. In the case of conforming finite elements, it is defined by its values at vertices $P_i \in \sigma_h$ of the FE mesh $\mathcal{T}_h$:

$$\begin{aligned}
w_{h1}^*(P_i, t) &= \rho_D(P_i, t), \quad P_i \in \sigma_h \cap \Gamma_I, \\
w_{h1}^*(P_i, t) &= 0, \quad P_i \in \sigma_h \setminus \Gamma_I, \\
w_{hj}^*(P_i, t) &= \rho_D(P_i, t) v_{D(j-1)}(P_i, t), \quad P_i \in \sigma_h \cap \Gamma_I, \\
w_{hj}^*(P_i, t) &= 0, \quad P_i \in \sigma_h \setminus \Gamma_I, \\
&\qquad j = 2, \dots, N+1, \\
w_{hm}^*(P_i, t) &= 0, \quad P_i \in \sigma_h,
\end{aligned} \tag{4.4.28}$$

$$t \in [0, T].$$

The approximate initial condition $\boldsymbol{w}_h^0 \in \boldsymbol{X}_h$ can be defined by

$$\boldsymbol{w}_h^0(P_i) = \boldsymbol{w}^0(P_i) \quad \forall P_i \in \sigma_h, \tag{4.4.29}$$

provided $\boldsymbol{w}^0 \in C(\overline{\Omega}_h)^m$.

In the case of nonconforming finite elements we define $\boldsymbol{w}_h^*$ and $\boldsymbol{w}_h^0$ in the same way, replacing σ_h by $\mathcal{Q}_h$.

Now we arrive at the formulation of the FV–FE *semidiscretization* of the problem:

Definition 4.53 *We define a finite volume–finite element approximate solution of the viscous compressible flow as a vector-valued function* $\boldsymbol{w}_h = \boldsymbol{w}_h(x,t)$ *defined for (a.a)* $x \in \overline{\Omega}_h$ *and all* $t \in [0,T]$ *satisfying the following conditions:*

$$\begin{aligned}
&a)\ \boldsymbol{w}_h \in C^1([0,T]; \boldsymbol{X}_h), \qquad\qquad (4.4.30)\\
&b)\ \boldsymbol{w}_h(t) - \boldsymbol{w}_h^*(t) \in \boldsymbol{V}_h,\\
&c)\ \left(\frac{\partial \boldsymbol{w}_h(t)}{\partial t}, \boldsymbol{\varphi}_h\right)_h + b_h(\boldsymbol{w}_h(t), \boldsymbol{\varphi}_h)\\
&\qquad + a_h(\boldsymbol{w}_h(t), \boldsymbol{\varphi}_h) - \beta_h(\boldsymbol{w}_h(t), \boldsymbol{\varphi}_h) = 0\\
&\qquad \forall \boldsymbol{\varphi}_h \in \boldsymbol{V}_h,\ \forall t \in (0,T),\\
&d)\ \boldsymbol{w}_h(0) = \boldsymbol{w}_h^0.
\end{aligned}$$

4.4.4 *Time discretization*

Problem (4.4.30) is equivalent to a large system of ordinary differential equations which is solved with the aid of a suitable time discretization. It is possible to use Runge–Kutta methods from Section 4.1.6.2.

For the solution of high speed flow, one usually uses the *Euler forward scheme.* Let $0 = t_0 < t_1 < t_2 \ldots$ be a partition of the time interval and let $\tau_k = t_{k+1} - t_k$. Then in (4.4.30), b), c) we use the approximations $\boldsymbol{w}_h^k \approx \boldsymbol{w}_h(t_k)$ and $(\partial \boldsymbol{w}_h/\partial t)(t_k) \approx (\boldsymbol{w}_h^{k+1} - \boldsymbol{w}_h^k)/\tau_k$ and obtain the scheme

$$\begin{aligned}
&\text{a)}\ \boldsymbol{w}_h^{k+1} - \boldsymbol{w}_h^*(t_{k+1}) \in \boldsymbol{V}_h, \qquad\qquad (4.4.31)\\
&\text{b)}\ (\boldsymbol{w}_h^{k+1}, \boldsymbol{\varphi}_h)_h = (\boldsymbol{w}_h^k, \boldsymbol{\varphi}_h)_h - \tau_k a_h(\boldsymbol{w}_h^k, \boldsymbol{\varphi}_h)\\
&\qquad - \tau_k b_k(\boldsymbol{w}_h^k, \boldsymbol{\varphi}_h) + \tau_k \beta_h(\boldsymbol{w}_h^k, \boldsymbol{\varphi}_h^k) \quad \forall \boldsymbol{\varphi}_h \in \boldsymbol{V}_h,\ k = 0, 1, \ldots.
\end{aligned}$$

Due to the use of the mass lumping, (4.4.31) represents explicit formulae for the computation of the values $\boldsymbol{w}_h^{k+1}(P_i)$, at nodes $P_i \in \sigma_h$ or $P_i \in \mathcal{Q}_h$.

Another possibility is the *semi-implicit Euler scheme*, implicit with respect to the diffusion terms represented by the form a_h:

$$\begin{aligned}
&\text{a)}\ \boldsymbol{w}_h^{k+1} - \boldsymbol{w}_h^*(t_{k+1}) \in \boldsymbol{V}_h, \qquad\qquad (4.4.32)\\
&\text{b)}\ (\boldsymbol{w}_h^{k+1}, \boldsymbol{\varphi}_h)_h + \tau_k a_h(\boldsymbol{w}_h^{k+1}, \boldsymbol{\varphi}_h) = (\boldsymbol{w}_h^k, \boldsymbol{\varphi}_h)_h\\
&\qquad - \tau_k b_h(\boldsymbol{w}_h^k, \boldsymbol{\varphi}_h) + \tau_k \beta_h(\boldsymbol{w}_h^k, \boldsymbol{\varphi}_h^k) \quad \forall \boldsymbol{\varphi}_h \in \boldsymbol{V}_h,\ k = 0, 1, \ldots\ .
\end{aligned}$$

Various other variants are possible.

4.4.5 *Realization of boundary conditions in the convective form b_h*

If $\Gamma_{ij}^\alpha \subset \partial\Omega_h$ (i.e. $j \in \gamma(i)$, $\alpha = 1$), then there is no finite volume adjacent to Γ_{ij}^α from the opposite side to D_i and it is necessary to interpret the value $L_h \boldsymbol{w}_h^k|_{D_j}$ in the definition (4.4.19) of the form b_h. This means that we need to determine a boundary state $\tilde{\boldsymbol{w}}_j^k$ which will be substituted for $L_h \boldsymbol{w}_h^k|_{D_j}$ in (4.4.19). For the explicit or semi-implicit schemes (4.4.31), (4.4.22), respectively, there are two possibilities:

a) Assuming the approximate solution is known at the time level t_k, the simplest way is to define the boundary state $\tilde{\boldsymbol{w}}_j^k$ by extrapolation:

$$\tilde{\boldsymbol{w}}_j^k := L_h \boldsymbol{w}_h^k|_{D_i}. \tag{4.4.33}$$

b) The second possibility is to apply the approach used in the FVM and explained in Section 3.3.6. This means that the individual components of the state vector $\tilde{\boldsymbol{w}}_j^k$ are either extrapolated (with the use of $L_h \boldsymbol{w}_h^k|_{D_i}$) or prescribed, according to the signs of eigenvalues of the matrix $\mathbb{P}(\boldsymbol{w}_h^k|_{D_i}, \boldsymbol{n}_{ij}^\alpha)$ – cf. 3.3.6. For example, in the case of the *subsonic outlet*, the auxiliary outlet pressure p_D is prescribed on Γ_O. If the flow is *supersonic on the inlet* Γ_I, it is suitable to prescribe the Dirichlet conditions for all components of the state vector on Γ_I. This means that beside conditions (4.3.8), a), b), the condition (4.3.8), c) is used instead of (4.3.8), c*). Then it is necessary to change the definition of the space $\boldsymbol{V}_h$ of test functions:

$$\begin{aligned}\boldsymbol{V}_h = \Big\{ \boldsymbol{\varphi}_h = (\varphi_{h1}, \dots, \varphi_{hm}) \in \boldsymbol{X}_h; \varphi_{hn}(P_i) = 0 \text{ for } n = 1, \dots, m, \\ \text{at } P_i \in \Gamma_{Ih}, \varphi_{hn}(P_i) = 0 \text{ for } n = 2, \dots, m-1 \text{ at } P_i \in \Gamma_{Wh} \Big\},\end{aligned} \tag{4.4.34}$$

where P_i are the vertices, i.e. $P_i \in \sigma_h$, or midpoint of faces, i.e. $P_i \in \mathcal{Q}_h$, for conforming or nonconforming finite elements, respectively. (Cf. (4.4.16).)

4.4.6 *Operator splitting*

One possible way for the discretization of the compressible Navier–Stokes equations by the combined FV–FE method is the *inviscid–viscous operator splitting.* We consider viscous terms as a perturbation of the inviscid system of the Euler equations and split the complete system (4.3.1) into the inviscid system and purely viscous system,

$$\frac{\partial \boldsymbol{w}}{\partial t} + \sum_{s=1}^{N} \frac{\partial \boldsymbol{f}_s(\boldsymbol{w})}{\partial x_s} = 0, \tag{4.4.35}$$

$$\frac{\partial \boldsymbol{w}}{\partial t} = \sum_{s=1}^{N} \frac{\partial \boldsymbol{R}_s(\boldsymbol{w}, \nabla \boldsymbol{w})}{\partial x_s} + \boldsymbol{F}(\boldsymbol{w}), \tag{4.4.36}$$

and discretize them separately. The Euler equations (4.4.35) are discretized by the FVM on a mesh $\mathcal{D}_h = \{D_i\}_{i \in J}$, whereas the viscous system (4.4.36) is discretized by the FEM on a mesh $\mathcal{T}_h$. We use the same notation as above.

In the cases a) and b) from Example 4.51, when the conforming P^1-finite elements are combined with dual finite volumes and the nonconforming P^1-finite elements are combined with barycentric finite volumes, the operator L_h is one-to-one and its inverse L_h^{-1} exists. In the case c) from Exercise 4.51, where triangular finite volumes are combined with conforming P^1-finite elements, the operator L_h is not one-to-one. It is therefore necessary to define a 'pseudoinverse' operator $\tilde{L}_h^{-1}$ to L_h in an appropriate way. It can be carried out in the following way. Let $\boldsymbol{w}_{h\mathrm{FV}} \in \boldsymbol{Z}_h$ be a finite volume piecewise constant function. Then we construct a finite element function $\boldsymbol{w}_{h\mathrm{FE}} = \tilde{L}_h^{-1}\boldsymbol{w}_{h\mathrm{FV}} \in \boldsymbol{X}_h$ determined by its values $\boldsymbol{w}_{h\mathrm{FE}}(P_i)$ at the vertices $P_i \in \sigma_h$ of the mesh $\mathcal{T}_h$: If $P_i \in \sigma_h \cap \Omega_h$, then P_i is the barycentre of D_i and we set

$$\boldsymbol{w}_{h\mathrm{FE}}(P_i) = \boldsymbol{w}_{h\mathrm{FV}}|_{D_i}. \tag{4.4.37}$$

For $P_i \in \sigma_h \cap \partial\Omega_h$, $\boldsymbol{w}_{h\mathrm{FE}}(P_i)$ is defined with aid of the Dirichlet boundary conditions or extrapolation using the weighted averages. For the component w_s of $\boldsymbol{w}$ with a prescribed value w_{Ds} at P_i we set

$$w_{hs\mathrm{FE}}(P_i) = w_{Ds}(P_i). \tag{4.4.38}$$

Otherwise,

$$w_{hs\mathrm{FE}}(P_i) = \frac{\sum_{j\in A_i} |D_j| w_{hs\mathrm{FV}}|_{D_j}}{\sum_{j\in A_i} |D_j|}, \tag{4.4.39}$$

where $A_i = \{j \in J; P_i \in \partial D_j\}$. Obviously,

$$L_h(\tilde{L}_h^{-1}\boldsymbol{w}_{h\mathrm{FV}}) = \boldsymbol{w}_{h\mathrm{FV}} \quad \forall \boldsymbol{w}_{h\mathrm{FV}} \in \boldsymbol{Z}_h, \tag{4.4.40}$$

but, in general, $\tilde{L}_h^{-1}(L_h\boldsymbol{w}_{h\mathrm{FE}}) \neq \boldsymbol{w}_{h\mathrm{FE}}$ for $\boldsymbol{w}_{h\mathrm{FE}} \in \boldsymbol{X}_h$.

For the sake of uniformity of notation, we set $\tilde{L}_h^{-1} = L_h^{-1}$ in the cases a) and b) from Example 4.51.

4.4.6.1 *The algorithm of the inviscid–viscous operator splitting* This algorithm can be written in the following way:

(0) We start from the FE initial approximation $\boldsymbol{w}^0_{h\mathrm{FE}} \in \boldsymbol{X}_h$ of the initial condition $\boldsymbol{w}^0$ and set

$$\boldsymbol{w}^0_{h\mathrm{FV}} = L_h\boldsymbol{w}^0_{h\mathrm{FE}}. \tag{4.4.41}$$

Let us assume that we have already obtained the FV approximate solution $\boldsymbol{w}^k_{h\mathrm{FV}}$ at time level t_k. The transition to the next time level is carried out in two *fractional steps*:

(1) Inviscid FV step:

(α) Set $\tilde{\boldsymbol{w}}_i^k = \boldsymbol{w}^k_{h\mathrm{FV}}|_{D_i}$ for all $D_i \in \mathcal{D}_h$ and compute

$$\tilde{\boldsymbol{w}}_i^{k+1/2} = \tilde{\boldsymbol{w}}_i^k - \frac{\tau_k}{|D_i|}\sum_{j\in S(i)}\sum_{\alpha=1}^{\beta_{ij}} \boldsymbol{H}(\tilde{\boldsymbol{w}}_i^k, \tilde{\boldsymbol{w}}_j^k, \boldsymbol{n}_{ij}^\alpha)|\Gamma_{ij}^\alpha|, \tag{4.4.42}$$

$D_i \in \mathcal{D}_h$, with the use of inviscid boundary conditions (cf. 3.3.6).

(β) Define $\boldsymbol{w}_{h\mathrm{FV}}^{k+1/2} \in \boldsymbol{Z}_h$ so that $\boldsymbol{w}_{h\mathrm{FV}}^{k+1/2}|_{D_i} = \tilde{\boldsymbol{w}}_i^{k+1/2}$ for $D_i \in \mathcal{D}_h$.

(2) Viscous FE step:

(α) Set $\boldsymbol{w}_{h\mathrm{FE}}^{k+1/2} = \tilde{L}_h^{-1} \boldsymbol{w}_{h\mathrm{FV}}^{k+1/2} \in \boldsymbol{X}_h$.

(β) Compute $\boldsymbol{w}_{h\mathrm{FE}}^{k+1} \in \boldsymbol{X}_h$ such that

a) $\boldsymbol{w}_{h\mathrm{FE}}^{k+1} - \boldsymbol{w}_h^*(t_{k+1}) \in \boldsymbol{V}_h$,

b) $$(\boldsymbol{w}_{h\mathrm{FE}}^{k+1}, \boldsymbol{\varphi}_h)_h = (\boldsymbol{w}_{h\mathrm{FE}}^{k+1/2}, \boldsymbol{\varphi}_h)_h - \tau_k a_h(\boldsymbol{w}_{h\mathrm{FE}}^{k+1/2}, \boldsymbol{\varphi}_h) + \tau_k \beta_h(\boldsymbol{w}_{h\mathrm{FE}}^{k+1/2}, \boldsymbol{\varphi}_h) \quad \forall \boldsymbol{\varphi}_h \in \boldsymbol{V}_h. \tag{4.4.43}$$

(γ) Set $\boldsymbol{w}_{h\mathrm{FV}}^{k+1} = L_h \boldsymbol{w}_{h\mathrm{FE}}^{k+1}$, $k := k+1$, and go to (1).

The above process with time step $\tau = \tau_k$ can be symbolically written as

$$\boldsymbol{w}_{h\mathrm{FV}}^{k+1} = \mathcal{L}_h(\tau) \boldsymbol{w}_{h\mathrm{FV}}^{k}, \tag{4.4.44}$$

with

$$\mathcal{L}_h(\tau) = L_h \mathcal{L}_{h\mathrm{FE}}(\tau) \tilde{L}_h^{-1} \mathcal{L}_{h\mathrm{FV}}(\tau). \tag{4.4.45}$$

Here $\mathcal{L}_{h\mathrm{FV}}$ and $\mathcal{L}_{h\mathrm{FE}}$ are the operators realizing the FV and FE fractional steps (1) and (2), respectively:

$$\begin{aligned} \boldsymbol{w}_{h\mathrm{FV}}^{k+1/2} &= \mathcal{L}_{h\mathrm{FV}}(\tau) \boldsymbol{w}_{h\mathrm{FV}}^{k}, \\ \boldsymbol{w}_{h\mathrm{FE}}^{k+1} &= \mathcal{L}_{h\mathrm{FE}}(\tau) \boldsymbol{w}_{h\mathrm{FE}}^{k+1/2}. \end{aligned} \tag{4.4.46}$$

With the aid of this notation, it is possible to define other variants of the fractional step method. Let us consider, for example, (4.4.44), where $\mathcal{L}_h(\tau)$ is replaced by

$$\hat{\mathcal{L}}_h(\tau) = \mathcal{L}_{h\mathrm{FV}}\left(\frac{\tau}{2}\right) L_h \mathcal{L}_{h\mathrm{FE}}(\tau) \tilde{L}_h^{-1} \mathcal{L}_{h\mathrm{FV}}\left(\frac{\tau}{2}\right) \tag{4.4.47}$$

or

$$\tilde{\mathcal{L}}_h(\tau) = L_h \mathcal{L}_{h\mathrm{FE}}\left(\frac{\tau}{2}\right) \tilde{L}_h^{-1} \mathcal{L}_{h\mathrm{FV}}(\tau) L_h \mathcal{L}_{h\mathrm{FE}}\left(\frac{\tau}{2}\right) \tilde{L}_h^{-1}. \tag{4.4.48}$$

4.4.6.2 *Stability of the combined FV–FE methods* Since schemes (4.4.31), (4.4.32) and the fractional step schemes from 4.4.6.1 are explicit or semi-implicit, it is necessary to apply some stability conditions. Unfortunately, there is no rigorous theory for the stability of schemes applied to the complete compressible Navier–Stokes system. We proceed heuristically. By virtue of the explicit FV discretization of inviscid terms, we apply the stability condition derived in Section 3.3.11 for the explicit FVM for the solution of the Euler equations:

$$\frac{\tau_k}{|D_i|} |\partial D_i| \max_{\substack{j \in S(i) \\ \alpha = 1, \ldots, \beta_{ij}}} \max_{\ell = 1, \ldots, m} \{ |\lambda_\ell(\boldsymbol{w}_i^k, \boldsymbol{n}_{ij}^\alpha)| \} \le \mathrm{CFL} \approx 0.85, \quad i \in J, \tag{4.4.49}$$

where $\lambda_\ell(\boldsymbol{w}_i^k, \boldsymbol{n}_{ij}^\alpha)$ are the eigenvalues of the matrix $\mathbb{P}(\boldsymbol{w}_i^k, \boldsymbol{n}_{ij}^\alpha)$ – see (3.1.14).

If the combined FV–FE scheme is fully explicit as (4.4.31), or as the operator splitting method 4.4.6.1, it is necessary to augment condition (4.4.49) taking into account the explicit discretization of the viscous (i.e. diffusion) terms. In this case, besides (4.4.49), we consider the stability condition

$$\frac{3}{4}\frac{h_K}{\rho_K}\frac{\tau_k}{|K|}\max(\mu, k) \leq \mathrm{CFL}, \quad K \in \mathcal{T}_h, \tag{4.4.50}$$

where h_K, ρ_K and $|K|$ are the diameter of K, the radius of the largest ball inscribed in K and the measure of K, respectively. This condition was derived as an L^∞-stability condition for the explicit P^1–finite element method for the solution of the heat conduction equation and applied heuristically to the flow problem under consideration. A theoretical background for these stability conditions will be given in Section 4.5 – see Remarks 4.60 and 4.69. Due to (4.4.49) and (4.4.50), $\tau_k = O(h)$ and $\tau_k = O\left(\min_{K\in\mathcal{T}_h} h_K^2/\max(\mu, k)\right)$, respectively. The second condition is rather restrictive for τ_k, if h_K is very small. (It is compensated a little by the small denominator $\max(\mu, k)$.)

4.4.7 *Applications of the combined FV–FE methods*

A number of numerical experiments ((Feistauer and Felcman, 1997), (Feistauer *et al.*, 1996), (Feistauer and Kliková, 2001), (Dolejší *et al.*, 2002), (Arminjon and Madrane, 1999*b*)) demonstrate the applicability, efficiency and robustness of the combined FV–FE methods. Here we present some computational results for viscous gas flow.

4.4.7.1 *GAMM channel* As a simple test problem demonstrating the application of the combined FV–FE method we use the viscous flow through the GAMM channel, represented by the domain Ω introduced in Section 3.7.2. The flow was computed for $\gamma = 1.4$, $\mu = 1.72 \cdot 10^{-5}\,\mathrm{kg\,m^{-1}\,s^{-1}}$, $\lambda = -\frac{2}{3}\mu = -1.15 \cdot 10^{-5}\,\mathrm{kg\,m^{-1}\,s^{-1}}$, $k = 2.4 \cdot 10^{-2}\,\mathrm{kg\,m\,s^{-3}\,K^{-1}}$, $c_v = 721.428\,\mathrm{J\,kg^{-1}\,K^{-1}}$.

On the boundary $\partial\Omega$ the boundary conditions (4.3.8), a), b), c*), (4.3.9), a), b*) and (4.3.10) are used. That is, at the inlet, the Mach number $M = 0.67$, the density $\rho = 1.5\,\mathrm{kg\,m^{-3}}$, the velocity component $v_1 = 206.9\,\mathrm{m\,s^{-1}}$ and the velocity component $v_2 = 0\,\mathrm{m\,s^{-1}}$. At the outlet the normal component of the viscous part of the stress tensor vanishes. On the walls the no-slip condition is used. On the walls and outlet the temperature satisfies the condition $\partial\theta/\partial n = 0$ and at the inlet condition (4.3.8), c*) with $g_N = 0$ holds. We neglect the volume force and heat sources. Thus, $\boldsymbol{F} = 0$ in equation (4.3.1). (See Remark 4.42.) The dimensionless form of the Navier–Stokes equations (4.3.4) is used for the numerical solution. As basic reference quantities (see Section 1.2.23) we choose $\rho^* = 1.5\,\mathrm{kg\,m^{-3}}$, $U^* = 206.9\,\mathrm{m\,s^{-1}}$, $L^* = 1\,\mathrm{m}$, $\mu^* = \mu$ and $k^* = k$. Using (1.2.135) and (1.2.68), we compute the Reynolds number $Re = 1.8 \cdot 10^7$ and the Prandtl number $Pr = 0.72$.

a) *Red–green refinement using the residual error indicator* We start from the weak formulation of the system of compressible Navier–Stokes equations

(4.3.4). Multiplying (4.3.4) by a test function $\varphi = (\varphi^1, \varphi^2, \ldots, \varphi^m)^{\mathrm{T}} \in \boldsymbol{H}_0^1(\Omega) = H_0^1(\Omega)^m$, integrating over Ω and the time interval $I = (t_k, t_{k+1})$ and using Green's theorem, we obtain the identity (we have omitted primes in the notation)

$$0 = \int_\Omega (\boldsymbol{w}(x, t_{k+1}) - \boldsymbol{w}(x, t_k))\varphi \, dx - \int_I \int_\Omega \sum_{s=1}^{N} \boldsymbol{f}_s(\boldsymbol{w}) \frac{\partial \varphi}{\partial x_s} \, dx \, dt \tag{4.4.51}$$
$$+ \int_I \int_\Omega \sum_{s=1}^{N} \boldsymbol{R}_s(\boldsymbol{w}, \nabla \boldsymbol{w}) \frac{\partial \varphi}{\partial x_s} \, dx \, dt =: \boldsymbol{a}_k(\boldsymbol{w}, \nabla \boldsymbol{w}, \varphi).$$

There exists an element $\boldsymbol{\mathcal{A}}_k(\boldsymbol{w}, \nabla \boldsymbol{w}) \in \boldsymbol{H}^{-1}(\Omega)$ (dual to $\boldsymbol{H}_0^1(\Omega)$) such that equation (4.4.51) is equivalent to the operator equation $\boldsymbol{\mathcal{A}}_k(\boldsymbol{w}, \nabla \boldsymbol{w}) = 0$. The FV–FE operator splitting method used from Section 4.4.6 yields at the time level t_k a piecewise constant solution $\boldsymbol{w}_{h\mathrm{FV}}^k$ of system (4.4.35). If the mesh $\mathcal{D}_h$ is N-simplicial, we denote by $\boldsymbol{w}_i^k$ the restriction $\boldsymbol{w}_{h\mathrm{FV}}^k | D_i$, $D_i \in \mathcal{D}_h$. For a dual or barycentric mesh $\mathcal{D}_h$, we denote by the symbol $\boldsymbol{w}_i^k$ the values on D_i of the piecewise constant recovery of $\boldsymbol{w}_{h\mathrm{FV}}^k$ on the N-simplicial mesh $\mathcal{T}_h$. (See Section 3.6.4.) Let us note that only triangular or tetrahedral meshes are adapted, as stated in the introductory part of Section 3.6. Further, by $\tilde{\boldsymbol{w}}_h^k$ we denote the piecewise linear reconstruction of $\boldsymbol{w}_{h\mathrm{FV}}^k$ (see Section 3.6.5) and set $\boldsymbol{r}_h^k = \boldsymbol{\mathcal{A}}_k(\boldsymbol{w}_h^k, \nabla \tilde{\boldsymbol{w}}_h^k)$ (= the approximation of the residual). We approximate the $\boldsymbol{H}^{-1}$-norm of the residual in the form $\|\boldsymbol{r}_h^k\|_{\boldsymbol{H}^{-1}} \approx (\sum_{D_i \in \mathcal{D}_h} g_{\mathrm{res}}^2(i))^{1/2}$, where g_{res} is called the *residual error indicator* (cf. Section 3.6.4.2). According to the operator splitting method, it is expressed in two parts corresponding to the inviscid and viscous fluxes:

$$\begin{aligned} g_{\mathrm{res}}^2(i) &= \beta^2(i) + \gamma^2(i), \\ \beta^2(i) &= \sum_{\ell=1}^{m} \left(\beta^\ell(i)\right)^2, \\ \gamma^2(i) &= \sum_{\ell=1}^{m} \left(\gamma^\ell(i)\right)^2, \end{aligned} \tag{4.4.52}$$

where, for $\ell = 1, \ldots, m$,

$$\beta^\ell(i) = \max_{j \in s(i)} \frac{|\Gamma ij|}{|\varphi_{ij}|_{H_0^1(\Omega)}} \left| \sum_{s=1}^{2} n_{ij}^s \left(f_s^\ell(\boldsymbol{w}_i^k) - f_s^\ell(\boldsymbol{w}_j^k) \right) \right|, \tag{4.4.53}$$

$$\gamma^\ell(i) = \max_{j \in s(i)} \frac{|\Gamma_{ij}|}{|\varphi_{ij}|_{H_0^1(\Omega)}} \left| \sum_{s=1}^{2} n_{ij}^s \left(R_s^\ell(\boldsymbol{w}_i^k, \nabla \tilde{\boldsymbol{w}}_i^k) - R_s^\ell(\boldsymbol{w}_j^k, \nabla \tilde{\boldsymbol{w}}_j^k) \right) \right|. \tag{4.4.54}$$

In (4.4.53) and (4.4.54) notation (3.3.11) is used. The test functions φ_{ij} are defined in Section 3.6.4.2. Finaly, we set

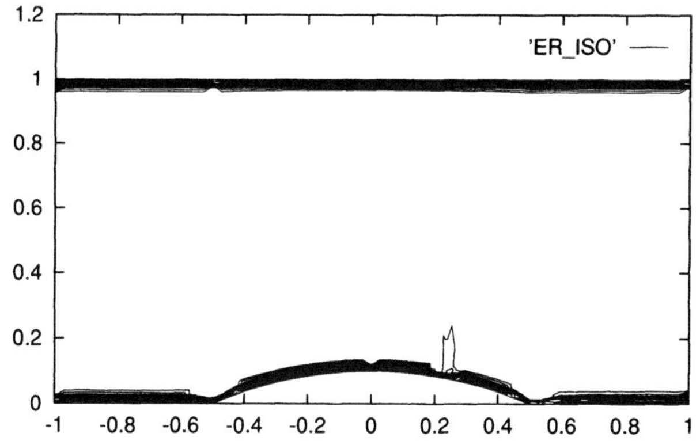

FIG. 4.4.3. Residual error isolines

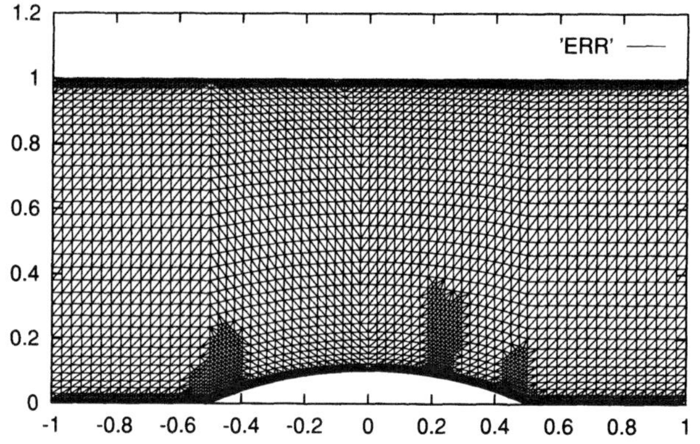

FIG. 4.4.4. Computational mesh

$$g_{\text{ind}} := g_{\text{res}} \tag{4.4.55}$$

and use the adaptation technique described in Section 3.6.3. Figures 4.4.3–4.4.5 show the computational results.

b) *User-defined refinement* This was used on the basis of a priori knowledge of the flow character in the case when triangular finite volumes were combined with adjoint triangular finite elements. In Fig. 4.4.6 and Fig. 4.4.7, the primary FV and adjoint FE meshes are presented. Figure 4.4.8 shows the convergence history to the steady state measured in the L^1-norm (the behaviour of the quantity $\log(e_k)$ with e_k defined in (3.7.4), depending on k) and Fig. 4.4.9 shows the dimensionless pressure distribution on the walls. In Fig. 4.4.10, Mach number isolines are plotted. We obtained identical results with the aid of the purely explicit scheme (4.4.31) as well as of the inviscid–viscous operator splitting method from Section 4.4.6.1. We can see in Fig. 4.4.10 that the boundary layer is thicker than in Fig. 4.4.5. The reason is that in this case the Reynolds number $Re = 1.8 \cdot 10^5$,

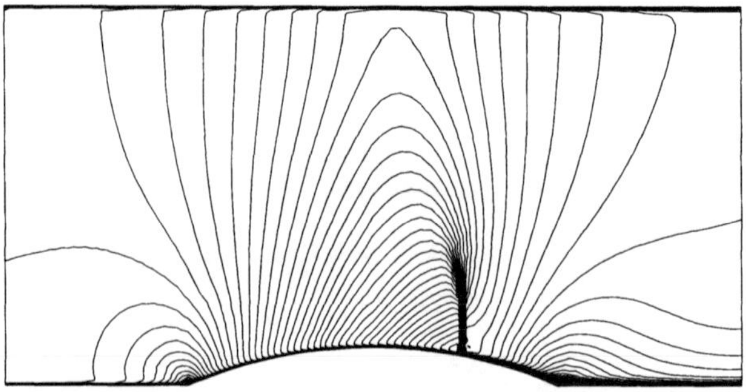

FIG. 4.4.5. Mach number isolines

whereas in the previous example the Reynolds number is a hundred times higher.

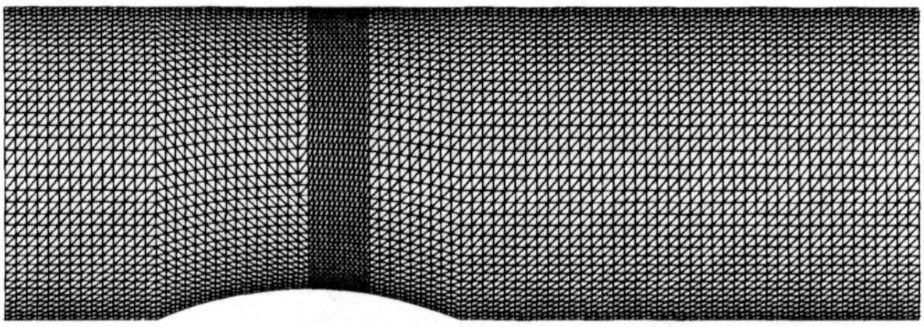

FIG. 4.4.6. Primary FV mesh

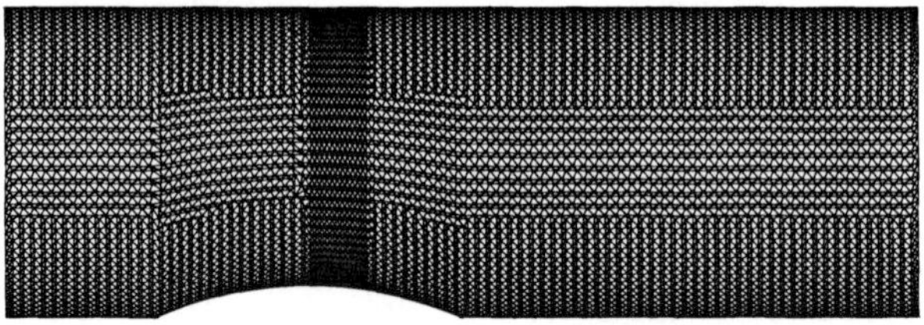

FIG. 4.4.7. Adjoint FE mesh

4.4.7.2 *Cascade of profiles* Flow past a cascade of profiles was solved by the inviscid–viscous operator splitting method. The computational results are compared with a wind tunnel experiment (by courtesy of the Institute of Thermomechanics of the Academy of Sciences of the Czech Republic in Prague, see (Šťastný and Šafařík, 1990)). The experiment and the computations were performed for the following data: angle of attack = $19^\circ\,18'$, inlet Mach number = 0.32, outlet Mach number = 1.18, $\gamma = 1.4$, $\mu = 1.72 \cdot 10^{-5}\,\mathrm{kg\,m^{-1}\,s^{-1}}$, $\lambda = -\frac{2}{3}\mu = -1.15 \cdot 10^{-5}\,\mathrm{kg\,m^{-1}\,s^{-1}}$, $k = 2.4 \cdot 10^{-2}\,\mathrm{kg\,m\,s^{-3}\,K^{-1}}$, $c_v = 721.428\,\mathrm{J\,kg^{-1}\,K^{-1}}$.

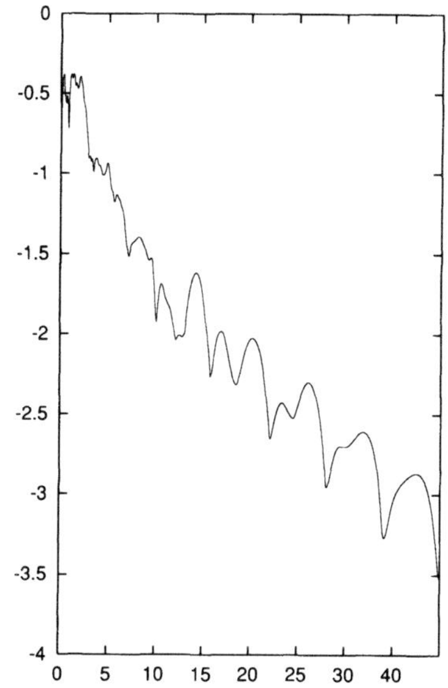

FIG. 4.4.8. Convergence history ($\log\|(\rho^{k+1}-\rho^k)/\tau\rho^k\|_{L^1(\Omega)}$ against the number of time steps $k/10^3$)

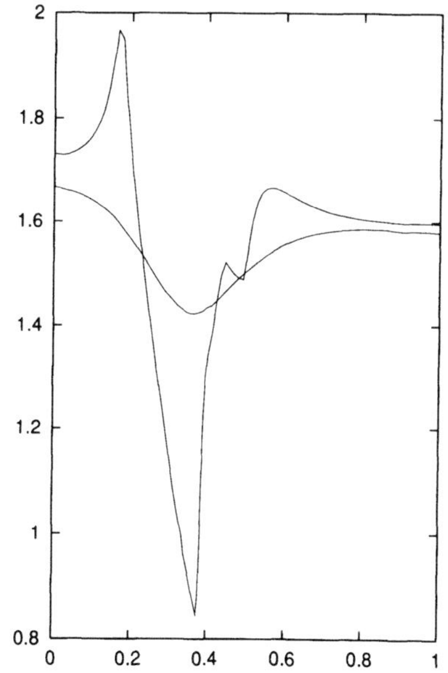

FIG. 4.4.9. Dimensionless pressure distribution on the walls against the relative arc length

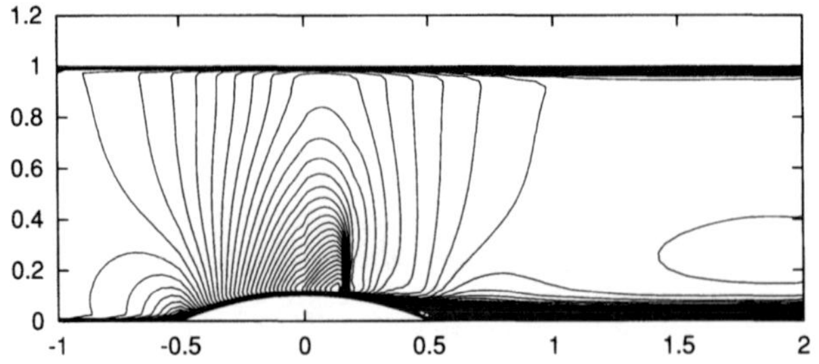

FIG. 4.4.10. Mach number isolines

At the inlet, the density $\rho = 1.5\,\mathrm{kg\,m^{-3}}$ and the magnitude of the velocity $|\boldsymbol{v}| = 98.2492\,\mathrm{m\,s^{-1}}$ were prescribed. Similarly as in Section 4.4.7.1 we neglect the term $\boldsymbol{F}$ in equation (4.3.1). The dimensionless form of the Navier–Stokes equations (4.3.4) is used for the numerical solution. As basic reference quantities (see Section 1.2.23) we choose $\rho^* = 1.5\,\mathrm{kg\,m^{-3}}$, $U^* = 98.2492\,\mathrm{m\,s^{-1}}$, $L^* = 0.1\,\mathrm{m}$, $\mu^* = \mu$ and $k^* = k$. Using (1.2.135) and (1.2.68), we compute the Reynolds number $Re = 8.5 \cdot 10^5$ and the Prandtl number $Pr = 0.72$. Two combined FV–FE methods were applied to the solution of the cascade flow. The computational mesh was constructed with the aid of the anisotropic mesh adaptation (AMA) from Section 3.6.7. The Mach number M was used as the quantity for the indication of the mesh adaptation.

a) *AMA in the framework of the combination of triangular finite volumes and barycentric finite elements* Figure 4.4.11 shows the barycentric FV mesh $\mathcal{D}_h$ constructed over a triangular mesh $\mathcal{T}_h$ from Fig. 4.4.12. In Fig. 4.4.13, density isolines are plotted. Compared with an interferogram (see Figs 3.7.58 and 3.7.59), an agreement between computational and experimental results is observed.

b) *AMA in the framework of the combination of triangular finite volumes and adjoint triangular finite elements* Figure 4.4.14 shows a detail of the primary FV mesh $\mathcal{D}_h$. A detail of the adjoint triangular FE mesh $\mathcal{T}_h$ is shown in Fig. 4.4.15. In Fig. 4.4.16, density isolines are plotted. Figure 4.4.17 shows a very good agreement between the computed and measured pressure distribution along the profile. Compare Figs 3.7.67 and 4.4.17 and notice the improvement in the numerical results with respect to the experiment when the model of viscous flow (Navier–Stokes equations) is used instead of the inviscid model (Euler equations). In both cases, the same finite volume mesh has been used.

c) *Shock indicator in the framework of the combination of triangular finite volumes and adjoint triangular finite elements* Figure 4.4.18 shows the primary FV mesh $\mathcal{D}_h$ obtained with the aid of adaptive mesh refinement in the vicinity of shock waves, based on the shock indicator from Section 3.6.4.1. Because of the resolution of a boundary layer, the mesh is refined a priori towards the profile. The adjoint triangular FE mesh $\mathcal{T}_h$ is plotted in Fig. 4.4.19. In Fig. 4.4.20 and Fig. 4.4.21, the density and Mach number isolines are shown, respectively. A

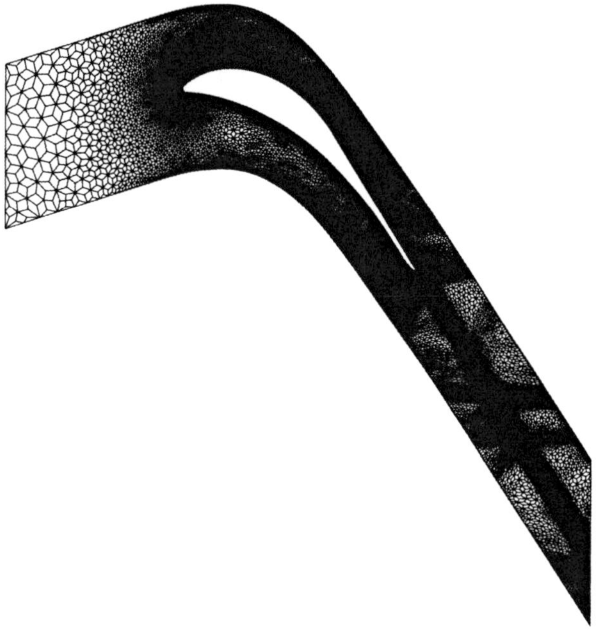

FIG. 4.4.11. Barycentric FV grid for viscous cascade flow

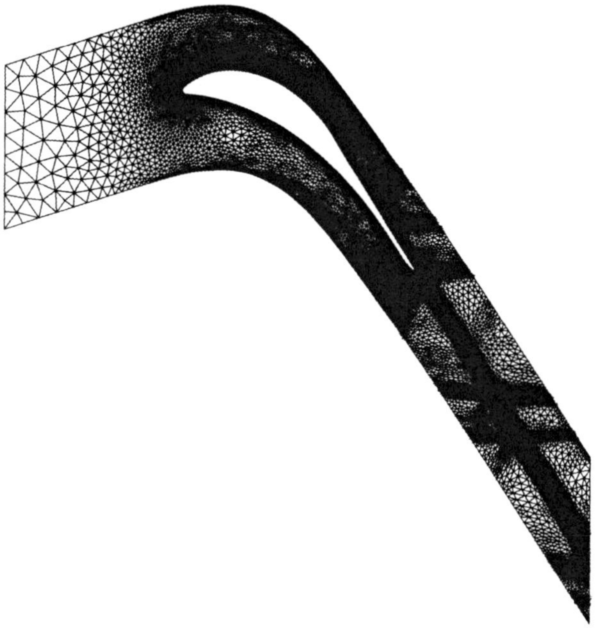

FIG. 4.4.12. Triangular FE grid for viscous cascade flow

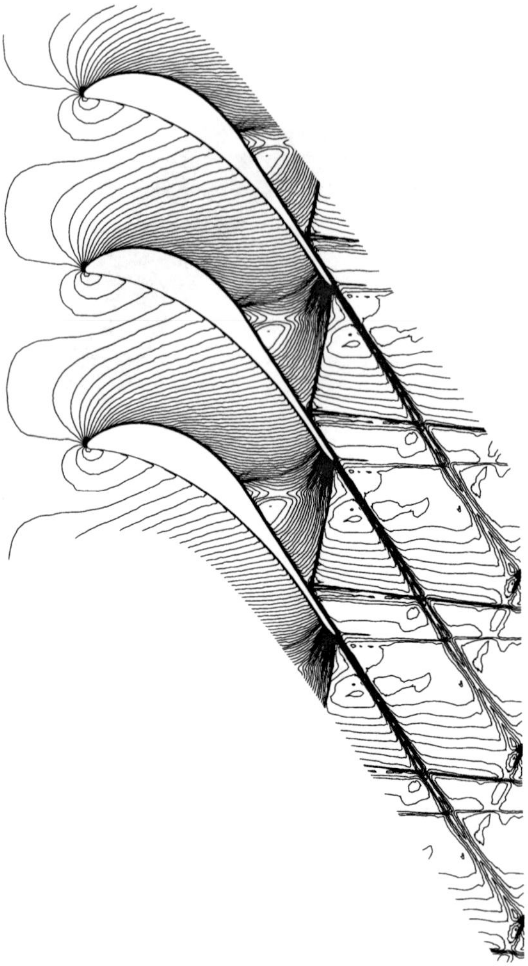

FIG. 4.4.13. Density isolines for viscous cascade flow

detail of the density isolines is plotted in Fig. 4.4.22 for the comparison with the interferogram in Fig. 3.7.59. The convergence history is shown in Fig. 4.4.23.

d) *The Osher–Solomon numerical flux versus the Godunov numerical flux* In this section we present computational results obtained with the aid of two different numerical fluxes, used in the framework of the anisotropic mesh adaptation. That is, we use the numerical flux defined by (3.3.37), where for $\boldsymbol{g}_{\mathrm{R}}$ we substitute either the Godunov flux defined by (3.2.71) and results from Section 3.1.7 – see Fig. 3.1.15 and Remark 3.28 – or the Osher–Solomon flux defined in Section 3.4. In Fig. 4.4.24 and Fig. 4.4.25, the Mach number isolines computed with the aid of the Osher–Solomon and Godunov numerical fluxes are plotted, respectively. The combination of triangular finite volumes with triangular finite elements was

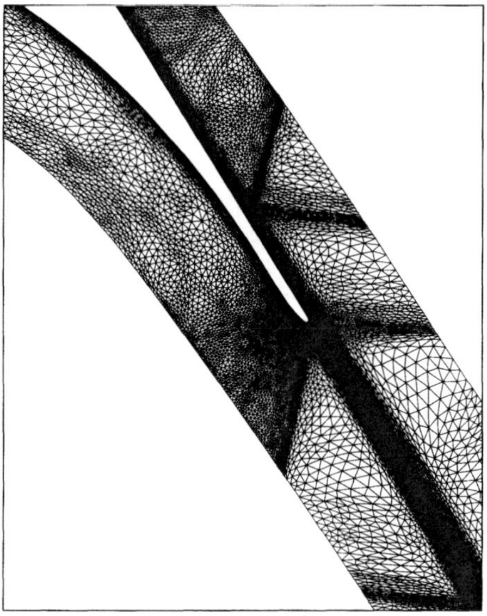

FIG. 4.4.14. Triangular finite volume mesh obtained using AMA

used. The differences in the graphical output are indistinguishable.

4.4.7.3 *NACA 0012* Similarly as in the previous section, we compare here the results obtained with the aid of the Osher–Solomon numerical flux and the Godunov numerical flux. As a test problem the viscous flow around the NACA 0012 profile Γ_{prof} is considered. We use the following data from (Bristeau *et al.*, 1987): inlet Mach number $M_{\mathrm{inlet}} = 0.85$, angle of attack $\alpha = 0^\circ$ and the Reynolds number $Re = 500$. The meshes used were generated by the AMA technique and they are slightly different for both solvers. We have compared the CPU time and the drag c_D and lift c_L coefficients for both solvers. The results can be seen in Table 4.4.1. Provided the chord of the profile Γ_{prof} is in the direction of the axis x_1, and the axis x_2 is orthogonal to the chord of Γ_{prof}, the drag and lift coefficients are defined as the first and the second components of the following vector, respectively:

$$-\frac{1}{\frac{1}{2}\rho_\infty |\boldsymbol{v}_\infty|^2} \int_{\Gamma_{\mathrm{prof}}} \{(-p + \lambda \operatorname{div} \boldsymbol{v})\mathbb{I} + 2\mu \mathbb{D}(\boldsymbol{v})\}\, \boldsymbol{n}\, dS, \tag{4.4.56}$$

where ρ_∞ and $\boldsymbol{v}_\infty$ denote the density and the velocity vector at infinity, respectively. The integral is calculated along the profile Γ_{prof}. See Section 4.4.8 for more details. Since the considered profile is symmetric and $\alpha = 0^\circ$, the lift coefficient c_L is expected to be zero in stationary flow. The triangulation and the Mach number isolines for both methods are shown in Figs 4.4.26–4.4.29.

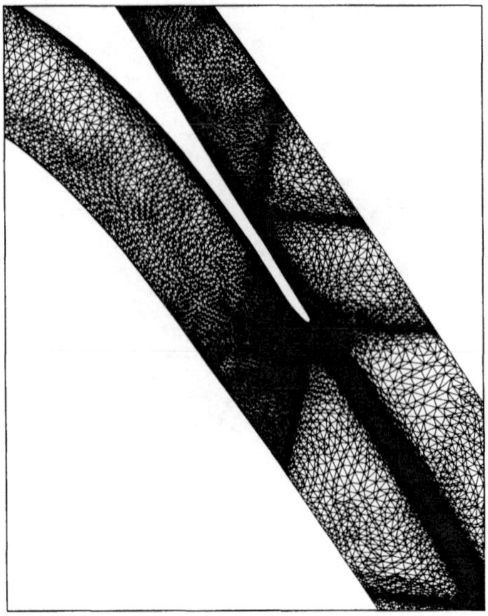

FIG. 4.4.15. Adjoint triangular finite element mesh

Table 4.4.1 *Comparison of the Osher–Solomon and Godunov numerical flux*

Solver	# triangles	CPU (s)	c_D	c_L	Figure
Osher–Solomon	2783	3345.7	0.254	0.000	Fig. 4.4.26
Godunov	2718	4255.3	0.258	0.000	Fig. 4.4.28

4.4.8 *Computation of the drag and lift*

Let us consider flow in a domain $\Omega \subset I\!R^2$ past a profile Γ_{prof}. This means that Γ_{prof} is an interior bounded component of $\partial\Omega$. The computation of the drag and lift (4.4.56) requires the evaluation of the force $\boldsymbol{F}_{\text{prof}}$ acting on the profile Γ_{prof}:

$$\boldsymbol{F}_{\text{prof}} = (F_{1\,\text{prof}}, F_{2\,\text{prof}}) = -\int_{\Gamma_{\text{prof}}} \{(-p + \lambda \operatorname{div} \boldsymbol{v})\,\mathbb{I} + 2\mu\mathbb{D}(\boldsymbol{v})\}\, \boldsymbol{n}\, \mathrm{d}S. \quad (4.4.57)$$

Here p is the pressure, $\boldsymbol{v} = (v_1, v_2)$ is the velocity vector, $\boldsymbol{n}$ is the unit outer normal to $\partial\Omega$ (pointing from Ω to Γ_{prof}) and μ, λ are the viscosity coefficients. The velocity deformation tensor $\mathbb{D}(\boldsymbol{v})$ has components $d_{ij}(\boldsymbol{v}) = (\partial v_i/\partial x_j + \partial v_j/\partial x_i)/2$. If the axis x_1 is parallel and x_2 is orthogonal to the chord of Γ_{prof}, then the components $F_{1\,\text{prof}}$ and $F_{2\,\text{prof}}$ are called the drag and lift of the profile, respectively.

There are two ways to evaluate the force $\boldsymbol{F}_{\text{prof}}$:

a) *Direct evaluation* This starts from formula (4.4.57). Of course, it makes sense provided the pressure p and the derivatives $\partial v_i/\partial x_j$ of the velocity com-

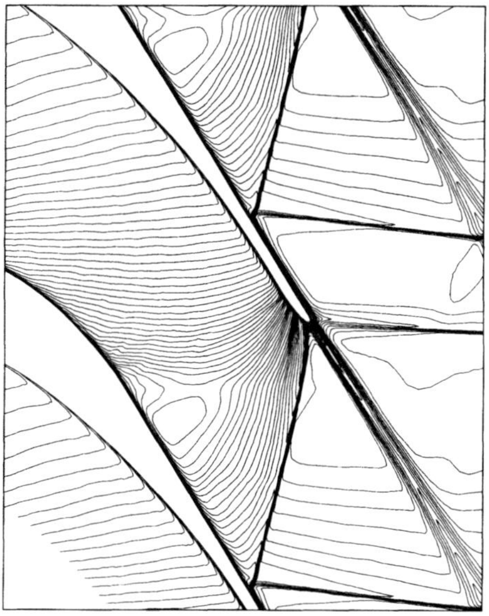

FIG. 4.4.16. Density isolines for viscous cascade flow

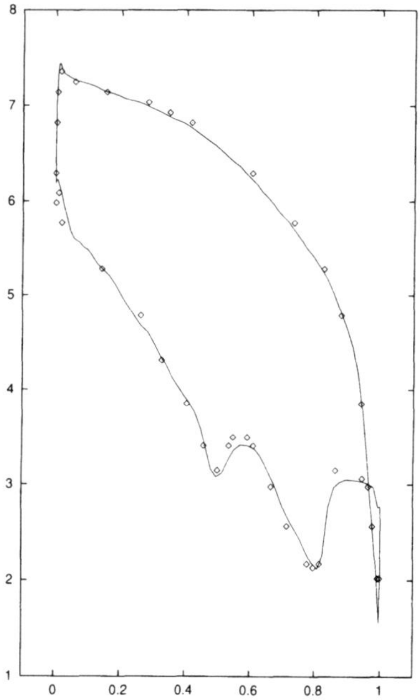

FIG. 4.4.17. Dimensionless pressure distribution along the profile (◇ experiment, — computation). Triangular FV mesh, triangular FE mesh, AMA

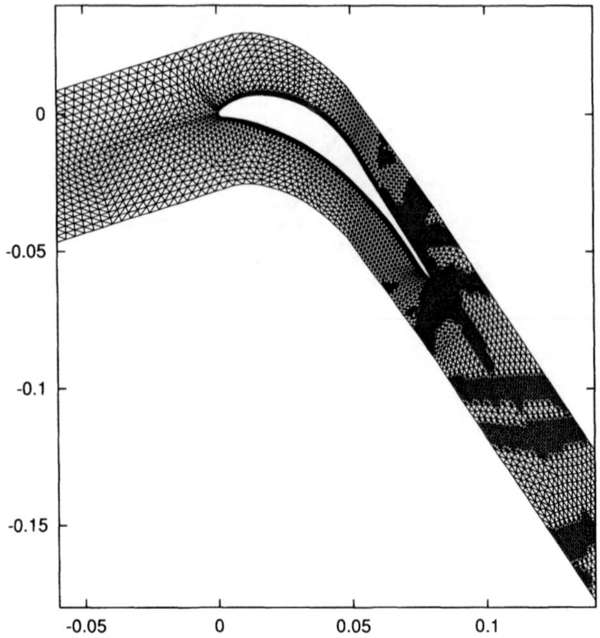

FIG. 4.4.18. Triangular finite volume mesh refined with the use of the shock indicator

ponents have integrable traces on Γ_{prof}. For example, in the classical sense we assume that p and $\partial v_i/\partial x_j$ can be continuously extended from Ω onto Γ_{prof}. In the framework of the discrete problem we proceed similarly. By p and $\boldsymbol{v}$ we now denote the approximation of the pressure and velocity, respectively, on the finite element mesh $\mathcal{T}_h$. If we write $\Gamma_{\text{prof}} = \bigcup_{j\in\gamma_{\text{prof}}} S_j$, where $S_j \subset \Gamma_{\text{prof}}$, $j \in \gamma_{\text{prof}}$, are faces of elements $K_j \in \mathcal{T}_h$, $j \in \gamma_{\text{prof}}$, adjacent to Γ_{prof}, then

$$\int_{\Gamma_{\text{prof}}} \cdots = \sum_{j\in\gamma_{\text{prof}}} \int_{S_j} \cdots . \tag{4.4.58}$$

The integrals along S_j are evaluated with the aid of suitable quadrature formulae and continuous extensions of p and $\boldsymbol{v}$ from K_j onto S_j, $j \in \gamma_{\text{prof}}$. For example, if a_j and b_j are the end points of S_j and $\boldsymbol{n}_j$ is the unit outer normal to $\partial\Omega$ on S_j, then we can set

$$\begin{aligned}\boldsymbol{F}_{\text{prof}} = -\frac{1}{2} \sum_{j\in\gamma_{\text{prof}}} \Big\{ & \left[(-p|_{K_j} + \lambda \operatorname{div}(\boldsymbol{v}|_{K_j}))\,\mathbb{I} + 2\mu\mathbb{D}(\boldsymbol{v}|_{K_j})\right](a_j) \\ & + \left[(-p|_{K_j} + \lambda \operatorname{div}(\boldsymbol{v}|_{K_j}))\,\mathbb{I} + 2\mu\mathbb{D}(\boldsymbol{v}|_{K_j})\right](b_j)\Big\}\boldsymbol{n}_j.\end{aligned} \tag{4.4.59}$$

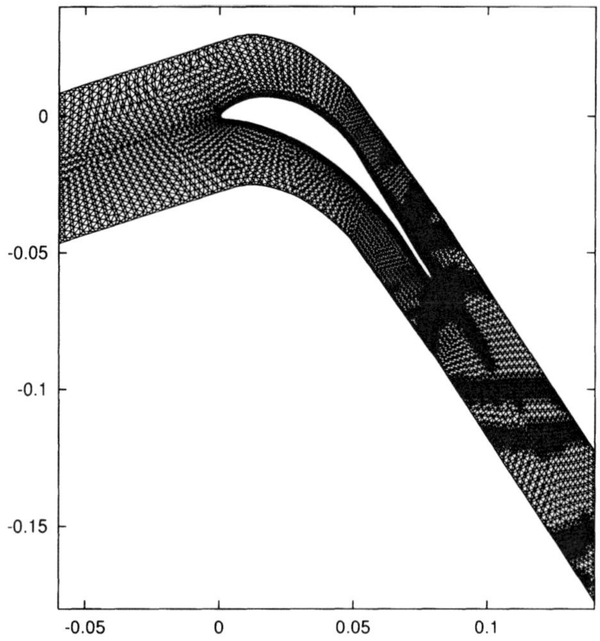

FIG. 4.4.19. Adjoint triangular finite element mesh

b) *Weak formulation* In the finite element or finite volume discrete problems, the starting point is usually the concept of a weak solution. Then, usually, p, $\partial v_i/\partial x_j \in L^2_{\text{loc}}(\Omega)$. Such functions have no traces on $\partial\Omega$ in general. In this case it is suitable to introduce a weak formulation of the force $\boldsymbol{F}_{\text{prof}}$, which can be used in the framework of the discrete problem in a natural way. First let us assume that ρ, p, $\boldsymbol{v}$ form a classical solution of the compressible Navier–Stokes equations written in the form

$$\frac{\partial(\rho v_i)}{\partial t} + \operatorname{div}(\rho v_i \boldsymbol{v}) = \rho f_i + \frac{\partial}{\partial x_i}(-p + \lambda \operatorname{div} \boldsymbol{v}) + \sum_{j=1}^{2} \frac{\partial}{\partial x_j}(2\mu d_{ij}(\boldsymbol{v})), \quad i = 1, 2. \tag{4.4.60}$$

Let $\varphi \in H^1(\Omega)$ be such a function that $\varphi|_{\Gamma_{\text{prof}}} = 1$ and $\varphi = 0$ outside some neighbourhood of Γ_{prof}. Multiplying (4.4.60) by φ, integrating over Ω and using Green's theorem, we get

$$\int_\Omega \left(\frac{(\rho v_i)}{\partial t} + \operatorname{div}(\rho v_i \boldsymbol{v}) - \rho f_i \right) \varphi \, dx$$

$$= \int_{\Gamma_{\text{prof}}} \left\{ (-p + \lambda \operatorname{div} \boldsymbol{v})\, n_i + 2\mu \sum_{j=1}^{2} d_{ij}(\boldsymbol{v})\, n_j \right\} \varphi \, dS$$

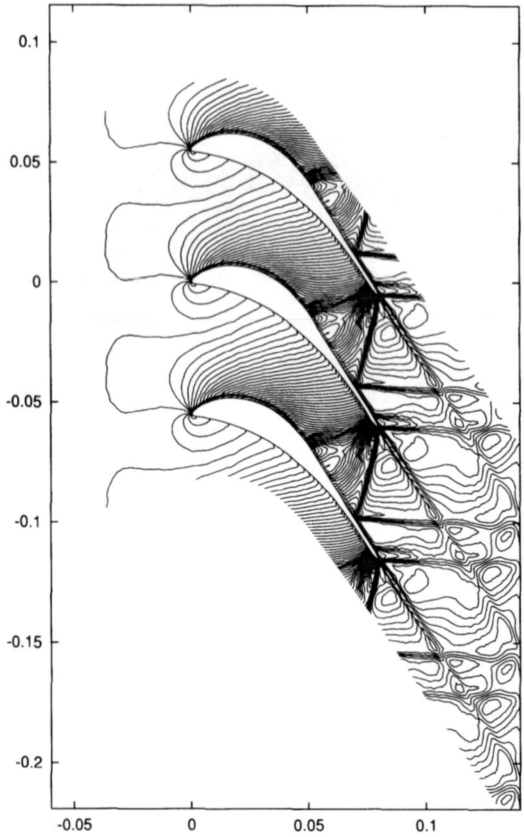

FIG. 4.4.20. Density isolines for viscous cascade flow

$$-\int_{\Omega}\left\{(-p+\lambda\,\mathrm{div}\,\boldsymbol{v})\,\frac{\partial\varphi}{\partial x_i}+2\mu\sum_{j=1}^{2}d_{ij}(\boldsymbol{v})\,\frac{\partial\varphi}{\partial x_j}\right\}dx,\qquad i=1,\,2.$$

Thus, since $\varphi|_{\Gamma_{\mathrm{prof}}}=1$, we see that the i-th component of $\boldsymbol{F}_{\mathrm{prof}}$ can be expressed as

$$\begin{aligned}F_{i\mathrm{prof}}=-\int_{\Omega}\Big\{\Big(&\frac{\partial(\rho v_i)}{\partial t}+\mathrm{div}(\rho v_i\boldsymbol{v})-\rho f_i\Big)\varphi \\ &+(-p+\lambda\,\mathrm{div}\,\boldsymbol{v})\,\frac{\partial\varphi}{\partial x_i}+2\mu\sum_{j=1}^{2}d_{ij}(\boldsymbol{v})\,\frac{\partial\varphi}{\partial x_j}\Big\}\,dx,\qquad i=1,\,2.\end{aligned}\tag{4.4.61}$$

Formula (4.4.61) makes sense even for the weak solution defined in Chapter 2. It can also be used in a natural way in the discrete problem. In this case, φ is chosen so that $\varphi\in X_h$, $\varphi|_{\Gamma_{\mathrm{prof}}}=1$ and $\varphi=0$ at all nodes (vertices) lying outside

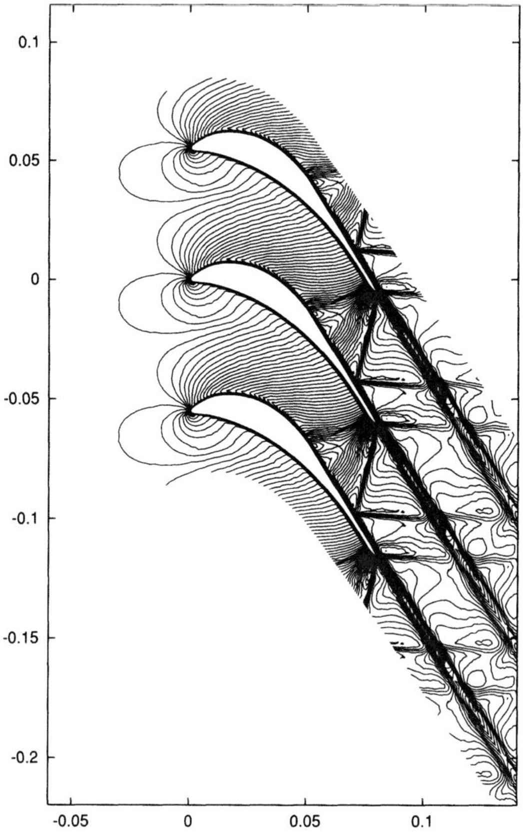

FIG. 4.4.21. Mach number isolines for viscous cascade flow

Γ_{prof}. Then the approximation of the components of the force acting on Γ_{prof} assumes the simple form

$$F_{i\text{prof}} = -\sum_{K \in \mathcal{T}_{\text{prof}}} \int_K \Big\{ \Big(\frac{\partial(\rho v_i)}{\partial t} + \operatorname{div}(\rho v_i \boldsymbol{v}) - \rho f_i \Big) \varphi \tag{4.4.62}$$

$$+ (-p + \lambda \operatorname{div} \boldsymbol{v}) \frac{\partial \varphi}{\partial x_i} + 2\mu \sum_{j=1}^{2} d_{ij}(\boldsymbol{v}) \frac{\partial \varphi}{\partial x_j} \Big\} \, dx, \quad i = 1, 2,$$

where $\mathcal{T}_{\text{prof}}$ denotes the set of all elements $K \in \mathcal{T}_h$ such that $K \cap \Gamma_{\text{prof}} \neq \emptyset$. In practical computations the integrals over $K \in \mathcal{T}_{\text{prof}}$ are evaluated with the aid of numerical integration and the time derivative is approximated by a finite difference.

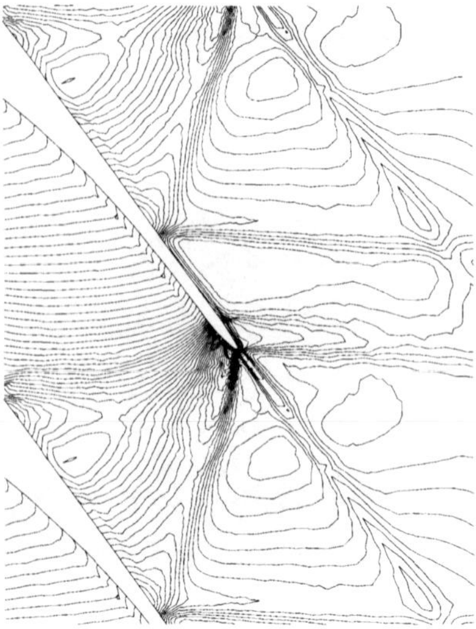

FIG. 4.4.22. Detail of the density isolines for viscous cascade flow

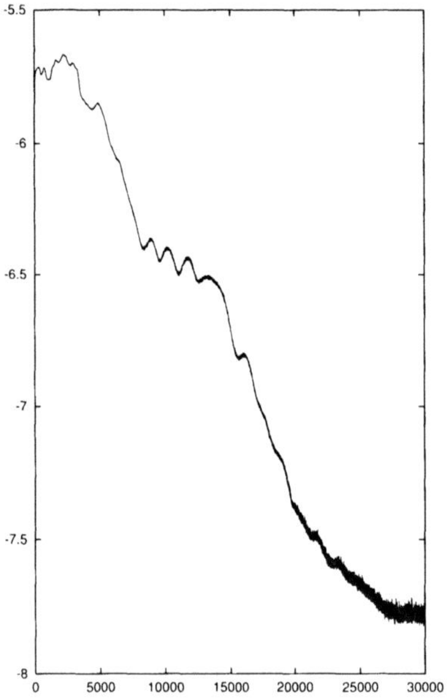

FIG. 4.4.23. Convergence history to the steady state

4.5 Theory of the combined FV–FE method

In order to support theoretically the combined FV–FE method for the numerical solution of viscous compressible flow treated in Section 4.4, we shall pay attention to the analysis of the semi-implicit and explicit combined FV–FE schemes applied to an initial-boundary value problem for a scalar 2D nonstationary, nonlinear

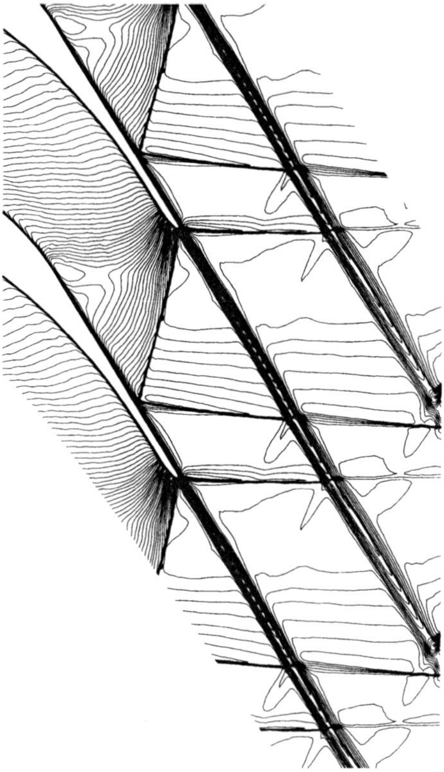

FIG. 4.4.24. Mach number isolines of the viscous cascade flow (Osher–Solomon numerical flux)

conservation law equation with a diffusion term. We shall briefly characterize some results from (Feistauer *et al.*, 1997), (Feistauer *et al.*, 1999*a*), (Feistauer *et al.*, 1999*b*), (Angot *et al.*, 1998) and (Dolejší *et al.*, 2002).

4.5.1 *Continuous problem*

Let $\Omega \subset I\!R^2$ be a bounded polygonal domain with a Lipschitz-continuous boundary $\partial\Omega$. In the space-time cylinder $Q_T = \Omega\times(0,T)$ $(0 < T < \infty)$ we shall consider the following initial-boundary value problem: find $u : Q_T \to I\!R$, $u = u(x,t)$, $x \in \Omega$, $t \in (0,T)$, such that

$$\frac{\partial u}{\partial t} + \sum_{s=1}^{2} \frac{\partial f_s(u)}{\partial x_s} - \nu\,\Delta\, u = g \quad \text{in } Q_T, \tag{4.5.1}$$

$$u|_{\partial\Omega\times(0,T)} = 0, \tag{4.5.2}$$

$$u(x,0) = u^0(x), \quad x \in \Omega, \tag{4.5.3}$$

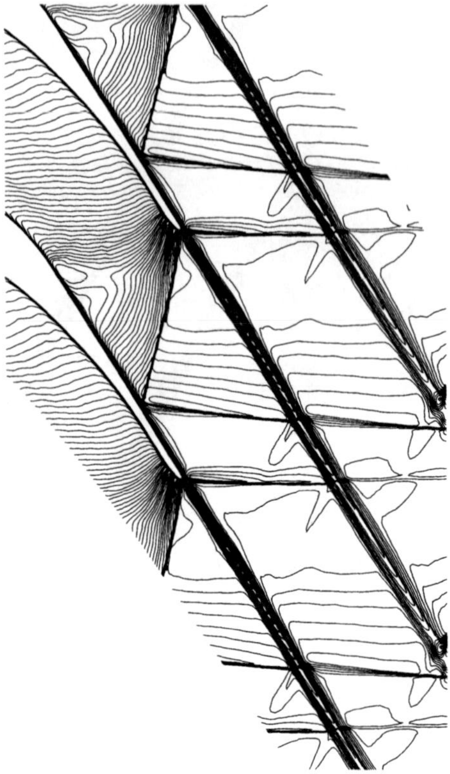

FIG. 4.4.25. Mach number isolines of the viscous cascade flow (Godunov numerical flux)

where $\nu > 0$ is a given constant and $f_s : \mathbb{R} \to \mathbb{R}$, $s = 1, 2$, $g : Q_T \to \mathbb{R}$, $u^0 : \Omega \to \mathbb{R}$ are given functions. In the theory of conservation laws the functions f_s are called the *fluxes of the quantity* u, g represents the *density of sources* and ν is the *diffusion coefficient.*

Provided the functions f_s, $s = 1, 2$, g and u^0 are sufficiently regular, the *classical solution* of this problem can be defined, for example, as a function $u \in C^2(\overline{Q}_T)$ satisfying (4.5.1)–(4.5.3).

In the sequel we shall be concerned with the concept of a weak solution. We shall use the notation introduced in Sections 1.3.3 and 1.3.4. Let us set

$$V = H_0^1(\Omega). \tag{4.5.4}$$

In the space $H^1(\Omega)$, besides its norm

$$\|u\|_{1,\Omega} = \left(\int_\Omega (|u|^2 + |\nabla u|^2) \, dx \right)^{1/2}, \tag{4.5.5}$$

we shall work with the seminorm

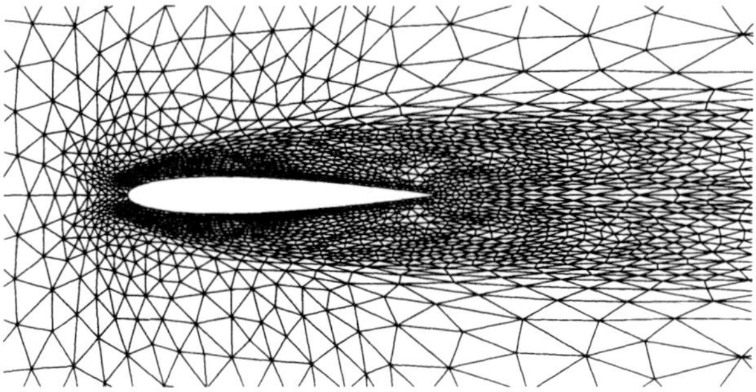

FIG. 4.4.26. Triangular FV mesh (Osher–Solomon numerical flux)

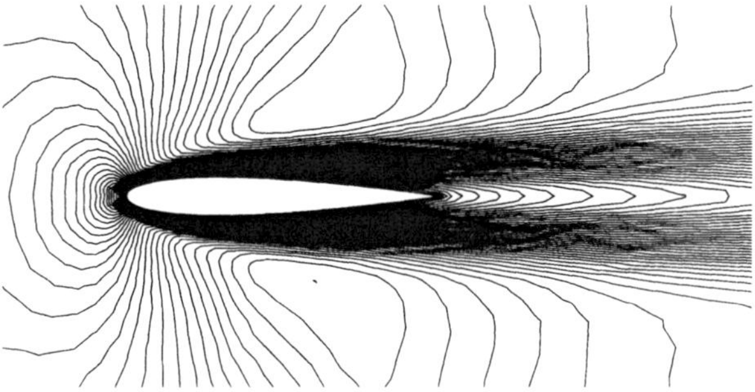

FIG. 4.4.27. Mach number isolines (Osher–Solomon numerical flux)

$$|u|_{1,\Omega} = \left(\int_\Omega |\nabla u|^2 \, dx \right)^{1/2}, \tag{4.5.6}$$

which is a norm on V equivalent to the norm $\|\cdot\|_{1,\Omega}$: there exist constants $c_1, c_2 > 0$ such that

$$c_1 \|v\|_{1,\Omega} \le |v|_{1,\Omega} \le c_2 \|v\|_{1,\Omega} \quad \forall\, v \in V. \tag{4.5.7}$$

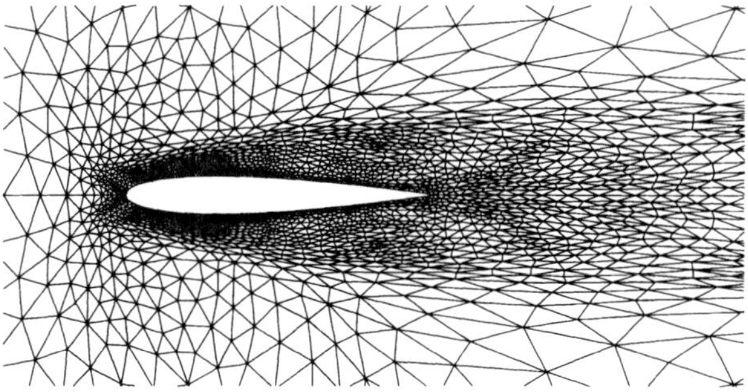

FIG. 4.4.28. Triangular FV mesh (Godunov numerical flux)

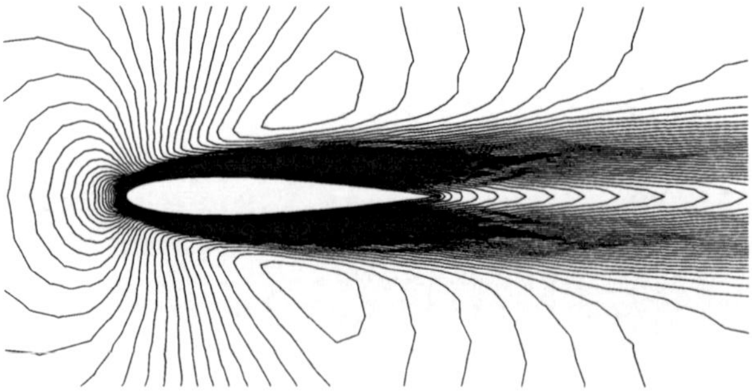

FIG. 4.4.29. Triangular FV mesh (Godunov numerical flux)

Further, we set

$$(u, v) = \int_\Omega uv\,dx, \quad u,\, v \in L^2(\Omega) \tag{4.5.8}$$

(scalar product in $L^2(\Omega)$) and

$$(\!(u, v)\!) = \int_\Omega \nabla u \cdot \nabla v\,dx, \quad u,\, v \in H^1(\Omega) \tag{4.5.9}$$

(scalar product in V inducing the norm $|\cdot|_{1,\Omega}$ in V).

In what follows, we shall consider the following *assumptions on the data*:

$$f_s \in C^1(I\!R), \quad f_s(0) = 0, \quad s = 1, 2, \tag{4.5.10}$$
$$g \in C([0,T]; W^{1,q}(\Omega)) \quad \text{for some } q > 2, \tag{4.5.11}$$
$$u^0 \in H^2(\Omega) \cap V. \tag{4.5.12}$$

With respect to the form of equation (4.5.1), the assumption that $f_s(0) = 0$ does not represent any limitation.

Now we derive the *weak formulation* of problem (4.5.1)–(4.5.3). Let us assume that u is a classical, sufficiently regular solution satisfying (4.5.1)–(4.5.3) pointwise. Multiplying equation (4.5.1) by an arbitrary $v \in V$, integrating over Ω, using Green's theorem and interchanging integration over Ω with differentiation with respect to t, we obtain the identity

$$\frac{d}{dt}\int_\Omega u(t)\,v\,dx - \int_\Omega \sum_{s=1}^{2} f_s(u(t))\,\frac{\partial v}{\partial x_s}\,dx + \nu \int_\Omega \nabla u(t) \cdot \nabla v\,dx \tag{4.5.13}$$
$$= \int_\Omega g(t)\,v\,dx \quad \text{for all } v \in V \text{ and all } t \in [0,T].$$

Let us set

$$b(\varphi, v) = -\int_\Omega \sum_{s=1}^{2} f_s(\varphi)\,\frac{\partial v}{\partial x_s}\,dx \quad \text{for } \varphi \in L^\infty(\Omega), \quad v \in V. \tag{4.5.14}$$

Identity (4.5.13) and the above notation leads us to the following concept:

Definition 4.54 *We say that a function u is a weak solution of problem* (4.5.1)–(4.5.3), *if it satisfies the following conditions:*

$$a)\ u \in L^2(0,T;V) \cap L^\infty(Q_T), \tag{4.5.15}$$
$$b)\ \frac{d}{dt}(u(t),v) + b(u(t),v) + \nu((u(t),v)) = (g(t),v) \quad \forall\, v \in V,$$
in the sense of distributions on $(0,T)$,
$$c)\ u(0) = u^0.$$

Identity (4.5.15), b), which is (4.5.13) rewritten with the aid of the above notation, means that

$$-\int_0^T (u(t),v)\,\psi'(t)\,dt + \nu \int_0^T ((u(t),v))\,\psi(t)\,dt + \int_0^T b(u(t),v)\,\psi(t)\,dt$$
$$= \int_0^T (g(t),v)\,\psi(t)\,dt \quad \forall\, v \in V,\ \forall\, \psi \in C_0^\infty((0,T)). \tag{4.5.16}$$

In view of (4.5.11), $g \in L^2(0,T;V^*)$. (V^* denotes the dual to V.) Using assumption (4.5.10) and conditions (4.5.15), a)–b), we find that u has the derivative u' defined a.e. in $(0,T)$ and $u' \in L^2(0,T;V^*)$. This implies that $u \in C([0,T],V^*)$ and we see that also condition (4.5.15), c) makes sense. (Cf., for example, (Feistauer, 1993), Chapter 8.)

With the aid of methods from (Lions, 1969), (Rektorys, 1982) and (Málek *et al.*, 1996) it is possible to prove that problem (4.5.15) has a unique solution.

4.5.2 *Semi-implicit method combining dual finite volumes with conforming finite elements*

The discretization of problem (4.5.1)–(4.5.3) will be carried out similarly as in Section 4.4. By $\mathcal{T}_h$ we denote a triangulation of the domain Ω with properties (4.1.15). Let $\mathcal{D}_h$ denote the mesh of *dual finite volumes* constructed in Example 4.51, a) over $\mathcal{T}_h$. Let $\sigma_h = \{P_i; i \in J\}$ be the set of all vertices of all $K \in \mathcal{T}_h$. (J is a suitable index set.) We set $\overset{\circ}{J} = \{i \in J; P_i \in \Omega\}$.

As usual, by h_K, ρ_K and ϑ_K we denote the length of the longest side of a triangle $K \in \mathcal{T}_h$, the radius of the largest circle inscribed in K and the magnitude of the smallest angle of K, respectively, and put

$$h = \max_{K\in\mathcal{T}_h} h_K, \quad \vartheta_h = \min_{K\in\mathcal{T}_h} \vartheta_K. \tag{4.5.17}$$

If the boundaries of two different finite volumes D_i and D_j contain a common straight segment, we call D_i and D_j *neighbours*. Then we write

$$\Gamma_{ij} = \bigcup_{\alpha=1}^{\beta_{ij}} \Gamma_{ij}^\alpha = \partial D_i \cap \partial D_j = \Gamma_{ji}, \tag{4.5.18}$$

where $\Gamma_{ij}^{\alpha} = \Gamma_{ji}^{\alpha}$ are straight segments. For $i \in J$, let

$$s(i) = \{j \in J; D_j \text{ is a neighbour of } D_i\}. \tag{4.5.19}$$

If $P_i \in \sigma_h \cap \partial\Omega$, then we denote by $\Gamma_{ij}^1, j \in \gamma(i)$, the segments that form $\partial D_i \cap \partial\Omega$. In this case we set $S(i) = s(i) \cup \gamma(i)$, otherwise for $P_i \in \sigma_h \cap \Omega$ we put $S(i) = s(i)$. Obviously, for every $D_i \in \mathcal{D}_h$ we have

$$\partial D_i = \bigcup_{j \in S(i)} \Gamma_{ij} = \bigcup_{j \in S(i)} \bigcup_{\alpha=1}^{\beta_{ij}} \Gamma_{ij}^{\alpha}. \tag{4.5.20}$$

Furthermore, we recall the following notation: $|D_i|$ is the area of $D_i \in \mathcal{D}_h$, $|K|$ is the area of $K \in \mathcal{T}_h$, $\boldsymbol{n}_{ij}^{\alpha} = (n_{1ij}^{\alpha}, n_{2ij}^{\alpha})$ is the unit outer normal to ∂D_i on Γ_{ij}^{α}, $|\Gamma_{ij}^{\alpha}|$ is the length of Γ_{ij}^{α}, $|\Gamma_{ij}|$ is the length of Γ_{ij}, and $|\partial D_i|$ is the length of ∂D_i. Moreover, let us consider a partition $0 = t_0 < t_1 < \ldots$ of the time interval $(0, T)$ and set $\tau_k = t_{k+1} - t_k$ for $k = 0, 1, \ldots$.

Let us define the following spaces over the grids $\mathcal{T}_h$ and $\mathcal{D}_h$:

$$\begin{aligned} X_h &= \{v_h \in C(\overline{\Omega}); v_h|_K \text{ is linear for each } K \in \mathcal{T}_h\} \\ &\qquad \text{(the space of linear conforming finite elements)}, \\ V_h &= \{v_h \in X_h; v_h = 0 \text{ on } \partial\Omega\}, \\ Z_h &= \{w \in L^2(\Omega); w|_{D_i} = \text{const for each } D_i \in \mathcal{D}_h\}, \\ Y_h &= \{w \in Z_h; w = 0 \text{ on } D_i \in \mathcal{D}_h \text{ for each } P_i \in \sigma_h \cap \partial\Omega\}. \end{aligned} \tag{4.5.21}$$

As mentioned in Section 4.1.2.1, $X_h \subset H^1(\Omega_h)$.

In the spaces from (4.5.21) we easily construct *simple bases.* The system $\{w_i; i \in J\}$ of functions $w_i \in X_h$ such that $w_i(P_j) = \delta_{ij}$ = Kronecker delta, $i, j \in J$, forms a basis in X_h. The system $\{w_i; i \in \overset{\circ}{J}\}$ $(= \{w_i; i \in J, \ P_i \in \sigma_h \cap \Omega\})$ is a basis in V_h. Furthermore, denoting by $d_i = \mathcal{X}_{D_i}$ the characteristic function of $D_i \in \mathcal{D}_h$, we have bases in Z_h and Y_h as the systems $\{d_i; i \in J\}$ and $\{d_i; i \in \overset{\circ}{J}\}$, respectively.

By r_h we denote the operator of the *Lagrange interpolation.* Hence, if $v : \sigma_h \to I\!R$, then

$$r_h v \in X_h, \quad (r_h v)(P_i) = v(P_i), \quad P_i \in \sigma_h. \tag{4.5.22}$$

By $L_h : C(\overline{\Omega}) \to Z_h$ we denote the *lumping operator:*

$$v_h \in C(\overline{\Omega}) \to L_h v_h = \sum_{i \in J} v_h(P_i)\, d_i \in Z_h. \tag{4.5.23}$$

Obviously, $L_h(V_h) = Y_h$.

In order to derive the discrete problem to (4.5.15), a)–c), we put

$$(u, v)_h = \int_{\Omega} r_h(uv)\, dx, \qquad u, v \in C(\overline{\Omega}), \tag{4.5.24}$$

$$\|u\|_h = (u,u)_h^{1/2}, \qquad u \in C(\overline{\Omega}),$$
$$\tilde{b}(u,v) = \sum_{s=1}^{2} \int_\Omega \frac{\partial f_s(u)}{\partial x_s} v\,dx, \quad u \in H^1(\Omega) \cap L^\infty(\Omega),\ v \in H^1(\Omega).$$
$$g^{k+1} = g(\cdot, t_{k+1}).$$

The forms b and $\tilde{b}$ are approximated by the FV approach explained in Section 4.4. Similarly as in (4.4.19), taking into account (4.5.23), for $u_h, v_h \in V_h$ we approximate the form b by

$$b_h(u_u, v_h) = \sum_{i \in J} v_h(P_i) \sum_{j \in s(i)} \sum_{\alpha=1}^{\beta_{ij}} H(u_u(P_i), u_h(P_j), \boldsymbol{n}_{ij}^\alpha)\,|\Gamma_{ij}^\alpha|, \quad u_h, v_h \in V_h, \tag{4.5.25}$$

where $H(\alpha, \beta, \boldsymbol{n})$ is a *numerical flux* defined on $I\!R \times \mathcal{S}_1$, where $\mathcal{S}_1 = \{\boldsymbol{n} \in I\!R^2; |\boldsymbol{n}| = 1\}$.

Now we come to the *semi-implicit discrete problem.*

Definition 4.55 *We define the approximate solution of* (4.5.1)–(4.5.3) *as functions* u_h^k, $t_k \in [0,T]$, *given by the conditions*

$$\begin{aligned} &a)\ u_h^0 = r_h u^0 \ (\in V_h), \\ &b)\ u_h^{k+1} \in V_h, \quad t_k \in [0,T), \\ &c)\ \frac{1}{\tau}(u_h^{k+1} - u_h^k, v_h)_h + b_h(u_h^k, v_h) + \nu(\!(u_h^{k+1}, v_h)\!)_h = (g^{k+1}, v_h)_h \\ &\qquad \forall v_h \in V_h,\ t_k \in [0,T). \end{aligned} \tag{4.5.26}$$

The function u_h^k *is the approximate solution at time* t_k.

In view of assumptions (4.5.11), (4.5.12) and the imbedding $W^{1,q}(\Omega) \subset C(\overline{\Omega})$ for $q > 2$, the expression $(g^{k+1}, v_h)_h$ and condition (4.5.26), c) make sense.

4.5.2.1 *Properties of the numerical flux* In the sequel we use the following assumptions:

a) $H = H(y, z, \boldsymbol{n})$ is *locally Lipschitz-continuous* with respect to y, z: for any $M \geq 0$ there exists $c(M) > 0$ such that

$$|H(y,z,\boldsymbol{n}) - H(y^*, z^*, \boldsymbol{n})| \leq c(M)\,(|y - y^*| + |z - z^*|)$$
$$\forall y, y^*, z, z^* \in [-M, M],\ \forall \boldsymbol{n} \in \mathcal{S}_1.$$

b) H is *consistent*:

$$H(u,u,\boldsymbol{n}) = \mathcal{F}(u,\boldsymbol{n}) = \sum_{s=1}^{2} f_s(u)\,n_s \quad \forall u \in I\!R,\ \forall \boldsymbol{n} = (n_1, n_2) \in \mathcal{S}_1.$$

c) H is *conservative:*

$$H(y, z, \boldsymbol{n}) = -H(z, y, -\boldsymbol{n}) \quad \forall y,\, z \in \mathbb{R},\ \forall \boldsymbol{n} \in \mathcal{S}_1.$$

d) H is *monotone* in the following sense: for a given fixed number $M > 0$ the function $H(y, z, \boldsymbol{n})$ is nonincreasing with respect to the second variable z on the set $\mathcal{M}_M = \{(y, z, \boldsymbol{n}); y,\, z \in [-M, M], \boldsymbol{n} \in \mathcal{S}_1\}$.

Example 4.56 We mention here two examples of numerical fluxes suitable for the solution of our problem.
a) The Lax–Friedrichs scheme has the numerical flux

$$H(u, v, \boldsymbol{n}) = \frac{1}{2}\left(\mathcal{F}(u, \boldsymbol{n}) + \mathcal{F}(v, \boldsymbol{n}) - \frac{1}{2\lambda}(v - u)\right),$$

where $\lambda > 0$ is in general different for different Γ_{ij} and is chosen so that assumption 4.5.2.1, d) is satisfied.
b) The Engquist–Osher scheme has the numerical flux

$$H(u, v, \boldsymbol{n}) = \frac{1}{2}\left(\mathcal{F}(u, \boldsymbol{n}) + \mathcal{F}(v, \boldsymbol{n}) - \int_u^v |F(q, \boldsymbol{n})|dq\right),$$

where $F(u, \boldsymbol{n}) = \sum_{s=1}^{2} f_s'(u)\, n_s$.

4.5.2.2 *Properties of the discrete problem* It is possible to prove the following results:

Lemma 4.57 *1) The bilinear forms $(\cdot,\cdot)_h$ and $(\!(\cdot,\cdot)\!)$ are scalar products on V_h.*

2) For each $u \in X_h$, $b_h(u, \cdot)$ is a linear form defined on V_h.

3) If $i \in J$ and $K \in \mathcal{T}_h$ is a triangle with the vertex $P_i \in \sigma_h$, then

$$|K \cap D_i| = \frac{1}{3}|K|. \tag{4.5.27}$$

4) The approximation $(\cdot,\cdot)_h$ of the L^2-scalar product can be defined with the aid of numerical integration using the vertices P_1^K, P_2^K, P_3^K of $K \in \mathcal{T}_h$ as the integration points or, equivalently, with the aid of the lumping operator L_h:

$$(u, v)_h = \sum_{K \in \mathcal{T}_h} |K| \sum_{n=1}^{3} u(P_n^K)\, v(P_n^K)/3 = \int_\Omega (L_h u)\,(L_h v)\, dx, \tag{4.5.28}$$
$$u,\, v \in C(\overline{\Omega}).$$

This implies that

$$\|u\|_h = \|L_h u\|_{0,\Omega}, \quad u \in C(\overline{\Omega}). \tag{4.5.29}$$

5) We have

$$(w_i, w_j)_h = |D_i|\,\delta_{ij}, \quad i, j \in J, \tag{4.5.30}$$
$$(u, w_i)_h = \frac{1}{3} \sum_{\{K \in \mathcal{T}_h;\, P_i \in K \cap \sigma_h\}} |K|\, u(P_i) = |D_i|\, u(P_i),$$
$$i \in J, \ u \in X_h,$$
$$(g^k, w_i)_h = |D_i|\, g(P_i, t_k), \quad i \in J, \ t_k \in [0, T].$$

6) Problem (4.5.26), *b)–c) has a unique solution* u_h^{k+1}.

7) Functions $z \in X_h$ *and* $y \in V_h$ *can be expressed in the form*

$$z = \sum_{j \in J} z(P_j)\, w_j \quad \textit{and} \quad y = \sum_{j \in \overset{\circ}{J}} y(P_j)\, w_j, \tag{4.5.31}$$

respectively.

8) Problem (4.5.26), *b)–c) is equivalent to the system of algebraic equations*

$$|D_i|\, u_h^{k+1}(P_i) + \tau\nu \sum_{j \in \overset{\circ}{J}} ((w_i, w_j))\, u_h^{k+1}(P_j) \tag{4.5.32}$$
$$= |D_i|\, u_h^k(P_i) - \tau\, b_h(u_h^k, w_i) + \tau |D_i|\, g(P_i, t_{k+1}), \quad i \in \overset{\circ}{J},$$

for unknown values $u_h^{k+1}(P_j)$, $j \in \overset{\circ}{J}$. *This system has a unique solution.*

Proof The proof of all these assertions is based on elementary considerations. Assertion 6) follows from the Lax–Milgram lemma 1.4.6. Assertion 8) is an immediate consequence of 6) and 7). □

4.5.3 *Convergence of the semi-implicit scheme*

Now we shall discuss the question whether the approximate solutions $\{u_h^k\}$, $t_k \in [0, T]$, obtained with the aid of scheme (4.5.26), a)–c), converge in some sense to the exact weak solution u of problem (4.5.1)–(4.5.3), provided the size of the mesh in Q_T tends to zero. To this end, we consider a system $\{\mathcal{T}_h\}_{h \in (0, h_0)}$ ($h_0 > 0$) of triangulations of the domain Ω, set $\tau = T/r$ for any integer $r > 1$ and define the partition of the interval $[0, T]$ formed by $t_k = k\tau$, $k = 0, \dots, r$. Hence, $\tau_k = \tau$ for $k = 0, \dots, r-1$.

We define the functions $u_{h\tau}$, $w_{h\tau} : (-\infty, +\infty) \to V_h$ associated with an approximate solution $\{u_h^k\}_{k=0}^r$:

$$u_{h\tau}(t) = u_h^0, \quad t \le 0, \tag{4.5.33}$$
$$u_{h\tau}(t) = u_h^k, \quad t \in (t_{k-1}, t_k],$$
$$k = 1, \dots, r,$$
$$u_{h\tau}(t) = u_h^r, \quad t \ge T;$$

$$
\begin{aligned}
&w_{h\tau} \text{ is a continuous, piecewise linear mapping of } [0,T] \text{ into } V_h, \qquad (4.5.34)\\
&w_{h\tau}(t_k) = u_h^k, \ k = 0, \ldots, r,\\
&w_{h\tau}(t) = 0 \quad \text{for} \quad t < 0 \text{ or } t > T.
\end{aligned}
$$

The goal is to prove the convergence (in suitable spaces) of the mappings $u_{h\tau}$ and $w_{h\tau}$ to u, if $h, \tau \to 0$ in an appropriate way.

4.5.3.1 *Assumptions* a) Let the system $\{\mathcal{T}_h\}_{h\in(0,h_0)}$ be *regular*, i.e. there exists $\vartheta_0 > 0$ such that

$$\vartheta_h \geq \vartheta_0 > 0 \quad \forall\, h \in (0, h_0). \tag{4.5.35}$$

b) The triangulations $\mathcal{T}_h$ are of *weakly acute type.* This means that the magnitude of all angles of all $K \in \mathcal{T}_h$, $h \in (0, h_0)$, is less than or equal to $\pi/2$.

In our further considerations we suppose that assumptions (4.5.10) – (4.5.12) and 4.5.3.1, a)–b) are satisfied and that the numerical flux H has properties 4.5.2.1. We shall proceed in several steps.

4.5.3.2 *L^∞-stability* By virtue of (4.5.11) and (4.5.12), there exist constants $\tilde{M}$ and $\tilde{C}$ such that

$$\|u^0\|_{0,\infty,\Omega} \leq \tilde{M}, \quad \|g\|_{L^\infty(Q_T)} \leq \tilde{C}. \tag{4.5.36}$$

Let us put

$$M = \tilde{M} + T\tilde{C}. \tag{4.5.37}$$

The main tool for proving the L^∞-stability is the *discrete maximum principle* represented here by the following result.

Lemma 4.58 *Let for $i \in \overset{\circ}{J}$ and $j \in J$ real numbers a_{ij}, b_{ij}, δ_i, φ_i, u_j, $\tilde{u}_j$, τ satisfy the following conditions:*

$$\tau > 0, \tag{4.5.38}$$

$$a_{ii} > 0 \text{ for } i \in \overset{\circ}{J}, \quad a_{ij} \leq 0 \text{ for } i \in \overset{\circ}{J},\ j \in J, i \neq j, \tag{4.5.39}$$

$$b_{ij} \geq 0 \text{ for } i \in \overset{\circ}{J},\ j \in J, \tag{4.5.40}$$

$$\sum_{j\in J} a_{ij} = \sum_{j\in J} b_{ij} = \delta_i > 0 \text{ for } i \in \overset{\circ}{J}, \tag{4.5.41}$$

$$\sum_{j\in J} a_{ij}\,\tilde{u}_j = \sum_{j\in J} b_{ij}\,u_j + \tau\,\delta_i\,\varphi_i \text{ for } i \in \overset{\circ}{J}, \tag{4.5.42}$$

$$\tilde{u}_i = u_i = 0 \text{ for } i \in J - \overset{\circ}{J}. \tag{4.5.43}$$

Then

$$\max_{j\in J} |\tilde{u}_j| \leq \max_{j\in J} |u_i| + \tau \max_{j\in \overset{\circ}{J}} |\varphi_i|. \tag{4.5.44}$$

Proof This follows from (Ikeda, 1983), Lemma 3.1.1, page 29. □

The following theorem is important for the derivation of a CFL stability condition for the numerical solution of the compressible Navier–Stokes equations by the semi-implicit scheme (4.4.32). Therefore, we give its detailed proof here.

Theorem 4.59 *If $\tau > 0$ and $h \in (0, h_0)$ satisfy the condition*

$$\tau\, c(M)\, |\partial D_i| \le |D_i|, \quad i \in J, \tag{4.5.45}$$

where $c(M)$ is the constant from 4.5.2.1, a), and if (4.5.36) and (4.5.37) hold, then

$$\begin{aligned} &\|u_h^k\|_{0,\infty,\Omega} \le M \quad \text{for each} \quad t_k \in [0,T], \\ &\|u_{h\tau}\|_{0,\infty,Q_T}, \quad \|w_{h\tau}\|_{0,\infty,Q_T} \le M. \end{aligned} \tag{4.5.46}$$

Proof By virtue of (4.5.32), identity (4.5.26), c) can be written in the form

$$\begin{aligned} &|D_i|\, u_h^{k+1}(P_i) + \tau\nu \sum_{j\in J} (\!(w_i, w_j)\!)\, u_h^{k+1}(P_j) \\ &= |D_i|\, u_h^k(P_i) - \tau\, b_h(u_h^k, w_i) + \tau |D_i|\, g(P_i, t_{k+1}), \quad i \in \mathring{J},\ k \ge 0. \end{aligned} \tag{4.5.47}$$

Using assumption 4.5.3.1, b), let us prove that

$$\begin{aligned} &(\!(w_i, w_i)\!) > 0, \quad i \in J, \\ &(\!(w_i, w_j)\!) \le 0, \quad i, j \in J,\ i \ne j, \\ &\sum_{j\in J} (\!(w_i, w_j)\!) = 0, \quad i \in J. \end{aligned} \tag{4.5.48}$$

Obviously,

$$(\!(w_i, w_j)\!) = \sum_{K\in\mathcal{T}_h} |K|\, (\nabla w_i \cdot \nabla w_j)\big|_K. \tag{4.5.49}$$

If $\nabla w_i|_K \cdot \nabla w_j|_K \ne 0$, then the vertices P_i, $P_j \in \sigma_h$ belong to the triangle K. So, let us assume that the element K has vertices $P_i = (x_1^i, x_2^i)$, $P_j = (x_1^j, x_2^j)$, $P_k = (x_1^k, x_2^k)$. It is clear that $(\!(w_i, w_i)\!) > 0$. Since $(w_i + w_j + w_k)|_K = 1$ we have $(\nabla w_i + \nabla w_j + \nabla w_k)|_K = 0$. This implies that $\nabla w_i|^2 + \nabla w_i \cdot \nabla w_j + \nabla w_i \cdot \nabla w_k = 0$ on K and, thus, from here and (4.5.49) we get the third relation in (4.5.48). Now let us prove the second inequality. Using the results from Exercise 4.5, we find that

$$\begin{aligned} \nabla w_i|_K &= \frac{1}{D}\left(x_2^j - x_2^k,\, x_1^k - x_1^j\right), \\ \nabla w_j|_K &= \frac{1}{D}\left(x_2^k - x_2^i,\, x_1^i - x_1^k\right), \end{aligned} \tag{4.5.50}$$

$$\nabla w_k|_K = \frac{1}{D}\left(x_2^i - x_2^j,\, x_1^j - x_1^i\right).$$

Further, using the well-known expression for the cosine of the angle between two vectors, denoting by α_k the angle in K at the vertex P_k, using (4.5.50) and assumption 4.5.3.1, we get

$$\begin{aligned}(\nabla w_i \cdot \nabla w_j)|_K &= -\frac{1}{D^2}(P_j - P_k)\cdot(P_i - P_k)\\ &= -\frac{1}{D^2}|P_j - P_k|\,|P_i - P_k|\cos\alpha_k \le 0.\end{aligned}$$

This and (4.5.49) yield what we wanted to prove.

Now, by induction with respect to k we shall prove that

$$\|u_h^k\|_{0,\infty,\Omega} \le \tilde{M} + k\,\tau\,\tilde{C} \le M, \quad t_k \in [0,T]. \tag{4.5.51}$$

Obviously, (4.5.51) holds for $k = 0$. Let us assume that (4.5.51) is valid for some $t_k \in [0,T)$.

Let us denote by L_i the left-hand side of (4.5.47) and set $u_i = u_h^k(P_i)$ and $\varphi_i = g(P_i, t_{k+1})$ (for simplicity we omit the superscript k). Then (4.5.47) reads

$$\begin{aligned}L_i &:= |D_i|\,u_i - \tau\, b_h(u_h^k, w_i) + \tau|D_i|\,\varphi_i\\ &= |D_i|\,u_i - \tau \sum_{j\in s(i)}\sum_{\alpha=1}^{\beta_{ij}} H(u_i, u_j, \boldsymbol{n}_{ij}^\alpha)\,|\Gamma_{ij}^\alpha| + \tau|D_i|\,\varphi_i\\ &= |D_i|\,u_i + \tau \sum_{j\in s(i)}\sum_{\alpha=1}^{\beta_{ij}} [(H(u_i, u_i, \boldsymbol{n}_{ij}^\alpha) - H(u_i, u_j, \boldsymbol{n}_{ij}^\alpha))\\ &\quad - H(u_i, u_i, \boldsymbol{n}_{ij}^\alpha)]\,|\Gamma_{ij}^\alpha| + \tau|D_i|\,\varphi_i, \quad i \in \mathring{J}.\end{aligned}$$

In view of the consistency of the numerical flux H and Green's theorem,

$$\sum_{j\in s(i)}\sum_{\alpha=1}^{\beta_{ij}} H(u_i, u_i, \boldsymbol{n}_{ij}^\alpha)\,|\Gamma_{ij}^\alpha| = \int_{\partial D_i}\sum_{s=1}^{2} f_s(u_i)\,n_s\,dS = 0. \tag{4.5.52}$$

Hence, setting

$$\mathcal{H}_{ij} = \begin{cases} 0, & u_i = u_j,\\ \displaystyle\sum_{\alpha=1}^{\beta_{ij}} \frac{H(u_i, u_i, \boldsymbol{n}_{ij}^\alpha) - H(u_i, u_j, \boldsymbol{n}_{ij}^\alpha)}{u_j - u_i}\,|\Gamma_{ij}^\alpha|, & u_i \ne u_j,\end{cases}$$

we can write

$$L_i = |D_i|\,u_i + \tau \sum_{j\in s(i)} \mathcal{H}_{ij}(u_j - u_i) + \tau|D_i|\,\varphi_i. \tag{4.5.53}$$

Due to the monotonicity of the numerical flux,

$$\mathcal{H}_{ij} \geq 0, \quad i \in \overset{\circ}{J},\ j \in s(i). \tag{4.5.54}$$

By virtue of the induction assumption, $|u_i| \leq \tilde{M} + k\,\tau\,\tilde{C} < M$ for all $i \in J$. This and the local Lipschitz-continuity of H imply that

$$0 \leq \mathcal{H}_{ij} \leq c(M) \sum_{\alpha=1}^{\beta_{ij}} |\Gamma_{ij}^{\alpha}| = c(M)\,|\Gamma_{ij}|.$$

Using (4.5.20), we find that

$$0 \leq \sum_{j \in s(i)} \mathcal{H}_{ij} \leq c(M) \sum_{j \in s(i)} |\Gamma_{ij}| \leq c(M)\,|\partial D_i|$$

and, hence, by (4.5.45),

$$|D_i| - \tau \sum_{j \in s(i)} \mathcal{H}_{ij} \geq 0, \quad i \in \overset{\circ}{J}\,. \tag{4.5.55}$$

From (4.5.53) it follows that (4.5.47) can be rewritten in the form

$$|D_i|\, u_h^{k+1}(P_i) + \nu\,\tau \sum_{j \in J} (\!(w_i, w_j)\!)\, u_h^{k+1}(P_j)$$

$$= \left(|D_i| - \tau \sum_{j \in s(i)} \mathcal{H}_{ij} \right) u_h^k(P_i) + \tau \sum_{j \in s(i)} \mathcal{H}_{ij}\, u_h^k(P_j) + \tau |D_i|\, g(P_i, t_{k+1}), \quad i \in \overset{\circ}{J}\,.$$

Taking into account (4.5.48), (4.5.54) and (4.5.55), we see that Lemma 4.58 can be applied. Inequality (4.5.38) and the fact that $u_h^{k+1} = 0$ on $\partial\Omega$ imply that

$$\|u_h^{k+1}\|_{0,\infty,\Omega} \leq \|u_h^k\|_{0,\infty,\Omega} + \tau \|g(\cdot, t_{k+1})\|_{0,\infty,\Omega}.$$

In view of the induction assumption and (4.5.36), we obtain

$$\|u_h^{k+1}\|_{0,\infty,\Omega} \leq \tilde{M} + (k+1)\,\tau\,\tilde{C} \leq M,$$

which we wanted to prove. □

Remark 4.60 Condition (4.5.45) represents a CFL stability condition guaranteeing an L^∞-bound for the approximate solution obtained by the semi-implicit combined FV–FE scheme. If we replace the constant $c(M)$ by the expression

$$\max_{\substack{j \in S(i) \\ \alpha = 1, \ldots, \beta_{ij}}} \max_{\ell = 1, \ldots, m} \{ |\lambda_\ell(\boldsymbol{w}_i^k, \boldsymbol{n}_{ij}^{\alpha})| \} \tag{4.5.56}$$

and write τ_k instead of τ, we obtain from (4.5.45) the CFL stability condition (4.4.49) of the combined FV–FE semi-implicit scheme (4.4.32) for the numerical solution of the compressible Navier–Stokes equations. This means that (4.4.49) can be considered as a generalization of condition (4.5.45).

If we introduce the additional inverse assumption $h \le c_{\text{inv}} h_K$ for $K \in \mathcal{T}_h$ and $h \in (0, h_0)$, then the stability condition (4.5.45) implies that

$$0 < \tau \le c\, c(M)^{-1} h \tag{4.5.57}$$

with a constant c depending on c_{inv} and ϑ_0. Hence, $\tau = O(h)$.

In what follows we shall give a brief survey of the convergence analysis without proofs which are rather technical and can be found in (Feistauer *et al.*, 1997). It will be necessary to use a number of various constants. By c we shall denote a generic constant independent of h, τ, $k, \ldots$ which attains in general different values at different places.

4.5.3.3 *Consistency* The consistency of the method is based on the following estimates:

Lemma 4.61 *There exist constants c, $\hat{c}_1$, $\hat{c}_2 > 0$ such that for any $h \in (0, h_0)$ we have*

a) $\hat{c}_1 \|v\|_{0,\Omega} \le \|L_h v\|_{0,\Omega} \le \hat{c}_2 \|v\|_{0,\Omega}, \quad v \in X_h,$

b) $\|v - L_h v\|_{0,\Omega} \le c h \|v\|_{1,\Omega}, \quad v \in X_h,$

c) $|(u, v) - (u, v)_h| \le c h^2 \|u\|_{1,\Omega} \|v\|_{1,\Omega}, \quad u, v \in X_h,$

d) $|(g^k, v) - (g^k, v)_h| \le c h \|g^k\|_{1,q,\Omega} \|v\|_{1,\Omega}, \quad v \in V_h.$

e) If $M > 0$, then there exists a constant $\tilde{c} = \tilde{c}(M)$ such that
$|b(u, v) - b_h(u, v)| \le \tilde{c} h \|u\|_{1,\Omega} \|v\|_{1,\Omega},$
$u \in V_h \cap L^\infty(\Omega), \ \|u\|_{0,\infty,\Omega} \le M, \ v \in V_h, \ h \in (0, h_0).$

f) If $M > 0$, then there exists a constant $c^* = c^*(M)$ such that
$|b_h(u, v)| \le c^* \|u\|_{0,\infty,\Omega} |v|_{1,\Omega},$
$v \in V_h \cap L^\infty(\Omega), \ \|u\|_{0,\infty,\Omega} \le M, \ v \in V_h, \ h \in (0, h_0).$

4.5.3.4 *A priori estimates* Besides the L^∞-bound, the following estimates of the approximate solution hold.

Theorem 4.62 *Let* (4.5.36) *and* (4.5.37) *hold. Then there exists a constant $c > 0$ independent of h and τ such that*

$$a) \max_{t_k \in [0,T]} \|u_h^k\|_{0,\Omega} \le c, \tag{4.5.58}$$

$$b) \sum_{k=1}^{r} \|u_h^k - u_h^{k-1}\|_{0,\Omega}^2 \le c,$$

$$c) \tau \sum_{k=0}^{m} \|u_h^k\|_{1,\Omega}^2 \le c,$$

for all τ, $h > 0$ satisfying the conditions $h \in (0, h_0)$ and (4.5.45).

Corollary 4.63 *Let* (4.5.36) *and* (4.5.37) *hold. Then there exists a constant* $c > 0$ *such that the functions* $u_{h\tau}$ *and* $w_{h\tau}$ *defined by* (4.5.33) *and* (4.5.34) *satisfy the estimates*

$$\begin{aligned} &a)\ \|u_{h\tau}\|_{L^2(-1,T;V)} \le c, \\ &b)\ \|w_{h\tau}\|_{L^2(0,T;V)} \le c \end{aligned} \tag{4.5.59}$$

for all $h \in (0, h_0)$ *and* τ *satisfying condition* (4.5.45). *Moreover, there exists a constant* $c > 0$ *such that*

$$\|u_{h\tau} - w_{h\tau}\|_{0,Q_T} \le c\tau \tag{4.5.60}$$

for all h *and* τ *with the above properties.*

(We consider estimate (4.5.59), a) on the interval $(-1, T)$ because of the delayed argument $t - \tau$ in equation (4.5.62).)

4.5.3.5 *Passage to limit* Since $L^2(0, T; V)$ is a reflexive Banach space and $L^\infty(Q_T)$ is the dual to the separable Banach space $L^1(Q_T)$, the above results imply the following:

Lemma 4.64 *There exist sequences* $h = h_n$, $\tau = \tau_n$, $n = 1, 2, \ldots$, *satisfying condition* (4.5.45) *and a function* $u \in L^2(0, T; V) \cap L^\infty(Q_T)$ *such that* $h = h_n \to 0$, $\tau = \tau_n \to 0$ *and*

$$\begin{aligned} u_{h\tau} &\to u \text{ weakly in } L^2(0,T;V), \\ u_{h\tau} &\to u \text{ weak-}* \text{ in } L^\infty(Q_T), \\ w_{h\tau} &\to u \text{ weakly in } L^2(0,T;V), \\ w_{h\tau} &\to u \text{ weak-}* \text{ in } L^\infty(Q_T), \text{ as } n \to \infty. \end{aligned} \tag{4.5.61}$$

(The definition of the weak and weak-$*$ convergence can be found in 1.4.3.5 and 1.4.3.8. For $t \le 0$ we set $u(t) = u^0$.)

Our goal now is to show that u is a weak solution of problem (4.5.1)–(4.5.3) and that the whole systems $\{w_{h\tau}\}$, $\{u_{h\tau}\}(h \in (0, h_0), \tau \in (0, T))$ converge to u, if h, τ approach zero so that (4.5.45) is satisfied. To this end, we rewrite scheme (4.5.26), c) in the equivalent form

$$\begin{aligned} &\frac{d}{dt}(w_{h\tau}(t), v_h)_h + \nu(\!(u_{h\tau}(t), v_h)\!) + b_h(u_{h\tau}(t-\tau), v_h) \\ &= (g_{h\tau}(t), v_h)_h \quad \text{for a.a. } t \in [0,T] \text{ and all } v_h \in V_h, \end{aligned} \tag{4.5.62}$$

where

$$g_{h\tau}(t) = g^{k+1}, \quad t \in (t_k, t_{k+1}). \tag{4.5.63}$$

However, (4.5.46), (4.5.59) and (4.5.61) are not sufficient for the passage to the limit in the nonlinear form b_h. We need a further compactness result. This is obtained with the aid of the *Fourier transform* $\hat{w}_{h\tau}$ of the function $w_{h\tau}$:

$$\hat{w}_{h\tau}(s) = \int_{-\infty}^{\infty} \exp(-2i\pi ts) w_{h\tau}(t)\, dt, \tag{4.5.64}$$

and the following theorem.

Theorem 4.65 *Let the following assumptions be satisfied:*

a) X_0, X, X_1 *are Hilbert spaces,*

b) $X_0 \hookrightarrow X \hookrightarrow X_1$ *(continuous imbeddings) and* $X_0 \hookrightarrow\hookrightarrow X$ *(compact imbedding),*

c) $\{v_n\}_{n=0}^{\infty}$ *is a bounded sequence in* $L^2(\mathbb{R}; X_0)$,

d) the sequence $\{|s|^\gamma \hat{v}_n(s)\}_{n=0}^{\infty}$ *of functions '*$s \in \mathbb{R} \to |s|^\gamma \hat{v}_n(s)$*' is bounded in* $L^2(\mathbb{R}; X_1)$ *for some* $\gamma > 0$,

e) $K \subset \mathbb{R}$ *is compact.*

Then $\{v_n\}$ *contains a subsequence strongly convergent in* $L^2(K; X)$.

Proof See (Temam, 1977), Chapter III, Theorem 2.2. □

Now we set $X_0 = V$, $X = X_1 = L^2(\Omega)$, $K = [0, T]$ and consider $h \in (0, h_0)$ and $\tau > 0$ satisfying (4.5.45). It is possible to prove that

$$\int_{\mathbb{R}} |s|^{2\gamma} \, \|\hat{w}_{h\tau}(s)\|^2_{L^2(\Omega)}\, ds \le c \quad \text{for } 0 < \gamma < 1/4 \tag{4.5.65}$$

with a constant c independent of h, τ.

From this, Theorem 4.65, Lemma 4.64 and Corollary 4.63 we conclude that there exist a function $u \in L^2(0, T; V) \cap L^\infty(Q_T)$ and sequences $h = h_n \to 0$, $\tau = \tau_n \to 0$ as $n \to \infty$ satisfying (4.5.45), such that (4.5.61) holds and

$$w_{h\tau} \to u \quad \textit{and} \quad u_{h\tau} \to u \quad \textit{strongly in } L^2(Q_T) \textit{ as } n \to \infty. \tag{4.5.66}$$

4.5.3.6 *Limit process* Let us consider sequences h_n, $\tau_n \to 0$ satisfying (4.5.45) and assume that the approximate solutions $w_{h\tau}$, $u_{h\tau}$ associated with $h = h_n$ and $\tau = \tau_n$ satisfy conditions (4.5.46), (4.5.59), (4.5.61) and (4.5.66). Then it is possible to carry out the limit process in the identity (4.5.62) and to prove that the limit function is a weak solution of problem (4.5.1)–(4.5.3), i.e. u satisfies (4.5.15), a)–c). On the basis of these considerations we come to the following conclusion.

Let us consider approximate solutions of problem (4.5.16), a)–c) obtained from (4.5.26), a)–c) with τ, $h > 0$ satisfying condition (4.5.45). Then the systems of functions $u_{h\tau}$, $w_{h\tau}$ defined by (4.5.33) and (4.5.34) can be split into sequences converging in the sense of (4.5.61) and (4.5.66). Every limit function of such a sequence is a solution of problem (4.5.16), a)–c). (As we see, we have proven the existence of the weak solution of problem (4.5.1)–(4.5.3), mentioned in Section 4.5.1.) Taking into account the uniqueness of the weak solution, we obtain the convergence of the whole systems $\{u_{h\tau}\}$, $\{w_{h\tau}\}$ to the weak solution u of problem (4.5.1)–(4.5.3). Thus, we come to the *main result*:

Theorem 4.66 *Let us assume that $\Omega \subset \mathbb{R}^2$ is a bounded polygonal domain with a Lipschitz-continuous boundary and polygonal and that conditions* (4.5.10)–(4.5.12), 4.5.2.1, *a)–d)*, 4.5.3.1, *a)–b) are satisfied. For $h \in (0, h_0)$ and $\tau = T/r$ let us construct approximate solutions with the aid of the FV–FE scheme* (4.5.26), *a)–c) and define the functions $u_{h\tau}$ and $w_{h\tau}$ by* (4.5.33) *and* (4.5.34). *Then*

$$\begin{aligned} u_{h\tau}, w_{h\tau} &\to u \quad \textit{weakly in } L^2(0,T;V), \\ u_{h\tau}, w_{h\tau} &\to u \quad \textit{weak-}* \textit{ in } L^\infty(Q_T), \\ u_{h\tau}, w_{h\tau} &\to u \quad \textit{strongly in } L^2(Q_T), \\ \textit{as } h,\, \tau &\to 0, \quad h,\, \tau \textit{ satisfy } (4.5.45), \end{aligned}$$

where u is the unique weak solution of problem (4.5.1)–(4.5.3).

4.5.3.7 *Error estimates* In (Feistauer *et al.*, 1999*a*), the error estimates of the combined FV–FE scheme (4.5.26) were derived under assumptions (4.5.10)–(4.5.12), 4.5.2.1, 4.5.3.1 and the assumptions on the regularity of the exact solution of problem (4.5.1)–(4.5.3):

$$\begin{aligned} &u \in L^\infty(0,T; H^{1+\gamma}(\Omega)), \\ &\partial u/\partial t \in L^\infty(0,T; L^2(\Omega)), \\ &\partial^2 u/\partial t^2 \in L^\infty(0,T; V^*), \end{aligned} \tag{4.5.67}$$

where V^* denotes the dual of the space V and $\gamma \in (1/4, 1]$. (Here $H^{1+\gamma}(\Omega)$ denotes the Sobolev–Slobodetskii space defined in Section 1.3.3.6.) Further, we consider the stability condition (4.5.45) and the *'inverse stability condition'*

$$h \le c\tau \tag{4.5.68}$$

with a suitable constant $c > 0$ independent of h and τ. (We can find similar nonstandard conditions in (Roos *et al.*, 1996), Sections 4.2, 5.1 or in (Kröner and Rokyta, 1994).)

In (Feistauer *et al.*, 1999*a*), the existence of such $\alpha \in (1/2, 1]$ is proven that under the above assumptions, the error $e_h^k = u_h^k - u(t_k)$ of the method can be estimated as follows:

$$\begin{aligned} &\max_{t_k \in [0,T]} \|e_h^k\|_{0,\Omega} \le (C_1 h + C_2 h^\varepsilon + C_3 h^{2\varepsilon - 1/2}) \exp(cT/\nu), \\ &\Big(\nu\tau \sum_{t_k \in [0,T]} |e_h^k|_{1,\Omega}^2\Big)^{1/2} \le (C_1 h + C_2 h^\varepsilon + C_3 h^{2\varepsilon - 1/2}) \exp(cT/\nu) \nu^{-1/2}, \end{aligned} \tag{4.5.69}$$

where $\varepsilon = \min(\gamma, \alpha)$ and C_1, C_2, C_3 are constants independent of h, τ, but depending on ν so that $C_1 = O(\nu^{-3/2}), C_2 = O(\nu), C_3 = O(\nu^{1/2})$. The constant c is independent of h, τ and ν. If the domain Ω is convex, then $\alpha = 1$. Hence, if also $\gamma = 1$, the error is of order $O(h)$.

As we can see, the above '$L^\infty(L^2)$' and '$L^2(H^1)$' estimates are rather pessimistic for $\nu \ll 1$. This is caused by the 'parabolic machinery' used in the proof of these estimates. The derivation of error estimates uniform with respect to the diffusion parameter $\nu \to 0$ for the combined FV–FE schemes (as well as for other methods) is a very challenging problem. One possibility might be to apply the 'hyperbolic machinery' used, for example, in (Eymard *et al.*, 2000), Section 29. However, to obtain optimal error estimates uniform with respect to ν, it is necessary to use a detailed knowledge of the behaviour of the exact solution. Up to now, this has been possible only for linear convection–diffusion equations with one space variable. See, for example, (Linss and Stynes, 2001), (Roos, 2002) and (Melenk and Schwab, 1999).

4.5.4 *Explicit method combining conforming finite elements with dual finite volumes*

Here we pay attention to the convergence analysis of the *fully explicit* combined FV–FE method using conforming piecewise linear finite elements and dual finite volumes.

4.5.4.1 *Explicit discretization* Under the above notation, we shall consider the following *explicit scheme* for the numerical solution of problem (4.5.1)–(4.5.3):

$$\begin{aligned}
&\text{a) } u_h^0 = r_h u^0 \ (\in V_h), \qquad (4.5.70)\\
&\text{b) } u_h^{k+1} \in V_h, \quad t_k \in [0, T),\\
&\text{c) } \frac{1}{\tau}(u_h^{k+1} - u_h^k, v_h)_h + b_h(u_h^k, v_h) + \nu(\!(u_h^k, v_h)\!) = (g^k, v_h)_h\\
&\qquad \forall v_h \in V_h,\ t_k \in [0, T).
\end{aligned}$$

Definition 4.67 *We define the approximate solution of* (4.5.1)–(4.5.3) *as functions* u_h^k, $t_k \in [0, T]$, *given by conditions* (4.5.70), *a)–c). The function* u_h^k *is the approximate solution at time* t_k.

It is obvious that problem (4.5.70), b)–c) for the unknown function u_h^{k+1} is equivalent to the system of algebraic equations

$$\begin{aligned}
&|D_i|\, u_h^{k+1}(P_i) + \tau\nu \sum_{j \in \overset{\circ}{J}} (\!(w_i, w_j)\!)\, u_h^k(P_j) \qquad (4.5.71)\\
&= |D_i|\, u_h^k(P_i) - \tau\, b_h(u_h^k, w_i) + \tau |D_i|\, g(P_i, t_k), \quad i \in \overset{\circ}{J},
\end{aligned}$$

for unknown values $u_h^{k+1}(P_j)$, $j \in \overset{\circ}{J}$. This system is uniquely solvable and its solution can be expressed explicitly.

In the sequel, we briefly characterize the analysis of the convergence of approximate solutions to the exact one.

4.5.4.2 *Assumptions* a) Let the system $\{\mathcal{T}_h\}_{h \in (0, h_0)}$ be regular and of weakly acute type. (See 4.5.3.1.)

b) Let the triangulations $\mathcal{T}_h$ satisfy the inverse assumption

$$h \leq c_{\text{inv}} h_K, \quad \forall K \in \mathcal{T}_h, \forall h \in (0, h_0), \tag{4.5.72}$$

with a constant $c_{\text{inv}} > 0$ independent of $K \in \mathcal{T}_h$ and h. Then the following inverse estimate holds (cf. (4.1.94)):

$$|v_h|_{1,\Omega} \leq c_4 h^{-1} \|v_h\|_{0,\Omega}, \quad v_h \in X_h, \, h \in (0, h_0). \tag{4.5.73}$$

Moreover, we have

$$h_K^2 \leq c_5 |K|, \quad K \in \mathcal{T}_h, \, h \in (0, h_0). \tag{4.5.74}$$

c) Let assumptions 4.5.2.1 be satisfied.

4.5.4.3 *L^∞-stability* By virtue of (4.5.11) and (4.5.12), we can consider (4.5.36) and (4.5.37), i.e.

$$\|u^0\|_{0,\infty,\Omega} \leq \tilde{M}, \quad \|g\|_{0,\infty,Q_T} \leq \tilde{C},$$

and

$$M = \tilde{M} + T\tilde{C}.$$

Let us set $A_i = \operatorname{supp} w_i = \bigcup_{\{K \in \mathcal{T}_h; P_i \in K\}} K$.

The application of the discrete maximum principle formulated in Lemma 4.58 yields the following theorem:

Theorem 4.68 *If $\tau > 0$ and $h \in (0, h_0)$ satisfy the condition*

$$|D_i| - \tau\, c(M)\, |\partial D_i| - \frac{1}{4} \tau \nu \sum_{K \subset A_i} \frac{h_K}{\rho_K} \geq 0, \quad i \in \mathring{J}, \tag{4.5.75}$$

where $c(M)$ is the constant from assumption 4.5.2.1, *a), and if* (4.5.36) *and* (4.5.37) *hold, then*

$$\|u_h^k\|_{0,\infty,\Omega} \leq M \quad \textit{for each } t_k \in [0, T]. \tag{4.5.76}$$

Remark 4.69 In view of (4.5.27), the condition

$$\frac{1}{3}|K| - \tau\, c(M)\, |K \cap \partial D_i| - \frac{1}{4} \tau \nu \frac{h_K}{\rho_K} \geq 0, \quad K \in \mathcal{T}_h, \, i \in \mathring{J}, \tag{4.5.77}$$

implies (4.5.75). It follows from (4.5.77) that $\tau = O\left(\frac{h^2}{h+\nu}\right)$.

Proof (of Theorem 4.68) For simplicity we set $u_i = u_h^k(P_i)$ and $\varphi_i = g(P_i, t_k)$. Further, we use the notation

$$\mathcal{H}_{ij} = \begin{cases} 0 & \text{for } u_i = u_j \\ \displaystyle\sum_{\alpha=1}^{\beta_{ij}} \frac{H(u_i, u_i, \boldsymbol{n}_{ij}^\alpha) - H(u_i, u_j, \boldsymbol{n}_{ij}^\alpha)}{u_j - u_i} \ell_{ij}^\alpha & \text{for } u_i \neq u_j. \end{cases} \tag{4.5.78}$$

By the proof of Theorem 4.59, we have

$$0 \leq \mathcal{H}_{ij} \leq c(M)\,|\Gamma_{ij}|, \quad i \in \overset{\circ}{J},\ j \in s(i), \tag{4.5.79}$$

and by virtue of (4.5.25) and (4.5.78), we can write

$$\begin{aligned} |D_i|\, u_h^{k+1}(P_i) = &\left\{ |D_i| - \tau \left[\sum_{j \in s(i)} \mathcal{H}_{ij} + \nu(\!(w_i, w_i)\!) \right] \right\} u_i \\ &+ \tau \sum_{j \in s(i)} (\mathcal{H}_{ij} - \nu(\!(w_i, w_j)\!))\, u_j + \tau |D_i|\, \varphi_i. \end{aligned} \tag{4.5.80}$$

By (4.5.48),

$$\begin{aligned} &(\!(w_i, w_i)\!) > 0, \quad i \in J, \\ &(\!(w_i, w_j)\!) \leq 0, \quad i, j \in J,\ i \neq j, \\ &\sum_{j \in s(i) \cup \{i\}} (\!(w_i, w_j)\!) = 0, \quad i \in \overset{\circ}{J}\,. \end{aligned} \tag{4.5.81}$$

Hence, the coefficients at u_j, $j \in s(i)$, in (4.5.80) are nonnegative:

$$\mathcal{H}_{ij} - \nu(\!(w_i, w_j)\!) \geq 0, \quad j \in s(i),\ i \in \overset{\circ}{J}\,. \tag{4.5.82}$$

Let us show also that the coefficient at u_i in (4.5.80) is nonnegative. In view of (4.5.79),

$$0 \leq \sum_{j \in s(i)} \mathcal{H}_{ij} \leq c(M) \sum_{j \in s(i)} |\Gamma_{ij}| = c(M)\,|\partial D_i|. \tag{4.5.83}$$

Further,

$$(\!(w_i, w_i)\!) = \int_\Omega |\nabla w_i|^2\, dx = \sum_{K \subset A_i} \int_K |\nabla w_i|^2\, dx, \tag{4.5.84}$$

$$\int_K |\nabla w_i|^2\, dx = |K|\, |\nabla w_i|_K|^2.$$

For $K \in \mathcal{T}_h$, $K \subset A_i$, we have $P_i \in K$. By $\mathcal{S}_i^K$ and V_i^K we denote the length of the side of K which does not contain the vertex P_i and the distance of P_i from this side, respectively. Then

$$|\nabla w_i|_K|^2 = \frac{1}{(V_i^K)^2}, \tag{4.5.85}$$

$$|K| = \frac{1}{2}\,\mathcal{S}_i^K\,V_i^K.$$

Taking into account that $\mathcal{S}_i^K \le h_K \le h$, $V_i^K \ge 2\rho_K$, we get from (4.5.84) and (4.5.85) the estimate

$$(\!(w_i, w_i)\!) = \frac{1}{2}\sum_{K\subset A_i} \mathcal{S}_i^K/V_i^K \le \frac{1}{4}\sum_{K\subset A_i} h_K/\rho_K. \tag{4.5.86}$$

Now, (4.5.83), (4.5.84), (4.5.86) and (4.5.75) imply that

$$\begin{aligned}&|D_i| - \tau\left[\sum_{j\in s(i)} \mathcal{H}_{ij} + \nu(\!(w_i, w_i)\!)\right] \qquad (4.5.87)\\ &\ge |D_i| - \tau\, c(M)\,|\partial D_i| - \frac{1}{4}\tau\,\nu \sum_{K\subset A_i} h_K/\rho_K \ge 0,\end{aligned}$$

which we wanted to prove.

We proceed further by induction. We have $\|u_h^0\|_{0,\infty,\Omega} \le \tilde{M}$. Let us assume that

$$\|u_h^k\|_{0,\infty,\Omega} \le \tilde{M} + \tau k\,\tilde{C} \quad (< M) \tag{4.5.88}$$

for some $t_k \in [0, T)$. Then, by virtue of (4.5.80), (4.5.81), (4.5.82), (4.5.87) and (4.5.88),

$$\begin{aligned}|D_i|\,|u_h^{k+1}(P_i)| &\le \left\{|D_i| - \tau\left[\sum_{j\in s(i)} \mathcal{H}_{ij} + \nu(\!(w_i, w_i)\!)\right]\right\}|u_h^k(P_i)|\\ &\quad + \tau\sum_{j\in s(i)} (\mathcal{H}_{ij} - \nu(\!(w_i, w_j)\!))\;|u_h^k(P_j)| + \tau|D_i|\,|g(P_i, t_k)|\\ &\le \left\{|D_i| - \tau\nu \sum_{j\in s(i)\cup\{i\}} (\!(w_i, w_j)\!)\right\}(\tilde{M} + \tau k\,\tilde{C}) + \tau\,\tilde{C}\,|D_i|\\ &= |D_i|\,(\tilde{M} + \tau(k+1)\,\tilde{C}) \le |D_i|\,M, \quad i \in \overset{\circ}{J}\,.\end{aligned}$$

Hence, $\|u_h^{k+1}\|_{0,\infty,\Omega} \le M$, which we wanted to prove. □

Remark 4.70 The above results represent a hint of how to choose a CFL stability condition of explicit numerical schemes for the solution of the compressible Navier–Stokes equations. In the case of the fully explicit scheme (4.4.31) we formulate the CFL condition in a heuristic way as condition (4.5.77), where we replace the constant $c(M)$ by expression (4.5.56) and set $\nu := \max(\mu, k)$, $\tau := \tau_k$. For the inviscid–viscous operator splitting algorithm 4.4.6.1 we propose the application of two CFL conditions in each time step consisting of fractional steps

(4.4.42) and (4.4.43). We request that the time step τ_k satisfies the 'inviscid' stability condition (4.4.49) (= generalization of (4.5.45)) and the 'viscous' stability condition (4.4.50). The latter is obtained from (4.5.77), if we set $c(M) := 0$ (we neglect convection terms), $\tau := \tau_k$ and $\nu := \max(\mu, k)$.

4.5.4.4 *Additional a priori estimates* These are formulated in the following theorem.

Theorem 4.71 *Let* (4.5.36) *and* (4.5.37) *hold. Then there exists a constant* c *such that*

$$a)\ \max_{t_k\in[0,T]} \|u_h^k\|_{0,\Omega} \le c, \tag{4.5.89}$$

$$b)\ \sum_{k=1}^{r} \|u_h^k - u_h^{k-1}\|_{0,\Omega}^2 \le c,$$

$$c)\ \tau \sum_{k=0}^{r} \|u_h^k\|_{1,\Omega}^2 \le c,$$

for all h, τ *satisfying the conditions* $h \in (0, h_0)$, (4.5.75) *and*

$$0 < \tau \le \frac{h^2}{\nu(h^2 + 1 + c_4^2 \hat{c}_1^{-2})}, \tag{4.5.90}$$

where $\hat{c}_1$ *and* c_4 *are constants from Lemma 4.61 and inequality* (4.5.73), *respectively.*

Now let us define the functions $u_{h\tau}, w_{h\tau} : I\!R \to V_h$ associated with the approximate solution u_h^k, $t_k \in [0, T]$:

$$\begin{aligned} u_{h\tau}(t) &= u_h^0, \quad t < t_1, \\ u_{h\tau}(t) &= u_h^k, \quad t_k \le t < t_{k+1},\ k = 1, \ldots, r-1, \\ u_{h\tau}(t) &= u_h^r, \quad t \ge T, \end{aligned} \tag{4.5.91}$$

$$\begin{aligned} &w_{h\tau}(t) = 0, \quad t < 0 \text{ or } t > T, \\ &w_{h\tau} \text{ is a continuous, piecewise linear mapping} \\ &\text{of } [0, T] \text{ into } V_h,\ w_{h\tau}(t_k) = u_h^k,\ k = 0, \ldots, r. \end{aligned} \tag{4.5.92}$$

On the basis of the above results we can derive estimates of these functions.

Lemma 4.72 *Let the assumptions of Theorems 4.68 and 4.71 be satisfied. Then*

$$\begin{aligned} &\|u_{h\tau}\|_{0,\infty,Q_T},\ \|w_{h\tau}\|_{0,\infty,Q_T} \le M, \\ &\|u_{h\tau}\|_{L^2(0,T;V)},\ \|w_{h\tau}\|_{L^2(0,T;V)} \le c \\ &\|u_{h\tau} - w_{h\tau}\|_{0,Q_T} \le c\sqrt{\tau}, \end{aligned} \tag{4.5.93}$$

for all $h \in (0, h_0)$ *and* $\tau > 0$ *satisfying conditions* (4.5.75) *and* (4.5.90).

4.5.4.5 *Convergence* The next step is to prove the convergence of the functions $u_{h\tau}$, $w_{h\tau}$ to the exact solution u of problem (4.5.14), a)–c), if h, $\tau \to 0$ in a suitable way. Condition (4.5.70), c) can be rewritten in the equivalent form

$$\frac{d}{dt}(w_{h\tau}(t), v_h)_h + \nu((u_{h\tau}(t), v_h)) + b_h(u_{h\tau}(t), v_h) = (g_{h\tau}(t), v_h) \quad (4.5.94)$$

for a.a. $t \in [0, T]$ and all $v_h \in V_h$,

where

$$g_{h\tau} = g^k = g(\cdot, t_k), \quad t \in [t_k, t_{k+1}). \quad (4.5.95)$$

In order to carry out the limit process in (4.5.94), we need an additional compactness for overcoming the nonlinearity of the form b_h. This is obtained similarly as in Section 4.5.2 with the aid of the Fourier transform leading to the following result:

Lemma 4.73 *There exist a function $u \in L^2(0, T; V) \cap L^\infty(Q_T)$ and sequences h_n, τ_n such that h_n, $\tau_n \to 0$ as $n \to \infty$, $h = h_n$, $\tau = \tau_n$ satisfy* (4.5.75) *and* (4.5.90), *and*

$$\begin{aligned} u_{h\tau}, w_{h\tau} &\to u \quad \textit{strongly in } L^2(Q_T), \\ u_{h\tau}, w_{h\tau} &\to u \quad \textit{weakly in } L^2(0, T; V), \\ u_{h\tau}, w_{h\tau} &\to u \quad \textit{weak-* in } L^\infty(Q_T), \end{aligned} \quad (4.5.96)$$

as $n \to \infty$.

If we carry out the limit process in (4.5.94), we come to the *main result* of this section:

Theorem 4.74 *Let $\Omega \subset I\!R^2$ be a bounded polygonal domain with a Lipschitz-continuous boundary and let conditions* (4.5.10)–(4.5.12) *and assumptions* 4.5.4.2, (4.5.36), (4.5.37) *be satisfied. For $h \in (0, h_0)$ and $\tau \in (0, T)$ let us construct approximate solutions of problem* (4.5.1)–(4.5.3) *with the aid of the FV–FE scheme* (4.5.70) *and define the functions $u_{h\tau}$ and $w_{h\tau}$ by* (4.5.91) *and* (4.5.92). *Then the systems $\{u_{h\tau}\}$, $\{w_{h\tau}\}$ converge to the solution u of problem* (4.5.15) *in the sense of* (4.5.96), *as h, $\tau \to 0$, so that h, τ satisfy the stability conditions* (4.5.75) *and* (4.5.90).

Error estimates for the explicit FV–FE scheme (4.5.70) have not yet been derived.

Remark 4.75 Application of scheme (4.5.70) requires the verification of the stability conditions (4.5.75) and (4.5.90). Condition (4.5.75), which is a consequence of (4.5.77), contains quantities depending on data only, namely the initial condition, right-hand side, the properties of the triangulation $\mathcal{T}_h$ and the local Lipschitz-continuity of the numerical flux. Its verification does not present any difficulty. In order to verify condition (4.5.75), it is necessary to evaluate the constants $\hat{c}_1$ from Lemma 4.61 and c_4 from the inverse estimate (4.5.73). By

(Houston and Süli, 2001), Appendix B, the constant c_4 can be expressed with the aid of parameters characterizing the triangulation $\mathcal{T}_h$:

$$c_4 = 6c_{\text{inv}}\, c_5, \tag{4.5.97}$$

where c_{inv} and c_5 are the constants from (4.5.72) and (4.5.74), respectively.

We show that $\hat{c}_1 = 1$, because

$$\|v_h\|_{0,\Omega} \leq \|L_h\, v_h\|_{0,\Omega} \quad \forall v_h \in X_h. \tag{4.5.98}$$

In order to establish (4.5.98), we consider an arbitrary $v_h \in X_h$. We can write

$$\|v_h\|_{0,\Omega}^2 = \sum_{K\in\mathcal{T}_h} \int_K |v_h(x)|^2\, dx, \tag{4.5.99}$$

$$\|L_h\, v_h\|_{0,\Omega}^2 = \frac{1}{3} \sum_{K\in\mathcal{T}_h} |K| \sum_{i=1}^{3} |v_h(P_i^K)|^2 \tag{4.5.100}$$

(see (4.5.28), where P_i^K, $i = 1, 2, 3$, are the vertices of $K \in \mathcal{T}_h$). Since $|v_h|^2$ is a quadratic function on K, the integrals in (4.5.99) can be evaluated exactly by the edge midpoint integration rule:

$$\int_K |v_h(x)|^2\, dx = \frac{1}{3}\left(\left(\frac{v_h(P_1^K) + v_h(P_2^K)}{2}\right)^2 + \left(\frac{v_h(P_2^K) + v_h(P_3^K)}{2}\right)^2 + \left(\frac{v_h(P_3^K) + v_h(P_1^K)}{2}\right)^2\right).$$

Now we use the inequality

$$\left(\frac{a+b}{2}\right)^2 \leq \frac{a^2}{2} + \frac{b^2}{2}$$

and from (4.5.99) and (4.5.100) obtain (4.5.98).

Remark 4.76 From the point of view of practical computations, assumption 4.5.3.1, b) (4.5.4.2, b)) is rather restrictive. It can be weakened by using a triangulation of the Delaunay type ((Frey and George, 2000), (Bern, 1999)), again allowing the application of the discrete maximum principle.

4.5.5 *Combination of barycentric finite volumes with nonconforming finite elements*

This section is devoted to an overview of theoretical results obtained in (Angot *et al.*, 1998) and (Dolejší *et al.*, 2002) for the combination of barycentric finite volumes with nonconforming piecewise linear finite elements. Numerical examples presented in Section 4.4.7 show that the application of this method to the compressible Navier–Stokes equations leads to very satisfactory results.

4.5.5.1 *Discretization* We assume that the domain $\Omega \subset \mathbb{R}^2$ is polygonal with a Lipschitz-continuous boundary. By $\mathcal{T}_h = \{K_i\}_{i\in I}$ we denote a triangulation of Ω with standard properties from Section 4.1.2 and construct the associated barycentric FV mesh $\mathcal{D}_h = \{D_j\}_{j\in J}$ as described in Example 4.51, b). These finite volumes are associated with midpoints Q_j of sides S_j of all elements $K \in \mathcal{T}_h$. We set $\mathcal{Q}_h = \{Q_j\}_{j\in J}$. We use the notation introduced in Section 4.4 and set $\overset{\circ}{J} = \{j \in J; Q_j \in \Omega\}$. Thus, $j \in J \setminus \overset{\circ}{J}$ if and only if $Q_j \in \partial\Omega$.

Let us define the following spaces over the grids $\mathcal{T}_h$ and $\mathcal{D}_h$:

$$
\begin{aligned}
X_h &= \left\{v_h \in L^2(\Omega); v_h|_K \text{ is linear } \forall K \in \mathcal{T}_h, v_h \text{ is continuous at } Q_j\ \forall j \in J\right\},\\
V_h &= \left\{v_h \in X_h; v_h(Q_i) = 0\ \forall i \in J - \overset{\circ}{J}\right\}, \qquad (4.5.101)\\
Z_h &= \left\{w_h \in L^2(\Omega); w_h|_{D_i} = \text{ const } \forall i \in J\right\},\\
Y_h &= \left\{w_h \in Z_h; w_h = 0 \text{ on } D_i \in \mathcal{D}_h\ \forall i \in J - \overset{\circ}{J}\right\}.
\end{aligned}
$$

Note that $X_h \not\subset H^1(\Omega)$ and $V_h \not\subset V = H_0^1(\Omega)$. Therefore, we speak of *nonconforming piecewise linear finite elements*. As mentioned in Section 4.1.4, the use of nonconforming FEs is one of the basic FE*variational crimes,* see (Strang, 1972).)

In the spaces from (4.5.101) we easily construct *simple bases.* The system $\{w_i;\ i \in J\}$ of functions $w_i \in X_h$ such that $w_i(Q_j) = \delta_{ij} =$ Kronecker delta, $i, j \in J$, forms a basis in X_h. The system $\{w_i,\ i \in \overset{\circ}{J}\}$ is a basis in V_h. Furthermore, denoting by $d_i = \chi_{D_i}$ the characteristic function of $D_i \in \mathcal{D}_h$, we have bases in Z_h and Y_h as the systems $\{d_i;\ i \in J\}$ and $\{d_i;\ i \in \overset{\circ}{J}\}$, respectively.

By I_h we denote the interpolation operator in the space of nonconforming finite elements. If $v \in H^1(\Omega)$, then

$$
I_h v \in X_h, \quad (I_h v)(Q_{ij}) = \frac{1}{|S_{ij}|}\int_{S_{ij}} v\, dS, \quad j = 1, 2, 3,\ i \in I. \qquad (4.5.102)
$$

This integral exists due to the theorem on traces in the space $H^1(K)$:

$$
\|\varphi\|_{0,\partial K} \le c\|\varphi\|_{1,K}, \quad \varphi \in H^1(K),\ K \in \mathcal{T}_h\ (c = c(K)). \qquad (4.5.103)
$$

The *lumping operator* $L_h : X_h \to Z_h$ is defined by

$$
L_h v_h = \sum_{i\in J} v_h(Q_i) d_i \in Z_h \quad \text{for } v_h : \mathcal{Q}_h \to \mathbb{R}. \qquad (4.5.104)
$$

Obviously, $L_h(V_h) = Y_h$.

In order to derive the discrete problem for (4.5.1)–(4.5.3), we define

$$
(u, v)_h = \int_\Omega (I_h u)(I_h v)\, dx, \quad u, v \in H^1(\Omega), \qquad (4.5.105)
$$

$$((u,v))_h = \sum_{i\in I} \int_{K_i} \nabla u \cdot \nabla v \, dx, \quad u,v \in L^2(\Omega), \quad u|_K, v|_K \in H^1(K) \; \forall K \in \mathcal{T}_h.$$

By $|\cdot|_h$ we denote the discrete L^2-norm induced by $(\cdot,\cdot)_h$. For $v_h \in X_h$ we set $I_h v_h = v_h$ and then

$$(u_h, v_h)_h = (u_h, v_h)_{0,\Omega}, \; |v_h|_h = \|v_h\|_{0,\Omega}, \quad u_h, v_h \in X_h. \tag{4.5.106}$$

For 'regular' functions we have

$$\begin{aligned} ((u,v))_h &= ((u,v)), && u,v \in H^1(\Omega), \\ \tilde{b}_h(u,v) &= b(u,v), && u \in H^1(\Omega) \cap L^\infty(\Omega), \; v \in L^2(\Omega). \end{aligned} \tag{4.5.107}$$

The form $((\cdot,\cdot))_h$ induces the seminorm

$$\|u_h\|_{X_h} = \left(\sum_{i\in I} \int_{K_i} |\nabla u_h|^2 \, dx \right)^{1/2}, \quad u_h \in X_h. \tag{4.5.108}$$

Under the notation

$$\|u_h\|_{X_h(K_i)} = \left(\int_{K_i} |\nabla u_h|^2 \, dx \right)^{1/2}, \quad i \in I, \; u_h \in X_h, \tag{4.5.109}$$

we have

$$\|u_h\|^2_{X_h} = \sum_{i\in I} \|u_h\|^2_{X_h(K_i)}, \quad u_h \in X_h. \tag{4.5.110}$$

The convective form $\tilde{b}_h$ is approximated with the aid of the FV approach explained in Section 4.4 (see (4.4.18) and (4.4.19)). Taking into account the form of the lumping operator L_h, we can write

$$b_h(u_h, v_h) = \sum_{i\in J} v_h(Q_i) \sum_{j\in s(i)} H\left(u_h(Q_i), u_h(Q_j), \boldsymbol{n}_{ij}\right) |\Gamma_{ij}|, \quad u_h, v_h \in V_h. \tag{4.5.111}$$

Here H is a numerical flux satisfying assumptions 4.5.2.1.

In what follows we shall be concerned with the *semi-implicit discrete problem*. We define the *approximate solution* of problem (4.5.1)–(4.5.3) as functions u_h^k, $t_k \in [0,T]$, given by the conditions

$$\begin{aligned} &\text{a)} \;\; u_h^0 = I_h u^0 \; (\in V_h), \\ &\text{b)} \;\; u_h^{k+1} \in V_h, \quad t_k \in [0,T), \\ &\text{c)} \;\; \frac{1}{\tau}(u_h^{k+1} - u_h^k, v_h)_h + b_h(u_h^k, v_h) + \nu((u_h^{k+1}, v_h))_h = (g^{k+1}, v_h)_h, \\ &\qquad \forall v_h \in V_h, \; t_k \in [0,T), \end{aligned} \tag{4.5.112}$$

where $g^k = g(\cdot, t_k)$. The function u_h^k is the approximate solution at time t_k.

4.5.5.2 *Convergence* Let us consider a system $\{\mathcal{T}_h\}_{h\in(0,h_0)}, h_0 > 0$ of triangulations of the domain Ω from 4.5.4.2. (That is, this system is regular, of weakly acute type, satisfying the inverse assumption.)

We define functions $u_{h\tau}, w_{h\tau} : (-\infty, \infty) \to V_h$ associated with an approximate solution $\{u_h^k\}_{k=0}^r$:

$$\begin{aligned} u_{h\tau}(t) &= u_h^0, \quad t \le 0, \\ u_{h\tau}(t) &= u_h^k, \quad t \in (t_{k-1}, t_k], \quad k = 1, \ldots, r, \\ u_{h\tau}(t) &= u_h^r, \quad t \ge T; \end{aligned} \tag{4.5.113}$$

$w_{h\tau}$ is a continuous, piecewise linear mapping of $[0, T]$ into V_h with values

$$\begin{aligned} w_{h\tau}(t_k) &= u_h^k, \quad k = 0, \ldots, r, \\ w_{h\tau}(t) &= 0 \text{ for } t < 0 \text{ or } t > T. \end{aligned}$$

The goal is to prove that the functions $u_{h\tau}, w_{h\tau}$ converge in some sense to the exact solution of problem (4.5.15), as $h, \tau \to 0$ in a suitable way. In the investigation of the convergence we proceed in a similar way as in Section 4.5.3, but it is necessary to deal with special technicalities due to the nonconformity of finite elements.

I. L^∞-*stability* Let us set

$$\tilde{M} = \|u^0\|_{0,\infty,\Omega}, \tilde{C} = \|g\|_{0,\infty,Q_T}, \tag{4.5.114}$$

$$M^* = \tilde{M} + T\tilde{C}, M = 3M^*. \tag{4.5.115}$$

Then, similarly as in Theorem 4.59, we can prove the estimates

$$\begin{aligned} \|u_h^k\|_{0,\infty,\Omega} &\le M, \quad t_k \in [0, T], \\ \|u_{h\tau}\|_{0,\infty,Q_T}, \|w_{h\tau}&\|_{0,\infty,Q_T} \le M, \end{aligned} \tag{4.5.116}$$

provided the *stability condition*

$$\tau c(M^*)|\partial D_i| \le |D_i|, \quad i \in J, \tag{4.5.117}$$

is satisfied.

Similarly as in previous sections, condition (4.5.117) can be heuristically generalized to a CFL stability condition of schemes for the solution of the compressible Navier–Stokes equations.

II. *Consistency of the method* This is based on various relations and estimates:

a) Discrete Friedrichs inequality:

$$\|u_h\|_{0,\Omega} \le C_F \|u_h\|_{X_h}, \quad u_h \in V_h, h \in (0, h_0). \tag{4.5.118}$$

b) Approximation properties of the operators I_h, L_h, discrete scalar L^2-product and the form b_h. Let $\varphi \in H^{k+1}(\Omega)$, where $k = 0$ or 1. Then for $h \in (0, h_0)$ we have

$$\|\varphi - I_h\varphi\|_{X_h} \leq ch^k\|\varphi\|_{k+1,\Omega}, \tag{4.5.119}$$

$$\|\varphi - I_h\varphi\|_{0,\Omega} \leq ch^{k+1}\|\varphi\|_{k+1,\Omega}, \tag{4.5.120}$$

$$\|I_h\varphi\|_{X_h} \leq c\|\varphi\|_{1,\Omega}, \tag{4.5.121}$$

$$\varphi \in H^1(\Omega) \Rightarrow \|\varphi - I_h\varphi\|_{X_h} \to 0 \text{ as } h \to 0 \tag{4.5.122}$$

with $c > 0$ independent of φ and h. Further,

$$\|v_h\|_{0,\Omega} = \|L_hv_h\|_{0,\Omega}, \quad v_h \in X_h, \tag{4.5.123}$$

$$\|v_h - L_hv_h\|_{0,\Omega} \leq ch\|v_h\|_{X_h}, \quad v_h \in X_h, \tag{4.5.124}$$

$$(u_h, v_h) = (u_h, v_h)_h, \quad u_h, v_h \in X_h, \tag{4.5.125}$$

$$|(g^k, v_h) - (g^k, v_h)_h| \leq ch\|g^k\|_{1,q,\Omega}\|v_h\|_{X_h}, \quad v_h \in V_h. \tag{4.5.126}$$

If $M > 0$ and $\kappa \in (0,1)$, then there exists a constant $\tilde{c} = \tilde{c}(M, \kappa)$ such that

$$|\tilde{b}_h(u_h, v_h) - b_h(u_h, v_h)| \leq \tilde{c}h^{1-\kappa}\left(\|u_h\|^2_{X_h} + \|u_h\|_{X_h}\right)\|v_h\|_{X_h}, \tag{4.5.127}$$
$$u_h \in V_h \cap L^\infty(\Omega),\ \|u_h\|_{0,\infty,\Omega} \leq M,\ v_h \in V_h,\ h \in (0, h_0).$$

If $M > 0$, then there exists a constant $c^* = c^*(M)$ such that

$$b_h(u_h, v_h) \leq c^*\|u_u\|_{0,\infty,\Omega}\|v_h\|_{X_h}, \tag{4.5.128}$$
$$u_h \in V_h \cap L^\infty(\Omega),\ \|u_h\|_{0,\infty,\Omega} \leq M,\ v_h \in V_h,\ h \in (0, h_0).$$

III. *A priori estimates* These have the form

$$\max_{t_k \in [0,T]} \|u_h^k\|_{0,\Omega} \leq c, \tag{4.5.129}$$

$$\sum_{k=1}^{r} \|u_h^k - u_h^{k-1}\|^2_{0,\Omega} \leq c,$$

$$\nu\tau \sum_{k=0}^{r} \|u_h^k\|^2_{X_h} \leq \hat{C},$$

for all $\tau,\ h > 0$ satisfying the conditions $h \in (0, h_0)$ and (4.5.117). This implies that

$$\|u_{h\tau}\|_{L^2(-1,T;L^2(\Omega))} \leq c, \tag{4.5.130}$$
$$\|w_{h\tau}\|_{L^2(0,T;L^2(\Omega))} \leq c,$$
$$\|u_{h\tau}\|_{L^2(-1,T;V_h)} \leq c,$$
$$\|w_{h\tau}\|_{L^2(0,T;V_h)} \leq c,$$
$$\|u_{h\tau} - w_{h\tau}\|_{0,Q_T} \leq c\sqrt{\tau},$$

for all $h \in (0, h_0)$ and $\tau > 0$ satisfying condition (4.5.117). (We consider some estimates on the interval $(-1, T)$ because of the delayed argument $t - \tau$ in equation (4.5.131).)

IV. *Auxiliary relations* Scheme (4.5.112) can be written in the form

$$\frac{d}{dt}(w_{h\tau}(t), v_h)_h + \nu((u_{h\tau}, v_h))_h + b_h(u_{h\tau}(t-\tau), v_h) = (g_{h\tau}(t), v_h)_h \quad (4.5.131)$$

for a.a. $t \in (0,T)$ and all $v_h \in V_h$,

where

$$g_{h\tau}(t) = g^{k+1} \quad \text{for} \quad t \in (t_k, t_{k+1}). \quad (4.5.132)$$

Since the distributional derivatives $v_h \in V_h$ are not elements of $L^2(\Omega)$, we define the discrete analogue $d_{ih}v_h$ of the derivatives $\partial v_h/\partial x_i$, $i = 1, 2$:

$$(d_{ih}v_h)(x) = \left(\frac{\partial v_h}{\partial x_i}\right)(x), \quad x \in K,\ K \in \mathcal{T}_h. \quad (4.5.133)$$

Obviously, $d_{ih}v \in L^2(\Omega)$. We introduce the space $F = \left[L^2(\Omega)\right]^3$ and the mapping $\omega : V \to F$ defined by

$$u \in V \longmapsto \omega u = \left(u, \frac{\partial u}{\partial x_1}, \frac{\partial u}{\partial x_2}\right) \in F. \quad (4.5.134)$$

The space F is equipped with the norm

$$\|\varphi\|_F = \left(\sum_{i=0}^{2} \|\varphi_i\|_{0,\Omega}^2\right)^{1/2} \quad \text{for } \varphi = (\varphi_0, \varphi_1, \varphi_2) \in F. \quad (4.5.135)$$

Further, we define the imbedding operator $J_h : V_h \to F$ by

$$v_h \in V_h \longmapsto J_h v_h = (v_h, d_{1h}v_h, d_{2h}v_h) \in F. \quad (4.5.136)$$

The operators J_h, $h \in (0, h_0)$, are *uniformly bounded:*

$$\|J_h\| = \sup_{0 \neq v_h \in V_h} \frac{\|J_h v_h\|_F}{\|v_h\|_{X_h}} \le c, \quad h \in (0, h_0). \quad (4.5.137)$$

Moreover, the following statements hold:

1) For each $v \in V$,

$$\lim_{h\to 0} J_h(I_h v) = \omega v \text{ strongly in } F. \quad (4.5.138)$$

2) If for a sequence $h_n \in (0, h_0)$, $n = 1, 2, \ldots$, we have $h = h_n \to 0$ as $n \to \infty$, $v_h \in V_h$ and

$$\lim_{h\to 0} J_h v_h = \phi \text{ weakly in } F, \quad (4.5.139)$$

then there exists $v \in V$ such that $\phi = \omega v$.

Remark 4.77 The family of triplets $\{V_h, J_h, I_h\}_{h\in(0,h_0)}$ together with $\{V, F, \omega\}$ is called the external approximation of the space V. If (4.5.137) holds, the external approximation of V is called stable. If the operators I_h, J_h have properties (4.5.138) and (4.5.139), then the external approximation of V is called convergent (cf. (Glowinski *et al.*, 1976), Chapter I, Section 5. or (Temam, 1977), Chapter I, Section 3).

The above results imply the existence of sequences $h = h_n,\ \tau = \tau_n \to 0$ as $n \to \infty$ satisfying (4.5.117) and of functions $u \in L^2(-1, T; V),\ \phi \in L_2(-1, T; F)$ such that

$$\begin{aligned} &u_{h\tau} \to u \text{ weakly in } L^2(-1, T; L^2(\Omega)), \\ &J_h u_{h\tau} \to \phi \text{ weakly in } L^2(-1, T; F), \\ &\phi = \omega u, \\ &w_{h\tau} \to u \text{ weakly in } L^2(0, T; L^2(\Omega)), \\ &J_h w_{h\tau} \to \omega u \text{ weakly in } L^2(0, T; F). \end{aligned} \tag{4.5.140}$$

V. *Compactness arguments* These are based on the Fourier transform $\hat{w}_{h\tau}$ of the function $w_{h\tau}$ with respect to time, to prove that for any sequences $h = h_n \to 0$ and $\tau = \tau_n \to 0$ satisfying (4.5.117) and $w_{h\tau} = w_{h_n\tau_n}$ we get

$$w_{h\tau} \to u \text{ strongly in } L^2(Q_T). \tag{4.5.141}$$

Summarizing the above results, we have

$$\begin{aligned} &J_h u_{h\tau} \to \omega u \quad \text{weakly in } L^2(-1, T; F), \\ &J_h w_{h\tau} \to \omega u \quad \text{weakly in } L^2(-1, T; F), \\ &u_{h\tau} \to u \quad \text{strongly in } L^2(\tilde{Q}_T), \\ &w_{h\tau} \to u \quad \text{strongly in } L^2(\tilde{Q}_T), \end{aligned} \tag{4.5.142}$$

and

$$\begin{aligned} &u_{h\tau} \to u \text{ weak-* in } L^\infty(Q_T), \\ &w_{h\tau} \to u \text{ weak-* in } L^\infty(Q_T). \end{aligned} \tag{4.5.143}$$

VI. *Limit process* On the basis of (4.5.142) and (4.5.143), we can pass to the limit in (4.5.131). As a *final result* we obtain the following *convergence theorem:*

Theorem 4.78 *Let assumptions* (4.5.10)–(4.5.12), 4.5.2.1 *and* 4.5.3.1 *be satisfied. Let us construct approximate solutions with the aid of the FV–FE scheme* (4.5.112) *and define functions $u_{h\tau}$ and $w_{h\tau}$ by* (4.5.113) *and* (4.5.114). *Then the systems $\{u_{h\tau}\}$, $\{w_{h\tau}\}$ converge to the unique weak solution of problem* (4.5.1)–(4.5.3) *in the following sense:*

$$\begin{aligned} &J_h u_{h\tau},\ J_h w_{h\tau} \to \omega u \quad \textit{weakly in } L^2(0, T; F), \\ &u_{h\tau},\ w_{h\tau} \to u \quad \textit{weak-* in } L^\infty(Q_T), \\ &u_{h\tau},\ w_{h\tau} \to u \quad \textit{strongly in } L^2(Q_T), \end{aligned} \tag{4.5.144}$$

as $h, \tau \to 0$ so that the stability condition (4.5.117) *is satisfied.*

4.5.5.3 *Error estimates* These were proven for the combined barycentric FV–nonconforming FE method were proven in (Dolejší *et al.*, 2002) under assumptions from Theorem 4.78 and the following assumptions on the regularity of the exact weak solution of problem (4.5.1)–(4.5.3):

$$\begin{aligned} & u \in L^\infty(0,T;H^2(\Omega)) \cap W^{1,\infty}(\Omega), \\ & \partial u/\partial t \in L^\infty(0,T;L^2(\Omega)), \\ & \partial^2 u/\partial t^2 \in L^\infty(0,T;L^2(\Omega)). \end{aligned} \tag{4.5.145}$$

Then the estimates of the error $e_h^k = u_h^k - u(t_k)$ were established in the form

$$\begin{aligned} \tau \sum_{k=0}^{r} \|e_h^k\|_{0,\Omega}^2 &\leq C_1 h^{2(1-\kappa)}, \\ \nu\tau \sum_{k=0}^{r-1} \|e_h^{k+1}\|_{X_h}^2 &\leq C_2 h^{1-2\kappa}, \end{aligned} \tag{4.5.146}$$

for all $h \in (0, h_0)$ and $\tau > 0$ satisfying the stability condition (4.5.117), the 'inverse stability condition'

$$h \leq \tilde{c}\tau \tag{4.5.147}$$

(see (4.5.68)) and the additional condition

$$c^*\tau \leq \nu \tag{4.5.148}$$

with suitable constants $\tilde{c}, c^* > 0$ independent of h, τ and ν.

Unfortunately, similar to 4.5.3.7, the behaviour of constants C_1, C_2 in dependence on the diffusion coefficient ν is rather pessimistic:

$$C_1 = O(\nu^{-6}\exp(cT/\nu)), \quad C_2 = O(\nu^{-7}\exp(cT\nu))$$

with a constant c independent of ν.

4.6 Discontinuous Galerkin finite element method

In the FEM, an important question arises: should we prefer to use *conforming* finite elements or *nonconforming* finite elements? Conforming (i.e. continuous) FE approximations are suitable for problems with sufficiently regular solutions. However, singularly perturbed problems or nonlinear conservation laws of fluid dynamics have solutions with steep gradients or discontinuities and their approximations by conforming finite elements suffer from the Gibbs phenomenon. One way to avoid this drawback is to use a suitable stabilization such as the streamline diffusion method or Galerkin least squares method and shock capturing stabilization, treated in Sections 4.1 and 4.3. As we can see there, the streamline diffusion and shock capturing terms look quite sophisticated. Moreover, the approximation of discontinuous solutions to conservation laws by continuous functions does not seem quite natural.

From this point of view, it seems that for conservation laws with a discontinuous solution, the FV method is more suitable, because the FV approximations are discontinuous on interelement interfaces, which allows better resolution of shock waves and contact discontinuities. On the other hand, as was shown in Section 4.5, the increase of accuracy in FV schemes seems to be problematic. A combination of ideas and techniques of the FV and FE methods yields the *discontinuous Galerkin finite element method* (DGFEM), using advantages of both approaches and allowing schemes to be obtained with a higher order of accuracy in a natural way. As was mentioned in Section 4.1.11, the DGFEM is based on the idea of approximating the solution of an initial-boundary value problem by piecewise polynomial functions over an FE mesh without any requirement on interelement continuity. The higher-order DGFEM was applied to nonlinear conservation laws as early as 1989 by Cockburn and Shu ((Cockburn and Shu, 1989)). It was used later for the numerical simulation of compressible flow by Bassi and Rebay in (Bassi and Rebay, 1997*b*) and (Bassi and Rebay, 1997*a*). In recent years the DGFE schemes have been extensively developed and become more and more popular. Some aspects of the DGFEM and applications to gas dynamics are also discussed in (Flaherty *et al.*, 2000), (Adjerid *et al.*, 2002), (Bejček *et al.*, 2002), (Dolejší *et al.*, 2002*a*), (Dolejší *et al.*, 2003), (Dolejší *et al.*, 2002*b*), (Dolejší and Feistauer, 2001). For a survey, see, for example, (Cockburn *et al.*, 2000) and (Cockburn, 1999).

In Section 4.1.11 we paid attention to the application and error analysis of the DGFEM for the numerical solution of a purely convective linear equation (with a reaction term) under the assumption that the exact solution is regular. Here we shall be concerned with the DGFEM for the solution of nonlinear conservation laws and compressible flow.

In order to explain the main ideas of the DGFEM applied to conservation laws, we start from a simple 1D problem.

4.6.1 *DGFEM for a scalar conservation law with one space variable*

Let us consider the scalar conservation law equation

$$\frac{\partial u}{\partial t} + \frac{\partial f(u)}{\partial x} = 0, \quad x \in \mathcal{I},\ t \in (0,T), \tag{4.6.1}$$

where $u = u(x,t)$, $\mathcal{I} = (0,1)$, $T > 0$, equipped with the initial condition

$$u(x,0) = u^0(x), \quad x \in \mathcal{I}, \tag{4.6.2}$$

and the periodic boundary condition

$$u(0,t) = u(1,t), \quad t \in (0,T). \tag{4.6.3}$$

We assume that $f \in C^1(\mathbb{R})$.

The *weak solution* of problem (4.6.1)–(4.6.3) is defined as a function $u \in L^\infty_{\mathrm{loc}}(\mathcal{I} \times (0,T))$ satisfying the identity

$$\int_0^1 \int_0^T \left(u(x,t)\, \frac{\partial v(x,t)}{\partial t} + f(u(x,t)) \frac{\partial v(x,t)}{\partial x} \right) dx dt + \int_0^1 u^0(x)\, v(x,0)\, dx = 0 \tag{4.6.4}$$

for all test functions $v \in C^\infty([0,1] \times [0,T])$ having a compact support in $[0,1] \times [0,T)$ and satisfying the periodicity condition (4.6.3). In order to guarantee the uniqueness of the solution, we require that the weak solution satisfies the entropy condition (cf. Section 2.3.4).

4.6.1.1 *Discrete problem* In the discretization of problem (4.6.1)–(4.6.3) we shall not start from the concept of the weak solution, but use a process leading to an integral identity with test functions depending on x only. Let us set

$$\mathcal{V} = \{v \in C^\infty([0,1]); v(0) = v(1)\} \tag{4.6.5}$$

and assume that $u \in C^1([0,1] \times [0,T])$ is a classical solution of (4.6.1)–(4.6.3). Then from (4.6.1) we obtain the identity

$$\int_0^1 \frac{\partial u(x,t)}{\partial t}\, v(x)\, dx - \int_0^1 f(u(x,t))\, \frac{dv(x)}{dx}\, dx = 0, \quad v \in \mathcal{V}, \tag{4.6.6}$$

which is the starting point for the *space semidiscretization* with the aid of conforming finite elements. In the DGFEM we proceed in a more modified way.

Let us construct a partition of the interval $\mathcal{I} = (0,1)$ formed by mesh points $x_{j+\frac{1}{2}}$, $j = 0, \ldots, n$, such that

$$0 = x_{\frac{1}{2}} < x_{\frac{3}{2}} < \cdots < x_{j-\frac{1}{2}} < x_{j+\frac{1}{2}} < \cdots < x_{n+\frac{1}{2}} = 1, \tag{4.6.7}$$

and set $\mathcal{I}_j = \left[x_{j-\frac{1}{2}}, x_{j+\frac{1}{2}}\right]$, $h_j = x_{j+\frac{1}{2}} - x_{j-\frac{1}{2}}$. Further, let

$$S_h = S_h^{p,-1} = \left\{v \in L^1(0,1); v|_{\mathcal{I}_j} \in P^p(\mathcal{I}_j),\, j = 1, \ldots, n\right\}, \tag{4.6.8}$$

where $p \geq 0$ is an integer and $P^p(\mathcal{I}_j)$ is the set of all polynomials on $\mathcal{I}_j$ of degree at most p.

Multiplying (4.6.1) by $v \in S_h$, integrating over $\mathcal{I}_j$ and using integration by parts, we get the relation

$$\begin{aligned} &\int_{\mathcal{I}_j} \frac{\partial u(x,t)}{\partial t}\, v(x)\, dx - \int_{\mathcal{I}_j} f(u(x,t))\, \frac{dv(x)}{dx}\, dx \\ &+ f(u(x_{j+\frac{1}{2}}-,t))\, v(x_{j+\frac{1}{2}}-) - f(u(x_{j-\frac{1}{2}}+,t)) v(x_{j-\frac{1}{2}}+) = 0, \\ &\qquad t \in (0,T),\ j = 1, \ldots, n. \end{aligned} \tag{4.6.9}$$

Here

$$v(x_{j+\frac{1}{2}}-) = \lim_{x \to x_{j+\frac{1}{2}}-} v(x), \tag{4.6.10}$$

$$v(x_{j-\frac{1}{2}}+) = \lim_{x \to x_{j-\frac{1}{2}}+} v(x).$$

Similarly we define the values $u(x_{j+\frac{1}{2}}-,t)$ and $u(x_{j-\frac{1}{2}}+,t)$.

Identity (4.6.9) is the basis for the derivation of the discrete problem for the approximate solution u_h which is in general discontinuous at $x = x_{j+\frac{1}{2}}$. This is the reason that the fluxes $f(u(x_{j+\frac{1}{2}}-,t))$ and $f(u(x_{j-\frac{1}{2}}+,t))$ will be approximated with the aid of a numerical flux g:

$$\begin{aligned} f(u(x_{j+\frac{1}{2}}-,t)) &\approx g(u_h(x_{j+\frac{1}{2}}-,t), u_h(x_{j+\frac{1}{2}}+,t)), \\ f(u(x_{j-\frac{1}{2}}+,t)) &\approx g(u_h(x_{j-\frac{1}{2}}-,t), u_h(x_{j-\frac{1}{2}}+,t)). \end{aligned} \tag{4.6.11}$$

Similarly as in Section 3.2 we assume that the function $g = g(\alpha, \beta)$ is continuous and consistent with the flux f:

$$g(\alpha, \alpha) = f(\alpha), \quad \alpha \in I\!R. \tag{4.6.12}$$

In this way we arrive at the formulation of the *discrete problem*: an *approximate solution* of problem (4.6.1)–(4.6.3) obtained by the DGFEM is defined as a function $u_h = u_h(x,t)$ such that

$$u_h(\cdot, t) \in S_h \quad \forall t \in [0,T], \tag{4.6.13}$$

$$\begin{aligned} &\int_{I_j} \frac{\partial u_h(x,t)}{\partial t} v_h(x)\, dx - \int_{I_j} f(u_h(x,t)) \frac{dv_h(x)}{dx}\, dx \\ &+ g(u_h(x_{j+\frac{1}{2}}-,t), u_h(x_{j+\frac{1}{2}}+,t)) v_h(x_{j+\frac{1}{2}}-) \\ &- g(u_h(x_{j-\frac{1}{2}}-,t), u_h(x_{j-\frac{1}{2}}+,t)) v_h(x_{j-\frac{1}{2}}+) = 0, \\ &\quad \forall v_h \in S_h,\ \forall j = 1, \dots, n,\ \forall t \in (0,T), \end{aligned} \tag{4.6.14}$$

$$u_h(x,0) = u_h^0(x), \quad x \in I, \tag{4.6.15}$$

$$\begin{aligned} u_h(x_{n+\frac{1}{2}}+,t) &= u_h(x_{\frac{1}{2}}+,t), \\ u_h(x_{\frac{1}{2}}-,t) &= u_h(x_{n+\frac{1}{2}}-,t), \end{aligned} \tag{4.6.16}$$

where u_h^0 is an approximation of u^0. Conditions (4.6.16) represent the periodicity condition (4.6.3).

The initial condition in the discrete problem can be defined in several ways. Let us mention two possibilities:

a) *L^2-projection* If $u^0 \in L^2(0,1)$, then we define $u_h^0 \in S_h$ in such a way that

$$(u_h^0, v_h) = (u^0, v_h) \quad \forall v_h \in S_h. \tag{4.6.17}$$

Here $(\cdot,\cdot)$ denotes the L^2-scalar product, i.e.

$$(u,v) = \int_0^1 uv dx, \quad u, v \in L^2(0,1). \tag{4.6.18}$$

b) *Lagrange interpolation* If $u^0 \in C([0,1])$, then $u_h^0 \in S_h$ is determined by the conditions

$$u_h^0(x_{i+\frac{1}{2}}-) = u_h^0(x_{i+\frac{1}{2}}+) = u^0(x_{i+\frac{1}{2}}), \quad i = 1, \dots, n. \tag{4.6.19}$$

The function g can be chosen as an arbitrary numerical flux of a three-point FVM from Section 3.2.

Let us note that the choice $p = 0$ (piecewise constant approximation of the solution) leads to the finite volume method. This means that the FVM method can be considered as a special case of the DGFEM.

4.6.2 *Realization of the discrete problem*

Let us choose a basis in the space $P^p(\mathcal{I}_j)$ formed by functions $\varphi_0^j, \dots, \varphi_p^j$. (It is possible to use, for instance, $\varphi_i^j = x^i$, $i = 0, \dots, p$.) Then we can write

$$u_h(x,t) = \sum_{\ell=0}^{p} u_\ell^j(t)\, \varphi_\ell^j(x), \quad x \in \mathcal{I}_j,\ j = 1, \dots, n. \tag{4.6.20}$$

The substitution into (4.6.14) and the use of (4.6.16) yield the system of ordinary differential equations which can be written in the form

$$\frac{d\boldsymbol{W}_h}{dt} = \boldsymbol{\Phi}_h(\boldsymbol{W}_h), \tag{4.6.21}$$

where $\boldsymbol{W}_h$ is an $n(p+1)$-dimensional vector with components u_ℓ^j, $\ell = 0, \dots, p$, $j = 1, \dots, n$, and $\boldsymbol{\Phi}_h$ is an $n(p+1)$-dimensional vector function depending on these arguments. System (4.6.21) is equipped with the initial condition

$$\boldsymbol{W}_h(0) = \boldsymbol{W}_h^0 \tag{4.6.22}$$

equivalent to (4.6.15).

In practical computations the integrals over the intervals $\mathcal{I}_j$ are evaluated with the aid of suitable quadrature formulae. The numerical integration is then reflected in the form of $\boldsymbol{\Phi}_h$.

The solution of the initial value problem (4.6.21)–(4.6.22) is carried out by a suitable numerical method. To this end, let $0 = t_0 < t_1 < \dots < T$ denote a partition of the time interval $(0, T)$ and let $\tau_k = t_{k+1} - t_k$. Denoting by $\boldsymbol{W}_h^k$ the approximation of $\boldsymbol{W}_h(t_k)$, the transition from $\boldsymbol{W}_h^k$ to $\boldsymbol{W}_h^{k+1}$ can be carried out, for example, with the aid of the Runge–Kutta methods from Section 3.2.2 (where we set $\tau := \tau_k$). Since these methods are explicit, it is necessary to use a suitable stability condition. By analogy to (3.2.55) we require that

$$\tau_k \le \mathrm{CFL}\, \frac{h_j}{\lambda_k}, \quad j = 1, \dots, n, \tag{4.6.23}$$

where

$$\lambda_k = \sup_{x \in (0,1)} |f'(u_h(x, t_k))| \tag{4.6.24}$$

and CFL > 0 is a suitable constant depending on the method.

4.6.3 *Investigation of the order of the DGFEM*

Numerical experiments carried out for the finite volume method (i.e. the DGFEM using piecewise constant approximations) applied to problems with sufficiently regular solutions show that the error measured in the L^1-norm is of the first order, i.e. $O(h)$. See, for example, Table 3.5.1. This leads us to the hypothesis that the DGFEM using polynomials of degree p is of order $p+1$, provided that the exact solution is sufficiently regular and a sufficiently accurate time discretization is applied.

In what follows we shall carry out some numerical experiments with the DGFEM using piecewise linear approximations, i.e. $p = 1$. Our goal is to show that this leads to a second order scheme. In all numerical experiments we use second order Runge–Kutta time stepping (3.2.20) and the following *modified Lax–Friedrichs numerical flux*:

$$g_{\mathrm{LF}}(\alpha, \beta) = \frac{1}{2}(f(\alpha) + f(\beta) - C(\beta - \alpha)), \qquad (4.6.25)$$

$$C = \max\{|f'(s)|; \inf u^0 \le s \le \sup u^0\}.$$

We set CFL $= 0.8$.

4.6.3.1 *Linear equation* Let us consider a linear equation

$$\frac{\partial u}{\partial t} + \frac{\partial u}{\partial x} = 0, \quad x \in \mathcal{I} = (0,1), \ t \in (0,T), \qquad (4.6.26)$$

with the initial condition

$$u(x,0) = \sin(2\pi x), \quad x \in \mathcal{I}, \qquad (4.6.27)$$

and the boundary condition (4.6.3). The exact solution has the form

$$u(x,t) = \sin(2\pi(x-t)), \quad x \in \mathcal{I}, \ t \in (0,T), \qquad (4.6.28)$$

and satisfies the periodic boundary condition (4.6.3).

Let us apply the DGFEM with $p = 1$ to this problem. We choose $\tau = 0.0001$ and a uniform partition of the space interval $\mathcal{I}$ with constant step h successively diminishing: $h = 1/12, 1/24, 1/48, 1/96$. The error $e_h^k = u_h^k - u(t_k)$ will be measured now in the L^2-norm over a subinterval $\tilde{\mathcal{I}} \subset \mathcal{I}$:

$$\|e_h^k\| = \|e_h^k\|_{L^2(\tilde{\mathcal{I}})} = \int_{\tilde{\mathcal{I}}} |e_h^k|^2 \, dx. \qquad (4.6.29)$$

The *experimental order of convergence* q (EOC) can be evaluated with the aid of the least squares method from Section 4.6.6. Here we use another approach.

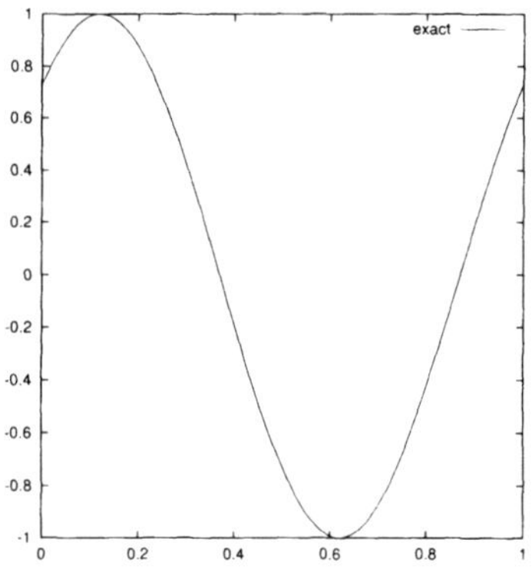

FIG. 4.6.1. Exact solution of a linear problem at $t = 3.87$

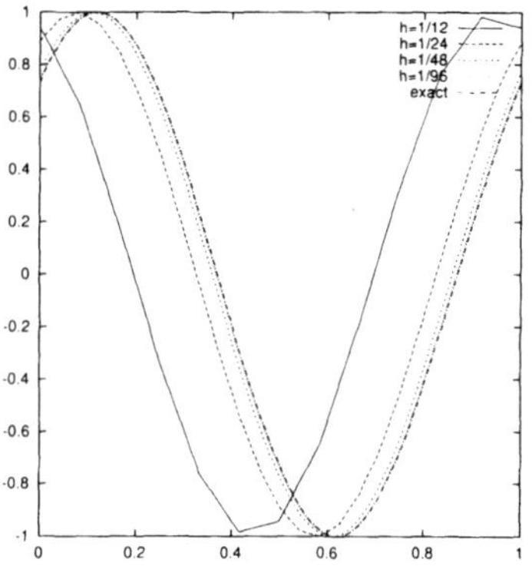

FIG. 4.6.2. 'Convergence' $u_h \to u$ for $h \to 0$

Assuming that the method is of order q in a norm $\|\cdot\|$, i.e. $\|e_h\| \approx Ch^q$, we also have $\|e_{h/2}\| \approx C(h/2)^q$. Then we obtain the formula

$$q \approx \log \frac{\|e_h^k\|}{\|e_{h/2}^k\|} \Big/ \log 2. \tag{4.6.30}$$

Figure 4.6.1 shows the exact solution in the interval $(0, 1)$ for $t_k := 3.87$. Figure 4.6.2 demonstrates the 'convergence' of the approximate solution u_h to the exact solution u as $h \to 0$. In Table 4.6.1 the error $\|e_h^k\|$ and the experimental order of accuracy q on the interval $\tilde{\mathcal{I}} = \mathcal{I}$ are given.

4.6.3.2 *Burgers equation.* Let us consider the nonlinear equation

$$\frac{\partial u}{\partial t} + \frac{\partial}{\partial x}\left(\frac{u^2}{2}\right) = 0, \quad x \in \mathcal{I},\ t \in (0, T), \tag{4.6.31}$$

Table 4.6.1 *Experimental order of convergence for a linear problem*

h	τ	$\|e_h\|$	$\|e_{h/2}\|$	EOC q
1/12	$\tau = 0.0001$	0.655749081	0.175056134	1.9053262
1/24	$\tau = 0.0001$	0.175056134	0.043759202	2.0001592
1/48	$\tau = 0.0001$	0.043759202	0.010652539	2.0383889
1/96	$\tau = 0.0001$	0.010652539	0.002363253	2.1723513

Table 4.6.2 *Experimental order of convergence for a regular solution of a nonlinear problem (Euler time stepping)*

h	τ	$\|e_h\|$	$\|e_{h/2}\|$	EOC q
1/12	$\tau = 0.001$	0.0643576329	0.0220072762	1.54813060
1/24	$\tau = 0.001$	0.0220072762	0.0059605949	1.88445228
1/48	$\tau = 0.001$	0.0059605949	0.0022604817	1.39882604
1/96	$\tau = 0.0001$	0.0017616627	0.0004170548	2.07862864

equipped with the periodic boundary condition (4.6.3) and the initial condition (4.6.27). In the entropy solution of this problem a discontinuity is developed in a finite time. The exact solution can be obtained by the process described in Exercise 4.80 – see later. In Fig. 4.6.3, the graph of the approximate solution at time $t = 0.1$ obtained with $\tau = 0.002$ and $h = 1/200$ is shown. Using $h = 1/12, 1/24, 1/48, 1/96$, we can follow the convergence history and the development of the EOC shown in Table 4.6.2. We see that the order of accuracy is 2 in the case of a regular solution.

For $t = 0.2$ the solution is discontinuous at $x = 0.5$. In Fig. 4.6.4 we see the 'overkilled' numerical solution at $t = 0.2$ obtained with $\tau = 10^{-5}$ and $h = 1/9600$. For $t = 0.2$, the numerical solution obtained with $\tau = 0.002$ and $h = 1/200$ suffers from spurious oscillations in a neighbourhood of the discontinuity. See Fig. 4.6.5.

In order to avoid spurious oscillations produced by higher order methods, suitable limiting procedures were used in Sections 3.2 and 3.5. They cause a decrease in the order of accuracy to one in the vicinity of discontinuities, due to the increase of numerical viscosity. This fact is the motivation for our strat-

Table 4.6.3 *Experimental order of convergence for a regular solution of a nonlinear problem (second order Runge–Kutta time stepping)*

h	τ	$\|e_h\|$	$\|e_{h/2}\|$	EOC q
1/12	$\tau = 0.001$	0.0647160483	0.0217834849	1.57088867
1/24	$\tau = 0.001$	0.0217834849	0.0051857532	2.07060924
1/48	$\tau = 0.001$	0.0051857532	0.0013997758	1.88935771
1/96	$\tau = 0.0001$	0.0016795505	0.0003075543	2.44916225

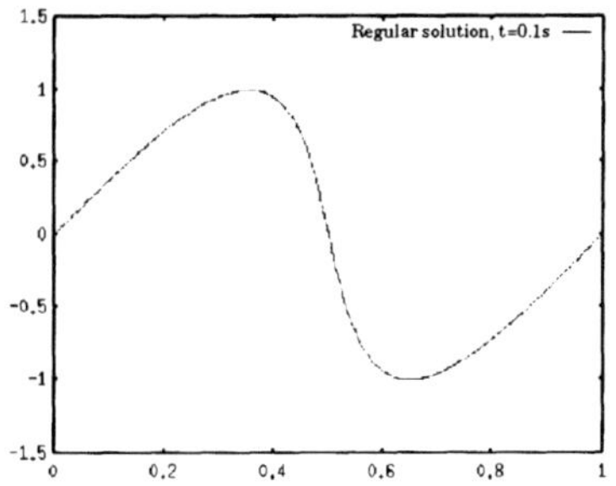

FIG. 4.6.3. Numerical solution of a nonlinear problem at $t = 0.1$ ($\tau = 10^{-3}$, $h = 1/200$)

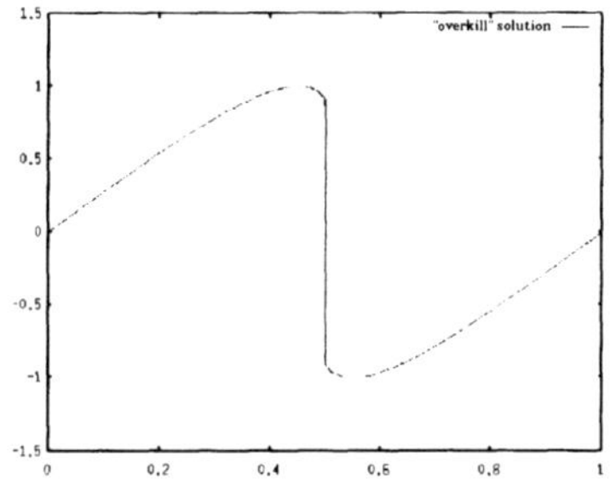

FIG. 4.6.4. 'Overkilled' numerical solution at $t = 0.2$ ($\tau = 10^{-5}$, $h = 1/9600$)

egy, based on the *order limiting* of the scheme in the vicinity of discontinuities. This means that instead of linear polynomials we shall use piecewise constant approximations in a neighbourhood of the discontinuity. Figure 4.6.6 shows the numerical solution with the order limiting. This idea will be developed in a more general way in Section 4.6.7 for multidimensional problems.

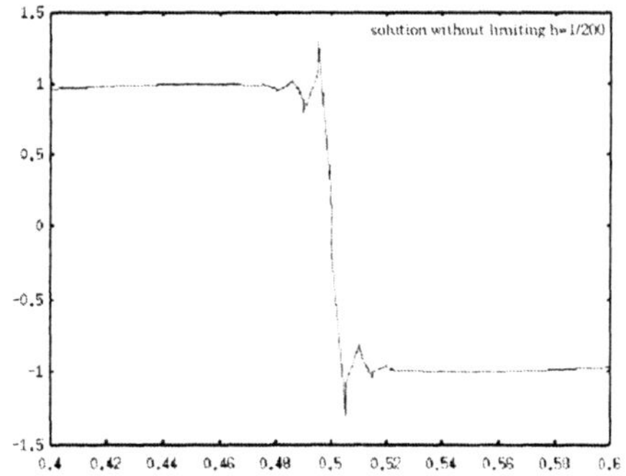

FIG. 4.6.5. Numerical solution of a nonlinear problem at $t = 0.2$ without limiting in the interval $(0.4, 0.6)$ ($\tau = 10^{-3}$, $h = 1/200$)

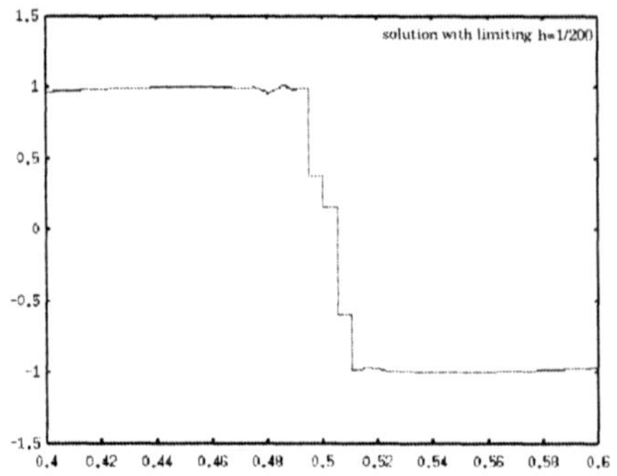

FIG. 4.6.6. Numerical solution of a nonlinear problem at $t = 0.2$ with limiting ($\tau = 10^{-3}$, $h = 1/200$)

Table 4.6.4 *Experimental order of convergence for a discontinuous solution (second order Runge–Kutta time stepping)*

h	τ	$\|e_h\|$	$\|e_{h/2}\|$	EOC q
1/12	$\tau = 0.001$	0.1110378991	0.0682359146	0.70244896
1/24	$\tau = 0.001$	0.0682359146	0.0445839462	0.61400693
1/48	$\tau = 0.001$	0.0445839462	0.0371368034	0.26367467
1/96	$\tau = 0.001$	0.0371368034	0.0272264644	0.44783997

The experimental order of convergence in the L^2-norm was investigated in the case of the discontinuous solution at time $t = 0.2$ in the whole space interval $(0, 1)$ including the discontinuity. In this case we obtain the EOC $= q < 1$, as can be seen in Table 4.6.4. However, in the set $(0,\ 0.4) \cup (0.6,\ 1)$, where the solution is regular, we obtain the EOC $= q \approx 2$, as follows from Table 4.6.5.

Exercise 4.79 Develop an algorithm for the evaluation of the exact entropy solution of problem (4.6.31), (4.6.3) and (4.6.27).
Hint: Let us extend the initial condition u^0 periodically on $I\!R$ with period 1. The exact solution is given by the implicit formula (2.2.28), where $a(u) = u$:

$$u(x, t) = u^0(x - u(x, t)t). \tag{4.6.32}$$

Table 4.6.5 *Experimental order of convergence for a discontinuous solution in a regularity region (second order Runge–Kutta time stepping)*

h	τ	$\|e_h\|$	$\|e_{h/2}\|$	EOC q
1/12	$\tau = 0.0001$	0.0443118136	0.0276906329	0.67849916
1/24	$\tau = 0.0001$	0.0276906329	0.0067115879	2.04467190
1/48	$\tau = 0.0001$	0.0067115878	0.0007476935	3.16613508
1/96	$\tau = 0.0001$	0.0007476935	0.0001181006	2.66243060

Moreover, due to the entropy condition, by virtue of Example 2.20, we require that the exact solution jumps down on a discontinuity, if we pass through the interval $\mathcal{I}$ in the direction x. The nonlinear equation (4.6.32) can be solved iteratively:

$$u^{(k+1)}(x,t) = \omega u^0(x - u^{(k)}(x,t)t) + (1-\omega)u^{(k)}(x,t), \qquad (4.6.33)$$
$$k \geq 0, \; x \in I\!R, \; t \geq 0.$$

Here $\omega \in (0,1)$ is a suitable relaxation parameter. It is necessary to choose the initial iteration $u^{(0)}(x,t)$ so that the iterative process converges to the entropy solution. This is obtained under the choice

$$u^{(0)}(x,t) = u^0(x) + C, \qquad (4.6.34)$$

where $C > 0$ is a sufficiently large constant.

4.6.4 *DGFEM for multidimensional problems*

In this section we shall discuss the discontinuous Galerkin finite element discretization of multidimensional initial-boundary value problems for conservation law equations and, in particular, for the Euler equations. Let $\Omega \subset I\!R^N$ be a bounded domain with a piecewise smooth Lipschitz-continuous boundary $\partial\Omega$ and let $T > 0$. In the space-time cylinder $Q_T = \Omega \times (0,T)$ we consider a system of m first order hyperbolic equations

$$\frac{\partial \boldsymbol{w}}{\partial t} + \sum_{s=1}^{N} \frac{\partial \boldsymbol{f}_s(\boldsymbol{w})}{\partial x_s} = 0. \qquad (4.6.35)$$

This system is equipped with the initial condition

$$\boldsymbol{w}(x,0) = \boldsymbol{w}^0(x), \quad x \in \Omega, \qquad (4.6.36)$$

where $\boldsymbol{w}^0$ is a given function, and with boundary conditions

$$B(\boldsymbol{w}) = 0, \qquad (4.6.37)$$

where B is a boundary operator. The choice of the boundary conditions is carried out similarly as in Section 3.3.6 in the framework of the discrete problem for the Euler equations describing gas flow.

4.6.4.1 *Discretization* By Ω_h we denote a polygonal or polyhedral approximation of the domain Ω, if $N = 2$ or $N = 3$, respectively. Let $\mathcal{T}_h$ $(h > 0)$ denote a partition of the closure $\overline{\Omega}_h$ of the domain Ω_h into a finite number of closed convex polygons (if $N = 2$) or polyhedra (if $N = 3$) K with mutually disjoint interiors. We call $\mathcal{T}_h$ a triangulation of Ω_h, but *do not* require the usual conforming properties from the FEM used in Section 4.1. In 2D problems we usually choose $K \in \mathcal{T}_h$ as triangles or quadrilaterals, in 3D, $K \in \mathcal{T}_h$ can be, for example,

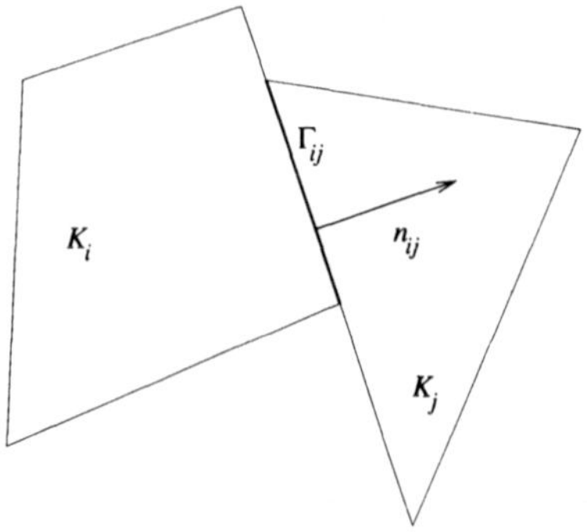

FIG. 4.6.7. Neighbouring elements K_i, K_j

tetrahedra, pyramids or hexahedra, but we can allow even more general convex elements K.

We set $h_K = \text{diam}(K)$, $h = \max_{K\in\mathcal{T}_h} h_K$. By $|K|$ we denote the N-dimensional Lebesgue measure of K. All elements of $\mathcal{T}_h$ will be numbered so that $\mathcal{T}_h = \{K_i\}_{i\in I}$, where $I \subset Z^+ = \{0,1,2,\dots\}$ is a suitable index set. If two elements K_i, $K_j \in \mathcal{T}_h$ contain a nonempty open face which is a part of an $(N-1)$-dimensional hyperplane (i.e. straight line in 2D or plane in 3D), we call them *neighbouring elements* or *neighbours*. In this case we set $\Gamma_{ij} = \partial K_i \cap \partial K_j$ and assume that the whole set Γ_{ij} is a part of an $(N-1)$-dimensional hyperplane. For $i \in I$ we set $s(i) = \{j \in I; K_j \text{ is a neighbour of } K_i\}$. The boundary $\partial\Omega$ is formed by a finite number of faces of elements K_i adjacent to $\partial\Omega$. We denote all these boundary faces by S_j, where $j \in I_b \subset Z^- = \{-1,-2,\dots\}$, and set $\gamma(i) = \{j \in I_b; S_j \text{ is a face of } K_i\}, \Gamma_{ij} = S_j$ for $K_i \in \mathcal{T}_h$ such that $S_j \subset \partial K_i$, $j \in I_b$. For K_i not containing any boundary face S_j we set $\gamma(i) = \emptyset$. Obviously, $s(i) \cap \gamma(i) = \emptyset$ for all $i \in I$. Now, if we write $S(i) = s(i) \cup \gamma(i)$, we have

$$\partial K_i = \bigcup_{j\in S(i)} \Gamma_{ij}, \qquad \partial K_i \cap \partial\Omega = \bigcup_{j\in\gamma(i)} \Gamma_{ij}. \tag{4.6.38}$$

Furthermore, we use the following notation: $\boldsymbol{n}_{ij} = ((n_{ij})_1, \dots, (n_{ij})_N)$ is the unit outer normal to ∂K_i on the face Γ_{ij} ($\boldsymbol{n}_{ij}$ is a constant vector on Γ_{ij}), $d(\Gamma_{ij}) = \text{diam}(\Gamma_{ij})$, and $|\Gamma_{ij}|$ is the $(N-1)$-dimensional Lebesgue measure of Γ_{ij}. See Fig. 4.6.7.

Over the triangulation $\mathcal{T}_h$ we define the *broken Sobolev space*

$$H^k(\Omega,\mathcal{T}_h) = \{v; v|_K \in H^k(K)\ \forall\, K \in \mathcal{T}_h\}. \tag{4.6.39}$$

For $v \in H^1(\Omega,\mathcal{T}_h)$ we introduce the following notation:

$$\begin{aligned} &v|_{\Gamma_{ij}} - \text{the trace of } \ v|_{K_i} \ \text{ on } \ \Gamma_{ij}, \\ &v|_{\Gamma_{ji}} - \text{the trace of } \ v|_{K_j} \ \text{ on } \ \Gamma_{ji} = \Gamma_{ij}. \end{aligned} \tag{4.6.40}$$

The approximate solution of problem (4.6.35)–(4.6.37) is sought in the space of discontinuous piecewise polynomial vector-valued functions $\boldsymbol{S}_h$ defined by

$$\begin{aligned} \boldsymbol{S}_h &= [S_h]^m, \\ S_h &= S^{p,-1}(\Omega, \mathcal{T}_h) = \{v; v|_K \in P^p(K)\ \forall\, K \in \mathcal{T}_h\}, \end{aligned} \tag{4.6.41}$$

where $p \in Z^+$ and $P^p(K)$ denotes the space of all polynomials on K of degree $\leq p$.

Let us assume that $\boldsymbol{w}$ is a classical C^1-solution of system (4.6.35). As usual, by $\boldsymbol{w}(t)$ we denote a function $\boldsymbol{w}(t) : \Omega \to I\!R^m$ such that $\boldsymbol{w}(t)(x) = \boldsymbol{w}(x,t)$ for $x \in \Omega$. In order to derive the discrete problem, we multiply (4.6.35) by a function $\boldsymbol{\varphi} \in H^1(\Omega, \mathcal{T}_h)^m$ and integrate over an element K_i, $i \in I$. With the use of Green's theorem, we obtain the integral identity

$$\begin{aligned} \frac{d}{dt}\int_{K_i} \boldsymbol{w}(t)\cdot\boldsymbol{\varphi}\, dx &- \int_{K_i}\sum_{s=1}^{N} \boldsymbol{f}_s(\boldsymbol{w}(t))\cdot\frac{\partial\boldsymbol{\varphi}}{\partial x_s}\, dx \\ &+ \sum_{j\in S(i)}\int_{\Gamma_{ij}}\sum_{s=1}^{N}\boldsymbol{f}_s(\boldsymbol{w}(t))\cdot\boldsymbol{\varphi}\, n_s\, dS = 0. \end{aligned} \tag{4.6.42}$$

Summing (4.6.42) over all $K_i \in \mathcal{T}_h$, we obtain the identity

$$\begin{aligned} \frac{d}{dt}\sum_{i\in I}\int_{K_i} \boldsymbol{w}(t)\cdot\boldsymbol{\varphi}\, dx &- \sum_{i\in I}\int_{K_i}\sum_{s=1}^{N} \boldsymbol{f}_s(\boldsymbol{w}(t))\cdot\frac{\partial\boldsymbol{\varphi}}{\partial x_s}\, dx \\ &+ \sum_{i\in I}\sum_{j\in S(i)}\int_{\Gamma_{ij}}\sum_{s=1}^{N}\boldsymbol{f}_s(\boldsymbol{w}(t))\cdot\boldsymbol{\varphi}\, n_s\, dS = 0. \end{aligned} \tag{4.6.43}$$

Under the notation

$$(\boldsymbol{w}, \boldsymbol{\varphi}) = \sum_{i\in I}\int_{K_i} \boldsymbol{w}\cdot\boldsymbol{\varphi}\, dx = \int_{\Omega_h} \boldsymbol{w}\cdot\boldsymbol{\varphi} dx \tag{4.6.44}$$

($[L^2]^m$-scalar product) and

$$\begin{aligned} b(\boldsymbol{w}, \boldsymbol{\varphi}) = &- \sum_{i\in I}\int_{K_i}\sum_{s=1}^{N} \boldsymbol{f}_s(\boldsymbol{w})\cdot\frac{\partial\boldsymbol{\varphi}}{\partial x_s}\, dx \\ &+ \sum_{i\in I}\sum_{j\in S(i)}\int_{\Gamma_{ij}}\sum_{s=1}^{N}\boldsymbol{f}_s(\boldsymbol{w})\cdot\boldsymbol{\varphi}\, n_s\, dS, \end{aligned} \tag{4.6.45}$$

(4.6.42) can be written in the form

$$\frac{d}{dt}(\boldsymbol{w}(t), \boldsymbol{\varphi}) + b(\boldsymbol{w}(t), \boldsymbol{\varphi}) = 0. \tag{4.6.46}$$

This equality represents a *weak form* of system (4.6.35) in the sense of the broken Sobolev space $H^1(\Omega, \mathcal{T}_h)$.

4.6.4.2 *Numerical solution* Now we shall introduce the discrete problem approximating identity (4.6.46). For $t \in [0,T]$, the exact solution $\boldsymbol{w}(t)$ will be approximated by an element $\boldsymbol{w}_h(t) \in \boldsymbol{S}_h$. It is not possible to replace $\boldsymbol{w}$ formally in the definition (4.6.45) of the form b, because $\boldsymbol{w}_h$ is discontinuous on Γ_{ij} in general. Similarly as in the FVM we use here the concept of the *numerical flux* $\boldsymbol{H} = \boldsymbol{H}(\boldsymbol{u}, \boldsymbol{v}, \boldsymbol{n})$ and write

$$\int_{\Gamma_{ij}} \sum_{s=1}^{N} \boldsymbol{f}_s(\boldsymbol{w}(t))\, n_s \cdot \boldsymbol{\varphi}\, dS \approx \int_{\Gamma_{ij}} \boldsymbol{H}(\boldsymbol{w}_h|_{\Gamma_{ij}}(t), \boldsymbol{w}_h|_{\Gamma_{ji}}(t), \boldsymbol{n}_{ij}) \cdot \boldsymbol{\varphi}|_{\Gamma ij}\, dS. \tag{4.6.47}$$

We assume that the numerical flux has the properties formulated in Section 3.3.3:

1) $\boldsymbol{H}(\boldsymbol{u}, \boldsymbol{v}, \boldsymbol{n})$ is defined and continuous on $D \times D \times \mathcal{S}_1$, where D is the domain of definition of the fluxes $\boldsymbol{f}_s$ and $\mathcal{S}_1$ is the unit sphere in $\mathbb{R}^N$: $\mathcal{S}_1 = \{\boldsymbol{n} \in \mathbb{R}^N; |\boldsymbol{n}| = 1\}$.
2) $\boldsymbol{H}$ is *consistent*:

$$\boldsymbol{H}(\boldsymbol{u}, \boldsymbol{u}, \boldsymbol{n}) = \mathcal{P}(\boldsymbol{u}, \boldsymbol{n}) = \sum_{s=1}^{N} \boldsymbol{f}_s(\boldsymbol{u})\, n_s, \quad \boldsymbol{u} \in D,\ \boldsymbol{n} \in \mathcal{S}_1. \tag{4.6.48}$$

3) $\boldsymbol{H}$ is *conservative*:

$$\boldsymbol{H}(\boldsymbol{u}, \boldsymbol{v}, \boldsymbol{n}) = -\boldsymbol{H}(\boldsymbol{v}, \boldsymbol{u}, -\boldsymbol{n}), \quad \boldsymbol{u}, \boldsymbol{v} \in D,\ \boldsymbol{n} \in \mathcal{S}_1. \tag{4.6.49}$$

The above considerations lead us to the definition of the approximation b_h of the convective form b:

$$b_h(\boldsymbol{w}, \boldsymbol{\varphi}) = -\sum_{i \in I} \int_{K_i} \sum_{s=1}^{N} \boldsymbol{f}_s(\boldsymbol{w}) \cdot \frac{\partial \boldsymbol{\varphi}}{\partial x_s}\, dx \tag{4.6.50}$$

$$+\sum_{i \in I} \sum_{j \in S(i)} \int_{\Gamma_{ij}} \boldsymbol{H}(\boldsymbol{w}|_{\Gamma_{ij}}, \boldsymbol{w}|_{\Gamma_{ji}}, \boldsymbol{n}_{ij}) \cdot \boldsymbol{\varphi}|_{\Gamma_{ij}} dS, \quad \boldsymbol{w}, \boldsymbol{\varphi} \in H^1(\Omega, \mathcal{T}_h)^m.$$

If $\Gamma_{ij} \subset \partial\Omega_h$, then there is no neighbour K_j of K_i adjacent to Γ_{ij} and the values of $\boldsymbol{w}|_{\Gamma_{ij}}$ must be determined on the basis of *boundary conditions.* In the case of the Euler equations we use the same approach as in the FVM, explained in Section 3.3.6.

By $\boldsymbol{w}_h^0$ we denote an $\boldsymbol{S}_h$-approximation of $\boldsymbol{w}^0$, e.g. the $[L^2]^m$-projection on $\boldsymbol{S}_h$ – cf. (4.6.17).

Now we come to the formulation of the *discrete problem.*

Definition 4.80 *We say that $\boldsymbol{w}_h$ is an approximate solution of (4.6.46), if it satisfies the conditions*

$$a)\ \boldsymbol{w}_h \in C^1([0,T]; \boldsymbol{S}_h), \tag{4.6.51}$$

$$b)\ \frac{d}{dt}(\boldsymbol{w}_h(t), \boldsymbol{\varphi}_h) + b_h(\boldsymbol{w}_h(t), \boldsymbol{\varphi}_h) = 0 \quad \forall\, \boldsymbol{\varphi}_h \in \boldsymbol{S}_h, \forall\, t \in (0,T),$$

$$c)\ \boldsymbol{w}_h(0) = \boldsymbol{w}_h^0.$$

The discrete problem (4.6.51) is equivalent to an initial value problem for a system of ordinary differential equations which can be solved by a suitable time stepping numerical method.

Remark 4.81 If we set $p = 0$ in the DGFEM for the solution of problem (4.6.35)–(4.6.37), which means that we use a piecewise constant approximation of the solution, then we get the finite volume method described in Section 3.3, where we use the notation $\mathcal{D}_h = \mathcal{T}_h$, $D_i = K_i$ and $J = I$, $J_b = I_b$. A comparison of the FVM and DGFEM is discussed in (Feistauer, 2002).

4.6.5 *An example of implementation*

Here we briefly describe the numerical realization of the DGFEM in the special case of piecewise linear elements over a triangular grid.

Let $\Omega \subset I\!R^2$ and let $\mathcal{T}_h$ be a triangulation of a polygonal approximation Ω_h of Ω satisfying the conditions from Section 4.1. (Thus, $\mathcal{T}_h$ is a regular triangulation.) On $\mathcal{T}_h$ we consider the DGFE method (4.6.51) with $p = 1$. Due to the regularity of $\mathcal{T}_h$, each interface Γ_{ij} is a whole side of neighbouring triangles K_i and K_j or $\Gamma_{ij} \subset \partial\Omega_h$ is a whole side of a triangle K_i adjacent to $\partial\Omega_h$. By Q_{ij} we denote the midpoint of Γ_{ij}.

In S_h we use the basis formed by functions associated with midpoints Q_{ij} of Γ_{ij}:

$$\{\phi_{ij} \in S_h; \phi_{ij}(Q_{kl}) = \delta_{ik}\delta_{jl}, \ j \in S(i), \, l \in S(k), \, i, k \in I\}. \tag{4.6.52}$$

Then the basis in the space $\boldsymbol{S}_h$ is formed by the system

$$\begin{aligned}\{\boldsymbol{\phi}_{ij}^1 = (\phi_{ij}, 0, \dots, 0)^{\mathrm{T}}, \boldsymbol{\phi}_{ij}^2 = (0, \phi_{ij}, \dots, 0)^{\mathrm{T}}, \dots,\\ \boldsymbol{\phi}_{ij}^m = (0, \dots, 0, \phi_{ij})^{\mathrm{T}} \in \boldsymbol{S}_h; i \in I, j \in S(i)\}.\end{aligned} \tag{4.6.53}$$

Obviously, the dimension of the space $\boldsymbol{S}_h$ is $3m \times \mathrm{card}(I)$, where $\mathrm{card}(I)$ is the number of elements of the set I.

The basis functions ϕ_{ij} are mutually L^2-orthogonal and, hence, also the functions $\boldsymbol{\phi}_{ij}^r$ are $[L^2]^m$-orthogonal. This implies that the *mass matrix* with entries $(\boldsymbol{\phi}_{ij}^r, \boldsymbol{\phi}_{k\ell}^s)$ is diagonal.

Now, if we substitute $\boldsymbol{\varphi}_h := \boldsymbol{\phi}_{ij}^r, r = 1, \dots, m, i \in I, j \in S(i)$, into (4.6.51), we obtain a system of ordinary differential equations for unknown scalar functions $W_{r;ij}(t) = (w_h)_{r;i}(Q_{ij}, t), r = 1, \dots, m, i \in I, t \in [0, T]$, where $(w_h)_{r;i}$ is the r-th component of the vector function $\boldsymbol{w}_h$ restricted on $K_i \times [0, T]$.

Obviously, we can write

$$\boldsymbol{w}_h(x, t) = \sum_{\substack{r=1,\dots,m\\ j \in S(i)}} W_{r;ij}(t)\, \boldsymbol{\phi}_{ij}^r(x), \quad x \in K_i, \ i \in I. \tag{4.6.54}$$

The integrals in the definition of the scalar product $(\cdot, \cdot)$ and the form b_h are computed with the aid of numerical integration. The *line integral* along Γ_{ij} is transformed to an integral over the interval $(-1, 1)$ with the use of the linear parametrization

$$X = Q_{ij} + \xi(B_{ij} - A_{ij})/2, \quad \xi \in (-1, 1), \tag{4.6.55}$$

where A_{ij} and B_{ij} are the end points of Γ_{ij}. Then $Q_{ij} = (A_{ij} + B_{ij})/2$. The resulting integral is evaluated by the two-point Gauss quadrature rule (see (Ueberhuber, 1997) or (Ralston, 1965))

$$\int_{-1}^{1} g(t)\,dt \approx g\Big(-\frac{1}{\sqrt{3}}\Big) + g\Big(\frac{1}{\sqrt{3}}\Big), \tag{4.6.56}$$

which is exact for polynomials of the third degree. The *volume integrals* are evaluated by the three-point integration rule

$$\int_{K_i} g(x)\,dx \approx \frac{1}{3}|K_i| \sum_{j\in S(i)} g(Q_{ij}), \tag{4.6.57}$$

which is exact for second degree polynomials (Ciarlet, 1979), Section 4.1. Therefore, the application of (4.6.57) to the evaluation of the scalar product (v_h, φ_h) is exact for $v_h, \varphi_h \in S_h = S_h^{1,-1}(\Omega, \mathcal{T}_h)$. The numerical integration in the form b_h gives its approximation $\tilde{b}_h$, which is reflected in the form of the resulting system of ordinary differential equations.

The initial value problem for this system can be written in the form (4.6.21)–(4.6.22) and solved by Runge–Kutta schemes from 3.2.2, under a suitable CFL stability condition. The determination of this condition is a difficult task and, therefore, analogy with the FVM (see 3.3.7) and theoretical results obtained in (Cockburn *et al.*, 1990) for a scalar conservation law is used:

$$\tau_k \frac{|\partial K_i|}{|K_i|} \lambda_{i,\max} \le CFL, \quad i \in I, \tag{4.6.58}$$

limiting the time step τ_k. Here

$$\lambda_{i,\max} = \max\nolimits_{r=1,\dots,m,j\in s(i)} |\lambda_r(\boldsymbol{w}_h^k|_{\Gamma_{ij}}(Q_{ij}), \boldsymbol{n}_{ij})|, \tag{4.6.59}$$

where $\lambda_r(\boldsymbol{w}, \boldsymbol{n})$, $r = 1, \dots, m$, are eigenvalues of the matrix $\mathbb{P}(\boldsymbol{w}, \boldsymbol{n})$ and $\boldsymbol{w}_h^k$ denotes the approximate solution at time t_k. In the case of the Euler equations we set

$$\lambda_{i,\max} = \max\nolimits_{j\in s(i)} (|\boldsymbol{v}_h^k| + a_h^k)|_{\Gamma_{ij}}(Q_{ij}). \tag{4.6.60}$$

Here $\boldsymbol{v}_h^k$ and a_h^k denote the approximations of the velocity vector and speed of sound, respectively, at the time level t_k.

Exercise 4.82 Prove the following assertions:

a) The basis functions ϕ_{ij}, $i \in I, j \in S(i)$, of the space S_h are mutually L^2-orthogonal.

b) The basis functions ϕ_{ij}^r, $r = 1, \dots, m, i \in I, j \in S(i)$, of the space $\boldsymbol{S}_h$ are mutually $[L^2]^m$-orthogonal.

c) Using the representation (4.6.54) of the approximate solution, construct a system of ordinary differential equations of the form (4.6.21) equivalent to identity (4.6.51), b).

4.6.6 *Problem with periodic boundary conditions*

A number of test problems for nonlinear conservation laws use, for simplicity, periodic boundary conditions. (For a 1D example, see Section 4.6.1.) Here we describe the formulation and DGFE discretization of system (4.6.35) considered in the space-time cylinder $Q_T = \Omega \times (0, T)$, where $\Omega = (-1, 1)^N \subset I\!R^N$ and equipped with the initial condition (4.6.36) and periodic boundary conditions:

$$u(x_1, \dots, x_{i-1}, -1, x_{i+1}, \dots, x_N, t) = u(x_1, \dots, x_{i-1}, 1, x_{i+1}, \dots, x_N, t), \tag{4.6.61}$$
$$x_1, \dots, x_{i-1}, x_{i+1}, \dots, x_n \in (-1, 1),\ t \in (0, T),\ i = 1, \dots, N.$$

For the discretization we use a partition $\mathcal{T}_h = \{K_i\}_{i \in I}$ of the domain Ω with properties from Section 4.6.4.1. Similarly as above, we denote by $\Gamma_{ij}, j \in s(i)$, the interfaces of K_i common with its neighbours K_j. In order to realize the periodic boundary conditions in the discrete problem, it is necessary to associate elements K_i adjacent to $\partial\Omega$ with suitable elements lying on 'the opposite side' of $\partial\Omega$. Let us denote $\boldsymbol{e} = (\delta_{1s}, \dots, \delta_{ss}, \dots, \delta_{Ns})$ (δ_{ij} = Kronecker delta) and let for some K_i, $K_j \in \mathcal{T}_h$ the sets

$$\begin{aligned} &S_i^+ \\ &= \partial K_i \cap \{(x_1, \dots, x_{s-1}, 1, x_{s+1}, \dots, x_n); x_1, \dots, x_{s-1}, x_{s+1}, \dots, x_N \in (-1, 1)\}, \\ &S_j^- \\ &= \partial K_j \cap \{(x_1, \dots, x_{s-1}, -1, x_{s+1}, \dots, x_n); x_1, \dots, x_{s-1}, x_{s+1}, \dots, x_N \in (-1, 1)\}, \end{aligned} \tag{4.6.62}$$

have positive $(N-1)$-dimensional measure. (Then S_i^+ and S_j^- are faces of elements K_i and K_j respectively, adjacent to $\partial\Omega$ and lying on 'the opposite sides' of $\partial\Omega$.) Now we set

$$\begin{aligned} \Gamma_{ij} &= S_i^+ \cap \{x + 2\boldsymbol{e}_s; x \in S_j^-\}, \\ \Gamma_{ji} &= \{x - 2\boldsymbol{e}_s; x \in \Gamma_{ij}\} \end{aligned} \tag{4.6.63}$$

and call K_i and K_j *boundary neighbours*, if Γ_{ij} and Γ_{ji} have positive $(N-1)$-dimensional measure. For any element K_i we denote

$$\begin{aligned} \gamma(i) &= \{j \in I; K_j \text{ is a boundary neighbour of } K_i\}, \\ S(i) &= s(i) \cup \gamma(i). \end{aligned} \tag{4.6.64}$$

Obviously, again we have

$$\partial K_i = \bigcup_{j \in S(i)} \Gamma_{ij} \tag{4.6.65}$$

and the notation from 4.6.4.1 makes sense.

Now, using the process described in Sections 4.6.4.1 and 4.6.4.2, we obtain the DGFE discretization of problem (4.6.35), (4.6.36) and (4.6.61).

Exercise 4.83 Construct the system of ordinary differential equations equivalent to the discrete problem (4.6.35), (4.6.36) and (4.6.61).

4.6.7 *Limiting of the order of accuracy*

Let us return to the discrete problem (4.6.51), equivalent to a system of ordinary differential equations. The simplest way to obtain a fully discrete problem is to use the *Euler forward method.* To this end, we consider a partition $0 = t_0 < t_1 < t_2 < \dots$ of the time interval $(0, T)$ and set $\tau_k = t_{k+1} - t_k$.

Using the approximations $\boldsymbol{w}_h^k \approx \boldsymbol{w}_h(t_k)$, $\frac{d}{dt}(\boldsymbol{w}_h(t), \boldsymbol{\varphi}_h) \approx (\boldsymbol{w}_h^{k+1} - \boldsymbol{w}_h^k, \boldsymbol{\varphi}_h)/\tau_k$, we obtain the *fully discrete problem*: for each $k \geq 0$ find $\boldsymbol{w}_h^{k+1}$ such that

$$\begin{aligned}
&\text{a)}\ \ \boldsymbol{w}_h^{k+1} \in \boldsymbol{S}_h, \\
&\text{b)}\ \ (\boldsymbol{w}_h^{k+1}, \boldsymbol{\varphi}_h) = (\boldsymbol{w}_h^k, \boldsymbol{\varphi}_h) - \tau_k b_h(\boldsymbol{w}_h^k, \boldsymbol{\varphi}_h) \quad \forall \boldsymbol{\varphi}_h \in \boldsymbol{S}_h.
\end{aligned} \tag{4.6.66}$$

More precise time discretization is obtained with the aid of the Runge–Kutta methods.

As we have mentioned several times, the disadvantage of higher order schemes is the rise of the Gibbs phenomenon manifested by nonphysical spurious oscillations, undershoots and overshoots in the approximate solution in the vicinity of discontinuities or steep gradients. In order to cure this undesirable feature, in the framework of higher order finite volume methods one uses suitable limiting procedures, described in Sections 3.2.21.3 and 3.5.4. They should preserve the higher order of accuracy of the method in regions, where the solution is regular, and decrease the order to 1 in a neighbourhood of discontinuities or steep gradients. As we mentioned in Chapter 3, there exists a rigorous method using the concept of a flux limiter for finite difference or FV schemes applied to scalar conservation laws with one space variable over uniform meshes. (See, for example, (Sweby, 1984).) In (Cockburn and Shu, 1989) and (Cockburn *et al.*, 1990) it was proposed to generalize the FV limiting procedures also to DGFEM. However, as numerical experiments show, the extension of this approach to FV schemes for problems with several space variables and/or nonuniform unstructured meshes can lead to an undesired decrease of accuracy in the whole computational domain. (Cf. Section 3.5.6.) There is naturally also a danger that a similar effect appears in the DGFEM applied on nonuniform unstructured meshes.

Here we present a different type of limiting proposed in (Dolejší *et al.*, 2003) which keeps the higher order of accuracy in the region of regularity of the solution, because the original higher order scheme remains unchanged in this region and is modified to a first order scheme only in the vicinity of steep gradients or discontinuities. We explain this approach for a 2D situation.

Let us consider a fully discrete problem (4.6.66) with $N = 2$, $\Omega \subset I\!R^2$, for problem (4.6.35)–(4.6.37) discretized by piecewise linear elements (i.e. $p = 1$). The use of the Euler forward time discretization (4.6.66) yields the formal accuracy of order $O(\tau + h^2)$, but the poor time approximation can be 'overkilled' by the choice $\tau = O(h^2)$. We assume, of course, that $h \to 0+$ and, thus, $0 < h < 1$.

Let us denote by u_h^k some scalar quantity characterizing the approximate solution $\boldsymbol{w}_h^k$. (For example, for the Euler equations we choose this quantity as the density ρ.) By $[u_h^k]_{\Gamma_{ij}}$ we denote the jump of u_h^k on Γ_{ij}, i.e. $[u_h^k]_{\Gamma_{ij}} = u_h^k|_{\Gamma_{ij}} - u_h^k|_{\Gamma_{ji}}$. Further, we define $[u_h^k]_{\partial K_i}$ as a function on ∂K_i such that $[u_h^k]_{\partial K_i}(x) = [u_h^k]_{\Gamma_{ij}}(x)$ for $x \in \partial K_i \cap \Gamma_{ij}$. Numerical experiments show that the interelement jumps in the approximate solution are of the order $O(1)$ on discontinuities, but $O(h^2)$ in the regions where the solution is regular. On general unstructured grids it appears to be suitable to measure the magnitude of interelement jumps in the integral form by

$$\int_{\partial K_i} [u_h^k]^2 \, dS, \quad K_i \in \mathcal{T}_h. \tag{4.6.67}$$

In areas where the solution is regular, we have

$$g^1(i) = \int_{\partial K_i} [u_h^k]^2 \, dS/h^5 \approx \int_{\partial K_i} (O(h^2))^2 \, dS/h^5 \approx O(1), \tag{4.6.68}$$

whereas

$$g^2(i) = \int_{\partial K_i} [u_h^k]^2 \, dS/h \approx \int_{\partial K_i} (O(1))^2 \, dS/h \approx O(1) \tag{4.6.69}$$

for very steep gradients (discontinuities). These results lead us to the idea that the switch between the higher order scheme and its first order modification should be tested with the aid of the indicator

$$g(i) = \int_{\partial K_i} [u_h^k]^2 \, dS/h^\alpha, \quad K_i \in \mathcal{T}_h, \tag{4.6.70}$$

where $\alpha \in (1,\, 5)$. We choose, for example, $\alpha = 5/2$. However, in general, using unstructured grids, it is suitable to define the indicator $g(i)$ in terms of h_{K_i} and $|K_i|$. For $\alpha = 5/2$ we set

$$g(i) = \int_{\partial K_i} [u_h^k]^2 \, dS/(h_{K_i}|K_i|^{3/4}), \quad K_i \in \mathcal{T}_h. \tag{4.6.71}$$

We call $g(i)$ a *discontinuity indicator.*

Now we define an *adaptive strategy* for an *automatic limiting* of the order of accuracy of scheme (4.6.66):

$$\begin{aligned} &\text{a)}\ \ \boldsymbol{w}_h^{k+1} \in \boldsymbol{S}_h = \boldsymbol{S}_h^{1,-1}(\Omega, \mathcal{T}_h), \\ &\text{b)}\ \ (\boldsymbol{w}_h^{k+1}, \boldsymbol{\varphi}) = (\tilde{\boldsymbol{w}}_h^k, \boldsymbol{\varphi}_h) - \tau_k\, b_h(\tilde{\boldsymbol{w}}_h^k, \boldsymbol{\varphi}_h) \quad \forall\, \boldsymbol{\varphi}_h \in \boldsymbol{S}_h, \end{aligned} \tag{4.6.72}$$

where $\tilde{\boldsymbol{w}}_h^k$ is the modification of $\boldsymbol{w}_h^k$ defined with the aid of our limiting strategy in the following way:

$$\text{a)}\ \ \text{Set}\ \tilde{\boldsymbol{w}}_h^k|_{K_i} := \boldsymbol{w}_h^k|_{K_i}, \quad \forall\, i \in I. \tag{4.6.73}$$

b) If $g(i) > 1$ for some $i \in I$, then $\tilde{\boldsymbol{w}}_h^k|_{K_i} := \pi_0 \boldsymbol{w}_h^k|_{K_i}$,

where π_0 is the $[L^2]^m$-projection operator $\pi_0 : [L^2(\Omega)]^m \to [S^{0,-1}(\Omega, \mathcal{T}_h)]^m$ (= the space of piecewise constant vector functions): if $\boldsymbol{v} \in [L^2(\Omega)]^m$, then $\pi_0 \boldsymbol{v} \in [S^{0,-1}(\Omega, \mathcal{T}_h)]^m$ is defined by

$$(\pi_0 \boldsymbol{v}, \boldsymbol{\varphi}_h) = (\boldsymbol{v}, \boldsymbol{\varphi}_h) \quad \forall \boldsymbol{\varphi}_h \in [S^{0,-1}(\Omega, \mathcal{T}_h)]^m. \tag{4.6.74}$$

Hence,

$$\pi_0 \boldsymbol{v}|_{K_i} = \int_{K_i} \boldsymbol{v} \, dx / |K_i|, \quad i \in I. \tag{4.6.75}$$

The described procedure means that in (4.6.73) the limiting of the order of the scheme is applied on elements lying on the discontinuity or in the area with a very steep gradient via the piecewise constant approximation of the numerical solution just on the chosen elements. In other areas, where the solution is regular, the numerical scheme is unchanged and the higher order of accuracy is preserved. (The extension to the case $N = 3$ is straightforward.)

The above approach to the *adaptive limiting* was developed in (Dolejší *et al.*, 2003) and (Dolejší *et al.*, 2002*b*). In (Dolejší *et al.*, 2002*b*), a detailed numerical investigation and verification of this algorithm was carried out. A theoretical justification is still missing.

Example 4.84 In order to demonstrate the applicability of the described limiting procedure, let us consider the scalar 2D Burgers equation

$$\frac{\partial u}{\partial t} + u \frac{\partial u}{\partial x_1} + u \frac{\partial u}{\partial x_2} = 0 \quad \text{in } \Omega \times (0, T), \tag{4.6.76}$$

where $\Omega = (-1, 1) \times (-1, 1)$, equipped with initial condition

$$u^0(x_1, x_1) = 0.25 + 0.5 \sin(\pi(x_1 + x_2)), \quad (x_1, x_2) \in \Omega, \tag{4.6.77}$$

and periodic boundary conditions (4.6.61). The exact entropy solution of this problem becomes discontinuous for $t \geq 0.3$ (see Exercise 4.87). In Fig. 4.6.8, the graph of the exact solution at time $t = 0.45$ is plotted. If we apply scheme (4.6.66) to this problem on the mesh from Fig. 4.6.9, with time step $\tau = 2.5 \cdot 10^{-4}$, we obtain the numerical solution shown in Fig. 4.6.10. It can be seen here that the numerical solution contains spurious oscillations near discontinuities.

If we apply the limiting procedure described in Section 4.6.7 to our problem, we obtain the numerical solution displayed in Fig. 4.6.11. We see that the adaptive limiting algorithm (4.6.72)–(4.6.73) avoids spurious oscillations and gives accurate approximate solution in areas where the exact solution is regular. In Fig. 4.6.12, the graphs of the solution along the diagonal $x_1 = x_2$ are plotted:

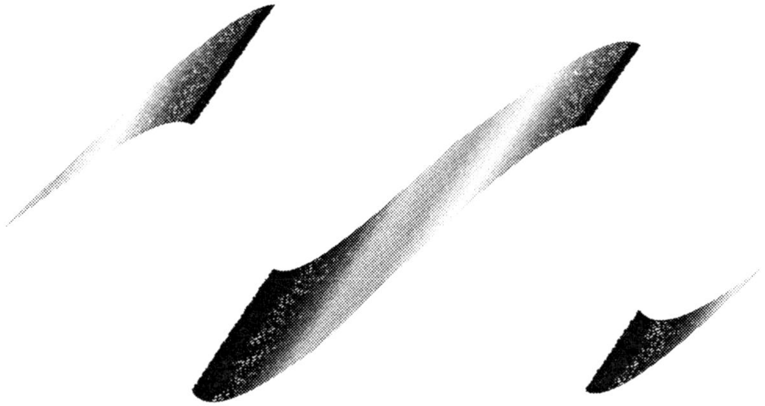

FIG. 4.6.8. Exact solution of the problem from Example 4.84 plotted at $t = 0.45$

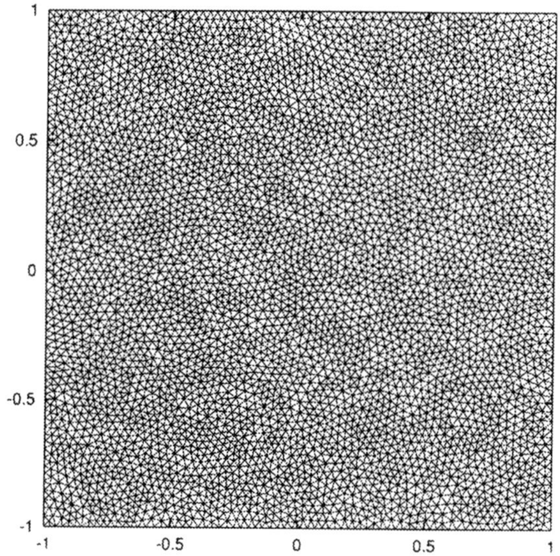

FIG. 4.6.9. Triangulation used for the numerical solution

DGFEM (left), exact solution (centre), DGFEM with limiting (right). In our numerical experiments we use the numerical flux of the form

$$H(u_1, u_2, \boldsymbol{n}) = \begin{cases} \sum_{s=1}^{2} f_s(u_1)\, n_s, \text{ if } A > 0, \\ \sum_{s=1}^{2} f_s(u_2)\, n_s, \text{ if } A \le 0, \end{cases} \tag{4.6.78}$$

where $f_s(u) = u^2/2$, $A = \sum_{s=1}^{2} f_s'((u_1+u_2)/2)\, n_s$. In Figs 4.6.13 and 4.6.14 three

FIG. 4.6.10. Numerical solution of the problem from Example 4.84 computed by DGFEM, plotted at $t = 0.45$

unstructured meshes are displayed, which were used for the numerical solution of problem (4.6.76), (4.6.77), (4.6.61). By + we denote the elements, where the order limiting was used for the computation of the approximate solution at time $t = 0.45$. It can be clearly seen that this limiting is applied in a narrow neighbourhood of discontinuities.

Remark 4.85 Note that the scheme with numerical flux (4.6.78) contains a partial upwinding (such as the Vijayasundaram method – see Sections 3.2.17, 3.3.4). Actually, if we apply the FVM with the numerical flux (4.6.78) to a linear scalar conservation law

$$\frac{\partial u}{\partial t} + a_1 \frac{\partial u}{\partial x_1} + a_2 \frac{\partial u}{\partial x_2} = 0, \quad a_1, a_2 \in \mathbb{R} \tag{4.6.79}$$

on a uniform quadrilateral mesh, we obtain a standard *upwind scheme* – see Exercise 4.86. Let us recall that upwinding used in FV schemes causes that their order of accuracy does not exceed one. There is an amazing fact that the use

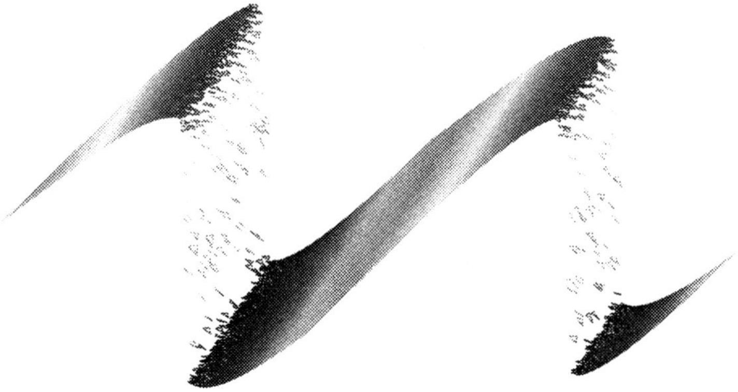

FIG. 4.6.11. Numerical solution of the problem from Example 4.84 computed by DGFEM with limiting, plotted at $t = 0.45$

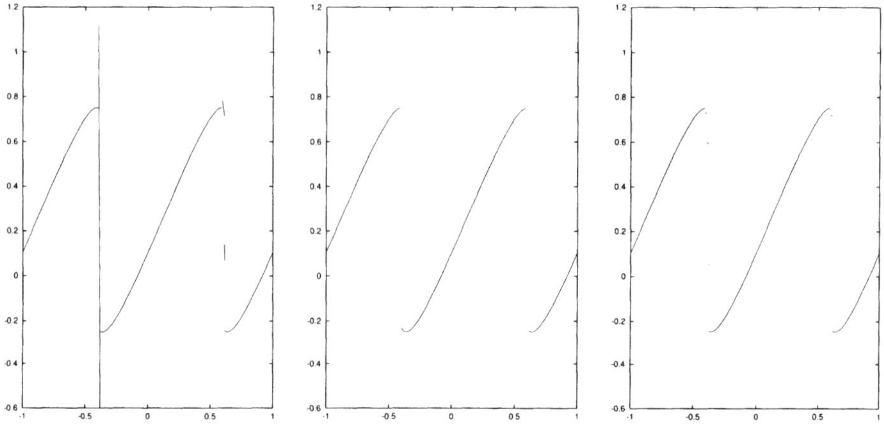

FIG. 4.6.12. Inviscid Burgers equation: graphs along the diagonal, DGFEM (left), exact solution (centre), DGFEM with limiting (right) plotted at $t = 0.45$

of an upwind numerical flux in a DG method with $p \geq 1$ yields a higher order scheme. We could have observed it already in Section 4.1.11, where some sort of upwinding was used for the derivation of a DGFEM applied to a linear first order equation.

Exercise 4.86 a) Verify that, provided $f_s \in C^1(\mathbb{R})$, $s = 1, 2$, the numerical flux (4.6.78) is continuous, consistent and conservative in the sense of (4.6.48)–(4.6.49).

b) Derive the FV scheme (i.e. DGFEM with $p = 0$) with the numerical flux

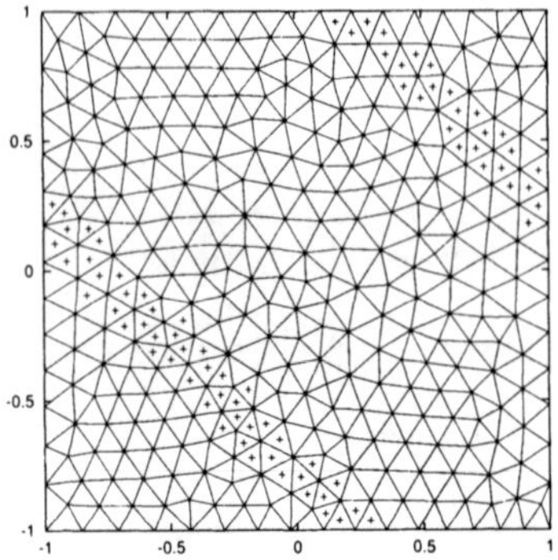

FIG. 4.6.13. Triangulation mesh.500 with marked triangles with $g(i) > 1$

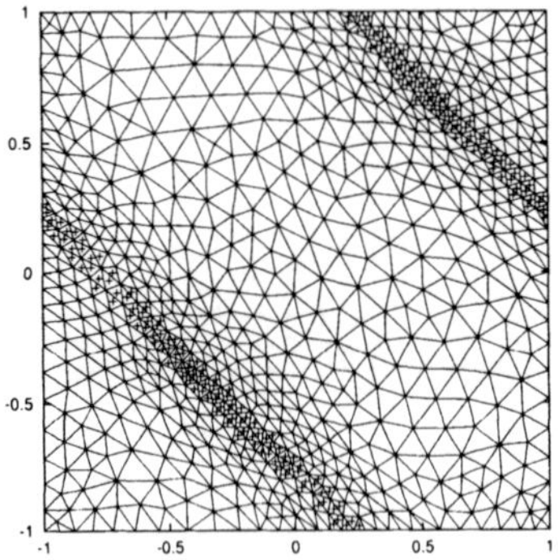

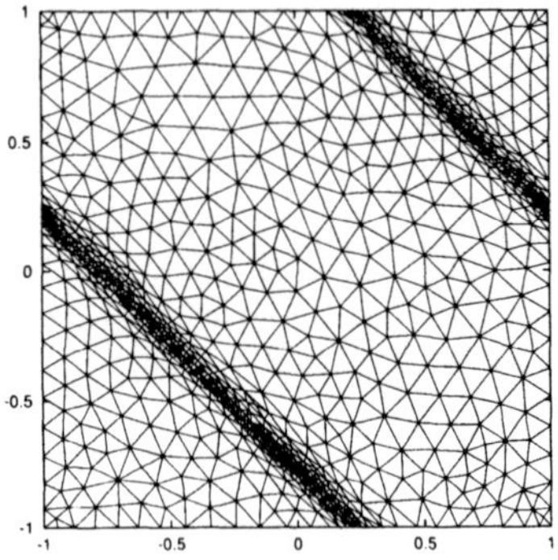

FIG. 4.6.14. Examples of locally refined triangulations with marked triangles satisfying the condition $g(i) > 1$

(4.6.78) applied to equation (4.6.79) on a uniform quadrilateral mesh and explain why we speak of upwinding here. (Cf. (3.3.82).)

Exercise 4.87 In analogy to Exercise 4.79 derive a method for obtaining the exact solution of problem (4.6.76), (4.6.77), (4.6.61).
Hint: The transformation of coordinates $\tilde{x}_1 = x_1 + x_2, \tilde{x}_2 = x_1 - x_2$ and the assumption that $\partial/\partial\tilde{x}_2 \equiv 0$ lead to the Burgers equation (4.6.31) with one space variable $\tilde{x}_1$. Now it is possible to apply the procedure from Exercise 4.79.

The convergence of DGFE approximate solutions to the unique weak entropy solution of a scalar conservation law and error estimates were analysed in (Jaffre *et al.*, 1995) and (Cockburn and Gremaud, 1996*b*), where stabilization artificial diffusion terms were introduced. This type of stabilization was applied in (Hartmann and Houston, 2001*b*) to the solution of the stationary Euler equations.

The stationary approximate solution is defined by the modification of scheme (4.6.51) in the form

$$b_h(\boldsymbol{w}_h, \varphi_h) = \sum_{K \in \mathcal{T}_h} \int_K C h^{2-\beta} \Big| \sum_{s=1}^{N} \frac{\partial \boldsymbol{f}_s(\boldsymbol{w}_h)}{\partial x_s} \Big| \sum_{i=1}^{N} \frac{\partial \boldsymbol{w}_h}{\partial x_i} \cdot \frac{\partial \varphi_h}{\partial x_i}\, dx, \quad (4.6.80)$$
$$\forall \varphi_h \in \boldsymbol{S}_h,$$

where $C > 0$ and $\beta \in (0, 1/2)$ are constants and h is considered here as a function of x such that $h|_K = h_K$ for $K \in \mathcal{T}_h$. (Compare this stabilization with those introduced in Section 4.3.5.6.)

4.6.8 *Approximation of the boundary*

In the FVM applied to conservation laws or in the FEM using piecewise linear approximations applied to elliptic or parabolic problems, it is sufficient to use a polygonal or polyhedral approximation Ω_h of the 2D or 3D domain Ω, respectively. However, numerical experiments show that in some cases the DGFEM does not give a good resolution in the neighbourhood of curved parts of the boundary $\partial\Omega$, if the mentioned approximations of Ω are used. In 1997, Bassi and Rebay (Bassi and Rebay, 1997*b*) showed the importance of a sufficiently accurate approximation of the boundary $\partial\Omega$. For example, if a polygonal approximation of a plane domain is used in the case of flow past a cylinder, then each of the vertices of the polygon introduces nonphysical entropy production and the approximate solution presents a nonphysical wake which does not disappear by further refining the grid.

In order to get a good quality numerical solution, it is necessary to use a sufficiently precise approximation of the boundary. We shall discuss this topic for a 2D problem and piecewise linear elements.

4.6.8.1 *Superparametric elements at a curved boundary* Let $\Omega \subset I\!R^2$ and $\mathcal{T}_h$ be its partition formed by triangles K_i, $i \in I$. Let $\{K_i,\ i \in I_c\}$ with $I_c \subset I$ be a set of triangles adjacent to a curved part of $\partial\Omega$, see Fig. 4.6.15. For $i \in I_c$ let P_i^k, $k = 0, 1, 2$, be the vertices of K_i and let $P_i^0 \in \Omega, P_i^1, P_i^2 \in \partial\Omega$. We suppose that the centre $P_i^{12} \in \partial\Omega$ of the curved side with end points P_i^1, P_i^2 is close to the centre of the straight segment $P_i^1 P_i^2$. Then there exists a unique *bilinear* mapping $F_i : \hat{K} \to K_i, F_i = (F_i^1, F_i^2)$ such that

$$F_i(\hat{P}^k) = P_i^k, \quad k = 0, 1, 2, \qquad (4.6.81)$$
$$F_i(\hat{P}^{12}) = P_i^{12},$$

where $\hat{K}$ is a reference triangle with vertices

$$\hat{P}^0 = [0; 0], \quad \hat{P}^1 = [1; 0], \quad \hat{P}^2 = [0; 1] \qquad (4.6.82)$$

and

$$\hat{P}^{12} = [1/2; 1/2], \qquad (4.6.83)$$

see Fig. 4.6.16. Then triangles K_i, $i \in I_c$, are replaced by the curved triangles

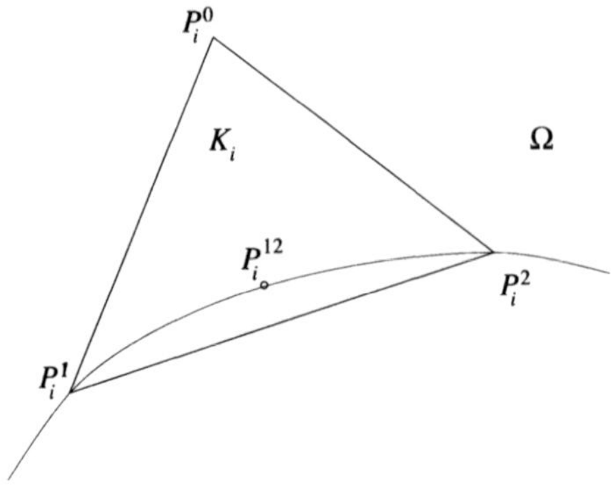

FIG. 4.6.15. Triangle K_i lying on a curved part of $\partial\Omega$

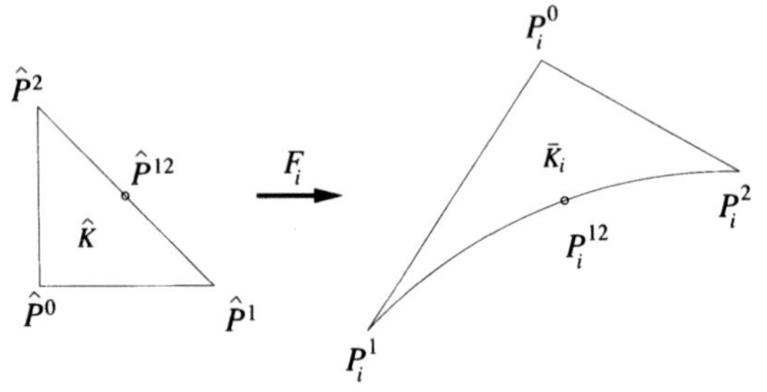

FIG. 4.6.16. Bilinear mapping $F_i : \hat{K} \to K_i$

defined by

$$\bar{K}_i = F_i(\hat{K}). \tag{4.6.84}$$

The set $\bar{K}_i$ is a plane figure having two straight sides and one curved side. If $i \notin I_c$ then F_i is a linear mapping and, therefore, $\bar{K}_i = K_i$.

Now let us describe how to evaluate the volume and boundary integrals over elements $\bar{K}_i$, $i \in I_c$, and their sides Γ_{ij}. We denote by

$$J_{F_i}(\hat{x}) \equiv \frac{D\,F_i}{D\,\hat{x}}(\hat{x}), \quad \hat{x} \in \hat{K}, \tag{4.6.85}$$

the Jacobi matrix of the mapping F_i. Since F_i is bilinear, we easily find that $\det J_{F_i}$ is a linear function of the variable $\hat{x}$. The vector-valued test functions $\boldsymbol{\varphi}_h$ from (4.6.51) are defined on the boundary elements $\bar{K}_i$ with the aid of the mapping F_i:

$$\boldsymbol{\varphi}_h(x) = \hat{\boldsymbol{\varphi}}(F_i^{-1}(x)), \quad x \in \bar{K}_i, \tag{4.6.86}$$

where $\hat{\boldsymbol{\varphi}} \in P^1(\hat{K})^m$. This means that the components of $\hat{\boldsymbol{\varphi}}$ are linear functions on $\hat{K}$. The approximate solution $\boldsymbol{w}_h(\cdot, t)$ is defined in a similar way on the element $\bar{K}_i$:

$$\boldsymbol{w}_h(x,t) = \hat{\boldsymbol{w}}_i(F_i^{-1}(x), t), \quad x \in \bar{K}_i, \ t \in [0,T], \tag{4.6.87}$$

where $\hat{\boldsymbol{w}}_i(\cdot, t) \in P^1(\hat{K})^m$.

Then the forms in (4.6.51) are evaluated in the following way.
The L^2-scalar product is expressed as

$$\int_{K_i} \boldsymbol{w}_h(x,t) \cdot \boldsymbol{\varphi}_h(x)\, dx = \int_{\hat{K}} \hat{\boldsymbol{w}}_i(\hat{x},t) \cdot \hat{\boldsymbol{\varphi}}(\hat{x}) \det J_{F_i}(\hat{x}) d\hat{x}, \quad i \in I. \tag{4.6.88}$$

The volume integrals in the form b_h in (4.6.50) are written in the form

$$\begin{aligned} &\int_{K_i} \sum_{s=1}^{2} \boldsymbol{f}_s(\boldsymbol{w}_h(x,t)) \cdot \frac{\partial \boldsymbol{\varphi}_h(x)}{\partial x_s}\, dx \qquad (4.6.89) \\ &= \int_{\hat{K}} \sum_{s=1}^{2} \boldsymbol{f}_s(\hat{\boldsymbol{w}}_i(\hat{x},t)) \cdot \sum_{j=1}^{2} \frac{\partial \hat{\boldsymbol{\varphi}}(\hat{x})}{\partial \hat{x}_j} \frac{\partial (F_i^{-1})^j(F_i(\hat{x}))}{\partial x_s} \det J_{F_i}(\hat{x})\, d\hat{x}, \quad i \in I, \end{aligned}$$

where $(F_i^{-1})^j$ denotes the j-th component of the inverse mapping F_i^{-1}. In order to compute the inverse mapping F_i^{-1}, we use the following relation written in the matrix form:

$$\frac{D F_i^{-1}}{Dx}(F_i(\hat{x})) = \left[\frac{D F_i}{D\hat{x}}(\hat{x})\right]^{-1} \tag{4.6.90}$$

following from the identity $x = F_i(F_i^{-1}(x))$. The computation of the inverse matrix in (4.6.90) is simpler than the evaluation of F_i^{-1}.

Further, the boundary integral over a curved side $\Gamma_{ij} \subset \partial K_i$ in the form b_h in (4.6.50) is computed with the aid of a suitable parametrization of Γ_{ij} and the side $\hat{\Gamma}$ of $\hat{K}$ corresponding to Γ_{ij} in the mapping F_i:

$$x = x(\xi) = F_i(\hat{x}(\xi)), \quad \xi \in [-1,1].$$

Then, if we put

$$u(x) = \boldsymbol{H}(\boldsymbol{w}_h|_{\Gamma_{ij}}(x,t), \boldsymbol{w}_h|_{\Gamma_{ji}}(x,t), \boldsymbol{n}_{ij}) \cdot \boldsymbol{\varphi}_h(x)$$

(for a fixed t), we get

$$\begin{aligned} &\int_{\Gamma_{ij}} u(x) dS = \int_{-1}^{1} u(x(\xi))|x'(\xi)| d\xi \qquad (4.6.91) \\ &= \int_{-1}^{1} u(F_i(\hat{x}(\xi))) \Big\{ \sum_{j=1}^{2} \Big(\sum_{k=1}^{2} \frac{\partial F_i^j(\hat{x}(\xi))}{\partial \hat{x}_k} \hat{x}_k'(\xi) \Big)^2 \Big\}^{1/2} d\xi. \end{aligned}$$

The parametrization $\hat{x} = \hat{x}(\xi)$ of $\hat{\Gamma}$ is expressed in the form

$$\hat{x}(\xi) = \hat{Q} + \xi(\hat{B} - \hat{A})/2,$$

where $\hat{Q}$ is the midpoint of $\hat{\Gamma}$ and $\hat{A}, \hat{B}$ are the end points of $\hat{\Gamma}$. The integrals over the reference edge $\hat{\Gamma}$ and the reference triangle $\hat{K}$ in (4.6.88), (4.6.89), (4.6.91) are evaluated with the aid of the quadrature formula (4.6.56) and the quadrature formula on $\hat{K}$ (similar to (4.6.57)), using midpoints of sides as integration points, respectively.

Since the polynomial mapping of the reference triangle $\hat{K}$ onto an element K_i has a higher degree than the degree of basis and trial functions on $\hat{K}$, we speak of *superparametric elements* in contrast to isoparametric elements mentioned in Section 4.1.4. The use of superparametric elements at a curved boundary essentially improves the quality of the solution in a neighbourhood of curved parts of the boundary as shown below in Section 4.6.10, Figs 4.6.18 and 4.6.19. If one uses higher degree elements, i.e. $p > 1$, then it is necessary to increase the polynomial degree in the geometric mapping F_i. Various combinations of polynomial approximations on the reference element $\hat{K}$ and the geometric transformations F_i of $\hat{K}$ onto $\bar{K}_i$ were investigated in (Bassi and Rebay, 1997*b*).

4.6.9 *DGFEM for convection–diffusion problems and viscous flow*

Now we shall be concerned with the application of the discontinuous Galerkin finite element method to convection–diffusion problems including the compressible Navier–Stokes equations.

4.6.9.1 *Example of a scalar problem* First let us consider a simple nonstationary nonlinear convection-diffusion problem to find $u : Q_T = \Omega \times (0,T) \to I\!R$ such that

$$\text{a)}\quad \frac{\partial u}{\partial t} + \sum_{s=1}^{N} \frac{\partial f_s(u)}{\partial x_s} = \nu \Delta u + g \quad \text{in } Q_T, \tag{4.6.92}$$

$$\text{b)}\quad u|_{\Gamma_D \times (0,T)} = u_D, \quad \text{c)}\quad \nu \frac{\partial u}{\partial n}|_{\Gamma_N \times (0,T)} = g_N,$$

$$\text{d)}\quad u(x,0) = u^0(x), \quad x \in \Omega.$$

We assume that $\Omega \subset I\!R^N$ is a bounded polygonal domain, if $N = 2$, or polyhedral domain, if $N = 3$, with a Lipschitz boundary $\partial\Omega = \overline{\Gamma}_D \cup \overline{\Gamma}_N$, $\Gamma_D \cap \Gamma_N = \emptyset$, and $T > 0$. The diffusion coefficient $\nu > 0$ is a given constant, $g : Q_T \to I\!R$, $u_D : \Gamma_D \times (0,T) \to I\!R$, $g_N : \Gamma_N \times (0,T) \to I\!R$ and $u^0 : \Omega \to I\!R$ are given functions, $f_s \in C^1(I\!R)$, $s = 1, \dots, N$, are given inviscid fluxes.

We define a *classical solution* of problem (4.6.92) as a sufficiently regular function in $\overline{Q}_T$ satisfying (4.6.92), a)–d) pointwise.

Note that in Definition 4.54 the concept of a *weak solution* of this problem was introduced provided $\Gamma_N = \emptyset$ and, thus, $\Gamma_D = \partial\Omega$. We leave to the reader the definition of a weak solution to problem (4.6.92) as an exercise.

4.6.9.2 *Discretization* The discretization of convective terms is carried out in the same way as in Section 4.6.4.1. There are several approaches to the discretization of the diffusion term. For example, in (Cockburn, 1999) the so-called *local discontinuous Galerkin FEM* is described. It is based on a mixed formulation introducing first-order derivatives of unknown functions as new dependent variables. This method is also used in (Karniadakis and Sherwin, 1999) in spectral methods and in (Bassi and Rebay, 1997*a*). Its theoretical analysis can be found in (Cockburn and Shu, 1998) or (Castillo *et al.*, 2001). However, the use of this approach leads to an extreme increase in the number of unknowns and, therefore, we shall apply another technique, used, for example, in (Oden *et al.*, 1998), (Babuška *et al.*, 1999).

We use the notation from Section 4.6.4.1. Moreover, for $i \in I$, by $\gamma_D(i)$ and $\gamma_N(i)$ we denote the subsets of $\gamma(i)$ formed by such indexes j that the faces Γ_{ij} approximate the parts Γ_D and Γ_N, respectively, of $\partial\Omega$. Thus, we suppose that

$$\gamma(i) = \gamma_D(i) \cup \gamma_N(i), \quad \gamma_D(i) \cap \gamma_N(i) = \emptyset. \tag{4.6.93}$$

For $v \in H^1(\Omega, \mathcal{T}_h)$ we set

$$\begin{aligned} \langle v \rangle_{\Gamma_{ij}} &= \frac{1}{2}\left(v\big|_{\Gamma_{ij}} + v\big|_{\Gamma_{oj}}\right), \\ [v]_{\Gamma_{ij}} &= v\big|_{\Gamma_{ij}} - v\big|_{\Gamma_{ji}}, \end{aligned} \tag{4.6.94}$$

denoting the *average* and *jump of the traces* of v on $\Gamma_{ij} = \Gamma_{ji}$ defined in (4.6.40). The approximate solution as well as test functions are supposed to be elements of the space $S_h = S^{p,-1}(\Omega, \mathcal{T}_h)$ introduced in (4.6.41). Obviously, $\langle v \rangle_{\Gamma_{ij}} = \langle v \rangle_{\Gamma_{ji}}$, $[v]_{\Gamma_{ij}} = -[v]_{\Gamma_{ji}}$ and $[v]_{\Gamma_{ij}} \boldsymbol{n}_{ij} = [v]_{\Gamma_{ji}} \boldsymbol{n}_{ji}$.

In order to derive the discrete problem, we assume that u is a classical solution of problem (4.6.92). The regularity of u implies that $u(\cdot, t) \in H^2(\Omega) \subset H^2(\Omega, \mathcal{T}_h)$ and

$$\begin{aligned} \langle u(\cdot,t)\rangle_{\Gamma_{ij}} &= u(\cdot,t)|_{\Gamma_{ij}}, \ [u(\cdot,t)]_{\Gamma_{ij}} = 0, \\ \langle \nabla u(\cdot,t)\rangle_{\Gamma_{ij}} &= \nabla u(\cdot,t)|_{\Gamma_{ij}} = \nabla u(\cdot,t)|_{\Gamma_{ji}}, \\ &\qquad \text{for each } t \in (0,T). \end{aligned} \tag{4.6.95}$$

We multiply equation (4.6.92), a) by any $\varphi \in H^2(\Omega, \mathcal{T}_h)$, integrate over $K_i \in \mathcal{T}_h$, apply Green's theorem and sum over all $K_i \in \mathcal{T}_h$. After some manipulation we obtain the identity

$$\begin{aligned} &\int_\Omega \frac{\partial u}{\partial t}\varphi\, dx + \sum_{i\in I}\sum_{j\in S(i)} \int_{\Gamma_{ij}} \sum_{s=1}^N f_s(u)\,(n_{ij})_s\, \varphi|_{\Gamma_{ij}}\, dS \\ &- \sum_{i\in I}\int_{K_i}\sum_{s=1}^N f_s(u)\,\frac{\partial\varphi}{\partial x_s}\, dx + \sum_{i\in I}\int_{K_i} \nu\,\nabla u\cdot\nabla\varphi\, dx \\ &- \sum_{i\in I}\sum_{\substack{j\in s(i)\\ j<i}} \int_{\Gamma_{ij}} \nu\langle\nabla u\rangle\cdot \boldsymbol{n}_{ij}[\varphi]\, dS \\ &- \sum_{i\in I}\sum_{j\in\gamma_D(i)} \int_{\Gamma_{ij}} \nu\,\nabla u\cdot \boldsymbol{n}_{ij}\,\varphi\, dS \end{aligned} \tag{4.6.96}$$

$$= \int_\Omega g\,\varphi\,dx + \sum_{i\in I}\sum_{j\in\gamma_N(i)} \int_{\Gamma_{ij}} \nu\,\nabla u \cdot \boldsymbol{n}_{ij}\,\varphi\,dS.$$

To the left-hand side of (4.6.96) we now add the terms

$$\pm \sum_{i\in I}\sum_{\substack{j\in s(i)\\ j<i}} \int_{\Gamma_{ij}} \nu\langle\nabla\varphi\rangle \cdot \boldsymbol{n}_{ij}[u]\,dS = 0, \tag{4.6.97}$$

as follows from (4.6.95). Further, to the left-hand side and the right-hand side we add the terms

$$\pm \sum_{i\in I}\sum_{j\in\gamma_D(i)} \int_{\Gamma_{ij}} \nu\,\nabla\varphi \cdot \boldsymbol{n}_{ij}\,u\,dS$$

and

$$\pm \sum_{i\in I}\sum_{j\in\gamma_D(i)} \int_{\Gamma_{ij}} \nu\,\nabla\varphi \cdot \boldsymbol{n}_{ij}\,u_D\,dS,$$

respectively, which are identical by the Dirichlet condition (4.6.92), b). We can add these terms equipped with the + sign (the so-called *nonsymmetric DG discretization* of diffusion terms) or with the – sign (*symmetric DG discretization* of diffusion terms). Both possibilities have their advantages and disadvantages – see, for example, (Prudhomme *et al.*, 2000). Here we shall use the nonsymmetric discretization.

In view of the Neumann condition (4.6.92), c), we replace the second term on the right-hand side of (4.6.96) by

$$\sum_{i\in I}\sum_{j\in\gamma_N(i)} \int_{\Gamma_{ij}} g_N\,\varphi\,dS. \tag{4.6.98}$$

Because of the stabilization of the scheme we introduce the *interior penalty*

$$\sum_{i\in I}\sum_{\substack{j\in s(i)\\ j<i}} \int_{\Gamma_{ij}} \sigma[u]\,[\varphi]\,dS \tag{4.6.99}$$

and the *boundary penalty*

$$\sum_{i\in I}\sum_{j\in\gamma_D(i)} \int_{\Gamma_{ij}} \sigma\,u\,\varphi\,dS = \sum_{i\in I}\sum_{j\in\gamma_D(i)} \int_{\Gamma_{ij}} \sigma u_D \varphi dS \tag{4.6.100}$$

where σ is a *weight* defined by

$$\sigma|_{\Gamma_{ij}} = \nu/d(\Gamma_{ij}). \tag{4.6.101}$$

On the basis of the above considerations we introduce the following forms defined for u, $\varphi \in H^2(\Omega, \mathcal{T}_h)$:

$$a_h(u,\varphi) = \sum_{i\in I} \int_{K_i} \nu\,\nabla u \cdot \nabla\varphi\,dx \tag{4.6.102}$$

$$
\begin{aligned}
&-\sum_{i\in I}\sum_{\substack{j\in s(i)\\ j<i}}\int_{\Gamma_{ij}}\nu\langle\nabla u\rangle\cdot\boldsymbol{n}_{ij}[\varphi]\,dS\\
&+\sum_{i\in I}\sum_{\substack{j\in s(i)\\ j<i}}\int_{\Gamma_{ij}}\nu\langle\nabla\varphi\rangle\cdot\boldsymbol{n}_{ij}[u]\,dS\\
&-\sum_{i\in I}\sum_{j\in\gamma_D(i)}\int_{\Gamma_{ij}}\nu\,\nabla u\cdot\boldsymbol{n}_{ij}\,\varphi\,dS\\
&+\sum_{i\in I}\sum_{j\in\gamma_D(i)}\int_{\Gamma_{ij}}\nu\,\nabla\varphi\cdot\boldsymbol{n}_{ij}\,u\,dS
\end{aligned}
$$

(nonsymmetric variant of the diffusion form),

$$
\begin{aligned}
J_h(u,\varphi)=&\sum_{i\in I}\sum_{\substack{j\in s(i)\\ j<i}}\int_{\Gamma_{ij}}\sigma[u]\,[\varphi]\,dS\\
&+\sum_{i\in I}\sum_{j\in\gamma_D(i)}\int_{\Gamma_{ij}}\sigma\,u\,\varphi\,dS
\end{aligned}
\tag{4.6.103}
$$

(interior and boundary penalty jump terms),

$$
\begin{aligned}
\ell_h(\varphi)(t)=&\int_\Omega g(t)\,\varphi\,dx+\sum_{i\in I}\sum_{j\in\gamma_N(i)}\int_\Gamma g_N(t)\,\varphi\,dS\\
&+\sum_{i\in I}\sum_{j\in\gamma_D(i)}\int_{\Gamma_{ij}}\nu\,\nabla\varphi\cdot\boldsymbol{n}_{ij}\,u_D(t)\,dS+\sum_{i\in I}\sum_{j\in\gamma_D(i)}\int_{\Gamma_{ij}}\sigma\,u_D(t)\,\varphi\,dS
\end{aligned}
\tag{4.6.104}
$$

(right-hand side form).

Finally, the convective terms are approximated with the aid of a numerical flux $H = H(u, v, \boldsymbol{n})$ by the form $b_h(u, \varphi)$ defined analogously as in Section 4.6.4.2:

$$
\begin{aligned}
b_h(u,\varphi)=&-\sum_{i\in I}\int_K\sum_{s=1}^N f_s(u)\,\frac{\partial\varphi}{\partial x_s}\,dx\\
&+\sum_{i\in I}\sum_{j\in S(i)}\int_{\Gamma_{ij}}H\left(u|_{\Gamma_{ij}},u|_{\Gamma_{ji}},\boldsymbol{n}_{ij}\right)\varphi|_{\Gamma_{ij}}\,dS,\quad u,\,\varphi\in H^2(\Omega,\mathcal{T}_h).
\end{aligned}
\tag{4.6.105}
$$

We assume that the numerical flux H is continuous, consistent and conservative – see Section 4.6.4.2.

Now we can introduce the *discrete problem.*

Definition 4.88 *We say that u_h is a DGFE solution of the convection-diffusion problem* (4.6.92), *if*

a) $u_h \in C^1([0,T]; S_h)$, (4.6.106)

b) $\frac{d}{dt}(u_h(t), \varphi_h) + b_h(u_h(t), \varphi_h) + a_h(u_h(t), \varphi_h) + J_h(u_h(t), \varphi_h) = \ell_h(\varphi_h)(t)$
$\forall \varphi_h \in S_h, \ \forall t \in (0,T),$

c) $u_h(0) = u_h^0.$

By u_h^0 we denote an S_h-approximation of the initial condition u^0.

This discrete problem has been obtained with the aid of the method of lines, i.e. the space semidiscretization. In practical computations suitable time discretization is applied (Euler forward or backward scheme, Runge–Kutta methods or discontinuous Galerkin time discretization) and integrals are evaluated with the aid of numerical integration – cf. Section 4.6.5. Let us note that in contrast to Sections 4.2–4.5, we do not require here that the approximate solution satisfies the essential Dirichlet boundary condition pointwise, e.g. at boundary nodes. In the DGFEM, this condition is represented in the framework of the 'integral identity' (4.6.106), b).

The above DGFE discrete problem was investigated theoretically in (Dolejší *et al.*, 2002*a*) and (Dolejší *et al.*, 2003), where error estimates were analysed.

4.6.9.3 *Application of the DGFEM to the Navier–Stokes equations* Now we shall extend the above approach to the Navier–Stokes problem (CFP) introduced in Section 4.3. Hence, we want to find a vector-valued function $\boldsymbol{w} : Q_T \to I\!R^m$ such that

$$\frac{\partial \boldsymbol{w}}{\partial t} + \sum_{s=1}^{N} \frac{\partial \boldsymbol{f}_s(\boldsymbol{w})}{\partial x_s} = \sum_{s=1}^{N} \frac{\partial \boldsymbol{R}_s(\boldsymbol{w}, \nabla \boldsymbol{w})}{\partial x_s} + \boldsymbol{F}(\boldsymbol{w}) \text{ in } Q_T; \tag{4.6.107}$$

$$\text{a) } \rho|_{\Gamma_I \times (0,T)} = \rho_D, \quad \text{b) } \boldsymbol{v}|_{\Gamma_I \times (0,T)} = \boldsymbol{v}_D = (v_{D1}, \dots, v_{DN})^{\mathrm{T}}, \tag{4.6.108}$$

$$\text{c) } \sum_{j=1}^{N} \left(\sum_{i=1}^{N} \tau_{ij} n_i \right) v_j + k \frac{\partial \theta}{\partial n} = 0 \quad \text{on } \Gamma_I \times (0,T);$$

$$\text{a) } \boldsymbol{v}|_{\Gamma_W \times (0,T)} = 0, \quad \text{b) } \frac{\partial \theta}{\partial n}|_{\Gamma_W \times (0,T)} = 0; \tag{4.6.109}$$

$$\sum_{i=1}^{N} \tau_{ij} n_i = 0, \quad j = 1, \dots, N, \quad \frac{\partial \theta}{\partial n} = 0 \quad \text{on } \Gamma_O \times (0,T); \tag{4.6.110}$$

$$\boldsymbol{w}(x,0) = \boldsymbol{w}^0(x), \quad x \in \Omega. \tag{4.6.111}$$

The state vector $\boldsymbol{w}$, the inviscid Euler fluxes $\boldsymbol{f}_s$, the viscous fluxes $\boldsymbol{R}_s$ and the right-hand side $\boldsymbol{F}(\boldsymbol{w})$ are defined by (4.3.2). The terms $\boldsymbol{R}_s$ can be expressed in the form (4.3.11) with matrices $\mathbb{K}_{ij}(\boldsymbol{w})$, $i, j = 1, \dots, N$, defined for $N = 2$ in (4.3.12).

In the discretization of problem (CFP) we proceed in a similar way as in Section 4.6.4.1. The approximate solution $\boldsymbol{w}_h$ of problem (CFP) as well as test functions φ_h are elements of the finite dimensional space $\boldsymbol{S}_h = \boldsymbol{S}^{p,-1}(\Omega, \mathcal{T}_h)$

introduced in (4.6.41). By $\gamma_D(i)$ we now denote the subset of such indexes $j \in \gamma(i)$ that for at least one component w_r of the sought solution $\boldsymbol{w}$ the Dirichlet condition is prescribed on the face $\Gamma_{ij} \subset \partial\Omega$.

Assuming that $\boldsymbol{w}$ is a classical, sufficiently regular solution of problem (CFP) and $\boldsymbol{\varphi} \in H^2(\Omega, \mathcal{T}_h)^m$, we multiply equation (4.6.107) by $\boldsymbol{\varphi}$, integrate over $K_i \in \mathcal{T}_h$, apply Green's theorem, sum over all $K_i \in \mathcal{T}_h$ and arrive at the identity

$$\begin{aligned}
&\int_\Omega \frac{\partial \boldsymbol{w}}{\partial t} \cdot \boldsymbol{\varphi}\, dx + \sum_{i\in I} \sum_{j\in s(i)} \int_{\Gamma_{ij}} \sum_{s=1}^N \boldsymbol{f}_s(\boldsymbol{w})\, (n_{ij})_s \cdot \boldsymbol{\varphi}|_{\Gamma_{ij}} dS \qquad (4.6.112)\\
&- \sum_{i\in I} \int_{K_i} \sum_{s=1}^N \boldsymbol{f}_s(\boldsymbol{w}) \cdot \frac{\partial \boldsymbol{\varphi}}{\partial x_s}\, dx + \sum_{i\in I} \int_{K_i} \sum_{s=1}^N \boldsymbol{R}_s(\boldsymbol{w}, \nabla \boldsymbol{w}) \cdot \frac{\partial \boldsymbol{\varphi}}{\partial x_s}\, dx\\
&- \sum_{i\in I} \sum_{\substack{j\in s(i)\\ j<i}} \int_{\Gamma_{ij}} \sum_{s=1}^N \langle \boldsymbol{R}_s(\boldsymbol{w}, \nabla \boldsymbol{w})\rangle\, (n_{ij})_s \cdot [\boldsymbol{\varphi}]\, dS\\
&- \sum_{i\in I} \sum_{j\in\gamma(i)} \int_{\Gamma_{ij}} \sum_{s=1}^N \boldsymbol{R}_s(\boldsymbol{w}, \nabla \boldsymbol{w})\, (n_{ij})_s \cdot \boldsymbol{\varphi}\, dS\\
&= \int_\Omega \boldsymbol{F}(\boldsymbol{w}) \cdot \boldsymbol{\varphi}\, dx.
\end{aligned}$$

The extension of the DGFEM from the scalar equation to the Navier–Stokes system is not quite straightforward. It is caused by the fact that the viscous (i.e. diffusion) terms $\boldsymbol{R}_s$ are nonlinear. Therefore, it is not possible to construct additional viscous (diffusion) terms by a simple exchange of $\boldsymbol{w}$ and $\boldsymbol{\varphi}$, because the resulting form would not be linear with respect to the test functions $\boldsymbol{\varphi}$. This problem is overcome by a *partial linearization* of the viscous fluxes $\boldsymbol{R}_s$. We consider two possibilities of the partial linearization of $\boldsymbol{R}_s$ from (Dolejší, 2002) and (Dolejší and Feistauer, 2003*a*):

1) Using the representation (4.3.11), we see that $\boldsymbol{R}_s$, $s = 1, \dots, N$, are linear with respect to $\nabla \boldsymbol{w}$. This leads us to add the following terms to the left-hand side of (4.6.112):

$$\begin{aligned}
&\sum_{i\in I} \sum_{\substack{j\in s(i)\\ j<i}} \int_{\Gamma_{ij}} \sum_{s=1}^N \langle \boldsymbol{R}_s(\boldsymbol{w}, \nabla \boldsymbol{\varphi})\rangle\, (n_{ij})_s \cdot [\boldsymbol{w}]\, dS \qquad (4.6.113)\\
&+ \sum_{i\in I} \sum_{j\in\gamma(i)} \int_{\Gamma_{ij}} \sum_{s=1}^N \boldsymbol{R}_s(\boldsymbol{w}, \nabla \boldsymbol{\varphi})\, (n_{ij})_s \cdot \boldsymbol{w}\, dS.
\end{aligned}$$

(As we shall see, this will yield a nonsymmetric variant of the diffusion form a_h.) In the second term we use the zero natural Neumann boundary conditions (4.6.109), c), (4.6.109), b) and (4.6.112). The Dirichlet conditions are taken into account with the aid of additional terms on the right-hand side of (4.6.112).

Moreover, we add to the left-hand side of (4.6.112) the vanishing *interior penalty* terms

$$\sum_{i \in I} \sum_{\substack{j \in s(i) \\ j<i}} \int_{\Gamma_{ij}} \sigma[\boldsymbol{w}] \cdot [\boldsymbol{\varphi}]\, dS \tag{4.6.114}$$

with $\sigma|_{\Gamma_{ij}} = \mu / \mathrm{d}(\Gamma_{ij})$ and boundary penalty terms balanced by additional right-hand side terms containing the Dirichlet boundary data.

We arrive at the definition of the following forms:

$$\begin{aligned} & a_h(\boldsymbol{w}, \boldsymbol{\varphi}) \\ &= \sum_{i \in I} \int_{K_i} \sum_{s=1}^{N} \boldsymbol{R}_s(\boldsymbol{w}, \nabla \boldsymbol{w}) \cdot \frac{\partial \boldsymbol{\varphi}}{\partial x_s}\, dx \\ & - \sum_{i \in I} \sum_{\substack{j \in s(i) \\ j<i}} \int_{\Gamma_{ij}} \sum_{s=1}^{N} \langle \boldsymbol{R}_s(\boldsymbol{w}, \nabla \boldsymbol{w}) \rangle\, (n_{ij})_s \cdot [\boldsymbol{\varphi}]\, dS \\ & + \sum_{i \in I} \sum_{\substack{j \in s(i) \\ j<i}} \int_{\Gamma_{ij}} \sum_{s=1}^{N} \langle \boldsymbol{R}_s(\boldsymbol{w}, \nabla \boldsymbol{\varphi}) \rangle\, (n_{ij})_s \cdot [\boldsymbol{w}]\, dS \\ & - \sum_{i \in I} \sum_{j \in \gamma_D(i)} \int_{\Gamma_{ij}} \sum_{s=1}^{N} \boldsymbol{R}_s(\boldsymbol{w}, \nabla \boldsymbol{w})\, (n_{ij})_s \cdot \boldsymbol{\varphi}\, dS \\ & + \sum_{i \in I} \sum_{j \in \gamma_D(i)} \int_{\Gamma_{ij}} \sum_{s=1}^{N} \boldsymbol{R}_s(\boldsymbol{w}, \nabla \boldsymbol{\varphi})\, (n_{ij})_s \cdot \boldsymbol{w}\, dS \end{aligned} \tag{4.6.115}$$

(nonsymmetric variant of the diffusion form),

$$\begin{aligned} J_h(\boldsymbol{w}, \boldsymbol{\varphi}) = & \sum_{i \in I} \sum_{\substack{j \in s(i) \\ j<i}} \int_{\Gamma_{ij}} \sigma[\boldsymbol{w}] \cdot [\boldsymbol{\varphi}] dS \\ & + \sum_{i \in I} \sum_{j \in \gamma_D(i)} \int_{\Gamma_{ij}} \sigma\, \boldsymbol{w} \cdot \boldsymbol{\varphi}\, dS \end{aligned} \tag{4.6.116}$$

(interior and boundary penalty jump terms),

$$\begin{aligned} \beta_h(\boldsymbol{w}, \boldsymbol{\varphi}) = & \int_{\Omega} \boldsymbol{F}(\boldsymbol{w}) \cdot \boldsymbol{\varphi} \\ & + \sum_{i \in I} \sum_{j \in \gamma_D(i)} \int_{\Gamma_{ij}} \sum_{s=1}^{N} \boldsymbol{R}_s(\boldsymbol{w}, \nabla \boldsymbol{\varphi})\, (n_{ij})_s \cdot \boldsymbol{w}_B\, dS \\ & + \sum_{i \in I} \sum_{j \in \gamma_D(i)} \int_{\Gamma_{ij}} \sigma\, \boldsymbol{w}_B \cdot \boldsymbol{\varphi}\, dS \end{aligned} \tag{4.6.117}$$

(right-hand side form). The convective terms are represented by the form $b_h(\boldsymbol{w}, \boldsymbol{\varphi})$ defined in (4.6.50).

The boundary state $\boldsymbol{w}_B = (w_{B1}, \dots, w_{Bm})^{\mathrm{T}}$ is determined in the following way. We set

$$w_{Br}|_{\Gamma_{ij}} = w_r^*|_{\Gamma_{ij}}, \tag{4.6.118}$$

if the r-th component w_r of $\boldsymbol{w}$ is prescribed on Γ_{ij}. Here w_r^* is the r-th component of the function $\boldsymbol{w}^*$ representing the Dirichlet boundary conditions and given by (4.4.2). Otherwise, we set

$$w_{Br}|_{\Gamma_{ij}} = w_r|_{\Gamma_{ij}}, \tag{4.6.119}$$

which means that we use 'extrapolation' of w_r onto Γ_{ij} from $K_i \in \mathcal{T}_h$. In particular, we have

$$\begin{aligned} \boldsymbol{w}_B &= (\rho_{ij}, 0, \dots, 0, \rho_{ij}\theta_{ij}) \quad \text{on } \Gamma_W, \\ \boldsymbol{w}_B &= \left(\rho_D, \rho_D v_{D1}, \dots, \rho_D v_{DN}, \rho_{ij}\theta_{ij} + \frac{1}{2}\rho_D|\boldsymbol{v}_D|^2\right) \quad \text{on } \Gamma_I, \end{aligned} \tag{4.6.120}$$

where ρ_D and $\boldsymbol{v}_D$ are the given density and velocity from the boundary conditions (4.6.108) and ρ_{ij}, θ_{ij} are the values of the density and absolute temperature extrapolated from K_i onto Γ_{ij}.

Now the *discrete DGFE Navier–Stokes problem* reads:

Definition 4.89 *An approximate DGFE solution of the compressible Navier–Stokes problem (CFP) is defined as a vector-valued function* $\boldsymbol{w}_h$ *such that*

$$\begin{aligned} &a)\ \boldsymbol{w}_h \in C^1([0,T]; \boldsymbol{S}_h), \\ &b)\ \frac{d}{dt}(\boldsymbol{w}_h(t), \boldsymbol{\varphi}_h) + b_h(\boldsymbol{w}_h(t), \boldsymbol{\varphi}_h) + a_h(\boldsymbol{w}_h(t), \boldsymbol{\varphi}_h) + J_h(\boldsymbol{w}_h(t), \boldsymbol{\varphi}_h) \\ &\quad = \beta_h(\boldsymbol{w}_h(t), \boldsymbol{\varphi}_h) \quad \forall \boldsymbol{\varphi}_h \in \boldsymbol{S}_h,\ t \in (0,T), \\ &c)\ \boldsymbol{w}_h(0) = \boldsymbol{w}_h^0, \end{aligned} \tag{4.6.121}$$

where $\boldsymbol{w}_h^0$ *is an* $\boldsymbol{S}_h$*-approximation of* $\boldsymbol{w}^0$.

2) Another partial linearization of the viscous terms $\boldsymbol{R}_s(\boldsymbol{w}, \nabla w)$ is obtained by the differentiation inside the definition of $\boldsymbol{R}_s(\boldsymbol{w}, \nabla \boldsymbol{w})$. For example, if $N = 2$, $\lambda = -2\mu/3$, then from (4.3.2) we obtain

$$\boldsymbol{R}_1(\boldsymbol{w}, \nabla \boldsymbol{w}) \tag{4.6.122}$$

$$= \begin{pmatrix} 0 \\ \frac{2}{3}\frac{\mu}{w_1}\left[2\left(\frac{\partial w_2}{\partial x_1} - \frac{w_2}{w_1}\frac{\partial w_1}{\partial x_1}\right) - \left(\frac{\partial w_3}{\partial x_2} - \frac{w_3}{w_1}\frac{\partial w_1}{\partial x_2}\right)\right] \\ \frac{\mu}{w_1}\left[\left(\frac{\partial w_3}{\partial x_1} - \frac{w_3}{w_1}\frac{\partial w_1}{\partial x_1}\right) + \left(\frac{\partial w_2}{\partial x_2} - \frac{w_2}{w_1}\frac{\partial w_1}{\partial x_2}\right)\right] \\ \frac{w_2}{w_1}\boldsymbol{R}_1^{(2)} + \frac{w_3}{w_1}\boldsymbol{R}_1^{(3)} + \frac{k}{c_v w_1}\left[\frac{\partial w_4}{\partial x_1} - \frac{w_4}{w_1}\frac{\partial w_1}{\partial x_1} - \frac{1}{w_1}\left(w_2\frac{\partial w_2}{\partial x_1} + w_3\frac{\partial w_3}{\partial x_1}\right)\right. \\ \qquad\qquad \left. + \frac{1}{w_1^2}\left(w_2^2 + w_3^2\right)\frac{\partial w_1}{\partial x_1}\right] \end{pmatrix},$$

$$
\boldsymbol{R}_2(\boldsymbol{w}, \nabla \boldsymbol{w}) = \begin{pmatrix} 0 \\ \frac{\mu}{w_1}\left[\left(\frac{\partial w_3}{\partial x_1} - \frac{w_3}{w_1}\frac{\partial w_1}{\partial x_1}\right) + \left(\frac{\partial w_2}{\partial x_2} - \frac{w_2}{w_1}\frac{\partial w_1}{\partial x_2}\right)\right] \\ \frac{2}{3}\frac{\mu}{w_1}\left[2\left(\frac{\partial w_3}{\partial x_2} - \frac{w_3}{w_1}\frac{\partial w_1}{\partial x_2}\right) - \left(\frac{\partial w_2}{\partial x_1} - \frac{w_2}{w_1}\frac{\partial w_1}{\partial x_1}\right)\right] \\ \frac{w_2}{w_1}\boldsymbol{R}_2^{(2)} + \frac{w_3}{w_1}\boldsymbol{R}_2^{(3)} + \frac{k}{c_v w_1}\left[\frac{\partial w_4}{\partial x_1} - \frac{w_4}{w_1}\frac{\partial w_1}{\partial x_2} - \frac{1}{w_1}\left(w_2\frac{\partial w_2}{\partial x_2} + w_3\frac{\partial w_3}{\partial x_2}\right) \right. \\ \left. + \frac{1}{w_1^2}\left(w_2^2 + w_3^2\right)\frac{\partial w_1}{\partial x_2}\right] \end{pmatrix},
$$

where $\boldsymbol{R}_s^{(r)} = \boldsymbol{R}_s^{(r)}(\boldsymbol{w}, \nabla \boldsymbol{w})$ denotes the r-th component of $\boldsymbol{R}_s$ ($s = 1, 2,\ r = 2, 3$).

Now for $\boldsymbol{w} = (w_1, \ldots, w_4)^{\mathrm{T}}$ and $\boldsymbol{\varphi} = (\varphi_1, \ldots, \varphi_4)^{\mathrm{T}}$ we define the vector-valued functions

$$
\boldsymbol{D}_1(\boldsymbol{w}, \nabla \boldsymbol{w}, \boldsymbol{\varphi}, \nabla \boldsymbol{\varphi}) = \begin{pmatrix} 0 \\ \frac{2}{3}\frac{\mu}{w_1}\left[2\left(\frac{\partial \varphi_2}{\partial x_1} - \frac{\varphi_2}{w_1}\frac{\partial w_1}{\partial x_1}\right) - \left(\frac{\partial \varphi_3}{\partial x_2} - \frac{\varphi_3}{w_1}\frac{\partial w_1}{\partial x_2}\right)\right] \\ \frac{\mu}{w_1}\left[\left(\frac{\partial \varphi_3}{\partial x_1} - \frac{\varphi_3}{w_1}\frac{\partial w_1}{\partial x_1}\right) + \left(\frac{\partial \varphi_2}{\partial x_2} - \frac{\varphi_2}{w_1}\frac{\partial w_1}{\partial x_2}\right)\right] \\ \frac{w_2}{w_1}\boldsymbol{D}_1^{(2)} + \frac{w_3}{w_1}\boldsymbol{D}_1^{(3)} + \frac{k}{c_v w_1}\left[\frac{\partial \varphi_4}{\partial x_1} - \frac{\varphi_4}{w_1}\frac{\partial w_1}{\partial x_1} - \frac{1}{w_1}\left(w_2\frac{\partial \varphi_2}{\partial x_1} + w_3\frac{\partial \varphi_3}{\partial x_1}\right) \right. \\ \left. + \frac{1}{w_1^2}\left(w_2\varphi_2 + w_3\varphi_3\right)\frac{\partial w_1}{\partial x_1}\right] \end{pmatrix}, \tag{4.6.123}
$$

$$
\boldsymbol{D}_2(\boldsymbol{w}, \nabla \boldsymbol{w}, \boldsymbol{\varphi}, \nabla \boldsymbol{\varphi}) = \begin{pmatrix} 0 \\ \frac{\mu}{w_1}\left[\left(\frac{\partial \varphi_3}{\partial x_1} - \frac{\varphi_3}{w_1}\frac{\partial w_1}{\partial x_1}\right) + \left(\frac{\partial \varphi_2}{\partial x_2} - \frac{\varphi_2}{w_1}\frac{\partial w_1}{\partial x_2}\right)\right] \\ \frac{2}{3}\frac{\mu}{w_1}\left[2\left(\frac{\partial \varphi_3}{\partial x_2} - \frac{\varphi_2}{w_1}\frac{\partial w_1}{\partial x_2}\right) - \left(\frac{\partial \varphi_2}{\partial x_1} - \frac{\varphi_2}{w_1}\frac{\partial w_1}{\partial x_1}\right)\right] \\ \frac{w_2}{w_1}\boldsymbol{D}_2^{(2)} + \frac{w_3}{w_1}\boldsymbol{D}_2^{(3)} + \frac{k}{c_v w_1}\left[\frac{\partial \varphi_4}{\partial x_2} - \frac{\varphi_4}{w_1}\frac{\partial w_1}{\partial x_2} - \frac{1}{w_1}\left(w_2\frac{\partial \varphi_2}{\partial x_2} + w_2\frac{\partial \varphi_3}{\partial x_2}\right) \right. \\ \left. + \frac{1}{w_1^2}\left(w_2 + \varphi_2 + w_3\varphi_2\right)\frac{\partial w_1}{\partial x_2}\right] \end{pmatrix},
$$

where $\boldsymbol{D}_s^{(r)}$ denotes the r-th component of $\boldsymbol{D}_s$ ($s = 1, 2,\ r = 2, 3$). Obviously, $\boldsymbol{D}_1$ and $\boldsymbol{D}_2$ are linear with respect to $\boldsymbol{\varphi}$ and $\nabla\boldsymbol{\varphi}$ and

$$
\boldsymbol{D}_s(\boldsymbol{w}, \nabla \boldsymbol{w}, \boldsymbol{w}, \nabla \boldsymbol{w}) = \boldsymbol{R}_s(\boldsymbol{w}, \nabla \boldsymbol{w}), \quad s = 1, 2. \tag{4.6.124}
$$

Now we define the diffusion form

$$
a_h(\boldsymbol{w}, \boldsymbol{\varphi}) \tag{4.6.125}
$$

$$
\begin{aligned}
&= \sum_{i\in I} \int_{K_i} \sum_{s=1}^{2} \boldsymbol{R}_s(\boldsymbol{w}, \nabla \boldsymbol{w}) \cdot \frac{\partial \boldsymbol{\varphi}}{\partial x_s}\, dx \\
&- \sum_{i\in I} \sum_{\substack{j\in s(i)\\ j<i}} \int_{\Gamma_{ij}} \sum_{s=1}^{2} \langle \boldsymbol{R}_s(\boldsymbol{w}, \nabla \boldsymbol{w})\rangle\, (n_{ij})_s \cdot [\boldsymbol{\varphi}]\, dS \\
&+ \sum_{i\in I} \sum_{\substack{j\in s(i)\\ j<i}} \int_{\Gamma_{ij}} \sum_{s=1}^{2} \langle \boldsymbol{D}_s(\boldsymbol{w}, \nabla \boldsymbol{w}, \boldsymbol{\varphi}, \nabla \boldsymbol{\varphi})\rangle\, (n_{ij})_s \cdot [\boldsymbol{w}]\, dS \\
&- \sum_{i\in I} \sum_{j\in \gamma_D(i)} \int_{\Gamma_{ij}} \sum_{s=1}^{2} \boldsymbol{R}_s(\boldsymbol{w}, \nabla \boldsymbol{w})\, (n_{ij})_s \cdot \boldsymbol{\varphi}\, dS \\
&+ \sum_{i\in I} \sum_{j\in \gamma_D(i)} \int_{\Gamma_{ij}} \sum_{s=1}^{2} \boldsymbol{D}_s(\boldsymbol{w}, \nabla \boldsymbol{w}, \boldsymbol{\varphi}, \nabla \boldsymbol{\varphi})\, (n_{ij})_s\, \boldsymbol{w}\, dS,
\end{aligned}
$$

linear with respect to $\boldsymbol{\varphi}$. The goal was to obtain a form $a_h(\boldsymbol{w}, \boldsymbol{\varphi})$ linear with respect to $\boldsymbol{\varphi}$, but independent of first order derivatives of the first component of $\boldsymbol{\varphi}$. This approach yields a scheme, in which the continuity equation is perturbed by penalty jump terms only. According to numerical experiments, this method behaves better than the method using the first type of the partial linearization leading to the form a_h defined by (4.6.115). Using the forms J_h, β_h and b_h introduced in (4.6.116), (4.6.117) and (4.6.50), respectively, we define an approximate solution of problem (CFP) again as in Definition 4.89.

Other formulations of the DGFE Navier–Stokes problem can be found in (Baumann, 1997), (Baumann and Oden, 1999*b*) and (Baumann and Oden, 1999*a*). Various types of the DGFE approximations of diffusion terms in the case of the scalar Poisson equation are analysed in (Babuška *et al.*, 1999), (Brezzi *et al.*, 2000). The DGFEM applied to linear scalar convection–diffusion problems is analyzed, for example, in (Wheeler, 1978), (Houston *et al.*, 2002), (Roos and Zarin, 2002). A nonlinear parabolic problem is treated in (Rivière and Wheeler, 2000). A detailed analysis of the DGFEM for a nonlinear convection-diffusion problem on general mixed meshes can be found in (Dolejší *et al.*, 2003).

Up to now we have assumed that the domain Ω is polygonal ($N = 2$) or polyhedral ($N = 3$). In the case of a curved boundary $\partial\Omega$ superparametric finite elements are used – see Section 4.6.8.

For transonic flow, when the solution contains steep gradients, the limiting of the order of the method explained in Section 4.6.7 is applied in order to avoid the Gibbs phenomenon.

4.6.10 *Numerical examples*

4.6.10.1 *Application of the DGFEM to the solution of inviscid compressible flow* The first numerical example deals with inviscid transonic flow through the GAMM channel with inlet Mach number = 0.67. (See Section 3.7.) In order

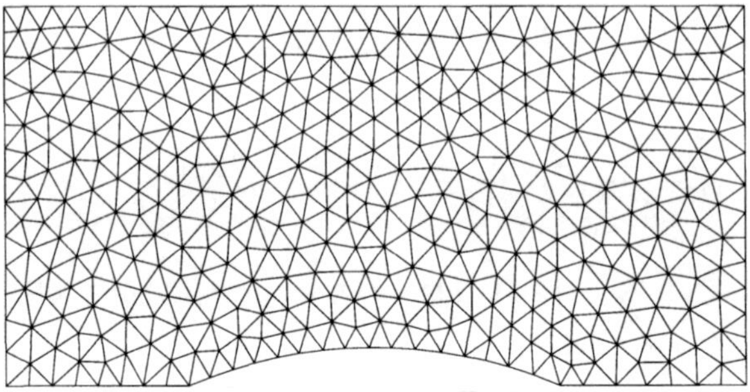

FIG. 4.6.17. Coarse triangular mesh (784 triangles) in the GAMM channel

to obtain a steady-state solution, the time stabilization method for $t \to \infty$ is applied.

We demonstrate the influence of the use of superparametric elements at the curved part of $\partial\Omega$, explained in Section 4.6.8. The computations were performed on a coarse grid shown in Fig. 4.6.17 having 784 triangles. Figures 4.6.18 and 4.6.19 show the density distribution in the computational domain and along the lower wall obtained by scheme (4.6.72) without and with the use of a bilinear mapping, respectively. One can see a difference in the quality of the approximate solutions.

Figure 4.6.20 shows the computational grid constructed with the aid of the anisotropic mesh adaptation (AMA) technique ((Dolejší, 1998*b*)) described in Section 3.6. Figure 4.6.21 shows the density distribution in Ω and along the lower wall obtained with the aid of the bilinear mapping on a refined mesh. As we can see, a very sharp shock wave and the Zierep singularity were obtained. (Cf. Section 3.7.)

The second example is concerned with the computation of inviscid compressible flow past the airfoil NACA 0012 with far-field Mach number = 0.8 and angle of attack = 1.25°. To achieve a sharp capturing of shock waves, the AMA method was applied. Figure 4.6.22 shows the triangulation used and computed Mach number isolines.

The isolines were computed without any postprocessing on each element separately. Although the approximate solution is, in general, discontinuous on interelement interfaces, the isolines are smooth curves due to a sufficiently fine mesh and negligible interelement jumps in regions where the exact solution is regular.

In the above examples, the explicit Euler time stepping and limiting of the order of accuracy from Section 4.6.7 were used. This method requires to satisfy the CFL stability condition representing a severe restriction of the time step. In (Dolejší and Feistauer, 2003*b*), an efficient semi-implicit time stepping scheme was developed for the numerical solution of the Euler equations, allowing to use a long time step in the DGFEM.

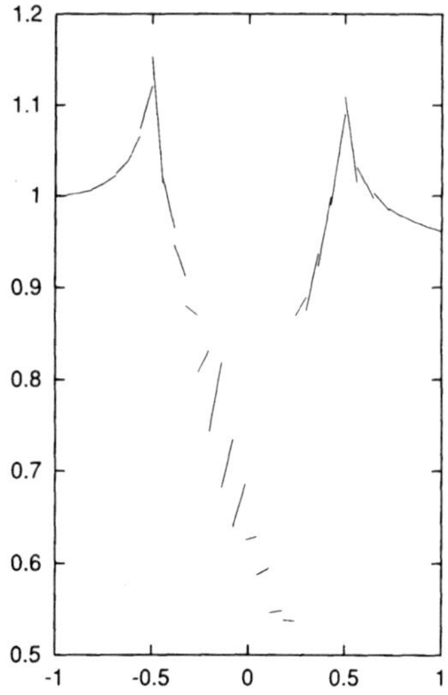

FIG. 4.6.18. Density distribution along the lower wall in the GAMM channel without the use of a bilinear mapping at $\partial\Omega$

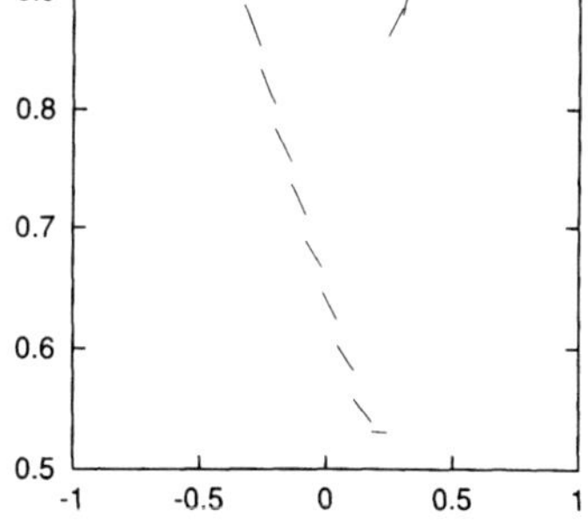

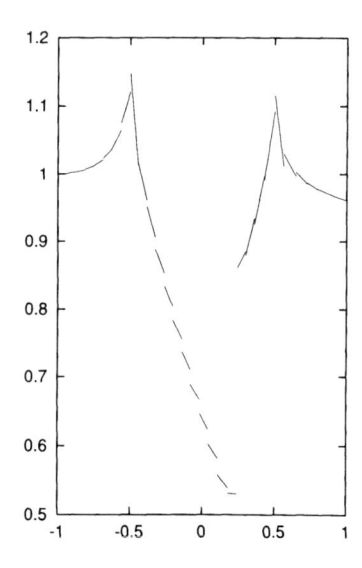

FIG. 4.6.19. Density distribution along the lower wall in the GAMM channel with the use of a bilinear mapping at $\partial\Omega$

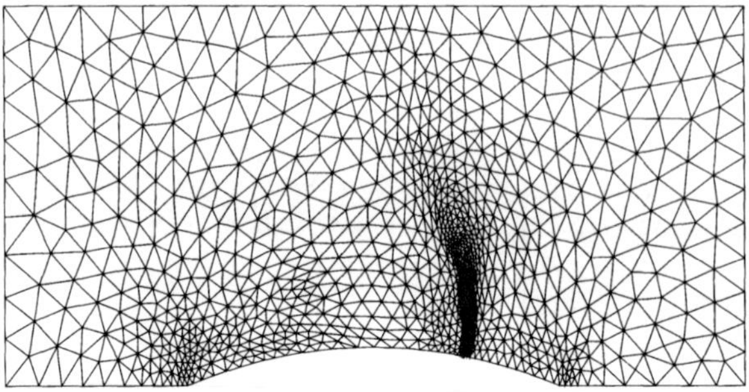

FIG. 4.6.20. Adapted triangular mesh (2131 triangles) in the GAMM channel

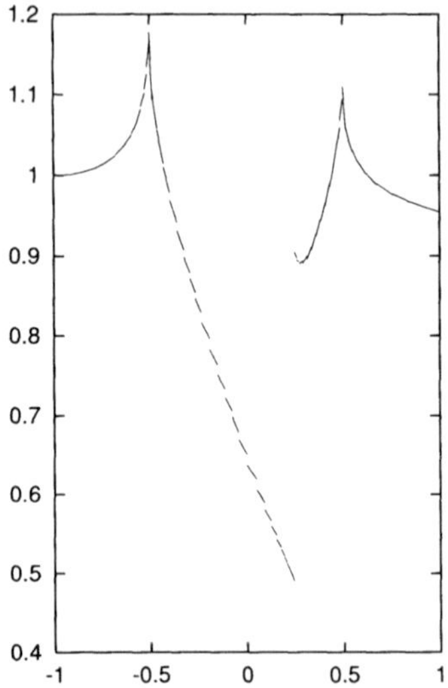

FIG. 4.6.21. Density distribution along the lower wall in the GAMM channel with the use of a bilinear mapping on an adapted mesh

4.6.10.2 *Application of the DGFEM to the solution of viscous compressible flow*
We present here results from (Dolejší, 2002) on the numerical solution of the viscous flow past the airfoil NACA 0012 by the DGFEM. The computation was performed for data from (Bristeau *et al.*, 1987): angle of attack $\alpha = 10°$, far-field Mach number $M_\infty = 0.8$, Reynolds number $Re = 500$. The mesh was constructed with the aid of adaptive anisotropic mesh refinement ((Dolejší, 2001)). Figure 4.6.23 shows the mesh used. In Figure 4.6.24, Mach number isolines for the stationary solution obtained with the aid of the Euler explicit time stepping and time stabilization for $t \to \infty$ are plotted. The results are comparable with those from (Bristeau *et al.*, 1987). The isolines are not smooth in areas where

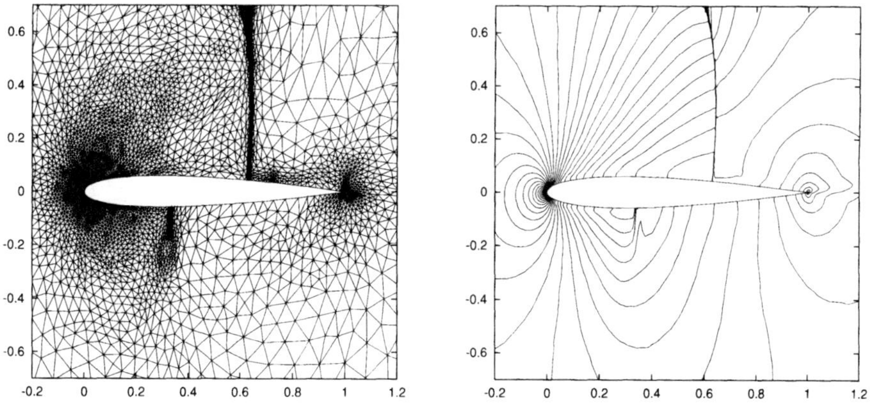

FIG. 4.6.22. Transonic flow past the profile NACA 0012: triangulation and Mach number isolines

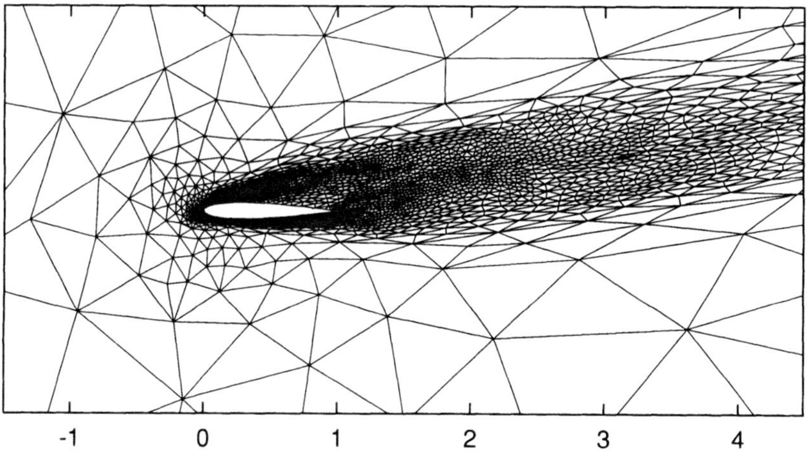

FIG. 4.6.23. Viscous flow past the profile NACA 0012: triangulation

the mesh is coarse. From Fig. 4.6.25 we see that the quality of isolines is better in a neighbourhood of the profile, where the mesh was refined.

Remark 4.90 Numerical experiments show that the stability of the Euler forward time stepping for the DGFEM requires rather severe time step limitation. From this point of view it appears that implicit methods are more efficient. They were applied in (Bassi and Rebay, 2000). In (Dolejší and Feistauer, 2003*b*) a semi-implicit DGFEM was developed, allowing to use long time steps. To obtain a steady-state solution it is suitable to solve directly the stationary system of gov-

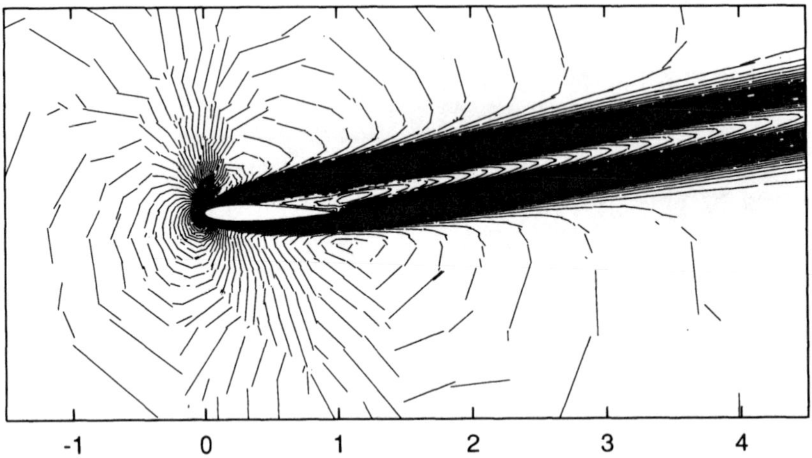

FIG. 4.6.24. Viscous flow past the profile NACA 0012: Mach number isolines

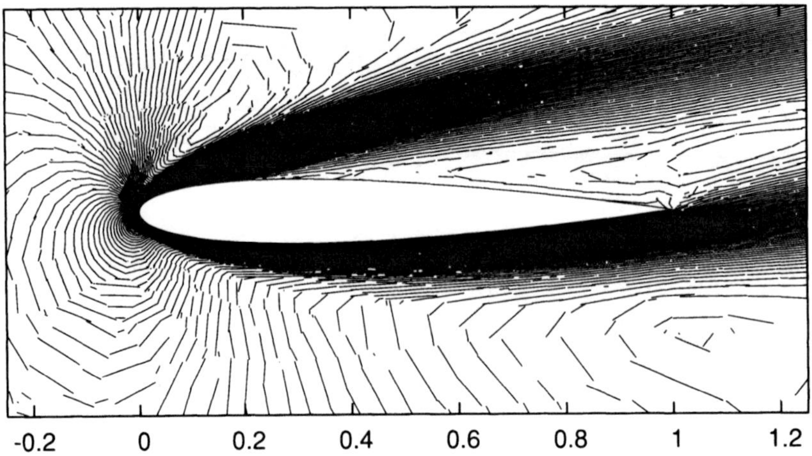

FIG. 4.6.25. Detail of Mach number isolines in a neighbourhood of the profile NACA 0012

erning equations, as for example in the papers (Hartmann and Houston, 2001*b*), (Hartmann and Houston, 2001*a*) concerned with the stationary Euler equations.

REFERENCES

Abgrall, R. (1994). On essentially non-oscillatory schemes on unstructured meshes: analysis and implementation. *J. Comput. Phys.*, **114**, 45–58.

Adjerid, S., Devine, D., Flaherty, J. E., and Krivodonova, L. (2002). A posteriori error estimation for discontinuous Galerkin solutions of hyperbolic problems. *Comput. Methods Appl. Mech. Eng.*, **191**, 1097–1112.

Aliabadi, S. K., Ray, S. E, and Tezduyar, T. E. (1993). SUPG finite element computation of compressible flows based on the conservation and entropy variable formulation. *Comput. Mech.*, **11**, 300–312.

Almeida, R. C. and Galeao, A. C. (1996). An adaptive Petrov–Galerkin formulation for the compressible Euler and Navier–Stokes equations. *Comput. Methods Appl. Mech. Eng.*, **129**, 157–176.

Angermann, L. (1991). Numerical solution of second–order elliptic equations on plane domains. *RAIRO Modél. Math. Anal. Numér.*, **25**, 165–191.

Angermann, L. (1993). Addendum to the paper: Numerical solution of second order equations on plane domains. *RAIRO Modél. Math. Anal. Numér.*, **7**, 1–7.

Angermann, L. (1995). Error estimates for the finite-element solution of an elliptic singularly perturbed problem. *IMA J. Numer. Anal.*, **15**, 161–196.

Angot, Ph., Debieve, J. F., Fürst, J., and Kozel, K. (1997). Two- and three-dimensional applications of TVD and ENO schemes. In *Numerical Modelling in Continuum Mechanics* (ed. M. Feistauer, R. Rannacher, and K. Kozel), pp. 103–111. MATFYZPRESS, Prague.

Angot, Ph., Dolejší, V., Feistauer, M., and Felcman, J. (1998). Analysis of a combined barycentric finite volume – nonconforming finite element method for nonlinear convection–diffusion problems. *Appl. Math.*, **43(4)**, 263–310.

Apel, T. (1999). *Anisotropic Finite Elements: Local Estimates and Applications.* Teubner, Stuttgart–Leipzig.

Apel, T. and Lube, G. (1996). Anisotropic mesh refinement in stabilized Galerkin methods. *Numer. Math.*, **74**, 261–282.

Apel, T. and Lube, G. (1998). Anisotropic mesh refinement for a singularly perturbed reaction diffusion model problem. *Appl. Numer. Math.*, **26**, 415–433.

Apel, T., Nicaise, S., and Schöberl, J. (2001). Finite element methods with anisotropic meshes near edges. In *Finite Element Methods for Three-dimensional Problems* (ed. P. Neittaanmäki and M. Křížek), Volume 15 of *GAKUTO Int. Ser., Math. Sci. Appl.*, Tokyo, pp. 1–8. GAKKOTOSHO.

Arminjon, P., Dervieux, A., Fezoui, L., Steve, H., and Stoufflet, B. (1989). Non-oscillatory schemes for multidimensional Euler calculations with unstructured grids. In *Nonlinear hyperbolic equations – theory, computational methods, and applications* (ed. J. Ballmann and R. Jeltsch), Volume 24, pp. 31–42. (Proceedings of the Second International Conference on Nonlinear Hyperbolic Problems Aachen 1988), Notes on Numerical Fluid Mechanics. Vieweg, Braunschweig.

Arminjon, P. and Madrane, A. (1999*a*). A mixed finite volume/finite element method for 2-dimensional compressible Navier–Stokes equations on unstructured grids. In *Hyperbolic Problems: Theory, Numerics, Applications* (ed. M. Fey and R. Jeltsch), Volume I, pp. 11–20. Birkhäuser, Basel.

Arminjon, P. and Madrane, A. (1999*b*). A mixed finite volume/finite element method for 2-dimensional compressible Navier–Stokes equations on unstructured grids. In *Hyperbolic Problems: Theory, Numerics, Applications* (ed. M. Fey and R. Jeltsch), Volume I, pp. 11–20. Birkhäuser, Basel.

Axelsson, O. (1977). Solution of linear systems of equations: Iterative methods. In *Sparse Matrix Techniques* (ed. V. A. Barker), Number 572 in Lecture Notes in Mathematics. Springer, Berlin.

Babuška, I., Oden, J. T., and Baumann, C. E. (1999). A discontinuous *hp* finite element method for diffusion problems: 1-D analysis. *Comput. Math. Appl.*, **37**, 103–122.

Babuška, I. and Strouboulis, T. (2001). *A Posteriori Error Estimators for the Finite Element Method.* Clarendon Press, Oxford.

Bank, R., Sherman, A., and Weiser, A. (1983). Refinement algorithm and data structures for regular local mesh refinement. In *Scientific Computing* (ed. R. Stepleman *et al.*), Volume 44, pp. 3–17. IMACS. North-Holland, Amsterdam.

Bardos, C., Le Roux, A.-Y., and Nedelec, J.-C. (1979). First order quasilinear equations with boundary conditions. *Commun. Partial. Differ. Equations*, **4**, 1017–1034.

Barth, T. J. and Jespersen, D. C. (1989). The design and application of upwind schemes on unstructured meshes. AIAA Paper, 89-0366.

Bassi, F. and Rebay, S. (1997*a*). A high-order accurate discontinuous finite element method for the numerical solution of the compressible Navier–Stokes equations. *J. Comput. Phys*, **131**, 267–279.

Bassi, F. and Rebay, S. (1997*b*). High-order accurate discontinuous finite element solution of the 2D Euler equations. *J. Comput. Phys.*, **138**, 251–285.

Bassi, F. and Rebay, S. (2000). A high order discontinuous Galerkin method for compressible turbulent flow. In *Discontinuous Galerkin Method: Theory, Computations and Applications* (ed. B. Cockburn, G. E. Karniadakis, and C. W. Shu). Springer, Berlin.

Baumann, C. E. (1997). *An hp-adaptive discontinuous Galerkin method for computational fluid dynamics.* Ph. D. thesis, The University of Texas at Austin.

Baumann, C. E. and Oden, J. T. (1999*a*). A discontinuous *hp* finite element method for convection–diffusion problems. *Comput. Methods Appl. Mech. Eng.*, **175**, 311–341.

Baumann, C. E. and Oden, J. T. (1999*b*). A discontinuous *hp* finite element method for the Euler and Navier–Stokes equations. *Int. J. Numer. Methods Fluids*, **31**, 79–95.

Bayliss, A. and Turkel, E. (1980). Radiation boundary conditions for wave-like equations. *Commun. Pure Appl. Math.*, **33**, 708–725.

Becker, R. and Rannacher, R. (1996). A feed-back approach to error control in finite element methods: Basic analysis and examples. *East–West J. Numer. Math.*, **4**, 237–264.

Beirão da Veiga, H. (1987). An L^p-theory for the n-dimensional, stationary, compressible Navier–Stokes equations, and incompressible limit for compressible fluids. The equilibrium solutions. *Commun. Math. Phys.*, **109**, 229–248.

Bejček, M., Dolejší, V., and Feistauer, M. (2002). On discontinuous Galerkin method for numerical solution of conservation laws and convection–diffusion problems. In *Proceedings of the XIVth Summer School Software and Algorithms of Numerical Mathematics*, pp. 7–32. University of West Bohemia, Pilsen.

Benharbit, S., Chalabi, A., and Villa, J. P. (1995). Numerical viscosity and convergence of finite volume methods for conservation laws with boundary conditions. *SIAM J. Numer. Anal.*, **32(3)**, 775–796.

Bern, M. (1999). Adaptive mesh generation. In *Error Estimation and Adaptive Discretization Methods in Computational Fluid Dynamics* (ed. T. J. Barth and H. Deconinck), Volume 25 of *Lecture Notes in Computational Science and Engineering*, pp. 1–46. Springer, Berlin.

Bijl, H. and Wesseling, P. (1998). A unified method for computing on incompressible and compressible flows in boundary-fitted coordinates. *J. Comput. Phys.*, **141**, 153–173.

Braack, M. (2001). Adaptive finite elements for stationary compressible flows at low Mach numbers. In *Hyperbolic Problems: Theory, Numerics, Applications* (ed. M. Freistühler and G. Warnecke), pp. 169–176. Birkhäuser, Basel.

Braack, M. (2002). A finite element method for subsonic viscous flows: I. The stationary case. Universität Heidelberg, SFB 359.

Brenner, S. and Scott, R. L. (1994). *The Mathematical Theory of Finite Element Methods*. Springer, New York.

Bressan, A. (2000). *Hyperbolic Systems of Conservation Laws. The One-dimensional Cauchy Problem.* Oxford Lecture Series. Oxford University Press, Oxford.

Brezzi, F., Manzini, G., Marini, D., Pietra, P., and Russo, A. (2000). Discontinuous Galerkin approximations for elliptic problems. *Numer. Methods Partial Differ. Equations*, **16**, 335–378.

Bristeau, M. O., Glowinski, R., Periaux, J., and Viviand, H. (ed.) (1987). *Numerical Simulation of Compressible Navier–Stokes Flows*, Number 18 in Notes on Numerical Fluid Mechanics. Vieweg, Braunschweig.

Buscaglia, G. C. and Dari, E. A. (1977). Anisotropic mesh optimization and its application in adaptivity. *Int. J. Numer. Methods Eng.*, **40**, 4119–4136.

Casper, J. and Atkins, H. (1993). A finite-volume high-order ENO scheme for two dimensional hyperbolic systems. *J. Comput. Phys.*, **106**, 62–76.

Castillo, P., Cockburn, B., Schötzau, D., and Schwab, C. (2001). Optimal a priori error estimates for the *hp*-version of the local discontinuous Galerkin method for convection–diffusion problems. *Math. Comput.*, **71**, 455–478.

Castro-Díaz, M. J., Borouchaki, H., George, P. L., Hecht, F., and Mohammadi, B. (1996). Anisotropic adaptive mesh generation in two dimensions for CFD. In *Computational Fluid Dynamics '96* (ed. J.-A. Désidéri, C. Hirsch, P. Le Tallec, M. Pandolfi, and J. Périaux), Paris, pp. 181–186. Wiley, Chichester.

Champier, S., Gallouet, T., and Herbin, R. (1993). Convergence of an upstream finite volume scheme for a nonlinear hyperbolic equation on a triangular mesh. *Numer. Math.*, **66(2)**, 139–157.

Chen, G.-Q. and Le Floch, Ph. (2000). Compressible Euler equations with general pressure law. *Arch. Ration. Mech. Anal.*, **153(3)**, 221–259.

Chen, G.-Q. and Liu, J. G. (1993). Convergence of second order schemes for isentropic gas dynamics. *Math. Comput.*, **61**, 607–627.

Chen, G.-Q. and Wang, D. (2001). The Cauchy problem for the Euler equations for compressible fluids. In *Handbook of Mathematical Fluid Dynamics*, Volume 1, pp. 421–543. North-Holland, Amsterdam.

Chiocchia, G. (1985). Exact solutions to transonic and supersonic flows. In *Test Cases for Inviscid Flow Field Methods*, AGARD Advisory Report 211.

Chorin, A. J. and Marsden, J. E. (1993). *A Mathematical Introduction to Fluid Mechanics.* Springer, Berlin.

Ciarlet, P. G. (1979). *The Finite Element Method for Elliptic Problems.* North-Holland, Amsterdam.

Cockburn, B. (1999). Discontinuous Galerkin methods for convection dominated problems. In *High-Order Methods for Computational Physics* (ed. T. J. Barth and H. Deconinck), Number 9 in Lecture Notes in Computational Science and Engineering, pp. 69–224. Springer, Berlin.

Cockburn, B., Coquel, F., and Le Floch, Ph. (1994). An error estimate for finite volume methods for multidimensional conservation laws. *Math. Comput.*, **63**, 77–103.

Cockburn, B., Coquel, F., and Le Floch, Ph. (1995). Convergence of finite volume method for multidimensional conservation laws. *SIAM J. Numer. Anal.*, **32(3)**, 687–706.

Cockburn, B. and Gremaud, P.-A. (1996*a*). A priori error estimates for numerical methods for scalar conservation laws. Part I: The general approach. *Math. Comput.*, **65**, 553–573.

Cockburn, B. and Gremaud, P.-A. (1996*b*). Error estimates for finite element methods for scalar conservation laws. *SIAM J. Numer. Anal.*, **33**, 522–554.

Cockburn, B., Hou, S., and Shu, C. W. (1990). TVB Runge–Kutta local projection discontinuous Galerkin finite element for conservation laws IV: The multi-dimensional case. *Math. Comput.*, **54**, 545–581.

Cockburn, B., Karniadakis, G. E., and Shu, C.-W. (ed.) (2000). *Discontinuous*

Galerkin Methods, Number 11 in Lecture Notes in Computational Science and Engineering. Springer, Berlin.

Cockburn, B. and Shu, C. W. (1998). The local discontinuous Galerkin finite element method for convection–diffusion systems. *SIAM J. Numer. Anal.*, **35**, 2440–2463.

Cockburn, B. and Shu, S. W. (1989). TVB Runge–Kutta local projection discontinuous Galerkin finite element method for conservation laws II: General framework. *Math. Comput.*, **52**, 411–435.

Coré, X., Angot, P., and Latché, J.-C. (2002). A multilevel FIC projection method for low Mach natural convection flows. In *Finite Volumes for Complex Applications III* (ed. R. Herbin and D. Kröner), pp. 317–324. Hermes Penton Science, London.

Cournede, P.-H., Debiez, Ch., and Dervieux, A. (1998). A positive MUSCL scheme for triangulations. Technical Report 3465, INRIA.

Crouzeix, M. and Raviart, P. A. (1973). Conforming and nonconforming finite element methods for solving the stationary Stokes equations. *RAIRO, Anal. Numér.*, **7**, 33–76.

Dafermos, C. (2000). *Hyperbolic Conservation Laws in Continuum Physics.* Springer, Heidelberg.

D'Azevedo, E. F. and Simpson, R. B. (1989). On optimal interpolation triangle incidences. *SIAM. J. Sci. Statist. Comput.*, **10**, 1063–1075.

D'Azevedo, E. F. and Simpson, R. B. (1991). On optimal triangular meshes for minimizing the gradient error. *Numer. Math.*, **59**, 321–348.

Delaunay, B. (1934). Sur la sphere vide. *Bull. Acad. Sci. USSR*, 793–800.

Demkowicz, L., Oden, J. T., and Rachowicz, W. (1990). A new finite element method for solving compressible Navier–Stokes equations based on an operator splitting method and *hp* adaptivity. *Comput. Methods Appl. Mech. Eng.*, **84**, 275–326.

Dervieux, A. and Vijayasundaram, G. (1983). On numerical schemes for solving Euler equations of gas dynamics. In *Numerical methods for Euler equations of fluid dynamics, Proceedings of INRIA workshop* (ed. F. Angrand *et al.*), Rocquencourt, France, pp. 121–144. SIAM, Philadelphia (1985).

Díaz, M. J. Castro, Hecht, F., and Mohammadi, B. (1995). New progress in anisotropic grid adaptation for inviscid and viscous flows simulations. In *Proceedings of the 4th Annual International Meshing Roundtable.* Sandia National Laboratories.

Dick, E. (1988). A multigrid method for steady Euler equations based on flux–difference splitting with respect to primitive variables. *Not. Numer. Fluid Mech.*, **23**, 69–85.

Dick, E. (1990). Multigrid formulation of polynomial flux–difference splitting for steady Euler equations. *J. Comput. Phys.*, **91**, 161–173.

Dick, E. (1991). *Multigrid methods for steady Euler- and Navier–Stokes equations based on polynomial flux-difference splitting*, pp. 1–20. Number 98 in International Series of Numerical Mathematics. Birkhäuser, Basel.

Dinkler, D., Dornberger, R., Grohmann, B., and Kröplin, B. (1997). An adaptive multigrid method for the finite space–time element discretization of the unsteady Euler equations. In *Numerical Modelling in Continuum Mechanics* (ed. M. Feistauer, R. Rannacher, and K. Kozel), pp. 236–245. MATFYZPRESS, Prague.

Dolejší, V. (1998*a*). Adaptive methods for the numerical solution of the compressible Navier–Stokes equations. In *Computational Fluid Dynamics '98* (ed. K. D. Papailiou, D. Tsahalis, J. Périaux, C. Hirsch, and M. Pandolfi), Volume 1, pp. 393–397. ECCOMAS, Wiley, Chichester.

Dolejší, V. (1998*b*). Anisotropic mesh adaptation for finite volume and finite element methods on triangular meshes. *Comput. Visualization Sci.*, **1(3)**, 165–178.

Dolejší, V. (1998*c*). *Sur des méthodes combinant des volumes finis et des éléments finis pour le calcul d'écoulements compressibles sur des maillages non structurés.* Ph. D. thesis, Charles Univeristy Prague and Université Mediterannée Aix–Marseille II.

Dolejší, V. (2000). Numerical simulation and adaptive methods for transonic flow. In *ECCOMAS 2000*. CD-ROM. ISBN 84-89925-70-4.

Dolejší, V. (2001). Anisotropic mesh adaptation technique for viscous flow simulation. *East–West J. Numer. Math.*, **9(1)**, 1–24.

Dolejší, V. (2002). On the discontinuous Galerkin method for the numerical solution of the Euler and Navier–Stokes equations. *Int. J. Numer. Meth. Fluids*. (submitted).

Dolejší, V. (2003). *Adaptive higher order methods for compressible flow.* Habilitation Thesis. Charles University, Prague.

Dolejší, V. and Angot, Ph. (1996). Finite volume methods on unstructured meshes for compressible flows. In *Finite Volumes for Complex Applications (Problems and Perspectives)* (ed. F. Benkhaldoun and R. Vilsmeier), pp. 667–674. Hermes, Rouen.

Dolejší, V. and Feistauer, M. (2001). On the discontinuous Galerkin method for the numerical solution of compressible high-speed flow. In *Proceedings of the Conference ENUMATH 2001*. Springer Italia.

Dolejší, V. and Feistauer, M. (2003*a*). Discontinuous Galerkin finite element method for convection–diffusion problems and the compressible Navier–Stokes equations. The Preprint Series of the School of Mathematics MATH-KNM-2003/1, Charles University, Prague.

Dolejší, V. and Feistauer, M. (2003*b*). A semiimplicit discontinuous Galerkin finite element method for the numerical solution of inviscid compressible flows. The Preprint Series of the School of Mathematics MATH-KNM-2003/2, Charles University, Prague.

Dolejší, V., Feistauer, M., Felcman, J., and Kliková, A. (2002). Error estimates for barycentric finite volumes combined with nonconforming finite elements applied to nonlinear convection–diffusion problems. *Appl. Math.*, **47(4)**, 301–340.

Dolejší, V., Feistauer, M., and Schwab, C. (2002*a*). A finite volume discontinuous Galerkin scheme for nonlinear convection–diffusion problems. *Calcolo*, **39**, 1–40.

Dolejší, V., Feistauer, M., and Schwab, C. (2002*b*). On discontinuous Galerkin methods for nonlinear convection–diffusion problems and compressible flow. *Math. Bohemica*, **127(2)**, 163–179.

Dolejší, V., Feistauer, M., and Schwab, C. (2003). On some aspects of the discontinuous Galerkin finite element method for conservation laws. *Math. Comput. Simul.*, **61**, 333–346.

Dolejší, V., Feistauer, M., Schwab, C., and Sobotíková, V. (2003). Analysis of the discontinuous Galerkin method for nonlinear convection–diffusion problems. The Preprint Series of the School of Mathematics MATH-KNM-2003/3, Charles University, Prague.

Dolejší, V. and Felcman, J. (2001). Anisotropic mesh adaptation and its application for scalar diffusion equations. The Preprint Series of the School of Mathematics MATH-KNM-2001/1, Charles University, Prague.

Durlofsky, L. J., Enquist, B., and Osher, S. (1992). Triangle based adaptive stencils for the solution of hyperbolic conservation laws. *J. Comput. Phys.*, **98**, 64–73.

Edwards, R. E. (1965). *Functional Analysis, Theory and Applications.* Holt, Rinehart and Winston, New York.

Eymard, R., Gallouët, T., and Herbin, R. (2000). *Handbook of Numerical Analysis*, Volume VII, Chapter Finite Volume Methods, pp. 717–1020. North-Holland-Elsevier, Amsterdam.

Feireisl, E., Novotný, A., and Petzeltová, H. (2001). On the existence of globally defined weak solutions to the Navier–Stokes equations of compressible isentropic fluids. *J. Math. Fluid Mech.*, **3**, 358–392.

Feistauer, M. (1993). *Mathematical Methods in Fluid Dynamics.* Longman Scientific & Technical, Harlow.

Feistauer, M. (1998). Analysis in compressible fluid mechanics. *ZAMM*, **78**, 579–596.

Feistauer, M. (2002). Discontinuous Galerkin method: Compromise between FV and FE schemes. In *Finite Volumes for Complex Applications III* (ed. R. Herbin and D. Kröner), pp. 81–96. Hermes Penton Science, London.

Feistauer, M. and Felcman, J. (1997). Theory and applications of numerical schemes for nonlinear convection–diffusion problems and compressible viscous flow. In *The Mathematics of Finite Elements and Applications, Highlights 1996* (ed. J. Whiteman), pp. 175–194. Wiley, Chichester.

Feistauer, M., Felcman, J., and Dolejší, V. (1996). Numerical simulation of compresssible viscous flow through cascades of profiles. *ZAMM*, **76**, 297–300.

Feistauer, M., Felcman, J., and Lukáčová-Medviďová, M. (1995). Combined finite element–finite volume solution of compressible flow. *J. Comput. Appl. Math.*, **63**, 179–199.

Feistauer, M., Felcman, J., and Lukáčová-Medviďová, M. (1997). On the convergence of a combined finite volume–finite element method for nonlinear convection–diffusion problems. *Numer. Methods Partial Differ. Equations*, **13**, 163–190.

Feistauer, M., Felcman, J., Lukáčová-Medviďová, M., and Warnecke, G. (1999*a*). Error estimates of a combined finite volume–finite element method for nonlinear convection–diffusion problems. *SIAM J. Numer. Anal.*, **36(5)**, 1528–1548.

Feistauer, M. and Kliková, A. (2001). Adaptive methods for the solution of compressible flow. In *Hyperbolic Problems: Theory, Numerics, Applications* (ed. H. Freistühler and G. Warnecke), Volume I, pp. 363–372. Birkhäuser, Basel.

Feistauer, M., Slavík, J., and Stupka, P. (1999*b*). On the convergence of a combined finite volume–finite element method for nonlinear convection–diffusion problems. Explicit schemes. *Numer. Methods Partial Differ. Equations*, **15**, 215–235.

Felcman, J., Dolejší, V., and Feistauer, M. (1994). Adaptive finite volume method for the numerical solution of the compressible Euler equations. In *Computational Fluid Dynamics '94* (ed. J. P. S. Wagner, E. H. Hirschel and R. Piva), pp. 894–901. Wiley, Stuttgart.

Felcman, J. and Šolín, P. (1998). On the construction of the Osher–Solomon scheme for 3D Euler equations. *East–West J. Numer. Math.*, **6(1)**, 43–64.

Fernandez, G. (1989). *Schemas conservativs implicites et decentres pour des ecoulements fortement non stationnaires.* INRIA, Sophia Antipolis.

Ferziger, J. H. and Perić, M. (1996). *Computational Methods for Fluid Dynamics.* Springer, Berlin.

Fey, M. (1995). Decompositions of the multidimensional Euler equations into advection equations. SAM Research Report 95-14, ETH Zürich.

Fey, M. and Jeltsch, R. (1992*a*). A new multidimensional Euler-scheme. In *Seminar für Angewandte Mathematik*, ETH Zürich.

Fey, M. and Jeltsch, R. (1992*b*). A simple multidimensional Euler-scheme. In *Seminar für Angewandte Mathematik*, ETH Zürich. Also in Proceedings of the 1st European Computational Fluid Dynamics Conference, Brussels, 7–11 September 1992.

Fezoui, L. and Stouffet, B. (1989). A class of implicit upwind schemes for Euler simulations with unstructured meshes. *J. Comput. Phys.*, **84**, 174–206.

Flaherty, J. E., Loy, R. M., Shephart, M. S., and Teresco, J. D. (2000). Software for the parallel adaptive solution of conservation laws by discontinuous Galerkin methods. In *Discontinuous Galerkin Methods: Theory, Computation and Applications* (ed. B. Cockburn, G. E. Karniadakis, and G. W. Shu), pp. 113–124. Springer, Berlin.

Fletcher, C. A. J. (1991). *Computational Techniques for Fluid Dynamics I, II.* Springer, Berlin.

Fortin, M., Vallet, M.-G., Dompierre, J., Bourgault, Y., and Habashi, W. G. (1996). Anisotropic mesh adaptation: Theory, validation and applications. In

Computational Fluid Dynamics '96 (ed. J.-A. Désidéri, C. Hirsch, P. Le Tallec, M. Pandolfi, and J. Périaux), Paris, pp. 174–180. Wiley, Chichester.

Frey, P. J. and George, P.-L. (2000). *Mesh Generation – Application to Finite Elements.* Hermes Science Publishing, Oxford & Paris.

Friedrich, O. (1998). Weighted essential non-oscillatory schemes for the interpolation of mean values on unstructured grids. *J. Comput. Phys.*, **144**, 194–212.

Fürst, J. and Kozel, K. (1999). Using TVD and ENO schemes for numerical solution of multidimensional system of Euler and Navier–Stokes equations. In *Navier–Stokes equations: theory and numerical methods* (ed. R. Salvi), Pitman Research Notes in Mathematics Series, pp. 264–276. Longman, London.

Fürst, J. and Kozel, K. (2001). Numerical solution of inviscid and viscous flows using modern schemes and quadrilateral or triangular mesh. *Math. Bohemica*, **126(2)**, 379–393.

Fürst, J. and Kozel, K. (2002*a*). Application of second order TVD and ENO schemes in internal aerodynamics. *J. Sci. Comput.*, **17**, 263–272.

Fürst, J. and Kozel, K. (2002*b*). Numerical solution of transonic flows through 2D and 3D turbine cascades. *Comput. Visualization Sci.*, **4(3)**, 183–189.

Ghidaglia, J. M. and Pascal, F. (1999). Footbridges finite volumes–finite elements. In *C. R. Acad. Sci., Ser. I, Math. 328*, Volume 8, pp. 711–716.

Giles, M. B. (1990). Non-reflecting boundary conditions for Euler equation calculations. *AIAA J.*, **42**, 2050–2058.

Girault, V. and Raviart, P. A. (1979). *Finite Element Approximation of the Navier–Stokes Equations.* Number 749 in Lecture Notes in Mathematics. Springer, Berlin.

Girault, V. and Raviart, P. A. (1986). *Finite Element Methods for the Navier–Stokes Equations.* Springer, Berlin.

Glimm, J. (1965). Solutions in the large for nonlinear hyperbolic systems of equations. *Commun. Pure Appl. Math.*, **18(4)**, 697–715.

Glowinski, R. (1984). *Numerical Methods for Nonlinear Variational Problems.* Springer, New York.

Glowinski, R., Lions, J. L., and Trémalières, R. (1976). *Analyse numérique des inéquations variationnelles.* Dunod, Paris.

Godlewski, E. and Raviart, P. A. (1991). *Hyperbolic Systems of Conservation Laws.* Number 3–4 in Mathematiques & Applications. Ellipses, Paris.

Godlewski, E. and Raviart, P. A. (1996). *Numerical Approximation of Hyperbolic Systems of Conservation Laws.* Number 118 in Applied Mathematical Sciences. Springer, New York.

Godunov, S. K. (1959). A difference schemes for numerical computation of discontinuous solutions of equations of fluid dynamics. *Math. Sb.*, **47(89)**, 271–306.

Göhner, U. and Warnecke, G. (1995). A second order finite difference error indicator for adaptive transonic flow computations. *Numer. Math.*, **70**, 129–161.

Gresho, P. M. and Sani, R. L. (2000). *Incompressible Flow and the Finite Element Method.* Wiley, Chichester.

Gustafsson, B. (1982). The choice of numerical boundary conditions for hyperbolic systems. *J. Comput. Phys.*, 270–283.

Gustafsson, B. and Fern, L. (1987). Far fields boundary conditions for steady state solutions to hyperbolic systems. In *Nonlinear hyperbolic problems, Proceedings, St Etienne 1986*, Number 1270 in Lecture Notes in Mathematics, pp. 238–252. Springer, Berlin.

Gustafsson, B. and Fern, L. (1988). Far fields boundary conditions for time-dependent hyperbolic systems. *SIAM J. Sci. Statist. Comput.*, **9**, 812–848.

Gustafsson, B. and Kreiss, H.-O. (1979). Boundary conditions for time-dependent problems with an artificial boundary. *J. Comput. Phys.*, **30**, 333–351.

Habashi, W. G., Fortin, M., Ait-ali-Yahia, D., Boivin, S., Y.Bourgault, Dompierre, J., Robichaud, M. P., Tam, A., and Vallet, M. G. (1996). Anisotropic mesh optimization: Towards a solver-independent and mesh-independent CFD. In *Compressible Fluid Dynamics*, VKI Lecture Series 1996-06. von Karman Institute for Fluid Dynamics - Concordia University.

Hackbusch, W. (1989). On first and second order box schemes. In *Computing 44*, pp. 277–296.

Hagstrom, T. and Hariharan, S. I. (1988). Accurate boundary conditions for exterior problems in gas dynamics. *Math. Comput.*, **51**, 581–597.

Hansbo, P. (1993). Explicit streamline diffusion finite element methods for compressible Euler equations in conservation variables. *J. Comput. Phys.*, **109**, 274–288.

Hansbo, P. and Johnson, C. (1991). Adaptive streamline diffusion methods for compressible flow using conservation variables. *Comput. Methods Appl. Mech. Eng.*, **87**, 267–280.

Harten, A. (1983). On the symmetric form of systems of conservation laws with entropy. *J. Comput. Phys.*, **49**, 151–164.

Harten, A. (1989). ENO schemes with subcell resolutions. *J. Comput. Phys.*, **83**, 148–184.

Harten, A. (1991). Multi-dimensional ENO schemes for general geometries. ICASE Report 91-76, Langley Research Center, Hampton.

Harten, A., Engquist, B., Osher, S., and Chakravarthy, S. (1987). Uniformly high order essentially non-oscillatory schemes III. *J. Comput. Phys.*, 231–303.

Harten, A., Hyman, J. M., and Lax, P. D. (1976). On finite difference approximations and entropy conditions for shocks. *Commun. Pure Appl. Math.*, **29**, 297–322.

Harten, A., Lax, P. D., and Van Leer, B. (1983). On upstream differencing and Godunov-type schemes for hyperbolic conservation laws. *SIAM Rev.*, **25**, 35–61.

Harten, A. and Osher, S. (1987). Uniformly high-order accurate non-oscillatory schemes I. *SIAM J. Numer. Anal.*, **24**, 279–309.

Hartmann, R. and Houston, P. (2001*a*). Adaptive discontinuous Galerkin finite element methods for nonlinear hyperbolic conservation laws. Technical Report 2001-20 (SFB 359), IWR Heidelberg.

Hartmann, R. and Houston, P. (2001*b*). Adaptive discontinuous Galerkin finite element methods for the compressible Euler equations. Technical Report 2001-42 (SFB 359), IWR Heidelberg.

Hedstrom, G. W. (1979). Nonreflecting boundary conditions for nonlinear hyperbolic systems. *J. Comput. Phys.*, **30**, 222–237.

Heinrich, B. (1987). *Finite Difference Methods on Irregular Networks.* Birkhäuser, Basel.

Hemker, P. W. and Spekreijse, S. (1986). Multiple and Osher's scheme for the efficient solution of the steady Euler equations. *Appl. Numer. Math.*, **2**, 475–493.

Hindmarsh, A. C. (1983). ODEPACK, a systematized collection of ODE solvers. In *Scientific Computing*, Number 1 in IMACS Transactions on Scientific Computation, pp. 55–64. North-Holland, Amsterdam.

Hinton, E. and Owen, D. R. J. (1977). *Finite Element Programming.* Academic Press, London.

Hirsch, Ch. (1988). *Numerical Computation of Internal and External Flows*, Volume 1. Wiley, New York.

Hoff, D. (1995). Global solution of the Navier–Stokes equations for multidimensional, compressible flow with discontinuous initial data. *J. Differ. Equations*, **120**, 215–254.

Hoff, D. (1997). Discontinuous solutions of the Navier–Stokes equations for multidimensional heat-conducting flows. *Arch. Ration. Mech. Anal.*, **139**, 303–354.

Hou, T. and Le Floch, Ph. (1994). Why nonconservative schemes converge to wrong solutions: Error analysis. *Math. Comput.*, **62**, 497–530.

Houston, P., Schwab, C., and Süli, E. (2002). Discontinuous *hp*-finite element method for advection–diffusion problems. *SIAM J. Numer. Anal.*, **39**, 2133–2163.

Houston, P. and Süli, E. (2001). Adaptive Lagrange–Galerkin method for unsteady convection-dominated diffusion problem. *Math. Comput.*, **70**, 77–106.

Hu, C. and Shu, C.-W. (1998). High order weighted ENO schemes for unstructured meshes: Preliminary results. In *Computational fluid dynamics '98* (ed. K. Papailiou, D. Tsahalis, J. Periaux, and D. Knorzer), Volume 2, pp. 356–362. Wiley, Chichester.

Hughes, T. J., Franca, L. P., and Hulbert, G. M. (1989). A new finite element formulation for computational fluid dynamics: VIII. The Galerkin/least squares method for advective–diffusive equations. *Comput. Methods Appl. Mech. Eng.*, **73**, 173–189.

Hughes, T. J., Franca, L. P., and Mallet, M. (1986). A new finite element formulation for computational fluid dynamics: I. Symmetric forms of the compressible Euler and Navier–Stokes equations and the second law of thermodynamics. *Comput. Methods Appl. Mech. Eng.*, **54**, 223–234.

Hughes, T. J., Franca, L. P., and Mallet, M. (1987). A new finite element formulation for computational fluid dynamics: IV. Convergence analysis of the generalized SUPG formulation for linear time-dependent multidimensional advective–diffusive systems. *Comput. Methods Appl. Mech. Eng.*, **63**, 97–112.

Hughes, T. J. and Mallet, M. (1986*a*). A new finite element formulation for computational fluid dynamics: III. The generalized streamline operator for multidimensional advective–diffusive systems. *Comput. Methods Appl. Mech. Eng.*, **58**, 305–328.

Hughes, T. J. and Mallet, M. (1986*b*). A new finite element formulation for computational fluid dynamics: IV. A discontinuity-capturing operator for multidimensional advective–diffusive systems. *Comput. Methods Appl. Mech. Eng.*, **58**, 329–336.

Hughes, T. J., Mallet, M., and Mizukami, A. (1986). A new finite element formulation for computational fluid dynamics: II. Beyond SUPG. *Comput. Methods Appl. Mech. Eng.*, **54**, 341–355.

Hughes, T. J. R. and Brooks, A. N. (1979). A multidimensional upwind scheme with no crosswind diffusion. In *Finite Element Methods for Convection Dominated Flow* (ed. T. J. R. Hughes), Number 34 in ADM, pp. 19–35. ASME, New York.

Hughes, T. J. R. and Shakib, F. (1988). Computational aerodynamics and the finite element method. Technical report, AIAA-88-0031.

Hughes, T. J. R. and Tezduyar, T. E. (1984). Finite element methods for first-order hyperbolic systems with particular emphasis on the compressible Euler equations. *Comput. Methods Appl. Mech. Eng.*, **45**, 217–284.

Ikeda, T. (1983). *Maximum Principle in Finite Element Methods for Convection–Diffusion Phenomena.* North-Holland, Amsterdam.

Iljin, A. M. (1969). A difference scheme for a differential equation with a small parameter affecting the highes derivative. *Mat. zametki*, **6(4)**, 237–248. (in Russian).

Isaacson, E. and Temple, B. (1995). Convergence of a 2×2 Godunov method for a general resonant nonlinear balance law. *SIAM J. Appl. Math.*, **55**, 625–640.

Jaffre, J., Johnson, C., and Szepessy, A. (1995). Convergence of the discontinuous Galerkin finite element method for hyperbolic conservation laws. *Math. Models Methods Appl. Sci.*, **5**, 367–386.

Jeltsch, R. and Botta, N. (1994). A numerical method for unsteady flows. SAM Research Report 94-11, ETH Zürich.

Jeng, Y. N. and Chen, J. L. (1992). Truncation error analysis of finite volume method for a model steady convective equation. *J. Comput. Phys.*, **100**, 64–76.

Jiang, G. and Shu, C.-W. (1996). Efficient implementation of weighted ENO schemes. *J. Comput. Phys.*, **126**, 202–228.

Johnson, C. (1987). *Numerical Solution of Partial Differential Equations by the Finite Element Method.* Cambridge University Press, New York.

Johnson, C. (1994). A new paradigm for adaptive finite element methods. In

The Mathematics of Finite Elements and Applications, Highlights 1993 (ed. J. Whiteman), pp. 105–120. Wiley, Chichester.

Johnson, C. and Hansbo, P. (1992). Adaptive finite element methods in computational mechanics. *Comput. Methods Appl. Mech. Eng.*, **101**, 143–181.

Johnson, C., Hansbo, P., and Szepessy, A. (1990). On the convergence of shock capturing streamline diffusion finite element methods for hyperbolic conservation laws. *Math. Comput.*, **54**, 107–129.

Johnson, C., Nävert, U., and Pitkäranta, J. (1984). Finite element methods for linear hyperbolic problem. *Comput. Methods Appl. Mech. Eng.*, **45**, 285–312.

Johnson, C. and Pitkäranta, J. (1986). An analysis of the discontinuous Galerkin method for a scalar hyperbolic equation. *Math. Comput.*, **46**, 1–26.

Johnson, C. and Szepessy, A. (1987). On the convergence of a finite element method for a nonlinear hyperbolic conservation laws. *Math. Comput.*, **49**, 427–444.

Karni, S. (1992). Accelerated convergence to steady state by gradual far-field damping. *AIAA J.*, **30**, 1220–1228.

Karniadakis, G. E. and Sherwin, S. J. (1999). *Spectral/hp Element Methods for CFD*. Oxford University Press, Oxford.

Kellog, R. B. and Liu, B. (1996). A finite element method for the compressible Stokes equations. *SIAM J. Numer. Anal.*, **33**, 780–788.

Kellog, R. B. and Liu, B. (1997). A finite element method for the compressible Stokes system. *SIAM J. Numer. Anal.*, **34**, 1093–1105.

Klainerman, S. and Majda, A. (1981). Singular limits of quasilinear hyperbolic systems with large parameters and the incompressible limit of compressible fluids. *Commun. Pure Appl. Math.*, **34**, 481–524.

Klein, R. (1995). Semi-implicit extension of a Godunov-type scheme based on low Mach number asymptotics I. *J. Comput. Phys.*, **121**, 213–237.

Klein, R., Botta, N., Schneider, T., Munz, C. D., Roller, S., Meister, L., Hoffmann, L., and Sonar, T. (2000). Asymptotic adaptive methods for multi-scale problems in fluid mechanics. *J. Eng. Math.*. Accepted for publication.

Klein, R. and Munz, C. D. (1994). The multiple pressure variable (MVP) approach for the numerical approximation of weakly compressible fluid flow. In *Proceedings of the International Conference on Numerical Modeling in Continuum Mechanics*, Volume II, Prague, pp. 151–161.

Klein, R. and Munz, C.-D. (1995). The multiple pressure variable (MPV) approach for the numerical approximation of weakly compressible fluid flow. In *Numerical Modelling in Continuum Mechanics* (ed. M. Feistauer, R. Rannacher, and K. Kozel), pp. 151–161. Faculty of Mathematics and Physics, Charles University, Prague.

Kliková, A. (2000). *Finite volume–finite element solution of compressible flow*. Ph. D. thesis, Charles Univeristy, Prague.

Koren, B. (1996). Improving Euler computations at low Mach numbers. *Comput. Fluid Dyn.*, **6**, 51–70.

Koren, B. and Hemker, P. W. (1991). Dampled, direction-dependent multigrid for hypersonic flow computations. *Appl. Numer. Math.*, **7**, 309–328.

Koren, B. and van Leer, B. (1995). Analysis of preconditioning and multigrid for Euler flows with low-subsonic regions. *Adv. Comput. Math.*, **4**, 127–144.

Kovenya, V. M., Tarnavskii, G. A., and Chornyi, S. G. (1990). *Application of Splitting-Up Method in Problems of Aerodynamics.* Nauka, Novosibirsk. (in Russian).

Kovenya, V. M. and Yanenko, N. N. (1981). *Splitting-Up Method in Problems for Gas Dynamics.* Nauka, Novosibirsk. (in Russian).

Křížek, M. (1991). On semiregular families of triangulations and linear interpolation. *Appl. Math.*, **36**, 223–232.

Křížek, M. and Neittaanmäki, P. (1990). *Finite Element Approximation of Variational Problems and Applications.* Longman, Harlow.

Kröner, D. (1991). Absorbing boundary conditions for the linearized Euler equations in 2-D. *Math. Comput.*, **57**, 153–167.

Kröner, D. (1997). *Numerical Schemes for Conservation Laws.* Wiley–Teubner, Stuttgart.

Kröner, D., Noelle, S., and Rokyta, M. (1995). Convergence of higher order upwind finite volume schemes on unstructured grids for scalar conservation laws in several space dimensions. *Numer. Math.*, **71**, 527–560.

Kröner, D. and Rokyta, M. (1994). Convergence of upwind finite volume schemes for scalar conservation laws in 2D. *SIAM J. Numer. Anal.*, **31**, 324–343.

Kröner, D., Rokyta, M., and Wierse, M. (1996). A Lax–Wendroff type theorem for upwind finite volume schemes in 2-D. *East–West J. Numer. Math.*, **4**, 279–292.

Kruzhkov, S. N. (1970). First order quasilinear equations in several independent variables. *Math. USSR Sb.*, **10**, 217–243.

Kufner, A., John, O., and Fučík, S. (1977). *Function Spaces.* Academia, Prague.

Kurzweil, J. (1986). *Ordinary Differential Equations.* Elsevier, Amsterdam.

Kuznetsov, N. N. (1976). Accuracy of some approximate methods for computing the weak solutions of a first-order quasi-linear equation. *USSR Comput. Math. Phys.*, **16**, 105–119.

Lallemand, M.-H., Steve, H., and Dervieux, A. (1992). Unstructured multi-grid by volume agglomeration: Current status. *Comput. Fluids*, **21(3)**, 397–433.

Landau, L. D. and Lifschitz, E. M. (1959). *Fluid Mechanics.* Pergamon Press, London.

Larrouturou, B. (1991). How to preserve the mass fractions positivity when computing compressible multicomponent flows. *J. Comput. Phys.*, **92**, 273–295.

Lax, P. and Wendroff, B. (1960). Systems of conservation laws. *Commun. Pure Appl. Math.*, **13**, 217–237.

Le Beau, G. J., Ray, S. E., Aliabadi, S. K., and Tezduyar, T. E. (1993). SUPG finite element computation of compressible flows with the entropy and conservation variables formulations. *Comput. Methods Appl. Mech. and Eng.*, **104**,

397–422.

Le Saint, P. and Raviart, P. A. (1974). On a finite element method for solving the neutron transport equation. In *Mathematical Aspects of Finite Elements in Partial Differential Equations* (ed. C. de Boor), pp. 89–145. Academic Press, New York.

Le Veque, R. J. (1992). *Numerical Methods for Conservation Laws.* Birkhäuser, Basel.

Linss, T. and Stynes, M. (2001). The SDFEM on Shishkin meshes for linear convection–diffusion problems. *Numer. Math.*, **87**, 457–484.

Lions, J. L. (1969). *Quelques méthodes de résolution des problémes aux limites non linéaires.* Dunod, Paris.

Lions, P. L. (1993). Existence globale de solutions pour les équations de Navier–Stokes compressibles isentropiques. *C.R. Acad. Sci. Paris*, **316**, 1335–1340.

Lions, P. L. (1998). *Mathematical Topics in Fluid Mechanics, Volume 2: Compressible Models.* Oxford Science Publications, Oxford.

Liu, B. (2000). The analysis of a finite element method with streamline diffusion for the compressible Navier–Stokes equations. *SIAM J. Numer. Anal.*, **38(1)**, 1–16.

Liu, T. P. (1975). The Riemann problem for general systems of conservation laws. *J. Differ. Equations*, **56**, 218–234.

Liu, X.-D. and Osher, S. (1998). Convex ENO high order multi-dimensional schemes without field by field decomposition or staggered grids. *J. Comput. Phys.*, **42**, 304–330.

Liu, X.-D., Osher, S., and Chan, T. (1994). Weighted essentially nonoscillatory schemes. *J. Comput. Phys.*, **115**, 200–212.

Ljusternik, L. A. and Sobolev, V. I. (1974). *Elements of Functional Analysis.* Hindustan Publishing, Delphi–New York.

Löhner, R., Morgan, K., and Zienkiewicz, O. C. (1984). The solutions of nonlinear hyperbolic equation systems by the finite element method. *Int. J. Numer. Methods Fluids*, **4**, 1043–1063.

Löhner, R., Morgan, K., and Zienkiewicz, O. C. (1985). An adaptive finite element procedure for compressible high speed flows. *Comput. Methods Appl. Mech. Eng.*, **51**, 411–465.

Loitsianskii, S. G. (1973). *Mechanics of Liquids and Gases.* Nauka, Moscow. (in Russian).

Lomtev, I., Quillen, C. W., and Karniadakis, G. E. (1998). Spectral/hp methods for viscous compressible flows on unstructured 2D meshes. *J. Comput. Phys.*, **144**, 325–357.

Lube, G. and Weiss, D. (1995). Stabilized finite element methods for singularly perturbed parabolic problems. *Appl. Numer. Math.*, **17**, 431–459.

Mackenzie, J. A., Süli, E., and Warnecke, G. (1994). A posteriori error estimates for the cell–vertex finite volume method. In *Proceedings of the 9th GAMM Seminar on Adaptive Methods* (ed. W. Hackbusch and G. Wittum), pp. 221–235. Vieweg, Braunschweig–Wiesbaden.

Majda, A. (1984). *Compressible Fluid Flow and Systems of Conservation Laws in Several Space Variables.* Springer, New York.

Málek, J., Nečas, J., Rokyta, M., and Růžička, M. (1996). *Weak and Measure-Valued Solutions to Evolutionary PDEs.* Chapman & Hall, London.

Matsumura, A. and Nishida, T. (1982). Initial boundary value problems for the equations of motion of general fluids. In *Computing Methods in Applied Sciences and Engineering* (ed. R. Glowinski and J. L. Lions), Volume V, pp. 389–406. North-Holland, Amsterdam.

Matsumura, A. and Nishida, T. (1983). Initial boundary value problems for the equations of motion of compressible viscous and heat-conductive fluids. *Commun. Math. Phys.*, **89**, 445–464.

Matsumura, A. and Padula, M. (1992). Stability of stationary flows of compressible fluids subject to large external potential forces. *Stabil. Appl. Anal. Contin. Media*, **2**, 183–202.

Matsumura, A. and Yamagata, N. (2001). Global weak solutions of the Navier–Stokes equations for multidimensional compressible flow subject to large external forces. *Osaka J. Math.*, **38**, 399–418.

Mavriplis, D. J. (1990). Accurate multigrid solutions of the Euler equation on unstructured and adaptive grids. *AIAA J.*, **28**, 2.

Mavriplis, J. D., Jameson, A., and Martinelli, L. (1989). Multigrid solution of the Navier–Stokes equations on triangular meshes. ICASE Report 89–11, NASA.

Mehlman, G. (1991). *Etude de quelques problèmes liés aux écoulements en déséquilibre chmique.* Ph. D. thesis, Ecole Polytechnique.

Meister, A. (1998). Comparison of different Krylov subspace methods embedded in an implicit finite volume scheme for the computation of viscous and inviscid flow fields on unstructured grids. *J. Comput. Phys.*, **140**, 311–345.

Meister, A. (1999). Asymptotic single and multiple scale expansions in the low Mach number limit. *SIAM J. Appl. Math.*, **60(1)**, 256–271.

Melenk, J. M. and Schwab, C. (1999). The *hp* streamline diffusion finite element method for convection dominated problems in one space dimension. *East–West J. Numer. Math.*, **7**, 31–60.

Morgan, K., Hassan, O., Manzari, M. T., Verhoeven, N., and Weatherill, N. P. (1997). Towards a finite element capability for the modelling of viscous compressible flows. In *The Mathematics of Finite Elements and Applications, Highlights 1996* (ed. J. Whiteman), pp. 195–222. Wiley, Chichester.

Morgan, K., Peraira, J., Vahdata, M., and Zienkiewicz, O. C. (1987). Adaptive remeshing for compressible flow computations. *J. Comput. Phys.*, **72**, 449–466.

Mulder, W. and van Leer, B. (1985). Experiments with implicit upwind methods for the Euler equations. *J. Comput. Phys.*, **59**, 232–246.

Nash, J. (1962). Le problème de Cauchy pour les équations différentielles d'un fluide général. *Bull. Soc. Math. France*, **90**, 487–497.

Nečas, J. (1967). *Les Méthodes Directes en Théorie des Équations Elliptiques.* Academia, Prague.

Nessyahu, H. and Tadmor, E. (1990). Non oscillatory central differencing for hyperbolic conservation laws. *J. Comput. Phys.*, **87**, 408–463.

Nessyahu, H., Tassa, T., and Tadmor, E. (1994). The convergence rate of Godunov type schemes. *SIAM J. Numer. Anal.*, **31**, 1–16.

Noelle, S. (1995). Convergence of higher order finite volume schemes on irregular grids. *Adv. Comput. Math.*, **3(III)**, 197–218.

Novo, S. and Novotný, A. (2002). On the existence of weak solutions to steady compressible Navier–Stokes equations when the density is not square integrable. *J. Math. Kyoto Univ.*, **42(3)**.

Novotný, A. (1993). Steady flows of a viscous compressible fluid-L^2-approach. In *Proceedings of EQUAM 92, Varenna (Italy), SAACM* (ed. R. Salvi and I. Straškraba), Volume 3(3), pp. 181–199.

Novotný, A. and Padula, M. (1993). Existence and uniqueness of stationary solutions for viscous compressible fluid with large potential and small nonpotential forces. *Sib. Math. J.*, **34(5)**, 120–146.

Novotný, A. and Straškraba, I. (2003). *Mathematical Theory of Compressible Flow.* Oxford University Press, Oxford.

Oden, J. T., Babuška, I., and Baumann, C. E. (1998). A discontinuous *hp* finite element method for diffusion problems. *J. Comput. Phys*, **146**, 491–519.

Ohmori, K. and Ushijima, T. (1984). A technique of upstream type applied to a linear nonconforming finite element approximation of convective diffusion equations. *RAIRO Anal. Numér.*, **18**, 309–322.

Osher, S. and Shu, C. W. (1988). Efficient implementation of essentially non-oscillatory shock-capturing schemes, I. *J. Comput. Phys.*, **77**, 439–471.

Osher, S. and Shu, C. W. (1989). Efficient implementation of essentially non-oscillatory shock-capturing schemes, II. *J. Comput. Phys.*, **83**, 32–78.

Osher, S. and Shu, C.-W. (1991). High-order essentially nonoscillatory schemes for Hamilton–Jacobi equations. *SIAM J. Numer. Anal.*, **28**, 907–922.

Osher, S. and Solomon, F. (1982). Upwind difference schemes for hyperbolic systems of conservation laws. *Math. Comput.*, **38**, 339–374.

Pascal, F. and Ghidaglia, J. M. (2001). Footbridges between finite volumes and finite elements with application to CFD. Technical report, University Paris-Sud. (preprint).

Persiano, R. M., Comba, J. L. D., and Barbalho, V. (1993). An adaptive triangulation refinement scheme and construction. In *Proceedings of the VI Sibgrapi (Brazilian Symposium on Computer Graphics and Image Processing).*

Perthame, B. and Qiu, Y. (1994). A variant of van Leer's method for multidimensional systems of conservation laws. *J. Comput. Phys.*, **112**, 370–381.

Peyret, R. and Taylor, T. D. (1983). *Computational Methods for Fluid Flow.* Springer, New York.

Pike, J. (1987). Grid adaptative algorithms for the solution of the Euler equation on irregular grids. *J. Comput. Phys.*, **71**, 194–223.

Pironneau, O. (1989). *Finite Element Methods for Fluids.* Wiley, Chichester.

Prudhomme, S., Pascal, F., Oden, J. T., and Romkes, A. (2000). Review of a priori error estimation for discontinuous Galerkin methods. Technical report, TICAM.

Quarteroni, A. and Valli, A. (1997). *Numerical Approximation of Partial Differential Equations.* Springer, Berlin.

Rachowicz, W. (1997). An anisotropic h-adaptive finite element method for compressible Navier–Stokes equations. *Comput. Methods Appl. Mech. Eng.*, **146**, 231–252.

Ralston, A. (1965). *A First Course in Numerical Analysis.* McGraw–Hill, New York.

Reed, W. H. and Hill, T. R. (1973). Triangular mesh methods for the neutron transport equation. Technical Report LA-UR-73-479, Los Alamos Scientific Laboratory.

Rektorys, K. (1982). *The Method of Discretization in Time and Partial Differential Equations.* Reidel, Dordrecht.

Riemslagh, K. and Dick, E. (1994). Mixed discretization multigrid methods for TVD-schemes on unstructured grids. In *Computational Fluid Dynamics '94* (ed. S. Wagner, E. H. Hirschel, J. Périaux, and R. Piva), pp. 317–324. Wiley, Stuttgart.

Ringleb, F. (1940). Exakte Lösungen der Differentialgleichungen einer adiabatischen Strömung. *ZAMM*, **20**, 185–198.

Rivara, M.-C. (1996). New mathematical tools and techniques for the refinement and/or improvement of unstructured triangulations. In *Proceedings of the 5th International Meshing Roundtable*, pp. 77–86.

Rivière, B. and Wheeler, M. F. (2000). A discontinuous Galerkin method applied to nonlinear parabolic equations. In *Discontinuous Galerkin Methods* (ed. B. Cockburn, G. E. Karniadakis, and C.-W. Shu), pp. 231–246. Springer, Berlin.

Roache, P. J. (1972). *Computational fluid dynamics.* Hermosa, Albuquerque, NM.

Roache, P. J. (1998*a*). *Fundamentals of Computational Fluid Dynamics.* Hermosa, Albuquerque, NM.

Roache, P. J. (1998*b*). *Verification and Validation in Computational Science and Engineering.* Hermosa, Albuquerque, NM.

Roe, P. L. (1981). Approximate Riemann solvers, parameter vectors and difference schemes. *J. Comput. Phys.*, **43**, 357–372.

Roe, P. L. (1989). Remote boundary conditions for unsteady multidimensional aerodynamic computations. *Comput. Fluids*, **17**, 221–231.

Roos, H.-G. (2002). Optimal convergence of basic schemes for elliptic boundary value problems with strong parabolic layers. *J. Math. Anal. Appl.*, **267**, 194–208.

Roos, H.-G., Stynes, M., and Tobiska, L. (1996). *Numerical Methods for Singularly Perturbed Differential Equations (Convection–Diffusion and Flow Problems).* Springer, Berlin.

Roos, H.-G. and Zarin, H. (2002). The discontinuous Galerkin finite element method for singularly perturbed problems. Technical Report MATH-NM-14-2002, Technische Universität Dresden.

Rudin, W. (1974). *Real and Complex Analysis.* McGraw-Hill, London–New York.

Saad, Y. (1996). *Iretative Methods for Sparse Linear Systems.* PWS, Boston.

Saad, Y. and Schultz, M. H. (1986). GMRES: A generalized minimal residual algorithm for solving non-symmetric linear systems. *SIAM J. Sci. Stat. Comput.*, **7**, 856–869.

Salvi, R. and Straškraba, I. (1993). Global existence for viscous compressible fluids and their behaviour as $t \to \infty$. *J. Fac. Sci. Univ. Tokyo Sect. IA Math.*, **40(1)**, 17–52.

Schatzman, M. (2002). *Numerical Analysis: A Mathematical Introduction.* Oxford University Press, Oxford.

Schieweck, F. and Tobiska, L. (1989). A nonconforming finite element method of upstream type applied to the stationary Navier–Stokes equations. *RAIRO Anal. Numér.*, **23**, 627–647.

Schneider, T., Botta, N., Geratz, K. J., and Klein, R. (1999). Extension of finite volume compressible flow solvers to multidimensional, variable density zero Mach number flows. *J. Comput. Phys.*, **155**, 248–286.

Schochet, S. (1986). The compressible Euler equations in a bounded domain: Existence of solutions and incompressible limit. *Commun. Math. Phys.*, **104**, 49–75.

Schwab, C. (1998). *p- and hp-Finite Element Methods.* Clarendon Press, Oxford.

Serre, D. (1997). Solutions globales des équations paraboliques de lois de conservations. *Ann. Inst. Fourier, Grenoble*, **48**, 1069–1091.

Sestertenn, J., Müller, B., and Thomann, H. (1993). Flux-vector splitting for compressible low Mach number flow. *Comput. Fluids*, **22**, 441–451.

Shakib, F., Hughes, T. J. R., and Johan, Z. (1991). A new finite element formulation for computational fluid dynamics: X. The compressible Euler and Navier–Stokes equations. *Comput. Methods Appl. Mech. Eng.*, **89**, 141–219.

Shinbrot, M. (1973). *Lectures on Fluid Mechanics.* Gordon and Breach, New York.

Shu, C.-W. (1990). Numerical experiments on the accuracy of ENO and modified ENO schemes. *SIAM J. Sci. Comput.*, **5**, 127–149.

Shu, C.-W. (1999). Higher order ENO and WENO schemes for computational fluid dynamics. In *High-Order Methods for Computational Physics* (ed. T. J. Barth and H. Deconinck), Volume 9 of *Lecture Notes in Computational Science and Engineering*, pp. 439–582. Springer, Berlin.

Shu, C.-W. and Osher, S. (1988). Efficient implementation of essentially non-oscillatory shock capturing schemes. *J. Comput. Phys.*, **77**, 439–471.

Shu, C.-W. and Osher, S. (1989). Efficient implementation of essentially non-oscillatory shock capturing schemes II. *J. Comput. Phys.*, **77**, 32–78.

Simpson, R. B. (1994). Anisotropic mesh transformations and optimal error control. *Appl. Numer. Math.*, **14**, 183–198.

Smoller, J. (1983). *Shock Waves and Reaction–Diffusion Equations.* Springer, New York.

Smoller, J. (1994). *Shock Waves and Reaction–Diffusion Equations.* Springer, New York.

Sod, G. A. (1985). *Numerical methods in fluid dynamics.* Cambridge University Press, Cambridge.

Solonnikov, V. A. (1980). Solvability of the initial-boundary value problem for the equations of a viscous compressible fluid. *J. Sov. Math.*, **14**, 1120–1133.

Sonar, Th. (1993). Strong and weak norm refinement indicators based on the finite element residual for compressible flow computation. *Impact Comput. Sci. Eng.*, **5**, 111–127.

Sonar, Th. (1997*a*). *Mehrdimensionale ENO–Verfahren.* Advances in Numerical Mathematics. Teubner, Stuttgart.

Sonar, T. (1997*b*). On the construction of essentially non-oscillatory finite volume approximations to hyperbolic conservation laws on general triangulations: polynomial recovery, accuracy and stencil selection. *Comput. Methods Appl. Mech. Eng.*, **140**, 157–181.

Sonar, Th. and Warnecke, G. (1996). On finite difference error indication for adaptive approximations of conservation laws. Reihe A Preprint 109, Universität Hamburg.

Spekreijse, S. P. (1988). *Multigrid Solution of the Steady Euler Equations.* Centre for Mathematics and Computer Science, Amsterdam.

Šťastný, M. and Šafařík, P. (1990). Experimental analysis data on the transonic flow past a plane turbine. Technical report, ASME Paper 90-GT-313, New York.

Steger, J. L. and Warming, R. F. (1981). Flux vector splitting of the inviscid gas dynamics equations with applications to finite difference methods. *J. Comput. Phys.*, **40**, 263–293.

Stiller, J. (1996). A finite element method for the compressible Navier–Stokes equations and its application to transonic nozzle flows. In *Aerothermodynamics of internal flows III, Proceedings of the third international symposium on experimental and computational aerothermodynamics of internal flows* (ed. S. Yu, N. Chen, and Y. Bai), Beijing, China, pp. 437–444. World Publishing Corporation.

Stouffet, B. (1983). Implicit finite element methods for the Euler equations. In *Numerical methods for the Euler equations of fluid dynamics, Proceedings of INRIA workshop* (ed. F. Angrand *et al.*), Rocquencourt, France, pp. 409–434. SIAM Philadelphia.

Stoufflet, B., Periaux, J., Fezoui, F., and Dervieux, A. (1987). 3-D hypersonic Euler numerical simulation around space vehicles using adapted finite elements. AIAA Paper, 86-0560. 25th AIAA Aerospace Meeting, Reno.

Strang, G. (1972). Variational crimes in the finite element method. In *The Mathematical Foundations of the Finite Element Method* (ed. A. K. Aziz), pp.

689–710. Academic Press, New York.

Strang, G. and Fix, G. J. (1973). *An Analysis of the Finite Element Method.* Prentice Hall, Englewood Cliffs, NJ.

Süli, E. (1999). A posteriori error analysis and adaptivity for finite element approximations of hyperbolic problems. In *An Introduction to Recent Developments in Theory and Numerics for Conservation Laws* (ed. D. Kröner, M. Ohlberger, and C. Rohde), Volume 5 of *Lecture Notes in Computational Science and Engineering*, pp. 123–194. Springer, Berlin.

Sweby, P. K. (1984). High resolution schemes using flux limiters for hyperbolic conservation laws. *SIAM J. Numer. Anal.*, **21**, 995–1011.

Szabo, B. A. and Babuška, I. (1991). *Finite Element Analysis.* Wiley, New York.

Tani, A. (1977). On the first initial boundary value problem of compressible viscous fluid motion. *Publ. Res. Math. Sci. Kyoto Univ.*, **13**, 193–253.

Taylor, A. E. (1967). *Introduction to Functional Analysis.* Wiley, New York.

Temam, R. (1977). *Navier–Stokes Equations. Theory and Numerical Analysis.* North-Holland, Amsterdam.

Thomée, V. (1997). *Galerkin Finite Element Methods for Parabolic Problems.* Springer, Berlin.

Toro, E. F. (1997). *Riemann Solvers and Numerical Methods for Fluid Dynamics.* Springer, Berlin.

Turek, S. (1999). *Efficient Solvers for Incompressible Flow Problems. An Algorithmic and Computational Approach.* Springer, Berlin.

Turkel, E. (1987). Preconditioned methods for solving the incompressible and low speed compressible equations. *J. Comput. Phys.*, **72**, 277–298.

Turkel, E. (1996). Preconditioned methods for low speed flows. *AIAA Paper AIAA-96-2460.*

Tveito, A. and Winther, R. (1993). An error estimate for a finite difference scheme approximating a hyperbolic system of conservation laws. *SIAM J. Numer. Anal.*, **30**, 401–424.

Ueberhuber, C. W. (1997). *Numerical Computation 2, Methods, Software, and Analysis.* Springer, Berlin.

Valli, A. (1992). Mathematical results for compressible flows. In *Mathematical topics in fluid mechanics* (ed. J. Rodrigues and A. Sequeira), Volume 274, pp. 193–229. Wiley, New York.

Valli, A. and Zajączkowski, W. M. (1986). Navier–Stokes equations for compressible fluids: Global existence and qualitative properties of the solutions in the general case. *Commun. Math. Phys.*, **103**, 256–296.

van der Maarel, H. T. M., Hemker, P. W., and Koren, B. (1993). Application of a solution-adaptive multigrid method to the Euler equations. *CWI Q.*, **6**, 49–75.

van der Vorst, H. A. (1992). Bi-CGSTAB: A fast and smoothly converging variant of Bi-CG for solution of non-symmetric linear systems. *SIAM J. Sci. Stat. Comput.*, **13**, 631–644.

Van Dyke, M. (1988). *An Album of Fluid Motion.* The Parabolic Press, Stanford, CA.

Van Leer, B. (1979). Towards the ultimate conservative difference scheme III. A second-order sequel to Godunov's method. *J. Comput. Phys.*, **32**, 101–136.

Varga, R. (1962). *Matrix Iterative Analyis.* Prentice Hall, Englewood Cliffs, NJ.

Verfürth, R. (1996). *A Review of a posteriori Error Estimation and Adaptive Mesh-Refinement Techniques.* Wiley–Teubner, New York.

Vijayasundaram, G. (1986). Transonic flow simulation using upstream centered scheme of Godunov type in finite elements. *J. Comput. Phys.*, **63**, 416–433.

Vila, J. P. (1994). Convergence and error estimates in finite volume schemes for general multidimensional scalar conservation laws I. Explicit monotone schemes. *RAIRO Math. Model. Numer. Anal.*, **28**, 267–295.

Weiss, R. (1996). *Parameter-Free Iterative Linear Solvers.* Akademie Verlag, Berlin.

Wesseling, P. (1992). *An Introduction to Multigrid Methods.* Wiley, Chichester.

Wesseling, P. (2001). *Principles of Computational Fluid Dynamics.* Springer, Berlin.

Wheeler, M. F. (1978). An elliptic collocation–finite element method with interior penalties. *SIAM J. Numer. Anal.*, **15**, 152–161.

Yosida, K. (1974). *Functional Analysis.* Springer, New York.

Zeidler, E. (1985-1989). *Nonlinear Functional Analysis and its Applications*, Volume 1-4. Springer, New York.

Ženíšek, A. (1990). *Nonlinear Elliptic and Evolution Problems and Their Finite Element Approximations.* Academic Press, London.

Zienkiewicz, O. C. and Morgan, K. (1983). *Finite Elements and Approximation.* Wiley, New York.

INDEX

Lightning Source UK Ltd.
Milton Keynes UK
29 March 2011

170008UK00001B/12/A

9 780198 505884